Image Processing, Analysis, and Machine Vision

Image Processing, Analysis, and Machine Vision

Second Edition

Milan Sonka
The University of Iowa, Iowa City

Vaclav Hlavac
Czech Technical University, Prague

Roger Boyle
University of Leeds, Leeds

PWS Publishing
An Imprint of Brooks/Cole Publishing Company
I(T)P® An International Thomson Publishing Company

Pacific Grove • Albany • Bonn • Boston • Cincinnati • Detroit • London • Madrid • Melbourne
Mexico City • New York • Paris • San Francisco • Singapore • Tokyo • Toronto • Washington

Project Development Editor: *Suzanne Jeans*
Marketing Team: *Nathan Wilbur, Michele Mootz*
Marketing Communications: *Jean Thompson*
Production Editor: *Kelsey McGee*
Proofreader: *Lynne Lackenbach*
Cover Design: *Laurie Albrecht*
Cover Printing/Printing and Binding:
R. R. Donnelley & Sons, Crawfordsville

For more information, contact PWS Publishing at Brooks/Cole Publishing Company:

BROOKS/COLE PUBLISHING COMPANY
511 Forest Lodge Road
Pacific Grove, CA 93950
USA

International Thomson Publishing Europe
Berkshire House 168-173
High Holborn
London WC1V 7AA
England

Thomas Nelson Australia
102 Dodds Street
South Melbourne, 3205
Victoria, Australia

Nelson Canada
1120 Birchmount Road
Scarborough, Ontario
Canada M1K 5G4

International Thomson Editores
Seneca 53
Col. Polanco
11560 México, D. F., México

International Thomson Publishing GmbH
Königswinterer Strasse 418
53227 Bonn
Germany

International Thomson Publishing Asia
60 Albert Street
#15-01 Albert Complex
Singapore 189969

International Thomson Publishing Japan
Hirakawacho Kyowa Building, 3F
2-2-1 Hirakawacho
Chiyoda-ku, Tokyo 102
Japan

Printed in the United States of America

10 9 8 7 6 5 4 3 2 1

Library of Congress Cataloging-in-Publication Data
Sonka, Milan
Image processing, analysis, and machine vision / Milan Sonka, Vaclav Hlavac, and Roger Boyle. – 2nd. ed.
p. cm.
Includes bibliographical references and index.
ISBN 0-534-95393-X
1. Image processing. 2. Computer vision. 3. Image analysis.
I. Hlavac, Vaclav. II. Boyle, Roger. III. Title.
TA1637.S66 1998
006.3'7—dc21 98-38396

Contents

List of algorithms

List of symbols and abbreviations

$\arg(x, y)$ — angle (in radians) from x axis to the point (x, y)
$\text{argmax}_i(\text{expr}(i))$ — the value of i that causes $\text{expr}(i)$ to be maximal
$\text{argmin}_i(\text{expr}(i))$ — the value of i that causes $\text{expr}(i)$ to be minimal
div — integer division
mod — remainder after integer division
$\text{round}(x)$ — largest integer which is not bigger than $x + 0.5$
$\emptyset$ — empty set
A^c — complement of the set A
$A \subset B$, $B \supset A$ — set A is included in set B
$A \cap B$ — intersection between sets A and B
$A \cup B$ — union of sets A and B
$A \mid B$ — difference between sets A and B
$\mathbf{A}$ — (uppercase bold) matrices
$\mathbf{x}$ — (lowercase bold) vectors
$|\mathbf{x}|$ — magnitude (or modulus) of the vector $\mathbf{x}$
$\mathbf{x} \cdot \mathbf{y}$ — scalar product between vectors $\mathbf{x}$ and $\mathbf{y}$
$\tilde{x}$ — estimate of the value x
$|x|$ — absolute value of a scalar
Δx — small finite interval of x, difference
$\partial f / \partial x$ — partial derivative of the function f with respect to x
$\nabla \mathbf{f}$, grad $\mathbf{f}$ — gradient of $\mathbf{f}$
$\nabla^2 \mathbf{f}$ — Laplace operator applied to $\mathbf{f}$
$f * g$ — convolution between functions f and g
D_E — Euclidean distance (see Section 2.3.1)
D_4 — city block distance (see Section 2.3.1)
D_8 — chessboard distance (see Section 2.3.1)
F — complex conjugate of the complex function F
$\text{rank}(A)$ — Rank of a matrix A
T^* — transformation dual to transformation T
$\delta(x)$ — Dirac function
$\mathcal{E}$ — mean value operator
$\mathcal{L}$ — linear operator

$\mathcal{O}$	origin of the coordinate system
$\#$	number of (e.g., pixels)
$\breve{B}$	point set symmetrical to point set B
$\oplus$	morphological dilation
$\ominus$	morphological erosion
$\circ$	morphological opening
$\bullet$	morphological closing
$\otimes$	morphological hit-or-miss transformation
$\oslash$	morphological thinning
$\odot$	morphological thickening

1D	one dimension(al)
2D	two dimension(al)
3D	three dimension(al)
AI	artificial intelligence
ASM	active shape model
B-rep	boundary representation
CAD	computer-aided design
CCD	charge-coupled device
CSG	constructive solid geometry
CT	computer tomography
dof	degrees of freedom
ECG	electro-cardiogram
EEG	electro-encephalogram
FFT	fast Fourier transform
FOE	focus of expansion
GA	genetic algorithm
HMM	hidden Markov model
IHS	intensity, hue, saturation
JPEG	Joint Photographic Experts Group
MR	magnetic resonance
MRI	magnetic resonance imaging
OCR	optical character recognition
OS	order statistics
PDM	point distribution model
PET	positron emission tomography
PMF	Pollard-Mayhew-Frisby (correspondence algorithm)
RGB	red, green, blue
SNR	signal-to-noise ratio
SVD	singular value decomposition
TV	television

Preface

Image processing, analysis and machine vision represent an exciting and dynamic part of cognitive and computer science. Following an explosion of interest during the 1970s, the 1980s and 1990s were characterized by the maturing of the field and the significant growth of active applications; remote sensing, technical diagnostics, autonomous vehicle guidance, medical imaging (2D and 3D) and automatic surveillance are the most rapidly developing areas. This progress can be seen in an increasing number of software and hardware products on the market, as well as in a number of digital image processing and machine vision courses offered at universities worldwide.

There are many texts available in the areas we cover—most (indeed, all of which we know) are referenced somewhere in this book. The subject suffers, however, from a shortage of texts which are 'complete' in the sense that they are accessible to the novice, of use to the educated, and up to date. Here we present the second edition of a text first published in 1993 in which we hope to include many of the very rapid developments that have taken and are still taking place, which quickly age some of the very good textbooks produced over the last two decades or so. The target audience is the undergraduate with negligible experience in the area through to the Master's and research student seeking an advanced springboard in a particular topic. Every section of this text has been updated since the first version (particularly with respect to references); additionally, wholly new sections are presented on: compression via JPEG and MPEG; fractals; fuzzy logic recognition; hidden Markov models; Kalman filters; point distribution models; three-dimensional vision; watershed segmentation; wavelets; and an entire chapter devoted to case studies. Additionally, each chapter now includes a concise Summary section. To help the reader to acquire practical understanding, newly added Exercise sections accompany each chapter; these are in the form of short-answer questions and problems of varying difficulty, frequently requiring practical usage of computer tools and/or development of application programs.

This book reflects the authors' experience in teaching one- and two-semester undergraduate and graduate courses in Digital Image Processing, Digital Image Analysis, Machine Vision, Pattern Recognition, and Intelligent Robotics at their respective institutions. We hope that this combined experience will give a thorough grounding to the beginner and provide material that is advanced enough to allow the more mature student to understand fully the relevant areas of the subject. We acknowledge that in a very short time the more active areas will have moved beyond this text.

This book could have been arranged in many ways. It begins with low-level processing and works its way up to higher levels of image interpretation; the authors have chosen this framework because they believe that image understanding originates from a common database of information. The book is formally divided into 16 chapters, beginning with low-level processing and working toward higher-level image representation, although this structure will be less apparent after Chapter 10, when we present transforms, compression, morphology, texture, and motion analysis which are very useful but often special-purpose approaches that may not always be included in the processing chain. The final chapter presents four live research projects which illustrate in practical use much of what has gone before.

Decimal section numbering is used, and equations and figures are numbered within each chapter. Each chapter is accompanied by an extensive list of references and exercises. A selection of algorithms is summarized formally in a manner that should aid implementation—not all the algorithms discussed are presented in this way (this might have doubled the length of the book); we have chosen what we regard as the key, or most useful or illustrative, examples for this treatment.

Each chapter presents material from an introductory level through to an overview of current work; as such, it is unlikely that the beginner will, at the first reading, expect to absorb all of a given topic. Often it has been necessary to make reference to material in later chapters and sections, but when this is done an understanding of material in hand will not depend on an understanding of that which comes later. It is expected that the more advanced student will use the book as a reference text and signpost to current activity in the field—we believe at the time of going to press that the reference list is full in its indication of current directions, but record here our apologies to any work we have overlooked. The serious reader will note that many references are very recent, and should be aware that before long more relevant work will have been published that is not listed here.

This is a long book and therefore contains material sufficient for much more than one course. Clearly, there are many ways of using it, but for guidance we suggest an ordering that would generate four distinct modules:

Digital Image Processing, an undergraduate course.

Digital Image Analysis, an undergraduate/graduate course, for which Digital Image Processing may be regarded as prerequisite.

Computer Vision I, an undergraduate/graduate course, for which Digital Image Processing may be regarded as prerequisite.

Computer Vision II, a graduate course, for which Computer Vision I may be regarded as prerequisite.

The important parts of a course, and necessary prerequisites, will naturally be specified locally; a suggestion for partitioning the contents follows this Preface.

Assignments should wherever possible make use of existing software; it is our experience that courses of this nature should not be seen as 'programming courses', but it is the case that the more direct practical experience the students have of the material discussed, the better is their understanding. Since the first edition was published, an explosion of World Wide Web-based material has been made available, permitting many of the exercises we present to be conducted without the necessity of implementing from scratch—we do not present explicit pointers to Web material, since they evolve so quickly; however, pointers to specific support materials for this book and others may be located via the publisher, *http://www.brookscole.com.*

The book has been prepared using the LaTeX text processing system. Its completion would have been impossible without extensive usage of the Internet computer network and electronic mail. We should like to acknowledge the University of Iowa, the Czech Technical University, and the School of Computer Studies at Leeds University for providing the environment in which this book was prepared.

Milan Sonka was a faculty member of the Department of Control Engineering, Faculty of Electrical Engineering, Czech Technical University, Prague, Czech Republic for ten years, and is now an Associate Professor at the Department of Electrical and Computer Engineering, the University of Iowa, Iowa City, Iowa, USA. His research interests include medical image analysis, knowledge-based image analysis, and machine vision. Vaclav Hlavac is an Associate Professor at the Department of Control Engineering, Czech Technical University, Prague. His research interests are knowledge-based image analysis and 3D model-based vision. Roger Boyle is a Senior Lecturer in Artificial Intelligence in the School of Computer Studies at the University of Leeds, England, where his research interests are in low-level vision and pattern recognition. The first two authors have worked together for some years, and have been co-operating with the third since 1991.

The authors have spent many hours in discussions with their teachers, colleagues, and students, from which many improvements to early drafts of this text resulted. Particular thanks are due to Tomáš Pajdla, Petr Kodl, Radim Šára at the Czech Technical University; Steve Collins at the University of Iowa; Jussi Parkkinen at the University of Lappeenranta; Guido Prause at the University of Bremen; David Hogg at the University of Leeds; and many others whose omission from this list does not diminish the value of their contribution. The continuous support and encouragement we received from our wives and families, while inexplicable, was essential to us throughout this project—once again, we promise that our next book will not be written outside standard office hours or during holidays (but this time we mean it).

All authors have contributed throughout—the ordering on the cover corresponds to the weight of individual contribution. Any errors of fact are the joint responsibility of all, while any errors of typography are the responsibility of Roger Boyle. Jointly, they will be glad to incorporate any corrections into future editions.

Milan Sonka (milan-sonka@uiowa.edu)
The University of Iowa, Iowa City, Iowa, USA

Vaclav Hlavac (hlavac@vision.felk.cvut.cz)
Czech Technical University, Prague, Czech Republic

Roger Boyle (roger@scs.leeds.ac.uk)
University of Leeds, Leeds, England

Thanks to the reviewers of this edition for their comments, critiques, and suggestions:

John M. Canning
University of Oklahoma

John G. Harris
University of Florida

Bruce Maxwell
University of North Dakota

Guido Prause
University of Bremen

Christian Roux
École Nationale Supérieure des Télécommunications de Bretagne

Course contents

In this section, one possible ordering of the material covered in the four courses proposed in the Preface is given. This coverage should not be considered the only possibility—on the contrary, the possibilities for organizing Image Processing and Analysis courses are practically endless. Therefore, what follows should only be regarded as suggestions, and instructors should tailor course content to fit the already acquired knowledge, abilities, and needs of the students enrolled.

Digital Image Processing. An undergraduate course.

1 Introduction

2 The digitized image and its properties

3 Data structures for image analysis

4 Image pre-processing (excluding 4.3.6–4.3.9, 4.4.3, limited coverage of 4.3.4, 4.3.5)

5 Segmentation

- 5.1 Thresholding (excluding 5.1.3, 5.1.4)
- 5.2 Edge-based segmentation (excluding 5.2.8, limited coverage of 5.2.4, 5.2.5)
- 5.3 Region growing segmentation (excluding 5.3.4)
- 5.4 Matching

12 Linear discrete image transforms

13 Image data compression

16 Case studies (selected topics)

Digital Image Analysis. An undergraduate/graduate course, for which Digital Image Processing may be regarded as prerequisite. Sections that were covered in the Digital Image Processing class and re-appear are intended to be discussed at more depth than is possible in the introductory course.

1 Introduction (brief review)

2 The digitized image and its properties (brief review)

5 Segmentation

- 5.1.3 Multi-spectral thresholding
- 5.1.4 Thresholding in hierarchical data structures

- 5.2.4 Edge following as graph searching
- 5.2.5 Edge following as dynamic programming
- 5.3.4 Watershed segmentation

6 Shape representation and description (excluding 6.2.7, 6.3.4–6.3.6, 6.4)

7 Object recognition

- 7.1 Knowledge representation
- 7.2 Statistical pattern recognition
- 7.3 Neural networks
- 7.4 Syntactic pattern recognition

11 Mathematical morphology

14 Texture

16 Case studies (selected topics)

Computer Vision I. An undergraduate/graduate course, for which Digital Image Processing may be regarded as prerequisite.

1 Introduction (brief review)

2 The digitized image and its properties (brief review)

4 Image pre-processing

- 4.3.3 Zero-crossings of the second derivative
- 4.3.4 Scale in image processing
- 4.3.5 Canny edge detection
- 4.3.6 Parametric edge models
- 4.3.7 Edges in multi-spectral images
- 4.3.8 Other local pre-processing operators
- 4.3.9 Adaptive neighborhood pre-processing

6 Shape representation and description

7 Object recognition

8 Image understanding

16 Case studies (selected topics)

Computer Vision II. A graduate course, for which Computer Vision I may be regarded as prerequisite.

5 Segmentation

- 5.2.4 Edge following as graph searching
- 5.2.5 Edge following as dynamic programming
- 5.5 Advanced border and surface detection approaches

9 3D Vision, geometry and radiometry

10 Use of 3D vision

15 Motion analysis

Practical 3D vision projects

Chapter 1

Introduction

Vision allows humans to perceive and understand the world surrounding them. Computer vision aims to duplicate the effect of human vision by electronically perceiving and understanding an image. Giving computers the ability to see is not an easy task—we live in a three-dimensional (3D) world, and when computers try to analyze objects in 3D space, the visual sensors available (e.g., TV cameras) usually give two-dimensional (2D) images, and this projection to a lower number of dimensions incurs an enormous loss of information. Dynamic scenes such as those to which we are accustomed, with moving objects or a moving camera, make computer vision even more complicated.

A particular image analysis problem will illustrate a few commonly encountered problems. Figure 1.1 shows the ozone layer hole over Antarctica, as captured by apparatus on board

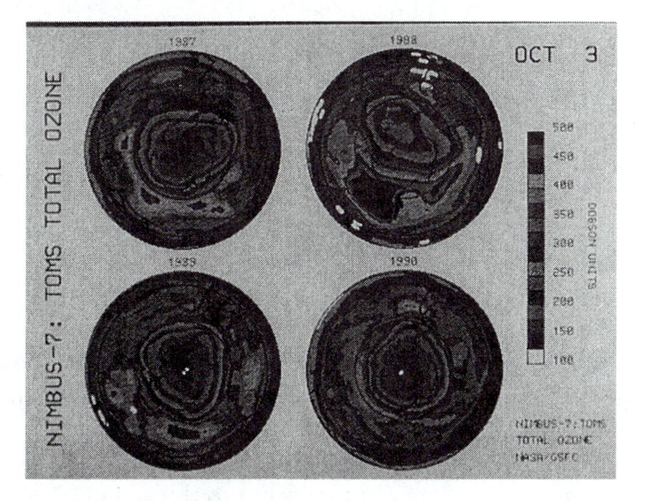

Figure 1.1: *The ozone layer hole. A color version of this picture may be seen in the color inset. Courtesy NASA, Goddard Space Flight Center, Greenbelt, MD.*

NASA's Nimbus 7 satellite over a period of years: 'TOMS' here stands for Total Ozone Mapping Spectrometer, and Dobson units are standard units (multiple of molecules per cubic centimeter) used in ozone mapping—the normal value is about 300 Dobson units and so depletion is obvious in these pictures. The ozone hole is observed to open each year between September and November; shortly after the last picture in this sequence was taken, the lowest ever level was recorded (110 Dobson units on 6th October 1991 [Appenzeller 91]). The hole has varied in seriousness and duration throughout the 1990s—109 Dobson units were measured in 1993.

It is natural to investigate ways in which pictures like these may be analyzed automatically—the *qualitative* conclusion that there is a trend towards ozone depletion should be available from the changes in colors between the successive pictures, and we might also hope for some *quantitative* conclusions saying exactly how much change is occurring. In fact, though, these pictures contain formidable quantities of information that make them far from trivial to analyze. In the picture presented, a great deal of computer processing has already been applied in the form of intensity normalization, geometric transformation, and a number of other filters which will be described in later chapters. Overlooking all the textual annotation, and concentrating on just the most recent shot, Figure 1.2 shows the result of some very early processing that we might perform in an attempt to process the information *beyond* require-

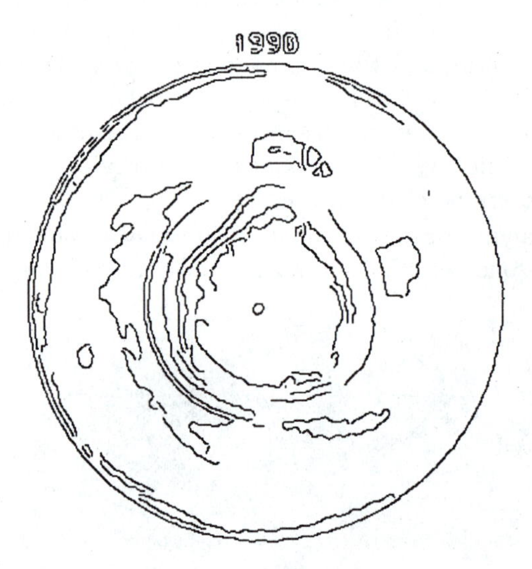

Figure 1.2: *Early processing of part of Figure 1.1.*

ments simply of display (that is, for human inspection)—this picture is the result of applying algorithms described fully in Chapter 4. The purpose of deriving this picture might be to delimit accurately the regions visible in the original, then to assign labels related to ozone concentration to each one, and then to draw conclusions about overall concentrations and trends. We see that, while some of the clear evidence in the original is still visible, not all of it is, and we would have a significant amount of further work to do on the early processing before being able to proceed. Several of the boundaries are incomplete, while some significant information has disappeared altogether. These are problems that will be discussed, and solutions presented, in later chapters. Following this stage would come higher-level analysis in which numbers derived from the scale on the right of Figure 1.1 would be attached to the regions of Figure 1.2—here we would make reference to the colors detected within the regions located, and would probably deploy high-level ('domain') knowledge, such as expecting ozone concentration to vary continuously across the surface of the globe. This sequence of operations—image capture, early processing, region extraction, region labeling, high-level identification, qualitative/quantitative conclusion—is characteristic of image understanding and computer vision problems.

This example is a relatively simple one, but many computer vision techniques use the results and methods of mathematics, pattern recognition, artificial intelligence (AI), psychophysiology, computer science, electronics, and other scientific disciplines. In order to simplify the task of computer vision understanding, two levels are usually distinguished: *Low-level* image processing and *high-level* image understanding.

Low-level methods usually use very little knowledge about the content of images. In the case of the computer knowing image content, it is usually provided by high-level algorithms or directly by a human who knows the problem domain. Low-level methods often include image compression, pre-processing methods for noise filtering, edge extraction, and image sharpening, all of which we shall discuss in this book. Low-level image processing uses data which resemble the input image; for example, an input image captured by a TV camera is 2D in nature, being described by an image function whose value is usually brightness depending on two parameters, the co-ordinates of the location in the image. If the image is to be processed using a computer it will be digitized first, after which it may be represented by a rectangular matrix with elements corresponding to the brightness at appropriate image locations. Such matrices are the inputs and outputs of low-level image processing.

High-level processing is based on knowledge, goals, and plans of how to achieve those goals, and artificial intelligence methods are widely applicable. High-level computer vision tries to imitate human cognition and the ability to make decisions according to the information contained in the image. In the ozone example described, high-level knowledge would be the continuity of the concentration figures (that is, they do not change too sharply between closely neighboring areas), and the fact that areas of similar concentration appear as (distorted) annuli centered at the polar area.

Computer vision is based on high-level processing, and the cognition process is tightly bound to prior knowledge about image content (semantics). An image is mapped into a formalized model of the world, but this model does not remain unchanged. Although the initial model may consist of some general a priori knowledge, high-level processing constantly extracts new information from the images, and updates and clarifies the knowledge.

High-level vision begins with some form of formal model of the world, and then the 'reality' perceived in the form of digitized images is compared to the model. A match is attempted, and when differences emerge, partial matches (or sub-goals) are sought that overcome the mismatches; the computer switches to low-level image processing to find information needed to update the model. This process is then repeated iteratively, and 'understanding' an image thereby becomes a co-operation between top-down and bottom-up processes. A feedback loop is introduced in which high-level partial results create tasks for low-level image processing, and the iterative image understanding process should eventually converge to the global goal.

Computer vision is expected to solve very complex tasks, the goal being to obtain similar results to those provided by biological systems. To illustrate the complexity of these tasks, consider Figure 1.3 in which a particular image representation is presented—the value on the vertical axis gives the brightness of its corresponding location in the image. Consider what this image might be before looking at Figure 1.4, which is a rather more common representation of the same image. Both representations contain exactly the same information, but for a human

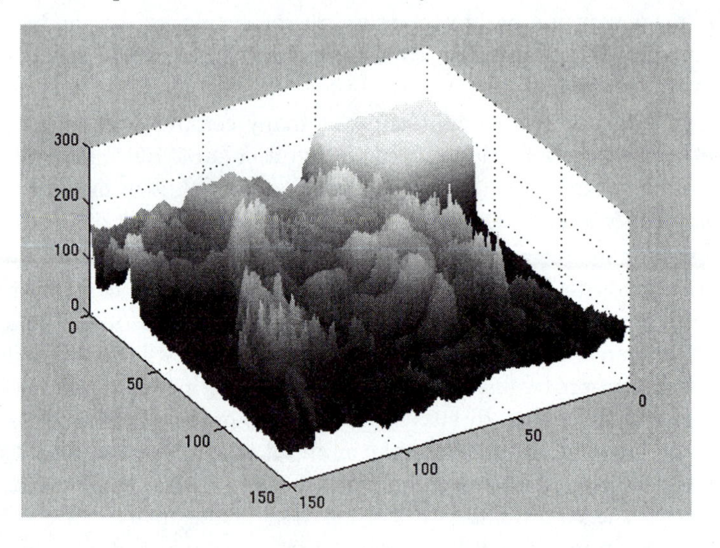

Figure 1.3: *An unusual image representation.*

observer it is very difficult to find a correspondence between them, and without the second, it is unlikely that one would recognize the face of a child. The point is that a lot of a priori knowledge is used by humans to interpret the images; the machine only begins with an array of numbers and so will be attempting to make identifications and draw conclusions from data that to us are more like Figure 1.3 than Figure 1.4. Internal image representations are not directly understandable—while the computer is able to process local parts of the image, it is difficult for it to locate global knowledge. General knowledge, domain-specific knowledge, and information extracted from the image will be essential in attempting to 'understand' these arrays of numbers.

Low-level computer vision techniques overlap almost completely with digital image processing, which has been practiced for decades. The following sequence of processing steps is commonly seen: An image is captured by a sensor (such as a TV camera) and digitized; then the computer suppresses noise (image pre-processing) and maybe enhances some object features which are relevant to understanding the image. Edge extraction is an example of processing carried out at this stage.

Image segmentation is the next step, in which the computer tries to separate objects from the image background and from each other. Total and partial segmentation may be distinguished; total segmentation is possible only for very simple tasks, an example being the recognition of dark non-touching objects from a light background. For example, in analyzing images of printed text (an early step in optical character recognition, OCR) even this superficially simple problem is very hard to solve without error. In more complicated problems (the general case), low-level image processing techniques handle the partial segmentation tasks, in which only the cues which will aid further high-level processing are extracted. Often, finding parts of object boundaries is an example of low-level partial segmentation.

Object description and classification in a totally segmented image are also understood as part of low-level image processing. Other low-level operations are image compression, and techniques to extract information from moving scenes.

Low-level image processing and high-level computer vision differ in the data used. Low-level data are comprised of original images represented by matrices composed of brightness values, while high-level data originate in images as well, but only those data which are relevant to high-level goals are extracted, reducing the data quantity considerably. High-level data represent knowledge about the image content—for example, object size, shape, and mutual relations between objects in the image. High-level data are usually expressed in symbolic form.

Most current low-level image processing methods were proposed in the 1970s or earlier. Recent research is trying to find more efficient and more general algorithms and is implementing them on more technologically sophisticated equipment—in particular, parallel machines are being used to ease the enormous computational load of operations conducted on image data sets. The requirement for better and faster algorithms is fueled by technology delivering larger images (better spatial resolution), and color.

A complicated and so far unsolved problem is how to order low-level steps to solve a specific task, and the aim of automating this problem has not yet been achieved. It is usually still a human operator who finds a sequence of relevant operations, and domain-specific knowledge and uncertainty cause much to depend on this operator's intuition and previous experience. It is not surprising that in the 1980s many projects focused on this problem using expert systems, but expert systems do not solve the problem by themselves. They only allow the use of conflicting opinions of experts in the field.

High-level vision tries to extract and order image processing steps using all available knowledge—image understanding is the heart of the method, in which feedback from high-level to low-level is used. Unsurprisingly this task is very complicated and compute intensive. David Marr's book [Marr 82], discussed in Section 9.1.1, influenced computer vision considerably throughout the 1980s; it described a new methodology and computational theory inspired by biological vision systems. Developments in the 1990s are moving away from dependence on this paradigm.

Computer vision uses pattern recognition and AI techniques and terminology, and where necessary, concepts from these disciplines will be explained in the appropriate chapters of this book. However, it is important to explain a few of them here in order to have a common vocabulary for our basic ideas.

The first concept is that of a **heuristic**. In physics, engineering, and other exact sciences, a deterministic approach is common, where for example the acceleration of a mass can be expressed exactly by a mathematical function. However, knowledge need not be a set of defined assertions with precisely specified domains; it may be both empirical and very vague in nature, and artificial intelligence calls this vague knowledge 'heuristic'. Problems with exponential complexity may often be solved using heuristics to reduce the solution search space. A heuristic solution might be found more quickly than by precise and exhaustive methods, and it is often the case that a good but sub-optimal solution is adequate. In contrast to exact knowledge, one often does not know under which conditions a heuristic is applicable. Heuristic methods are also used in the case when exact algorithms do not exist. The heuristic domain is not defined precisely, so in some cases a solution may exist but a heuristic method may not find it. A good example of a heuristic approach may often be found

Figure 1.4: *Another representation of Figure 1.3.*

in transport scheduling, where the problem space (a rail or bus timetable for a large district, maybe) is huge, but a good working solution (which may be sub-optimal) will be of great use.

A priori knowledge (information) is a second useful concept. Similarly to a priori probability in probability theory, this refers to knowledge available before the search for a problem solution starts—for example, in analysis of music scores we have foreknowledge of the existence of sets of five long parallel lines (see Section 16.1). A priori knowledge, if available, may ease the image understanding task.

Third there are **syntactic** and **semantic** concepts, which come from the theory of formal languages and logic (see Chapter 7 for details). Syntax and semantics allow us to study formal relationships between form and content: 'Syntax' is defined as a set of rules that enable formulae to be created from symbols, and 'semantics' are concerned with formulae interpretation, i.e., finding a mapping from formulae to some model. For example, in the analysis of engineering drawings there may be definite rules—syntax—governing which symbol may connect to any other, while any putative interpretation of the scene or some part of it would inform the semantics. When we use the expression 'semantic information', we will be referring to the information originating from image content.

Computer vision is a relatively young discipline, and until recently the books available have been limited in number and often specialized. This is no longer the case, and libraries can provide many texts. Some address an audience similar to this one; others will aim at a particular area such as image processing, or search, or geometry. Orthogonally, some will aim at junior students and others at the research community. A full library would be extensive, and we have chosen in this text to refer to relevant texts at specific points, with per-chapter summaries. The reader in search of wider or deeper knowledge should search these references, which include a number of highly authoritative works.

Another important source of information of particular importance in this fast-moving science are the research journals; the major current ones are: *Artificial Intelligence*; *Computer Vision, Graphics and Image Processing [CVGIP]* (to 1990); *CVGIP—Graphical Models and Image Processing* (1991 onward); *CVGIP—Image Understanding* (1991 onward); *IEE Proceedings, Part I: Communications, Speech and Vision*; *IEEE Transactions on Acoustics, Speech, and Signal Processing*; *IEEE Transactions on Image Processing*; *IEEE Transactions on Medical Imaging*; *IEEE Transactions on Pattern Analysis and Machine Intelligence*; *IEEE Transactions on Remote Sensing*; *IEEE Transactions on Systems, Man and Cybernetics*; *Image and Vision Computing*; *Information Sciences*; *International Journal of Computer Vision*; *International Journal on Neural Networks*; *International Journal on Pattern Recognition and Artificial Intelligence*; *International Journal of Remote Sensing*; *International Journal of Robotics Research*; *Medical Image Analysis*; *Machine Vision and Applications*; *Neural Computing and Applications*; *Neural Networks*; *Pattern Recognition*; *Pattern Recognition Letters*; *Perception*; *Radiology*; *Robotics*; *Signal Processing: Image Communication*; *Spatial Vision* and *Vision Research*. Attempting to keep this list up to date is a difficult task, particularly since the advent of electronic publication. Some journals are now available primarily or exclusively via the Internet, and the researcher is advised, as with many other disciplines, to open a search via the World Wide Web (WWW) search engines; several sections of this book owe their most recent additions and citations to research initiated by WWW data.

Recent developments and results are published in the proceedings of regularly held image processing and computer vision conferences. Each year an excellent survey of published material is given by Professor A. Rosenfeld of the University of Maryland in the May volume of *CVGIP (Image Understanding)*. In addition, there is an increasing number of more popular magazines that cover some part of the subject, in particular the 'trade' and related journals that document commercially available hardware and software systems. We do not in this book attempt to present up-to-date hardware information, since it would age so very quickly—in common with many other areas of computer science, vision technology is improving rapidly and often dropping in price simultaneously.

Computer vision is growing in theory and application; since the first edition of this book (1993), applications of image analysis and understanding, and their availability over the Internet, have grown beyond possible prediction. Additionally, applications that bring vision into contact with the wider public are increasingly evident—obvious examples are teleconferencing and video phones, and the burgeoning use of imaging of many modalities in medical diagnosis. Each of these makes use of algorithms and theories that we cover here; examples are image representation, filtering, compression, noise reduction, and segmentation. We foresee that before any future edition, the use for the material we present will double and redouble.

This book is intended for those interested in computer vision and image processing; it may serve as an introductory textbook for university students and as a handbook for experts in the field. Knowledge of elementary calculus is assumed. The outline and content of the book is motivated by practical needs, and stress has been put on algorithms, but even so it is not a collection of algorithms usable in all cases. Algorithms should be studied, understood, and their range of applicability carefully taken into account before applying them to specific cases. The authors have tried to give the reader a key to easy access to the literature and have provided references to original papers, and later developments are referenced wherever possible. Both computer vision and image processing are rapidly expanding fields, with more than a thousand papers published each year—despite our best efforts, the reader engaged in research would always be well advised to search the recent literature.

1.1 Summary

- Human vision is natural and seems easy; computer mimicry of this is difficult.
- We might hope to examine pictures, or sequences of pictures, for quantitative and qualitative analysis.
- Many standard and advanced AI techniques are relevant.
- 'High' and 'low' levels of computer vision can be identified.
- Processing moves from digital manipulation, through pre-processing, segmentation, and recognition to understanding.
- An understanding of the notions of heuristics, a priori knowledge, syntax, and semantics is necessary.
- The vision literature is large and growing; books may be specialized, elementary, or advanced.
- A knowledge of the research literature is necessary to stay up to date with the topic.
- Developments in electronic publishing and the Internet are making access to vision simpler.

1.2 Exercises

1. Walking around your local town, locate five instances in which digital images are now publicly visible or in use where they would not have been 10 years ago.
2. For some or all of the instances identified in Exercise 1.1, estimate the economics of the application—the size of market implied and the cost of the unit.
3. Determine the current cost of domestic video cameras and recorders, and image-capture hardware for home computers. Try to determine the cost of comparable equipment 1, 5, and 10 years ago. Make some predictions about the power and price of comparable equipment in 1, 5, and 10 years time.
4. In your own, or a local, university, determine how long image processing, image analysis, or computer vision have been taught. Try to determine the major syllabus changes over the last 1, 5, and (if possible) 10 years.
5. Locate a topic area of computer science outside vision in which *heuristics* are commonly used.
6. Locate a topic area of computer science outside vision in which *a priori information* is commonly used.
7. Locate a topic area of computer science outside vision in which *syntax* is commonly used.
8. Locate a topic area of computer science outside vision in which *semantics* are commonly used.

9. In a technical or academic library, locate the section on computer vision. Examine the texts there, noting the date of publication and chapter titles; compare them with those of this text. Draw some conclusions about the topics (or, at least, their titles) that are static, and those which have developed recently.

10. Select at random some journal references from the listings at the end of each chapter in this book. Construct a histogram of publication year.

11. Locate in a library some of the references selected in Exercise 1.10; study the publication dates of references listed there, and draw another histogram of years. What do this exercise and Exercise 1.10 tell you about the development of computer vision?

1.3 References

[Appenzeller 91] T Appenzeller. Ozone loss hits us where we live. *Science*, 254:645, November 1991.

[Marr 82] D Marr. *Vision—A Computational Investigation into the Human Representation and Processing of Visual Information.* Freeman, San Francisco, 1982.

Chapter 2

The digitized image and its properties

2.1 Basic concepts

We will introduce here some useful concepts and mathematical tools which will be used throughout the book. Readers without a thorough mathematical background might find some parts difficult to follow; in this case, skip the mathematical details and concentrate on the intuitive meaning of the basic concepts, which are emphasized in the text and summarized at the end of the chapter. This approach will not affect an understanding of the book.

Mathematical models are often used to describe images and other signals. A signal is a function depending on some variable with physical meaning; it can be one-dimensional (e.g., dependent on time), two-dimensional (e.g., dependent on two co-ordinates in a plane), three-dimensional (e.g., describing an object in space), or higher-dimensional. A scalar function might be sufficient to describe a monochromatic image, while vector functions are used in image processing to represent, for example, color images consisting of three component colors.

Functions we shall work with may be categorized as **continuous**, **discrete**, or **digital**. A continuous function has continuous domain and range; if the domain set is discrete, then we get a discrete function; if the range set is also discrete, then we have a digital function.

2.1.1 Image functions

By **image**, we shall understand the usual intuitive meaning—an example might be the image on the human eye retina or the image captured by a TV camera. The image can be modeled by a continuous function of two or three variables; in the simple case arguments are co-ordinates (x, y) in a plane, while if images change in time a third variable t might be added.

The image function values correspond to the brightness at image points. The function value can express other physical quantities as well (temperature, pressure distribution, distance from the observer, etc.). **Brightness** integrates different optical quantities—using brightness as a basic quantity allows us to avoid the description of the very complicated process of image formation.

The image on the human eye retina or on a TV camera sensor is intrinsically two-dimensional (2D). We shall call such a 2D image bearing information about brightness points an **intensity image**.

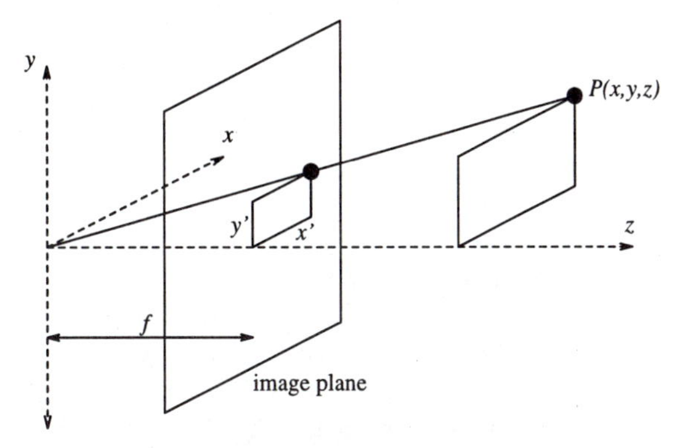

Figure 2.1: *Perspective projection geometry.*

The real world which surrounds us is intrinsically three-dimensional (3D). The 2D intensity image is the result of a **perspective projection** of the 3D scene, which is modeled by the image captured by a pin-hole camera, illustrated in Figure 2.1. In this figure, the image plane has been reflected with respect to the xy plane in order not to get a mirrored image with negative co-ordinates; the quantities x, y, and z are co-ordinates of the point P in a 3D scene in world co-ordinates, and f is the focal length of the lens. The projected point has co-ordinates (x', y') in the 2D image plane, where

$$x' = \frac{x\,f}{z} \qquad y' = \frac{y\,f}{z} \tag{2.1}$$

A non-linear perspective projection is often approximated by a linear **parallel** (or **orthographic) projection**, where $f \to \infty$. Implicitly, $z \to \infty$ too—orthographic projection is a limiting case of perspective projection for faraway objects.

When 3D objects are mapped into the camera plane by perspective projection, a lot of information disappears because such a transformation is not one-to-one. Recognizing or reconstructing objects in a 3D scene from one image is an ill-posed problem.

In due course, in Chapter 9, we shall consider more elaborate representations that attempt to recapture information about the 3D original scene that an image depicts. As may be expected, this is not a simple business and involves intermediate representations that try to establish the **depth** of points in the image. The aim is to recover a full 3D representation such as may be used in computer graphics—that is, a representation that is independent of viewpoint, and expressed in the co-ordinate system of the object rather than of the viewer. If such a representation can be recovered, then any intensity image view of the object(s) may be synthesized by standard computer graphics techniques.

Recovering information lost by perspective projection is only one, mainly geometric, problem of computer vision—a second problem is understanding image brightness. The only information available in an intensity image is the brightness of the appropriate pixel, which is dependent on a number of independent factors such as object surface reflectance properties (given by the surface material, microstructure, and marking), illumination properties, and object surface orientation with respect to a viewer and light source. It is a non-trivial and again ill-posed problem to separate these components when trying to recover the 3D geometry of an object from the intensity image.

Some scientific and technical disciplines work with 2D images directly—for example, an image of a flat specimen viewed by a microscope with transparent illumination, a character drawn on a sheet of paper, the image of a fingerprint, etc. Many basic and useful methods used in digital image analysis do not therefore depend on whether the object was originally 2D or 3D. Much of the material in this book restricts itself to the study of such methods—the problem of 3D understanding is addressed explicitly in Chapters 9 and 10.

The image formation process is described in [Horn 86]. Related disciplines are **photometry** (see Section 9.3), which is concerned with brightness measurement, and **colorimetry**, which studies light reflectance or emission depending on wavelength. Colorimetry is considered in the domain of image processing in [Pratt 78, Pratt 91].

Image processing often deals with **static** images, in which time t is constant. A monochromatic static image is represented by a continuous image function $f(x, y)$ whose arguments are two co-ordinates in the plane. Most images considered in this book will be monochromatic, static images, unless the contrary is explicitly stated. It is often the case that the extension of the techniques we will develop to the multi-spectral case is obvious.

Computerized image processing uses digital image functions which are usually represented by matrices, so co-ordinates are integer numbers. The domain of the image function is a region R in the plane

$$R = \{(x, y),\ 1 \le x \le x_m,\ 1 \le y \le y_n\} \tag{2.2}$$

where x_m, y_n represent maximal image co-ordinates. The image function has a limited domain—infinite summation or integration limits can be used, as it is assumed that the image function value is zero outside the domain R. The customary orientation of co-ordinates in an image is in the normal Cartesian fashion (horizontal x axis, vertical y axis), although the (*row*, *column*) orientation used in matrices is also often used in digital image processing.

The range of image function values is also limited; by convention, in monochromatic images the lowest value corresponds to black and the highest to white. Brightness values bounded by these limits are **gray-levels**.

The quality of a digital image grows in proportion to the spatial, spectral, radiometric, and time resolutions. The **spatial resolution** is given by the proximity of image samples in the image plane; **spectral resolution** is given by the bandwidth of the light frequencies captured by the sensor; **radiometric resolution** corresponds to the number of distinguishable gray-levels; and **time resolution** is given by the interval between time samples at which images are captured. The question of time resolution is important in dynamic image analysis, where time sequences of images are processed.

Images $f(x, y)$ can be treated as deterministic functions or as realizations of stochastic processes. Mathematical tools used in image description have roots in linear system theory, integral transformations, discrete mathematics, and the theory of stochastic processes. In this section only an overview of mathematical tools used in later explanations is presented; precise descriptions of background mathematics can be found in supporting references given to individual problems. If the reader intends to study image processing mathematics, the recommended books with which to start are [Pavlidis 82, Rosenfeld and Kak 82].

Mathematical transforms assume that the image function $f(x, y)$ is 'well-behaved', meaning that the function is integrable, has an invertible Fourier transform, etc. Existence problems of the Fourier transform of special signals (constant, impulses, non-periodic functions) [Papoulis 62] are not discussed; the Fourier transform always exists for discrete images.

2.1.2 The Dirac distribution and convolution

An ideal impulse is an important input signal that enables the use of linear mathematical theory in the domain of continuous image functions. The ideal impulse in the image plane is defined using the **Dirac distribution**, $\delta(x, y)$,

$$\int_{-\infty}^{\infty} \int_{-\infty}^{\infty} \delta(x, y) \, dx \, dy = 1 \tag{2.3}$$

and $\delta(x, y) = 0$ for all $x, y \neq 0$.

The following equation (2.4) is called the 'sifting property' of the Dirac distribution; it provides the value of the function $f(x, y)$ at the point λ, μ:

$$\int_{-\infty}^{\infty} \int_{-\infty}^{\infty} f(x, y) \, \delta(x - \lambda, y - \mu) \, dx \, dy = f(\lambda, \mu) \tag{2.4}$$

The sifting equation can be used to describe the sampling process of a continuous image function $f(x, y)$. We may express the image function as a linear combination of Dirac pulses located at the points a, b that cover the whole image plane; samples are weighted by the image function $f(x, y)$,

$$\int_{-\infty}^{\infty} \int_{-\infty}^{\infty} f(a, b) \, \delta(a - x, b - y) \, da \, db = f(x, y) \tag{2.5}$$

Convolution is an important operation in the linear approach to image analysis. The convolution g of two-dimensional functions f and h is denoted by $f * h$, and is defined by the integral

$$\begin{aligned} g(x, y) &= \int_{-\infty}^{\infty} \int_{-\infty}^{\infty} f(a, b) \, h(x - a, y - b) \, da \, db \\ &= \int_{-\infty}^{\infty} \int_{-\infty}^{\infty} f(x - a, y - b) \, h(a, b) \, da \, db \\ &= (f * h)(x, y) = (h * f)(x, y) \end{aligned} \tag{2.6}$$

Convolution is a very useful linear, translation-invariant operation. A digital image has a limited domain on the image plane, and so translational invariance is valid only for small translations—convolution is thus often used locally. The convolution expresses a linear filtering process using the filter h; linear filtering is often used in local image pre-processing and image restoration.

2.1.3 The Fourier transform

An image is a function of two parameters in a plane. One possible way to investigate its properties is to decompose the image function using a linear combination of orthonormal functions. The **Fourier transform** uses harmonic functions for the decomposition [Papoulis 62, Rosenfeld and Kak 82]. The two-dimensional Fourier transform is defined by the integral

$$F(u,v) = \int_{-\infty}^{\infty} \int_{-\infty}^{\infty} f(x,y)\, e^{-2\pi i(xu+yv)}\, dx\, dy \tag{2.7}$$

The existence conditions of the Fourier transform can be found in [Papoulis 62], but for image processing purposes it is reasonable to assume that the Fourier transform of periodic functions always exists. An inverse Fourier transform is defined by

$$f(x,y) = \int_{-\infty}^{\infty} \int_{-\infty}^{\infty} F(u,v)\, e^{2\pi i(xu+yv)}\, du\, dv \tag{2.8}$$

Parameters (x, y) denote image co-ordinates, and co-ordinates (u, v) are called **spatial frequencies**. The function $f(x, y)$ on the left-hand side of equation (2.8) can be interpreted as a linear combination of simple periodic patterns $e^{2\pi i(xu+yv)}$. The real and imaginary components of the pattern are sine and cosine functions, and the function $F(u, v)$ is a weight function which represents the influence of the elementary patterns.

Denoting the Fourier transform by an operator $\mathcal{F}$, equation (2.7) can be abbreviated to

$$\mathcal{F}\{f(x,y)\} = F(u,v)$$

Then the following properties of the Fourier transform are interesting from the image processing point of view:

- Linearity

$$\mathcal{F}\{a\, f_1(x,y) + b\, f_2(x,y)\} = a\ F_1(u,v) + b\, F_2(u,v) \tag{2.9}$$

- Shift of the origin in the image domain

$$\mathcal{F}\{f(x-a,\ y-b)\} = F(u,v)\, e^{-2\pi i(au+bv)} \tag{2.10}$$

- Shift of the origin in the frequency domain

$$\mathcal{F}\{f(x,y)\, e^{2\pi i(u_0 x+v_0 y)}\} = F(u-u_0, v-v_0) \tag{2.11}$$

- Symmetry: If $f(x, y)$ is real valued,

$$F(-u,-v) = F^*(u,v) \tag{2.12}$$

  where * denotes complex conjugate. An image function is always real valued and we can thus use the results of its Fourier transform in the first quadrant, i.e., $u \geq 0$, $v \geq 0$, without loss of generality. If in addition the image function is symmetrical, $f(x, y) = f(-x, -y)$, then the result of the Fourier transform $F(u, v)$ is a real function.

- Duality of the convolution: Convolution [equation (2.6)], and its Fourier transform are related by

$$\begin{aligned} \mathcal{F}\{(f*h)(x,y)\} &= F(u,v)\, H(u,v) \\ \mathcal{F}\{f(x,y)\, h(x,y)\} &= (F*H)(u,v) \end{aligned} \tag{2.13}$$

  This is the **Convolution theorem.**

These remarks about convolution and the Fourier transform in the continuous function domain are also valid for discrete functions (images). Integration is changed to summation in the respective equations.

The use of Fourier transforms in image analysis is pervasive. We will see in Chapter 4 how it can assist in the detection of edges by locating high frequencies (sharp changes) in the image function; it also has applications in restoring images from corruption (Section 4.4.2), fast matching using the convolution theorem (Section 5.4.1), boundary characterization (Section 6.2.3), image compression (Chapter 12), and several other areas.

2.1.4 Images as a stochastic process

Images are statistical in nature due to random changes and noise, and it is sometimes of advantage to treat image functions as realizations of a stochastic process [Papoulis 65, Rosenfeld and Kak 82]. Then questions regarding image information content and redundancy can be answered using probability distributions and correlation functions. If a probability distribution is known, we can measure image information content using **entropy**, H. Let $A = \{a_1, \ldots, a_n\}$ be the set of symbols and $p(a_k)$ be the probability of a symbol a_k, so the sum of all such probabilities is 1. Entropy is calculated as

$$H = -\sum_{k=1}^{n} p(a_k) \log_2 p(a_k) \tag{2.14}$$

The base of the logarithm in this formula determines the unit in which entropy is measured. If this base is 2 then the entropy is given in bits.

A **stochastic process** (random process, random field) is a generalization of the random variable concept. Denote a stochastic process by $f(x, y, \omega_i)$, where the event ω_i is an element from a set of all events $\Omega = \{\omega_1, \omega_2, \ldots, \omega_n\}$. In our case the event may represent an image picked from a given set of images. The probability distribution p_i corresponds to each event ω_i, and the function $f(x, y, \omega_i)$ is the random variable for fixed image co-ordinates (x, y). If an event is fixed, then the function $f(x, y, \omega_i)$ is a deterministic function which is called a **realization** of the stochastic process.

A stochastic process f is entirely described by a collection of k-dimensional distribution functions P_k, $k = 1, 2, \ldots$. The distribution function of k arguments $z_1, \ldots, z_k$ is

$$\begin{aligned} & P_k(z_1, \ldots, z_k; x_1, y_1, \ldots, x_k, y_k) \\ = \; & \mathcal{P}\{f(x_1, y_1, \omega_{i_1}) < z_1, f(x_2, y_2, \omega_{i_2}) < z_2, \ldots, f(x_k, y_k, \omega_{i_k}) < z_k\} \end{aligned} \tag{2.15}$$

where $\mathcal{P}$ denotes the probability of the conjunction of events listed in the brackets in equation (2.15).

The k-order probability density is defined by

$$\begin{aligned} & p_k(z_1, \ldots, z_k; x_1, y_1, \omega_{i_1}, \ldots, x_k, y_k, \omega_{i_k}) \\ = \; & \frac{\partial P_k(z_1, \ldots, z_k; x_1, y_1, \omega_{i_1}, \ldots, x_k, y_k, \omega_{i_k})}{\partial z_1 \ldots \partial z_k} \end{aligned} \tag{2.16}$$

The k-order distribution function or k-order probability distribution is often not known in practice—it expresses a complex relation among many events. The second-order distribution function or second-order probability density is used to relate pairs of events. Even simpler is the first-order probability density, $p_1(z; x, y)$, which can quite often be modeled if it is known how an image was obtained.

Simpler characteristics are used to describe the stochastic processes. The **mean** of the stochastic process f is defined using first-order probability density by the equation

$$\mu_f(x, y, \omega_i) = E\{f(x, y, \omega_i)\} = \int_{-\infty}^{\infty} z\, p_1(z; x, y, \omega_i)\, dz \tag{2.17}$$

(E is the mathematical expectation operator).

The **autocorrelation** and **cross correlation** functions [Papoulis 65] are often used in searching for similarities in images or image parts. The autocorrelation function R_{ff} of the random process f is defined as a mean of the product of the random variables $f(x_1, y_1, \omega_i)$ and $f(x_2, y_2, \omega_i)$,

$$R_{ff}(x_1, y_1, \omega_i, x_2, y_2, \omega_i) = E\{f(x_1, y_1, \omega_i)\, f(x_2, y_2, \omega_i)\} \tag{2.18}$$

The **autocovariance** function C_{ff} is defined as

$$\begin{aligned} &C_{ff}(x_1, y_1, \omega_i, x_2, y_2, \omega_i) \\ &= R_{ff}(x_1, y_1, \omega_i, x_2, y_2, \omega_i) - \mu_f(x_1, y_1, \omega_i)\, \mu_f(x_2, y_2, \omega_i) \end{aligned} \tag{2.19}$$

The cross correlation function R_{fg} and **cross covariance** function C_{fg} use similar definitions to equations (2.18) and (2.19). The only difference is that a point from one image (process) $f(x_1, y_1, \omega_i)$ is related to a point from another image (process) $g(x_2, y_2, \omega_j)$. Two stochastic processes are **uncorrelated** if their cross covariance function equals zero for any two points (x_1, y_1, ω_i), (x_2, y_2, ω_j).

The **stationary process** is a special stochastic process. Its properties are independent of absolute position in the image plane. The mean μ_f of the stationary process is a constant.

The autocorrelation function R_{ff} of a stationary stochastic process is translation invariant and depends only on the difference between co-ordinates $a = x_1 - x_2$, $b = y_1 - y_2$:

$$R_{ff}(x_1, y_1, \omega_i, x_2, y_2, \omega_i) = R_{ff}(a, b, \omega_i, 0, 0, \omega_i) \equiv R_{ff}(a, b, \omega_i)$$

$$R_{ff}(a, b, \omega_i) = \int_{-\infty}^{\infty} \int_{-\infty}^{\infty} f(x + a, y + b, \omega_i)\, f(x, y, \omega_i)\, dx\, dy \tag{2.20}$$

Similarly, the cross correlation function between samples of processes $f(x_1, y_1, \omega_i)$ and $g(x_2, y_2, \omega_i)$ is defined as

$$R_{fg}(x_1, y_1, \omega_i, x_2, y_2, \omega_i) = R_{fg}(a, b, \omega_i, 0, 0, \omega_i) \equiv R_{fg}(a, b, \omega_i)$$

$$R_{fg}(a, b, \omega_i) = \int_{-\infty}^{\infty} \int_{-\infty}^{\infty} f(x + a, y + b, \omega_i)\, g(x, y, \omega_i)\, dx\, dy \tag{2.21}$$

The properties of correlation functions are interesting after transformation into the frequency domain. The Fourier transform of the cross correlation function of the stationary stochastic process can be expressed as the product of the Fourier transforms of processes (images),

$$\mathcal{F}\{R_{fg}(a, b, \omega_i)\} = F^*(u, v)\, G(u, v) \tag{2.22}$$

Similarly, the autocorrelation function can be written as

$$\mathcal{F}\{R_{ff}(a,b,\omega_i)\} = F^*(u,v)\ F(u,v) = \mid F(u,v) \mid^2 \tag{2.23}$$

The Fourier transform of the autocorrelation function, equation (2.20), is given by the following expression—the result is called the **power spectrum**[1] or **spectral density**.

$$S_{ff}(u,v) = \int_{-\infty}^{\infty} \int_{-\infty}^{\infty} R_{ff}(a,b,\omega_i)\, e^{-2\pi i(au+bv)} da\, db \tag{2.24}$$

where u, v are spatial frequencies. Power spectral density communicates how much power the corresponding spatial frequency of the signal has.

Note that infinitely many functions have the same correlation function and therefore the same power spectrum as well. If an image is shifted, its power spectrum remains unchanged.

Let $g(x,y)$ be the result of the convolution of the functions $f(x,y)$ and $h(x,y)$ [equation (2.6)]. Assume that $f(x,y)$, $g(x,y)$ are stationary stochastic processes and S_{ff}, S_{gg} are their corresponding power spectral densities. If the mean of the process $f(x,y)$ is zero, then

$$S_{gg}(u,v) = S_{ff}(u,v)\ \mid H(u,v) \mid^2 \tag{2.25}$$

where $H(u,v)$ is the Fourier transform of the function $h(x,y)$. Equation (2.25) is used to describe spectral properties of a linear image filter h.

A special class of stochastic processes are the **ergodic processes** [Rosenfeld and Kak 82]. This is a stationary process for which the mean calculated from realizations is equal to the mean computed according to spatial variables. The mean from realization is calculated according to equation (2.17), when there are not usually enough data to evaluate in the domain of real images. This calculation is often replaced by calculation of the mean in image spatial co-ordinates (x,y). Be aware that this replacement is valid from the theoretical point of view only for ergodic processes.

2.1.5 Images as linear systems

Images and their processing can be modeled as a superposition of point spread functions which are represented by Dirac pulses δ [equation (2.3)]. If this image representation is used, well-developed linear system theory can be employed.

A linear system is an operator $\mathcal{L}$ which maps two (or more) input functions into an output function according to the following rule:

$$\mathcal{L}\{af_1 + bf_2\} = a\,\mathcal{L}\{f_1\} + b\,\mathcal{L}\{f_2\} \tag{2.26}$$

An image f can be expressed as a linear combination of point spread functions represented by Dirac pulses δ. Assume that the input image $f(x,y)$ is given by equation (2.5) and that the operation is translation invariant. The response $g(x,y)$ of the linear system $\mathcal{L}$ to the

[1] The concept of power spectrum can also be defined for functions for which the Fourier transform is not defined.

input image $f(x,y)$ is given by

$$\begin{aligned} g(x,y) &= \mathcal{L}\{f(x,y)\} \\ &= \int_{-\infty}^{\infty}\int_{-\infty}^{\infty} f(a,b)\,\mathcal{L}\{\delta(x-a,y-b)\}\,da\,db \\ &= \int_{-\infty}^{\infty}\int_{-\infty}^{\infty} f(a,b)\,h(x-a,y-b)\,da\,db \\ &= f(*h)\,(x,y) \end{aligned} \tag{2.27}$$

where $h(x,y)$ is the impulse response of the linear system $\mathcal{L}$. In other words the output of the linear system $\mathcal{L}$ is expressed as the convolution of the input image f with an impulse response h of the linear system $\mathcal{L}$. If the Fourier transform is applied to equation (2.27), the following equation is obtained:

$$G(u,v) = F(u,v)\,H(u,v) \tag{2.28}$$

Equation (2.28) is often used in image pre-processing to express the behavior of smoothing or sharpening operations, and is considered further in Chapter 4.

It is important to remember that real images are not in fact linear—both the image co-ordinates and values of the image function (brightness) are limited. Real images always have limited size, and the number of brightness levels is also finite. Nevertheless, images can be approximated by linear systems in many cases.

2.2 Image digitization

An image to be processed by computer must be represented using an appropriate discrete data structure, for example, a matrix. An image captured by a sensor is expressed as a continuous function $f(x,y)$ of two co-ordinates in the plane. Image digitization means that the function $f(x,y)$ is **sampled** into a matrix with M rows and N columns. Image **quantization** assigns to each continuous sample an integer value—the continuous range of the image function $f(x,y)$ is split into K intervals. The finer the sampling (i.e., the larger M and N) and quantization (the larger K), the better the approximation of the continuous image function $f(x,y)$ achieved.

Two questions should be answered in connection with image function sampling. First, the sampling period should be determined—this is the distance between two neighboring sampling points in the image. Second, the geometric arrangement of sampling points (sampling grid) should be set.

2.2.1 Sampling

A continuous image function $f(x,y)$ can be sampled using a discrete grid of sampling points in the plane. A second possibility is to expand the image function using some orthonormal function as a base—the Fourier transform is an example. The coefficients of this expansion then represent the digitized image. Here we consider only the first possibility, but a detailed description of both approaches can be found in [Rosenfeld and Kak 82].

The image is sampled at points $x = j\,\Delta x$, $y = k\,\Delta y$, for $j = 1,\ldots,M$, $k = 1,\ldots,N$. Two neighboring sampling points are separated by distance Δx along the x axis and Δy along the

y axis. Distances Δx and Δy are called the **sampling interval** (on the x or y axis), and the matrix of samples $f(j\,\Delta x, k\,\Delta y)$ constitutes the discrete image. The ideal sampling $s(x,y)$ in the regular grid can be represented using a collection of Dirac distributions δ

$$s(x,y) = \sum_{j=1}^{M} \sum_{k=1}^{N} \delta(x - j\,\Delta x,\; y - k\,\Delta y) \tag{2.29}$$

The sampled image $f_s(x,y)$ is the product of the continuous image $f(x,y)$ and the sampling function $s(x,y)$

$$\begin{aligned} f_s(x,y) &= f(x,y)s(x,y) \\ &= f(x,y) \sum_{j=1}^{M} \sum_{k=1}^{N} \delta(x - j\,\Delta x, y - k\,\Delta y) \end{aligned} \tag{2.30}$$

The collection of Dirac distributions in equation (2.30) can be regarded as periodic with period Δx, Δy and expanded into a Fourier series. To fulfill the periodicity condition of the Fourier transform, assume for a moment that the sampling grid covers the whole plane (infinite limits).

$$\mathcal{F}\{\sum_{j=-\infty}^{\infty} \sum_{k=-\infty}^{\infty} \delta(x - j\,\Delta x, y - k\,\Delta y)\} = \sum_{m=-\infty}^{\infty} \sum_{n=-\infty}^{\infty} a_{mn}\, e^{2\pi i(\frac{mx}{\Delta x} + \frac{ny}{\Delta y})} \tag{2.31}$$

Coefficients of the Fourier expansion a_{mn} can be calculated as

$$a_{mn} = \frac{1}{\Delta x\,\Delta y} \int_{-\frac{\Delta x}{2}}^{\frac{\Delta x}{2}} \int_{-\frac{\Delta y}{2}}^{\frac{\Delta y}{2}} \sum_{j=-\infty}^{\infty} \sum_{k=-\infty}^{\infty} \delta(x - j\,\Delta x, y - k\,\Delta y)\, e^{-2\pi i(\frac{mx}{\Delta x} + \frac{ny}{\Delta y})} dx\, dy \tag{2.32}$$

Then, noting that only the term for $j = 0$ and $k = 0$ in the sum is non-zero in the range of integration, we see that

$$a_{mn} = \frac{1}{\Delta x\,\Delta y} \int_{-\frac{\Delta x}{2}}^{\frac{\Delta x}{2}} \int_{-\frac{\Delta y}{2}}^{\frac{\Delta y}{2}} \delta(x,y)\, e^{-2\pi i(\frac{mx}{\Delta x} + \frac{ny}{\Delta y})} dx\, dy \tag{2.33}$$

Noting that the integral in equation (2.33) is uniformly 1 [Rosenfeld and Kak 82], we see that the coefficients a_{mn} can be expressed as

$$a_{mn} = \frac{1}{\Delta x\,\Delta y} \tag{2.34}$$

Thus, as $M, N \to \infty$, equation (2.30) can be rewritten using the derived value of the coefficients a_{mn}:

$$f_s(x,y) = f(x,y)\, \frac{1}{\Delta x\,\Delta y} \sum_{m=-\infty}^{\infty} \sum_{n=-\infty}^{\infty} e^{2\pi i(\frac{mx}{\Delta x} + \frac{ny}{\Delta y})} \tag{2.35}$$

Equation (2.35) can be expressed in the frequency domain using equation (2.11):

$$F_s(u,v) = \frac{1}{\Delta x\,\Delta y} \sum_{j=-\infty}^{\infty} \sum_{k=-\infty}^{\infty} F(u - \frac{j}{\Delta x}, v - \frac{k}{\Delta y}) \tag{2.36}$$

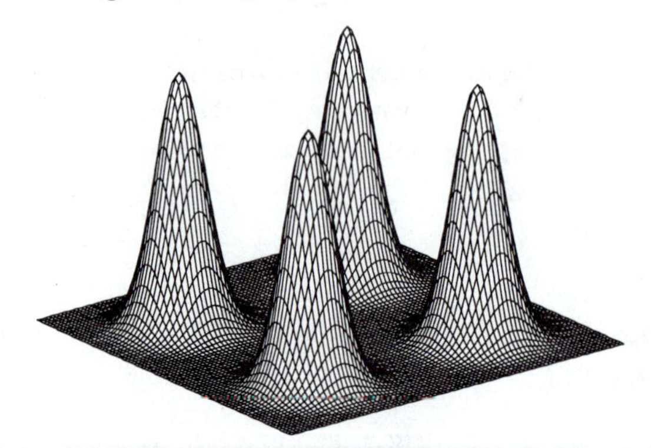

Figure 2.2: *2D spectrum of sampled image function.*

Thus the Fourier transform of the sampled image is the sum of periodically repeated Fourier transforms $F(u,v)$ of the image (see Figure 2.2).

Periodic repetition of the Fourier transform result $F(u,v)$ may under certain conditions cause distortion of the image, which is called **aliasing**; this happens when individual digitized components $F(u,v)$ overlap. The situation without spectra overlapping is shown in Figure 2.2, which assumes that the image function $f(x,y)$ has a **band-limited** spectrum, meaning that its Fourier transform $F(u,v)$ is zero outside a certain interval of frequencies $|u| > U$, $|v| > V$.

Overlapping of the periodically repeated results of the Fourier transform $F(u,v)$ of an image with a band-limited spectrum can be prevented if the sampling interval is chosen such that

$$\Delta x < \frac{1}{2U}, \qquad \Delta y < \frac{1}{2V} \tag{2.37}$$

This is the **Shannon sampling theorem**, known from signal processing theory or control theory. The theorem has a simple physical interpretation in image analysis: The sampling interval should be chosen in size such that it is less than half of the smallest interesting detail in the image.

The sampling function is not the Dirac distribution in real digitizers—limited impulses (quite narrow ones with limited amplitude) are used instead. Assume a rectangular sampling grid which consists of $M \times N$ such equal and non-overlapping impulses $h_s(x,y)$ with sampling period Δx, Δy; this function simulates realistically the real image sensors. Outside the sensitive area of the sensor, element $h_s(x,y) = 0$. Values of image samples are obtained by integration of the product $f(x,y)\, h_s(x,y)$—in reality this integration is done on the surface of the sensor sensitive element. The sampled image is then given by the convolution computed in discrete co-ordinates $j\,\Delta x$, $k\,\Delta y$,

$$f_s(x,y) = \sum_{j=1}^{M} \sum_{k=1}^{N} f(x,y)\, h_s(x - j\,\Delta x, y - k\,\Delta y) \tag{2.38}$$

The sampled image f_s is distorted by the convolution of the original image f and the limited impulse h_s. The distortion of the frequency spectrum of the function F_s can be expressed using the Fourier transform,

$$F_s(u,v) = \frac{1}{\Delta x\, \Delta y} \sum_{j=1}^{M} \sum_{k=1}^{N} F(u - \frac{j}{\Delta x}, v - \frac{k}{\Delta y})\, H_s(u - \frac{j}{\Delta x}, v - \frac{k}{\Delta y}) \tag{2.39}$$

In real image digitizers, a sampling interval about ten times smaller than that indicated by the Shannon sampling theorem [equation (2.37)] is used. The reason is that algorithms which reconstruct the continuous image on a display from the digitized image function use only a step function [Pavlidis 82]; i.e., a line is created from pixels represented by individual squares.

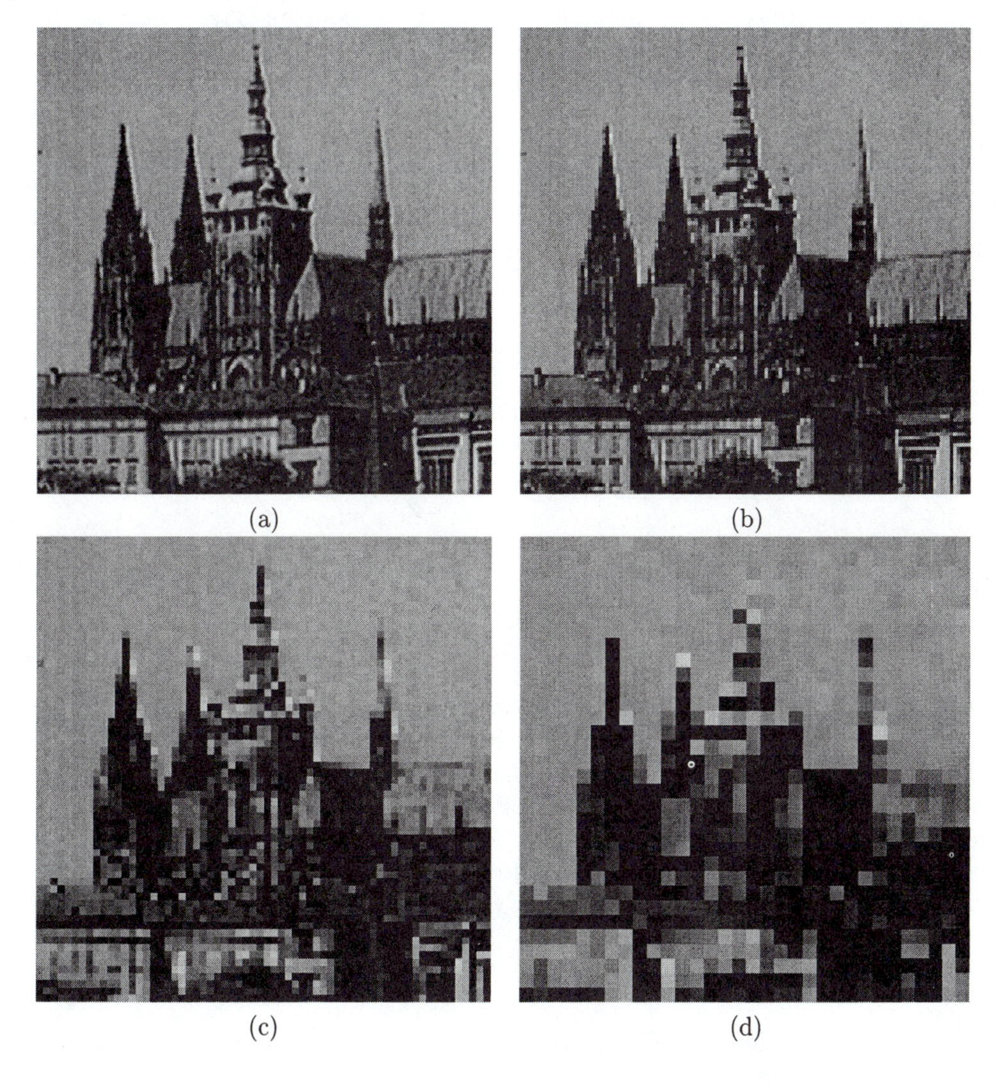

Figure 2.3: *Digitizing: (a) 256 × 256, (b) 128 × 128, (c) 64 × 64, (d) 32 × 32. Images have been enlarged to the same size to illustrate the loss of detail.*

A demonstration with an image of 256 gray-levels will illustrate the effect of sparse sampling. Figure 2.3a shows a monochromatic image with 256 × 256 pixels; Figure 2.3b shows the same scene digitized into a reduced grid of 128 × 128 pixels, Figure 2.3c into 64 × 64 pixels, and Figure 2.3d into 32 × 32 pixels. Decline in image quality is clear from Figures 2.3a to 2.3d. Quality may be improved by viewing from a distance and with screwed-up eyes, implying that the undersampled images still hold substantial information. Much of this visual degradation is caused by aliasing in the reconstruction of the continuous image function for display. This display can be improved by the reconstruction algorithm interpolating brightness values in neighboring pixels; this technique is called **anti-aliasing** and is often used in computer graphics [Rogers 85]; if anti-aliasing is used, the sampling interval can be brought near to the theoretical value of Shannon's theorem [equation (2.37)]. In real image processing devices, anti-aliasing is rarely used because of its computational requirements.

If quality comparable to an ordinary television image is required, sampling into a 512×512 grid is used; this is the reason most image frame grabbers use this (or higher) resolution.

A continuous image is digitized at **sampling points**. These sampling points are ordered in the plane, and their geometric relation is called the **grid**. The digital image is then a data structure, usually a matrix. Grids used in practice are usually square (Figure 2.4a) or hexagonal (Figure 2.4b).

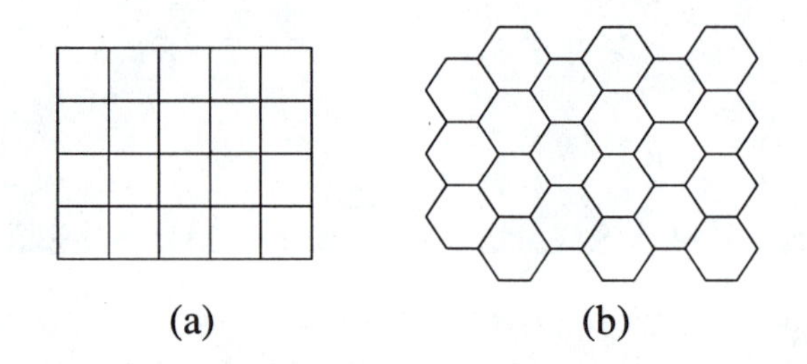

Figure 2.4: *(a) Square grid; (b) hexagonal grid.*

It is important to distinguish the grid from the raster; the **raster** is the grid on which a neighborhood relation between points is defined[2].

One infinitely small sampling point in the grid corresponds to one picture element (**pixel**) in the digital image. The set of pixels together covers the entire image; however, the pixel captured by a real digitization device has finite size, since the sampling function is not a collection of ideal Dirac impulses but a collection of limited impulses [equation (2.38)]. The pixel is a unit which is not further divisible[3] from the image analysis point of view. We shall often refer to a pixel as a 'point' as well.

2.2.2 Quantization

A value of the sampled image $f_s(j\,\Delta x, k\,\Delta y)$ is expressed as a digital value in image processing. The transition between continuous values of the image function (brightness) and its digital equivalent is called **quantization**. The number of quantization levels should be high enough to permit human perception of fine shading details in the image.

[2]E.g., if 4-neighborhoods are used on the square grid, the square raster is obtained. Similarly, if 8-neighborhoods are used on the the same square grid, then the octagonal raster is obtained. These 4-neighborhood and 8-neighborhood concepts are introduced in Section 2.3.1.

[3]In some case, the properties of an image at sub-pixel resolution can be computed. This is achieved by approximating the image function by a continuous function.

Most digital image processing devices use quantization into k equal intervals. If b bits are used to express the values of the pixel brightness, then the number of brightness levels is $k = 2^b$. Eight bits per pixel are commonly used, although some systems use six or four bits. A binary image pixel, which is either black or white, can be represented by one bit. Specialized measuring devices use 12 and more bits per pixel, although these are becoming more common.

The occurrence of false contours is the main problem in images which have been quantized with insufficient brightness levels. This effect arises when the number of brightness levels is lower than that which humans can easily distinguish. This number is dependent on many factors—for example, the average local brightness—but displays which avoid this effect will normally provide a range of at least 100 intensity levels [Gonzalez and Wintz 87]. This problem can be reduced when quantization into intervals of unequal length is used; the size of intervals corresponding to less probable brightnesses in the image is enlarged. These gray-scale transformation techniques are considered in Section 4.1.2.

An efficient representation of brightness values in digital images requires that eight bits, four bits, or one bit are used per pixel, meaning that one, two, or eight pixel brightnesses can be stored in one byte of computer memory.

Figures 2.3a and 2.5a-d demonstrate the effect of reducing the number of brightness levels in an image. An original image with 256 brightness levels is in Figure 2.3a. If the number of brightness levels is reduced to 64 (Figure 2.5a), no degradation is perceived. Figure 2.5b uses 16 brightness levels and false contours begin to emerge, and this becomes clearer in Figure 2.5c with four brightnesses and in Figure 2.5d with only two.

2.2.3 Color images

Color is a property of enormous importance to human visual perception, but historically it has not been particularly used in digital image processing. The reason for this has been the cost of suitable hardware, but since the 1980s this has declined sharply, and color images are now routinely accessible via TV cameras or scanners. The related problems of internal storage of large arrays associated with multi-spectral data have also receded with the reduction in cost of memory. Color display is, of course, the default in most computer systems. For our purposes this is useful, since a monochromatic image may not contain enough information for many applications, while color or **multi-spectral images** can often help.

Color is connected with the ability of objects to reflect electromagnetic waves of different wavelengths; the chromatic spectrum spans the electromagnetic spectrum from approximately 400 nm to 700 nm. Humans detect colors as combinations of the **primary colors** red, green, and blue, which for the purpose of standardization have been defined as 700 nm, 546.1 nm, and 435.8 nm respectively [Pratt 78], although this standardization does not imply that all colors can be synthesized as combinations of these three.

Hardware will generally deliver or display color via an **RGB model** (referring to red, green, and blue); thus a particular pixel may have associated with it a three-dimensional vector (r, g, b) which provides the respective color intensities, where $(0, 0, 0)$ is black, (k, k, k) is white, $(k, 0, 0)$ is 'pure' red, and so on—k here is the quantization granularity for each primary (256 is common). This implies a color space of k^3 distinct colors (2^{24} if $k = 256$) which not all displays, particularly older ones, can accommodate. For this reason, for display purposes, it is common to specify a subset of this space that is actually used; this subset is often called a palette.

(a) (b) (c) (d)

Figure 2.5: *Brightness levels: (a) 64, (b) 16, (c) 4, (d) 2.*

The RGB model may be thought of as a 3D co-ordinatization of color space (see Figure 2.6); note the secondary colors which are combinations of two pure primaries.

Most image sensors will provide data according to this model; the image can be captured by several sensors, each of which is sensitive to a rather narrow band of wavelengths, and the image function at the sensor output is just as in the simple case (see Section 2.2.1): Each spectral band is digitized independently and is represented by an individual digital image function as if it were a monochromatic image. Sometimes images are delivered using a similar approach, but with different spectral components—for instance, the LANDSAT 4

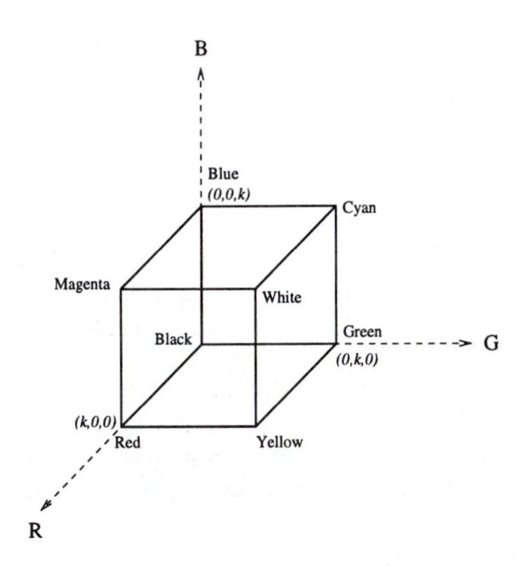

Figure 2.6: *Color space co-ordinatized as red, green, blue.*

satellite transmits digitized images in five spectral bands from near-ultraviolet to infrared.

Other color models turn out to be equally important, if less intuitive. The simplest to explain is **CMY**—Cyan, Magenta, Yellow—which is based on the secondaries and is used to construct a subtractive color scheme. For example, yellow subtracts blue from pure white light, while a combination of yellow and magenta subtracts blue and green from white (to provide red). This **pigment** approach is used in combining inks in color hard-copy devices.

The **YIQ** model (sometimes referred to as IYQ) is useful in color TV broadcasting, and is a simple linear transform of an RGB representation:

$$\begin{pmatrix} Y \\ I \\ Q \end{pmatrix} = \begin{pmatrix} 0.299 & 0.587 & 0.114 \\ 0.596 & -0.275 & -0.321 \\ 0.212 & -0.523 & 0.311 \end{pmatrix} \begin{pmatrix} R \\ G \\ B \end{pmatrix}$$

(with the inverse computed in the obvious manner). This model is useful since the Y component provides all that is necessary for a monochrome display; further, it exploits advantage of properties of the human visual system, in particular our sensitivity to **luminance**, the perceived energy of a light source. Details of this model and its use may be found in relevant texts [Pritchard 77, Smith 78].

The alternative model of most relevance to image processing is **HSI** (sometimes referred to as IHS) - Hue, Saturation, and Intensity. Hue refers to the perceived color (technically, the dominant wavelength), for example, 'purple' or 'orange', and saturation measures its dilution by white light, giving rise to 'light purple', 'dark purple', etc. HSI decouples the intensity information from the color, while hue and saturation correspond to human perception, thus making this representation very useful for developing image processing algorithms. This will become clearer as we proceed to describe image enhancement algorithms (for example, equalization Algorithm 4.1), which if applied to each component of an RGB model would corrupt the human sense of color, but which would work more or less as expected if applied to the intensity component of HSI (leaving the color information unaffected).

To convert from an RGB representation (r, g, b) to HSI, assume we have normalized the primary measurements so that

$$0 \leq r, g, b \leq 1 \tag{2.40}$$

Then set

$$\begin{aligned} i &= \frac{r+g+b}{3} \\ h &= \cos^{-1}\left\{\frac{\frac{1}{2}[(r-g)+(r-b)]}{[(r-g)^2+(r-b)(g-b)]^{1/2}}\right\} \\ s &= 1 - \frac{3}{r+g+b} min(r,g,b) \end{aligned}$$

If $b/i > g/i$, then set $h := 2\pi - h$. These measurements are normalized to the range $[0, 1]$ if we also set $h := h/2\pi$. Note that if $r = g = b$, then h is undefined; and if $i = 0$, s is undefined.

To convert from HSI to RGB, there are three cases to consider. Writing $H = 2\pi h$:

Case $0 < H \leq \frac{2\pi}{3}$

$$\begin{aligned} r &= i\left[1 + \frac{s\cos H}{\cos(\pi/3 - H)}\right] \\ b &= i(1-s) \\ g &= 3i\left(1 - \frac{r+b}{3i}\right) \end{aligned}$$

Case $\frac{2\pi}{3} < H \leq 2 \times \frac{2\pi}{3}$

$$\begin{aligned} H &= h - \frac{2\pi}{3} \\ g &= i\left[1 + \frac{s\cos H}{\cos(\pi/3 - H)}\right] \\ r &= i(1-s) \\ b &= 3i\left(1 - \frac{r+g}{3i}\right) \end{aligned}$$

Case $2 \times \frac{2\pi}{3} < H \leq 2\pi$

$$\begin{aligned} H &= h - \frac{4\pi}{3} \\ b &= i\left[1 + \frac{s\cos H}{\cos(\pi/3 - H)}\right] \\ g &= i(1-s) \\ r &= 3i\left(1 - \frac{g+b}{3i}\right) \end{aligned}$$

These derivations are based on consideration of a particular, and well-known, chromaticity triangle. A full derivation is given in [Gonzalez and Woods 92].

A fuller account of color and its applications in computer vision may be found in [Luong 93].

2.3 Digital image properties

A digital image has several properties, both metric and topological, which are somewhat different from those of continuous two-dimensional functions with which we are familiar from basic calculus. Another feature of difference is human perception of images, since judgment of image quality is also important.

2.3.1 Metric and topological properties of digital images

A digital image consists of picture elements with finite size—these pixels carry information about the brightness of a particular location in the image. Usually (and we assume this hereafter) pixels are arranged into a rectangular sampling grid. Such a digital image is represented by a two-dimensional matrix whose elements are integer numbers corresponding to the quantization levels in the brightness scale.

Some intuitively clear properties of continuous images have no straightforward analogy in the domain of digital images [Pavlidis 77, Ballard and Brown 82]. **Distance** is an important example. The distance between points with co-ordinates (i,j) and (h,k) may be defined in several different ways; the **Euclidean distance** D_E known from classical geometry and everyday experience is defined by

$$D_E[(i,j),(h,k)] = \sqrt{(i-h)^2 + (j-k)^2} \qquad (2.41)$$

The advantage of the Euclidean distance is the fact that it is intuitively obvious. The disadvantages are costly calculation due to the square root, and its non-integer value.

The distance between two points can also be expressed as the minimum number of elementary steps in the digital grid which are needed to move from the starting point to the end point. If only horizontal and vertical moves are allowed, the distance D_4 is obtained. D_4 is also called 'city block' distance because of the analogy with the distance between two locations in a city with a rectangular grid of streets and closed blocks of houses.

$$D_4[(i,j),(h,k)] = | i-h | + | j-k | \qquad (2.42)$$

If moves in diagonal directions are allowed in the digitization grid, we obtain the distance D_8, often called 'chessboard' distance. The distance D_8 is equal to the number of moves of the king on the chessboard from one part to another.

$$D_8[(i,j),(h,k)] = \max\{| i-h |, | j-k |\} \qquad (2.43)$$

Any of these metrics may be used as the basis of **chamfering**, in which the distance of pixels from some image subset (perhaps describing some feature) is generated. The resulting 'image' has pixel values of 0 for elements of the relevant subset, low values for close pixels, and then high values for pixels remote from it—the appearance of this array gives the name to the technique. Chamfering is of value in **chamfer matching**, which is described in Section 5.4. The following two-pass algorithm is derived from one based on a simplification of the Euclidean metric [Barrow et al. 77], and is originally described in [Rosenfeld and Pfalz 68].

AL	AL	BR
AL	AL BR	BR
AL	BR	BR

Figure 2.7: *Pixel neighborhoods used in chamfering—pixel p is the central one.*

Algorithm 2.1: Chamfering

1. To chamfer a subset S of an image of dimension $M \times N$ with respect to a distance metric D, where D is one of D_4 or D_8, construct an $M \times N$ array F with elements corresponding to the set S to 0, and each other element to infinity.

2. Pass through the image row by row, from top to bottom and left to right. For each neighboring pixel above and to the left, illustrated in Figure 2.7 by the set AL, set

$$F(p) = \min_{q \epsilon AL}[F(p), D(p,q) + F(q)]$$

3. Pass through the image row by row, from bottom to top and right to left. For each neighboring pixel below and to the right, illustrated in Figure 2.7 by the set BR, set

$$F(p) = \min_{q \epsilon BR}[F(p), D(p,q) + F(q)]$$

4. The array F now holds a chamfer of the subset S.

This algorithm needs obvious adjustments at the image boundaries, where the sets AL and BR are truncated.

Pixel **adjacency** is another important concept in digital images. Any two pixels are called **4-neighbors** if they have distance $D_4 = 1$ from each other. Analogously, **8-neighbors** are two pixels with $D_8 = 1$. 4-neighbors and 8-neighbors are illustrated in Figure 2.8.

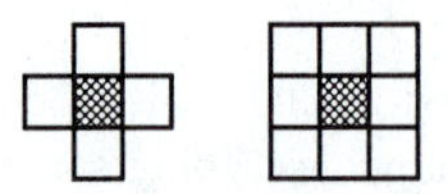

Figure 2.8: *Pixel neighborhoods.*

It will become necessary to consider important sets consisting of several adjacent pixels—**regions**. For those familiar with set theory, we can simply say that a region is a contiguous set. More descriptively, we can define a **path** from pixel P to pixel Q as a sequence of points $A_1, A_2, \ldots, A_n$, where $A_1 = P$, $A_n = Q$, and A_{i+1} is a neighbor of A_i, $i = 1, \ldots, n-1$; then a **region** is a set of pixels in which there is a path between any pair of its pixels, all of whose pixels also belong to the set.

If there is a path between two pixels in the image, these pixels are called **contiguous**. Alternatively, we can say that a region is the set of pixels where each pair of pixels in the set is contiguous. The relation *to be contiguous* is reflexive, symmetric, and transitive and therefore defines a decomposition of the set (in our case image) into equivalence classes (regions).

Assume that R_i are disjoint regions in the image which were created by the relation *to be contiguous*, and further assume (to avoid special cases) that these regions do not touch the image limits (meaning the rows or columns in the image matrix with minimum and maximum indices). Let region R be the union of all regions R_i; it then becomes sensible to define a set R^C which is the set complement of region R with respect to the image. The subset of R^C which is contiguous with the image limits is called **background**, and the rest of the complement R^C is called **holes**[4]. If there are no holes in a region we call it a **simply contiguous** region. A region with holes is called **multiply contiguous**.

Note that the concept of *region* uses only the property *to be contiguous*. Secondary properties can be attached to regions which originate in image data interpretation. It is common to call some regions in the image **objects**; a process which determines which regions in an image correspond to objects in the world is part of image **segmentation** and is discussed in Chapter 5.

The brightness of a pixel is a very simple property which can be used to find objects in some images; if, for example, a pixel is darker than some predefined value (threshold), then it belongs to the object. All such points which are also contiguous constitute one object. A hole consists of points which do not belong to the object and are surrounded by the object, and all other points constitute the background.

An example is the black printed text on this white sheet of paper, in which individual letters are objects. White areas surrounded by the letter are holes, for example, the area inside a letter 'o'. Other white parts of the paper are background.

These neighborhood and contiguity definitions on the square grid create some paradoxes. Figure 2.9 shows two digital line segments with 45° slope. If 4-connectivity is used, the lines are not contiguous at each of their points. An even worse conflict with intuitive understanding of line properties is also illustrated; two perpendicular lines do intersect in one case (upper right intersection) and do not intersect in another case (lower left), as they do not have any common point (i.e., their set intersection is empty).

It is known from Euclidean geometry that each closed curve (e.g., a circle) divides the plane into two non-contiguous regions. If images are digitized in a square grid using 8-connectivity, we can draw a line from the inner part of a closed curve into the outer part which does not intersect the curve (Figure 2.10). This implies that the inner and outer parts of the curve constitute only one region because all pixels of the line belong to only one region. This is another paradox.

One possible solution to contiguity paradoxes is to treat objects using 4-neighborhoods and background using 8-neighborhoods (or vice versa). More exact treatment of digital images

[4]Some image processing literature does not distinguish holes and background and calls both of them background.

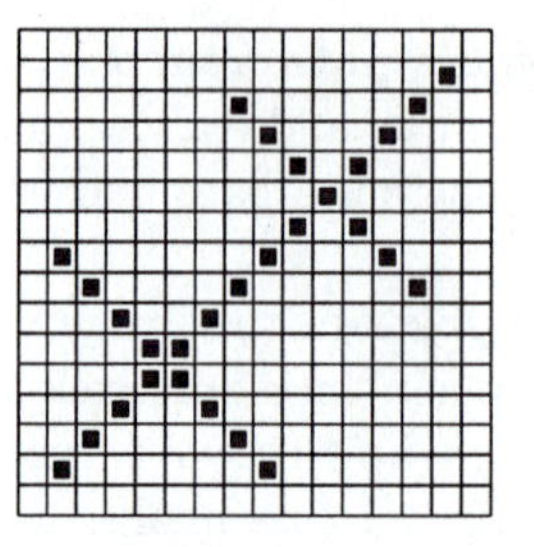

Figure 2.9: *Digital line.*

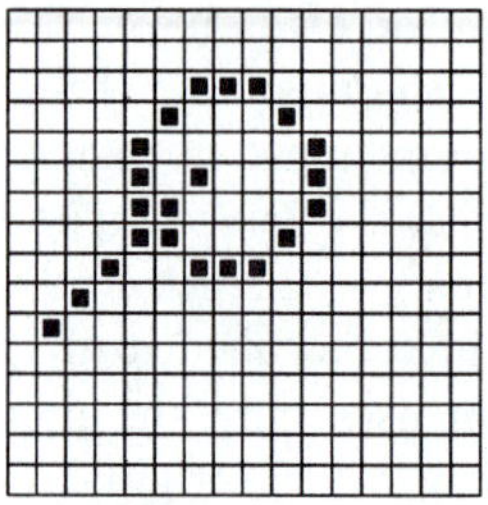

Figure 2.10: *Closed curve paradox.*

paradoxes and their solution for binary images and images with more brightness levels can be found in [Pavlidis 77, Horn 86].

These problems are typical on square grids—a hexagonal grid (see Figure 2.4), however, solves many of them. Any point in the hexagonal raster has the same distance to all its six neighbors. There are some problems peculiar to the hexagonal raster as well—for instance, it is difficult to express a Fourier transform on it.

An alternative approach to the connectivity problems is to use discrete topology based on cell complexes [Kovalevsky 89]. This approach develops a complete strand of image encoding and segmentation that deals with many issues we shall come to later, such as the representation of boundaries and regions. The idea, first proposed by Riemann in the nineteenth century, considers families of sets of different dimensions; points, which are 0-dimensional sets, may then be assigned to sets containing higher-dimensional structures (such as pixel arrays), which permits the removal of the paradoxes we have seen.

For reasons of simplicity and ease of processing, most digitizing devices use a square grid despite the stated drawbacks.

The **border** of a region is another important concept in image analysis. The border of a region R is the set of pixels within the region that have one or more neighbors outside R. The definition corresponds to an intuitive understanding of the border as a set of points at the limit of the region. This definition of border is sometimes referred to as **inner border** to distinguish it from the **outer border**, that is, the border of the background (i.e., its complement) of the region.

An **edge** is a further concept. This is a local property of a pixel and its immediate neighborhood—it is a vector given by a magnitude and direction. Images with many brightness levels are used for edge computation, and the gradient of the image function is used to compute edges. The edge direction is perpendicular to the gradient direction which points in the direction of image function growth. Edges are considered in detail in Section 4.3.2.

Note that there is a difference between *border* and *edge*. The border is a global concept related to a region, while *edge* expresses local properties of an image function. The border and edges are related as well. One possibility for finding boundaries is chaining the significant edges (points with high gradient of the image function). Methods of this kind are described in Section 5.2.

The edge property is attached to one pixel and its neighborhood—sometimes it is of advantage to assess properties between pairs of neighboring pixels, and the concept of the **crack edge** comes from this idea. Four crack edges are attached to each pixel, which are defined by its relation to its 4-neighbors. The direction of the crack edge is that of increasing brightness, and is a multiple of 90°, while its magnitude is the absolute difference between the brightness of the relevant pair of pixels. Crack edges are illustrated in Figure 2.11 and will be used later on in image segmentation, Chapter 5.

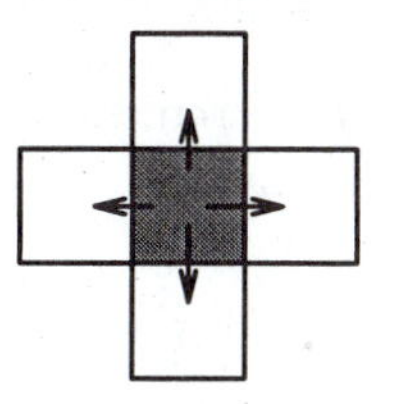

Figure 2.11: *Crack edges.*

Topological properties of images are invariant to *rubber sheet transformations.* Imagine a small rubber balloon with an object painted on it; topological properties of the object are those which are invariant to the arbitrary stretching of the rubber sheet. Stretching does not change contiguity of the object parts and does not change the number of holes in regions. One such image property is the **Euler–Poincaré characteristic**, defined as the difference between the number of regions and the number of holes in them. Further properties that are not rubber sheet invariant are described in Section 6.3.1.

A **convex hull** is a concept used to describe topological properties of objects. The convex hull is the smallest region which contains the object, such that any two points of the region can be connected by a straight line, all points of which belong to the region. For example, consider an object whose shape resembles a letter 'R' (see Figure 2.12). Imagine a thin rubber band pulled around the object; the shape of the rubber band provides the convex hull of the object. Calculation of the convex hull is described in Section 6.3.3.

Figure 2.12: *Description using topological components: An 'R' object, its convex hull, and the associated lakes and bays.*

An object with non-regular shape can be represented by a collection of its topological components. The set inside the convex hull which does not belong to an object is called the **deficit of convexity**; this can be split into two subsets. First, **lakes** (Figure 2.12 and hatched) are fully surrounded by the object; and second, **bays** are contiguous with the border of the convex hull of the object.

The convex hull, lakes, and bays are sometimes used for object description; these features are used in Chapter 6 (object description) and in Chapter 11 (mathematical morphology).

2.3.2 Histograms

The **brightness histogram** $h_f(z)$ of an image provides the frequency of the brightness value z in the image—the histogram of an image with L gray-levels is represented by a one-dimensional array with L elements.

Algorithm 2.2: Computing the brightness histogram

1. Assign zero values to all elements of the array h_f.

2. For all pixels (x, y) of the image f, increment $h_f[f(x, y)]$ by 1.

Recalling that an image can be analyzed as the realization of a stochastic process, we might want to find a first-order density function $p_1(z, x, y)$ to indicate that pixel (x, y) has brightness z. If the position of the pixel is not of interest, we obtain a density function $p_1(z)$, and the brightness histogram is its estimate.

The histogram is often displayed as a bar graph. The histogram of the image from Figure 2.3 is given in Figure 2.13.

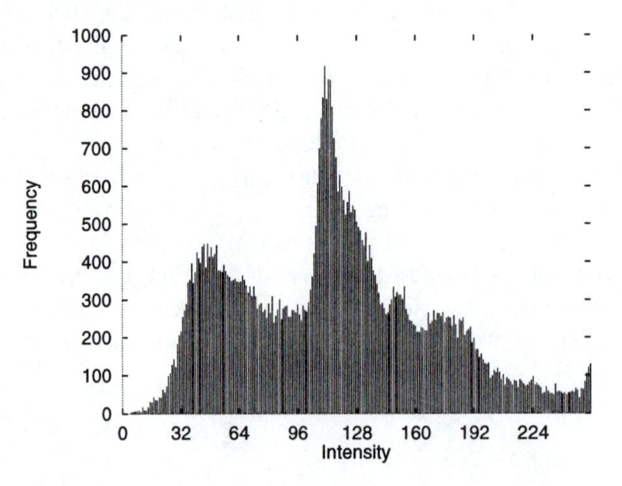

Figure 2.13: *A brightness histogram.*

The histogram is usually the only global information about the image which is available. It is used when finding optimal illumination conditions for capturing an image, gray-scale transformations, and image segmentation to objects and background. Note that one histogram may correspond to several images; for instance, change of object position on a constant background does not affect the histogram.

The histogram of a digital image typically has many local minima and maxima, which may complicate its further processing. This problem can be avoided by local smoothing of the histogram; this may be done, for example, using local averaging of neighboring histogram elements as the base, so that a new histogram $h'_f(z)$ is calculated according to

$$h'_f(z) = \frac{1}{2K+1} \sum_{i=-K}^{K} h_f(z+i) \tag{2.44}$$

where K is a constant representing the size of the neighborhood used for smoothing. This algorithm would need some boundary adjustment, and carries no guarantee of removing all local minima. Other techniques for smoothing exist, notably Gaussian blurring; in the case of a histogram, this would be a one-dimensional simplification of the 2D Gaussian blur, equation (4.52), which will be introduced in Section 4.3.3.

2.3.3 Visual perception of the image

Anyone who creates or uses algorithms or devices for digital image processing should take into account the principles of human image perception. If an image is to be analyzed by a human the information should be expressed using variables which are easy to perceive; these are psycho-physical parameters such as contrast, border, shape, texture, color, etc. Humans will find objects in images only if they may be distinguished effortlessly from the background. A detailed description of the principles of human perception can be found in [Cornsweet 70, Winston 75, Marr 82, Levine 85]. Human perception of images provokes many illusions, the understanding of which provides valuable clues about visual mechanisms. Some of the better-known illusions will be mentioned here—the topic is covered exhaustively from the point of view of computer vision in [Frisby 79].

The situation would be relatively easy if the human visual system had a linear response to composite input stimuli—i.e., a simple sum of individual stimuli. A decrease of some stimulus, e.g., area of the object in the image, could be compensated by its intensity, contrast, or duration. In fact, the sensitivity of human senses is roughly logarithmically proportional to the intensity of an input signal. In this case, after an initial logarithmic transformation, response to composite stimuli can be treated as linear.

Contrast

Contrast is the local change in brightness and is defined as the ratio between average brightness of an object and the background brightness. The human eye is logarithmically sensitive to brightness, implying that for the same perception, higher brightness requires higher contrast.

Apparent brightness depends very much on the brightness of the local background; this effect is called conditional contrast. Figure 2.14 illustrates this with two small squares of the same brightness on a dark and light background. Humans perceive the brightness of the small squares as different.

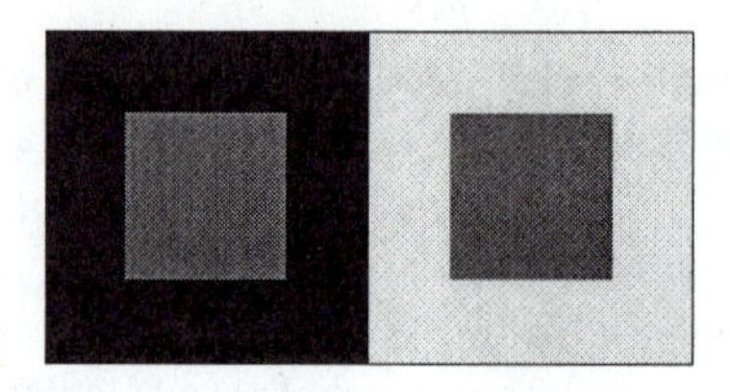

Figure 2.14: *Conditional contrast.*

Acuity

Acuity is the the ability to detect details in the image. The human eye is less sensitive to slow and fast changes in brightness in the image plane but is more sensitive to intermediate changes. Acuity also decreases with increasing distance from the optical axis.

Resolution in an image is firmly bounded by the resolution ability of the human eye; there is no sense in representing visual information with higher resolution than that of the viewer. Resolution in optics is defined as the inverse value of a maximum viewing angle between the viewer and two proximate points which humans cannot distinguish, and so fuse together.

Human vision has the best resolution for objects which are at a distance of about 250 mm from an eye under illumination of about 500 lux; this illumination is provided by a 60W bulb from a distance of 400 mm. Under these conditions the distance between two distinguishable points is approximately 0.16 mm.

Object border

Object borders carries a lot of information [Marr 82]. Boundaries of objects and simple patterns such as blobs or lines enable adaptation effects similar to conditional contrast, mentioned above. The Ebbinghaus illusion is a well known example—two circles of the same diameter in the center of images appear to have different diameters (Figure 2.15).

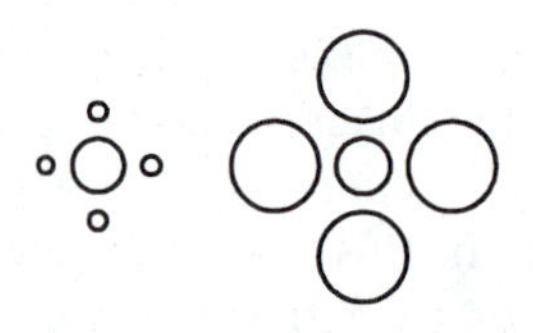

Figure 2.15: *The Ebbinghaus illusion.*

Color

Color is very important for perception, since under normal illumination conditions the human eye is more sensitive to color than to brightness. Color quantization and representation are described in Section 2.2.3; color can be expressed as a combination of the three basic components, red, green, and blue (RGB), but color **perception** is better expressed in the alternative HSI co-ordinate system.

Color perception is burdened with similar adaptation illusions as other psycho-physical quantities.

2.3.4 Image quality

An image might be degraded during capture, transmission, or processing, and measures of image quality can be used to assess the degree of degradation. The quality required naturally depends on the purpose for which an image is used.

Methods for assessing image quality can be divided into two categories: subjective and objective. Subjective methods are often used in television technology, where the ultimate criterion is the perception of a selected group of professional and lay viewers. They appraise an image according to a list of criteria and give appropriate marks. Details about subjective methods may be found in [Pratt 78].

Objective quantitative methods measuring image quality are more interesting for our purposes. Ideally such a method also provides a good subjective test, and is easy to apply; we might then use it a criterion in parameter optimization. The quality of the image $f(x, y)$ is usually estimated by comparison with a known reference image $g(x, y)$ [Rosenfeld and Kak 82]. A synthesized image is often used for this purpose. One class of methods uses simple measures such as the mean quadratic difference $\sum\sum(g - f)^2$. The problem here is that it is not possible to distinguish a few big differences from a lot of small differences. Instead of the mean quadratic difference, the mean absolute difference or simply maximal absolute difference may be used. Correlation between images f and g is another alternative.

Another class measures the resolution of small or proximate objects in the image. An image consisting of parallel black and white stripes is used for this purpose; then the number of black and white pairs per millimeter gives the resolution.

Measures of image similarity are becoming more important since they may be used in assisting retrieval from image databases. Picture information measures of described in [Chang 89].

2.3.5 Noise in images

Real images are often degraded by some random errors—this degradation is usually called **noise**. Noise can occur during image capture, transmission, or processing, and may be dependent on, or independent of, image content.

Noise is usually described by its probabilistic characteristics. Idealized noise, called **white noise**, which has constant power spectrum $S = c$ [see equation (2.24)], meaning that its intensity does not decrease with increasing frequency, is often used. White noise is frequently applied as the worst approximation of degradation, its advantage being that its use simplifies calculations. A special case of white noise is **Gaussian noise**. A random variable with Gaussian (normal) distribution has its probability density given by the Gaussian curve. In the 1D case the density function is

$$p(x) = \frac{1}{\sigma\sqrt{2\pi}} e^{\frac{-(x-\mu)^2}{2\sigma^2}} \tag{2.45}$$

where μ is the mean and σ the standard deviation of the random variable. Gaussian noise is a very good approximation to noise that occurs in many practical cases.

When an image is transmitted through some channel, noise which is usually independent of the image signal occurs. Similar noise arises in old-fashioned vidicon cameras. This signal-independent degradation is called **additive noise** and can be described by the model

$$f(x,y) = g(x,y) + \nu(x,y) \tag{2.46}$$

where the noise ν and the input image g are independent variables. The following algorithm will generate zero mean additive Gaussian noise in an image—this can often be of use in testing or demonstrating many other algorithms in this book which are designed to remove noise, or to be noise resistant.

Algorithm 2.3: Generate additive, zero mean Gaussian noise

1. Select a value for σ; low values generate less noise effect.
2. If the image gray-level range is $[0, G-1]$, calculate

$$p[i] = \frac{1}{\sigma\sqrt{2\pi}} e^{\frac{-i^2}{2\sigma^2}} \qquad i = 0, 1, \ldots, G-1$$

3. For each pixel (x,y), of intensity $g(x,y)$, generate a random number q_1 in the range $[0,1]$. Determine

$$j = \text{argmin}_i (q_1 - p[i])$$

4. Generate a random number (sign) q_2 from the set $\{-1,1\}$. Set $f^*(x,y) = g(x,y) + q_2 j$.
5. Set

$$\begin{aligned} f(x,y) &= 0 && \text{if } f^*(x,y) < 0 \\ f(x,y) &= G-1 && \text{if } f^*(x,y) > G-1 \\ f(x,y) &= f^*(x,y) && \text{otherwise} \end{aligned} \tag{2.47}$$

6. Go to 3 until all pixels have been scanned.

The truncation (2.47) will attenuate the Gaussian nature of the noise; this will become more marked for values of σ that are high relative to G. Other algorithms for noise generation may be found in [Pitas 93].

Equation (2.46) leads to a definition of **signal-to-noise ratio** (SNR); computing the total square value of the noise contribution

$$E = \sum_{(x,y)} \nu^2(x,y)$$

we compare this with the total square value of the observed signal,

$$F = \sum_{(x,y)} f^2(x,y)$$

The signal-to-noise ratio is then

$$\mathrm{SNR} = \frac{F}{E} \tag{2.48}$$

(strictly, we are measuring the *mean* observation with the *mean* error—the computation is obviously the same). SNR represents a measure of image quality, with high values being 'good'.

The noise magnitude depends in many cases on the signal magnitude itself. If the noise magnitude is much higher in comparison with the signal, we can write

$$f = g + \nu g = g(1 + \nu) \approx g\nu \tag{2.49}$$

This model describes **multiplicative noise**. An example of multiplicative noise is television raster degradation, which depends on TV lines; in the area of a line this noise is maximal, and between two lines it is minimal. Another example of multiplicative noise is the degradation of film material caused by the finite size of silver grains used in photosensitive emulsion.

Quantization noise occurs when insufficient quantization levels are used, for example, 50 levels for a monochromatic image. In this case false contours appear. Quantization noise can be eliminated simply, see Section 2.2.2.

Impulsive noise means that an image is corrupted with individual noisy pixels whose brightness differs significantly from that of the neighborhood. The term **salt-and-pepper noise** is used to describe saturated impulsive noise—an image corrupted with white and/or black pixels is an example. Salt-and-pepper noise can corrupt binary images.

The problem of suppressing noise in images is addressed in Chapter 4. If nothing is known a priori about noise properties, local pre-processing methods are appropriate (Section 4.3). If the noise parameters are known in advance, image restoration techniques can be used (Section 4.4).

2.4 Summary

- **Basic concepts**
  - An image is formed as the result of projection of a 3D scene into 2D.
  - Images are co-ordinatized in 2D; discrete points will have an associated discrete *gray-level* or *brightness.*
  - The discretization of images from a continuous domain can be achieved by convolutions with Dirac functions.
  - The Fourier transform can be a useful way of decomposing image data.
  - Images are statistical in nature, and it can be natural to represent them as a stochastic process.
- **Image digitization**
  - Discretization of an image can be seen as a product of a sampling function and a continuous representation.

- Fourier analysis of a digital sampling explains the *aliasing* effect; this is avoided in band-limited images if the sampling is fine enough. This is the Shannon sampling theorem.
- A pixel represents the elemental part of an image.
- Gray-level quantization governs the appearance of shading and false contour. At least 100 intensity levels are desirable.
- Color representations are now common; these are usually a combination of three signals: red, green, and blue.
- Equivalent color representations exist that are more useful for image manipulation. These include YIQ and HSI.

- **Digital image properties**
  - Images have metric and topological properties.
  - Various metrics are available to measure distance in images. The Euclidean is common but expensive.
  - Distance measurements may be used as the basis of various manipulations; *chamfering* is one that describes proximity to a subset of interest.
  - Neighborhood relations on rectangular grids are usually 4- or 8-based. Both of these give rise to paradoxes. Hexagonal grids overcome these but cause other problems of representation.
  - Sets of pixels may define *regions* which have *borders.*
  - Borders may have *inner* or *outer* properties, depending on whether the interior or exterior of the region is considered.
  - Images have topological properties that are invariant to rubber sheet transformations.
  - The *convex hull* of a region may be described as the minimal convex subset containing it.
  - The brightness histogram is a global descriptor of image intensity.
  - Human visual perception is essential understanding for design of image display.
  - Human visual perception is sensitive to contrast, acuity, border perception, and color. Each of these may provoke visual paradoxes.
  - Live images are always prone to noise. It may be possible to measure its extent quantitatively.
  - White, Gaussian, impulsive, and salt-and-pepper noise are common models.
  - Signal-to-noise ratio is a measure of image quality.

2.5 Exercises

Short-answer questions

1. Define *perspective projection.*

2. Define *orthographic projection.*
3. Define *photometry.*
4. Define *colorimetry.*
5. Define
   - *Spatial* resolution
   - *Spectral* resolution
   - *Radiometric* resolution
   - *Time* resolution
6. Give the formulae for the Fourier transform and its inverse.
7. State the convolution theorem.
8. Determine the Fourier transform of the Dirac $\delta(x, y)$ function.
9. What do you understand by *aliasing*?
10. Define
    - Additive noise
    - Multiplicative noise
    - Gaussian noise
    - Impulsive noise
    - Salt-and-pepper noise

Problems

1. Discuss the various factors that influence the *brightness* of a pixel in an image.
2. Why might high frequencies of an image function be of interest? Why would the Fourier transform be of use in their analysis?
3. Suppose a convolution of two finite digital functions is to be performed; determine how many elemental operations (additions and multiplications) are required for given sized domains. How many operations are required if the convolution theorem is exploited (excluding the cost of the Fourier transform).
4. Explain the aliasing effect in terms of Fourier frequency overlaps.
5. Develop a program that reads an input image and manipulates its resolution in the spatial and gray domains; for a range of images (synthetic, of man-made objects, of natural scenes ...) conduct experiments on the minimum resolution that leaves the image recognizable.

   Conduct such experiments on a range of subjects.
6. Acquire some RGB images. Develop software to convert them into YIQ and HSI representations. Subject them to various degrees of noise (by, for example, Algorithm 2.3) and convert back to RGB for display
7. Implement chamfering on a rectangular grid, and test it on a synthetic image consisting of a (black) subset of specified shape on a (white) background. Display the results for a range of shapes, basing the chamfering on;
   - The Euclidean metric

- The city block metric
- The chessboard metric

8. Implement chamfering on a hexagonal grid and display the results.
9. One solution to digitization paradoxes is to mix connectivities. Using 8-neighborhoods for foreground and 4-neighborhoods for background, examine the paradoxes cited in the text (Figures 2.9 and 2.10). Do new paradoxes occur? What might be the disadvantages of such an approach?
10. For each uppercase printed letter of the alphabet, determine the number of lakes and bays it has. Derive a look-up table that lists the candidate letters, given the number of lakes and bays. Comment on this quality of this *feature* as an identifier of letters.
11. Write a program that computes an image histogram; plot the histogram of a range of images. Plot also the histogram of the three components of a color image when represented as
    - RGB
    - YIQ
    - HSI
12. Implement histogram smoothing; determine how much smoothing is necessary to suppress turning points in the histogram due to what you consider to be noise, or small-scale image effects.
13. Implement Algorithm 2.3. For a range of images, plot the distribution of $f(x,y) - g(x,y)$ for various values of σ. Measure the deviation of this distribution from a 'perfect' zero mean Gaussian.
14. For a range of images and a range of noise corruption, compute the signal-to-noise ratio [equation (2.48)]. Draw some subjective conclusions about what 'bad' noise is.

2.6 References

[Ballard and Brown 82] D H Ballard and C M Brown. *Computer Vision.* Prentice-Hall, Englewood Cliffs, NJ, 1982.

[Barrow et al. 77] H G Barrow, J M Tenebaum, R C Bolles, and H C Wolf. Parametric correspondence and chamfer matching: Two new techniques for image matching. In *5th International Joint Conference on Artificial Intelligence,* Cambridge, CA, pages 659–663. Carnegie-Mellon University, 1977.

[Chang 89] S K Chang. *Principles of Pictorial Information Design.* Prentice-Hall, Englewood Cliffs, NJ, 1989.

[Cornsweet 70] T N Cornsweet. *Visual Perception.* Academic Press, New York, 1970.

[Frisby 79] J P Frisby. *Seeing—Illusion, Brain and Mind.* Oxford University Press, Oxford, 1979.

[Gonzalez and Wintz 87] R C Gonzalez and P Wintz. *Digital Image Processing.* Addison-Wesley, Reading, MA, 2nd edition, 1987.

[Gonzalez and Woods 92] R C Gonzalez and R E Woods. *Digital Image Processing.* Addison-Wesley, Reading, MA, 1992.

[Horn 86] B K P Horn. *Robot Vision.* MIT Press, Cambridge, MA, 1986.

[Kovalevsky 89] V A Kovalevsky. Finite topology as applied to image analysis. *Computer Vision, Graphics, and Image Processing,* 46:141–161, 1989.

[Levine 85] M D Levine. *Vision in Man and Machine.* McGraw-Hill, New York, 1985.

[Luong 93] Q T Luong. Color in computer vision. In C H Chen, L F Pau, and P S P Wang, editors, *Handbook of Pattern Recognition and Computer Vision,* pages 311–368. World Scientific, 1993.

[Marr 82] D Marr. *Vision—A Computational Investigation into the Human Representation and Processing of Visual Information.* Freeman, San Francisco, 1982.

[Papoulis 62] A Papoulis. *The Fourier Integral and Its Application.* McGraw-Hill, New York, 1962.

[Papoulis 65] A Papoulis. *Probability, Random Variables, and Stochastic Processes.* McGraw-Hill, New York, 1965.

[Pavlidis 77] T Pavlidis. *Structural Pattern Recognition.* Springer Verlag, Berlin, 1977.

[Pavlidis 82] T Pavlidis. *Algorithms for Graphics and Image Processing.* Computer Science Press, New York, 1982.

[Pitas 93] I Pitas. *Digital Image Processing Algorithms.* Prentice-Hall, Hemel Hempstead, UK, 1993.

[Pratt 78] W K Pratt. *Digital Image Processing.* Wiley, New York, 1978.

[Pratt 91] W K Pratt. *Digital Image Processing.* Wiley, New York, 2nd edition, 1991.

[Pritchard 77] D H Pritchard. US Color Television Fundamentals. *IEEE Transactions on Consumer Electronics,* CE-23(4):467–478, 1977.

[Rogers 85] D F Rogers. *Procedural Elements of Computer Graphics.* McGraw-Hill, New York, 1985.

[Rosenfeld and Kak 82] A Rosenfeld and A C Kak. *Digital Picture Processing.* Academic Press, New York, 2nd edition, 1982.

[Rosenfeld and Pfalz 68] A Rosenfeld and J L Pfalz. Distance functions on digital pictures. *Pattern Recognition,* 1(1):33–62, 1968.

[Smith 78] A R Smith. Color gamut transform pairs. *Computer Graphics,* 12(3):12–19, 1978.

[Winston 75] P H Winston, editor. *The Psychology of Computer Vision.* McGraw-Hill, New York, 1975.

Chapter 3

Data structures for image analysis

Data and an algorithm are the two basic related parts of any program. Data organization often considerably affects the simplicity of the selection and the implementation of an algorithm, and the choice of data structures is therefore a fundamental question when writing a program [Wirth 76]. Information about the representation of image data, and the data which can be deduced from them, will be introduced here before explaining different image processing methods. Relations between different types of representations of image data will then be clearer.

First we shall deal with basic levels of representation of information in image analysis tasks; then with traditional data structures such as matrices, chains, and relational structures. Lastly we consider hierarchical data structures such as pyramids and quadtrees.

3.1 Levels of image data representation

The aim of computer visual perception is to find a relation between an input image and models of the real world. During the transition from the raw input image to the model, image information becomes denser and semantic knowledge about the interpretation of image data is used more and more. Several levels of visual information representation are defined on the way between the input image and the model; computer vision then comprises a design of the

- Intermediate representations (data structures)
- Algorithms used for the creation of representations and introduction of relations between them

The representations can be stratified in four levels [Ballard and Brown 82]—however, there are no strict borders between them and a more detailed classification of the representational levels is used in some applications. These four representational levels are ordered from signals at a low level of abstraction to the description that a human can perceive. The information flow between the levels may be bi-directional, and for some specific uses, some representations can be omitted.

The first, lowest representational level—**iconic images**—consists of images containing original data: integer matrices with data about pixel brightness. Images of this kind are also outputs of pre-processing operations (e.g., filtration or edge sharpening) used for highlighting some aspects of the image important for further treatment.

The second level of representation is **segmented images**. Parts of the image are joined into groups that probably belong to the same objects. For instance, the output of the segmentation of a scene with polyhedra is either line segments coinciding with borders or two-dimensional regions corresponding with faces of bodies. It is useful to know something about the application domain while doing image segmentation; it is then easier to deal with noise and other problems associated with erroneous image data.

The third representational level is **geometric representations** holding knowledge about 2D and 3D shapes. The quantification of a shape is very difficult but also very important. Geometric representations are useful while doing general and complex simulations of the influence of illumination and motion in real objects. We need them also for the transition between natural raster images (gained, for example, by a TV camera) and data used in computer graphics (CAD—computer-aided design, DTP – desktop publishing).

The fourth level of representation of image data is **relational models**. They give us the ability to treat data more efficiently and at a higher level of abstraction. A priori knowledge about the case being solved is usually used in processing of this kind. AI techniques are often explored; the information gained from the image may be represented by semantic nets or frames [Nilsson 82].

An example will illustrate a priori knowledge. Imagine a satellite image of a piece of land, and the task of counting planes standing at an airport; the a priori knowledge is the position of the airport, which can be deduced, for instance, from a map. Relations to other objects in the image may help as well, e.g., to roads, lakes, or urban areas. Additional a priori knowledge is given by geometric models of planes for which we are searching.

3.2 Traditional image data structures

Traditional image data structures such as matrices, chains, graphs, lists of object properties, and relational databases are important not only for the direct representation of image information, but also as a basis for more complex hierarchical methods of image representation.

3.2.1 Matrices

A matrix is the most common data structure for low-level representation of an image. Elements of the matrix are integer numbers corresponding to brightness, or to another property of the corresponding pixel of the sampling grid. Image data of this kind are usually the direct output of the image-capturing device, e.g., a scanner. Pixels of both rectangular and hexagonal sampling grids can be represented by a matrix; the correspondence between data and matrix elements is obvious for a rectangular grid; with a hexagonal grid every even row in the image is shifted half a pixel to the right.

Image information in the matrix is accessible through the co-ordinates of a pixel that correspond with row and column indices. The matrix is a full representation of the image, independent of the contents of image data—it implicitly contains **spatial relations** among semantically important parts of the image. The space is two-dimensional in the case of an image—a plane. One very natural spatial relation is the **neighborhood relation**. A representation of a segmented image by a matrix usually saves more memory than an explicit list of all spatial relations between all objects, although sometimes we need to record other relations between objects.

Some special images that are represented by matrices are:

- A **binary image** (an image with two brightness levels only) is represented by a matrix containing only zeros and ones.

- Several matrices can contain information about one **multispectral image**. Each of these matrices contains one image corresponding to one spectral band.

- Matrices of different resolution are used to obtain **hierarchical image data structures**. This hierarchical representation of the image can be very convenient for parallel computers with the 'processor array' architecture.

Most programming languages use a standard array data structure to represent a matrix, and most modern machines provide adequate physical memory to accommodate image data structures. If they do not, they are usually provided with virtual memory to make storage transparent. Historically, memory limitations were a significant obstacle to image applications, requiring individual image parts to be retrieved from disk independently.

There is much image data in the matrix, so processing takes a long time. algorithms can be speeded up if global information is derived from the original image matrix first—global information is more concise and occupies less memory. We have already mentioned the most popular example of global information—the histogram – in Section 2.3.2. Looking at the image from a probabilistic point of view, the normalized histogram is an estimate of the probability density of a phenomenon: that an image pixel has a certain brightness.

Another example of global information is the **co-occurrence matrix** [Pavlidis 82], which represents an estimate of the probability that a pixel (i_1, j_1) has intensity z and a pixel (i_2, j_2) has intensity y. Suppose that the probability depends only on a certain spatial relation r between a pixel of brightness z and a pixel of brightness y; then information about the relation r is recorded in the square co-occurrence matrix C_r, whose dimensions correspond to the number of brightness levels of the image. To reduce the number of matrices C_r, introduce some simplifying assumptions; first consider only direct neighbors, and then treat relations as symmetrical (without orientation). The following algorithm calculates the co-occurrence matrix C_r from the image $f(i, j)$.

Algorithm 3.1: Co-occurrence matrix $C_r(z, y)$ for the relation r

1. Assign $C_r(z, y) = 0$ for all z, $y \in [0, L]$, where L is the maximum brightness.

2. For all pixels (i_1, j_1) in the image, determine (i_2, j_2) which has the relation r with the pixel (i_1, j_1), and perform

$$C_r[f(i_1, j_1), f(i_2, j_2)] = C_r[f(i_1, j_1), f(i_2, j_2)] + 1$$

If the relation r is *to be a southern or eastern 4-neighbor of the pixel* (i_1, j_1), *or identity*[1], elements of the co-occurrence matrix have some interesting properties. Values of the elements at the diagonal of the co-occurrence matrix $C_r(k,k)$ are equal to the area of the regions in the image with brightness k. Thus the diagonal elements correspond to the histogram. The values of elements off the diagonal of the matrix $C_r(k,j)$ are equal to the length of the border dividing regions with brightnesses k and j, $k \neq j$. For instance, in an image with low contrast, the elements of the co-occurrence matrix that are far from the diagonal are equal to zero or are very small. For high-contrast images the opposite is true.

The main reason for considering co-occurrence matrices is their ability to describe texture. This approach to texture analysis is introduced in Chapter 14.

3.2.2 Chains

Chains are used for the description of object borders in computer vision. One element of the chain is a basic symbol; this approach permits the application of formal language theory for computer vision tasks. Chains are appropriate for data that can be arranged as a sequence of symbols, and the neighboring symbols in a chain usually correspond to the neighborhood of primitives in the image. The primitive is the basic descriptive element that is used in syntactic pattern recognition (see Chapter 7).

This rule of proximity (neighborhood) of symbols and primitives has exceptions—for example, the first and the last symbol of the chain describing a closed border are not neighbors, but the corresponding primitives in the image are. Similar inconsistencies are typical of image description languages [Shaw 69], too. Chains are linear structures, which is why they cannot describe spatial relations in the image on the basis of neighborhood or proximity.

Chain codes (and Freeman codes) [Freeman 61] are often used for the description of object borders, or other one-pixel-wide lines in images. The border is defined by the co-ordinates of its reference pixel and the sequence of symbols corresponding to the line of the unit length in several pre-defined orientations. Notice that a chain code is of a relative nature; data are expressed with respect to some reference point. An example of a chain code is shown in Figure 3.1, where 8-neighborhoods are used—it is possible to define chain codes using 4-neighborhoods as well. An algorithm to extract a chain code may be implemented as an obvious simplification of Algorithm 5.8; chain codes and their properties are described in more detail in Chapter 6.

If local information is needed from the chain code, then it is necessary to search through the whole chain systematically. For instance, if we want to know whether the border turns somewhere to the left by 90°, we must just find a sample pair of symbols in the chain—it is simple. On the other hand, a question about the shape of the border near the pixel (i_0, j_0) is not trivial. It is necessary to investigate all chains until the pixel (i_0, j_0) is found and only then we can start to analyze a short part of the border that is close to the pixel (i_0, j_0).

The description of an image by chains is appropriate for syntactic pattern recognition that is based on formal language theory methods. When working with real images, the problem of how to deal with uncertainty caused by noise arises, which is why several syntactic analysis techniques with deformation correction have arisen [Lu and Fu 78]. Another way to deal with

[1]For the purpose of co-occurrence matrix creation we need to consider the identity relation $(i_1, j_1) = (i_2, j_2)$, or individual pixels would not contribute to the histogram.

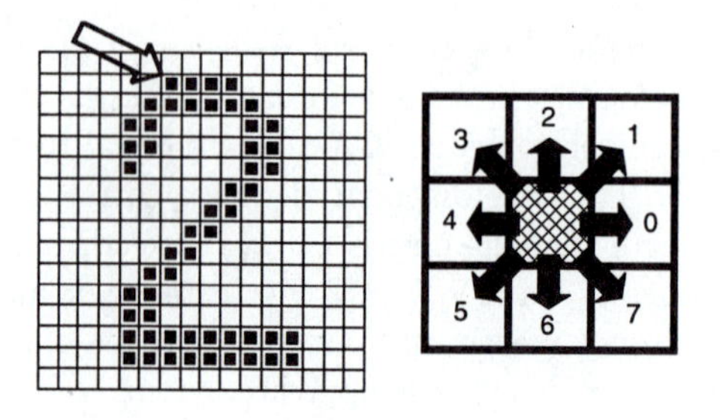

Figure 3.1: *An example chain code; the reference pixel is marked by an arrow: 00077665555556600000006444444442221111112234445652211.*

noise is to smooth the border or to approximate it by another curve. This new border curve is then described by chain codes [Pavlidis 77].

Run length coding is quite often used to represent strings of symbols in an image matrix (for instance, FAX machines use run length coding). For simplicity, consider a binary image first. Run length coding records only areas that belong to the object in the image; the area is then represented as a list of lists. Various schemes exist which differ in detail - a representative one describes each row of the image by a sublist, the first element of which is the row number. Subsequent terms are co-ordinate pairs; the first element of a pair is the beginning of a run and the second is the end. There can be several such sequences in the row. Run length coding is illustrated in Figure 3.2. The main advantage of run length coding is the existence of simple algorithms for intersections and unions of regions in the image.

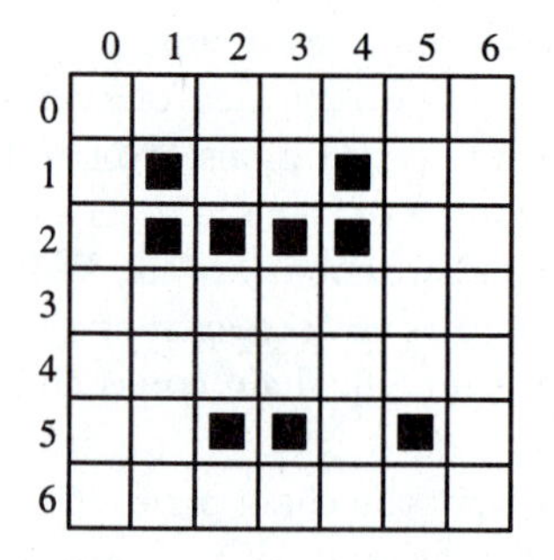

Figure 3.2: *Run length coding; the code is ((11144)(214)(52355)).*

Run length coding can be used for an image with multiple brightness levels as well—in this case sequences of neighboring pixels in a row that has constant brightness are considered. In the sublist we must record not only the beginning and the end of the sequence, but its brightness, too.

From the implementational point of view, chains can be represented using static data structures (e.g., 1D arrays); their size is the longest length of the chain expected. This might be too memory consuming, and so dynamic data structures are more advantageous. Lists from the LISP language are an example.

3.2.3 Topological data structures

Topological data structures describe the image as a set of elements and their relations; these relations are often represented using graphs. A **graph** $G = (V, E)$ is an algebraic structure which consists of a set of nodes $V = \{v_1, v_2, \ldots, v_n\}$ and a set of arcs $E = \{e_1, e_2, \ldots, e_m\}$. Each arc e_k is incident to an unordered pair of nodes $\{v_i, v_j\}$ which are not necessarily distinct [Even 79]. The degree of the node is equal to the number of incident arcs of the node.

An **evaluated graph** is a graph in which values are assigned to arcs, to nodes, or to both—these values may, for example, represent weights, or costs.

The **region adjacency graph** is typical of this class of data structures, in which nodes correspond to regions and neighboring regions are connected by an arc. The segmented image (see Chapter 5) consists of regions with similar properties (brightness, texture, color, ...) that correspond to some entities in the scene, and the neighborhood relation is fulfilled when the regions have some common border. An example of an image with areas labeled by numbers and the corresponding region adjacency graph is shown in Figure 3.3; the label 0 denotes pixels out of the image. This value is used to indicate regions that touch borders of the image in the region adjacency graph.

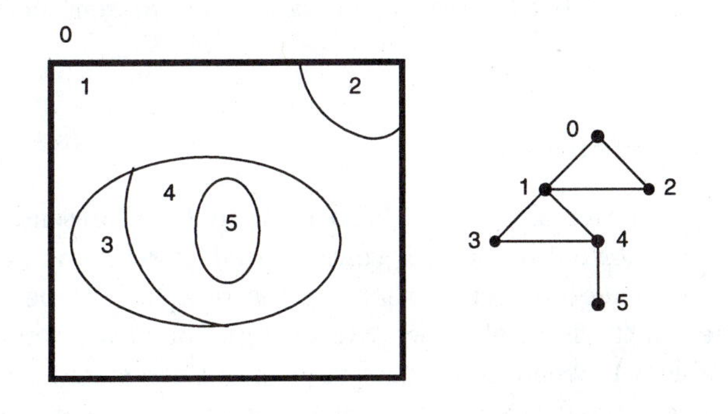

Figure 3.3: *An example region adjacency graph.*

The region adjacency graph has several attractive features. If a region encloses other regions, then the part of the graph corresponding with the areas inside can be separated by a cut in the graph. Nodes of degree 1 represent simple holes.

Arcs of the region adjacency graph can include a description of the relations between neighboring regions—the relations *to be to the left* or *to be inside* are common. The region adjacency graph can be used for matching with a stored pattern for recognition purposes.

The region adjacency graph is usually created from the **region map**, which is a matrix of the same dimensions as the original image matrix whose elements are identification labels of the regions. To create the region adjacency graph, borders of all regions in the image are traced, and labels of all neighboring regions are stored. The region adjacency graph can easily be created from an image represented by a quadtree as well (see Section 3.3.2).

The region adjacency graph stores information about the neighbors of all regions in the image explicitly. The region map contains this information as well, but it is much more difficult to recall from there. If we want to relate the region adjacency graph to the region map quickly, it is sufficient for a node in the region adjacency graph to be marked by the identification label of the region and some representative pixel (e.g., the top left pixel of the region).

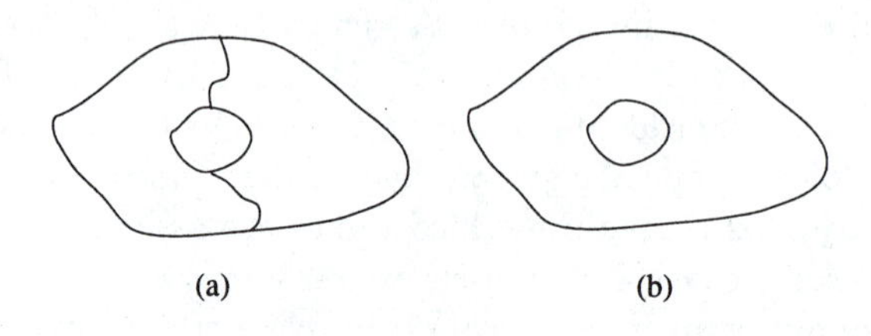

Figure 3.4: *Region merging may create holes: (a) Before a merge; (b) After.*

Construction of the boundary data structures that represent regions is not trivial, and is considered in Section 5.2.3. Region adjacency graphs can be used to approach region merging (where, for instance, neighboring regions thought to have the same image interpretation are merged into one region)—this topic is considered in Section 8.6. In particular, note that merging representations of regions that may border each other more than once can be intricate, for example, with the creation of 'holes' not present before the merge—see Figure 3.4.

3.2.4 Relational structures

Relational databases [Kunii et al. 74] can also be used for representation of information from an image; all the information is then concentrated in relations between semantically important parts of the image—objects – that are the result of segmentation (Chapter 5). Relations are recorded in the form of tables. An example of such a representation is shown in Figure 3.5 and Table 3.1, where individual objects are associated with their names and other features, e.g., the top-left pixel of the corresponding region in the image. Relations between objects are expressed in the relational table as well. In Figure 3.5 and Table 3.1, such a relation is *to be inside*; for example, the object 7 (pond) is situated inside the object 6 (hill).

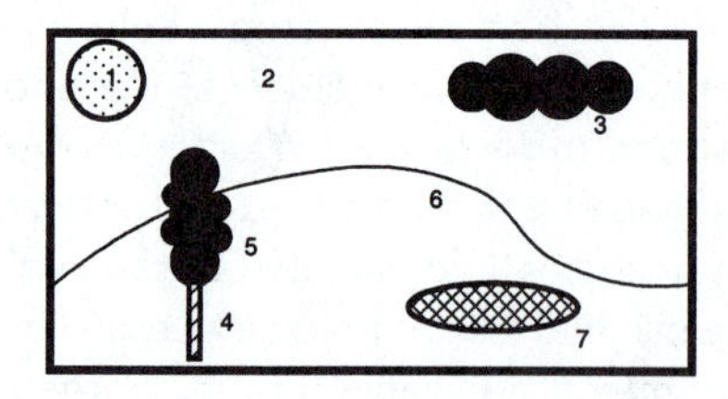

Figure 3.5: *Description of objects using relational structure.*

Description by means of relational structures is appropriate for higher levels of image understanding. In this case searches using keys, similar to database searches, can be used to speed up the whole process.

No.	Object name	Color	Min. row	Min. col.	Inside
1	sun	white	5	40	2
2	sky	blue	0	0	-
3	cloud	gray	20	180	2
4	tree trunk	brown	95	75	6
5	tree crown	green	53	63	-
6	hill	light green	97	0	-
7	pond	blue	100	160	6

Table 3.1: *Relational table*

3.3 Hierarchical data structures

Computer vision is by its nature very computationally expensive, if for no other reason than the large amount of data to be processed. Systems which we might call sophisticated must process considerable quantities of image data—hundreds of kilobytes to tens of megabytes. Usually a very quick response is expected because interactive systems are desirable. One of the solutions is to use parallel computers (in other words brute force). Unfortunately there are many computer vision problems that are very difficult to divide among processors, or decompose in any way. Hierarchical data structures make it possible to use algorithms which decide a strategy for processing on the basis of relatively small quantities of data. They work at the finest resolution only with those parts of the image for which it is essential, using knowledge instead of brute force to ease and speed up the processing. We are going to introduce two typical structures, pyramids and quadtrees.

3.3.1 Pyramids

Pyramids are among the simplest hierarchical data structures. We distinguish between **M-pyramids** (matrix-pyramids) and **T-pyramids** (tree-pyramids).

A **Matrix-pyramid** (M-pyramid) is a sequence $\{M_L, M_{L-1}, \ldots, M_0\}$ of images, where M_L has the same dimensions and elements as the original image, and M_{i-1} is derived from the M_i by reducing the resolution by one-half. When creating pyramids, it is customary to work with square matrices having dimensions equal to powers of 2—then M_0 corresponds to one pixel only.

M-pyramids are used when it is necessary to work with an image at different resolutions simultaneously. An image having one degree smaller resolution in a pyramid contains four times less data, so it is processed approximately four times as quickly.

Often it is advantageous to use several resolutions simultaneously rather than choose just one image from the M-pyramid. For such algorithms we prefer to use **tree-pyramids**, a tree structure. Let 2^L be the size of an original image (the highest resolution). A tree-pyramid (T-pyramid) is defined by:

1. A set of nodes $P = \{P = (k, i, j)$ such that level $k \in [0, L]$; $i, j \in [0, 2^k - 1]\}$.

2. A mapping F between subsequent nodes P_{k-1}, P_k of the pyramid,

$$F(k,i,j) = (k-1, i \text{ div } 2, j \text{ div } 2)$$

   where 'div' denotes whole-number division.

3. A function V that maps a node of the pyramid P to Z, where Z is the subset of the whole numbers corresponding to the number of brightness levels, for example, $Z = \{0, 1, 2, \ldots, 255\}$.

Nodes of a T-pyramid correspond for a given k with image points of an M-pyramid; elements of the set of nodes $P = \{(k,i,j)\}$ correspond with individual matrices in the M-pyramid—k is called the level of the pyramid. An image $P = \{(k,i,j)\}$ for a specific k constitutes an image at the kth level of the pyramid. F is the so-called parent mapping, which is defined for all nodes P_k of the T-pyramid except its root $(0,0,0)$. Every node of the T-pyramid has four child nodes except leaf nodes, which are nodes of level L that correspond to the individual pixels in the image.

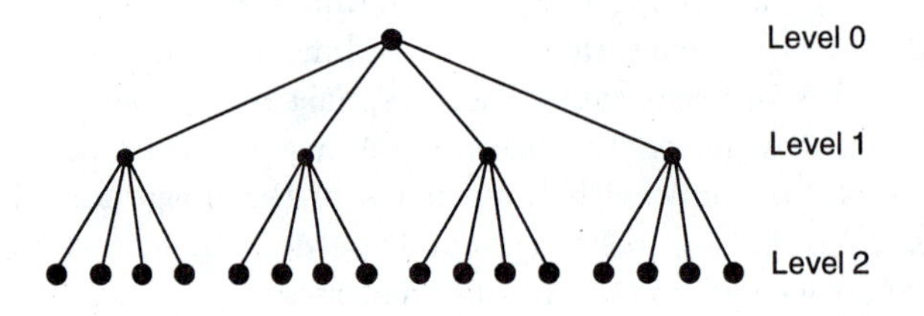

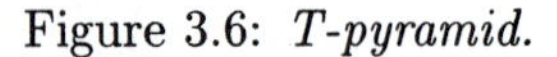

Figure 3.6: *T-pyramid.*

Values of individual nodes of the T-pyramid are defined by the function V. Values of leaf nodes are the same as values of the image function (brightness) in the original image at the finest resolution; the image size is 2^{L-1}. Values of nodes in other levels of the tree are either an arithmetic mean of four child nodes or they are defined by coarser sampling, meaning that the value of one child (e.g., top left) is used. Figure 3.6 shows the structure of a simple T-pyramid.

The number of image pixels used by an M-pyramid for storing all matrices is given by

$$N^2 \left(1 + \frac{1}{4} + \frac{1}{16} + \ldots\right) \approx 1.33\, N^2 \tag{3.1}$$

where N is the dimension of the original matrix (the image of finest resolution)—usually a power of two, 2^L.

The T-pyramid is represented in memory similarly. Arcs of the tree need not be recorded because addresses of the both child and parent nodes are easy to compute due to the regularity of the structure. An algorithm for the effective creation and storing of a T-pyramid is given in [Pavlidis 82].

3.3.2 Quadtrees

Quadtrees are modifications of T-pyramids. Every node of the tree except the leaves has four children (NW, north-western; NE, north-eastern; SW, south-western; SE, south-eastern). Similarly to T-pyramids, the image is divided into four quadrants at each hierarchical level; however, it is not necessary to keep nodes at all levels. If a parent node has four children of the same value (e.g., brightness), it is not necessary to record them. This representation is less expensive for an image with large homogeneous regions; Figure 3.7 is an example of a simple quadtree.

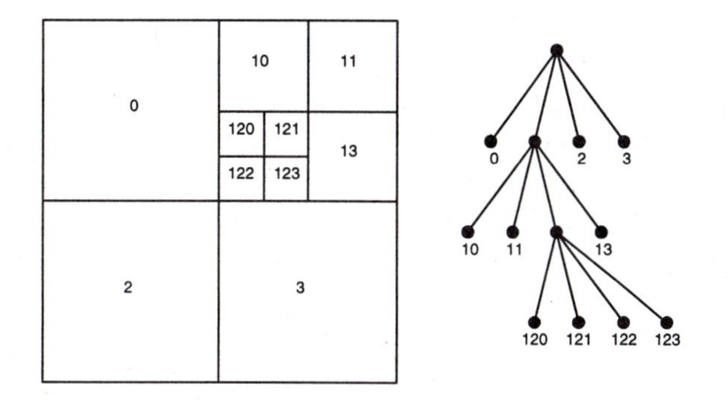

Figure 3.7: *Quadtree.*

An advantage of image representation by means of quadtrees is the existence of simple algorithms for addition of images, computing object areas, and statistical moments. The main disadvantage of quadtrees and pyramid hierarchical representations is their dependence on the position, orientation, and relative size of objects. Two similar images with just very small differences can have very different pyramid or quadtree representations. Even two images depicting the same, slightly shifted scene, can have entirely different representations.

These disadvantages can be overcome using a normalized shape of quadtree in which we do not create the quadtree for the whole image, but for its individual objects. Geometric features of objects such as the center of gravity and principal axis are used (see Chapter 6); the center of gravity and principal axis of every object are derived first and then the smallest enclosing square centered at the center of gravity having sides parallel with the principal axes is located. The square is then represented by a quadtree. An object described by a normalized quadtree and several additional items of data (co-ordinates of the center of gravity, angle of main axes) is invariant to shifting, rotation, and scale.

Quadtrees are usually represented by recording the whole tree as a list of its individual nodes, every node being a record with several items characterizing it. An example is given in Figure 3.8. In the item *Node type* there is information about whether the node is a leaf or inside the tree. Other data can be the level of the node in the tree, position in the picture, code of the node, etc. This kind of representation is expensive in memory. Its advantage is easy access to any node because of pointers between parents and children.

It is possible to represent a quadtree with less demand on memory by means of a **leaf**

Node type
Pointer to the NW son
Pointer to the NE son
Pointer to the SW son
Pointer to the SE son
Pointer to the father
Other data

Figure 3.8: *Record describing a quadtree node.*

code. Any point of the picture is coded by a sequence of digits reflecting successive divisions of the quadtree; zero means the NW (north-west) quadrant, and likewise for other quadrants: 1-NE, 2-SW, 3-SE. The most important digit of the code (on the left) corresponds to the division at the highest level, the least important one (on the right) with the last division. The number of digits in the code is the same as the number of levels of the quadtree. The whole tree is then described by a sequence of pairs—the leaf code and the brightness of the region. Programs creating quadtrees can use recursive procedures to advantage.

T-pyramids are very similar to quadtrees, but differ in two basic respects. A T-pyramid is a balanced structure, meaning that the corresponding tree divides the image regardless of the contents, which is why it is regular and symmetric. A quadtree is not balanced. The other difference is in the interpretation of values of the individual nodes.

Quadtrees have seen widespread application, particularly in the area of Geographic Information Systems (GIS) where, along with their three-dimensional generalization *octrees*, they have proved very useful in hierarchical representation of layered data [Samet 89, Samet 90].

3.3.3 Other pyramidical structures

The pyramid structure is widely used, and has seen several elaborations. Recalling that a (simple) M-pyramid was defined as a sequence of images $\{M_L, M_{L-1}, \ldots, M_0\}$ in which M_i is a 2×2 reduction of M_{i+1}, we can define the notion of a *reduction window*; for every cell c of M_i, the reduction window is its set of children in M_{i+1}, $w(c)$. If the images are constructed such that all interior cells have the same number of neighbors (e.g., a square grid, as is customary), and they all have the same number of children, the pyramid is called *regular*.

A taxonomy of regular pyramids may be constructed by considering the reduction window together with the *reduction factor* λ, which defines the rate at which the image area decreases between levels;

$$\lambda \leq \frac{|M_{i+1}|}{|M_i|}, i = 0, 1, \ldots, L-1$$

In the simple case, in which reduction windows do not overlap and are 2×2, we have $\lambda = 4$; if we choose to let the reduction windows overlap, the factor will reduce. The notation used to describe this characterization of regular pyramids is $(reduction\ window)/(reduction\ factor)$. Figure 3.9 illustrates some simple examples.

The reduction window of a given cell at level i may be propagated down to higher resolution than level $i+1$. For a cell c_i at level i, we can write $w^0(c_i) = w(c_i)$, and then recursively

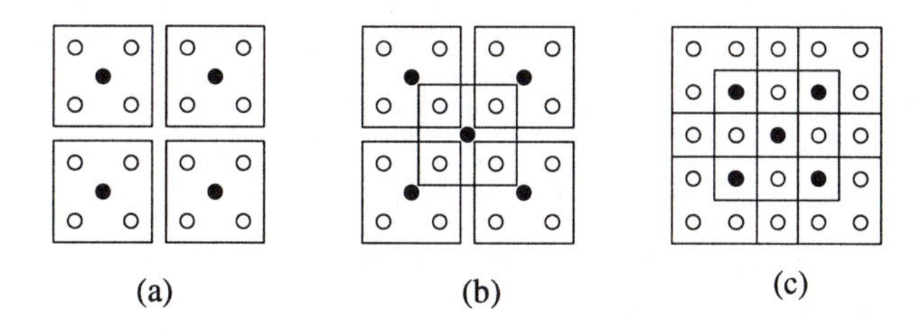

Figure 3.9: *Some regular pyramid definitions: (a)* $2 \times 2/4$, *(b)* $2 \times 2/2$, *(c)* $3 \times 3/2$. *(Solid dots are at the higher, lower-resolution, level.)*

define

$$w^{k+1}(c_i) = \bigcup_{q \in w(c_i)} w^k(q) \tag{3.2}$$

$w^k(c_i)$ is the *equivalent window* that covers all cells at level $i + k + 1$ that link to the cell c_i. Note that the shape of this window is going to depend on the type of pyramid—for example, an $n \times n/2$ pyramid will generate octagonal equivalent windows, while for an $n \times n/4$ pyramid they will be square. Use of this feature prevents square features dominating, as they will in the case of simple $2 \times 2/4$ pyramids.

The $2 \times 2/4$ pyramid is widely used and is what is usually interpreted as an 'image pyramid'; the $2 \times 2/2$ structure is often referred to as an 'overlap pyramid'. $5 \times 5/2$ pyramids have been used [Burt and Adelson 83] in compact image coding, where the image pyramid is augmented by a *Laplacian* pyramid of differences. Here, the Laplacian at a given level is computed as the per-pixel difference between the image at that level, and the image derived by 'expanding' the image at the next lowest resolution. The Laplacian may be expected to have zero (or close) values in areas of low contrast, and therefore be amenable to compression.

Irregular pyramids are derived from contractions of graphical representations of images (for example, region adjacency graphs). Here, a graph may be reduced to a smaller one by selective removal of edges and vertices. Depending on how these selections are made, important structures in the parent graph may be retained while reducing its overall complexity [Kropatsch 95]. The pyramid approach is quite general and lends itself to many developments—for example, the reduction algorithm need not be deterministic [Meer 89]. A brief survey of segmentation algorithms may be found in [Bister et al. 90].

3.4 Summary

- **Level of image data representation**
  - Data structures together with algorithms are used to devise solutions to computational tasks.
  - Data structures for vision may be loosely classified as
    * Iconic
    * Segmented
    * Geometric

    * Relational

  Boundaries between these layers may not be well defined.

- **Traditional image data structures**
  - The matrix (2D array) is the most common data structure used for low-level representations, implemented as an array.
  - Matrices hold image data explicitly. Spatial characteristics are implicitly available.
  - Binary images are represented by binary matrices; multispectral images are represented by binary matrices; Hierarchical image structures are represented by matrices of different dimensions;
  - The *co-occurrence matrix* is an example of global information derived from an image matrix; it is useful in describing texture.
  - Chains may be used to describe pixel paths, especially borders.
  - Chain codes are useful for recognition based on syntactic approaches.
  - Run length codes are useful for simple image compression.
  - Graph structures may be used to describe regions and their adjacency. These may be derived from a region map, a matrix of the same size as the image.
  - Relational structures may be used to describe semantic relationships between image regions.
- **Hierarchical data structures**
  - Hierarchical structures can be used to extract large-scale features, which may be used to initialize analysis. They can provide significant computational efficiency.
  - M-pyramids and T-pyramids provide data structures to describe multiple image resolutions.
  - Quadtrees are a variety of T-pyramid in which selected areas of an image are stored at higher resolution than others, permitting selective extraction of detail.
  - Many algorithms for manipulation of quadtrees are available. Quadtrees are prone to great variation from small image differences.
  - Leaf codes provide a more efficient form of quadtree.
  - Many ways of deriving pyramids exist, dependent on choice of reduction window.

3.5 Exercises

Short-answer questions

1. Determine the chain codes of the regions shown in Figure 3.10.
2. Define *run length encoding.*
3. Determine the run length encoding of the images shown in Figure 3.10.
4. Define a *region adjacency graph.*

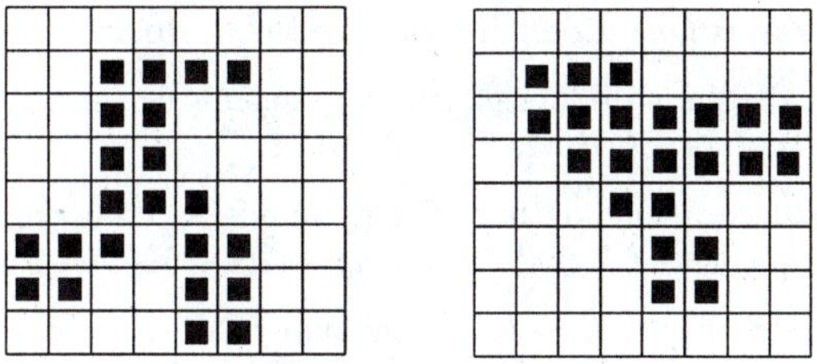

Figure 3.10: *Exercise 3.1*

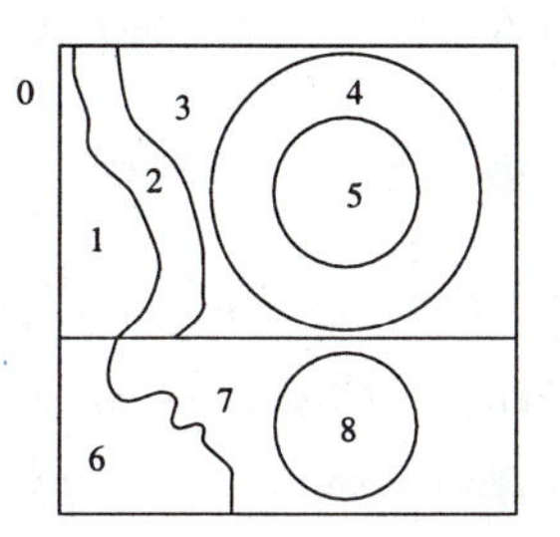

Figure 3.11: *Exercise 3.5*

5. Draw the region adjacency graph for the image depicted in Figure 3.11.
6. Define *M-pyramid.*
7. Define *T-pyramid.*

Problems

1. Implement Algorithm 3.1. Run it on a variety of images for a variety of neighborhood relations.
2. Implement run length encoding, as described in Section 3.2.2. Run it on a range of binary images and determine the compression ratio achieved (this is mostly usefully done with the most compact possible representation of the run length code).
3. Adapt the program written for Problem 3.2 to work for images which are not binary. Test it on a range of synthetic and real images, computing each time the compression ration it provides.
4. Write a program that computes the T-pyramid of an image.
5. Write a program that derives the quadtree representation of an image using the homogeneity criterion of equal intensity

3.6 References

[Ballard and Brown 82] D H Ballard and C M Brown. *Computer Vision.* Prentice-Hall, Englewood Cliffs, NJ, 1982.

[Bister et al. 90] M Bister, J Cornelius, and A Rosenfeld. A critical view of pyramid segmentation algorithms. *Pattern Recognition Letters*, 11(9):605–617, 1990.

[Burt and Adelson 83] P J Burt and E H Adelson. The Laplacian pyramid as a compact image code. *IEEE Transactions on Computers*, COM-31(4):532–540, 1983.

[Even 79] S Even. *Graph Algorithms.* Computer Science Press, Rockville, MD, 1979.

[Freeman 61] H Freeman. On the encoding of arbitrary geometric configuration. *IRE Transactions on Electronic Computers,* EC–10(2):260–268, 1961.

[Kropatsch 95] W G Kropatsch. Building irregular pyramids by dual graph contraction. *IEE Proceedings: Vision, Image and Signal Processing,* 142(6):366–374, 1995.

[Kunii et al. 74] T L Kunii, S Weyl, and I M Tenenbaum. A relation database schema for describing complex scenes with color and texture. In *Proceedings of the 2nd International Joint Conference on Pattern Recognition,* pages 310–316, Copenhagen, Denmark, 1974.

[Lu and Fu 78] S Y Lu and K S Fu. A syntactic approach to texture analysis. *Computer Graphics and Image Processing,* 7:303–330, 1978.

[Meer 89] P Meer. Stochastic image pyramids. *Computer Vision, Graphics, and Image Processing,* 45(3):269–294, 1989.

[Nilsson 82] N J Nilsson. *Principles of Artificial Intelligence.* Springer Verlag, Berlin, 1982.

[Pavlidis 77] T Pavlidis. *Structural Pattern Recognition.* Springer Verlag, Berlin, 1977.

[Pavlidis 82] T Pavlidis. *Algorithms for Graphics and Image Processing.* Computer Science Press, New York, 1982.

[Samet 89] H Samet. *The Design and Analysis of Spatial Data Structures.* Addison-Wesley, Reading, MA, 1989.

[Samet 90] H Samet. *Applications of Spatial Data Structures.* Addison-Wesley, Reading, MA, 1990.

[Shaw 69] A C Shaw. A formal picture description schema as a basis for picture processing systems. *Information and Control,* 14:9–52, 1969.

[Wirth 76] N Wirth. *Algorithms + Data Structures = Programs.* Prentice-Hall, Englewood Cliffs, NJ, 1976.

Chapter 4

Image pre-processing

Pre-processing is the name used for operations on images at the lowest level of abstraction—both input and output are intensity images. These iconic images are of the same kind as the original data captured by the sensor, with an intensity image usually represented by a matrix of image function values (brightnesses).

It is necessary to realize that pre-processing does not increase image information content. If information is measured using entropy (Section 2.1.4), then pre-processing typically decreases image information content. From the information-theoretic viewpoint it can thus be concluded that the best pre-processing is no pre-processing, and without question, the best way to avoid (elaborate) pre-processing is to concentrate on high-quality image acquisition. Nevertheless, pre-processing is very useful in a variety of situations since it helps to suppress information that is not relevant to the specific image processing or analysis task. Therefore, the aim of pre-processing is an improvement of the image data that suppresses undesired distortions or enhances some image features important for further processing, although geometric transformations of images (e.g., rotation, scaling, translation) are also classified as pre-processing methods here since similar techniques are used.

Image pre-processing methods are classified here into four categories according to the size of the pixel neighborhood that is used for the calculation of a new pixel brightness. Section 4.1 deals with pixel brightness transformations, Section 4.2 describes geometric transformations, Section 4.3 considers pre-processing methods that use a local neighborhood of the processed pixel, and Section 4.4 briefly characterizes image restoration that requires knowledge about the entire image.

Some authors [Moik 80, Rosenfeld and Kak 82] classify image pre-processing methods differently into **image enhancement**, covering pixel brightness transformations (local pre-processing in our sense), and **image restoration**.

Image pre-processing methods use the considerable redundancy in images. Neighboring pixels corresponding to one object in real images have essentially the same or similar brightness value, so if a distorted pixel can be picked out from the image, it can usually be restored as an average value of neighboring pixels.

If pre-processing aims to correct some degradation in the image, the nature of a priori information is important:

- A first group of methods uses no knowledge about the nature of the degradation; only very general properties of the degradation are assumed.

- A second group assumes knowledge about the properties of the image acquisition device, and the conditions under which the image was obtained. The nature of noise (usually its spectral characteristics) is sometimes known.

- A third approach uses knowledge about objects that are sought in the image, which may simplify the pre-processing considerably. If knowledge about objects is not available in advance, it can be estimated during the processing. The following strategy is possible. First the image is coarsely processed to reduce data quantity and to find image objects. The image information derived is used to create a hypothesis about image object properties, and this hypothesis is then verified in the image at finer resolution. Such an iterative process can be repeated until the presence of knowledge is verified or rejected. This feedback may span more than pre-processing, since segmentation also yields semantic knowledge about objects—thus feedback can be initiated after the object segmentation.

4.1 Pixel brightness transformations

A brightness transformation modifies pixel brightness—the transformation depends on the properties of a pixel itself. There are two classes of pixel brightness transformations: **brightness corrections** and **gray-scale transformations**. Brightness correction modifies the pixel brightness, taking into account its original brightness and its position in the image. Gray-scale transformations change brightness without regard to position in the image.

4.1.1 Position-dependent brightness correction

Ideally, the sensitivity of image acquisition and digitization devices should not depend on position in the image, but this assumption is not valid in many practical cases. The lens attenuates light more if it passes farther from the optical axis, and the photosensitive part of the sensor (vacuum-tube camera, CCD camera elements) is not of identical sensitivity. Uneven object illumination is also a source of degradation.

If degradation is of a systematic nature, it can be suppressed by brightness correction. A multiplicative error coefficient $e(i,j)$ describes the change from the ideal identity transfer function. Assume that $g(i,j)$ is the original undegraded image (or desired or true image) and $f(i,j)$ is the image containing degradation. Then

$$f(i,j) = e(i,j)\, g(i,j) \tag{4.1}$$

The error coefficient $e(i,j)$ can be obtained if a reference image $g(i,j)$ with known brightnesses is captured, the simplest being an image of constant brightness c. The degraded result is the image $f_c(i,j)$. Then systematic brightness errors can be suppressed by

$$g(i,j) = \frac{f(i,j)}{e(i,j)} = \frac{c\, f(i,j)}{f_c(i,j)} \tag{4.2}$$

This method can be used only if the image degradation process is stable. If we wish to suppress this kind of error in the image capturing process, we should perhaps calibrate the device [find error coefficients $e(i,j)$] from time to time.

This method implicitly assumes linearity of the transformation, which is not true in reality because the brightness scale is limited to some interval. The calculation according to equation (4.1) can overflow, and the limits of the brightness scale are used instead, implying that the best reference image has brightness that is far enough from both limits. If the gray-scale has 256 brightness levels, the ideal image[1] has constant brightness values of 128.

4.1.2 Gray-scale transformation

Gray-scale transformations do not depend on the position of the pixel in the image. A transformation $\mathcal{T}$ of the original brightness p from scale $[p_0, p_k]$ into brightness q from a new scale $[q_0, q_k]$ is given by

$$q = \mathcal{T}(p) \tag{4.3}$$

The most common gray-scale transformations are shown in Figure 4.1; the straight line a denotes the negative transformation; the piecewise linear function b enhances the image contrast between brightness values p_1 and p_2. The function c is called **brightness thresholding** and results in a black-and-white image.

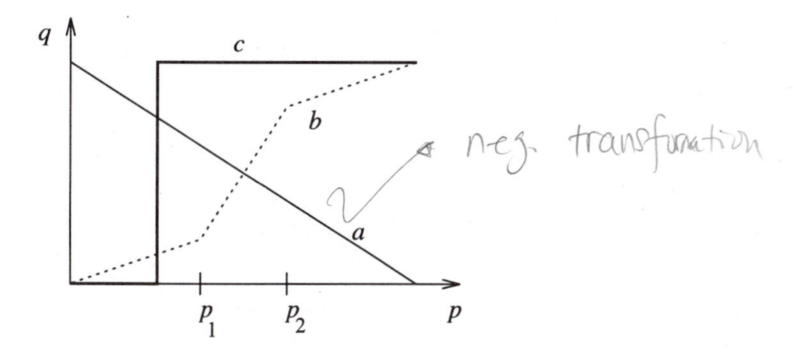

Figure 4.1: *Some gray-scale transformations.*

Digital images have a very limited number of gray-levels, so gray-scale transformations are easy to realize both in hardware and software. Often only 256 bytes of memory (called a **look-up table**) are needed. The original brightness is the index to the look-up, and the table content gives the new brightness. The image signal usually passes through a look-up table in image displays, enabling simple gray-scale transformation in real time.

The same principle can be used for color displays. A color signal consists of three components—red, green, and blue; three look-up tables provide all possible color scale transformations. These tables are called the **palette** in personal computer terminology (see Section 2.2.3 for more detail on color representation).

Gray-scale transformations are used mainly when an image is viewed by a human observer, and a transformed image might be more easily interpreted if the contrast is enhanced. For instance, an X-ray image can often be much clearer after transformation.

[1]Most TV cameras have automatic control of the gain, which allows them to operate under changing illumination conditions. If systematic errors are suppressed using error coefficients, this automatic gain control should be switched off first.

A gray-scale transformation for contrast enhancement is usually found automatically using the **histogram equalization** technique. The aim is to create an image with equally distributed brightness levels over the whole brightness scale (see Figure 4.2). Histogram equalization enhances contrast for brightness values close to histogram maxima, and decreases contrast near minima.

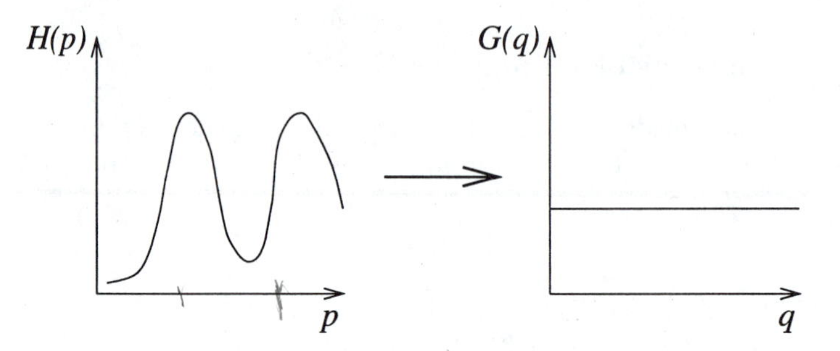

Figure 4.2: *Histogram equalization.*

Denote the input histogram by $H(p)$ and recall that the input gray-scale is $[p_0, p_k]$. The intention is to find a monotonic pixel brightness transformation $q = \mathcal{T}(p)$ such that the desired output histogram $G(q)$ is uniform over the whole output brightness scale $[q_0, q_k]$.

The histogram can be treated as a discrete probability density function. The monotonic property of the transform $\mathcal{T}$ implies

$$\sum_{i=0}^{k} G(q_i) = \sum_{i=0}^{k} H(p_i) \tag{4.4}$$

The sums in equation (4.4) can be interpreted as discrete distribution functions. Assume that the image has N rows and columns; then the equalized histogram $G(q)$ corresponds to the uniform probability density function f whose function value is a constant:

$$f = \frac{N^2}{q_k - q_0} \tag{4.5}$$

The value from equation (4.5) replaces the left side of equation (4.4). The equalized histogram can be obtained precisely only for the 'idealized' continuous probability density, in which case equation (4.4) becomes

$$N^2 \int_{q_0}^{q} \frac{1}{q_k - q_0} \, ds = \frac{N^2 (q - q_0)}{q_k - q_0} = \int_{p_0}^{p} H(s) \, ds \tag{4.6}$$

The desired pixel brightness transformation $\mathcal{T}$ can then be derived as

$$q = \mathcal{T}(p) = \frac{q_k - q_0}{N^2} \int_{p_0}^{p} H(s) \, ds + q_0 \tag{4.7}$$

The integral in equation (4.7) is called the **cumulative histogram**, which is approximated by a sum in digital images, so the resulting histogram is not equalized ideally. The discrete approximation of the continuous pixel brightness transformation from equation (4.7) is

$$q = \mathcal{T}(p) = \frac{q_k - q_0}{N^2} \sum_{i=p_0}^{p} H(i) + q_0 \tag{4.8}$$

Formally, the algorithm to perform equalization is as follows.

Algorithm 4.1: Histogram equalization

1. For an $N \times M$ image of G gray-levels (often 256), create an array H of length G initialized with 0 values.

2. Form the image histogram: Scan every pixel and increment the relevant member of H—if pixel p has intensity g_p, perform

$$H[g_p] = H[g_p] + 1$$

3. Form the cumulative image histogram H_c:

$$\begin{aligned} H_c[0] &= H[0] \\ H_c[p] &= H_c[p-1] + H[p] \qquad p = 1, 2, \ldots, G-1 \end{aligned}$$

4. Set

$$T[p] = \text{round}\left(\frac{G-1}{NM} H_c[p]\right)$$

   (This step obviously lends itself to more efficient implementation by constructing a look-up table of the multiples of NM, and making comparisons with the values in H_c, which are monotonic increasing).

5. Rescan the image and write an output image with gray-levels g_q, setting

$$g_q = T[g_p]$$

(This presentation assumes that the intensity range of source and destination images is [0,G-1]—the adjustment if this is not the case is trivial.)

These results can be demonstrated on an image of a lung. An input image and its equalization are shown in Figure 4.3; their respective histograms are shown in Figure 4.4.

The **logarithmic** gray-scale transformation function is another frequently used technique. It simulates the logarithmic sensitivity of the human eye to the light intensity.

Also belonging to the group of pixel brightness transformations are **adaptive neighborhood histogram modification** and **adaptive neighborhood contrast enhancement**. These methods are discussed later in Section 4.3.9 in the context of other adaptive neighborhood pre-processing approaches.

Pseudo-color is yet another kind of gray-scale transform. The individual brightnesses in the input monochromatic image are coded to some color. Since the human eye is much more sensitive to change in color than to change in brightness, much more detail can be perceived in pseudo-colored images.

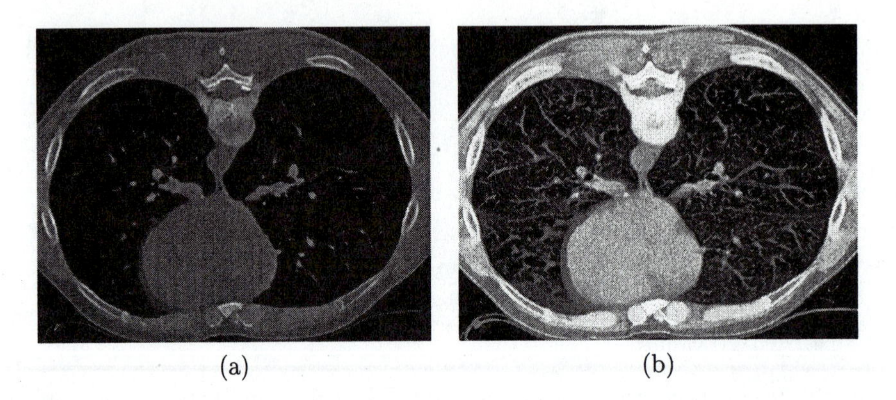

Figure 4.3: *Histogram equalization: (a) original image; (b) equalized image.*

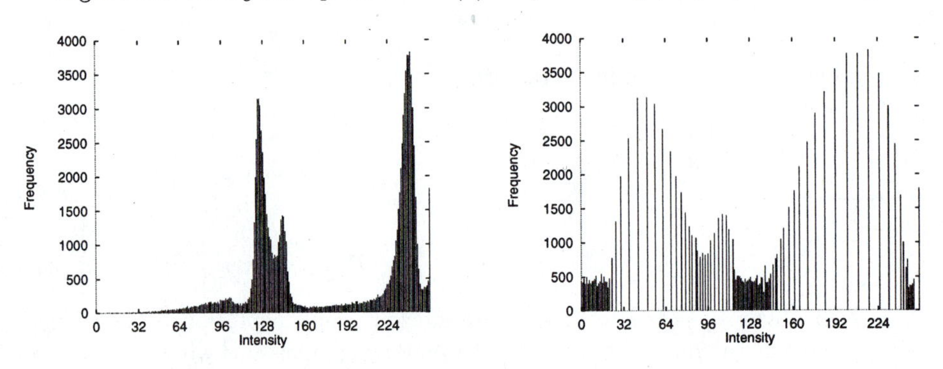

Figure 4.4: *Histogram equalization: Original and equalized histograms corresponding to Figure 4.3a,b.*

4.2 Geometric transformations

Geometric transforms are common in computer graphics, and are often used in image analysis as well. They permit elimination of the geometric distortion that occurs when an image is captured. If one attempts to match two different images of the same object, a geometric transformation may be needed. We consider geometric transformations only in 2D, as this is sufficient for most digital images. One example is an attempt to match remotely sensed images of the same area taken after one year, when the more recent image was probably not taken from precisely the same position. To inspect changes over the year, it is necessary first to execute a geometric transformation, and then subtract one image from the other. Another example, commonly encountered in document image processing applications, is correcting for document skew, which occurs when an image with an obvious orientation (for example, a printed page) is scanned, or otherwise captured, at a different orientation. This difference may be very small, but can be critical if the orientation is exploited in subsequent processing—this is usually the case in optical character recognition (OCR). An example application of this is described in Section 16.1 (see Figures 16.1 and 16.2).

A geometric transform is a vector function $\mathbf{T}$ that maps the pixel (x, y) to a new posi-

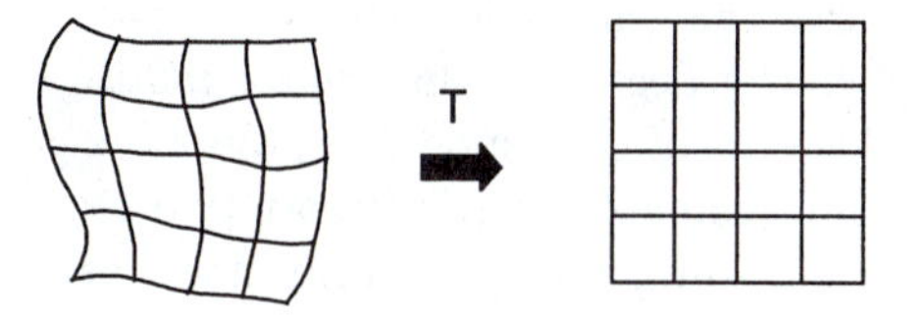

Figure 4.5: *Geometric transform on a plane.*

tion (x', y')—an illustration of the whole region transformed on a point-to-point basis is in Figure 4.5. **T** is defined by its two component equations,

$$x' = T_x(x, y) \qquad y' = T_y(x, y) \tag{4.9}$$

The transformation equations T_x and T_y are either known in advance—for example, in the case of rotation, translation, scaling—or can be determined from known original and transformed images. Several pixels in both images with known correspondences are used to derive the unknown transformation.

A geometric transform consists of two basic steps. First is the **pixel co-ordinate transformation**, which maps the co-ordinates of the input image pixel to the point in the output image. The output point co-ordinates should be computed as continuous values (real numbers), as the position does not necessarily match the digital grid after the transform. The second step is to find the point in the digital raster which matches the transformed point and determine its brightness value. The brightness is usually computed as an **interpolation** of the brightnesses of several points in the neighborhood.

This idea enables the classification of geometric transforms among other pre-processing techniques, the criterion being that only the neighborhood of a processed pixel is needed for the calculation. Geometric transforms are on the boundary between point and local operations.

4.2.1 Pixel co-ordinate transformations

Equation (4.9) shows the general case of finding the co-ordinates of a point in the output image after a geometric transform. It is usually approximated by a polynomial equation,

$$x' = \sum_{r=0}^{m} \sum_{k=0}^{m-r} a_{rk}\, x^r\, y^k \qquad y' = \sum_{r=0}^{m} \sum_{k=0}^{m-r} b_{rk}\, x^r\, y^k \tag{4.10}$$

This transform is linear with respect to the coefficients a_{rk}, b_{rk} and so if pairs of corresponding points (x, y), (x', y') in both images are known, it is possible to determine a_{rk}, b_{rk} by solving a set of linear equations. More points than coefficients are usually used to provide robustness; the mean square method is often used.

In the case where the geometric transform does not change rapidly depending on position in the image, low-order approximating polynomials, $m = 2$ or $m = 3$, are used, needing at least 6 or 10 pairs of corresponding points. The corresponding points should be distributed in the image in a way that can express the geometric transformation—usually they are spread uniformly. In general, the higher the degree of the approximating polynomial, the more sensitive to the distribution of the pairs of corresponding points the geometric transform is.

Equation (4.9) is in practice approximated by a **bilinear transform** for which four pairs of corresponding points are sufficient to find the transformation coefficients:

$$\begin{aligned} x' &= a_0 + a_1 x + a_2 y + a_3 xy \\ y' &= b_0 + b_1 x + b_2 y + b_3 xy \end{aligned} \tag{4.11}$$

Even simpler is the **affine transformation**, for which three pairs of corresponding points are sufficient to find the coefficients:

$$\begin{aligned} x' &= a_0 + a_1 x + a_2 y \\ y' &= b_0 + b_1 x + b_2 y \end{aligned} \tag{4.12}$$

The affine transformation includes typical geometric transformations such as rotation, translation, scaling, and skewing.

A geometric transform applied to the whole image may change the co-ordinate system, and a **Jacobian** J provides information about how the co-ordinate system changes:

$$J = \left| \frac{\partial(x', y')}{\partial(x, y)} \right| = \left| \begin{matrix} \frac{\partial x'}{\partial x} & \frac{\partial x'}{\partial y} \\ \frac{\partial y'}{\partial x} & \frac{\partial y'}{\partial y} \end{matrix} \right| \tag{4.13}$$

If the transformation is singular (has no inverse), then $J = 0$. If the area of the image is invariant under the transformation, then $J = 1$.

The Jacobian for the bilinear transform (4.11) is

$$J = a_1 b_2 - a_2 b_1 + (a_1 b_3 - a_3 b_1)x + (a_3 b_2 - a_2 b_3)y \tag{4.14}$$

and for the affine transformation (4.12) it is

$$J = a_1 b_2 - a_2 b_1 \tag{4.15}$$

Some important geometric transformations are:

- **Rotation** by the angle ϕ about the origin

$$\begin{aligned} x' &= x \cos\phi + y \sin\phi \\ y' &= -x \sin\phi + y \cos\phi \\ J &= 1 \end{aligned} \tag{4.16}$$

- **Change of scale** a in the x axis and b in the y axis

$$\begin{aligned} x' &= ax \\ y' &= bx \\ J &= ab \end{aligned} \tag{4.17}$$

- **Skewing by the angle ϕ,** given by

$$\begin{aligned} x' &= x + y \tan \phi \\ y' &= y \\ J &= 1 \end{aligned} \tag{4.18}$$

It is possible to approximate complex geometric transformations (distortion) by partitioning an image into smaller rectangular subimages; for each subimage, a simple geometric transformation, such as the affine, is estimated using pairs of corresponding pixels. The geometric transformation (distortion) is then repaired separately in each subimage.

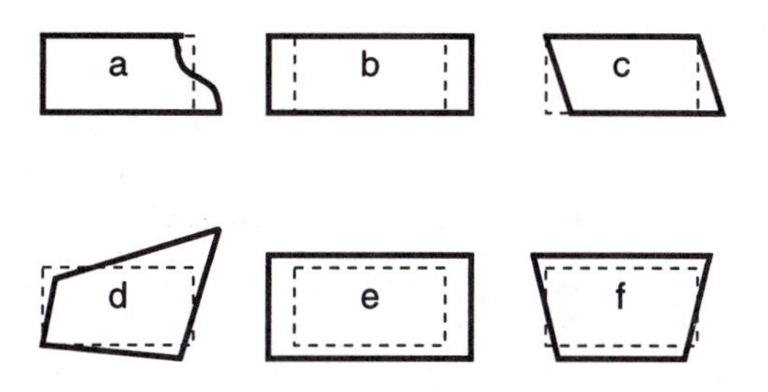

Figure 4.6: *Geometric distortion types.*

There are some typical geometric distortions which have to be overcome in remote sensing. Errors may be caused by distortion of the optical systems, by the non-linearities in row-by-row scanning and a non-constant sampling period. Wrong position or orientation of the sensor (or the satellite) with respect to the object is the main cause of rotation, skew, and line non-linearity distortions. Panoramic distortion (Figure 4.6b) appears in line scanners with the mirror rotating at constant speed. Line non-linearity distortion (Figure 4.6a) is caused by variable distance of the object from the scanner mirror. The rotation of the Earth during image capture in a mechanical scanner generates skew distortion (Figure 4.6c). Change of distance from the sensor induces change-of-scale distortion (Figure 4.6e). Perspective projection causes perspective distortion (Figure 4.6f).

4.2.2 Brightness interpolation

Assume that the planar transformation given by equation (4.9) has been accomplished, and new point co-ordinates (x', y') obtained. The position of the point does not in general fit the discrete raster of the output image, and the collection of transformed points gives the samples of the output image with non-integer co-ordinates. Values on the integer grid are needed, and each pixel value in the output image raster can be obtained by **brightness interpolation** of some neighboring non-integer samples [Moik 80].

Brightness interpolation influences image quality. The simpler the interpolation, the greater is the loss in geometric and photometric accuracy, but the interpolation neighborhood is often reasonably small due to computational load. The three most common interpolation methods are nearest neighbor, linear, and bi-cubic.

The brightness interpolation problem is usually expressed in a dual way by determining the brightness of the original point in the input image that corresponds to the point in the output image lying on the discrete raster. Assume that we wish to compute the brightness value of the pixel (x', y') in the output image where x' and y' lie on the discrete raster (integer numbers, illustrated by solid lines in figures). The co-ordinates of the point (x, y) in the original image can be obtained by inverting the planar transformation in equation (4.9):

$$(x, y) = \mathbf{T}^{-1}(x', y') \tag{4.19}$$

In general, the real co-ordinates after inverse transformation (dashed lines in figures) do not fit the input image discrete raster (solid lines), and so brightness is not known. The only information available about the originally continuous image function $f(x, y)$ is its sampled version $g_s(l\,\Delta x, k\,\Delta y)$. To get the brightness value of the point (x, y), the input image is resampled.

Denote the result of the brightness interpolation by $f_n(x, y)$, where n distinguishes different interpolation methods. The brightness can be expressed by the convolution equation

$$f_n(x, y) = \sum_{l=-\infty}^{\infty} \sum_{k=-\infty}^{\infty} g_s(l\,\Delta x,\, k\,\Delta y)\, h_n(x - l\,\Delta x,\, y - k\,\Delta y) \tag{4.20}$$

The function h_n is called the **interpolation kernel**. Usually, only a small neighborhood is used, outside which h_n is zero. [The same idea was used in continuous image sampling—recall that in equation (2.38) the function h_s represented the limited impulse.]

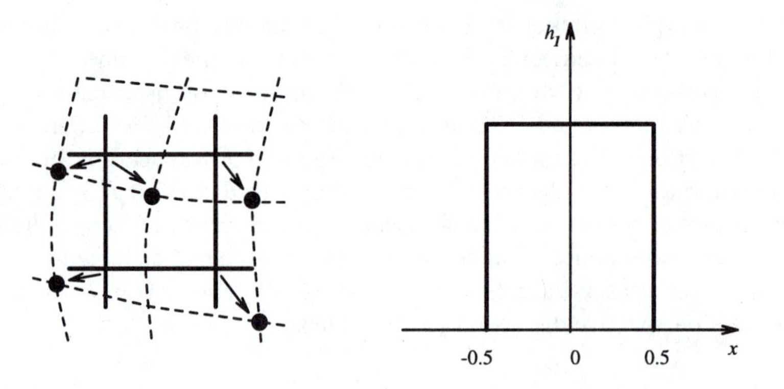

Figure 4.7: *Nearest-neighborhood interpolation. The discrete raster of the original image is depicted by the solid line.*

Nearest-neighborhood interpolation assigns to the point (x, y) the brightness value of the nearest point g in the discrete raster; this is demonstrated in Figure 4.7. On the right side is the interpolation kernel h_1 in the 1D case. The left side of Figure 4.7 shows how the new brightness is assigned. Dashed lines show how the inverse planar transformation maps the raster of the output image into the input image; full lines show the raster of the input image.

Nearest-neighborhood interpolation is given by

$$f_1(x,y) = g_s[\text{round}\,(x), \text{round}\,(y)] \tag{4.21}$$

The position error of the nearest-neighborhood interpolation is at most half a pixel. This error is perceptible on objects with straight-line boundaries that may appear step-like after the transformation.

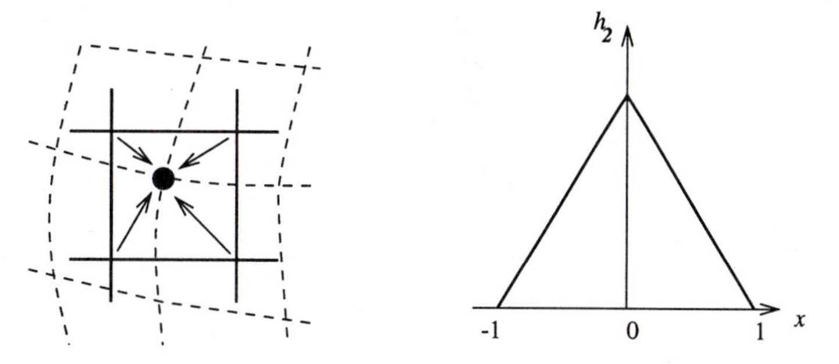

Figure 4.8: *Linear interpolation. The discrete raster of the original image is depicted by the solid line.*

Linear interpolation explores four points neighboring the point (x,y), and assumes that the brightness function is linear in this neighborhood. Linear interpolation is demonstrated in Figure 4.8, the left-hand side of which shows which points are used for interpolation. Linear interpolation is given by the equation

$$\begin{aligned} f_2(x,y) = \; & (1-a)(1-b)\, g_s(l,k) \\ & + a(1-b)\, g_s(l+1,k) \\ & + b(1-a)\, g_s(l,k+1) \\ & + ab\, g_s(l+1,k+1) \end{aligned} \tag{4.22}$$

$$\begin{aligned} l &= \text{round}\,(x), \quad & a = x - l \\ k &= \text{round}\,(y), \quad & b = y - k \end{aligned}$$

Linear interpolation can cause a small decrease in resolution, and blurring due to its averaging nature. The problem of step-like straight boundaries with the nearest-neighborhood interpolation is reduced.

Bi-cubic interpolation improves the model of the brightness function by approximating it locally by a bi-cubic polynomial surface; 16 neighboring points are used for interpolation. The one-dimensional interpolation kernel ('Mexican hat') is shown in Figure 4.9 and is given by

$$h_3 = \begin{cases} 1 - 2|x|^2 + |x|^3 & \text{for } 0 \le |x| < 1 \\ 4 - 8|x| + 5|x|^2 - |x|^3 & \text{for } 1 \le |x| < 2 \\ 0 & \text{otherwise} \end{cases} \tag{4.23}$$

Bi-cubic interpolation does not suffer from the step-like boundary problem of nearest-neighborhood interpolation, and copes with linear interpolation blurring as well. Bi-cubic interpolation is often used in raster displays that enable zooming with respect to an arbitrary point. If the nearest-neighborhood method were used, areas of the same brightness would increase. Bi-cubic interpolation preserves fine details in the image very well.

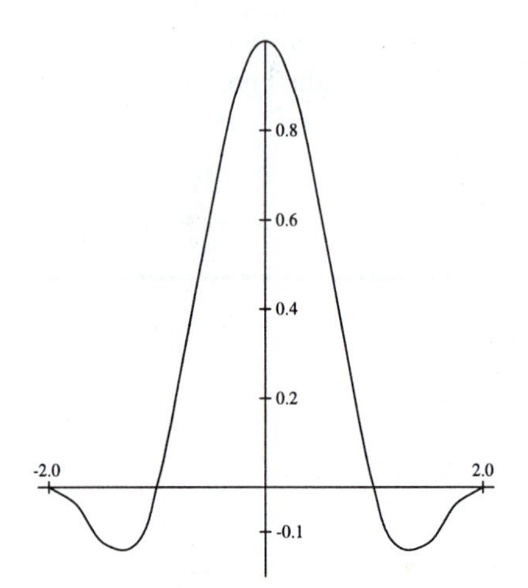

Figure 4.9: *Bi-cubic interpolation kernel.*

4.3 Local pre-processing

The object of interest in this section is pre-processing methods that use a small neighborhood of a pixel in an input image to produce a new brightness value in the output image. Such pre-processing operations are called also **filtration** (or **filtering**) if signal processing terminology is used.

Local pre-processing methods can be divided into two groups according to the goal of the processing. First, **smoothing** aims to suppress noise or other small fluctuations in the image; it is equivalent to the suppression of high frequencies in the Fourier transform domain. Unfortunately, smoothing also blurs all sharp edges that bear important information about the image. Second, **gradient operators** are based on local derivatives of the image function. Derivatives are bigger at locations of the image where the image function undergoes rapid changes, and the aim of gradient operators is to indicate such locations in the image. Gradient operators have a similar effect to suppressing low frequencies in the Fourier transform domain. Noise is often high frequency in nature; unfortunately, if a gradient operator is applied to an image, the noise level increases simultaneously.

Clearly, smoothing and gradient operators have conflicting aims. Some pre-processing algorithms solve this problem and permit smoothing and edge enhancement simultaneously.

Another classification of local pre-processing methods is according to the transformation properties; **linear** and **non-linear** transformations can be distinguished. Linear operations calculate the resulting value in the output image pixel $g(i,j)$ as a linear combination of brightnesses in a local neighborhood $\mathcal{O}$ of the pixel $f(i,j)$ in the input image. The contribution of the pixels in the neighborhood $\mathcal{O}$ is weighted by coefficients h:

$$f(i,j) = \sum_{(m,n) \in \mathcal{O}} \sum h(i-m,\, j-n)\, g(m,n) \tag{4.24}$$

Equation (4.24) is equivalent to discrete convolution with the kernel h, which is called a **convolution mask**. Rectangular neighborhoods $\mathcal{O}$ are often used with an odd number of pixels in rows and columns, enabling specification of the central pixel of the neighborhood.

Local pre-processing methods typically use very little a priori knowledge about the image contents. It is very difficult to infer this knowledge while an image is being processed, as the known neighborhood $\mathcal{O}$ of the processed pixel is small. Smoothing operations will benefit if some general knowledge about image degradation is available; this might, for instance, be statistical parameters of the noise.

The choice of the local transformation, size, and shape of the neighborhood $\mathcal{O}$ depends strongly on the size of objects in the processed image. If objects are rather large, an image can be enhanced by smoothing of small degradations.

Convolution-based operations (filters) can be used for smoothing, gradient operators, and line detectors. There are methods that enable the speed-up of calculations to ease implementation in hardware—examples are recursive filters or separable filters [Yaroslavskii 87].

4.3.1 Image smoothing

Image smoothing is the set of local pre-processing methods whose predominant use is the suppression of image noise—it uses redundancy in the image data. Calculation of the new value is based on the averaging of brightness values in some neighborhood $\mathcal{O}$. Smoothing poses the problem of blurring sharp edges in the image, and so we shall concentrate on smoothing methods which are **edge preserving**. They are based on the general idea that the average is computed only from those points in the neighborhood which have similar properties to the point being processed.

Local image smoothing can effectively eliminate impulse noise or degradations appearing as thin stripes, but does not work if degradations are large blobs or thick stripes. The solution for complicated degradations may be to use image restoration techniques, described in Section 4.4.

Averaging

Assume that the noise value ν at each pixel is an independent random variable with zero mean and standard deviation σ. We can obtain such an image by capturing the same static scene several times. The result of smoothing is an average of the same n points in these images $g_1, \ldots, g_n$ with noise values $\nu_1, \ldots, \nu_n$:

$$\frac{g_1 + \ldots + g_n}{n} + \frac{\nu_1 + \ldots + \nu_n}{n} \tag{4.25}$$

The second term here describes the effect of the noise, which is again a random value with zero mean and standard deviation $\sigma/\sqrt{n}$; the standard deviation is decreased by a factor $\sqrt{n}$.

Thus, if n images of the same scene are available, the smoothing can be accomplished without blurring the image by

$$f(i,j) = \frac{1}{n} \sum_{k=1}^{n} g_k(i,j) \tag{4.26}$$

$$h = \frac{1}{9} \begin{bmatrix} 1 & 1 & 1 \\ 1 & 1 & 1 \\ 1 & 1 & 1 \end{bmatrix} \tag{4.27}$$

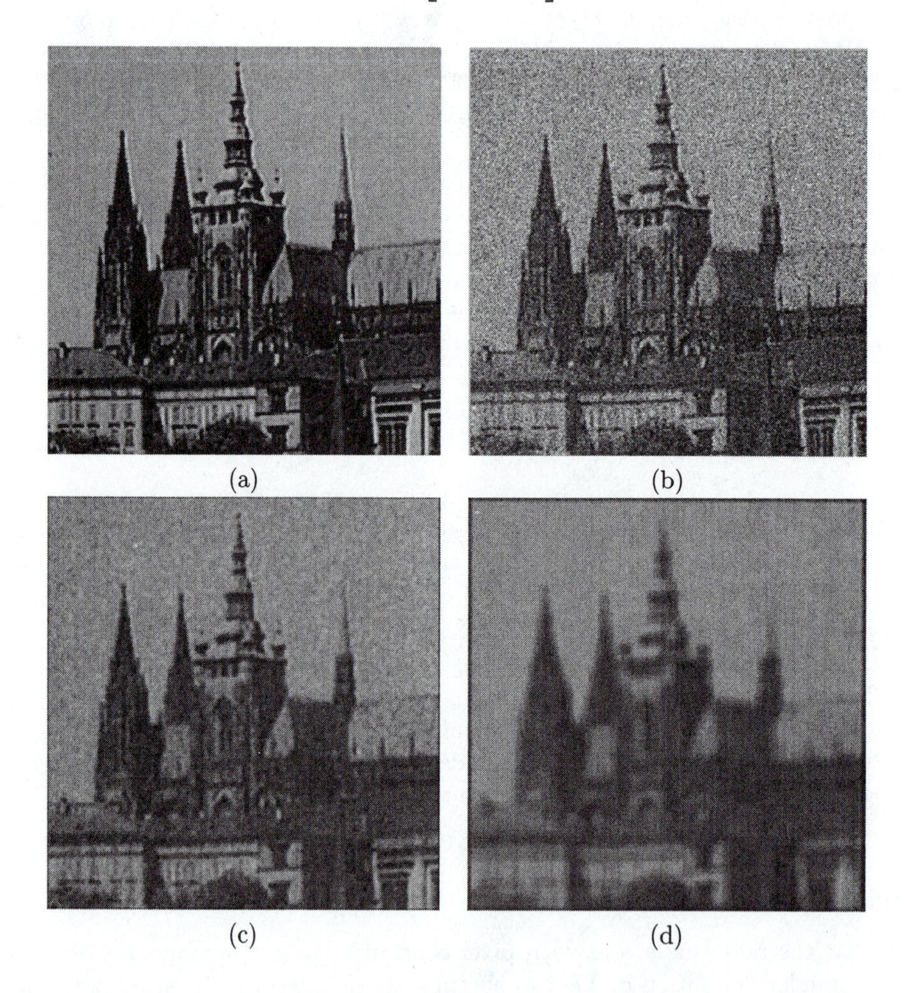

Figure 4.10: *Noise with Gaussian distribution and averaging filters: (a) original image; (b) superimposed noise (random Gaussian noise characterized by zero mean and standard deviation equal to one-half of the gray-level standard deviation of the original image); (c) 3×3 averaging; (d) 7×7 averaging. Compare with noise reduction results shown in Figures 12.9 and 12.10.*

In many cases only one image with noise is available, and averaging is then realized in a local neighborhood. Results are acceptable if the noise is smaller in size than the smallest objects of interest in the image, but blurring of edges is a serious disadvantage. In the case of smoothing within a single image, one has to assume that there are no changes in the gray-levels of the underlying image data. This assumption is clearly violated at locations of image edges, and edge blurring is a direct consequence of violating the assumptions. Averaging is a special case of discrete convolution [equation (4.24)]. For a 3×3 neighborhood, the convolution mask h is

$$\frac{1}{9}\begin{bmatrix} 1 & 1 & 1 \\ 1 & 1 & 1 \\ 1 & 1 & 1 \end{bmatrix} \tag{4.28}$$

The significance of the pixel in the center of the convolution mask h or its 4-neighbors is sometimes increased, as it better approximates the properties of noise with a Gaussian probability distribution (Gaussian noise, see Section 2.3.5).

$$h = \frac{1}{10}\begin{bmatrix} 1 & 1 & 1 \\ 1 & 2 & 1 \\ 1 & 1 & 1 \end{bmatrix} \qquad h = \frac{1}{16}\begin{bmatrix} 1 & 2 & 1 \\ 2 & 4 & 2 \\ 1 & 2 & 1 \end{bmatrix} \tag{4.29}$$

Larger convolution masks for averaging are created analogously according to the Gaussian distribution formula [equation (4.52)] and the mask coefficients are normalized to have a unit sum.

An example will illustrate the effect of this noise suppression. Images with low resolution (256×256) were chosen deliberately to show the discrete nature of the process. Figure 4.10a shows an original image of Prague castle with 256 brightnesses; Figure 4.10b shows the same image with superimposed additive noise with Gaussian distribution; Figure 4.10c shows the result of averaging with a 3×3 convolution mask (4.29)—noise is significantly reduced and the image is slightly blurred. Averaging with a larger mask (7×7) is demonstrated in Figure 4.10d, where the blurring is much more serious.

Alternative techniques, which are mostly non-linear, will now be discussed. These attempt not to blur sharp edges by avoiding averaging across edges.

Averaging with limited data validity

Methods that average with limited data validity [McDonnell 81] try to avoid blurring by averaging only those pixels which satisfy some criterion, the aim being to prevent involving pixels that are part of a separate feature.

A very simple criterion is to apply image averaging only to pixels in the original image with brightness in a pre-defined interval of invalid data [min, max] that typically corresponds to the gray-level interval of noise or other image faults. Considering the point (m, n) in the image, the convolution mask is calculated in the neighborhood $\mathcal{O}$ from the non-linear formula

$$h(i,j) = \begin{cases} 1 & \text{for } g(m+i, n+j) \notin [\text{min,max}] \\ 0 & \text{otherwise} \end{cases} \tag{4.30}$$

where (i, j) specify the mask element. Therefore, only values of pixels with invalid gray-levels are replaced with an average of their neighborhoods, and only valid data contribute to the neighborhood averages. The power of this approach is illustrated in Figure 4.11—with the exception of slight local blurring of the towers, the method removes the significant image corruptions successfully.

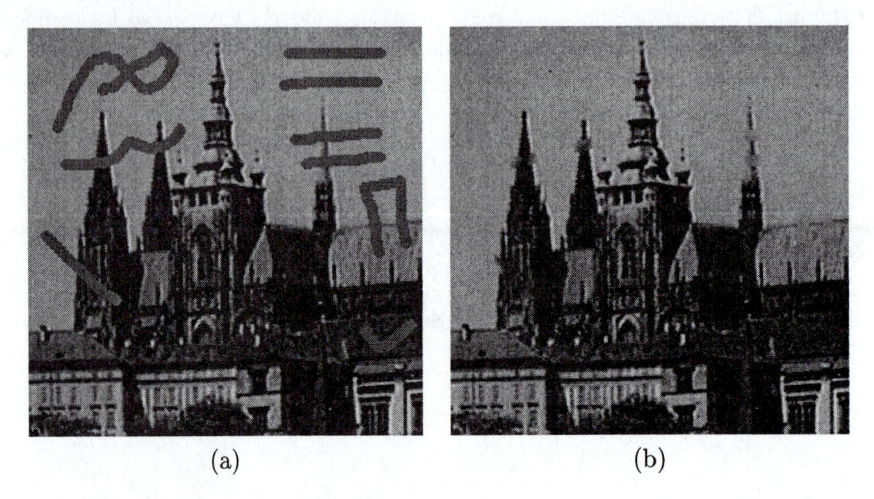

(a) (b)

Figure 4.11: *Averaging with limited data validity: (a) original corrupted image; (b) result of corruption removal.*

A second method performs averaging only if the computed brightness change of a pixel is in some pre-defined interval. This method permits repair to large-area errors resulting from slowly changing brightness of the background without affecting the rest of the image.

A third method uses edge strength (i.e., magnitude of a gradient) as a criterion. The magnitude of some gradient operator (see Section 4.3.2) is first computed for the entire image, and only pixels in the input image with a gradient magnitude smaller than a pre-defined threshold are used in averaging. This method effectively rejects averaging at edges and therefore suppresses blurring, but setting of the threshold is laborious.

Averaging according to inverse gradient

The convolution mask is calculated at each pixel according to the inverse gradient [Wang and Vagnucci 81], the idea being that the brightness change within a region is usually smaller than between neighboring regions. Let pixel location (m,n) correspond to that of a central pixel of a convolution mask with odd size; the inverse gradient δ at the point (i,j) with respect to (m,n) is then

$$\delta(i,j) = \frac{1}{|g(m,n) - g(i,j)|} \tag{4.31}$$

If $g(m,n) = g(i,j)$, then we define $\delta(i,j) = 2$; the inverse gradient δ is then in the interval $(0,2]$, and δ is smaller on the edge than in the interior of a homogeneous region. Weight coefficients in the convolution mask h are normalized by the inverse gradient, and the whole term is multiplied by 0.5 to keep brightness values in the original range. The constant 0.5 has the effect of assigning half the weight to the central pixel (m,n), and the other half to its neighborhood.

$$h(i,j) = 0.5 \frac{\delta(i,j)}{\sum_{(m,n)\in \mathcal{O}} \delta(i,j)} \tag{4.32}$$

The convolution mask coefficient corresponding to the central pixel is defined as $h(i,j) = 0.5$.

This method assumes sharp edges. When the convolution mask is close to an edge, pixels from the region have larger coefficients than pixels near the edge, and it is not blurred. Isolated noise points within homogeneous regions have small values of the inverse gradient; points from the neighborhood take part in averaging and the noise is removed.

Averaging using a rotating mask

Averaging using a rotating mask is a method that avoids edge blurring by searching for the homogeneous part of the current pixel neighborhood and the resulting image is in fact sharpened [Nagao and Matsuyama 80]. The brightness average is calculated only within this region; a brightness dispersion σ^2 is used as the region homogeneity measure. Let n be the number of pixels in a region R and g be the input image. Dispersion σ^2 is calculated as

$$\sigma^2 = \frac{1}{n} \left\{ \sum_{(i,j)\in R} \left[g(i,j) - \frac{1}{n} \sum_{(i,j)\in R} g(i,j) \right]^2 \right\} \tag{4.33}$$

The computational complexity (number of multiplications) of the dispersion calculation can be reduced if equation (4.33) is expressed another way:

$$\begin{aligned} \sigma^2 &= \frac{1}{n} \sum_{(i,j)\in R} \left\{ [g(i,j)]^2 - 2\,g(i,j) \frac{\sum_{(i,j)\in R} g(i,j)}{n} + \left[\frac{\sum_{(i,j)\in R} g(i,j)}{n} \right]^2 \right\} \\ &= \frac{1}{n} \left\{ \sum_{(i,j)\in R} [g(i,j)]^2 - 2\,\frac{\left[\sum_{(i,j)\in R} g(i,j)\right]^2}{n} + n \left[\frac{\sum_{(i,j)\in R} g(i,j)}{n} \right]^2 \right\} \\ &= \frac{1}{n} \left\{ \sum_{(i,j)\in R} [g(i,j)]^2 - \frac{\left[\sum_{(i,j)\in R} g(i,j)\right]^2}{n} \right\} \end{aligned} \tag{4.34}$$

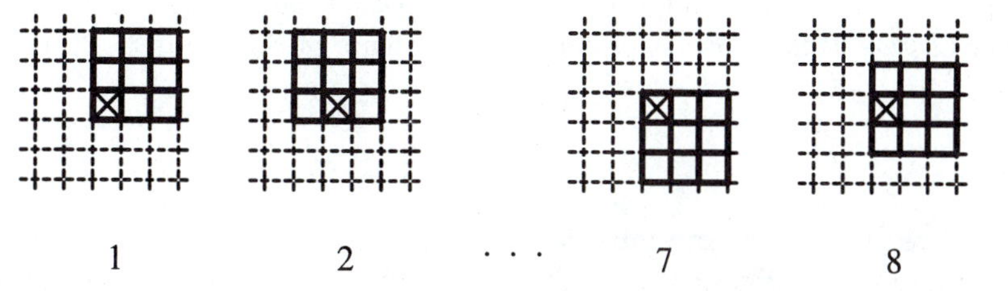

Figure 4.12: *Eight possible rotated* 3×3 *masks.*

Having computed region homogeneity, we consider its shape and size. The eight possible 3 $\times$ 3 masks that cover a 5×5 neighborhood of a current pixel (marked by the small cross) are shown in Figure 4.12. The ninth mask is the 3×3 neighborhood of the current pixel itself. Other mask shapes can also be used. Figure 4.13 shows another set of eight masks covering a 5×5 neighborhood of the current pixel. Again the ninth mask is the 3×3 neighborhood of the current pixel. Another possibility is to rotate a small 2×1 mask to cover the 3×3

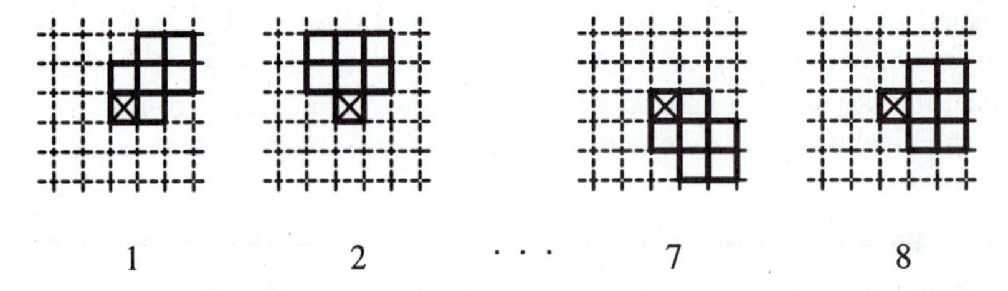

Figure 4.13: *Alternative shape of eight possible rotated masks.*

neighborhood of the current pixel.

Image smoothing using the rotating mask technique uses the following algorithm.

Algorithm 4.2: Smoothing using a rotating mask

1. Consider each image pixel (i, j).
2. Calculate dispersion in the mask for all possible mask rotations about pixel (i, j) according to equation (4.33).
3. Choose the mask with minimum dispersion.
4. Assign to the pixel $f(i, j)$ in the output image f the average brightness in the chosen mask.

Algorithm 4.2 can be used iteratively; the iterative process converges quite quickly to the stable state (that is, the image does not change any more). The size and shape of masks influence the convergence—the smaller the mask, the smaller are the changes and more iterations are needed. A larger mask suppresses noise faster and the sharpening effect is stronger. On the other hand, information about details smaller than the mask may be lost. The number of iterations is also influenced by the shape of regions in the image and noise properties.

Median filtering

In a set of ordered values, the **median**[2] is the central value.

Median filtering is a non-linear smoothing method that reduces the blurring of edges [Tyan 81], in which the idea is to replace the current point in the image by the median of the brightnesses in its neighborhood. The median of the brightnesses in the neighborhood is not

[2]The median is defined in probability theory. For a random variable x, the median M is the value for which the probability of the outcome $x < M$ is 0.5.

affected by individual noise spikes and so median smoothing eliminates impulse noise quite well. Further, as median filtering does not blur edges much, it can be applied iteratively. Clearly, performing a sort on pixels within a (possibly large) rectangular window at every pixel position may become very expensive. A more efficient approach [Huang et al. 79, Pitas and Venetsanopoulos 90] is to notice that as the window moves across a row by one column, the only change to its contents is to lose the leftmost column and replace it with a new right column—for a median window of m rows and n columns, $mn - 2 * m$ pixels are unchanged and do not need re-sorting. The algorithm is as follows,

Algorithm 4.3: Efficient median filtering

1. Set
$$th = \frac{mn}{2}$$

2. Position the window at the beginning of a new row, and sort its contents. Construct a histogram H of the window pixels, determine the median med, and record lt_med, the number of pixels with intensity less than or equal to med.

3. For each pixel p in the leftmost column of intensity p_g, perform
$$H[p_g] = H[p_g] - 1$$
Further, if $p_g < med$, set
$$lt\_med = lt\_med - 1$$

4. Move the window one column right. For each pixel p in the rightmost column of intensity p_g, perform
$$H[p_g] = H[p_g] + 1$$
If $p_g < med$, set
$$lt\_med = lt\_med + 1$$

5. If $lt\_med > th$ then go to 6.

   Repeat
$$\begin{aligned} lt\_med &= lt\_med + H[med] \\ med &= med + 1 \end{aligned}$$
   until $lt\_med \geq th$. Go to 7.

6. Repeat
$$\begin{aligned} med &= med - 1 \\ lt\_med &= lt\_med - H[med] \end{aligned} \tag{4.35}$$
   until $lt\_med \leq th$.

7. If the right-hand column of the window is not at the right-hand edge of the image, go to step 3.

8. If the bottom row of the window is not at the bottom of the image, go to step 2.

The effect of median filtering is shown in Figure 4.14.

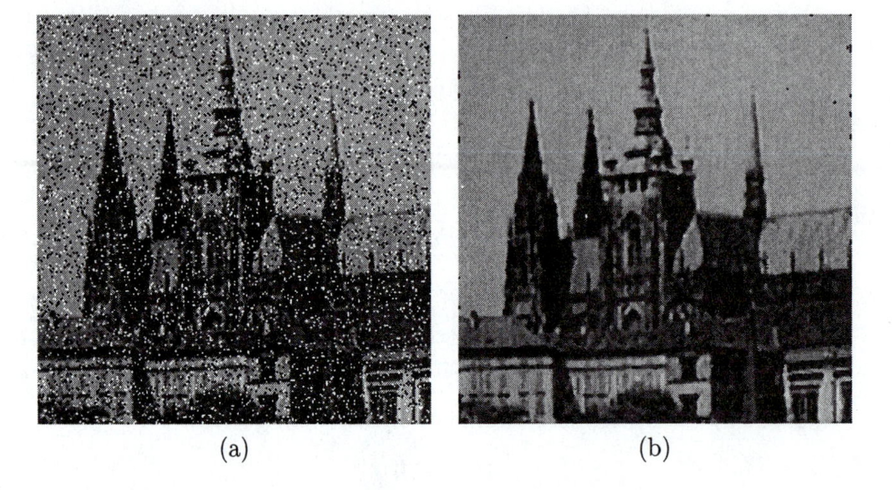

Figure 4.14: *Median filtering: (a) image corrupted with impulse noise (14% of image area covered with bright and dark dots); (b) result of* 3×3 *median filtering.*

The main disadvantage of median filtering in a rectangular neighborhood is its damaging of thin lines and sharp corners in the image—this can be avoided if another shape of neighborhood is used. For instance, if horizontal/vertical lines need preserving, a neighborhood such as that in Figure 4.15 can be used.

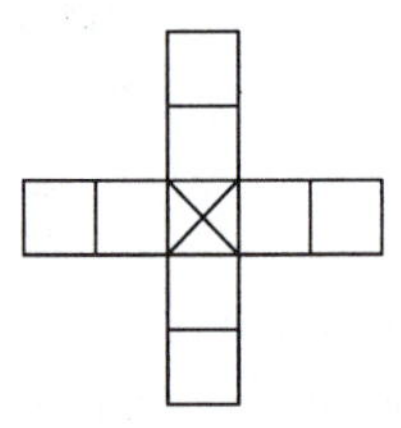

Figure 4.15: *Horizontal/vertical line preserving neighborhood for median filtering.*

Median smoothing is a special instance of more general **rank filtering** techniques [Rosenfeld and Kak 82, Yaroslavskii 87], the idea of which is to order pixels in some neighborhood into sequence. The results of pre-processing are some statistics over this sequence, of which the median is one possibility. Another variant is the maximum or the minimum values of the sequence. This defines generalizations of dilation and erosion operators (Chapter 11) in images with more brightness values.

A similar generalization of median techniques is given in [Borik et al. 83]. Their method is called **order statistics** (OS) filtering. Values in the neighborhood are again ordered into sequence, and a new value is given as a linear combination of the values of this sequence. Median smoothing, and minimum or maximum filters, are a special case of OS filtering.

Non-linear mean filter

The non-linear mean filter is another generalization of averaging techniques [Pitas and Venetsanopulos 86]; it is defined by

$$f(m,n) = u^{-1}\left\{\frac{\sum_{(i,j)\in\mathcal{O}} a(i,j)\, u[g(i,j)]}{\sum_{(i,j)\in\mathcal{O}} a(i,j)}\right\} \tag{4.36}$$

where $f(m,n)$ is the result of the filtering, $g(i,j)$ is the pixel in the input image, and $\mathcal{O}$ is a local neighborhood of the current pixel (m,n). The function u of one variable has an inverse function u^{-1}; the $a(i,j)$ are weight coefficients.

If the weights $a(i,j)$ are constant, the filter is called **homomorphic**. Some homomorphic filters used in image processing are

- Arithmetic mean, $u(g) = g$
- Harmonic mean, $u(g) = 1/g$
- Geometric mean, $u(g) = \log g$

Yet another approach to image pre-processing performed in homogeneous pixel neighborhoods is discussed in Section 4.3.9 in the context of several other adaptive-neighborhood pre-processing methods.

4.3.2 Edge detectors

Edge detectors are a collection of very important local image pre-processing methods used to locate changes in the intensity function; edges are pixels where this function (brightness) changes abruptly.

Calculus describes changes of continuous functions using derivatives; an image function depends on two variables—co-ordinates in the image plane—and so operators describing edges are expressed using partial derivatives. A change of the image function can be described by a gradient that points in the direction of the largest growth of the image function.

An edge is a property attached to an individual pixel and is calculated from the image function behavior in a neighborhood of that pixel. It is a **vector variable** with two components, **magnitude** and **direction**. The edge magnitude is the magnitude of the gradient, and the edge direction ϕ is rotated with respect to the gradient direction ψ by $-90°$. The gradient direction gives the direction of maximum growth of the function, e.g., from black $[f(i,j) = 0]$ to white $[f(i,j) = 255]$. This is illustrated in Figure 4.16, in which closed lines are lines of equal brightness. The orientation $0°$ points east.

Edges are often used in image analysis for finding region boundaries. Provided that the region has homogeneous brightness, its boundary is at the pixels where the image function varies and so in the ideal case without noise consists of pixels with high edge magnitude. It can be seen that the boundary and its parts (edges) are perpendicular to the direction of the gradient.

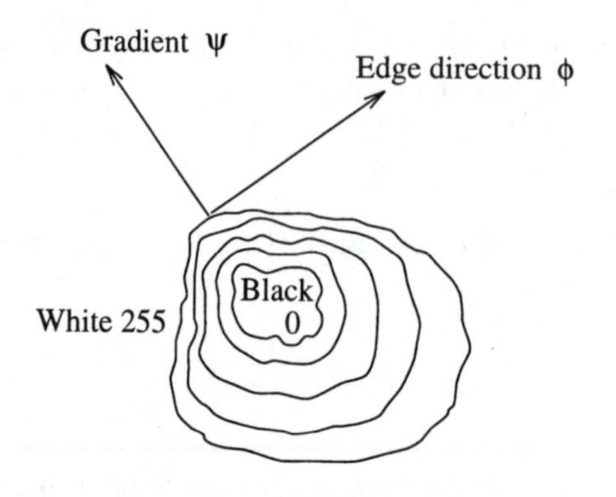

Figure 4.16: *Gradient direction and edge direction.*

The edge profile in the gradient direction (perpendicular to the edge direction) is typical for edges, and Figure 4.17 shows examples of several standard edge profiles. Roof edges are typical for objects corresponding to thin lines in the image. Edge detectors are usually tuned for some type of edge profile.

The gradient magnitude $|\text{grad } g(x,y)|$ and gradient direction ψ are continuous image functions calculated as

$$|\text{grad } g(x,y)| = \sqrt{\left(\frac{\partial g}{\partial x}\right)^2 + \left(\frac{\partial g}{\partial y}\right)^2} \tag{4.37}$$

$$\psi = \arg\left(\frac{\partial g}{\partial x}, \frac{\partial g}{\partial y}\right) \tag{4.38}$$

where $\arg(x,y)$ is the angle (in radians) from the x axis to the point (x,y).

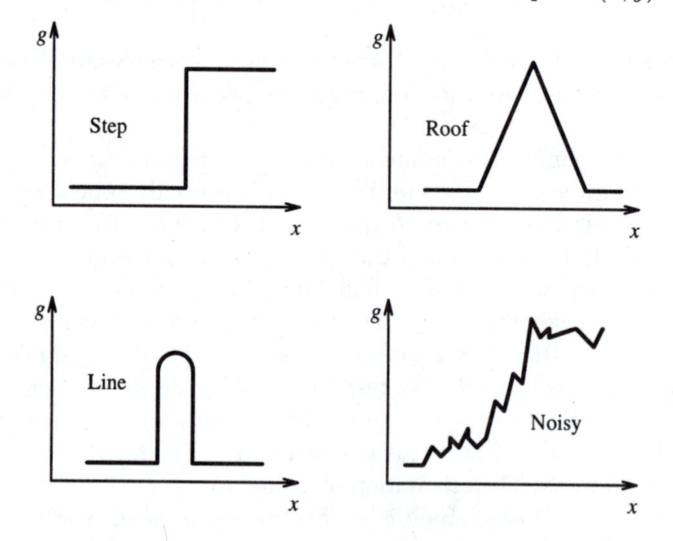

Figure 4.17: *Typical edge profiles.*

Sometimes we are interested only in edge magnitudes without regard to their orientations—a linear differential operator called the **Laplacian** may then be used. The Laplacian has the same properties in all directions and is therefore invariant to rotation in the image. It is defined as

$$\nabla^2 g(x,y) = \frac{\partial^2 g(x,y)}{\partial x^2} + \frac{\partial^2 g(x,y)}{\partial y^2} \tag{4.39}$$

Image **sharpening** [Rosenfeld and Kak 82] has the objective of making edges steeper—the sharpened image is intended to be observed by a human. The sharpened output image f is obtained from the input image g as

$$f(i,j) = g(i,j) - C S(i,j) \tag{4.40}$$

where C is a positive coefficient which gives the strength of sharpening and $S(i,j)$ is a measure of the image function sheerness, calculated using a gradient operator. The Laplacian is very often used for this purpose. Figure 4.18 gives an example of image sharpening using a Laplacian.

Image sharpening can be interpreted in the frequency domain as well. We already know that the result of the Fourier transform is a combination of harmonic functions. The derivative of the harmonic function $\sin(nx)$ is $n\cos(nx)$; thus the higher the frequency, the higher the magnitude of its derivative. This is another explanation of why gradient operators enhance edges.

A similar image sharpening technique to that given in equation (4.40), called **unsharp masking**, is often used in printing industry applications [Jain 89]. A signal proportional to an unsharp image (e.g., heavily blurred by a smoothing operator) is subtracted from the original image.

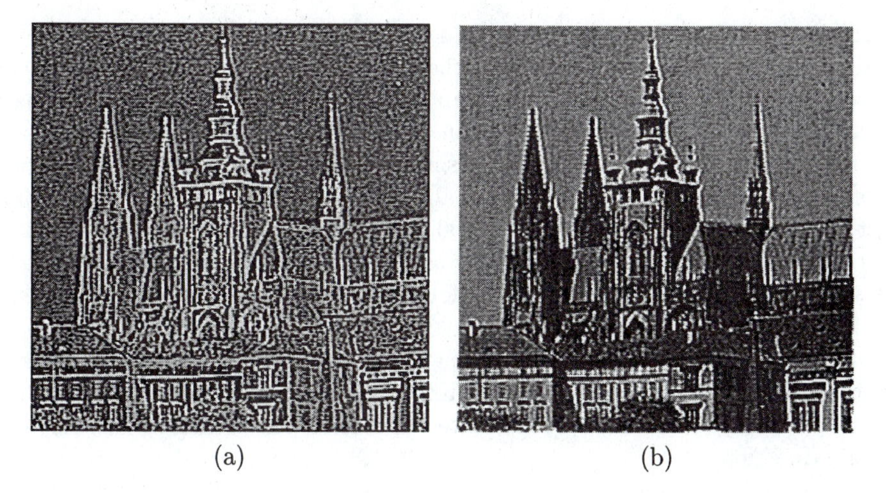

(a) (b)

Figure 4.18: *Laplace gradient operator: (a) Laplace edge image using the 8-connectivity mask; (b) sharpening using the Laplace operator [equation (4.40), $C = 0.7$]. Compare the sharpening effect with the original image in Figure 4.10a.*

A digital image is discrete in nature and so equations (4.37) and (4.38), containing derivatives, must be approximated by **differences**. The first differences of the image g in the vertical direction (for fixed i) and in the horizontal direction (for fixed j) are given by

$$\begin{aligned} \Delta_i\, g(i,j) &= g(i,j) - g(i-n,j) \\ \Delta_j\, g(i,j) &= g(i,j) - g(i,j-n) \end{aligned} \tag{4.41}$$

where n is a small integer, usually 1. The value n should be chosen small enough to provide a good approximation to the derivative, but large enough to neglect unimportant changes in the image function. Symmetric expressions for the difference,

$$\begin{aligned} \Delta_i\, g(i,j) &= g(i+n,j) - g(i-n,j) \\ \Delta_j\, g(i,j) &= g(i,j+n) - g(i,j-n) \end{aligned} \tag{4.42}$$

are not usually used because they neglect the impact of the pixel (i,j) itself.

Gradient operators as a measure of edge sheerness can be divided into three categories:

1. Operators approximating derivatives of the image function using differences. Some of them are rotationally invariant (e.g., Laplacian) and thus are computed from one convolution mask only. Others, which approximate first derivatives, use several masks. The orientation is estimated on the basis of the best matching of several simple patterns.

2. Operators based on the zero-crossings of the image function second derivative (e.g., Marr-Hildreth or Canny edge detectors).

3. Operators which attempt to match an image function to a parametric model of edges.

The remainder of this section will consider some of the many operators which fall into the first category, and the next section will consider the second. The last category is briefly outlined in Section 4.3.6. Edge detection represents an extremely important step facilitating higher-level image analysis and therefore remains an area of active research, with new approaches continually being developed. Recent examples include edge detectors using fuzzy logic, neural networks, or wavelets [Law et al. 96, Wang et al. 95, Sun and Sclabassi 95, Ho and Ohnishi 95, Aydin et al. 96, Vrabel 96, Hebert and Kim 96, Bezdek et al. 96] It may be difficult to select the most appropriate edge detection strategy; a comparison of edge detection approaches and an assessment of their performance may be found in [Ramesh and Haralick 94, Demigny et al. 95].

Individual gradient operators that examine small local neighborhoods are in fact convolutions [cf. equation (4.24)], and can be expressed by convolution masks. Operators which are able to detect edge direction are represented by a collection of masks, each corresponding to a certain direction.

Roberts operator

The Roberts operator is one of the oldest operators [Roberts 65]. It is very easy to compute as it uses only a 2×2 neighborhood of the current pixel. Its convolution masks are

$$h_1 = \begin{bmatrix} 1 & 0 \\ 0 & -1 \end{bmatrix} \quad h_2 = \begin{bmatrix} 0 & 1 \\ -1 & 0 \end{bmatrix} \tag{4.43}$$

so the magnitude of the edge is computed as

$$|g(i,j) - g(i+1,j+1)| + |g(i,j+1) - g(i+1,j)| \tag{4.44}$$

The primary disadvantage of the Roberts operator is its high sensitivity to noise, because very few pixels are used to approximate the gradient.

Laplace operator

The Laplace operator ∇^2 is a very popular operator approximating the second derivative which gives the gradient magnitude only. The Laplacian, equation (4.39), is approximated in digital images by a convolution sum. A 3×3 mask h is often used; for 4-neighborhoods and 8-neighborhoods it is defined as

$$h = \begin{bmatrix} 0 & 1 & 0 \\ 1 & -4 & 1 \\ 0 & 1 & 0 \end{bmatrix} \qquad h = \begin{bmatrix} 1 & 1 & 1 \\ 1 & -8 & 1 \\ 1 & 1 & 1 \end{bmatrix} \tag{4.45}$$

A Laplacian operator with stressed significance of the central pixel or its neighborhood is sometimes used. In this approximation it loses invariance to rotation:

$$h = \begin{bmatrix} 2 & -1 & 2 \\ -1 & -4 & -1 \\ 2 & -1 & 2 \end{bmatrix} \qquad h = \begin{bmatrix} -1 & 2 & -1 \\ 2 & -4 & 2 \\ -1 & 2 & -1 \end{bmatrix} \tag{4.46}$$

The Laplacian operator has a disadvantage—it responds doubly to some edges in the image.

Prewitt operator

The Prewitt operator, similarly to the Sobel, Kirsch, Robinson (as discussed later), and some other operators, approximates the first derivative. The gradient is estimated in eight (for a 3 $\times$ 3 convolution mask) possible directions, and the convolution result of greatest magnitude indicates the gradient direction. Larger masks are possible.

Operators approximating the first derivative of an image function are sometimes called **compass operators** because of their ability to determine gradient direction. We present only the first three 3×3 masks for each operator; the others can be created by simple rotation.

$$h_1 = \begin{bmatrix} 1 & 1 & 1 \\ 0 & 0 & 0 \\ -1 & -1 & -1 \end{bmatrix} \quad h_2 = \begin{bmatrix} 0 & 1 & 1 \\ -1 & 0 & 1 \\ -1 & -1 & 0 \end{bmatrix} \quad h_3 = \begin{bmatrix} -1 & 0 & 1 \\ -1 & 0 & 1 \\ -1 & 0 & 1 \end{bmatrix} \tag{4.47}$$

The direction of the gradient is given by the mask giving maximal response. This is also the case for all the following operators approximating the first derivative.

Sobel operator

$$h_1 = \begin{bmatrix} 1 & 2 & 1 \\ 0 & 0 & 0 \\ -1 & -2 & -1 \end{bmatrix} \quad h_2 = \begin{bmatrix} 0 & 1 & 2 \\ -1 & 0 & 1 \\ -2 & -1 & 0 \end{bmatrix} \quad h_3 = \begin{bmatrix} -1 & 0 & 1 \\ -2 & 0 & 2 \\ -1 & 0 & 1 \end{bmatrix} \tag{4.48}$$

The Sobel operator is often used as a simple detector of horizontality and verticality of edges, in which case only masks h_1 and h_3 are used. If the h_1 response is y and the h_3 response x, we might then derive edge strength (magnitude) as

$$\sqrt{x^2 + y^2} \quad \text{or} \quad |x| + |y| \tag{4.49}$$

and direction as $\tan^{-1}(y/x)$.

Figure 4.19: *First-derivative edge detection using Prewitt compass operators: (a) north direction (the brighter the pixel value, the stronger the edge); (b) east direction; (c) strong edges from (a); (d) strong edges from (b).*

Robinson operator

$$h_1 = \begin{bmatrix} 1 & 1 & 1 \\ 1 & -2 & 1 \\ -1 & -1 & -1 \end{bmatrix} \quad h_2 = \begin{bmatrix} 1 & 1 & 1 \\ -1 & -2 & 1 \\ -1 & -1 & 1 \end{bmatrix} \quad h_3 = \begin{bmatrix} -1 & 1 & 1 \\ -1 & -2 & 1 \\ -1 & 1 & 1 \end{bmatrix} \tag{4.50}$$

Kirsch operator

$$h_1 = \begin{bmatrix} 3 & 3 & 3 \\ 3 & 0 & 3 \\ -5 & -5 & -5 \end{bmatrix} \quad h_2 = \begin{bmatrix} 3 & 3 & 3 \\ -5 & 0 & 3 \\ -5 & -5 & 3 \end{bmatrix} \quad h_3 = \begin{bmatrix} -5 & 3 & 3 \\ -5 & 0 & 3 \\ -5 & 3 & 3 \end{bmatrix} \tag{4.51}$$

To illustrate the application of gradient operators on real images, consider again the image given in Figure 4.10a. The Laplace edge image calculated is shown in Figure 4.18a; the value of the operator has been histogram equalized to enhance its visibility.

The properties of an operator approximating the first derivative are demonstrated using the Prewitt operator—results of others are similar. The original image is again given in Figure 4.10a; Prewitt approximations to the directional gradients are in Figures 4.19a,b, in which north and east directions are shown. Significant edges (those with above-threshold magnitude) in the two directions are given in Figures 4.19c,d.

4.3.3 Zero-crossings of the second derivative

In the 1970s, Marr's theory (see Section 9.1.1) concluded from neurophysiological experiments that object boundaries are the most important cues that link an intensity image with its interpretation. Edge detection techniques existing at that time (e.g., the Kirsch, Sobel, and Pratt operators) were based on convolution in very small neighborhoods and worked well only for specific images. The main disadvantage of these edge detectors is their dependence on the size of the object and sensitivity to noise.

An edge detection technique based on the **zero-crossings** of the second derivative (in its original form, the **Marr-Hildreth** edge detector [Marr and Hildreth 80] or the same paper in a more recent collection, [Marr and Hildreth 91]) explores the fact that a step edge corresponds to an abrupt change in the image function. The first derivative of the image function should have an extremum at the position corresponding to the edge in the image, and so the second derivative should be zero at the same position; however, it is much easier and more precise to find a zero-crossing position than an extremum. In Figure 4.20 this principle is illustrated in 1D for the sake of simplicity. Figure 4.20a shows step edge profiles of the original image function with two different slopes, Figure 4.20b depicts the first derivative of the image function, and Figure 4.20c illustrates the second derivative; notice that this crosses the zero level at the same position as the edge. Considering a step-like edge in 2D, the 1D profile of Figure 4.20a corresponds to a cross section through the 2D step. The steepness of the profile will change if the orientation of the cutting plane changes—the maximum steepness is observed when the plane is perpendicular to the edge direction.

The crucial question is how to compute the second derivative robustly. One possibility is to smooth an image first (to reduce noise) and then compute second derivatives. When choosing a smoothing filter, there are two criteria that should be fulfilled [Marr and Hildreth 80]. First, the filter should be smooth and roughly band limited in the frequency domain to

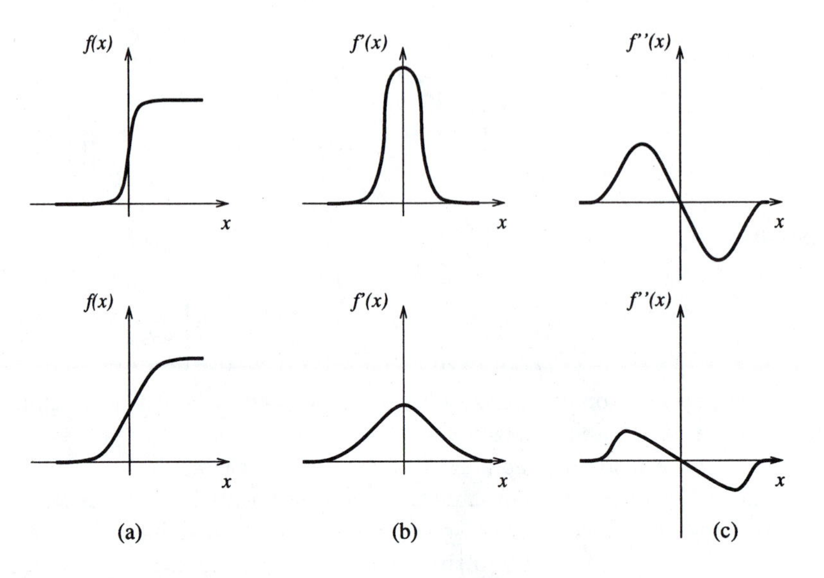

Figure 4.20: *1D edge profile of the zero-crossing.*

reduce the possible number of frequencies at which function changes can take place. Second, the constraint of spatial localization requires the response of a filter to be from nearby points in the image. These two criteria are conflicting, but they can be optimized simultaneously using a Gaussian distribution. In practice, one has to be more precise about what is meant by the localization performance of an operator, and the Gaussian may turn out to be sub-optimal. We shall consider this in the next section.

The 2D Gaussian smoothing operator $G(x, y)$ (also called a Gaussian filter, or simply a Gaussian) is given by

$$G(x, y) = e^{-\frac{x^2+y^2}{2\sigma^2}} \tag{4.52}$$

where x, y are the image co-ordinates and σ is a standard deviation of the associated probability distribution. Sometimes this is presented with a normalizing factor:

$$G(x, y) = \frac{1}{2\pi\sigma^2} e^{-\frac{x^2+y^2}{2\sigma^2}} \quad \text{or} \quad G(x, y) = \frac{1}{\sqrt{2\pi}\sigma} e^{-\frac{x^2+y^2}{2\sigma^2}}$$

The standard deviation σ is the only parameter of the Gaussian filter—it is proportional to the size of the neighborhood on which the filter operates. Pixels more distant from the center of the operator have smaller influence, and pixels farther than 3σ from the center have negligible influence.

Our goal is to obtain a second derivative of a smoothed 2D function $f(x, y)$. We have already seen that the Laplace operator ∇^2 gives the second derivative, and is non-directional (isotropic). Consider then the Laplacian of an image $f(x, y)$ smoothed by a Gaussian (expressed using a convolution $*$). The operation is abbreviated by some authors as **LoG**, from **Laplacian of Gaussian**.

$$\nabla^2[G(x, y, \sigma) * f(x, y)] \tag{4.53}$$

The order of performing differentiation and convolution can be interchanged because of the linearity of the operators involved:

$$[\nabla^2 G(x, y, \sigma)] * f(x, y) \tag{4.54}$$

The derivative of the Gaussian filter $\nabla^2 G$ can be pre-computed analytically, since it is independent of the image under consideration. Thus, the complexity of the composite operation is reduced. For simplicity, we use the substitution $r^2 = x^2 + y^2$, where r measures distance from the origin; this is reasonable, as the Gaussian is circularly symmetric. This substitution converts the 2D Gaussian [equation (4.52)] into a 1D function that is easier to differentiate:

$$G(r) = e^{-\frac{r^2}{2\sigma^2}} \tag{4.55}$$

The first derivative $G'(r)$ is then

$$G'(r) = -\frac{1}{\sigma^2}\, r\, e^{-\frac{r^2}{2\sigma^2}} \tag{4.56}$$

and the second derivative $G''(r)$, the Laplacian of a Gaussian, is

$$G''(r) = \frac{1}{\sigma^2} \left(\frac{r^2}{\sigma^2} - 1 \right) e^{-\frac{r^2}{2\sigma^2}} \tag{4.57}$$

After returning to the original co-ordinates x, y and introducing a normalizing multiplicative coefficient c, we get a convolution mask of a LoG operator:

$$h(x, y) = c \left(\frac{x^2 + y^2 - \sigma^2}{\sigma^4} \right) e^{-\frac{x^2+y^2}{2\sigma^2}} \tag{4.58}$$

where c normalizes the sum of mask elements to zero. Because of its shape, the inverted LoG operator is commonly called a **Mexican hat**. Examples of discrete approximations of 5×5 and 17×17 LoG operators $\nabla^2 G$ follow [Jain et al. 95]:

$$\begin{bmatrix} 0 & 0 & -1 & 0 & 0 \\ 0 & -1 & -2 & -1 & 0 \\ -1 & -2 & 16 & -2 & -1 \\ 0 & -1 & -2 & -1 & 0 \\ 0 & 0 & -1 & 0 & 0 \end{bmatrix}$$

$$
\begin{bmatrix}
0 & 0 & 0 & 0 & 0 & 0 & -1 & -1 & -1 & -1 & -1 & 0 & 0 & 0 & 0 & 0 & 0 \\
0 & 0 & 0 & 0 & -1 & -1 & -1 & -1 & -1 & -1 & -1 & -1 & -1 & 0 & 0 & 0 & 0 \\
0 & 0 & -1 & -1 & -1 & -2 & -3 & -3 & -3 & -3 & -3 & -2 & -1 & -1 & -1 & 0 & 0 \\
0 & 0 & -1 & -1 & -2 & -3 & -3 & -3 & -3 & -3 & -3 & -3 & -2 & -1 & -1 & 0 & 0 \\
0 & -1 & -1 & -2 & -3 & -3 & -3 & -2 & -3 & -2 & -3 & -3 & -3 & -2 & -1 & -1 & 0 \\
0 & -1 & -2 & -3 & -3 & -3 & 0 & 2 & 4 & 2 & 0 & -3 & -3 & -3 & -2 & -1 & 0 \\
-1 & -1 & -3 & -3 & -3 & 0 & 4 & 10 & 12 & 10 & 4 & 0 & -3 & -3 & -3 & -1 & -1 \\
-1 & -1 & -3 & -3 & -2 & 2 & 10 & 18 & 21 & 18 & 10 & 2 & -2 & -3 & -3 & -1 & -1 \\
-1 & -1 & -3 & -3 & -3 & 4 & 12 & 21 & 24 & 21 & 12 & 4 & -3 & -3 & -3 & -1 & -1 \\
-1 & -1 & -3 & -3 & -2 & 2 & 10 & 18 & 21 & 18 & 10 & 2 & -2 & -3 & -3 & -1 & -1 \\
-1 & -1 & -3 & -3 & -3 & 0 & 4 & 10 & 12 & 10 & 4 & 0 & -3 & -3 & -3 & -1 & -1 \\
0 & -1 & -2 & -3 & -3 & -3 & 0 & 2 & 4 & 2 & 0 & -3 & -3 & -3 & -2 & -1 & 0 \\
0 & -1 & -1 & -2 & -3 & -3 & -3 & -2 & -3 & -2 & -3 & -3 & -3 & -2 & -1 & -1 & 0 \\
0 & 0 & -1 & -1 & -2 & -3 & -3 & -3 & -3 & -3 & -3 & -3 & -2 & -1 & -1 & 0 & 0 \\
0 & 0 & -1 & -1 & -1 & -2 & -3 & -3 & -3 & -3 & -3 & -2 & -1 & -1 & -1 & 0 & 0 \\
0 & 0 & 0 & 0 & -1 & -1 & -1 & -1 & -1 & -1 & -1 & -1 & -1 & 0 & 0 & 0 & 0 \\
0 & 0 & 0 & 0 & 0 & 0 & -1 & -1 & -1 & -1 & -1 & 0 & 0 & 0 & 0 & 0 & 0
\end{bmatrix}
$$

Finding second derivatives in this way is very robust. Gaussian smoothing effectively suppresses the influence of the pixels that are up to a distance 3σ from the current pixel; then the Laplace operator is an efficient and stable measure of changes in the image.

After the image convolution with $\nabla^2 G$, the locations in the convolved image where the zero level is crossed correspond to the positions of edges. The advantage of this approach compared to classical edge operators of small size is that a larger area surrounding the current pixel is taken into account; the influence of more distant points decreases according to the σ of the Gaussian. In the ideal case of an isolated step edge, the σ variation does not affect the location of the zero-crossing.

Convolution masks become large for larger σ; for example, $\sigma = 4$ needs a mask about 40 pixels wide. Fortunately, there is a separable decomposition of the $\nabla^2 G$ operator [Huertas and Medioni 86] that can speed up computation considerably.

The practical implication of Gaussian smoothing is that edges are found reliably. If only globally significant edges are required, the standard deviation σ of the Gaussian smoothing filter may be increased, having the effect of suppressing less significant evidence.

The $\nabla^2 G$ operator can be very effectively approximated by convolution with a mask that is the difference of two Gaussian averaging masks with substantially different σ—this method is called the **difference of Gaussians**, abbreviated as **DoG**. The correct ratio of the standard deviations σ of the Gaussian filters is discussed in [Marr 82].

Even coarser approximations to $\nabla^2 G$ are sometimes used—the image is filtered twice by an averaging operator with smoothing masks of different sizes.

When implementing a zero-crossing edge detector, trying to detect *zeros* in the LoG or DoG image will inevitably fail, while naive approaches of thresholding the LoG/DoG image and defining the zero-crossings in some interval of values close to zero give piecewise disconnected edges at best. To end up with a well-functioning second-derivative edge detector, it is necessary to implement a true zero-crossing detector. A simple detector may identify a zero crossing in a moving 2×2 window, assigning an edge label to any one corner pixel, say the upper left, if LoG/DoG image values of both polarities occur in the 2×2 window; no edge label would be given if values within the window are either all positive or all negative. Another post-processing step to avoid detection of zero-crossings corresponding to nonsignificant edges in regions of almost constant gray-level would admit only those zero-crossings for which there is sufficient edge evidence from a first-derivative edge detector. Figure 4.21 provides several examples of edge detection using zero crossings of the second derivative.

Many other approaches improving zero-crossing performance can be found in the literature [Qian and Huang 94, Mehrotra and Shiming 96]; some of them are used in pre-processing [Hardie and Boncelet 95] or post-processing steps [Alparone et al. 96].

Figure 4.21: *Zero-crossings of the second derivative, see Figure 4.10a for the original image: (a) DoG image ($\sigma_1 = 0.10, \sigma_2 = 0.09$), dark pixels correspond to negative DoG values, bright pixels represent positive DoG values; (b) zero-crossings of the DoG image; (c) DoG zero-crossing edges after removing edges lacking first-derivative support; (d) LoG zero-crossing edges ($\sigma = 0.20$) after removing edges lacking first-derivative support—note different scale of edges due to different Gaussian smoothing parameters.*

The traditional second-derivative zero-crossing technique has disadvantages as well. First, it smoothes the shape too much; for example, sharp corners are lost. Second, it tends to create closed loops of edges (nicknamed the 'plate of spaghetti' effect). Although this property was highlighted as an advantage in original papers, it has been seen as a drawback in many applications.

Neurophysiological experiments [Marr 82, Ullman 81] provide evidence that the human eye retina in the form of the **ganglion cells** performs operations very similar to the $\nabla^2 G$ operations. Each such cell responds to light stimuli in a local neighborhood called the **receptive field**, which has a center-surround organization of two complementary types, off-center and on-center. When a light stimulus occurs, activity of on-center cells increases and that of off-center cells is inhibited. The retinal operation on the image can be described analytically as the convolution of the image with the $\nabla^2 G$ operator.

4.3.4 Scale in image processing

Many image processing techniques work locally, theoretically at the level of individual pixels—edge detection methods are an example. The essential problem in such computation is **scale**. Edges correspond to the gradient of the image function, which is computed as a difference between pixels in some neighborhood. There is seldom a sound reason for choosing a particular size of neighborhood, since the 'right' size depends on the size of the objects under investigation. To know what the objects are assumes that it is clear how to interpret an image, and this is not in general known at the pre-processing stage. The solution to the problem formulated above is a special case of a general paradigm called the **system approach**. This methodology is common in cybernetics or general system theory to study complex phenomena.

The phenomenon under investigation is expressed at different resolutions of the description, and a formal model is created at each resolution. Then the qualitative behavior of the model is studied under changing resolution of the description. Such a methodology enables the deduction of meta-knowledge about the phenomenon that is not seen at the individual description levels.

Different description levels are easily interpreted as different scales in the domain of digital images. The idea of scale is fundamental to Marr's edge detection technique, introduced in Section 4.3.3, where different scales are provided by different sizes of Gaussian filter masks. The aim was not only to eliminate fine scale noise but also to separate events at different scales arising from distinct physical processes [Marr 82].

Assume that a signal has been smoothed with several masks of variable sizes. Every setting of the scale parameters implies a different description, but it is not known which one is correct; for many tasks, no one scale is categorically correct. If the ambiguity introduced by the scale is inescapable, the goal of scale-independent description is to reduce this ambiguity as much as possible.

Many publications tackle scale-space problems, e.g., [Hummel and Moniot 89, Perona and Malik 90, Williams and Shah 90, Mokhtarian and Mackworth 92, Mokhtarian 95, Morrone et al. 95, Elder and Zucker 96, Aydin et al. 96, Lindeberg 96]. A symbolic approach to constructing a multi-scale primitive shape description to 2D binary (contour) shape images is presented in [Saund 90], and the use of a scale-space approach for object recognition is in [Topkar et al. 90]. Here we shall consider just three examples of the application of multiple scale description to image analysis.

The first approach [Lowe 89] aims to process planar noisy curves at a range of scales—the segment of curve that represents the underlying structure of the scene needs to be found. The problem is illustrated by an example of two noisy curves; see Figure 4.22. One of these may be interpreted as a closed (perhaps circular) curve, while the other could be described as two intersecting straight lines. Local tangent direction and curvature of the curve are significant

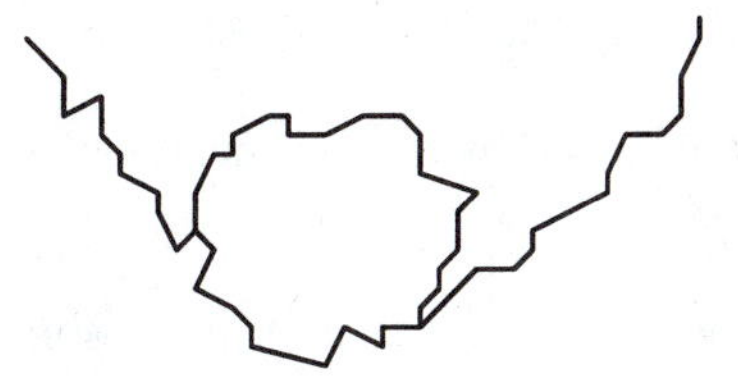

Figure 4.22: *Curves that may be analyzed at multiple scales.*

only with some idea of scale after the curve is smoothed by a Gaussian filter with varying standard deviations.

A second approach [Witkin 83], called **scale-space filtering**, tries to describe signals qualitatively with respect to scale. The problem was formulated for 1D signals $f(x)$, but it can easily be generalized for 2D functions as images. The original 1D signal $f(x)$ is smoothed by convolution with a 1D Gaussian,

$$G(x, \sigma) = e^{-\frac{x^2}{2\sigma^2}} \tag{4.59}$$

If the standard deviation σ is slowly changed, the function

$$F(x, \sigma) = f(x) * G(x, \sigma) \tag{4.60}$$

represents a surface on the (x, σ) plane that is called the **scale-space image**. Inflection points of the curve $F(x, \sigma_0)$ for a distinct value σ_0,

$$\frac{\partial^2 F(x, \sigma_0)}{\partial x^2} = 0 \quad \frac{\partial^3 F(x, \sigma_0)}{\partial x^3} \neq 0 \tag{4.61}$$

describe the curve $f(x)$ qualitatively. The positions of inflection points can be drawn as a set of curves in (x, σ) co-ordinates (see Figure 6.15). Coarse to fine analysis of the curves corresponding to inflection points, i.e., in the direction of decreasing value of the σ, localizes large-scale events.

The qualitative information contained in the scale-space image can be transformed into a simple **interval tree** that expresses the structure of the signal $f(x)$ over all observed scales. The interval tree is built from the root that corresponds to the largest scale ($\sigma_{\max}$), and then the scale-space image is searched in the direction of decreasing σ. The interval tree branches at those points where new curves corresponding to inflection points appear (see Chapter 6 and Section 6.2.4).

The third example of the application of scale is that used by the popular **Canny edge detector**. Since the Canny detector is a significant and widely used contribution to edge detection techniques, its principles will be explained in detail.

4.3.5 Canny edge detection

Canny proposed a new approach to edge detection [Canny 83, Brady 84, Canny 86] that is optimal for step edges corrupted by white noise. The optimality of the detector is related to three criteria.

- The **detection** criterion expresses the fact that important edges should not be missed and that there should be no spurious responses.
- The **localization** criterion says that the distance between the actual and located position of the edge should be minimal.
- The **one response** criterion minimizes multiple responses to a single edge. This is partly covered by the first criterion, since when there are two responses to a single edge, one of them should be considered as false. This third criterion solves the problem of an edge corrupted by noise and works against non-smooth edge operators [Rosenfeld and Thurston 71].

Canny's derivation of a new edge detector is based on several ideas.

1. The edge detector was expressed for a 1D signal and the first two optimality criteria. A closed-form solution was found using the calculus of variations.
2. If the third criterion (multiple responses) is added, the best solution may be found by numerical optimization. The resulting filter can be approximated effectively with error less than 20% by the first derivative of a Gaussian smoothing filter with standard deviation σ [Canny 86]; the reason for doing this is the existence of an effective implementation. There is a strong similarity here to the Marr-Hildreth edge detector [Marr and Hildreth 80], which is based on the Laplacian of a Gaussian—see Section 4.3.3.
3. The detector is then generalized to two dimensions. A step edge is given by its position, orientation, and possibly magnitude (strength). It can be shown that convolving an image with a symmetric 2D Gaussian and then differentiating in the direction of the gradient (perpendicular to the edge direction) forms a simple and effective directional operator (recall that the Marr-Hildreth zero-crossing operator does not give information about edge direction, as it uses a Laplacian filter).

   Suppose G is a 2D Gaussian [equation (4.52)] and assume we wish to convolve the image with an operator G_n which is a first derivative of G in the direction $\mathbf{n}$.

$$G_n = \frac{\partial G}{\partial \mathbf{n}} = \mathbf{n} \cdot \nabla G \tag{4.62}$$

   The direction $\mathbf{n}$ should be oriented perpendicular to the edge. Although this direction is not known in advance, a robust estimate of it based on the smoothed gradient direction is available. If f is the image, the normal to the edge $\mathbf{n}$ is estimated as

$$\mathbf{n} = \frac{\nabla(G * f)}{|\nabla(G * f)|} \tag{4.63}$$

The edge location is then at the local maximum of the image f convolved with the operator G_n in the direction $\mathbf{n}$.

$$\frac{\partial}{\partial \mathbf{n}} G_n * f = 0 \tag{4.64}$$

Substituting in equation (4.64) for G_n from equation (4.62), we get

$$\frac{\partial^2}{\partial \mathbf{n}^2} G * f = 0 \tag{4.65}$$

This equation (4.65) illustrates how to find local maxima in the direction perpendicular to the edge; this operation is often referred to as **non-maximal suppression** (see also Algorithm 5.5).

As the convolution and derivative are associative operations in equation (4.65), we can first convolve an image f with a symmetric Gaussian G and then compute the directional second-derivative using an estimate of the direction $\mathbf{n}$ computed according to equation (4.63). The strength of the edge (magnitude of the gradient of the image intensity function f) is measured as

$$|G_n * f| = |\nabla(G * f)| \tag{4.66}$$

A different generalization of this optimal detector into two dimensions was proposed by Spacek [Spacek 86], and the problem of edge localization is revisited in [Tagare and deFigueiredo 90].

4. Spurious responses to the single edge caused by noise usually create a 'streaking' problem that is very common in edge detection in general. The output of an edge detector is usually thresholded to decide which edges are significant, and streaking means the breaking up of the edge contour caused by the operator fluctuating above and below the threshold. Streaking can be eliminated by **thresholding with hysteresis**. If any edge response is above a *high threshold*, those pixels constitute definite edge output of the detector for a particular scale. Individual weak responses usually correspond to noise, but if these points are connected to any of the pixels with strong responses, they are more likely to be actual edges in the image. Such connected pixels are treated as edge pixels if their response is above a *low threshold*. The low and high thresholds are set according to an estimated signal-to-noise ratio [Canny 86] (see also Algorithm 5.6).

5. The correct scale for the operator depends on the objects contained in the image. The solution to this unknown is to use multiple scales and aggregate information from them. Different scales for the Canny detector are represented by different standard deviations σ of the Gaussians. There may be several scales of operators that give significant responses to edges (i.e., signal-to-noise ratio above the threshold); in this case the operator with the smallest scale is chosen, as it gives the best localization of the edge.

   Canny proposed a **feature synthesis** approach. All significant edges from the operator with the smallest scale are marked first, and the edges of a hypothetical operator with larger σ are synthesized from them (i.e., a prediction is made of how the large σ should perform on the evidence gleaned from the smaller σ—see also Section 4.3.4 and

Figure 6.15). Then the synthesized edge response is compared with the actual edge response for larger σ. Additional edges are marked only if they have a significantly stronger response than that predicted from synthetic output.

This procedure may be repeated for a sequence of scales, a cumulative edge map being built by adding those edges that were not identified at smaller scales.

Algorithm 4.4: Canny edge detector

1. Convolve an image f with a Gaussian of scale σ.
2. Estimate local edge normal directions $\mathbf{n}$ using equation (4.63) for each pixel in the image.
3. Find the location of the edges using equation (4.65) (non-maximal suppression).
4. Compute the magnitude of the edge using equation (4.66).
5. Threshold edges in the image with hysteresis to eliminate spurious responses.
6. Repeat steps (1) through (5) for ascending values of the standard deviation σ.
7. Aggregate the final information about edges at multiple scale using the 'feature synthesis' approach.

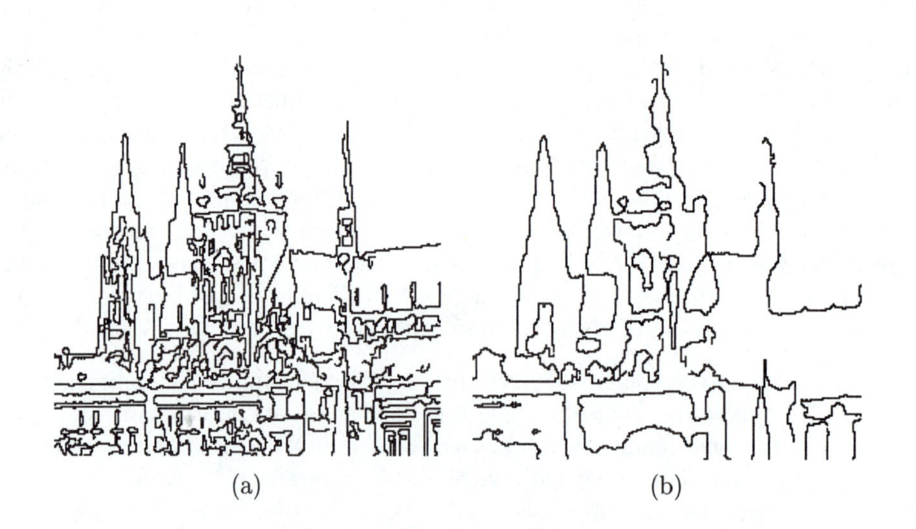

(a) (b)

Figure 4.23: *Canny edge detection at two different scales.*

Figure 4.23a shows the edges of Figure 4.10a detected by a Canny operator with $\sigma = 1.0$. Figure 4.23b shows the edge detector response for $\sigma = 2.8$ (feature synthesis has not been applied here).

Canny's detector represents a complicated but major contribution to edge detection. Its full implementation is unusual, it being common to find implementations that omit feature synthesis—that is, just steps 1–5 of Algorithm 4.4.

Recent developments of the optimal border detection concept have focused on optimal detection of more general edges than the step and ramp edges considered in Canny's original, and other early, approaches. Examples of Canny edge detector extensions can be found in [Laligant et al. 94, Jalali and Boyce 95, Sorrenti 95, Demigny et al. 95, Mehrotra and Shiming 96].

4.3.6 Parametric edge models

Parametric models are based on the idea that the discrete image intensity function can be considered a sampled and noisy approximation of the underlying continuous or piecewise continuous image intensity function [Nevatia 77]. While the continuous image intensity function is not known, it can be estimated from the available discrete image intensity function and image properties can be determined from this continuous estimate, possibly with sub-pixel precision. Since modeling the image as a single continuous function leads to high-order intensity functions in x and y, it is practically impossible to represent image intensities using a single continuous function. Instead, piecewise continuous function estimates called **facets** are used to represent (a neighborhood of) each image pixel. Such image representation is called the **facet model** [Haralick and Watson 81, Haralick 84, Haralick and Shapiro 92].

The intensity function in a pixel neighborhood can be estimated using models of different complexity. The simplest one is the flat facet model that uses piecewise constants and each pixel neighborhood is represented by a flat function of constant intensity. The sloped model uses piecewise linear functions forming a sloped plane fitted to the image intensities in the pixel neighborhood. Quadratic and bi-cubic facet models employ correspondingly more complex functions.

A thorough treatment of facet models and their modifications for peak noise removal, segmentation into constant-gray-level regions, determination of statistically significant edges, gradient edge detection, directional second-derivative zero-crossing edge detection, and line and corner detection is given in [Haralick and Shapiro 92]. Importantly, techniques for facet model parameter estimation are given there.

To provide an edge detection example, consider a bi-cubic facet model

$$g(i,j) = c_1 + c_2 x + c_3 y + c_4 x^2 + c_5 xy + c_6 y^2 + c_7 x^3 + c_8 x^2 y + c_9 x y^2 + c_{10} y^3 \qquad (4.67)$$

the parameters of which are estimated from a pixel neighborhood [the co-ordinates of the central pixel are (0,0)]. To determine the model parameters, a least-squares method with singular-value decomposition (see Section 9.2.4) may be used. Alternatively, when using a 5×5 pixel neighborhood, coefficients c_i can be computed directly using a set of ten 5×5 kernels that are provided in [Haralick and Shapiro 92].

Once the facet model parameters are available for each image pixel, edges can be detected as extrema of the first directional derivative and/or zero-crossings of the second directional derivative of the local continuous facet model functions.

Edge detectors based on parametric models describe edges more precisely than convolution-based edge detectors. Additionally, they carry the potential for sub-pixel edge localization.

However, their computational requirements are much higher. Promising extensions combine facet model with Canny's edge detection criteria (Section 4.3.5) and relaxation labeling (Section 5.2.2) [Matalas et al. 97].

4.3.7 Edges in multi-spectral images

One pixel in a multi-spectral image is described by an n-dimensional vector, and brightness values in n spectral bands are the vector components.

There are several possibilities for the detection of edges in multi-spectral images [Faugeras 93]. The first is to detect edges separately in individual image spectral components using the ordinary local gradient operators mentioned in Section 4.3.2. Individual images of edges can be combined to get the resulting image, with the value corresponding to edge magnitude and direction being the maximal edge value from all spectral components. A linear combination of edge spectral components can also be used, and other combination techniques are possible [Nagao and Matsuyama 80].

A second possibility is to use the brightness difference of the same pixel in two different spectral components. This is a very informative feature for classification based on properties of the individual pixel. The ratio instead of the difference can be used as well [Pratt 78], although it is necessary to assume that pixel values are not zero in this case.

A third possibility is to create a multi-spectral edge detector which uses brightness information from all n spectral bands; this approach is also applicable to multi-dimensional images forming three- or higher-dimensional data volumes. An edge detector of this kind is proposed in [Cervenka and Charvat 87]. The neighborhood used has size $2 \times 2 \times n$ pixels, where the 2×2 neighborhood is similar to that of the Roberts gradient, equation (4.43). The coefficients weighting the influence of the component pixels are similar to the correlation coefficients. Let $\overline{f}(i,j)$ denote the arithmetic mean of the brightnesses corresponding to the pixels with the same co-ordinates (i,j) in all n spectral component images, and f_r be the brightness of the rth spectral component. The edge detector result in pixel (i,j) is given as the minimum of the following expressions:

$$\frac{\sum_{r=1}^{n}[f_r(i,j)-\overline{f}(i,j)]\,[f_r(i+1,j+1)-\overline{f}(i+1,j+1)]}{\sqrt{\sum_{r=1}^{n}[f_r(i,j)-\overline{f}(i,j)]^2\,\sum_{r=1}^{n}[f_r(i+1,j+1)-\overline{f}(i+1,j+1)]^2}}\,.$$

$$\cdot\frac{\sum_{r=1}^{n}[f_r(i+1,j)-\overline{f}(i+1,j)]\,[f_r(i,j+1)-\overline{f}(i,j+1)]}{\sqrt{\sum_{r=1}^{n}[f_r(i+1,j)-\overline{f}(i+1,j)]^2\,\sum_{r=1}^{n}[f_r(i,j+1)-\overline{f}(i,j+1)]^2}} \tag{4.68}$$

This multi-spectral edge detector gives very good results on remotely sensed images.

4.3.8 Other local pre-processing operators

Several other local operations exist which do not belong to the taxonomy given in Section 4.3, as they are used for different purposes. Line finding, line thinning, line filling, and interest point operators are among them. Another class of local operators, mathematical morphology techniques, is mentioned in Chapter 11.

Line finding operators aim to find very thin curves in the image; it is assumed that curves do not bend sharply. Such curves and straight lines are called **lines** for the purpose of describing this technique. If a cross section perpendicular in direction to the tangent of a line is examined, we get a roof profile (see Figure 4.17) when examining edges. We assume that the width of the lines is approximately one or two pixels; such lines may correspond, for example, to roads in satellite images or to dimension lines in engineering drawings.

Lines in the image can be detected [Cervenka and Charvat 87] by a number of local convolution operators h_k. The output value of the line finding detector in pixel (i,j) is given by

$$f(i,j) = \max[0, \max_k(f * h_k)] \tag{4.69}$$

where $f * h_k$ denotes convolution of the kth mask with the neighborhood of a pixel (i,j) in the input image f.

One possibility is a convolution mask of size 5×5. There are 14 possible orientations of the line finding convolution mask of this size; we shall show only the first eight of them, as the others are obvious by rotation.

$$h_1 = \begin{bmatrix} 0 & 0 & 0 & 0 & 0 \\ 0 & -1 & 2 & -1 & 0 \\ 0 & -1 & 2 & -1 & 0 \\ 0 & -1 & 2 & -1 & 0 \\ 0 & 0 & 0 & 0 & 0 \end{bmatrix} \qquad h_2 = \begin{bmatrix} 0 & 0 & 0 & 0 & 0 \\ 0 & 0 & -1 & 2 & -1 \\ 0 & -1 & 2 & -1 & 0 \\ 0 & -1 & 2 & -1 & 0 \\ 0 & 0 & 0 & 0 & 0 \end{bmatrix}$$

$$h_3 = \begin{bmatrix} 0 & 0 & 0 & 0 & 0 \\ 0 & 0 & -1 & 2 & -1 \\ 0 & -1 & 2 & -1 & 0 \\ -1 & 2 & -1 & 0 & 0 \\ 0 & 0 & 0 & 0 & 0 \end{bmatrix} \qquad h_4 = \begin{bmatrix} 0 & 0 & 0 & 0 & 0 \\ 0 & -1 & 2 & -1 & 0 \\ 0 & -1 & 2 & -1 & 0 \\ -1 & 2 & -1 & 0 & 0 \\ 0 & 0 & 0 & 0 & 0 \end{bmatrix}$$

$$h_5 = \begin{bmatrix} 0 & 0 & 0 & 0 & 0 \\ -1 & 2 & -1 & 0 & 0 \\ 0 & -1 & 2 & -1 & 0 \\ 0 & -1 & 2 & -1 & 0 \\ 0 & 0 & 0 & 0 & 0 \end{bmatrix} \qquad h_6 = \begin{bmatrix} 0 & 0 & 0 & 0 & 0 \\ 0 & -1 & 2 & -1 & 0 \\ 0 & -1 & 2 & -1 & 0 \\ 0 & 0 & -1 & 2 & -1 \\ 0 & 0 & 0 & 0 & 0 \end{bmatrix}$$

$$h_7 = \begin{bmatrix} 0 & 0 & 0 & 0 & 0 \\ -1 & 2 & -1 & 0 & 0 \\ 0 & -1 & 2 & -1 & 0 \\ 0 & 0 & -1 & 2 & -1 \\ 0 & 0 & 0 & 0 & 0 \end{bmatrix} \qquad h_8 = \begin{bmatrix} 0 & 0 & 0 & 0 & 0 \\ 0 & -1 & -1 & -1 & 0 \\ 0 & 2 & 2 & 2 & 0 \\ 0 & -1 & -1 & -1 & 0 \\ 0 & 0 & 0 & 0 & 0 \end{bmatrix} \tag{4.70}$$

The line detector of equation (4.69) with masks similar to (4.70) sometimes produces more lines than needed. Some other non-linear constraints may be added to reduce this number. More sophisticated approaches determine lines in images as ridges and ravines using the facet model [Haralick and Shapiro 92]. Line detection is frequently used in remote sensing and in document processing; examples include [Venkateswar and Chellappa 92, Tang et al. 97].

Local information about edges is the basis of a class of image segmentation techniques that are discussed in Chapter 5. Edges which are likely to belong to object boundaries are usually found by simple thresholding of the edge magnitude—such edge thresholding does not provide ideal contiguous boundaries that are one pixel wide. Sophisticated segmentation techniques that are dealt with in the next chapter serve this purpose. Here, much simpler edge thinning and filling methods are described. These techniques are based on knowledge of small local neighborhoods and are very similar to other local pre-processing techniques.

Thresholded edges are usually wider than one pixel, and **line thinning** techniques may give a better result. One line thinning method uses knowledge about edge orientation and in this case edges are thinned before thresholding. Edge magnitudes and directions provided by some gradient operator are used as input, and the edge magnitudes of two neighboring pixels perpendicular to the edge direction are examined for each pixel in the image. If at least one of these pixels has edge magnitude higher than the edge magnitude of the examined pixel, then the edge magnitude of the examined pixel is assigned a zero value. This technique is called **non-maximal suppression** and is similar to the idea mentioned in conjunction with the Canny edge detector.

A second line thinning method [Cervenka and Charvat 87] does not explore information about edge orientation. A binary image with edges that have magnitude higher than a specified threshold is used as input; ones denote edge pixels and zeros the rest of the image. Such edges are then thinned by a local operator. Example 3×3 masks are shown in equation (4.71), where the letter x denotes an arbitrary value (0 or 1). The match of each mask at each position of the image is checked and if the mask matches, the edge is thinned by replacing the one in the center of the mask by zero.

$$\begin{bmatrix} 1 & x & 0 \\ 1 & 1 & 0 \\ x & 0 & 0 \end{bmatrix} \quad \begin{bmatrix} x & 1 & 1 \\ 0 & 1 & x \\ 0 & 0 & 0 \end{bmatrix} \quad \begin{bmatrix} x & 1 & x \\ 1 & 1 & x \\ x & x & 0 \end{bmatrix} \quad \begin{bmatrix} x & 1 & x \\ x & 1 & 0 \\ 0 & x & 0 \end{bmatrix}$$
$$\begin{bmatrix} 0 & 0 & 0 \\ x & 1 & 0 \\ 1 & 1 & x \end{bmatrix} \quad \begin{bmatrix} 0 & 0 & x \\ 0 & 1 & 1 \\ 0 & x & 1 \end{bmatrix} \quad \begin{bmatrix} x & x & 0 \\ 1 & 1 & x \\ x & 1 & x \end{bmatrix} \quad \begin{bmatrix} 0 & x & x \\ x & 1 & 1 \\ x & 1 & x \end{bmatrix} \tag{4.71}$$

Another procedure permits a more reliable extraction of a set of edge points. Edge points after thresholding do not create contiguous boundaries, and the **edge filling** method tries to recover edge pixels on the potential object boundary which are missing. We present here a very simple local edge filling technique, but more complicated methods based on edge relaxation are mentioned in Chapter 5.

The local edge filling procedure [Cervenka and Charvat 87] checks whether the 3×3 neighborhood of the current pixel matches one of the following situations:

$$\begin{bmatrix} 0 & 1 & 0 \\ 0 & 0 & 0 \\ 0 & 1 & 0 \end{bmatrix} \quad \begin{bmatrix} 0 & 0 & 0 \\ 1 & 0 & 1 \\ 0 & 0 & 0 \end{bmatrix} \quad \begin{bmatrix} 1 & 0 & 0 \\ 0 & 0 & 0 \\ 0 & 1 & 0 \end{bmatrix} \quad \begin{bmatrix} 0 & 0 & 1 \\ 1 & 0 & 0 \\ 0 & 0 & 0 \end{bmatrix}$$
$$\begin{bmatrix} 0 & 1 & 0 \\ 0 & 0 & 0 \\ 0 & 0 & 1 \end{bmatrix} \quad \begin{bmatrix} 0 & 0 & 0 \\ 0 & 0 & 1 \\ 1 & 0 & 0 \end{bmatrix} \quad \begin{bmatrix} 1 & 0 & 0 \\ 0 & 0 & 0 \\ 0 & 0 & 1 \end{bmatrix} \quad \begin{bmatrix} 0 & 0 & 1 \\ 0 & 0 & 0 \\ 1 & 0 & 0 \end{bmatrix} \tag{4.72}$$

If so, the central pixel of the mask is changed from zero to one.

These methods for edge thinning and filling do not guarantee that the width of the lines will be equal to one, and the contiguity of the lines is not certain either. Note that local thinning and filling operators can be treated as special cases of mathematical morphology operators which are described in Chapter 11.

In many cases it is of advantage to find pairs of corresponding points in two similar images; we came across this fact in Section 4.2 when considering geometric transforms. Knowing the position of corresponding points enables the estimation of mathematical formulae describing geometric transforms from live data. The same transformation usually holds for all pixels of the image. The necessary number of corresponding pairs of points is usually rather small and is equal to the number of parameters of the transform. We shall see later on that finding corresponding points is also a core problem in the analysis of moving images (Chapter 15), and for recovering depth information from pairs of stereo images (Section 9.2.5). In general, all possible pairs of points should be examined to solve this **correspondence problem**, and this is very computationally expensive. If two images have n pixels each, the complexity is $\mathcal{O}(n^2)$. This process might be simplified if the correspondence is examined among a much smaller number of points, called **interest points**. An interest point should have some typical local property [Ballard and Brown 82]. For example, if square objects are present in the image, then **corners** are very good interest points.

Corners in images can be located using local detectors; input to the corner detector is the gray-level image, and output is the image $f(i,j)$ in which values are proportional to the likelihood that the pixel is a corner. The simplest corner detector is the **Moravec detector** [Moravec 77] which is maximal in pixels with high contrast. These points are on corners and sharp edges. The Moravec operator MO is given by

$$\text{MO}(i,j) = \frac{1}{8} \sum_{k=i-1}^{i+1} \sum_{l=j-1}^{j+1} |f(k,l) - f(i,j)| \tag{4.73}$$

Better results are produced by computationally more expensive corner operators such as those proposed by Zuniga–Haralick [Zuniga and Haralick 83, Haralick and Shapiro 92] or Kitchen–Rosenfeld [Huang 83] which are based on the facet model (Section 4.3.6). The image function f is approximated in the neighborhood of the pixel (i,j) by a cubic polynomial with coefficients c_k:

$$f(i,j) = c_1 + c_2 x + c_3 y + c_4 x^2 + c_5 xy + c_6 y^2 + c_7 x^3 + c_8 x^2 y + c_9 x y^2 + c_{10} y^3 \tag{4.74}$$

The Zuniga–Haralick operator ZH is given by

$$\text{ZH}(i,j) = \frac{-2\,(c_2^2 c_6 - c_2 c_3 c_5 - c_3^2 c_4)}{(c_2^2 + c_3^2)^{\frac{3}{2}}} \tag{4.75}$$

The Kitchen-Rosenfeld KR operator has the same numerator as equation (4.75), but the denominator is $(c_2^2 + c_3^2)$. The ZH operator has been shown to outperform the KR corner detector in test images [Haralick and Shapiro 92].

Interest points are obtained by thresholding the result of the corner detector.

A corner detection technique that defines a corner as two half-edges and uses a more recent edge detection approach based on derivatives of Gaussian smoothing operators is given in [Mehrotra and Nichani 90]. Two fast corner detectors using dissimilarity along the contour and estimates of the contour curvature are presented in [Cooper et al. 93].

4.3.9 Adaptive neighborhood pre-processing

The importance of scale has been presented in Section 4.3.4 together with possible solutions. Nevertheless, the majority of pre-processing operators work in neighborhoods of fixed sizes in the whole image, of which square windows (3×3, 5×5, or 7×7) are most common. Further, pre-processing operators of variable sizes and shapes exist and bring improved pre-processing results. Often, they are based on detection of the most homogeneous neighborhood of each pixel. However, they are not widely used, mostly because of computational demands and the non-existence of a unifying approach.

A recent approach to image pre-processing introduces the concept of an adaptive neighborhood which is determined for each image pixel [Gordon and Rangayyan 84, Morrow and Rangayyan 90, Morrow et al. 92]. The neighborhood size and shape are dependent on characteristics of image data and on parameters which define measures of homogeneity of a pixel neighborhood. Therefore, a significant property of the neighborhood for each pixel is the ability to self-tune to contextual details in the image.

Neighborhood

An adaptive neighborhood is constructed for each pixel, this pixel being called a **seed pixel** of the neighborhood. The adaptive neighborhood consists of all the 8-connected pixels which satisfy a property of similarity with the seed pixel. The pixel property may represent a gray-level, or some more complex image properties such as texture, local motion parameters, etc. Consider gray-level as a basic pixel property—the adaptive neighborhood for gray-level image pre-processing is based on an additive or multiplicative tolerance interval; all the pixels which are 8-connected with the seed pixel and which have their gray-levels in a tolerance interval become members of the adaptive neighborhood. Specifically, let $f(i,j)$ represent the seed pixel, and $f(k,l)$ represent pixels 8-connected to the seed pixel. Then, the adaptive neighborhood of the pixel $f(i,j)$ is defined as a set of pixels $f(k,l)$ 8-connected to the seed pixel and either satisfying the additive tolerance property

$$|f(k,l) - f(i,j)| \leq T_1 \tag{4.76}$$

or satisfying a multiplicative property,

$$\frac{|f(k,l) - f(i,j)|}{f(i,j)} \leq T_2 \tag{4.77}$$

where T_1, T_2 are parameters of the adaptive neighborhood and represent the maximum allowed dissimilarity of a neighborhood pixel from the seed pixel. Note that each pixel is assigned one adaptive neighborhood, and therefore adaptive neighborhoods may overlap. This specification defines the first layer of the adaptive neighborhood (called the foreground layer) which is used in all adaptive neighborhood pre-processing operations. Sometimes not only the foreground layer but also a background layer must be used to represent more diverse contextual information. The second (background) layer is molded to the outline of the first layer and has a thickness of s pixels, s being a parameter of the adaptive neighborhood.

The foreground layer has gray-levels similar to the gray-level of the seed pixel, but the background-layer values are quite different from the seed pixel. The adaptive neighborhood definition may result in a neighborhood with many interior holes, and these holes or parts thereof may represent a portion of the background layer. The foreground layer may be constructed using a region-growing approach (see Section 5.3), and the background layer may result from a dilation-like operation applied to the foreground layer (see Section 11.3.1), the number of dilation steps being specified by the parameter s.

Although a neighborhood is constructed for each pixel, all pixels in the given foreground layer that have the same gray-level as the seed pixel construct the same adaptive neighborhood. These pixels are called **redundant seed pixels**. Using redundant seed pixels significantly decreases the complexity of adaptive neighborhood construction in the image (see Figure 4.24).

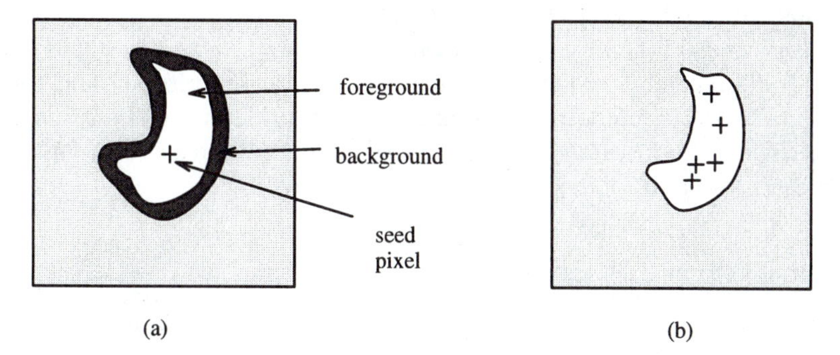

Figure 4.24: *Adaptive neighborhood: (a) construction; (b) redundant seed pixels—8-connected seed pixels of the same gray-level construct the same adaptive neighborhood.*

Many fixed-neighborhood pre-processing techniques may be implemented by applying the adaptive neighborhood concept. We shall demonstrate the power of adaptive neighborhood pre-processing in noise suppression, histogram modification, and contrast enhancement [Paranjape et al. 92b, Paranjape et al. 92a].

Noise suppression

An adaptive neighborhood is not formed across region boundaries; therefore, noise suppression will not blur image edges as often happens with other techniques. If noise suppression is the goal, only the foreground neighborhood layer is used. Having constructed the adaptive neighborhood for each pixel, the rest is straightforward: Each seed pixel is assigned a new value computed as a mean, median, etc., of all the pixels in the adaptive neighborhood. If the noise is additive, the additive criterion for neighborhood construction should be applied; if the noise is multiplicative, the multiplicative tolerance criterion is appropriate. Adaptive neighborhood noise suppression may be applied several times in a sequence with good results, and the edges will not be blurred. If the median is used to compute a new seed pixel value, the operation does not destroy corners and thin lines as is typical in fixed-size neighborhoods (see Section 4.3.1). Adaptive neighborhood smoothing does not work well for impulse noise, because large gray-level differences between the noise pixel and other pixels in the neighborhood cause the adaptive neighborhood to consist of only the noise pixel. A solution may be to apply a fixed-size averaging or median filtering pre-processing step prior to the adaptive neighborhood operations (a small size of fixed neighborhood should be used in order not to blur the edges too much). Examples of adaptive neighborhood noise suppression are given in Figure 4.25.

Histogram modification

Full-frame histogram equalization was discussed in Section 4.1.2; its main disadvantage is that the global image properties may not be appropriate under a local context. Local area histogram equalization computes a new gray-level for each pixel based on the equalization of a histogram acquired in a local fixed-size neighborhood [Pizer et al. 87]. Adaptive neighborhood histogram modification is based on the same principle—the local histogram is computed from a neighborhood which reflects local contextual image properties.

The adaptive neighborhood consists of both foreground and background layers in this application. The foreground variance is based on the additive criterion and the parameter T_1 should be chosen relatively large. The background portion of the adaptive neighborhood provides a mechanism for mediating the introduced gray-level change. Very good results from this method can be seen in Figure 4.26.

Contrast enhancement

Contrast is a property based on human perception abilities. An approximate definition of contrast is [Gordon and Rangayyan 84]

$$c = \frac{F - B}{F + B} \tag{4.78}$$

where F and B are the mean gray-levels of two regions whose contrast is evaluated. Standard contrast enhancement techniques such as sharpening (Section 4.3.2) do not enhance the contrast of regions, only local edge perception. Moreover, the larger the contrast between image parts, the larger is the enhancement. In other words, the most serious enhancement is achieved where the contrast is sufficient anyway. Conversely, the adaptive neighborhood is associated with objects, and therefore it is feasible to enhance contrast in regions by modifying gray-levels in regions and not only along their borders. Further, the contrast enhancement may be non-linear:

- No enhancement for very small gray-level differences between neighborhoods (caused probably by quantization noise or very small gray-level variance)
- Moderate to strong enhancement applied if the contrast between regions is small but outside the range of quantization contrast
- No contrast enhancement is applied if the contrast is already sufficient

The contrast modification curve is shown in Figure 4.27. For contrast enhancement, both foreground and background adaptive neighborhood layers are used, the background size being comparable in size to the foreground size. For each seed pixel and corresponding adaptive neighborhood, the original contrast c is computed using equation (4.78). The new desired contrast c' is obtained from the applied contrast curve (Figure 4.27). The new gray value $f'(i,j)$ to be assigned to the seed pixel (i,j) is computed as

$$f'(i,j) = \frac{B(1+c')}{1-c'} \tag{4.79}$$

where B is the original mean gray-level of the background adaptive neighborhood layer.

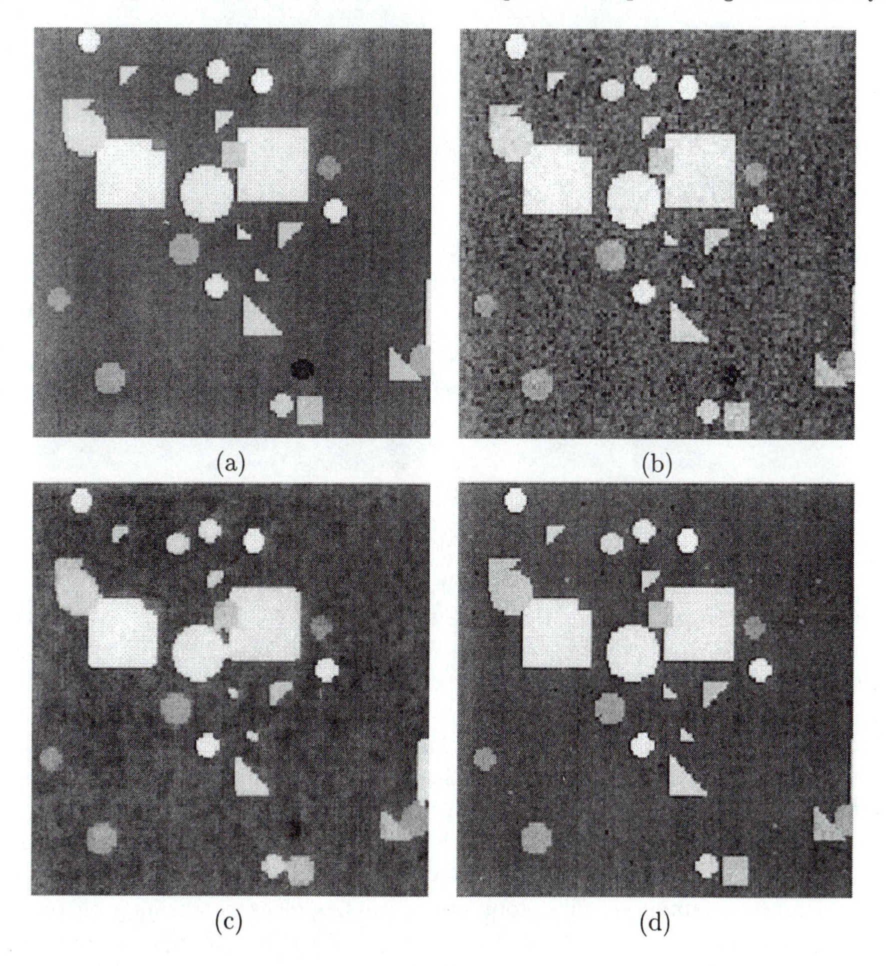

Figure 4.25: *Adaptive-neighborhood noise suppression: (a) original image; (b) noisy image (superimposed noise); (c) fixed neighborhood median filter 3×3; (d) adaptive neighborhood median filter, $T_1 = 16$. Compare corners, thin lines, and thin gaps. Courtesy R. Paranjape, R. Rangayyan, University of Calgary.*

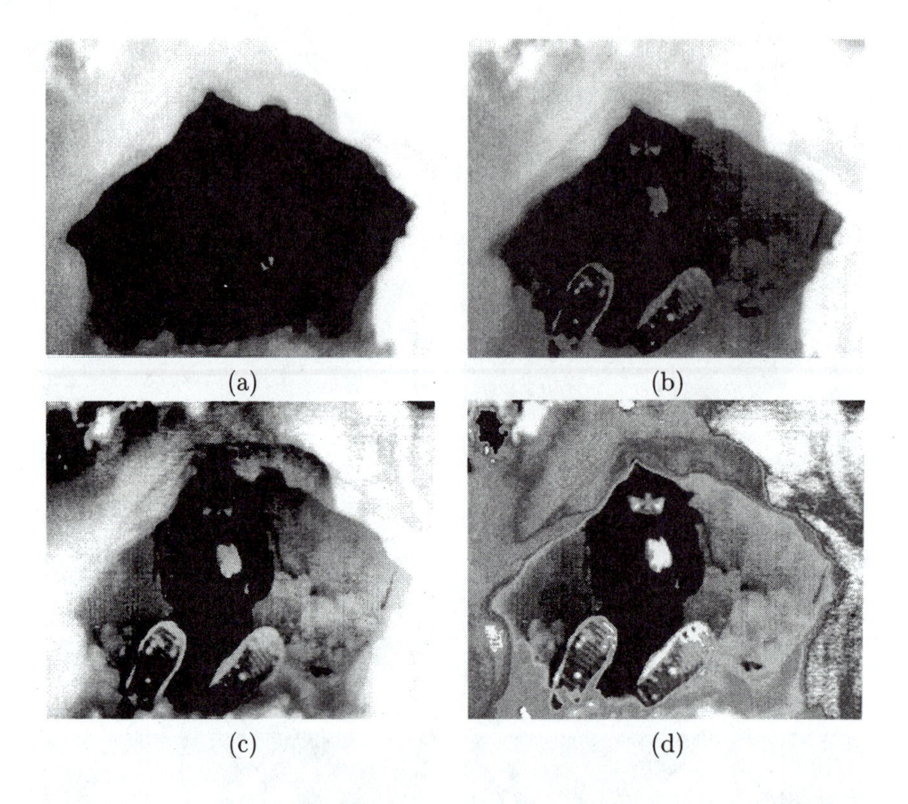

Figure 4.26: *Adaptive-neighborhood histogram modification: (a) original low-contrast image of a person in a snow igloo; (b) full-frame histogram equalization; (c) fixed-neighborhood adaptive histogram equalization; (d) adaptive-neighborhood histogram equalization. Courtesy R. Paranjape, R. Rangayyan, University of Calgary, and Academic Press.*

Note that, in addition to the contrast enhancement, the gray-level variations inside regions decrease. The multiplicative neighborhood construction criterion should be used to avoid dependence of the resulting contrast on the mean gray-level of the neighborhood. The adaptive contrast enhancement results may be compared with the fixed-size neighborhood contrast enhancement in Figure 4.28, and the contrast improvement is clearly visible.

The principle of adaptive neighborhood pre-processing gives significantly better results in many images, but large computational load is the price to pay for this improvement. Nevertheless, taking advantage of redundant seed pixels decreases the computational demands; also, feasibility of implementing these methods in parallel may soon make these methods as standard as fixed neighborhood methods are today.

4.4 Image restoration

Pre-processing methods that aim to suppress degradation using knowledge about its nature are called **image restoration**. Most image restoration methods are based on convolution applied globally to the whole image.

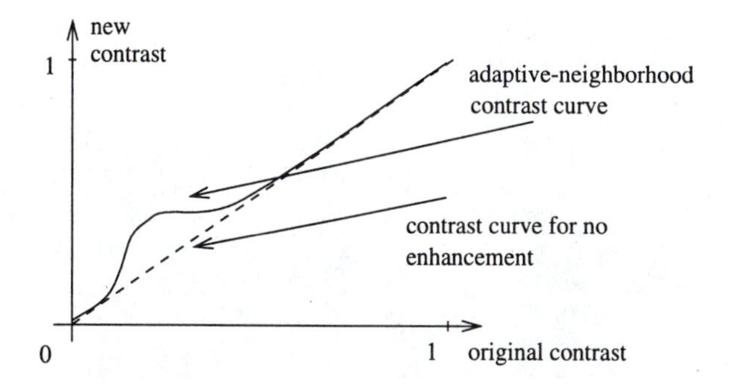

Figure 4.27: *Adaptive-neighborhood contrast curves.*

Degradation of images can have many causes: defects of optical lenses, non-linearity of the electro-optical sensor, graininess of the film material, relative motion between an object and camera, wrong focus, atmospheric turbulence in remote sensing or astronomy, scanning of photographs, etc. [Jain 89, Pratt 91, Gonzalez and Woods 92, Tekalp and Pavlovic 93, Sid-Ahmed 95]. The objective of image restoration is to reconstruct the original image from its degraded version.

Image restoration techniques can be classified into two groups: deterministic and stochastic. **Deterministic** methods are applicable to images with little noise and a known degradation function. The original image is obtained from the degraded one by a transformation inverse to the degradation. **Stochastic** techniques try to find the best restoration according to a particular stochastic criterion, e.g., a least-squares method. In some cases the degradation transformation must be estimated first.

It is advantageous to know the degradation function explicitly. The better this knowledge is, the better are the results of the restoration. There are three typical degradations with a simple function: relative constant speed movement of the object with respect to the camera, wrong lens focus, and atmospheric turbulence.

In most practical cases, there is insufficient knowledge about the degradation, and it must be estimated and modeled. The estimation can be classified into two groups according to the information available: a priori and a posteriori. If degradation type and/or parameters need to be estimated, this step is the most crucial one, being responsible for image restoration success or failure. It is also the most difficult part of image restoration.

A priori knowledge about degradation is either known in advance or can be obtained before restoration. For example, if it is known in advance that the image was degraded by relative motion of an object with respect to the sensor, then the modeling determines only the speed and direction of the motion. An example of the second case is an attempt to estimate parameters of a capturing device such as a TV camera or digitizer, whose degradation remains unchanged over a period of time and can be modeled by studying a known sample image and its degraded version.

A posteriori knowledge is that obtained by analyzing the degraded image. A typical example is to find some interest points in the image (e.g., corners, straight lines) and guess how they looked before degradation. Another possibility is to use spectral characteristics of the regions in the image that are relatively homogeneous.

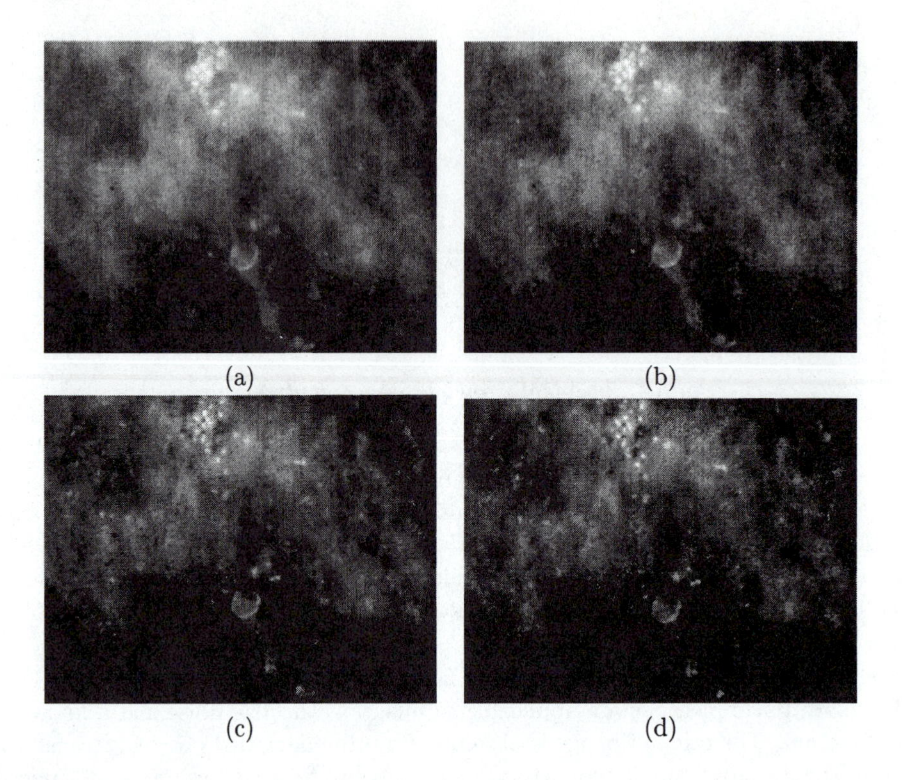

Figure 4.28: *Adaptive-neighborhood contrast enhancement: (a) original mammogram image; (b) unsharp masking 3×3; (c) adaptive contrast enhancement,* $T_2 = 0.03, s = 3$*; (d) adaptive contrast enhancement,* $T_2 = 0.05, s = 3$*.* T_2 *and* s *are parameters of the adaptive neighborhood; see equation (4.77) and the accompanying text. Courtesy R. Paranjape, R. Rangayyan, University of Calgary.*

Image restoration is considered in more detail in [Pratt 78, Rosenfeld and Kak 82, Bates and McDonnell 86, Pratt 91, Gonzalez and Woods 92, Castleman 96] and only the basic principles of the restoration and three typical degradations are considered here.

A degraded image g can arise from the original image f by a process which can be expressed as

$$g(i,j) = s\left[\int\int_{(a,b)\in\mathcal{O}} f(a,b)\, h(a,b,i,j)\, da\, db\right] + \nu(i,j) \tag{4.80}$$

where s is some non-linear function and ν describes the noise. The degradation is very often simplified by neglecting the non-linearity and by assuming that the function h is invariant with respect to position in the image. Degradation can be then expressed as convolution:

$$g(i,j) = (f * h)\,(i,j) + \nu(i,j) \tag{4.81}$$

If the degradation is given by equation (4.81) and the noise is not significant, then image restoration equates to inverse convolution (also called deconvolution). If noise is not negligible, then the inverse convolution is solved as an overdetermined system of linear equations.

Methods based on minimization of the least square error such as Wiener filtering (off-line) or Kalman filtering (recursive, on-line; see Section 15.4) are examples [Bates and McDonnell 86].

4.4.1 Degradations that are easy to restore

We mentioned that there are three types of degradations that can be easily expressed mathematically and also restored simply in images. These degradations can be expressed by convolution, equation (4.81); the Fourier transform H of the convolution function is used. In the absence of noise, the relationship between the Fourier representations F, G, H of the undegraded image f, the degraded image g, and the degradation convolution kernel h, respectively, is

$$G = H\,F \tag{4.82}$$

Therefore, not considering image noise ν, knowledge of the degradation function fully facilitates image restoration by inverse convolution (Section 4.4.2). We first discuss several degradation functions.

Relative motion of the camera and object

Assume an image is acquired with a camera with a mechanical shutter. Relative motion of the camera and the photographed object during the shutter open time T causes smoothing of the object in the image. Suppose V is the constant speed in the direction of the x axis; the Fourier transform $H(u,v)$ of the degradation caused in time T is given by [Rosenfeld and Kak 82]

$$H(u,v) = \frac{\sin(\pi V T u)}{\pi V u} \tag{4.83}$$

Wrong lens focus

Image smoothing caused by imperfect focus of a thin lens can be described by the following function [Born and Wolf 69]:

$$H(u,v) = \frac{J_1(a\,r)}{a\,r} \tag{4.84}$$

where J_1 is the Bessel function of the first order, $r^2 = u^2 + v^2$, and a is the displacement—the model is not space invariant.

Atmospheric turbulence

Atmospheric turbulence is degradation that needs to be restored in remote sensing and astronomy. It is caused by temperature non-homogeneity in the atmosphere that deviates passing light rays. The mathematical model is derived in [Hufnagel and Stanley 64] and is expressed as

$$H(u,v) = e^{-c(u^2+v^2)^{5/6}} \tag{4.85}$$

where c is a constant that depends on the type of turbulence which is usually found experimentally. The power $5/6$ is sometimes replaced by 1.

4.4.2 Inverse filtration

An obvious approach to image restoration is inverse filtration based on properties of the Fourier transforms [Sondhi 72, Andrews and Hunt 77, Rosenfeld and Kak 82]. Inverse filtering uses the assumption that degradation was caused by a linear function $h(i,j)$ [cf. equation (4.81)] and considers the additive noise ν as another source of degradation. It is further assumed that ν is independent of the signal. After applying the Fourier transform to equation (4.81), we get

$$G(u,v) = F(u,v)\, H(u,v) + N(u,v) \tag{4.86}$$

The degradation can be eliminated using the restoration filter with a transfer function that is inverse to the degradation h. The Fourier transform of the inverse filter is then expressed as $H^{-1}(u,v)$.

We derive the original undegraded image F (its Fourier transform to be exact) from its degraded version G [equation (4.86)], as follows:

$$F(u,v) = G(u,v)\, H^{-1}(u,v) - N(u,v)\, H^{-1}(u,v) \tag{4.87}$$

This equation shows that inverse filtration works well for images that are not corrupted by noise [not considering possible computational problems if $H(u,v)$ gets close to zero at some location of the u,v space—fortunately, such locations can be neglected without perceivable effect on the restoration result]. However, if noise is present, several problems arise. First, the noise influence may become significant for frequencies where $H(u,v)$ has small magnitude. This situation usually corresponds to high frequencies u,v. In reality, $H(u,v)$ usually decreases in magnitude much more rapidly than $N(u,v)$ and thus the noise effect may dominate the entire restoration result. Limiting the restoration to a small neighborhood of the u,v origin in which $H(u,v)$ is sufficiently large overcomes this problem, and the results are usually quite acceptable. The second problem deals with the spectrum of the noise itself—we usually do not have enough information about the noise to determine $N(u,v)$ sufficiently well.

4.4.3 Wiener filtration

Based on the preceding discussion, it is no surprise that inverse filtration gives poor results in pixels suffering from noise, since the information about noise properties is not taken into account. Wiener (least mean square) filtration [Wiener 42, Helstrom 67, Slepian 67, Pratt 72, Rosenfeld and Kak 82, Gonzalez and Woods 92, Castleman 96] incorporates a priori knowledge about the noise properties in the image restoration formula.

Restoration by the Wiener filter gives an estimate $\hat{f}$ of the original uncorrupted image f with minimal mean square error e^2:

$$e^2 = \mathcal{E}\left\{[f(i,j) - \hat{f}(i,j)]^2\right\} \tag{4.88}$$

where $\mathcal{E}$ denotes the mean operator. If no constraints are applied to the solution of equation (4.88), then an optimal estimate $\hat{f}$ is the conditional mean value of the ideal image f under the condition g. This approach is complicated from the computational point of view. Moreover, the conditional probability density between the optimal image f and the corrupted image g is not usually known. The optimal estimate is in general a non-linear function of the image g.

Minimization of equation (4.88) is easy if the estimate $\hat{f}$ is a linear combination of the values in the image g; the estimate $\hat{f}$ is then close (but not necessarily equal) to the theoretical optimum. The estimate is equal to the theoretical optimum only if the stochastic processes describing images f, g, and the noise ν are homogeneous, and their probability density is Gaussian [Andrews and Hunt 77]. These conditions are not usually fulfilled for typical images.

Denote the Fourier transform of the Wiener filter by H_W. Then, the estimate $\hat{F}$ of the Fourier transform F of the original image f can be obtained as

$$\hat{F}(u,v) = H_W(u,v)\; G(u,v) \tag{4.89}$$

The function H_W is not derived here, but may be found elsewhere [Papoulis 65, Rosenfeld and Kak 82, Bates and McDonnell 86, Gonzalez and Woods 92]. The result is

$$H_W(u,v) = \frac{H^*(u,v)}{|H(u,v)|^2 + [S_{\nu\nu}(u,v)/S_{ff}(u,v)]} \tag{4.90}$$

where H is the transform function of the degradation, $*$ denotes complex conjugate, $S_{\nu\nu}$ is the spectral density of the noise, and S_{ff} is the spectral density of the undegraded image.

If Wiener filtration is used, the nature of degradation H and statistical parameters of the noise need to be known. Wiener filtration theory solves the problem of optimal a posteriori linear mean square estimates—all statistics (for example, power spectrum) should be available in advance. Note the term $S_{ff}(u,v)$ in equation (4.90), which represents the spectrum of the undegraded image. This information may be difficult to obtain considering the goal of image restoration, to determine the undegraded image.

Note that the ideal inverse filter is a special case of the Wiener filter in which noise is absent, i.e., $S_{\nu\nu} = 0$.

Restoration is illustrated in Figures 4.29 and 4.30. Figure 4.29a shows an image that was degraded by 5 pixels motion in the direction of the x axis, and Figure 4.29b shows the result of restoration where Wiener filtration was used. Figure 4.30a shows an image degraded by wrong focus and Figure 4.30b is the result of restoration using Wiener filtration.

Despite its unquestionable power, Wiener filtration suffers several substantial limitations. First, the criterion of optimality is based on minimum mean square error and weights all errors equally, a mathematically fully acceptable criterion that unfortunately does not perform well if an image is restored for human viewing. The reason is that humans perceive the restoration errors more seriously in constant-gray-level areas and in bright regions, while they are much less sensitive to errors located in dark regions and in high-gradient areas. Second, spatially variant degradations cannot be restored using the standard Wiener filtration approach, and these degradations are common. Third, most images are highly non-stationary, containing large homogeneous areas separated by high-contrast edges. Wiener filtration cannot handle non-stationary signals and noise. To deal with real-life image degradations, more sophisticated approaches may be needed. Examples include **power spectrum equalization** and **geometric mean filtration**. These and other specialized restoration techniques can be found in higher-level texts devoted to this topic; [Castleman 96] is well suited for such a purpose.

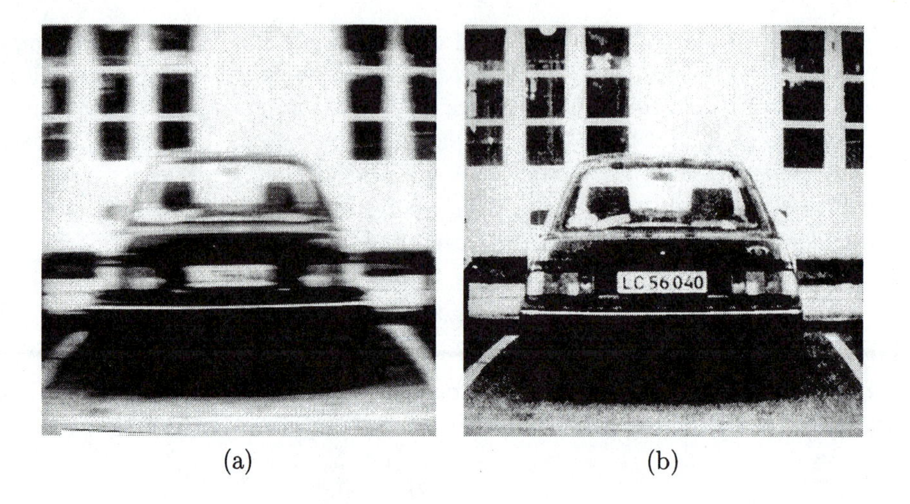

(a) (b)

Figure 4.29: *Restoration of motion blur using Wiener filtration. Courtesy P. Kohout, Criminalistic Institute, Prague.*

4.5 Summary

- **Image pre-processing**
  - Operations with images at the lowest level of abstraction—both input and output are intensity images—are called *pre-processing.*
  - The aim of pre-processing is an improvement of the image data that suppresses unwilling distortions or enhances some image features important for further processing.
  - Four basic types of pre-processing methods exist:
    * Brightness transformations
    * Geometric transformations
    * Local neighborhood pre-processing
    * Image restoration
- **Pixel brightness transformations**
  - There are two classes of pixel brightness transformations:
    * *Brightness corrections*
    * *Gray-scale transformations*
  - Brightness corrections modify pixel brightness taking into account its original brightness and its position in the image.
  - Gray-scale transformations change brightness without regard to position in the image.
  - Frequently used brightness transformations include:

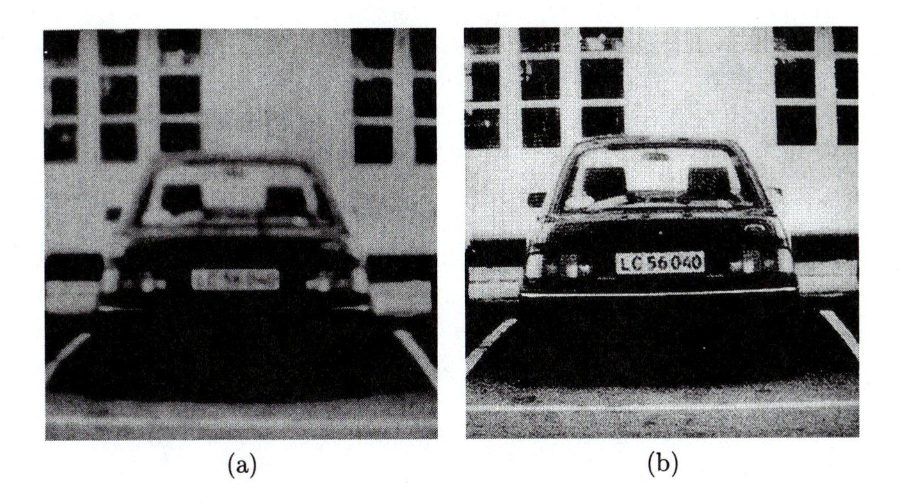

(a) (b)

Figure 4.30: *Restoration of wrong focus blur using Wiener filtration. Courtesy P. Kohout, Criminalistic Institute, Prague.*

  * Brightness thresholding
  * Histogram equalization
  * Logarithmic gray-scale transforms
  * Look-up table transforms
  * Pseudo-color transforms
- The goal of histogram equalization is to create an image with equally distributed brightness levels over the whole brightness scale.

- **Geometric transformations**
  - Geometric transforms permit the elimination of the geometric distortions that occur when an image is captured.
  - A geometric transform typically consists of two basic steps:
    * *Pixel co-ordinate transformation*
    * *Brightness interpolation*
  - Pixel co-ordinate transformations map the co-ordinates of the input image pixel to a point in the output image; *affine* and *bilinear* transforms are frequently used.
  - The output point co-ordinates do not usually match the digital grid after the transform and interpolation is employed to determine brightnesses of output pixels; *nearest-neighbor*, *linear*, and *bi-cubic* interpolations are frequently used.

- **Local pre-processing**
  - Local pre-processing methods use a small neighborhood of a pixel in an input image to produce a new brightness value in the output image.
  - For the pre-processing goal, two groups are common: *smoothing* and *edge detection.*

- Smoothing aims to suppress noise or other small fluctuations in the image; it is equivalent to suppressing high frequencies in the Fourier transform domain.
- Smoothing approaches based on direct averaging blur image edges. More sophisticated approaches reduce blurring by averaging in homogeneous local neighborhoods.
- *Median* smoothing is a non-linear operation; it reduces the blurring of edges by replacing the current point in the image by the median of the brightnesses in its neighborhood.
- *Gradient operators* determine edges—locations in which the image function undergoes rapid changes; they have a similar effect to suppressing low frequencies in the Fourier transform domain.
- *Edge* is a property attached to an individual pixel and has two components, *magnitude* and *direction.*
- Most gradient operators can be expressed using *convolution masks*; examples include Roberts, Laplace, Prewitt, Sobel, Robinson, and Kirsch operators.
- The main disadvantage of convolution edge detectors is their scale dependence and noise sensitivity. There is seldom a sound reason for choosing a particular size of a local neighborhood operator.
- *Zero-crossings* of the second derivative are more robust than small-size gradient detectors and can be calculated as a Laplacian of Gaussians (LoG) or as a difference of Gaussians (DoG).
- The *Canny* edge detector is optimal for step edges corrupted by white noise. The optimality criterion is based on requirements of *detecting* important edges, small *localization* error, and *single-edge response.* Canny edge detection starts with convolving an image with a symmetric 2D Gaussian and then differentiating in the direction of the gradient; further steps include *non-maximal edge suppression*, *hysteresis thresholding*, and *feature synthesis.*
- Edges can also be detected in multi-spectral images.
- Other local pre-processing operations include *line finding*, *line thinning*, *line filling*, and *interest point detection.*
- In *adaptive neighborhood pre-processing*, the neighborhood sizes and shapes are dependent on characteristics of image data and on parameters defining measures of homogeneity of a neighborhood.

- **Image restoration**

  - Image restoration methods aim to suppress degradation using knowledge about its nature. Most image restoration methods are based on *deconvolution* applied globally to the entire image.
  - Relative-constant-speed movement of the object with respect to the camera, wrong lens focus, and atmospheric turbulence are three typical image degradations with simple degradation functions.
  - *Inverse filtration* assumes that degradation was caused by a linear function.

- *Wiener filtration* gives an estimate of the original uncorrupted image with minimal mean square error; the optimal estimate is in general a non-linear function of the corrupted image.

4.6 Exercises

Short-answer questions

1. What is the main aim of image pre-processing?
2. Give examples of situations in which brightness transformations, geometric transformations, smoothing, edge detection, and/or image restorations are typically applied.
3. What is the main difference between brightness correction and gray-scale transformation?
4. Explain the rationale of histogram equalization.
5. Explain why the histogram of a discrete image is not flat after histogram equalization.
6. Consider the image given in Figure 4.3a. After histogram equalization (Figure 4.3b), much more detail is visible. Does histogram equalization increase the amount of information contained in image data? Explain.
7. What is a cumulative histogram?
8. What are the two main steps of geometric transforms?
9. What is the minimum number of corresponding pixel pairs that must be determined if a

    (a) Bilinear transform

    (b) Affine transform

    is used to perform a geometric correction?
10. Give a geometric transformation equation for

    (a) Rotation

    (b) Change of scale

    (c) Skewing by an angle
11. Consider brightness interpolation—explain why it is better to perform brightness interpolation using brightness values of neighboring points in the input image than interpolating in the output image.
12. Explain the principles of nearest-neighbor interpolation, linear interpolation, and bi-cubic interpolation.
13. Explain why smoothing and edge detection have conflicting aims.
14. Explain why Gaussian filtering is often the preferred averaging method.
15. Explain why smoothing typically blurs image edges.
16. Name several smoothing methods that try to avoid image blurring. Explain their main principles.
17. Explain why median filtering performs well in images corrupted by impulse noise.
18. Give convolution masks for the following edge detectors:

(a) Roberts
(b) Laplace
(c) Prewitt
(d) Sobel
(e) Robinson
(f) Kirsch

Which ones can serve as compass operators? List several applications in which determining edge direction is important.

19. Explain why subtraction of a second derivative of the image function from the original image results in the visual effect of image sharpening.
20. What are LoG and DoG? How do you compute them? How are they used?
21. Propose a robust way of detecting significant image edges using zero-crossings.
22. Explain why LoG is a better edge detector than Laplace edge detector.
23. Explain the notion of scale in image processing.
24. Explain the importance of hysteresis thresholding and non-maximal suppression in the Canny edge detection process. How do these two concepts influence the resulting edge image?
25. Explain the principles of noise suppression, histogram modification, and contrast enhancement performed in adaptive neighborhoods.
26. Explain the principles of image restoration based on

    (a) Inverse convolution
    (b) Inverse filtration
    (c) Wiener filtration

    List the main differences among the above methods.

27. Give image distortion functions for

    (a) Relative camera motion
    (b) Out-of-focus lens
    (c) Atmospheric turbulence

Problems

1. Consider calibrating a TV camera for non-homogeneous lighting. Develop a program that determines camera calibration coefficients after an image of a constant-gray-level surface is acquired with this camera. After calibration, the program should perform appropriate brightness correction to remove the effects of non-homogeneous lighting on other images acquired using the same camera under the same lighting conditions. Test the program's functionality under several non-homogeneous lighting conditions.
2. Determine a gray-scale transformation that maps the darkest 5% of image pixels to black (0), the brightest 10% of pixels to white (255), and linearly transforms the gray-levels of all remaining pixels between black and white.

3. Develop a program for gray-scale transformations as described in Problem 4.2. Develop it in such a way that the percentages of dark and bright pixels mapped to pure black and white are program parameters and can be modified by the operator.
4. Develop programs for the three gray-scale image transforms given in Figure 4.1. Apply them to several images and make a subjective judgment about the usefulness of the transforms.
5. Implement histogram equalization as described in Algorithm 4.1. Select several images with a variety of gray-level histograms to test the method's performance, include over- and under-exposed images, low-contrast images, and images with large dark or bright background regions. Compare the results.
6. Apply histogram equalization to an already equalized image; compare and explain the results of 1-step and 2-step histogram equalization.
7. Write a program that performs histogram equalization on HSI images (see Section 2.2.3). Verify visually that equalizing the I component alone has the desired effect, while equalizing the others does not.
8. Develop programs for the following geometric transforms:
   (a) Rotation
   (b) Change of scale
   (c) Skewing
   (d) Affine transform calculated from three pairs of corresponding points
   (e) Bilinear transform calculated from four pairs of corresponding points

   To avoid writing a program for solving systems of linear equations, use a mathematical calculation software package (e.g., Matlab, Mathematica, Maple) to determine transformation coefficients for the affine and bilinear transforms (d) and (e). For each of the above transforms, implement the following three brightness interpolation approaches:

   - Nearest-neighbor interpolation
   - Linear interpolation
   - Bi-cubic interpolation

   To implement all possible combinations efficiently, design your programs in a modular fashion with substantial code reuse. Compare the subjective image quality resulting from the three brightness interpolation approaches.
9. Develop a program for image convolution using a rectangular convolution mask of any odd size. The mask should be input as an ASCII text file. Test your program using the following convolution kernels:
   (a) 3×3 averaging
   (b) 7×7 averaging
   (c) 11×11 averaging
   (d) 5×5 Gaussian filtering [modification of equation (4.29)]
10. An imperfect camera is used to capture an image of a static scene:
   (a) The camera is producing random noise with zero mean. The single images look quite noisy.

(b) The camera has a dark spot in the middle of the image - the image is visible there, but it is darker than the rest of the image.

What approaches would you choose to obtain the best possible image quality? You can capture as many frames of the static scene as you may want to and/or you may capture an image of a constant gray-level and/or you may capture any other image with known gray-level properties. Give complete step-by-step procedures including the associated mathematics for both cases.

11. Implement image averaging using a rotating mask as described in Algorithm 4.2. Use the masks specified in

    (a) Figure 4.12

    (b) Figure 4.13

    Assess the amount of image blurring and sharpening in comparison to standard image averaging.

12. As an extension of Problem 4.11, consider iterative application of averaging using a rotating mask until convergence. Assess the smoothing/sharpening effect of the iterative approach in comparison to the single-step approach developed in Problem 4.11.

13. Show the linear character of Gaussian averaging and the non-linear character of median filtering. In other words, show that $\text{med}[f_1(x) + f_2(x)] \neq \text{med}[f_1(x)] + \text{med}[f_2(x)]$ for an arbitrary region of pixels x and two image brightness functions f_1 and f_2.

14. Consider the binary image given in Figure 4.31. Show the result of 3×3 median filtering if the following masks are used (a 'zero' in a mask position means that the corresponding pixel is not used for median calculation):

$$(a) \begin{bmatrix} 1 & 1 & 1 \\ 1 & 1 & 1 \\ 1 & 1 & 1 \end{bmatrix} \quad (b) \begin{bmatrix} 0 & 1 & 0 \\ 1 & 1 & 1 \\ 0 & 1 & 0 \end{bmatrix} \quad (c) \begin{bmatrix} 0 & 0 & 0 \\ 1 & 1 & 1 \\ 0 & 0 & 0 \end{bmatrix} \quad (d) \begin{bmatrix} 1 & 1 & 1 \\ 0 & 0 & 0 \\ 0 & 0 & 0 \end{bmatrix}$$

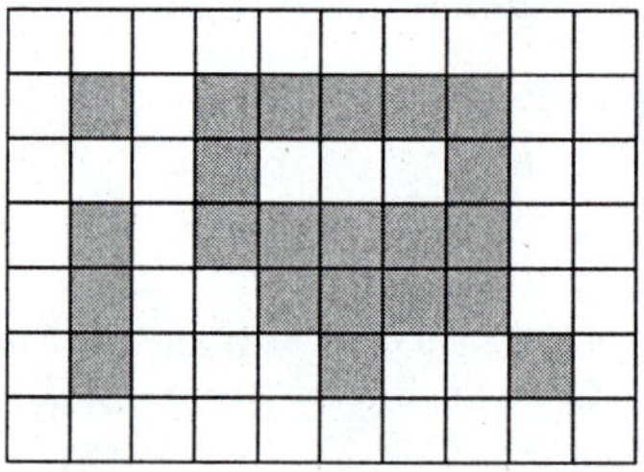

Figure 4.31: *Problem 4.14*

15. Implement efficient median filtering as described in Algorithm 4.3. Compare the processing efficiency in comparison with a 'naive' median filtering implementation. Use median filter sizes ranging from 3×3 to 15×15 (odd sizes) for comparison.

16. Median filtering that uses a 3×3 mask,

$$\begin{bmatrix} 1 & 1 & 1 \\ 1 & 1 & 1 \\ 1 & 1 & 1 \end{bmatrix}$$

    is damaging to thin lines and sharp corners. Give a 3×3 mask that can be used for median filtering and does not exhibit this behavior.

17. As a continuation of Problem 4.16, develop a program performing median filtering in neighborhoods of any size and shape. To test the behavior of different sizes and shapes of the median mask, corrupt input images with:

    (a) Impulse noise of varying severity
    (b) Horizontal lines of different width
    (c) Vertical lines of different width
    (d) Combinations of lines of different width and different direction

    For each of the above situations, determine the mask providing the subjectively best pre-processing performance.

18. As further continuation of Problem 4.17, consider the option of iteratively repeating median filtering several times in a sequence. For each of the situations given in Problem 4.17, assess the performance of the mask considered the best in Problem 4.17 in comparison to some other mask applied iteratively several times. Determine the mask and the number of iterations providing the subjectively best pre-processing performance. Consider the extent of removing the image corruption as well as the amount of image blurring introduced.

19. Develop a program performing averaging with limited data validity, as described by equation (4.30). The program must allow averaging with square masks from 3×3 to 15×15 (odd sizes), the convolution kernel coefficients must be calculated in the program, not listed as kernel values for each filter size. Averaging should only be done for pixels (i, j) with gray values $g(i, j)$ from the interval of invalid data ($\min\{invalid\} \leq g(i, j) \leq \max\{invalid\}$) and only valid data should contribute to the average calculated. Test your program on images corrupted by impulse noise and images corrupted by narrow (several pixels wide) elongated objects of gray-levels from a narrow gray-level interval. How does the effectiveness of your method compare to that of simple averaging and median filtering?

20. Create a set of noisy images by corrupting an image with

    (a) Additive Gaussian noise of five different severity levels
    (b) Multiplicative Gaussian noise of five different severity levels
    (c) Impulse noise of five different severity levels

    Apply

    - Averaging filters of different sizes [equation (4.27)]
    - Gaussian filters of different standard deviations
    - Median filters of different sizes and/or numbers of iterations
    - Averaging with limited data validity
    - Averaging according to inverse gradient
    - Rotating mask averaging

    to remove the superimposed noise as much as possible. Quantitatively compare the efficiency of individual approaches by calculating a mean square error between the original and pre-processed images. Formulate a general recommendation about applicability of pre-processing techniques for removing specific types of noise.

21. Using the program developed in Problem 4.9, implement the following edge detectors:

    (a) Laplace in 4-neighborhood
    (b) Laplace in 8-neighborhood

22. Consider the binary image given in Figure 4.32. Show the resulting edge images (magnitude and direction images where applicable) if the following edge detectors are used:

    (a) Laplace in 4-neighborhood

    (b) Prewitt

    (c) Sobel

    (d) Kirsch

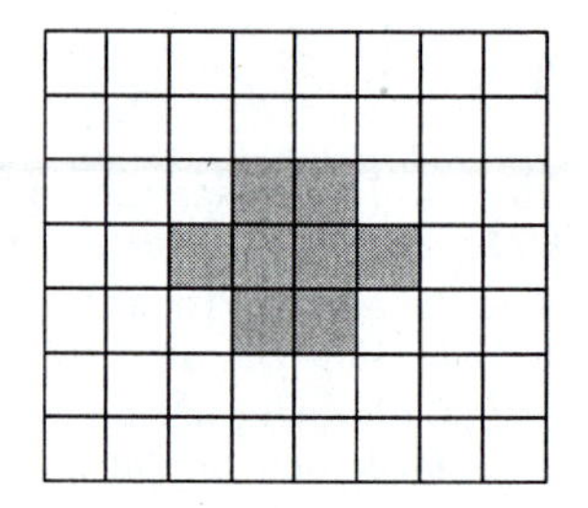

Figure 4.32: *Problem 4.22*

23. Develop programs for determining magnitude and direction edge image pairs using the following compass edge detectors:

    (a) Prewitt

    (b) Sobel

    (c) Robinson

    (d) Kirsch

    The programs must display:

    - Magnitude edge image
    - Direction edge image
    - Magnitude edge image for edges of a specified direction

    To show that your edge detection gives correct results, use a circle image to test your programs.

24. Develop a program for image sharpening as specified by equation (4.40). Use

    (a) Non-directional Laplacian to approximate $S(i,j)$

    (b) Unsharp masking

    Experiment with the value of the parameter C for both approaches and with the extent of smoothing for unsharp masking. Compare the sharpening effects of the two approaches.

25. Develop a program determining zero-crossing of the second derivative. Use the

    (a) LoG definition

    (b) DoG definition

    Explain why borders are disconnected if only zero pixels are used for zero-crossing definition. Propose and implement a modification providing contiguous borders. Is it possible to use zero crossings to determine edge positions with sub-pixel accuracy? If yes, how?

26. Apply the LoG edge detector developed in Problem 4.25 with several values of the smoothing parameter σ. Explain the relationship between the smoothing parameter and the scale of the resulting edge image.

27. Consider the double-step edge shown in Figure 4.33. Show that the locations of the zero-crossing of the second derivative depend on σ. Discuss the behavior of the zero-crossings as σ increases.

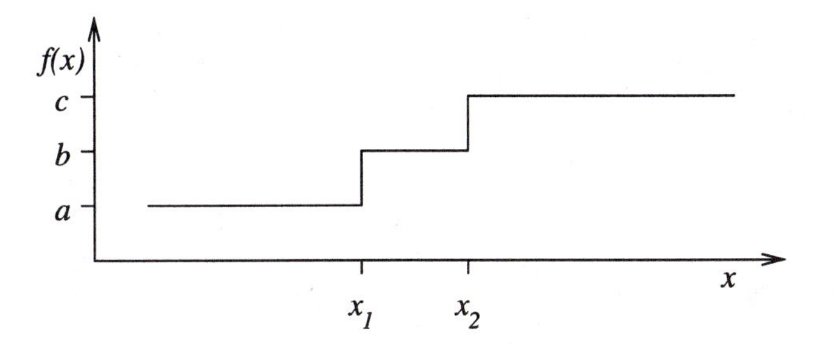

Figure 4.33: *Problem 4.27*

28. Implement the Canny edge detector as described by steps 1-5 of Algorithm 4.4 (implementing feature synthesis is hard, therefore we recommend skipping step 7).

29. Based on your theoretical understanding of Canny edge detector parameters of non-maximal suppression and hysteresis thresholding, generate hypotheses of how changes of these parameters will influence the resulting edge images. Using the Canny edge detector program developed in Problem 4.28 or using one of several implementations of the Canny edge detector freely available on the Web, prove the validity of your hypotheses by experimenting with the non-maximal suppression and hysteresis thresholding parameters. Summarize your observations.

30. Design and implement at least one operator (different from those given in text) for each of the following tasks:

    (a) Line finding
    (b) Line thinning
    (c) Line filling
    (d) Corner detection

    Test your operators in artificially generated images using the program developed in Problem 4.9

31. Develop a program for image restoration using inverse convolution. Use the program to restore images after the following degradations:

    (a) Relative motion of the camera
    (b) Wrong camera focus

    Assume that the parameters of the degradation are known. Test the programs on artificially degraded images. (Although not corresponding exactly to equation (4.83), camera motion distortion can be modeled using a simple sinusoidal filter. Create a sinusoidal image of the same size as your input image and use it as a sinusoidal filter in the frequency domain. By changing the numbers of waves along width and height, you can create a 'double exposure' image that may have resulted from an abrupt camera motion.]

4.7 References

[Alparone et al. 96] L Alparone, S Baronti, and A Casini. A novel approach to the suppression of false contours. In *International Conference on Image Processing*, pages 825–828, IEEE, Los Alamitos, CA, 1996.

[Andrews and Hunt 77] H C Andrews and B R Hunt. *Digital Image Restoration.* Prentice-Hall, Englewood Cliffs, NJ, 1977.

[Aydin et al. 96] T Aydin, Y Yemez, E Anarim, and B Sankur. Multidirectional and multiscale edge detection via M-band wavelet transform. *IEEE Transactions on Image Processing*, 5:1370–1377, 1996.

[Ballard and Brown 82] D H Ballard and C M Brown. *Computer Vision.* Prentice-Hall, Englewood Cliffs, NJ, 1982.

[Bates and McDonnell 86] R H T Bates and M J McDonnell. *Image Restoration and Reconstruction.* Clarendon Press, Oxford, 1986.

[Bezdek et al. 96] J C Bezdek, R Chandrasekhar, and Y Attikiouzel. A new fuzzy model for edge detection. In *Applications of Fuzzy Logic Technology III, Proc. SPIE Vol. 2761*, pages 11–28, SPIE, Bellingham, WA, 1996.

[Borik et al. 83] A C Borik, T S Huang, and D C Munson. A generalization of median filtering using combination of order statistics. *IEEE Proceedings*, 71(31):1342–1350, 1983.

[Born and Wolf 69] M Born and E Wolf. *Principles of Optics.* Pergamon Press, New York, 1969.

[Brady 84] M Brady. Representing shape. In M Brady, L A Gerhardt, and H F Davidson, editors, *Robotics and Artificial Intelligence*, pages 279–300. Springer + NATO, Berlin, 1984.

[Canny 83] J F Canny. Finding edges and lines in images. Technical Report AI-TR-720, MIT, Artificial Intelligence Laboratory, Cambridge, MA, 1983.

[Canny 86] J F Canny. A computational approach to edge detection. *IEEE Transactions on Pattern Analysis and Machine Intelligence*, 8(6):679–698, 1986.

[Castleman 96] K R Castleman. *Digital Image Processing.* Prentice-Hall, Englewood Cliffs, NJ, 1996.

[Cervenka and Charvat 87] V Cervenka and K Charvat. Survey of the image processing research applicable to the thematic mapping based on aerocosmic data (in Czech). Technical Report A 12–346–811, Geodetic and Carthographic Institute, Prague, Czechoslovakia, 1987.

[Cooper et al. 93] J Cooper, S Venkatesh, and L Kitchen. Early jump-out corner detectors. *IEEE Transactions on Pattern Analysis and Machine Intelligence*, 15:823–828, 1993.

[Demigny et al. 95] D Demigny, F G Lorca, and L Kessal. Evaluation of edge detectors performances with a discrete expression of Canny's criteria. In *International Conference on Image Processing*, pages 169–172, IEEE, Los Alamitos, CA, 1995.

[Elder and Zucker 96] J H Elder and S W Zucker. Local scale control for edge detection and blur estimation. In B Buxton and R Cipolla, editors, *4th European Conference on Computer Vision,* Cambridge, England, pages 57–69, Springer Verlag, Berlin, 1996.

[Faugeras 93] O D Faugeras. *Three-Dimensional Computer Vision: A Geometric Viewpoint.* MIT Press, Cambridge, MA, 1993.

[Gonzalez and Woods 92] R C Gonzalez and R E Woods. *Digital Image Processing.* Addison-Wesley, Reading, MA, 1992.

[Gordon and Rangayyan 84] R Gordon and R M Rangayyan. Feature enhancement of film mammograms using fixed and adaptive neighborhoods. *Applied Optics*, 23:560–564, 1984.

[Haralick 84] R M Haralick. Digital step edges from zero crossing of second directional derivatives. *IEEE Transactions on Pattern Analysis and Machine Intelligence*, 6:58–68, 1984.

[Haralick and Shapiro 92] R M Haralick and L G Shapiro. *Computer and Robot Vision, Volume I.* Addison-Wesley, Reading, MA, 1992.

[Haralick and Watson 81] R M Haralick and L Watson. A facet model for image data. *Computer Graphics and Image Processing*, 15:113–129, 1981.

[Hardie and Boncelet 95] R C Hardie and C G Boncelet. Gradient-based edge detection using nonlinear edge enhancing prefilters. *IEEE Transactions on Image Processing*, 4:1572–1577, 1995.

[Hebert and Kim 96] D J Hebert and H Kim. A fast-wavelet compass edge detector. In *Wavelet Applications in Signal and Image Processing IV, Proc. SPIE Vol. 2825*, pages 432–442, SPIE, Bellingham, WA, 1996.

[Helstrom 67] C W Helstrom. Image restoration by the method of least squares. *Journal of the Optical Society of America*, 57:297–303, 1967.

[Ho and Ohnishi 95] K H L Ho and N Ohnishi. FEDGE—fuzzy edge detection by fuzzy categorization and classification of edges. In *Fuzzy Logic in Artificial Intelligence. Towards Intelligent Systems. IJCAI '95 Workshop. Selected Papers*, pages 182–196, 1995.

[Huang 83] T S Huang, editor. *Image Sequence Processing and Dynamic Scene Analysis.* Springer Verlag, Berlin, 1983.

[Huang et al. 79] T S Huang, G J Yang, and G Y Tang. A fast two-dimensional median filtering algorithm. *IEEE Transactions on Acoustics, Speech and Signal Processing*, ASSP-27(1):13–18, 1979.

[Huertas and Medioni 86] A Huertas and G Medioni. Detection of intensity changes with subpixel accuracy using Laplacian-Gaussian masks. *IEEE Transactions on Pattern Analysis and Machine Intelligence*, 8:651–664, 1986.

[Hufnagel and Stanley 64] R E Hufnagel and N R Stanley. Modulation transfer function associated with image transmission through turbulent media. *Journal of the Optical Society of America*, 54:52–61, 1964.

[Hummel and Moniot 89] R Hummel and R Moniot. Reconstructions from zero crossings in scale space. *IEEE Transactions on Acoustics, Speech and Signal Processing*, 37(12):2111–2130, 1989.

[Jain 89] A K Jain. *Fundamentals of Digital Image Processing.* Prentice-Hall, Englewood Cliffs, NJ, 1989.

[Jain et al. 95] R Jain, R Kasturi, and B G Schunck. *Machine Vision.* McGraw-Hill, New York, 1995.

[Jalali and Boyce 95] S Jalali and J F Boyce. Determination of optimal general edge detectors by global minimization of a cost function. *Image and Vision Computing*, 13:683–693, 1995.

[Laligant et al. 94] O Laligant, F Truchete, and J Miteran. Edge detection by multiscale merging. In *Proceedings of the IEEE-SP International Symposium on Time-Frequency and Time-Scale Analysis*, pages 237–240, IEEE, Los Alamitos, CA, 1994.

[Law et al. 96] T Law, H Itoh, and H Seki. Image filtering, edge detection, and edge tracing using fuzzy reasoning. *IEEE Transactions on Pattern Analysis and Machine Intelligence*, 18:481–491, 1996.

[Lindeberg 96] T Lindeberg. Edge detection and ridge detection with automatic scale selection. In *Computer Vision and Pattern Recognition*, pages 465–470, IEEE, Los Alamitos, CA, 1996.

[Lowe 89] D G Lowe. Organization of smooth image curves at multiple scales. *International Journal of Computer Vision*, 1:119–130, 1989.

[Marr 82] D Marr. *Vision—A Computational Investigation into the Human Representation and Processing of Visual Information.* Freeman, San Francisco, 1982.

[Marr and Hildreth 80] D Marr and E Hildreth. Theory of edge detection. *Proceedings of the Royal Society*, B 207:187–217, 1980.

[Marr and Hildreth 91] D Marr and E Hildreth. Theory of edge detection. In R Kasturi and R C Jain, editors, *Computer Vision*, pages 77–107. IEEE, Los Alamitos, CA, 1991.

[Matalas et al. 97] L Matalas, R Benjamin, and R Kitney. An edge detection technique using the facet model and parameterized relaxation labeling. *IEEE Transactions on Pattern Analysis and Machine Intelligence*, 19:328–341, 1997.

[McDonnell 81] M J McDonnell. Box filtering techniques. *Computer Graphics and Image Processing*, 17(3):65–70, 1981.

[Mehrotra and Nichani 90] R Mehrotra and S Nichani. Corner detection. *Pattern Recognition Letters*, 23(11):1223–1233, 1990.

[Mehrotra and Shiming 96] R Mehrotra and Z Shiming. A computational approach to zero-crossing-based two-dimensional edge detection. *Graphical Models and Image Processing*, 58:1–17, 1996.

[Moik 80] J G Moik. *Digital Processing of Remotely Sensed Images.* NASA SP–431, Washington, DC, 1980.

[Mokhtarian 95] F Mokhtarian. Silhouette-based object recognition through curvature scale space. *IEEE Transactions on Pattern Analysis and Machine Intelligence*, 17:539–544, 1995.

[Mokhtarian and Mackworth 92] F Mokhtarian and A K Mackworth. A theory of multiscale, curvature-based shape representation for planar curves. *IEEE Transactions on Pattern Analysis and Machine Intelligence*, 14:789–805, 1992.

[Moravec 77] H P Moravec. Towards automatic visual obstacle avoidance. In *Proceedings of the 5th International Joint Conference on Artificial Intelligence*, Carnegie-Mellon University, Pittsburgh, PA, August 1977.

[Morrone et al. 95] M C Morrone, A Navangione, and D Burr. An adaptive approach to scale selection for line and edge detection. *Pattern Recognition Letters*, 16:667–677, 1995.

[Morrow and Rangayyan 90] W M Morrow and R M Rangayyan. Feature-adaptive enhancement and analysis of high-resolution digitized mammograms. In *Proceedings of 12th IEEE Engineering in Medicine and Biology Conference*, pages 165–166, IEEE, Piscataway, NJ, 1990.

[Morrow et al. 92] W M Morrow, R B Paranjape, R M Rangayyan, and J E L Desautels. Region-based contrast enhancement of mammograms. *IEEE Transactions on Medical Imaging*, 11(3):392–406, 1992.

[Nagao and Matsuyama 80] M Nagao and T Matsuyama. *A Structural Analysis of Complex Aerial Photographs.* Plenum Press, New York, 1980.

[Nevatia 77] R Nevatia. Evaluation of simplified Hueckel edge-line detector. *Computer Graphics and Image Processing*, 6(6):582–588, 1977.

[Papoulis 65] A Papoulis. *Probability, Random Variables, and Stochastic Processes.* McGraw-Hill, New York, 1965.

[Paranjape et al. 92a] R B Paranjape, R N Rangayyan, Morrow W M, and H N Nguyen. Adaptive neighborhood image processing. In *Proceedings of Visual Communications and Image Processing,* Boston, pages 198–207, SPIE, Bellingham, WA, 1992.

[Paranjape et al. 92b] R B Paranjape, R N Rangayyan, W M Morrow, and H N Nguyen. Adaptive neighborhood image processing. *CVGIP – Graphical Models and Image Processing,* 54(3):259–267, 1992.

[Perona and Malik 90] P Perona and J Malik. Scale-space and edge detection using anisotropic diffusion. *IEEE Transactions on Pattern Analysis and Machine Intelligence,* 12(7):629–639, 1990.

[Pitas and Venetsanopoulos 90] I Pitas and A N Venetsanopoulos. *Nonlinear Digital Filters: Principles and Applications.* Kluwer, Boston, 1990.

[Pitas and Venetsanopulos 86] I Pitas and A N Venetsanopulos. Nonlinear order statistic filters for image filtering and edge detection. *Signal Processing,* 10(10):573–584, 1986.

[Pizer et al. 87] S M Pizer, E P Amburn, J D Austin, R Cromartie, A Geselowitz, T Greer, B Haar-Romeny, J B Zimmerman, and K Zuiderveld. Adaptive histogram equalization and its variations. *Computer Vision, Graphics, and Image Processing,* 39:355–368, 1987.

[Pratt 72] W K Pratt. Generalized Wiener filter computation techniques. *IEEE Transactions on Computers,* 21:636–641, 1972.

[Pratt 78] W K Pratt. *Digital Image Processing.* Wiley, New York, 1978.

[Pratt 91] W K Pratt. *Digital Image Processing.* Wiley, New York, 2nd edition, 1991.

[Qian and Huang 94] R J Qian and T S Huang. Optimal edge detection in two-dimensional images. In *ARPA Image Understanding Workshop,* Monterey, CA, pages 1581–1588, ARPA, Los Altos, CA, 1994.

[Ramesh and Haralick 94] V Ramesh and R M Haralick. An integrated gradient edge detector. Theory and performance evaluation. In *ARPA Image Understanding Workshop,* Monterey, CA, pages 689–702, ARPA, Los Altos, CA, 1994.

[Roberts 65] L G Roberts. Machine perception of three-dimensional solids. In J T Tippett, editor, *Optical and Electro-Optical Information Processing,* pages 159–197. MIT Press, Cambridge, MA, 1965.

[Rosenfeld and Kak 82] A Rosenfeld and A C Kak. *Digital Picture Processing.* Academic Press, New York, 2nd edition, 1982.

[Rosenfeld and Thurston 71] A Rosenfeld and M Thurston. Edge and curve detection for visual scene analysis. *IEEE Transactions on Computers,* 20(5):562–569, 1971.

[Saund 90] E Saund. Symbolic construction of a 2D scale-space image. *IEEE Transactions on Pattern Analysis and Machine Intelligence,* 12:817–830, 1990.

[Sid-Ahmed 95] M A Sid-Ahmed. *Image Processing.* McGraw-Hill, New York, 1995.

[Slepian 67] D Slepian. Linear least-squares filtering of distorted images. *Journal of the Optical Society of America,* 57:918–922, 1967.

[Sondhi 72] M M Sondhi. Image restoration: The removal of spatially invariant degradations. *Proceedings IEEE,* 60:842–853, 1972.

[Sorrenti 95] D G Sorrenti. A proposal on local and adaptive determination of filter scale for edge detection. In *Image Analysis and Processing. ICIAP'95*, pages 405–410, Springer Verlag, Berlin, 1995.

[Spacek 86] L Spacek. Edge detection and motion detection. *Image and Vision Computing*, 4:43–52, 1986.

[Sun and Sclabassi 95] M Sun and R J Sclabassi. Symmetric wavelet edge detector of the minimum length. In *International Conference on Image Processing*, pages 177–180, IEEE, Los Alamitos, CA, 1995.

[Tagare and deFigueiredo 90] H D Tagare and R J P deFigueiredo. On the localization performance measure and optimal edge detection. *IEEE Transactions on Pattern Analysis and Machine Intelligence*, 12(12):1186–1190, 1990.

[Tang et al. 97] Y Y Tang, B F Li, and D Xi. Multiresolution analysis in extraction of reference lines from documents with gray level background. *IEEE Transactions on Pattern Analysis and Machine Intelligence*, 19:921–926, 1997.

[Tekalp and Pavlovic 93] A M Tekalp and G Pavlovic. Digital restoration of images scanned from photographic paper. *Journal of Electronic Imaging*, 2(1):19–27, 1993.

[Topkar et al. 90] V Topkar, B Kjell, and A Sood. Object detection using scale-space. In *Proceedings of the Applications of Artificial Intelligence VIII Conference, The International Society for Optical Engineering*, pages 2–13, SPIE, Orlando, FL, April 1990.

[Tyan 81] S G Tyan. Median filtering, deterministic properties. In T S Huang, editor, *Two-Dimensional Digital Signal Processing*, volume II. Springer Verlag, Berlin, 1981.

[Ullman 81] S Ullman. Analysis of visual motion by biological and computer systems. *IEEE Computer*, 14(8):57–69, August 1981.

[Venkateswar and Chellappa 92] V Venkateswar and R Chellappa. Extraction of straight lines in aerial images. *IEEE Transactions on Pattern Analysis and Machine Intelligence*, 14:1111–1114, 1992.

[Vrabel 96] M J Vrabel. Edge detection with a recurrent neural network. In *Applications and Science of Artificial Neural Networks II, Proc. SPIE Vol. 2760*, pages 365–371, SPIE, Bellingham, WA, 1996.

[Wang and Vagnucci 81] D C C Wang and A H Vagnucci. Gradient inverse weighting smoothing schema and the evaluation of its performace. *Computer Graphics and Image Processing*, 15, 1981.

[Wang et al. 95] D Wang, V Haese-Coat, and J Ronsin. Shape decomposition and representation using a recursive morphological operation. *Pattern Recognition*, 28:1783–1792, 1995.

[Wiener 42] N Wiener. *Extrapolation, Interpolation and Smoothing of Stationary Time Series.* MIT Press, Cambridge, MA, 1942.

[Williams and Shah 90] D J Williams and M Shah. Edge contours using multiple scales. *Computer Vision, Graphics, and Image Processing*, 51:256–274, September 1990.

[Witkin 83] A P Witkin. Scale-space filtering. In *Proceedings of the 8th Joint Conference on Artificial Intelligence*, pages 1019–1022, W Kaufmann, Karlsruhe, Germany, 1983.

[Yaroslavskii 87] L P Yaroslavskii. *Digital Signal Processing in Optics and Holography* (in Russian). Radio i svjaz, Moscow, 1987.

[Zuniga and Haralick 83] O Zuniga and R M Haralick. Corner detection using the facet model. In *Computer Vision and Pattern Recognition*, pages 30–37, IEEE, Los Alamitos, CA, 1983.

Chapter 5

Segmentation

Image segmentation is one of the most important steps leading to the analysis of processed image data—its main goal is to divide an image into parts that have a strong correlation with objects or areas of the real world contained in the image. We may aim for **complete segmentation**, which results in a set of disjoint regions corresponding uniquely with objects in the input image, or for **partial segmentation**, in which regions do not correspond directly with image objects. To achieve a complete segmentation, cooperation with higher processing levels which use specific knowledge of the problem domain is necessary. However, there is a whole class of segmentation problems that can be solved successfully using lower-level processing only. In this case, the image commonly consists of contrasted objects located on a uniform background—simple assembly tasks, blood cells, printed characters, etc. Here, a simple global approach can be used and the complete segmentation of an image into objects and background can be obtained. Such processing is context independent; no object-related model is used, and no knowledge about expected segmentation results contributes to the final segmentation.

If partial segmentation is the goal, an image is divided into separate regions that are homogeneous with respect to a chosen property such as brightness, color, reflectivity, texture, etc. If an image of a complex scene is processed, for example, an aerial photograph of an urban scene, a set of possibly overlapping homogeneous regions may result. The partially segmented image must then be subjected to further processing, and the final image segmentation may be found with the help of higher-level information.

Totally correct and complete segmentation of complex scenes usually cannot be achieved in this processing phase, although substantial reduction in data volume offers an immediate gain. A reasonable aim is to use partial segmentation as an input to higher-level processing.

Image data ambiguity is one of the main segmentation problems, often accompanied by information noise. Segmentation methods can be divided into three groups according to the dominant features they employ: First is **global knowledge** about an image or its part; the knowledge is usually represented by a histogram of image features. **Edge-based** segmentations form the second group, and **region-based** segmentations the third—many different characteristics may be used in edge detection or region growing, for example, brightness, texture, velocity field, etc. The second and the third groups solve a dual problem. Each region can be represented by its closed boundary, and each closed boundary describes a region. Because of the different natures of the various edge- and region-based algorithms, they may be expected to give somewhat different results and consequently different information. The segmentation results of these two approaches can therefore be combined in a single description structure. A common example of this is a region adjacency graph, in which regions are represented by nodes and graph arcs represent adjacency relations based on detected region borders (Section 3.2.3).

5.1 Thresholding

Gray-level thresholding is the simplest segmentation process. Many objects or image regions are characterized by constant reflectivity or light absorption of their surfaces; a brightness constant or **threshold** can be determined to segment objects and background. Thresholding is computationally inexpensive and fast—it is the oldest segmentation method and is still widely used in simple applications; thresholding can easily be done in real time using specialized hardware.

A complete segmentation of an image R is a finite set of regions $R_1, \ldots, R_S$,

$$R = \bigcup_{i=1}^{S} R_i \qquad R_i \cap R_j = \emptyset \qquad i \neq j \tag{5.1}$$

Complete segmentation can result from thresholding in simple scenes. Thresholding is the transformation of an input image f to an output (segmented) binary image g as follows:

$$\begin{aligned} g(i,j) &= 1 \quad \text{for } f(i,j) \geq T \\ &= 0 \quad \text{for } f(i,j) < T \end{aligned} \tag{5.2}$$

where T is the threshold, $g(i,j) = 1$ for image elements of objects, and $g(i,j) = 0$ for image elements of the background (or vice versa).

Algorithm 5.1: Basic thresholding

1. Search all the pixels $f(i,j)$ of the image f. An image element $g(i,j)$ of the segmented image is an object pixel if $f(i,j) \geq T$, and is a background pixel otherwise.

If objects do not touch each other, and if their gray-levels are clearly distinct from background gray-levels, thresholding is a suitable segmentation method. Such an example is found in Figure 5.1a, the threshold segmentation result for which is shown in Figure 5.1b. Figures 5.1c and 5.1d show segmentation results for different threshold values.

Correct threshold selection is crucial for successful threshold segmentation; this selection can be determined interactively or it can be the result of some threshold detection method that will be discussed in the next section. Only under very unusual circumstances can thresholding be successful using a single threshold for the whole image (global thresholding) since even in very simple images there are likely to be gray-level variations in objects and background; this variation may be due to non-uniform lighting, non-uniform input device parameters or a number of other factors. Segmentation using variable thresholds (also called **adaptive**

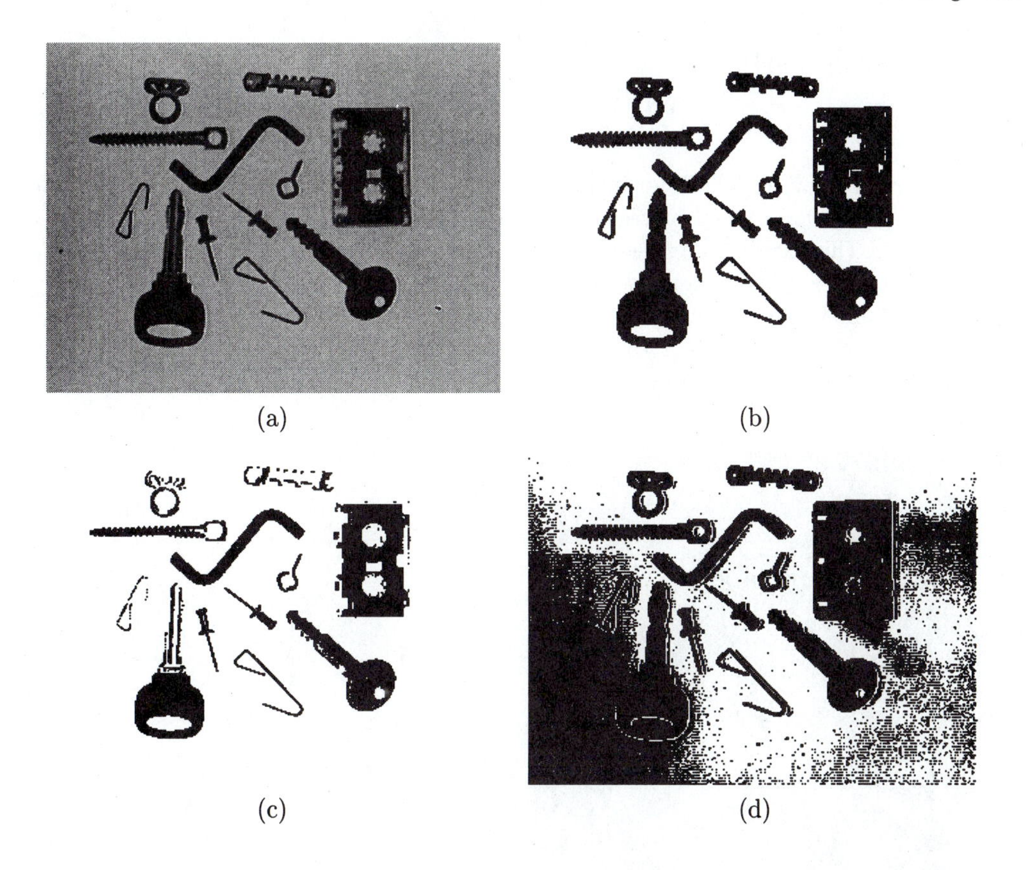

Figure 5.1: *Image thresholding: (a) original image; (b) threshold segmentation; (c) threshold too low; (d) threshold too high.*

thresholding), in which the threshold value varies over the image as a function of local image characteristics, can produce the solution in these cases.

A global threshold is determined from the whole image f:

$$T = T(f) \tag{5.3}$$

On the other hand, local thresholds are position dependent:

$$T = T(f, f_c) \tag{5.4}$$

where f_c is that image part in which the threshold is determined. One option is to divide the image f into subimages f_c and determine a threshold independently in each subimage; then if a threshold cannot be determined in some subimage, it can be interpolated from thresholds determined in neighboring subimages. Each subimage is then processed with respect to its local threshold.

Basic thresholding as defined by equation (5.2) has many modifications. One possibility is to segment an image into regions of pixels with gray-levels from a set D and into background otherwise (band thresholding):

$$\begin{aligned} g(i,j) &= 1 \quad \text{for } f(i,j) \in D \\ &= 0 \quad \text{otherwise} \end{aligned} \tag{5.5}$$

This thresholding can be useful, for instance, in microscopic blood cell segmentations, where a particular gray-level interval represents cytoplasma, the background is lighter, and the cell kernel darker. This thresholding definition can serve as a border detector as well; assuming dark objects on a light background, some gray-levels between those of objects and background can be found only in the object borders. If the gray-level set D is chosen to contain just these object-border gray-levels, and if thresholding according to equation (5.5) is used, object borders result as shown in Figure 5.2. Isolines of gray can be found using this appropriate gray-level set D.

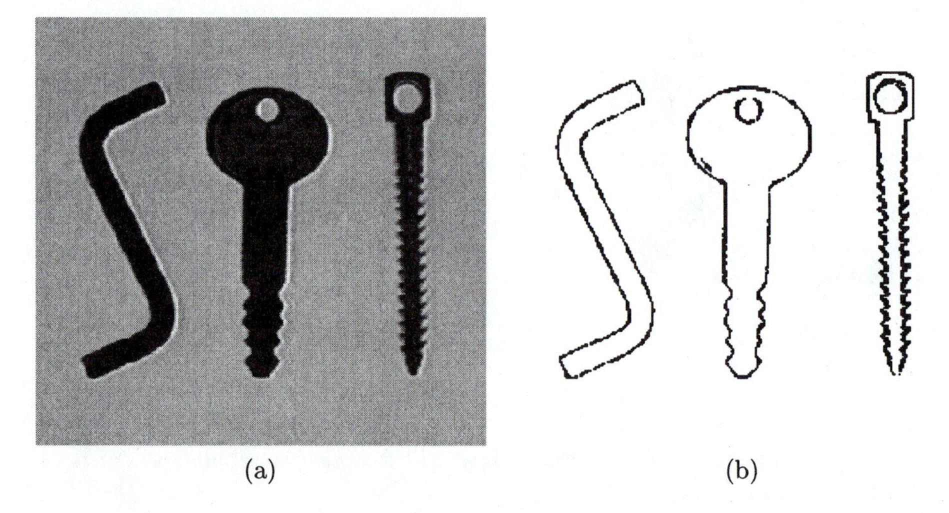

(a) (b)

Figure 5.2: *Image thresholding modification: (a) original image; (b) border detection using band-thresholding.*

There are many modifications that use multiple thresholds, after which the resulting image is no longer binary, but rather an image consisting of a very limited set of gray-levels:

$$\begin{aligned} g(i,j) &= 1 \quad \text{for } f(i,j) \in D_1 \\ &= 2 \quad \text{for } f(i,j) \in D_2 \\ &= 3 \quad \text{for } f(i,j) \in D_3 \\ &\dots \\ &= n \quad \text{for } f(i,j) \in D_n \\ &= 0 \quad \text{otherwise} \end{aligned} \tag{5.6}$$

where each D_i is a specified subset of gray-levels.

Another special choice of gray-level subsets D_i defines **semi-thresholding**, which is sometimes used to make human-assisted analysis easier:

$$\begin{aligned} g(i,j) &= f(i,j) \quad \text{for } f(i,j) \geq T \\ &= 0 \quad \text{for } f(i,j) < T \end{aligned} \tag{5.7}$$

This process aims to mask out the image background, leaving gray-level information present in the objects.

Thresholding has been presented relying only on gray-level image properties. Note that this is just one of many possibilities; thresholding can be applied if the values $f(i,j)$ do not represent gray-levels, but instead represent gradient, a local texture property (Chapter 14), or the value of any other image decomposition criterion.

5.1.1 Threshold detection methods

If some property of an image after segmentation is known a priori, the task of threshold selection is simplified, since the threshold is chosen to ensure that this property is satisfied. A printed text sheet may be an example if we know that characters of the text cover $1/p$ of the sheet area. Using this prior information about the ratio between the sheet area and character area, it is very easy to choose a threshold T (based on the image histogram) such that $1/p$ of the image area has gray values less than T and the rest has gray values larger than T. This method is called ***p*-tile thresholding**. Unfortunately, we do not usually have such definite prior information about area ratios. This information can sometimes be substituted by knowledge of another property, for example, the average width of lines in drawings, etc. The threshold can be determined to provide the required line width in the segmented image.

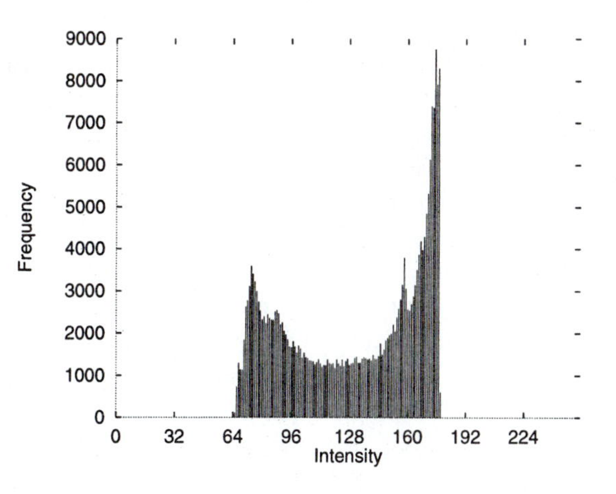

Figure 5.3: *A bi-modal histogram.*

More complex methods of threshold detection are based on histogram shape analysis. If an image consists of objects of approximately the same gray-level that differs from the gray-level of the background, the resulting histogram is bi-modal. Pixels of objects form one of its peaks, while pixels of the background form the second peak—Figure 5.3 shows a typical example. The histogram shape illustrates the fact that the gray values between the two peaks are not common in the image, and probably result from border pixels between objects and background. The chosen threshold must meet minimum segmentation error requirements; it makes intuitive sense to determine the threshold as the gray-level that has a minimum histogram value between the two mentioned maxima; see Figure 5.3. If the histogram is multi-modal, more thresholds may be determined at minima between any two maxima. Each threshold gives different segmentation results, of course. Multi-thresholding as given in equation (5.6) is another option.

To decide if a histogram is bi-modal or multi-modal may not be so simple in reality, it often being impossible to interpret the significance of local histogram maxima [Rosenfeld and de la Torre 83]. Bi-modal histogram threshold detection algorithms usually find the highest local maxima first and detect the threshold as a minimum between them; this technique is called the **mode method**. To avoid detection of two local maxima belonging to the same global maximum, a minimum distance in gray-levels between these maxima is usually required, or techniques to smooth histograms (see Section 2.3.2) are applied. Note that histogram bi-modality itself does not guarantee correct threshold segmentation—even if the histogram is bi-modal, correct segmentation may not occur with objects located on a background of different gray-levels. A two-part image with one half white and the second half black actually has the same histogram as an image with randomly spread white and black pixels (i.e., a salt-and-pepper noise image, see Section 2.3.5). This is one example showing the need to check threshold segmentation results whenever the threshold has been determined from a histogram only, using no other image characteristics.

A more general approach takes gray-level occurrences inside a local neighborhood into consideration when constructing a gray-level histogram, the goal being to build a histogram with a better peak-to-valley ratio. One option is to weight histogram contributions to suppress the influence of pixels with a high image gradient. This means that a histogram will consist mostly of the gray values of objects and background, and that border gray-levels (with higher gradient) will not contribute. This will produce a deeper histogram valley and allow an easier determination of the threshold. Another method based on the same idea uses only high-gradient pixels to form the gray-level histogram, meaning that the histogram will consist mostly of border gray-levels and should be unimodal in which the peak corresponds to the gray-level of borders between objects and background. The segmentation threshold can be determined as the gray value of this peak, or as a mean of a substantial part of the peak. Many modifications of **histogram transformation** methods can be found in [Weszka et al. 76, Weszka and Rosenfeld 79, Herman and Liu 78, Nagao and Matsuyama 80].

Thresholding is a very popular tool in image segmentation, and a large variety of threshold detection techniques exist in addition to the main techniques which have been discussed. The survey [Sahoo et al. 88] gives a good overview of existing methods: **histogram concavity analysis**, **entropic** methods, **relaxation** methods, **multi-thresholding** methods, and others can be found there, together with an extensive list of references. High processing speed has always been typical for threshold segmentations, and images can easily be thresholded in real time—real-time threshold detection is a current research effort [Hassan 89, Lavagetto 90].

5.1.2 Optimal thresholding

Methods based on approximation of the histogram of an image using a weighted sum of two or more probability densities with normal distribution represent a different approach called **optimal thresholding**. The threshold is set as the closest gray-level corresponding to the minimum probability between the maxima of two or more normal distributions, which results in minimum error segmentation (the smallest number of pixels is mis-segmented) [Chow and Kaneko 72, Rosenfeld and Kak 82, Gonzalez and Wintz 87]; see Figure 5.4 (and compare maximum-likelihood classification methods, Section 7.2.2). The difficulty with these methods is in estimating normal distribution parameters together with the uncertainty that the distribution may be considered normal. These difficulties may be overcome if an optimal threshold is sought that maximizes gray-level variance between objects and background. Note that this approach can be applied even if more than one threshold is needed [Otsu 79, Reddi et al. 84, Kittler and Illingworth 86, Mardia and Hainsworth 88, Cho et al. 89]. Alternatively, minimization of variance of the histogram, sum of square errors, spatial entropy, average clustering, or other optimization approaches may be used [Ramesh et al. 95, Brink 95, Beghdadi et al. 95, Venkateswarlu and Boyle 95]. For an overview of alternative approaches, 11 common algorithms for global threshold detection and relationships between them can be found in [Glasbey 93].

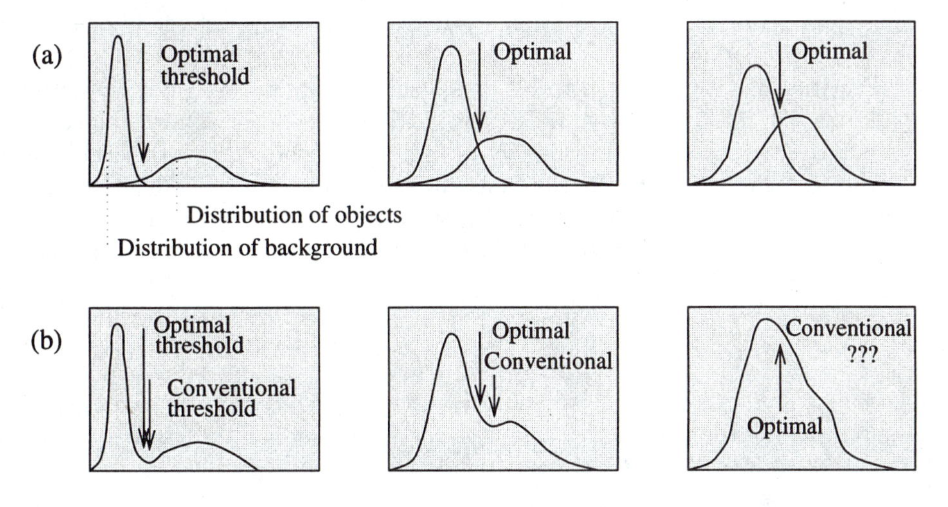

Figure 5.4: *Gray-level histograms approximated by two normal distributions— the threshold is set to give minimum probability of segmentation error: (a) probability distributions of background and objects; (b) corresponding histograms and optimal threshold.*

The following algorithm represents a simpler version that shows a rationale for this approach [Ridler and Calvard 78] and works well even if the image histogram is not bi-modal. This method assumes that regions of two main gray-levels are present in the image, thresholding of printed text being an example. The algorithm is iterative, four to ten iterations usually being sufficient.

Algorithm 5.2: Iterative (optimal) threshold selection

1. Assuming no knowledge about the exact location of objects, consider as a first approximation that the four corners of the image contain background pixels only and the remainder contains object pixels.

2. At step t, compute μ_B^t and μ_O^t as the mean background and object gray-level, respectively, where segmentation into background and objects at step t is defined by the

threshold value T^t determined in the previous step [equation 5.9];

$$\mu_B^t = \frac{\sum_{(i,j)\in\text{background}} f(i,j)}{\#\text{background\_pixels}} \qquad \mu_O^t = \frac{\sum_{(i,j)\in\text{objects}} f(i,j)}{\#\text{object\_pixels}} \tag{5.8}$$

3. Set

$$T^{(t+1)} = \frac{\mu_B^t + \mu_O^t}{2} \tag{5.9}$$

$T^{(t+1)}$ now provides an updated background—object distinction.

4. If $T^{(t+1)} = T^{(t)}$, halt; otherwise return to step 2.

The method performs well under a large variety of image contrast conditions; it is illustrated in use in Section 16.1.

A combination of optimal and adaptive thresholding [equation (5.4)] demonstrating practical applicability of these approaches is described in [Frank et al. 95]. The method determines optimal gray-level segmentation parameters in local sub-regions for which local histograms are constructed. The gray-level distributions corresponding to n individual (possibly non-contiguous) regions are fitted to each local histogram h_{region} which is modeled as a sum h_{model} of n Gaussian distributions so that the difference between the modeled and the actual histograms is minimized.

$$h_{\text{model}}(g) = \sum_{i=1}^{n} a_i e^{[-(g-\mu_i)^2/2\sigma_i^2]} \tag{5.10}$$

Variable g represents gray-level values from the set G of image gray-levels, a_i; μ_i and σ_i denote parameters of the Gaussian distribution for the region i. The optimal parameters of the Gaussian distributions are determined by minimizing the *fit function* F:

$$F = \sum_{g\in G} [h_{\text{model}}(g) - h_{\text{region}}(g)]^2 \tag{5.11}$$

In [Frank et al. 95], Levenberg-Marquardt [Marquardt 63, Press et al. 92] minimization was successfully used for segmentation of three-dimensional T1-weighted images from a magnetic resonance scanner into regions of white matter (WM), gray matter (GM), and cerebro-spinal fluid (CSF) (see Section 8.4 for a description of a different approach to the same problem using multi-band image data). A nine-parameter model ($n = 3$) was first fitted to the entire volume histogram and the parameters σ_i and μ_{CSF} were determined from the global histogram. (The μ_{CSF} parameter was determined globally, since CSF regions are relatively small and localized.) An example of a global histogram, fitted Gaussian distributions, and the three distributions corresponding to WM, GM, and CSF is given in Figure 5.5. The remaining five parameters are determined locally by minimizing the function F in each of the overlapping $45 \times 45 \times 45$ pixel (volume picture elements—voxels) 3D sub-regions located 10 voxels apart in all three dimensions. Non-convergent solutions ($F_{\text{local}} > 10F_{\text{global}}$) are neglected and interpolation of neighboring parameters is used instead. Then, Gaussian distribution parameters are tri-linearly interpolated for voxels between the minimized fit locations. Thus, the optimal thresholds can be determined for each voxel and used for segmentation. In [Frank et al. 95, Santago and Gage 93], the partial volume effect was also considered (in brain MR images, the finite-size voxels can consist of a combination of, e.g., gray and white matter) and a volume percentage corresponding to WM, GM, and CSF was calculated for each voxel. Figure 5.6 gives an example of such brain segmentation. The brighter the voxel location in individual segmented images, the higher the volume percentage of the GM, WM, or CSF in that particular voxel. In each voxel, the sum of partial volume percentages is 100%.

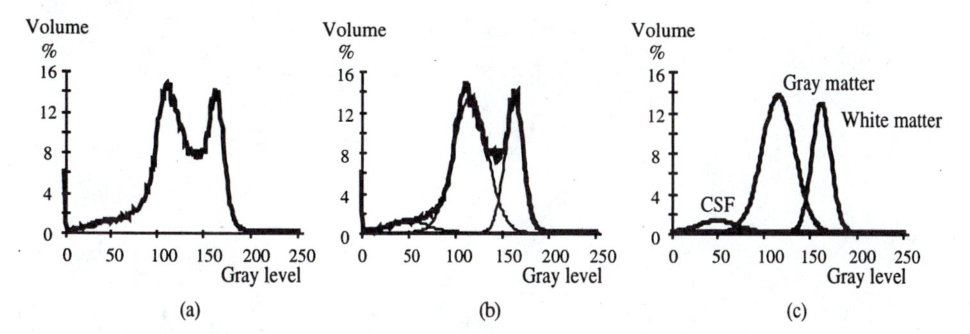

Figure 5.5: *Segmentation of 3D T1-weighted MR brain image data using optimal thresholding: (a) local gray-level histogram; (b) fitted Gaussian distributions, global 3D image fit; (c) Gaussian distributions corresponding to WM, GM, and CSF. Courtesy R.J. Frank, T.J. Grabowski, Human Neuroanatomy and Neuroimaging Laboratory, Department of Neurology, The University of Iowa.*

5.1.3 Multi-spectral thresholding

Many practical segmentation problems need more information than is contained in one spectral band. Color images are a natural example, in which information is coded in three spectral bands, for example, red, green, and blue; multi-spectral remote sensing images or meteorological satellite images may have even more spectral bands. One segmentation approach determines thresholds independently in each spectral band and combines them into a single segmented image.

Algorithm 5.3: Recursive multi-spectral thresholding

1. Initialize the whole image as a single region.

2. Compute a smoothed histogram (see Section 2.3.2) for each spectral band. Find the most significant peak in each histogram and determine two thresholds as local minima on either side of this maximum. Segment each region in each spectral band into sub-regions according to these thresholds. Each segmentation in each spectral band is projected into a multi-spectral segmentation—see Figure 5.7. Regions for the next processing steps are those in the multi-spectral image.

3. Repeat step 2 for each region of the image until each region's histogram contains only one significant peak.

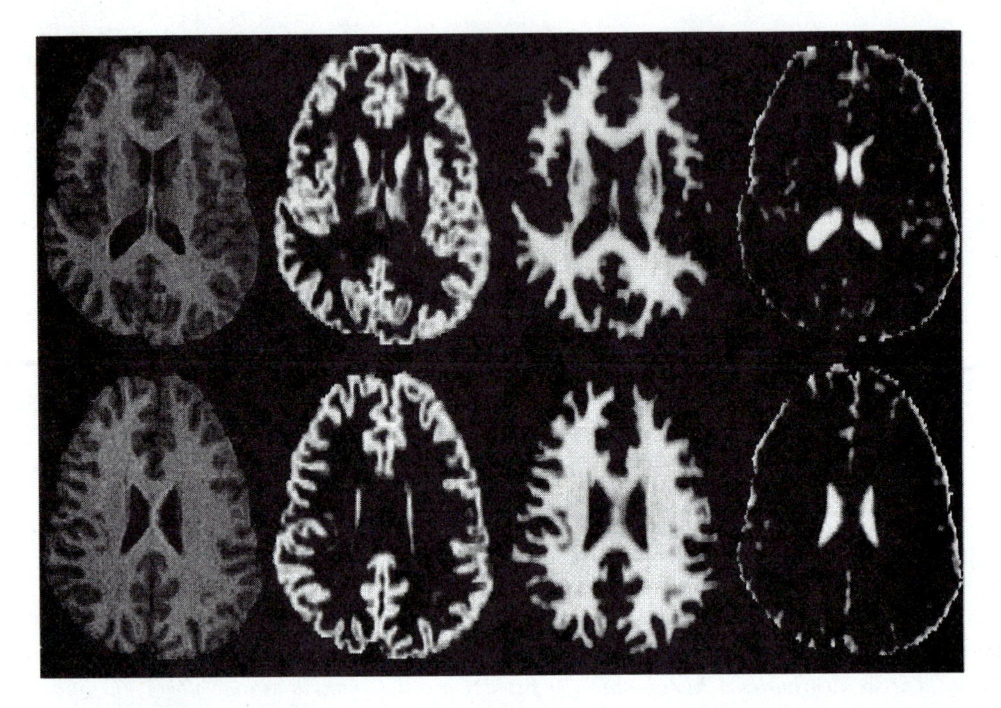

Figure 5.6: *Optimal MR brain image segmentation. Left column: original T1-weighted MR images, two of 120 slices of the 3D volume. Middle left: Partial-volume maps of gray matter. The brighter the voxel, the higher is the partial volume percentage of gray matter in the voxel. Middle right: Partial-volume maps of white matter. Right column: Partial-volume maps of cerebro-spinal fluid. Courtesy R.J. Frank, T.J. Grabowski, Human Neuroanatomy and Neuroimaging Laboratory, Department of Neurology, The University of Iowa.*

Region shapes can be adjusted during recursive pre-processing—for instance, boundary stretching, etc. (see Section 5.2.7). Better segmentation results can be achieved by analyzing multi-dimensional histograms [Hanson and Riseman 78b] instead of histograms for each spectral band in step 2 of the previous algorithm.

Multi-spectral segmentations are often based on n-dimensional vectors of gray-levels in n spectral bands for each pixel or small pixel neighborhood. This segmentation approach, widely used in remote sensing, results from a classification process which is applied to these n-dimensional vectors. Generally speaking, regions are formed from pixels with similar properties in all spectral bands, with similar n-dimensional description vectors; see Chapter 7 and [Narendra and Goldberg 77, Ohta et al. 80, Kittler and Illingworth 85]. Segmentation and region labeling based on supervised, unsupervised, and contextual classification is discussed in more detail in Section 8.4.

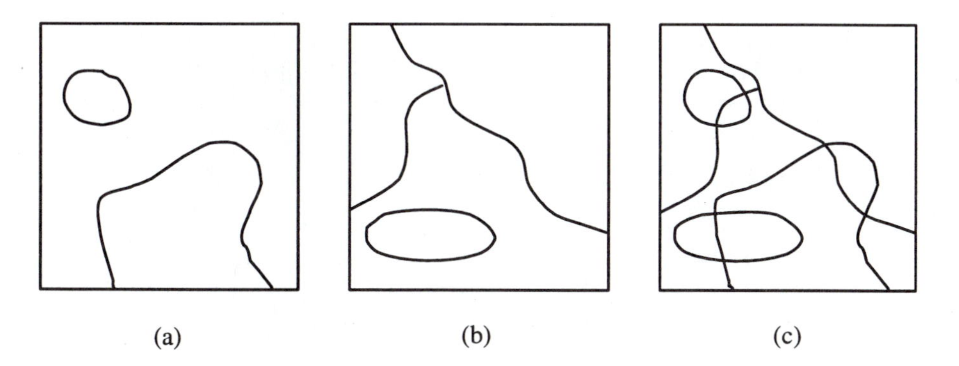

Figure 5.7: *Recursive multi-spectral thresholding: (a) band 1 thresholding; (b) band 2 thresholding; (c) multi-spectral segmentation.*

5.1.4 Thresholding in hierarchical data structures

The general idea of thresholding in hierarchical data structures is based on local thresholding methods [equation (5.4)], the aim being to detect the presence of a region in a low-resolution image, and to give the region more precision in images of higher to full resolution. Certain methods work in pre-computed pyramid data structures (Chapter 3), where low-resolution images are computed from higher-resolution images using averaging of gray values. The simplest method starts in the lowest-resolution image (the highest pyramid level), applying any of the segmentation methods discussed so far. The next step yields better segmentation precision—pixels close to boundaries are re-segmented into either object or background regions (Figure 5.8). This increase in precision is repeated for each pair of pyramid levels up to the full resolution level at which the final segmentation is obtained. A big advantage of this method is the significantly lower influence of image noise on the segmentation results, since segmentations at the lower resolution are based on smoothed image data, in which noise is suppressed. The imprecise borders that result from segmenting smoothed data are corrected by re-segmentation in areas close to borders using one-step-higher-resolution data [Tanimoto and Pavlidis 75, Tanimoto 78, Hartley 82, Rosenfeld 84, Baugher and Rosenfeld 86, Gross and Rosenfeld 87, Song et al. 90].

Another approach looks for a significant pixel in image data and segments an image into regions of any appropriate size. The pyramid data structure is used again—either 2×2 or 4×4 averaging is applied to construct the pyramid. If 4×4 averaging is used, the construction tiles overlap in the pyramid levels. A significant pixel detector is applied to all pixels of a pyramid level—this detector is based on 3×3 detection, which responds if the center pixel of a 3×3 neighborhood differs in gray-level from most other pixels in the neighborhood. It is assumed that the existence of this 'significant' pixel is caused by the existence of a different gray-level region in the full-resolution image. The corresponding part of the full-resolution image is then thresholded (that is, the part which corresponds to the size of a 3×3 neighborhood in the pyramid level where a significant pixel was found). The segmentation threshold is set to be between the gray-level of the detected significant pixel (which represents the average gray-level of the predicted region) and the average of the remaining 8-neighbors of the neighborhood (which represent the average of gray-levels of the predicted background).

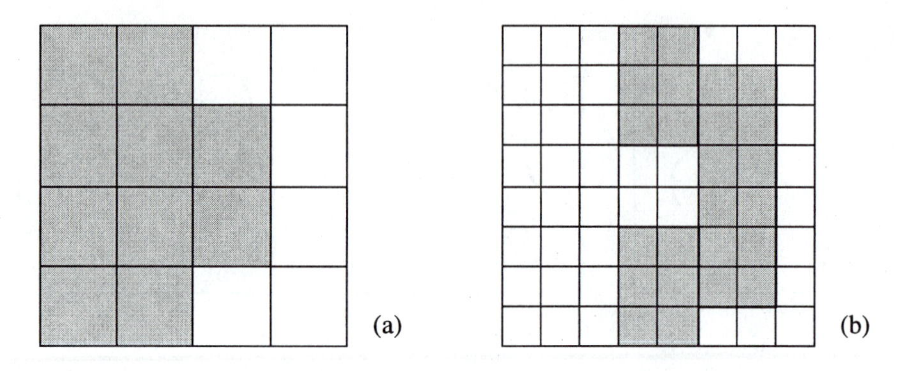

Figure 5.8: *Hierarchical thresholding: (a) pyramid level n, segmentation to objects and background; (b) pyramid level $n-1$, showing where the thresholding must be repeated for better precision.*

Algorithm 5.4: Hierarchical thresholding

1. Represent the image as a pyramid data structure.

2. Search sequentially for significant pixels in all pixels of all pyramid levels starting in the uppermost level. If a significant pixel is detected, set a threshold T

$$T = \frac{c + \frac{1}{8}\sum_i n_i}{2} \tag{5.12}$$

   where c is the gray value of the significant pixel, and n_i are gray values of its 3×3 neighbors at the current pyramid level. Apply the threshold to the corresponding part of the image at full resolution.

3. Continue the pyramid search as described in step 2.

5.2 Edge-based segmentation

Edge-based segmentation represents a large group of methods based on information about edges in the image; it is one of the earliest segmentation approaches and still remains very important. Edge-based segmentations rely on edges found in an image by edge detecting operators—these edges mark image locations of discontinuities in gray-level, color, texture, etc. A variety of edge detecting operators was described in Section 4.3.2, but the image resulting from edge detection cannot be used as a segmentation result. Supplementary processing steps must follow to combine edges into edge chains that correspond better with borders in the image. The final aim is to reach at least a partial segmentation—that is, to group local edges into an image where only edge chains with a correspondence to existing objects or image parts are present.

We will discuss several edge-based segmentation methods which differ in strategies leading to final border construction, and also differ in the amount of prior information that can be incorporated into the method. The more prior information that is available to the segmentation process, the better the segmentation results that can be obtained. Prior knowledge can be included in the confidence evaluation of the resulting segmentation as well. Prior information affects segmentation algorithms; if a large amount of prior information about the desired result is available, the boundary shape and relations with other image structures are specified very strictly and the segmentation must satisfy all these specifications. If little information about the boundary is known, the segmentation method must take more local information about the image into consideration and combine it with specific knowledge that is general for an application area. If little prior information is available, it cannot be used to evaluate the confidence of segmentation results, and therefore no basis for feedback corrections of segmentation results is available.

The most common problems of edge-based segmentation, caused by image noise or unsuitable information in an image, are an edge presence in locations where there is no border, and no edge presence where a real border exists. Clearly both these cases have a negative influence on segmentation results.

First, we will discuss simple edge-based methods requiring minimum prior information, and the necessity for prior knowledge will increase during the section. Construction of regions from edge-based partial segmentations is discussed at the end of the section.

5.2.1 Edge image thresholding

Almost no zero-value pixels are present in an edge image, but small edge values correspond to non-significant gray-level changes resulting from quantization noise, small lighting irregularities, etc. Simple thresholding of an edge image can be applied to remove these small values. The approach is based on an image of edge magnitudes [Kundu and Mitra 87] processed by an appropriate threshold. Figure 5.9a shows an original image, an edge image [as produced by a non-directional Sobel edge detector, see Section 4.3.2, equation (4.49)] is in Figure 5.9b, an 'over-thresholded' image is in Figure 5.9c, and an 'under-thresholded' image is in Figure 5.9d. Selection of an appropriate global threshold is often difficult and sometimes impossible; p-tile thresholding can be applied to define a threshold, and a more exact approach using orthogonal basis functions is described in [Flynn 72] which, if the original data has good contrast and is not noisy, gives good results.

A problem with simple detectors is the thickening that is evident in Figure 5.9b where there should only be a simple boundary. This can be partially rectified if edges carry directional information (as they do with the Sobel) by performing some form of non-maximal suppression (see Section 4.3.5) to suppress multiple responses in the neighborhood of single boundaries. The following algorithm generates Figure 5.11a from Figure 5.9b.

Algorithm 5.5: Non-maximal suppression of directional edge data

1. Quantize edge directions eight ways according to 8-connectivity (cf. Figures 2.8 and 3.1).

2. For each pixel with non-zero edge magnitude, inspect the two adjacent pixels indicated by the direction of its edge (see Figure 5.10).

3. If the edge magnitude of either of these two exceeds that of the pixel under inspection, mark it for deletion.

4. When all pixels have been inspected, re-scan the image and erase to zero all edge data marked for deletion.

This algorithm is based on 8-connectivity and may be simplified for 4-connectivity; it is also open to more sophisticated measurement of edge direction.

It is probable that such data will still be cluttered by noise (as in this case). The hysteresis approach outlined in Section 4.3.5 is also generally applicable if suitable thresholds can be determined. Supposing that edge magnitudes exceeding t_1 can be taken as certain (i.e., not due to noise), and that edge magnitudes less than t_0 may be assumed to be noise induced, the following algorithm may be defined.

Figure 5.9: *Edge image thresholding: (a) original image; (b) edge image (low contrast edges enhanced for display); (c) edge image thresholded at 30; (d) edge image thresholded at 10.*

Algorithm 5.6: Hysteresis to filter output of an edge detector

1. Mark all edges with magnitude greater than t_1 as correct.

2. Scan all pixels with edge magnitude in the range $[t_0, t_1]$.

3. If such a pixel borders another already marked as an edge, then mark it too. 'Bordering' may be defined by 4- or 8-connectivity.

4. Repeat from step 2 until stability.

Canny [Canny 86] reports choosing t_1/t_0 to be in the range 2 to 3; clearly, if a p-tile approach is available, this would guide the choice of t_1. Figure 5.11b illustrates the application of this algorithm. Hysteresis is a generally applicable technique that may be deployed when evidence is generated with 'strength' that is not susceptible to simple thresholding.

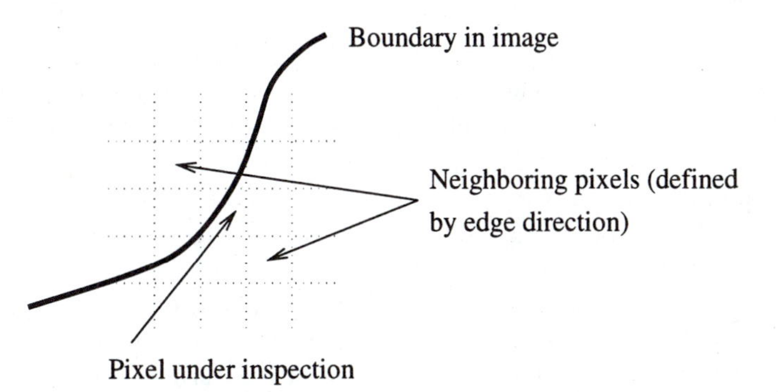

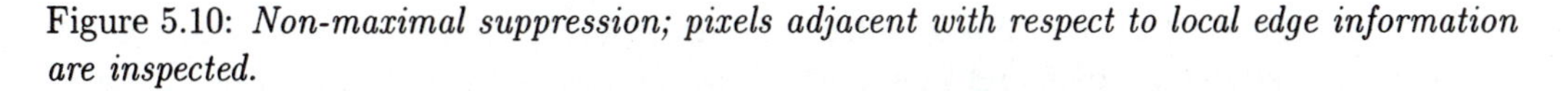

Figure 5.10: *Non-maximal suppression; pixels adjacent with respect to local edge information are inspected.*

Other forms of post-processing are also available, for example, to remove all border segments with length less than a specified value, or to consider the average edge strength of a (partial) border.

5.2.2 Edge relaxation

Borders resulting from the previous method are strongly affected by image noise, often with important parts missing. Considering edge properties in the context of their mutual neighbors can increase the quality of the resulting image. All the image properties, including those of further edge existence, are iteratively evaluated with more precision until the edge context is totally clear—based on the strength of edges in a specified local neighborhood, the confidence of each edge is either increased or decreased [Rosenfeld et al. 76, Zucker 76, Riseman and Arbib 77, Hancock and Kittler 90]. A weak edge positioned between two strong edges provides an example of context; it is highly probable that this inter-positioned weak edge should be a part of a resulting boundary. If, on the other hand, an edge (even a strong one) is positioned by itself with no supporting context, it is probably not a part of any border.

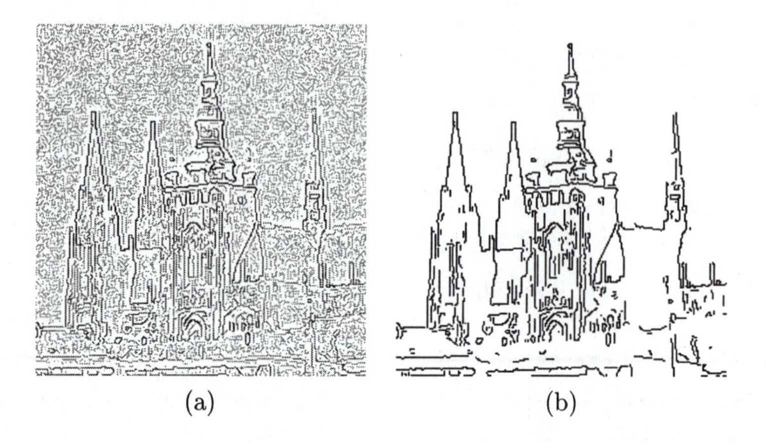

(a) (b)

Figure 5.11: *(a) Non-maximal suppression of the data in Figure 5.9a; (b) hysteresis applied to (a); high threshold 70, low threshold 10.*

A method we are going to discuss here [Hanson and Riseman 78b, Hanson and Riseman 78a, Prager 80] is a classical example of edge context evaluation. This method uses crack edges (edges located between pixels), which produce some favorable properties (see Section 2.3), although the method can work with other edge representations as well. Edge context is considered at both ends of an edge, giving the minimal edge neighborhood shown in Figure 5.12. All three possible edge positions at the end of the edge e must be included to cover all the possible ways the border can continue from both ends of e. Furthermore, two edge positions parallel with the edge e can be included in the local neighborhood—these parallel positions compete with the edge e in the placement of the border. Edge relaxation aims for continuous border construction, so we discuss the edge patterns that can be found in the local neighborhood. The central edge e has a vertex at each of its ends, and three possible border continuations can be found from both of these vertices. Let each vertex be evaluated according to the number of edges emanating from the vertex, not counting the edge e; call this number the vertex type. The type of edge e can then be represented using a number pair $i - j$ describing edge patterns at each vertex, where i and j are the vertex types of the edge e. For example, it assigns type $0 - 0$ for the edges shown in Figure 5.13a, type $3 - 3$ in for Figure 5.13d, etc. By symmetry, we need only consider the cases where $i \leq j$. The following context situations are possible:

- **0-0** isolated edge—negative influence on the edge confidence
- **0-1** uncertain—weak positive, or no influence on edge confidence
- **0-2, 0-3** dead end—negative influence on edge confidence
- **1-1** continuation—strong positive influence on edge confidence
- **1-2, 1-3** continuation to border intersection—medium positive influence on edge confidence

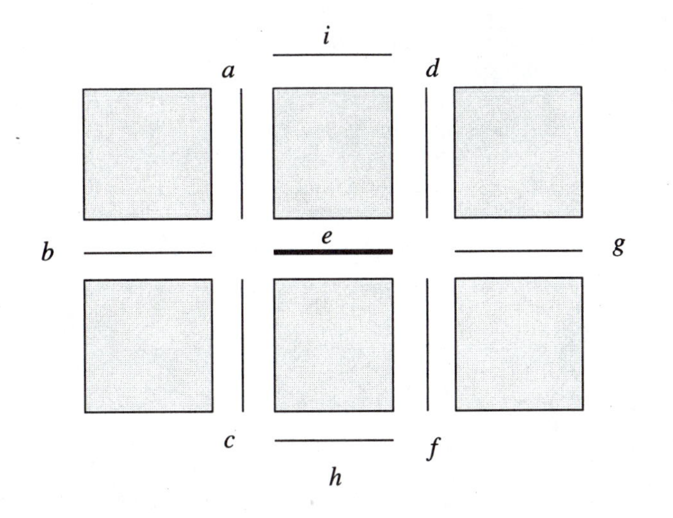

Figure 5.12: *Crack edges surrounding central edge e.*

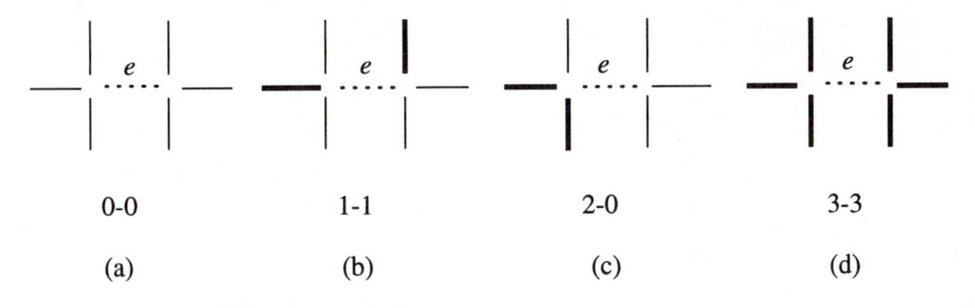

Figure 5.13: *Edge patterns and corresponding edge types.*

- **2-2, 2-3, 3-3** bridge between borders—not necessary for segmentation, no influence on edge confidence

An edge relaxation can be defined from the given context rules that may be considered as a production system [Nilsson 82], Section 7.1. Edge relaxation is an iterative method, with edge confidences converging either to zero (edge termination) or one (the edge forms a border). The confidence $c^{(1)}(e)$ of each edge e in the first iteration can be defined as a normalized magnitude of the crack edge, with normalization based on either the global maximum of crack edges in the whole image, or on a local maximum in some large neighborhood of the edge, thereby decreasing the influence of a few very high values of edge magnitude in the image.

Algorithm 5.7: Edge relaxation

1. Evaluate a confidence $c^{(1)}(e)$ for all crack edges e in the image.

2. Find the edge type of each edge based on edge confidences $c^{(k)}(e)$ in its neighborhood.

3. Update the confidence $c^{(k+1)}(e)$ of each edge e according to its type and its previous confidence $c^{(k)}(e)$.

4. Stop if all edge confidences have converged either to 0 or 1. Repeat steps 2 and 3 otherwise.

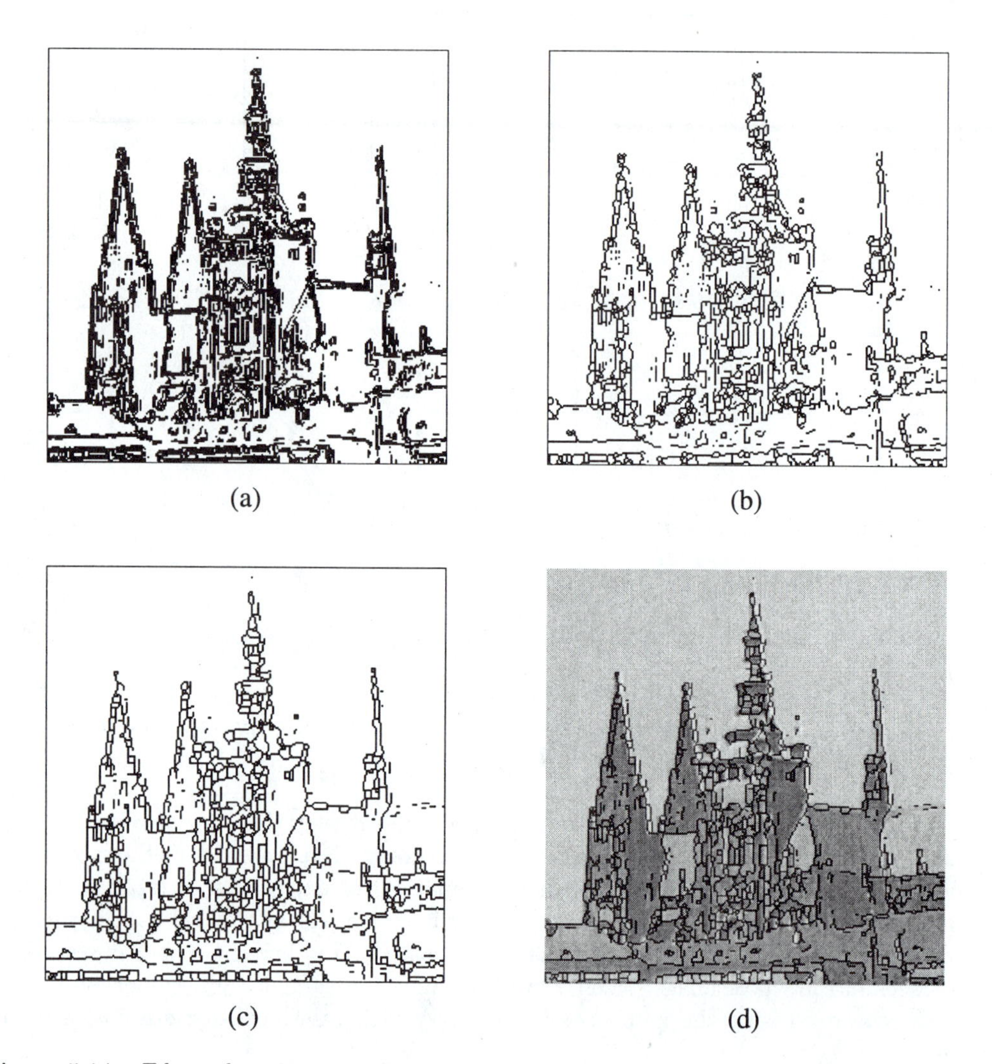

Figure 5.14: *Edge relaxation, see Figure 2.3a for original: (a) resulting borders after 10 iterations; (b) borders after thinning; (c) borders after 100 iterations, thinned; (d) borders after 100 iterations overlaid over original.*

The main steps of Algorithm 5.7 are evaluation of vertex types followed by evaluation of edge types, and the manner in which the edge confidences are modified. A vertex is considered to be of type i if

$$type(i) = \max_k [type(k)] \qquad k = 0, 1, 2, 3 \tag{5.13}$$

$$\begin{aligned} type(0) &= (m - a)(m - b)(m - c) \\ type(1) &= a(m - b)(m - c) \\ type(2) &= ab(m - c) \\ type(3) &= abc \end{aligned}$$

where a, b, c are the normalized values of the other incident crack edges, and without loss of generality we can assume $a \geq b \geq c$; q is a constant, for which a value of approximately 0.1 seems to be appropriate, and $m = \max(a, b, c, q)$ [Ballard and Brown 82]. Note that the introduction of the quantity q ensures that $type(0)$ is non-zero for small values of a.

For example, choosing $q = 0.1$, a vertex $(a, b, c) = (0.5, 0.05, 0.05)$ is a type 1 vertex, while a vertex $(0.3, 0.2, 0.2)$ is a type 3 vertex. Similar results can be obtained by simply counting the number of edges emanating from the vertex above a threshold value. Edge type is found as a simple concatenation of vertex types, and edge confidences are modified as follows:

$$\text{Confidence increase:} \qquad c^{(k+1)}(e) = \min[1, c^{(k)}(e) + \delta] \tag{5.14}$$

$$\text{Confidence decrease:} \qquad c^{(k+1)}(e) = \max[0, c^{(k)}(e) - \delta] \tag{5.15}$$

where δ is an appropriate constant, usually in the range 0.1 to 0.3.

Edge confidence modification rules can be simplified and just one value of δ can be used, not including the weak, moderate, or strong confidence increase/decrease options. Further, vertex types 2 and 3 can be considered the same in implementation because they result in the same production rules.

Edge relaxation, as described above, rapidly improves the initial edge labeling in the first few iterations. Unfortunately, it often slowly drifts, giving worse results than expected after larger numbers of iterations. A theoretical explanation, convergence proof, and practical solutions are given in [Levy 88]. The reason for this strange behavior is in searching for the global maximum of the edge consistency criterion over all the image, which may not give locally optimal results. A solution is found in setting edge confidences to zero under a certain threshold, and to one over another threshold which increases the influence of original image data. Therefore, one additional step must be added to the edge confidence computation [equations (5.14) and (5.15)]:

$$\text{If} \quad c^{(k+1)}(e) > T_1 \quad \text{then assign} \quad c^{(k+1)}(e) = 1 \tag{5.16}$$

$$\text{If} \quad c^{(k+1)}(e) < T_2 \quad \text{then assign} \quad c^{(k+1)}(e) = 0 \tag{5.17}$$

where T_1 and T_2 are parameters controlling the edge relaxation convergence speed and resulting border accuracy. Moreover, this method makes multiple labelings possible; the existence of two edges at different directions in one pixel may occur in corners, crosses, etc.

Edge relaxation results are shown in Figure 5.14, where edges parallel with the central edge were not considered in the relaxation process. The relaxation method can easily be implemented in parallel, with surprisingly good speedup. Implementation on a 16-processor hypercube showed almost linear speedup if 2, 4, 8, or 16 processors were used; in other words, using 16 processors the processing was almost 16 times faster than using one processor (note that linear speedup is uncommon in parallel implementations). All figures presented in this section were obtained running the parallel version of the algorithm [Clark 91]. More recent approaches to edge relaxation use edge and border information derived from image data. A method to determine probabilistic distribution of possible edge neighborhoods is given in [Sher 92]. In [Kim and Cho 94], fuzzy logic is used to assess neighborhood edge patterns; neural networks are employed as a means for fuzzy rule training.

5.2.3 Border tracing

If a region border is not known but regions have been defined in the image, borders can be uniquely detected. First, let us assume that the image with regions is either binary or that regions have been labeled (see Section 6.1). The first goal is to determine **inner** region borders. As defined earlier, an inner region border is a subset of the region—conversely, the **outer** border is not a subset of the region. The following algorithm covers inner boundary tracing in both 4-connectivity and 8-connectivity.

Algorithm 5.8: Inner boundary tracing

1. Search the image from top left until a pixel of a new region is found; this pixel P_0 then has the minimum column value of all pixels of that region having the minimum row value. Pixel P_0 is a starting pixel of the region border. Define a variable *dir* which stores the direction of the previous move along the border from the previous border element to the current border element. Assign

   (a) $dir = 3$ if the border is detected in 4-connectivity (Figure 5.15a)

   (b) $dir = 7$ if the border is detected in 8-connectivity (Figure 5.15b)

2. Search the 3×3 neighborhood of the current pixel in an anti-clockwise direction, beginning the neighborhood search in the pixel positioned in the direction

   (a) $(dir + 3) \mod 4$ (Figure 5.15c)

   (b) $(dir + 7) \mod 8$ if *dir* is *even* (Figure 5.15d)
   $(dir + 6) \mod 8$ if *dir* is *odd* (Figure 5.15e)

   The first pixel found with the same value as the current pixel is a new boundary element P_n. Update the *dir* value.

3. If the current boundary element P_n is equal to the second border element P_1, and if the previous border element P_{n-1} is equal to P_0, stop. Otherwise repeat step 2.

4. The detected inner border is represented by pixels $P_0 \dots P_{n-2}$.

Algorithm 5.8 works for all regions larger than one pixel (looking for the border of a single-pixel region is a trivial problem). This algorithm is able to find region borders but

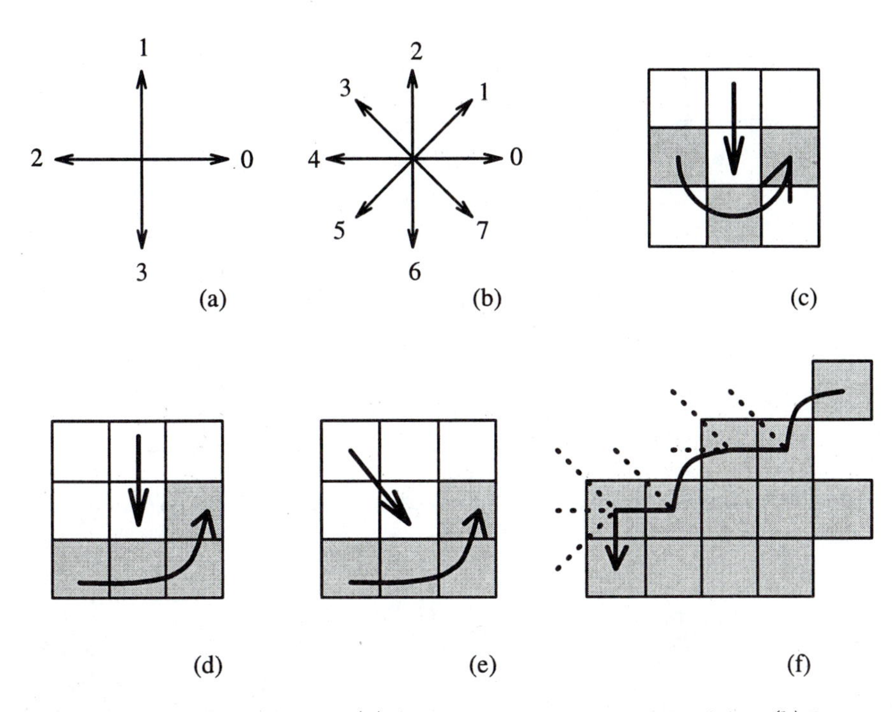

Figure 5.15: *Inner boundary tracing: (a) direction notation, 4-connectivity; (b) 8-connectivity; (c) pixel neighborhood search sequence in 4-connectivity; (d),(e) search sequence in 8-connectivity; (f) boundary tracing in 8-connectivity (dotted lines show pixels tested during the border tracing).*

does not find borders of region holes. To search for hole borders as well, the border must be traced starting in each region or hole border element if this element has never been a member of any border previously traced. The search for border elements always starts after a currently traced border is closed, and the search for 'unused' border elements can continue in the same way as the search for the first border element was done. Note that if objects are of unit width, more conditions must be added.

If the goal is to detect an outer region border, the given algorithm may still be used based on 4-connectivity.

Algorithm 5.9: Outer boundary tracing

1. Trace the inner region boundary in 4-connectivity until done.
2. The outer boundary consists of all non-region pixels that were tested during the search process; if some pixels were tested more than once, they are listed more than once in the outer boundary list.

Note that some border elements may be repeated in the outer border up to three times—see Figure 5.16. The outer region border is useful for deriving properties such as perimeter,

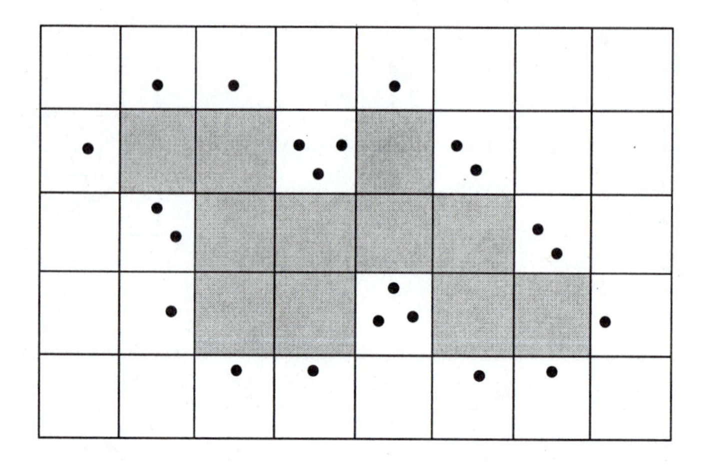

Figure 5.16: *Outer boundary tracing; • denotes outer border elements. Note that some pixels may be listed several times.*

compactness, etc., and is consequently often used—see Chapter 6.

The inner border is always part of a region but the outer border never is. Therefore, if two regions are adjacent, they never have a common border, which causes difficulties in higher processing levels with region description, region merging, etc. The inter-pixel boundary extracted, for instance, from crack edges is common to adjacent regions; nevertheless, its position cannot be specified in single pairs of pixel co-ordinates (compare the supergrid data structure in Figure 5.43). Boundary properties better than those of inner and outer borders may be found in **extended** borders [Pavlidis 77]. The main advantage of the extended boundary definition is that it defines a single common border between adjacent regions, and it may be specified using standard pixel co-ordinates (see Figure 5.17). All the useful properties of the outer border still remain; in addition, the boundary shape is exactly equal to the inter-pixel shape but is shifted one half-pixel down and one half-pixel right. The existence of a common border between regions makes it possible to incorporate into the boundary tracing a boundary description process. An evaluated graph consisting of border segments and vertices may result directly from the boundary tracing process; also, the border between adjacent regions may be traced only once and not twice as in conventional approaches.

The extended boundary is defined using 8-neighborhoods, and the pixels are coded according to Figure 5.18a [e.g., $P_4(P)$ denotes the pixel immediately to the left of pixel P]. Four kinds of inner boundary pixels of a region R are defined; if Q denotes pixels outside the region R, then a pixel $P \in R$ is

a LEFT pixel of R	if $P_4(P) \in Q$
a RIGHT pixel of R	if $P_0(P) \in Q$
an UPPER pixel of R	if $P_2(P) \in Q$
a LOWER pixel of R	if $P_6(P) \in Q$

Let LEFT(R), RIGHT(R), UPPER(R), LOWER(R) represent the corresponding subsets of R. The extended boundary EB is defined as a set of points P, P_0, P_6, P_7 satisfying the

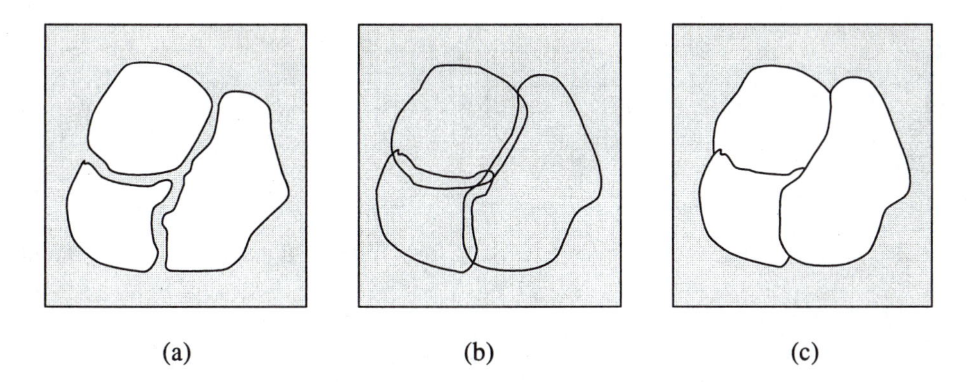

Figure 5.17: *Boundary locations for inner, outer, and extended boundary definition: (a) inner; (b) outer; (c) extended.*

following conditions [Pavlidis 77, Liow 91]:

$$\begin{aligned} EB = \; & \{P : P \in \mathrm{LEFT}(R)\} \cup \{P : P \in \mathrm{UPPER}(R)\} \; \cup \\ & \{P_6(P) : P \in \mathrm{LOWER}(R)\} \cup \{P_6(P) : P \in \mathrm{LEFT}(R)\} \; \cup \\ & \{P_0(P) : P \in \mathrm{RIGHT}(R)\} \cup \{P_7(P) : P \in \mathrm{RIGHT}(R)\} \end{aligned} \tag{5.18}$$

Figure 5.18 illustrates the definition.

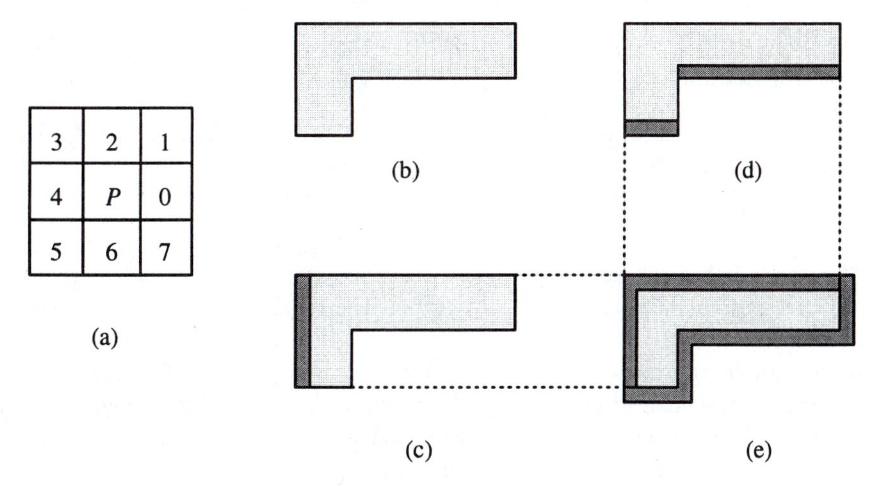

Figure 5.18: *Extended boundary definition: (a) pixel coding scheme; (b) region R; (c) LEFT(R); (d) LOWER(R); (e) extended boundary.*

The extended boundary can easily be constructed from the outer boundary. Using an intuitive definition of RIGHT, LEFT, UPPER, and LOWER outer boundary points, the extended boundary may be obtained by shifting all the UPPER outer boundary points one pixel down and right, shifting all the LEFT outer boundary points one pixel to the right, and shifting all the RIGHT outer boundary points one pixel down. The LOWER outer boundary point positions remain unchanged; see Figure 5.19.

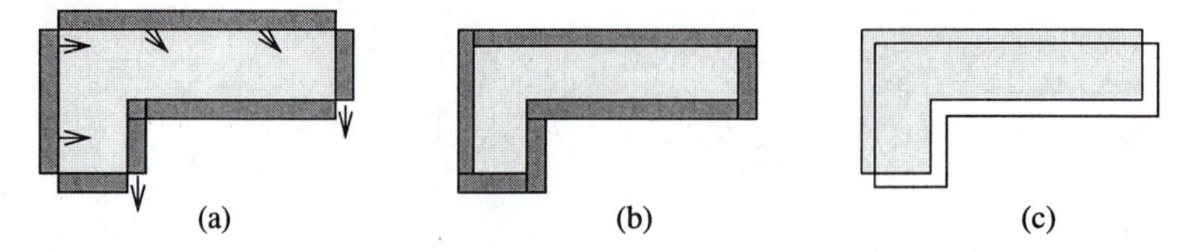

Figure 5.19: *Constructing the extended boundary from outer boundary: (a) outer boundary; (b) extended boundary construction; (c) extended boundary has the same shape and size as the natural object boundary.*

A more sophisticated method for extended boundary tracing was introduced together with an efficient algorithm in [Liow 91]. The approach is based on detecting common boundary segments between adjacent regions and vertex points in boundary segment connections. The detection process is based on a look-up table, which defines all 12 possible situations of the local configuration of 2×2 pixel windows, depending on the previous detected direction of boundary, and on the status of window pixels which can be inside or outside a region.

Algorithm 5.10: Extended boundary tracing

1. Define a starting pixel of an extended boundary in a standard way (the first region pixel found in a left-to-right and top-to-bottom line-by-line image search).
2. The first move along the traced boundary from the starting pixel is in direction $dir = 6$ (down), corresponding to the situation (i) in Figure 5.20.
3. Trace the extended boundary using the look-up table in Figure 5.20 until a closed extended border results.

Note that no hole-border tracing is included in the algorithm. The holes are considered separate regions and therefore the borders between the region and its hole are traced as a border of the hole.

The look-up table approach makes the tracing more efficient than conventional methods and makes parallel implementation possible. A pseudo-code description of algorithmic details is given in [Liow 91], where a solution to the problems of tracing all the borders in an image in an efficient way is given. In addition to extended boundary tracing, it provides a description of each boundary segment in chain code form together with information about vertices. This method is very suitable for representing borders in higher-level segmentation approaches including methods that integrate edge-based and region-based segmentation results. Moreover, in the conventional approaches, each border between two regions must be traced twice. The algorithm can trace each boundary segment only once, storing the information about what has already been done in double-linked lists.

A more difficult situation is encountered if the borders are traced in gray-level images where regions have not yet been defined [Dudani 76]. Therefore, the border is represented by a *simple path* of high-gradient pixels in the image (see Section 2.3.1). Border tracing should

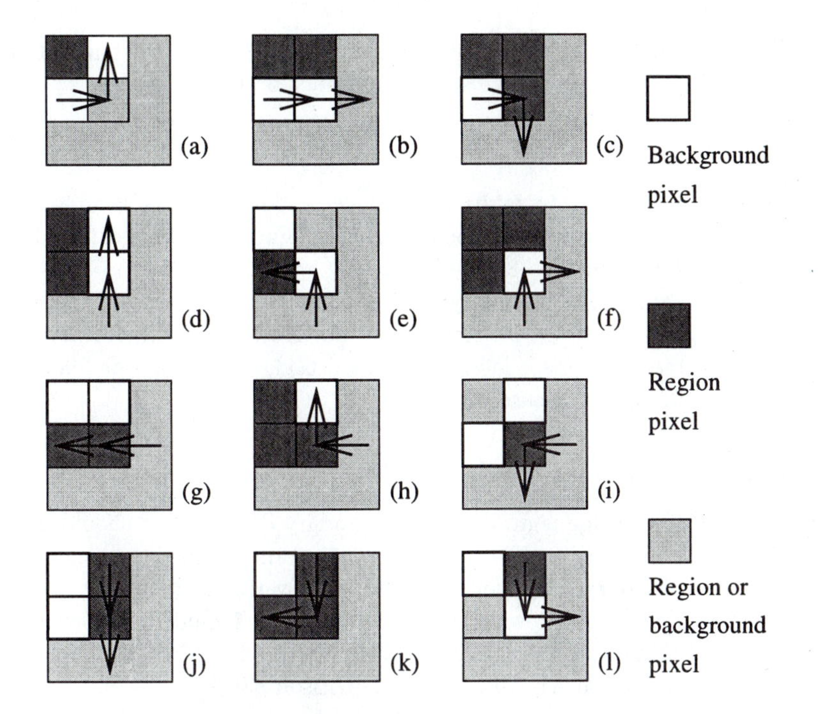

Figure 5.20: *Look-up table defining all 12 possible situations that can appear during extended border tracing. Current position is in the central pixel. The direction of the next move depends on the local configuration of background and region points, and on the direction of approach to the current pixel. Adapted from [Liow 91].*

be started in a pixel with a high probability of being a border element, and then border construction is based on the idea of adding the next elements which are in the most probable direction. To find the following border elements, edge gradient magnitudes and directions are usually computed in pixels of probable border continuation [Ballard and Brown 82].

Algorithm 5.11: Border tracing in gray-level images

1. Suppose the border has been determined up to the border element $\mathbf{x}_i$.
2. Define an element $\mathbf{x}_j$ as a pixel adjacent to $\mathbf{x}_i$ in the direction $\phi(\mathbf{x}_i)$. If the gradient magnitude in $\mathbf{x}_j$ is larger than the preset threshold, $\mathbf{x}_j$ is considered a border element; return to step 1. Otherwise proceed to step 3.
3. Compute the average gray-level value in the 3×3 neighborhood of the pixel $\mathbf{x}_j$. Compare the result with some preset gray-level value and decide whether $\mathbf{x}_j$ is positioned inside or outside the region. Proceed to step 4.
4. Try to continue the border tracing in pixel $\mathbf{x}_k$ which is adjacent to $\mathbf{x}_i$ in direction $[\phi(\mathbf{x}_i) \pm \pi/4]$, the sign being determined according to the result of step 3. If a border

continuation is found, $\mathbf{x}_k$ is a new border element, and return to step 1. If $\mathbf{x}_k$ is not a border element, start the border tracing at another promising pixel.

This algorithm can be applied to multi-spectral or dynamic images (temporal image sequences) as well, based on multi-dimensional gradients. Further details can be found in, for example, [Liu 77, Herman and Liu 78].

An algorithm for crack-edge boundary tracing that uses a concept of finite topological spaces and cell complexes is given in [Kovalevsky 90, Kovalevsky 92, Kovalevsky 94]. Its application to connected component labeling (Section 6.1), object interior filling, and shape description (Chapter 6) is also discussed.

5.2.4 Border detection as graph searching

Whenever additional knowledge is available for boundary detection, it should be used. One example of prior knowledge is a known starting point and a known ending point of the border, even if the precise border location is not known. Even some relatively weak additional requirements such as smoothness, low curvature, etc., may be included as prior knowledge. If this kind of supporting information is available in the border detection task, general problem-solving methods widely used in AI can be applied [Nilsson 82].

A graph is a general structure consisting of a set of nodes n_i and arcs between the nodes (n_i, n_j) (see Section 3.2.3). We consider oriented and numerically weighted arcs, these weights being called **costs**. The border detection process is transformed into a search for the optimal path in the weighted graph, the aim being to find the best path that connects two specified nodes, the starting and ending nodes. While cost minimization is considered throughout this section, the approach works equally well if a maximum cost path is sought.

Assume that both edge magnitude $s(\mathbf{x})$ and edge direction $\phi(\mathbf{x})$ information is available in an edge image. Each image pixel corresponds to a graph node weighted by a value $s(\mathbf{x})$. Two nodes n_i and n_j corresponding to two 8-connected adjacent pixels $\mathbf{x}_i$ and $\mathbf{x}_j$ are connected by an arc if the edge directions $\phi(\mathbf{x}_i)$ and $\phi(\mathbf{x}_j)$ match the local border direction. We can apply the following rules to construct the graph: To connect a node n_i representing the pixel $\mathbf{x}_i$ with a node n_j representing the pixel $\mathbf{x}_j$, pixel $\mathbf{x}_j$ must be one of three existing neighbors of $\mathbf{x}_i$ in the direction $d \in [\phi(\mathbf{x}_i) - \pi/4, \phi(\mathbf{x}_i) + \pi/4]$. Further, $s(\mathbf{x}_i)$ and $s(\mathbf{x}_j)$ must be greater than T, where T is some preset threshold of edge significance. Another common requirement is to connect two nodes only if the difference of their edge directions is less than $\pi/2$.

These conditions can be modified in specific edge detection problems. Figure 5.21a shows an image of edge directions, with only significant edges according to their magnitudes listed. Figure 5.21b shows an oriented graph constructed in accordance with the presented principles.

The application of graph search to edge detection was first published in [Martelli 72], in which Nilsson's A-algorithm [Nilsson 82] applies. Let $\mathbf{x}_A$ be the starting border element, and $\mathbf{x}_B$ be the end border element. To use graph search for region border detection, a method of oriented weighted-graph expansion must first be defined (one possible method was described earlier). A cost function $f(\mathbf{x}_i)$ must also be defined that is a cost estimate of the path between nodes n_A and n_B (pixels $\mathbf{x}_A$ and $\mathbf{x}_B$) which goes through an intermediate node n_i

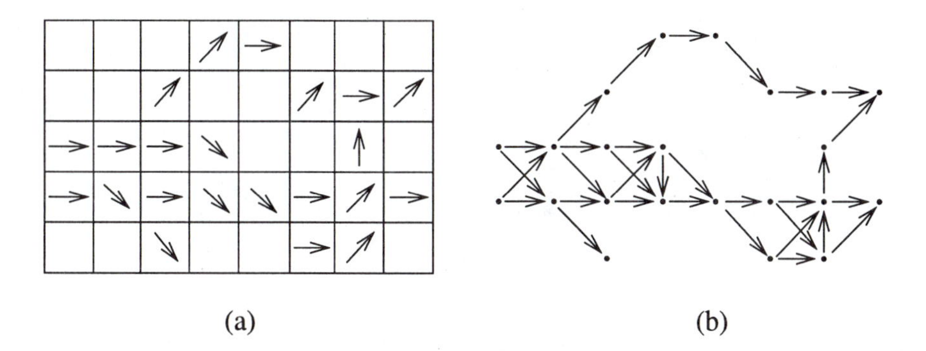

Figure 5.21: *Graph representation of an edge image: (a) edge directions of pixels with above-threshold edge magnitudes, (b) corresponding graph.*

(pixel $\mathbf{x}_i$). The cost function $f(\mathbf{x}_i)$ typically consists of two components; an estimate $\tilde{g}(\mathbf{x}_i)$ of the minimum path cost between the starting border element $\mathbf{x}_A$ and $\mathbf{x}_i$, and an estimate $\tilde{h}(\mathbf{x}_i)$ of the minimum path cost between $\mathbf{x}_i$ and the end border element $\mathbf{x}_B$. The cost $\tilde{g}(\mathbf{x}_i)$ of the path from the starting point to the node n_i is usually a sum of costs associated with the arcs or nodes that are in the path. The cost function must be separable and monotonic with respect to the path length, and therefore the local costs associated with arcs are required to be non-negative. A simple example of $\tilde{g}(\mathbf{x}_i)$ satisfying the given conditions is to consider the path length from $\mathbf{x}_A$ to $\mathbf{x}_i$. An estimate $\tilde{h}(\mathbf{x}_i)$ may be the length of the border from $\mathbf{x}_i$ to $\mathbf{x}_B$, it making sense to prefer shorter borders between $\mathbf{x}_A$ and $\mathbf{x}_B$ as the path with lower cost. This implies that the following graph search algorithm (Nilsson's A-algorithm) can be applied to the border detection.

Algorithm 5.12: A-algorithm graph search

1. Expand the starting node n_A and put all its successors into an OPEN list with pointers back to the starting node n_A. Evaluate the cost function f for each expanded node.

2. If the OPEN list is empty, fail. Determine the node n_i from the OPEN list with the lowest associated cost $f(n_i)$ and remove it. If $n_i = n_B$, then trace back through the pointers to find the optimum path and stop.

3. If the option to stop was not taken in step 2, expand the specified node n_i, and put its successors on the OPEN list with pointers back to n_i. Compute their costs f. Go to step 2.

An example of this algorithm is given in Figure 5.22. Here, nodes currently on the OPEN list are shaded and the minimum-cost node on the OPEN list is shaded and outlined. In Figure 5.22c, note that the node with a cumulative cost of 7 is also expanded; however no successors are found. In Figure 5.22e, since an expansion of a node in the final graph layer was attempted, the search is over.

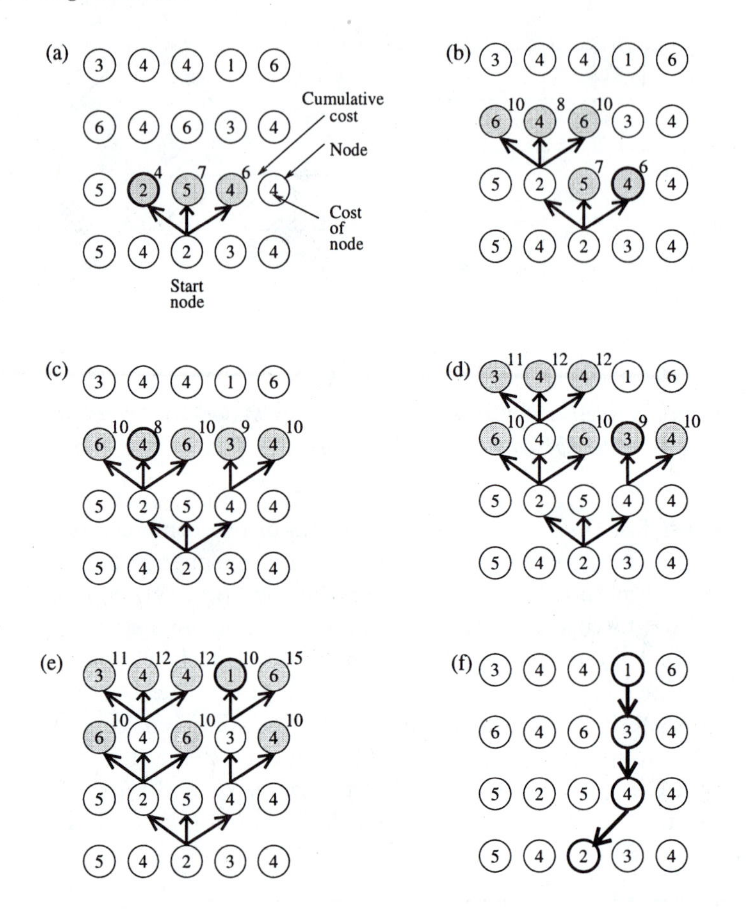

Figure 5.22: *Example of a graph searching sequence using the A-algorithm (see text for description of progress of algorithm). (a) Step 1, expansion of the start node; (b) step 2; (c) steps 3 and 4; (d) step 5; (e) step 6; (f) the optimal path is defined by back-tracking.*

If no additional requirements are set on the graph construction and search, this process can easily result in an infinite loop (see Figure 5.23). To prevent this behavior, no node expansion is allowed that puts a node on the OPEN list if this node has already been visited and put on the OPEN list in the past. A simple solution to the *loop* problem is not to allow searching in a backward direction. This approach can be used if a priori information about the boundary location and its local direction is available. In this case, it may be possible to straighten the processed image (and the corresponding graph) as shown in Figure 5.24. The edge image is geometrically warped by re-sampling the image along profile lines perpendicular to the approximate position of the sought border. The pre-processing step that straightens the image data provides a substantial computational convenience. No backward searches may be allowed to trace boundaries represented this way. This approach can be extremely useful if the borders of thin, elongated objects such as roads, rivers, vessels, etc., are to be detected [Collins and Skorton 86, Fleagle et al. 89]. On the other hand, prohibiting the backward search may limit the shapes of borders that can be successfully identified. In [van der Zwet

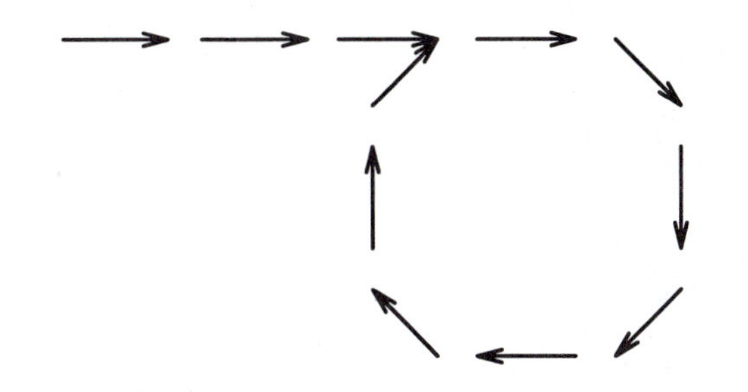

Figure 5.23: *Example of following a closed loop in image data.*

and Reiber 92, van der Zwet and Reiber 94], a graph-search based method called the **gradient field transform** is introduced that allows searching to proceed successfully in any direction.

The estimate of the cost of the path from the current node n_i to the end node n_B has a substantial influence on the search behavior. If this estimate $\tilde{h}(n_i)$ of the true cost $h(n_i)$ is not considered, so $(\tilde{h}(n_i) = 0)$, no heuristic is included in the algorithm and a breadth-first search is done. Because of this, the detected path will always be optimal according to the criterion used, and thus the minimum-cost path will always be found. Applying heuristics, the detected cost does not always have to be optimal but the search can often be much faster.

Given natural conditions for estimates $\tilde{g}$, the minimum-cost path result can be guaranteed if $\tilde{h}(n_i) \leq h(n_i)$ and if the true cost of any part of the path $c(n_p, n_q)$ is larger than the estimated cost of this part $\tilde{c}(n_p, n_q)$. The closer the estimate $\tilde{h}(n_i)$ is to $h(n_i)$, the lower the number of nodes expanded in the search. The problem is that the exact cost of the path from the node n_i to the end node n_B is not known beforehand. In some applications, it may be more important to get a quick rather than an optimal solution. Choosing $\tilde{h}(n_i) > h(n_i)$, optimality is not guaranteed but the number of expanded nodes will typically be smaller because the search can be stopped before the optimum is found.

A comparison of optimal and heuristic graph search border detection is given in Figure 5.25. The raw cost function [called the inverted edge image, inversion defined according to equation (5.19)] can be seen in Figure 5.25a; Figure 5.25b shows the optimal borders resulting from the graph search when $\tilde{h}(n_i) = 0$; 38% of nodes were expanded during the search, and expanded nodes are shown as white regions. When a heuristic search was applied [$\tilde{h}(n_i)$ was about 20% overestimated], only 2% of graph nodes were expanded during the search and the border detection was 15 times faster (Figure 5.25c). Comparing resulting borders in Figures 5.25b and 5.25c, it can be seen that despite a very substantial speedup, the resulting borders do not differ significantly [Tadikonda et al. 92].

We can summarize:

- If $\tilde{h}(n_i) = 0$, the algorithm produces a minimum-cost search.
- If $\tilde{h}(n_i) > h(n_i)$, the algorithm may run faster, but the minimum-cost result is not guaranteed.
- If $\tilde{h}(n_i) \leq h(n_i)$, the search will produce the minimum-cost path if and only if

$$c(n_p, n_q) \geq \tilde{h}(n_p) - \tilde{h}(n_q)$$

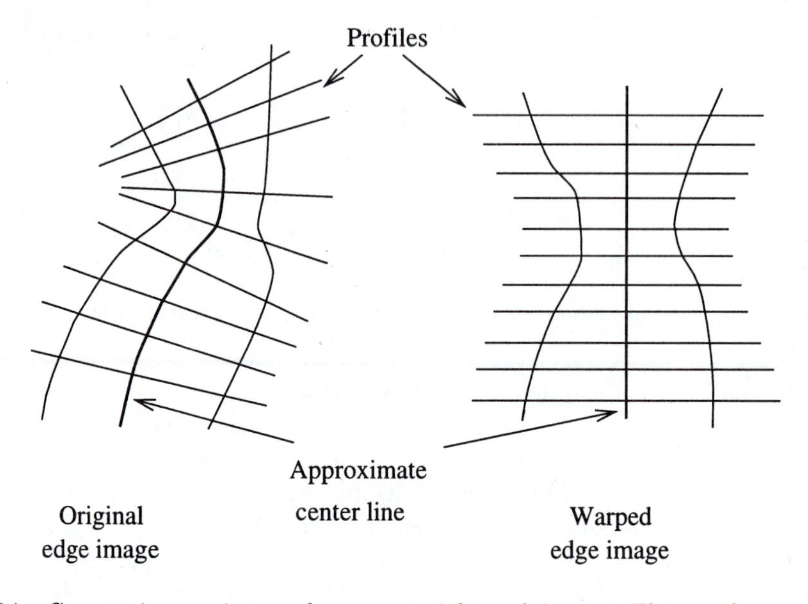

Figure 5.24: *Geometric warping produces a straightened image: The graph constructed requires (and allows) searches in one main direction only (e.g., top-down). Adapted from [Fleagle et al. 89].*

for any p, q, where $c(n_p, n_q)$ is the true minimum cost of getting from n_p to n_q, which is not easy to fulfill for a specific $f(x)$.

- If $h(n_i) = \tilde{h}(n_i)$, the search will always produce the minimum-cost path with a minimum number of expanded nodes.
- The better the estimate of $h(n)$, the smaller the number of nodes that must be expanded.

In image segmentation applications, the existence of a path between a starting pixel x_A and an ending pixel x_B is not guaranteed because of possible discontinuities in the edge image, and so more heuristics must often be applied to overcome these problems. For example, if there is no node in the OPEN list which can be expanded, it may be possible to expand nodes with non-significant edge-valued successors—this can build a bridge to pass these small discontinuities in border representations.

A crucial question is how to choose the evaluation cost functions for graph-search border detection. A good cost function should have elements common to most edge detection problems and also specific terms related to the particular application. Some generally applicable cost functions are

- Strength of edges forming a border: The heuristic 'the stronger the edges that form the border, the higher the probability of the border' is very natural and almost always gives good results. Note that if a border consists of strong edges, the cost of that border is small. The cost of adding another node to the border will be

$$(\max_{\text{image}} s(\mathbf{x}_k)) - s(\mathbf{x}_i) \tag{5.19}$$

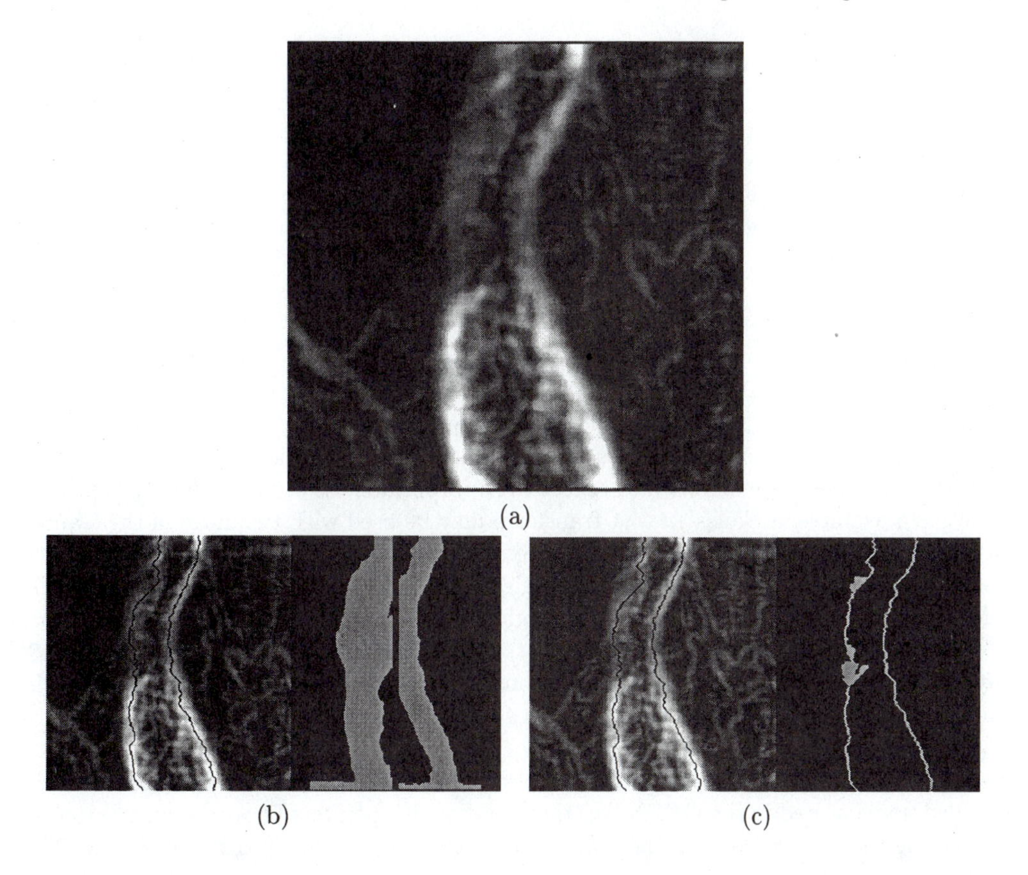

Figure 5.25: *Comparison of optimal and heuristic graph search performance: (a) raw cost function (inverted edge image of a vessel); (b) optimal graph search, resulting vessel borders are shown adjacent to the cost function, expanded nodes shown (38%); (c) heuristic graph search, resulting borders and expanded nodes (2%).*

where the maximum edge strength is obtained from all pixels in the image.

- Border curvature: Sometimes, borders with a small curvature are preferred. If this is the case, the total border curvature can be evaluated as a monotonic function of local curvature increments:

$$\text{diff}[\phi(\mathbf{x}_i) - \phi(\mathbf{x}_j)] \tag{5.20}$$

where *diff* is some suitable function evaluating the difference in edge directions in two consecutive border elements.

- Proximity to an approximate border location: If an approximate boundary location is known, it is natural to support the paths that are closer to the known approximation than others. When included into the border, a border element value can be weighted by the distance '*dist*' from the approximate boundary, the distance having either additive or multiplicative influence on the cost:

$$\text{dist}(\mathbf{x}_i, \text{approximate\_boundary}) \tag{5.21}$$

- Estimates of the distance to the goal (end point): If a border is reasonably straight, it is natural to support expansion of those nodes that are located closer to the goal node than other nodes:

$$\tilde{h}(\mathbf{x}_i) = \text{dist}(\mathbf{x}_i, \mathbf{x}_B) \tag{5.22}$$

Since the range of border detection applications is quite wide, cost functions may need some modification to be relevant to a particular task. For example, if the aim is to determine a region that exhibits a border of a moderate strength, a closely adjacent border of high strength may incorrectly attract the search if the cost given in equation (5.19) is used. Clearly, functions may have to be modified to reflect the appropriateness of individual costs properly. In the given example, a Gaussian cost transform may be used with the mean of the Gaussian distribution representing the desired edge strength and the standard deviation reflecting the interval of acceptable edge strengths. Thus, edge strengths close to the expected value will be preferred in comparison to edges of lower or higher edge strength. A variety of such transforms may be developed; a set of generally useful cost transforms can be found in [Falcao et al. 95]. Overall, a good cost function will very often consist of several components combined together; examples of application-specific complex cost functions can be found in Sections 16.2.1 and 16.2.2.

Graph-based border detection methods very often suffer from extremely large numbers of expanded nodes stored in the OPEN list, these nodes with pointers back to their predecessors representing the searched part of the graph. The cost associated with each node in the OPEN list is a result of all the cost increases on the path from the starting node to that node. This implies that even a good path can generate a higher cost in the current node than costs of the nodes on worse paths which did not get so far from the starting node. This results in expansion of these 'bad' nodes representing shorter paths with lower total costs, even with the general view that their probabilities are low. An excellent way to solve this problem is to incorporate a heuristic estimate $\tilde{h}(\mathbf{x}_i)$ into the cost evaluation, but unfortunately, a good estimate of the path cost from the current node to the goal is not usually available. Some modifications which make the method more practically useful, even if some of them no longer guarantee the minimum-cost path, are available.

- Pruning the solution tree: The set of nodes in the OPEN list can be reduced during the search. Deleting those paths that have high average cost per unit length, or deleting paths that are too short whenever the total number of nodes in the OPEN list exceeds a defined limit, usually gives good results (see also Section 7.4.2).

- Least maximum cost: The strength of a chain may be given by the strength of the weakest element—this idea is included in cost function computations. The cost of the current path is then set as the cost of the most expensive arc in the path from the starting node to the current node [Lester 78], whatever the sum of costs along the path. The path cost does not therefore necessarily grow with each step, and this is what favors expansion of good paths for a longer time.

- Branch and bound: This modification is based on maximum allowed cost of a path, no path being allowed to exceed this cost [Chien and Fu 74]. This maximum path cost is either known beforehand or it is computed and updated during the graph search. All the paths that exceed the allowed maximum path cost are deleted from the OPEN list.

- Lower bound: Another way to increase the search speed is to reduce the number of poor edge candidate expansions. Poor edge candidates are always expanded if the cost of the best current path exceeds that of any worse but shorter path in the graph. If the cost of the best successor is set to zero, the total cost of the path does not grow after the node expansion and the good path will be expanded again. The method developed by Collins et al. [Collins et al. 91, Sonka et al. 93] assumes that the path is searched in a straightened graph resulting from a warped image as discussed earlier. The cost of the minimum-cost node on each profile is subtracted from each node on the profile (lower bound). In effect, this shifts the range of costs from

$$\min(\text{profile\_node\_costs}) \leq \text{node\_cost} \leq \max(\text{profile\_node\_costs})$$

to

$$\begin{aligned} 0 &\leq \text{new\_node\_cost} \\ &\leq [\max(\text{profile\_node\_costs}) - \min(\text{profile\_node\_cost})] \end{aligned}$$

Note that the range of the costs on a given profile remains the same, but the range is translated such that at least one node for each profile is assigned a zero cost. Because the costs of the nodes for each profile are translated by different amounts, the graph is expanded in an order that supports expansion of good paths. For graph searching in the straightened image, the lower bound can be considered heuristic information when expanding nodes and assigning costs to subpaths in the graph. By summing the minimum value for each profile, the total is an estimate of the minimum-cost path through the graph. Obviously, the minimum cost nodes for each profile may not form a valid path, i.e., they may not be neighbors as required. However, the total cost will be the lower limit of the cost of any path through the graph. This result allows the heuristic to be admissible, thus guaranteeing the success of the algorithm in finding the optimal path. The assignment of a heuristic cost for a given node is implemented in a pre-processing step through the use of the lower bound.

- Multi-resolution processing: The number of expanded nodes can be decreased if a sequence of two graph search processes is applied. The first search is done in lower resolution, therefore a smaller number of graph nodes is involved in the search and a smaller number is expanded, compared to full resolution. The low-resolution search detects an approximate boundary. The second search is done in full resolution using the low-resolution results as a model, and the full-resolution costs are weighted by a factor representing the distance from the approximate boundary acquired in low resolution [equation (5.21)]. The weighting function should increase with the distance in a non-linear way. This approach assumes that the approximate boundary location can be detected from the low-resolution image [Sonka et al. 93, Sonka et al. 94].

- Incorporation of higher-level knowledge: Including higher-level knowledge into the graph search may significantly decrease the number of expanded nodes. The search may be directly guided by a priori knowledge of approximate boundary position. Another possibility is to incorporate a boundary shape model into the cost function computation. Both these approaches together with additional specific knowledge and the multi-resolution approach applied to coronary border detection are discussed in detail in Chapter 8 (see Figure 5.26 and Section 8.1.5).

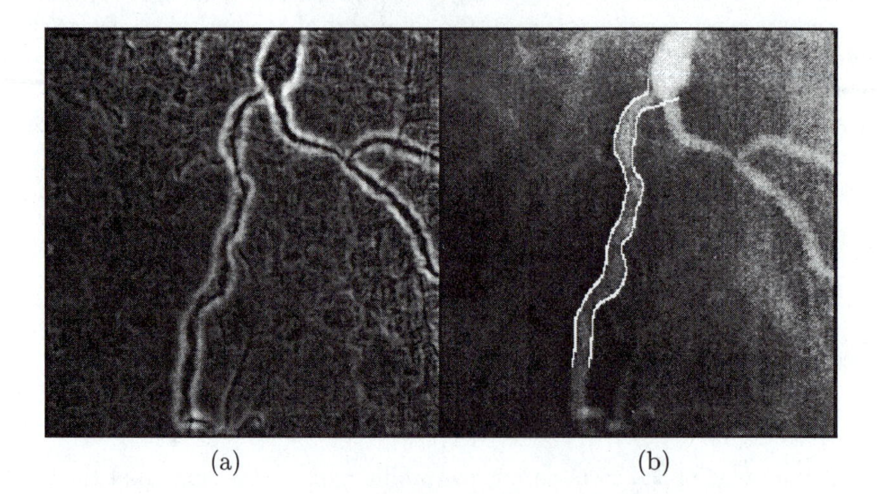

Figure 5.26: *Graph search applied to coronary vessel border detection: (a) edge image (see Figure 16.7 for the original angiographic image); (b) determined vessel borders.*

Graph searching techniques offer a convenient way to ensure global optimality of the detected contour. This technique has often been applied to the detection of approximately straight contours [Wang and Howarth 87, Wang and Howarth 89, Fleagle et al. 89]. The detection of closed structure contours would involve geometrically transforming the image using a polar-to-rectangular co-ordinate transformation in order to 'straighten' the contour, but this may prevent the algorithm from detecting the non-convex parts of the contour. To overcome this problem, the image may be divided into two segments (iteratively, if necessary) and separate, simultaneous searches can be conducted in each segment [Philip et al. 90]. The searches are independent and proceed in opposite directions from a start point until they meet at the dividing line between the two image segments.

The approaches discussed above search for optimal borders between two specific image points, but searching for all the borders in the image without knowledge of the start and end points is more complex. In an approach based on magnitudes and directions of edges in the image, edges are merged into edge chains (i.e., partial borders) [Ramer 75]. Edge chains are constructed by applying a bi-directional heuristic search in which half of each 8-neighborhood expanded node is considered as lying in front of the edge, the second half as lying behind the edge (see Figure 5.27). Partial borders are grouped together using other heuristics which are similar to the approach previously described in edge relaxation (Section 5.2.2), and final region borders result. The following algorithm describes the ideas above in more detail and is an example of applying a bottom-up control strategy (see Chapter 8).

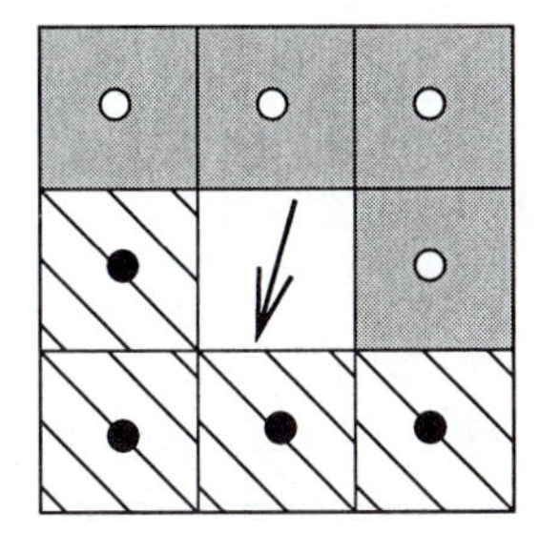

Figure 5.27: *Bidirectional heuristic search: Edge predecessors (marked ∘) and successors (marked •).*

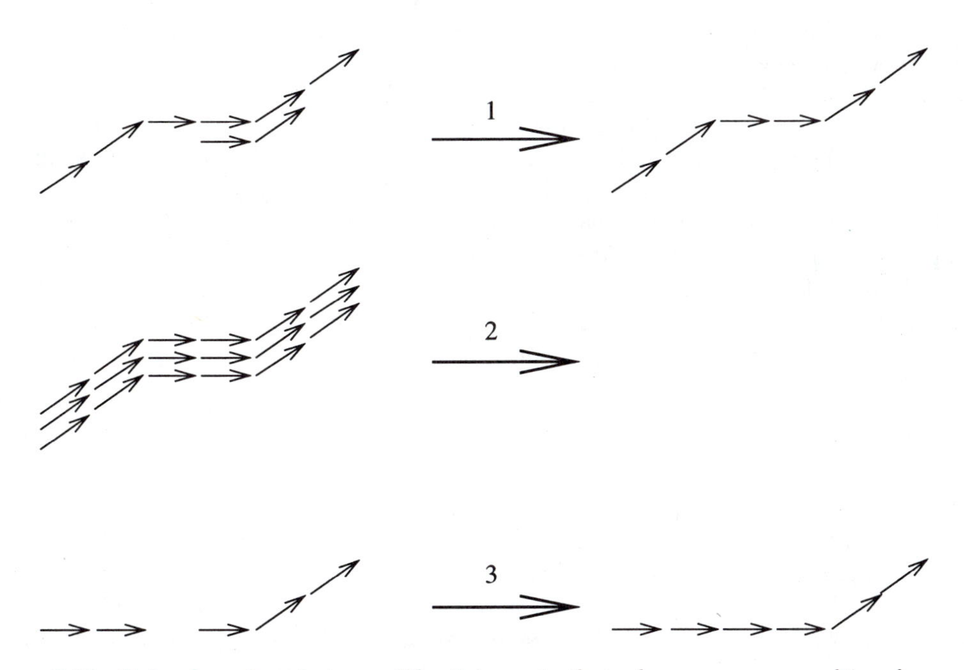

Figure 5.28: *Rules for edge chain modification; note that edge responses resulting from continuous changes of illumination are removed (case 2).*

Algorithm 5.13: Heuristic search for image borders

1. Search for the strongest edge in the image, not considering edges previously marked or edges that already are part of located borders. Mark the located edge. If the edge magnitude is less than the preset threshold or if no image edge was found, go to step 5.

2. Expand all the image edges positioned in front of the specified starting edge until no new successors can be found.

3. Expand all the image edges positioned behind the specified starting edge until no new predecessors can be found. In steps 2 and 3 do not include edges that are already part of any existing edge chain.

4. If the resulting edge chain consists of at least three edges, it is stored in the chain list, otherwise it is deleted. Proceed to step 1.

5. Modify edge chains according to the rules given in Figure 5.28.

6. Repeat step 5 until the resulting borders do not change substantially from step to step.

The rules given in Figure 5.28 (used in step 5 of Algorithm 5.13) solve three standard situations. First, thinner edge responses to a single border are obtained. Second, edge responses resulting from changes of lighting, where no real border exists, are removed. Third, small gaps in boundary-element chains are bridged. Detailed behavior given by these general rules can be modified according to the particular problem.

A good overview of recent border detection and edge linking methods can be found in [van der Heijden 95]. An interesting approach to border detection based on image filtering, detection of edges, sharp corners and border junctions (corners of at least two adjacent regions) is presented in [Law et al. 96]. The steps other than filtering use fuzzy reasoning to incorporate local image characteristics in the edge detection and edge linking processes. An approach to border detection that uses enhancements of corners and junctions can be found in [Demi 96]. Yet another approach, based on effectively searching for closed contours of edge elements using an optimal graph search approach to data clustering, is presented in [Wu and Leahy 93].

5.2.5 Border detection as dynamic programming

Dynamic programming is an optimization method based on the **principle of optimality** [Bellmann 57, Pontriagin 62, Pontriagin 90]. It searches for optima of functions in which not all variables are simultaneously interrelated.

Consider the following simple boundary-tracing problem (Figure 5.29). The aim is to find the best path (minimum cost) between one of the possible start points A, B, C and one of the possible ending points G, H, I. The boundary must be contiguous in 8-connectivity. The graph representing the problem, together with assigned partial costs, is shown in Figure 5.29a,b. As can be seen, there are three ways to get to the node E. Connecting A–E gives the cost $g(A, E) = 2$; connecting B–E, cost $g(B, E) = 6$; connecting C–E, cost $g(C, E) = 3$.

The **main idea** of the principle of optimality is: *Whatever the path to the node E was, there exists an optimal path between E and the end point.* In other words, *if the optimal path start point–endpoint goes through E, then both its parts start point–E and E–end point, are also optimal.*

In our case, the optimal path between the start point and E is the partial path A–E (see Figure 5.29c). Only the following information need be stored for future use; to get to E, the optimal path is A-E, cost $C(E) = 2$. Using the same approach, to get to D the optimal path is B–D, cost $C(D) = 2$; the best path to F is B–F, cost $C(F) = 1$ (see Figure 5.29d). The

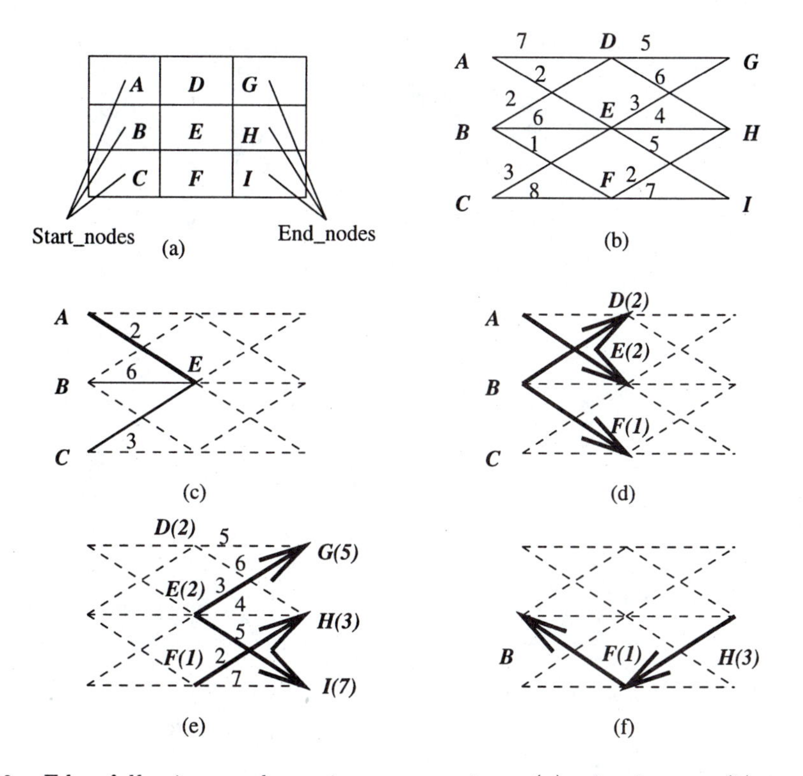

Figure 5.29: *Edge following as dynamic programming: (a) edge image; (b) corresponding graph, partial costs assigned; (c) possible paths from any start point to E, $A - E$ is optimal; (d) optimal partial paths to nodes D, E, F; (e) optimal partial paths to nodes G, H, I; (f) back-tracking from H defines the optimal boundary.*

path may get to node G from either D or E. The cost of the path through the node D is a sum of the cumulative cost $C(D)$ of the node D and the partial path cost $g(D, G)$. This cost $C(G_D) = 7$ represents the path B–D–G because the best path to D is from B. The cost to get to G from E is $C(G_E) = 5$, representing the path A–E–G. It is obvious that the path going through node E is better, the optimal path to G is the path A–E–G with cost $C(G) = 5$ (see Figure 5.29e). Similarly, cost $C(H) = 3$ (B–F–H) and cost $C(I) = 7$ (A–E–I). Now, the end point with the minimum path cost represents the optimum path; node H is therefore the optimal boundary end point, and the optimal boundary is B–F–H (see Figure 5.29f). Figure 5.30 gives an example in which node costs are used (not arc costs as in Figure 5.29). Note that the graph, the cost function, and the resulting path are identical to those used in Figure 5.22.

If the graph has more layers (in Figure 5.29 just three layers were present), the process is repeated until one of the end points is reached. Each repetition consists of a simpler optimization as shown in Figure 5.31a

$$C(x_k^{m+1}) = \min_i [C(x_i^m) + g^m(i,k)] \tag{5.23}$$

where $C(x_k^{m+1})$ is the new cost assigned to the node x_k^{m+1}, and $g^m(i,k)$ is the partial path

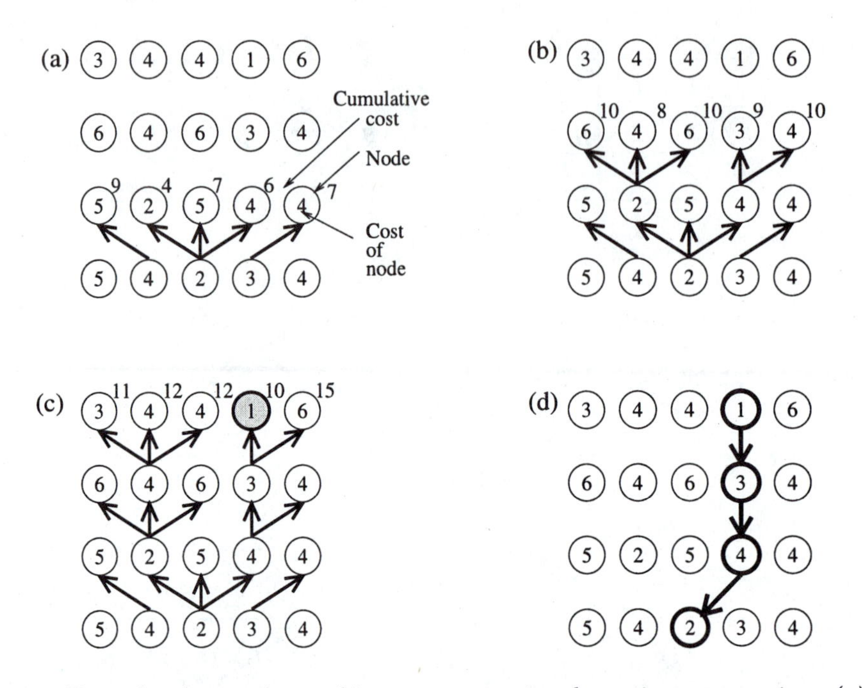

Figure 5.30: *Example of a graph searching sequence using dynamic programming: (a) step 1, expansion of the first graph layer; (b) step 2; (c) step 3—the minimum-cost node in the last layer marked; (d) the optimal path is defined by back-tracking.*

cost between nodes x_i^m and x_k^{m+1}. For the complete optimization problem,

$$\min[C(x^1, x^2, \dots, x^M)] = \min_{k=1,\dots,n} [C(x_k^M)] \tag{5.24}$$

where x_k^M are the end point nodes, M is the number of graph layers between start points and end points (see Figure 5.31b), and $C(x^1, x^2, \dots, x^M)$ denotes the cost of a path between the first and the last (M^{th}) graph layer. Requiring an 8-connected border and assuming n nodes x_i^m in each graph layer m, $3n$ cost combinations must be computed for each layer, $3n(M-1)+n$ being the total number of cost combination computations. Compared to the brute-force enumerative search, where $n(3^{M-1})$ combinations must be computed, the improvement is obvious. The final optimal path results from back-tracking through the searched graph. Note that the number of neighbors depends on the definition of contiguity and definition of the searched graph, and is not limited to three.

The complete graph must be constructed to apply dynamic programming, and this may follow general rules given in the previous section. The objective function must be separable and monotonic (as for the A-algorithm); evaluation functions presented in the previous section may also be appropriate for dynamic programming. Figure 5.32 shows the result of dynamic programming applied to detection of pulmonary fissures from X-ray CT images. Since the fissures are visualized as bright pixels, the cost function simply reflects inverted pixel gray values. Despite the fact that vessels are substantially brighter (their pixels correspond to locally lower costs), the path of the lowest overall cost reflects the continuous character of

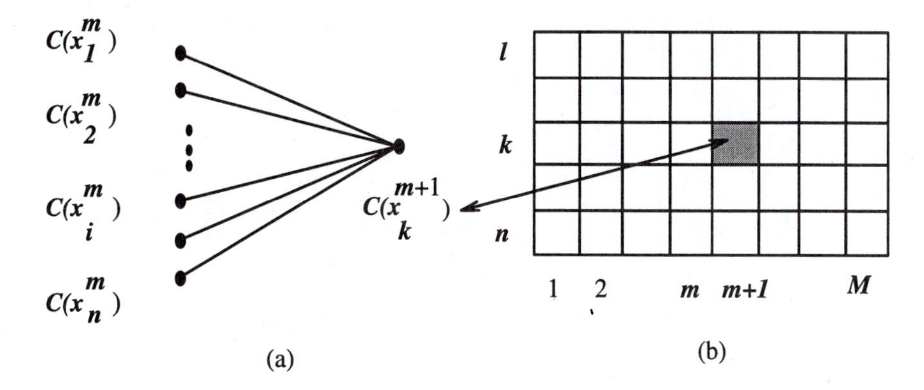

Figure 5.31: *Dynamic programming: (a) one step of the cost calculation; (b) graph layers, node notation.*

the fissure.

Algorithm 5.14: Boundary tracing as dynamic programming

1. Specify initial costs $C(x_i^1)$ of all nodes in the first graph layer, $i = 1, \ldots, n$ and partial path costs $g^m(i,k)$, $m = 1, \ldots, M-1$.

2. Repeat step 3 for all $m = 1, \ldots, M-1$.

3. Repeat step 4 for all nodes $k = 1, \ldots, n$ in the graph layer m.

4. Let
$$C(x_k^{m+1}) = \min_{i=-1,0,1} [C(x_{k+i}^m) + g^m(i,k)] \tag{5.25}$$
Set pointer from node x_k^{m+1} back to node x_i^{m*}; where $*$ denotes the optimal predecessor.

5. Find an optimal node x_k^{M*} in the last graph layer M and obtain an optimal path by back-tracking through the pointers from x_k^{M*} to x_i^{1*}.

It has been shown that heuristic search may be more efficient than dynamic programming for finding a path between two nodes in a graph [Martelli 76]. Further, an A-algorithm-based graph search does not require explicit definition of the graph. However, dynamic programming presents an efficient way of simultaneously searching for optimal paths from multiple starting and ending points. If these points are not known, dynamic programming is probably a better choice, especially if computation of the partial costs $g^m(i,k)$ is simple. Nevertheless, which approach is more efficient for a particular problem depends on evaluation functions and on the quality of heuristics for an A-algorithm. A comparison between dynamic programming and heuristic search efficiency can be found in [Ney 92]; dynamic programming was found to be faster and less memory demanding for a word recognition problem.

Dynamic programming was found to be more computationally efficient than edge relaxation (Section 5.2.2) [Bruel 88]. Another comparison [Wood 85] found dynamic programming

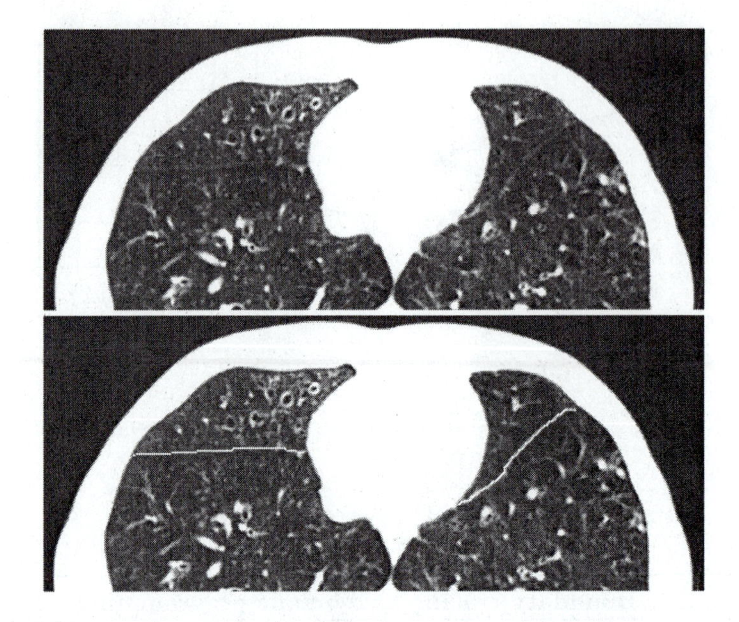

Figure 5.32: *Detection of pulmonary fissures using dynamic programming. Top: Sub-region of an original cross-sectional X-ray CT image of a human lung. Bottom: Detected fissures shown in white.*

more flexible and less restrictive than the Hough transform (Section 5.2.6), and it is a powerful tool in the presence of noise and in textured images [Furst 86, Derin and Elliot 87, Gerbrands 88, Cristi 88]. To increase processing speed, parallel implementations of dynamic programming are studied [Ducksbury 90]. Tracing borders of elongated objects such as roads and rivers in aerial photographs, and vessels in medical images, represents typical applications of dynamic programming in image segmentation [Pope et al. 85, Derin and Elliot 87, Degunst 90].

Practical border detection using two-dimensional dynamic programming was developed by Udupa et al. [Mortensen et al. 92, Udupa et al. 92, Falcao et al. 95, Barrett and Mortensen 96]. An interactive real-time border detection method called the *live wire* combines automated border detection with manual definition of the boundary start point and interactive positioning of the end point. In dynamic programming, the graph that is searched is always completely constructed at the beginning of the search process; therefore, interactive positioning of the end point invokes no time-consuming recreation of the graph as would be the case in heuristic graph searching (Section 5.2.4). Thus, after construction of the complete graph and associated node costs, optimal borders connecting the fixed start point and the interactively changing end point can be determined in real time. In the case of large or more complicated regions, the complete region border is usually constructed from several border segments. After definition of the initial start point, the operator interactively steers the end point so that the calculated optimal border is visually correct. If the operator is satisfied with the current border, and if further movement of the end point causes the border to diverge from the desired location, the end point is fixed and becomes a new start point for the next border segment detection. A new complete graph is calculated and the operator interactively defines the next end point. In many cases, a closed region contour can be formed from just two segments. While the border detection in response to the interactive modification of the end point is very fast, the initial construction of a complete graph needed for each optimal border segment search is computationally demanding, since the graph is of the size of the entire image. Clearly, substantial computational power is needed for real-time performance.

To overcome the computational needs of the live wire method, a modification called the *live lane* was developed [Falcao et al. 95]. In this approach, an operator defines a region of interest by approximately tracing the border by moving a square window. The size of the window is either pre-selected or is adaptively defined from the speed and acceleration of the manual tracing. When the border is of high quality, the manual tracing is fast and the live lane method is essentially identical to the live wire method applied to a sequence of rectangular windows. If the border is less obvious, manual tracing is usually slower and the window size adaptively decreases. If the window size reduces to a single pixel, the method degenerates to manual tracing. A flexible method results that combines the speed of automated border detection with the robustness of manual border detection whenever needed. Since the graph is constructed using an image portion comparable in size to the size of the moving window, the computational demands of the live lane method are much less than those of the live wire method.

Several additional features of the two *live* methods are worth mentioning. As was stressed earlier, design of border-detection cost functions often requires substantial experience and experimentation. To facilitate the method's use by non-experts, an automated approach has been developed that determines optimal border features from examples of the correct borders. Another automated step is available to specify optimal parameters of cost transforms to create a powerful cost function (Section 5.2.4). Consequently, the resultant optimal cost function is specifically designed for a particular application and can be conveniently obtained by presenting a small number of example border segments during the method's training stage. Additionally, the method can easily be applied to three-dimensional image data by incorporation of a cost element comparing the border positions in adjacent image slices.

5.2.6 Hough transforms

If an image consists of objects with known shape and size, segmentation can be viewed as a problem of finding this object within an image. Typical tasks are to locate circular pads in printed circuit boards, or to find objects of specific shapes in aerial or satellite data, etc. One of many possible ways to solve these problems is to move a mask with an appropriate shape and size along the image and look for correlation between the image and the mask, as discussed in Section 5.4. Unfortunately, the specified mask often differs too much from the object's representation in the processed data, because of shape distortions, rotation, zoom, etc. One very effective method that can solve this problem is the **Hough transform**, which can even be used successfully in segmentation of overlapping or semi-occluded objects.

To introduce the main concepts of the Hough transform, consider an example of circle detection. Let the task be to detect a dark circle of a known radius r in an image with a uniform bright background (shown in Figure 5.33a). The method starts with a search for dark image pixels; after such a pixel is found, a locus of potential center points of the circle associated with it can be determined. Such a locus of potential center points forms a circle

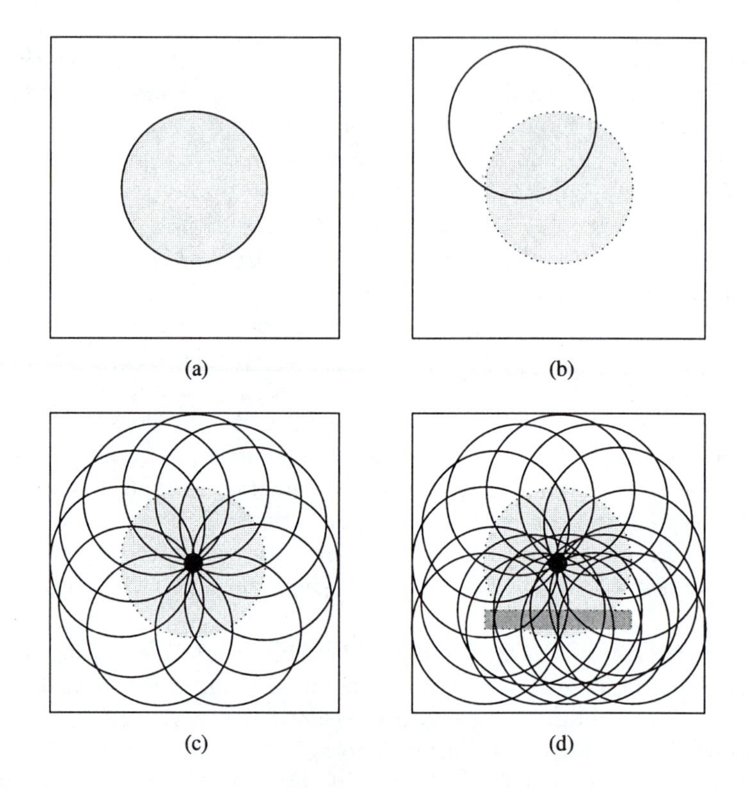

Figure 5.33: *Hough transform - example of circle detection: (a) original image of a dark circle (known radius r) on a bright background. (b) For each dark pixel, a potential circle-center locus is defined by a circle with radius r and center at that pixel. (c) The frequency with which image pixels occur in the circle-center loci is determined—the highest-frequency pixel represents the center of the circle (marked by •). (d) The Hough transform correctly detects the circle (marked by •) in the presence of incomplete circle information and overlapping structures. (See Figure 5.38 for a real-life example.)*

with the radius r as demonstrated in Figure 5.33b. If the loci of potential circle centers are constructed for all dark pixels identified in the original image, the frequency can be determined with which each pixel of the image space occurs as an element of the circle-center loci. As seen from Figure 5.33c, the true center of the circle being sought is represented by the pixel with the highest frequency of occurrence in the circle-center loci. Thus, the center of the searched circle is determined. With the known circle radius, the image segmentation is complete. Figure 5.33d presents intuitive proof that the Hough transform can be successfully applied to images with incomplete information about the searched objects (a circle in our case) and/or in the presence of additional structures and noise. The remainder of this section describes the Hough transform methodology in detail.

The original Hough transform was designed to detect straight lines and curves [Hough 62], and this original method can be used if analytic equations of object borderlines are known—no prior knowledge of region position is necessary. A big advantage of this approach is robustness of segmentation results; that is, segmentation is not too sensitive to imperfect data or noise. Nevertheless, it is often impossible to get analytic expressions describing borders. Later, a generalized Hough transform will be described that can find objects even if an analytic expression of the border is not known.

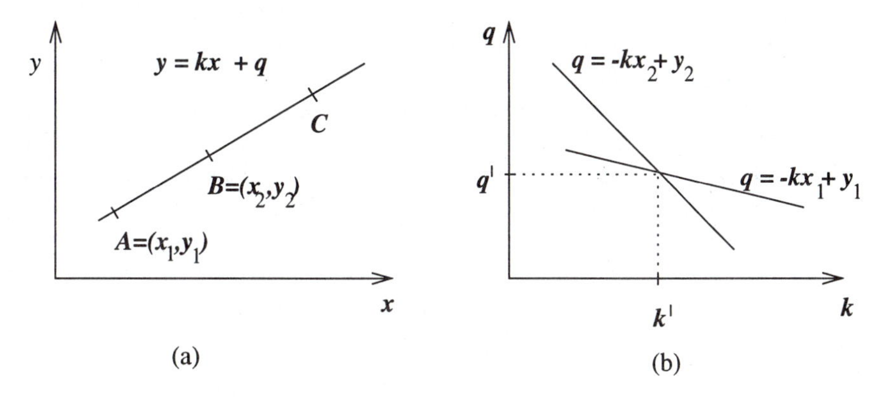

Figure 5.34: *Hough transform principles: (a) image space; (b) k, q parameter space.*

The basic idea of the method can be seen from the simple problem of detecting a straight line in an image [Duda and Hart 72, Duda and Hart 73]. A straight line is defined by two points $A = (x_1, y_1)$ and $B = (x_2, y_2)$ (shown in Figure 5.34a). All straight lines going through the point A are given by the expression $y_1 = kx_1 + q$ for some values of k and q. This means that the same equation can be interpreted as an equation in the parameter space k, q; all the straight lines going through the point A are then represented by the equation $q = -x_1 k + y_1$ (see Figure 5.34b). Straight lines going through the point B can likewise be represented as $q = -x_2 k + y_2$. The only common point of both straight lines in the k, q parameter space is the point which in the original image space represents the only existing straight line connecting points A and B.

This means that any straight line in the image is represented by a single point in the k, q parameter space and any part of this straight line is transformed into the same point. The main idea of line detection is to determine all the possible line pixels in the image, to transform all lines that can go through these pixels into corresponding points in the parameter space, and to detect the points (a, b) in the parameter space that frequently resulted from the Hough transform of lines $y = ax + b$ in the image.

These main steps will be described in more detail. Detection of all possible line pixels in the image may be achieved by applying an edge detector to the image; then, all pixels with edge magnitude exceeding some threshold can be considered possible line pixels (referred to as edge pixels below). In the most general case, nothing is known about lines in the image, and therefore lines of any direction may go through any of the edge pixels. In reality, the number of these lines is infinite; however, for practical purposes, only a limited number of line directions may be considered. The possible directions of lines define a discretization of the parameter k. Similarly, the parameter q is sampled into a limited number of values. The parameter space is not continuous any more, but rather is represented by a rectangular structure of cells. This array of cells is called the **accumulator array** A, whose elements are **accumulator cells** $A(k, q)$. For each edge pixel, parameters k, q are determined which represent lines of allowed directions going through this pixel. For each such line, the values of line parameters k, q are used to increase the value of the accumulator cell $A(k, q)$. Clearly, if a line represented by an equation $y = ax + b$ is present in the image, the value of the accumulator cell $A(a, b)$ will be increased many times—as many times as the line $y = ax + b$ is detected as a line possibly going through any of the edge pixels. For any pixel P, lines going through it may have any direction k (from the set of allowed directions), but the second parameter q is constrained by the image co-ordinates of the pixel P and the direction k. Therefore, lines existing in the image will cause large values of the appropriate accumulator cells in the image, while other lines possibly going through edge pixels, which do not correspond to lines existing in the image, have different k, q parameters for each edge pixel, and therefore the corresponding accumulator cells are increased only rarely. In other words, lines existing in the image may be detected as high-valued accumulator cells in the accumulator array, and the parameters of the detected line are specified by the accumulator array co-ordinates. As a result, line detection in the image is transformed to detection of local maxima in the accumulator space.

It has been noted that an important property of the Hough transform is its insensitivity to missing parts of lines, to image noise, and to other non-line structures co-existing in the image. Insensitivity to data imprecision and noise can be seen in Figure 5.35. This is caused by the robustness of transformation from the image space into the accumulator space—a missing part of the line will cause only a lower local maximum because a smaller number of edge pixels contributes to the corresponding accumulator cell. A noisy or only approximately straight line will not be transformed into a point in the parameter space, but rather will result in a cluster of points, and the cluster center of gravity can be considered the straight line representation.

Note that the parametric equation of the line $y = kx + q$ is appropriate only for explanation of the Hough transform principles—it causes difficulties in vertical line detection ($k \to \infty$) and in non-linear discretization of the parameter k. If a line is represented as

$$s = x \cos\theta + y \sin\theta \tag{5.26}$$

the Hough transform does not suffer from these limitations. Again, the straight line is transformed to a single point (see Figure 5.36). A practical example showing the segmentation of an MR image of the brain into the left and right hemispheres is given in Figure 5.37.

Discretization of the parameter space is an important part of this approach [Yuen and Hlavac 91]; also, detecting the local maxima in the accumulator array is a non-trivial problem. In reality, the resulting discrete parameter space usually has more than one local maximum per line existing in the image, and smoothing the discrete parameter space may be a solution. All these remarks remain valid if more complex curves are sought in the image using the Hough transform, the only difference being the dimensionality of the accumulator array.

Generalization to more complex curves that can be described by an analytic equation is straightforward. Consider an arbitrary curve represented by an equation $f(\mathbf{x}, \mathbf{a}) = 0$, where $\mathbf{a}$ is the vector of curve parameters.

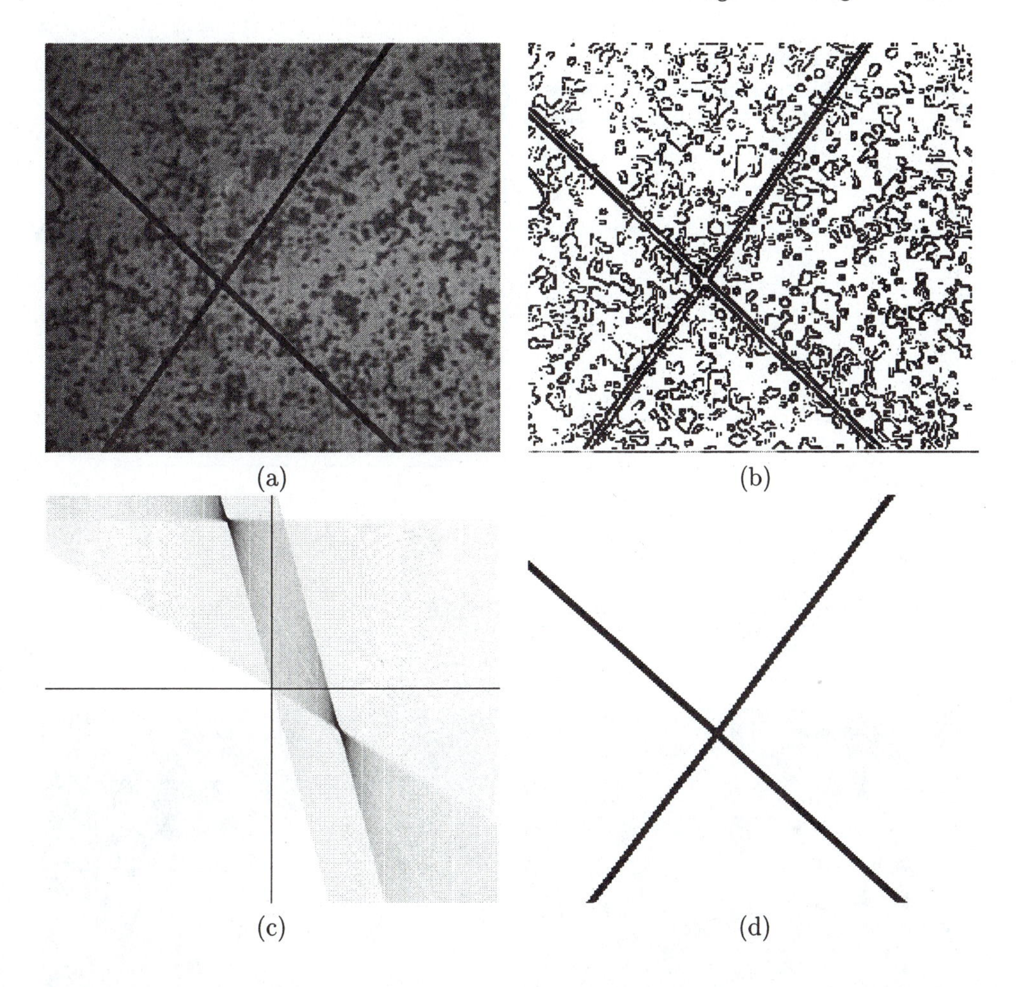

Figure 5.35: *Hough transform—line detection: (a) original image; (b) edge image (note many edges, which do not belong to the line); (c) parameter space; (d) detected lines.*

Algorithm 5.15: Curve detection using the Hough transform

1. Quantize parameter space within the limits of parameters $\mathbf{a}$. The dimensionality n of the parameter space is given by the number of parameters of the vector $\mathbf{a}$.

2. Form an n-dimensional accumulator array $A(\mathbf{a})$ with structure matching the quantization of parameter space; set all elements to zero.

3. For each image point (x_1, x_2) in the appropriately thresholded gradient image, increase all accumulator cells $A(\mathbf{a})$ if $f(\mathbf{x}, \mathbf{a}) = 0$

$$A(\mathbf{a}) = A(\mathbf{a}) + \Delta A$$

for all $\mathbf{a}$ inside the limits used in step 1.

4. Local maxima in the accumulator array $A(\mathbf{a})$ correspond to realizations of curves $f(\mathbf{x}, \mathbf{a})$ that are present in the original image.

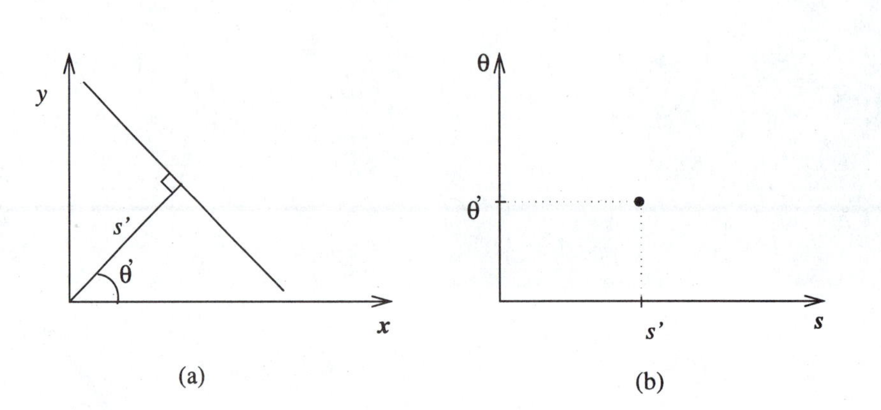

Figure 5.36: *Hough transform in s, θ space: (a) straight line in image space; (b) s, θ parameter space.*

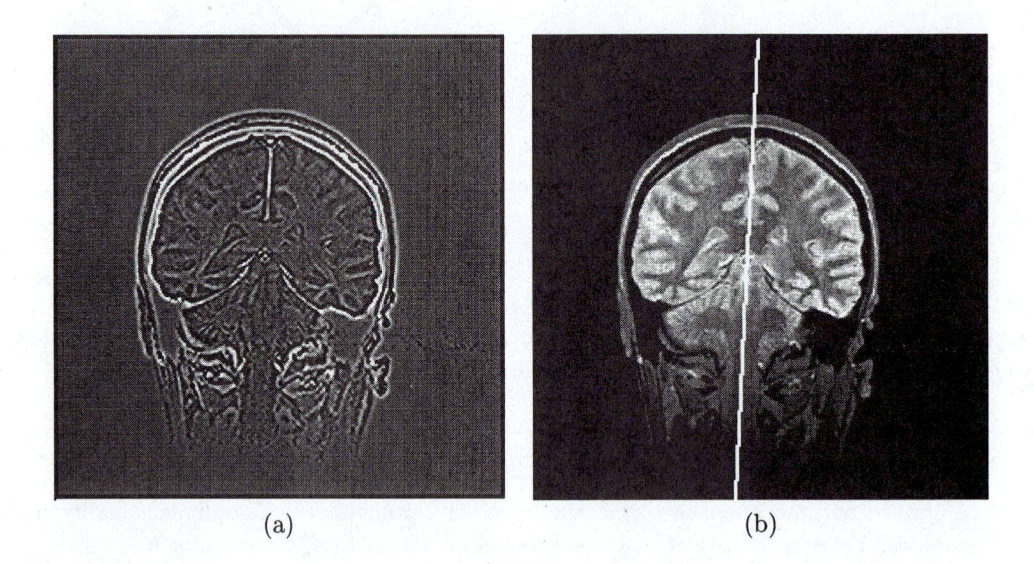

Figure 5.37: *Hough transform line detection used for MRI brain segmentation to the left and right hemispheres: (a) edge image; (b) segmentation line in original image data.*

If we are looking for circles, the analytic expression $f(\mathbf{x}, \mathbf{a})$ of the desired curve is

$$(x_1 - a)^2 + (x_2 - b)^2 = r^2 \tag{5.27}$$

where the circle has center (a, b) and radius r. Therefore, the accumulator data structure must be three-dimensional. For each pixel $\mathbf{x}$ whose edge magnitude exceeds a given threshold, all accumulator cells corresponding to potential circle centers (a, b) are incremented in step 3 of the given algorithm. The accumulator cell $A(a, b, r)$ is incremented if the point (a, b) is at distance r from point $\mathbf{x}$, and this condition is valid for all triplets (a, b, r) satisfying equation (5.27). If some potential center (a, b) of a circle of radius r is frequently found in the parameter space, it is highly probable that a circle with radius r and center (a, b) really exists in the processed data.

The processing results in a set of parameters of desired curves $f(\mathbf{x}, \mathbf{a}) = 0$ that correspond to local maxima of accumulator cells in the parameter space; these maxima best match the desired curves and processed data. Parameters may represent unbounded analytic curves (e.g., line, ellipse, parabola, etc.), but to look for finite parts of these curves, the end points must be explicitly defined and other conditions must be incorporated into the algorithm. Even though the Hough transform is a very powerful technique for curve detection, exponential growth of the accumulator data structure with the increase of the number of curve parameters restricts its practical usability to curves with few parameters.

If prior information about edge directions is used, computational demands can be decreased significantly. Consider the case of searching the circular boundary of a dark region, letting the circle have a constant radius $r = R$ for simplicity. Without using edge direction information, all accumulator cells $A(a, b)$ are incremented in the parameter space if the corresponding point (a, b) is on a circle with center $\mathbf{x}$. With knowledge of direction, only a small number of the accumulator cells need be incremented. For example, if edge directions are quantized into eight possible values, only one-eighth of the circle need take part in incrementing of accumulator cells. Of course, estimates of edge direction are unlikely to be precise—if we anticipate edge direction errors of $\pi/4$, three-eighths of the circle will require accumulator cell incrementing. Using edge directions, candidates for parameters a and b can be identified from the following formulae:

$$\begin{aligned} a &= x_1 - R\cos(\psi(\mathbf{x})) \\ b &= x_2 - R\sin(\psi(\mathbf{x})) \\ \psi(\mathbf{x}) &\in [\phi(\mathbf{x}) - \Delta\phi\,,\ \phi(\mathbf{x}) + \Delta\phi] \end{aligned} \tag{5.28}$$

where $\phi(\mathbf{x})$ refers to the edge direction in pixel $\mathbf{x}$ and $\Delta\phi$ is the maximum anticipated edge direction error. Accumulator cells in the parameter space are then incremented only if (a, b) satisfy equation (5.28). Another heuristic that has a beneficial influence on the curve search is to weight the contributions to accumulator cells $A(\mathbf{a})$ by the edge magnitude in pixel $\mathbf{x}$; thus the increment ΔA in step 3 of Algorithm 5.15 $[A(\mathbf{a}) = A(\mathbf{a}) + \Delta A]$ will be greater if it results from the processing of a pixel with larger edge magnitude. Figure 5.38 demonstrates circle detection when circular objects of known radius overlap and the image contains many additional structures causing the edge image to be very noisy. Note that the parameter space with three local maxima corresponding to centers of three circular objects.

The randomized Hough transform offers a different approach to achieve increased efficiency [Xu and Oja 93]; it randomly selects n pixels from the edge image and determines n parameters of the detected curve followed by incrementing a single accumulator cell only. Recent extensions to the randomized Hough transform use local information about the edge image and apply the Hough transform process to a neighborhood of the edge pixel [Kalviainen et. al. 95].

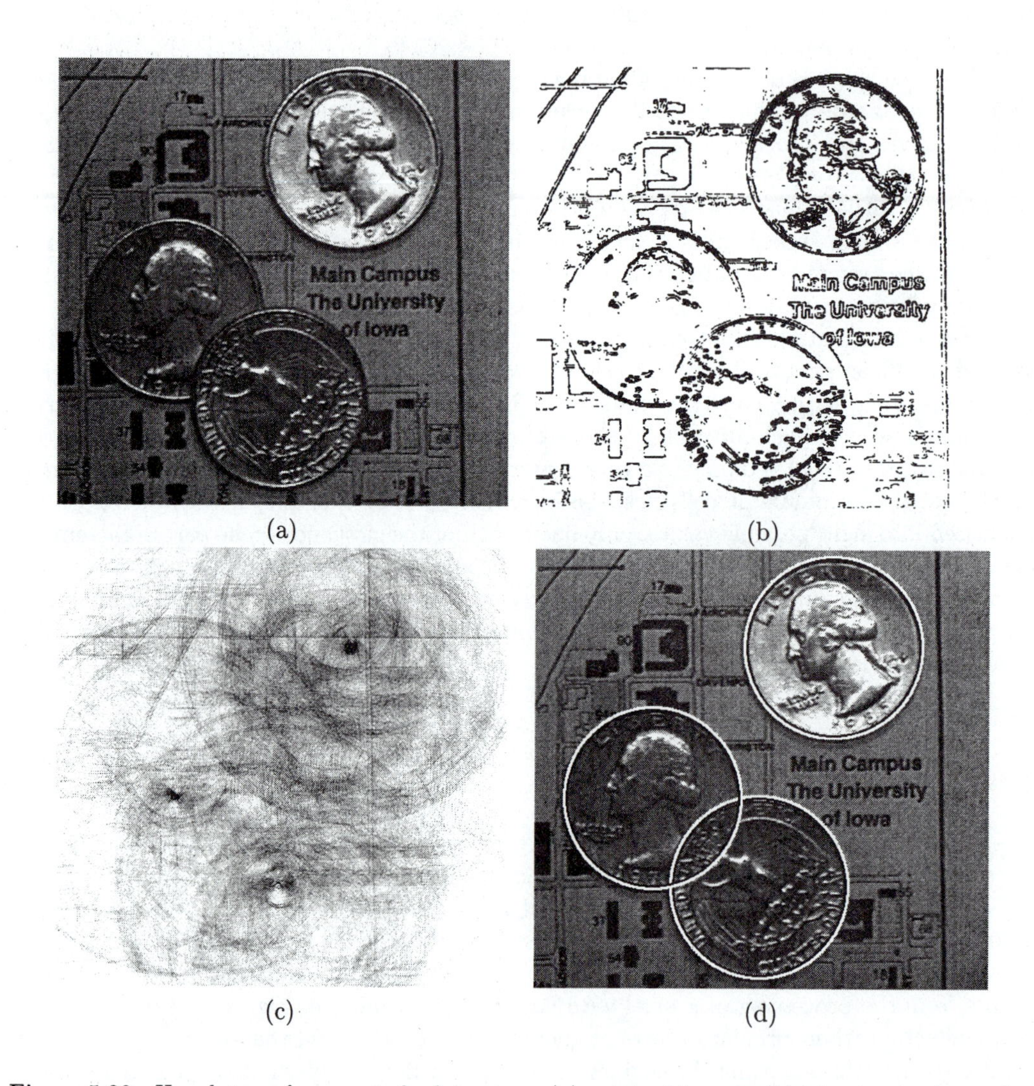

Figure 5.38: *Hough transform—circle detection: (a) original image; (b) edge image (note that the edge information is far from perfect); (c) parameter space; (d) detected circles.*

If the parametric representations of the desired curves or region borders are known, this method works very well, but unfortunately this is not often the case. The desired region borders can rarely be described using a parametric boundary curve with a small number of parameters; in this case, a generalized Hough transform [Ballard 81, Davis 82, Illingworth and Kittler 87] can offer the solution. This method constructs a parametric curve (region border) description based on sample situations detected in the learning stage. Assume that shape, size, and rotation of the desired region are known. A reference point $\mathbf{x}^R$ is chosen at any location inside the sample region, then an arbitrary line can be constructed starting at this reference point aiming in the direction of the region border (see Figure 5.39). The border direction (edge direction) is found at the intersection of the line and the region border. A reference table (referred to as the R-table in [Ballard 81]) is constructed, and intersection parameters are stored as a function of the border direction at the intersection point; using different lines aimed from the reference point, all the distances of the reference point to region borders and the border directions at the intersections can be found. The resulting table can be ordered according to the border directions at the intersection points. As Figure 5.39 makes clear, different points $\mathbf{x}$ of the region border can have the same border direction, $\phi(\mathbf{x}) = \phi(\mathbf{x}')$. This implies that there may be more than one (r, α) pair for each ϕ that can determine the co-ordinates of a potential reference point.

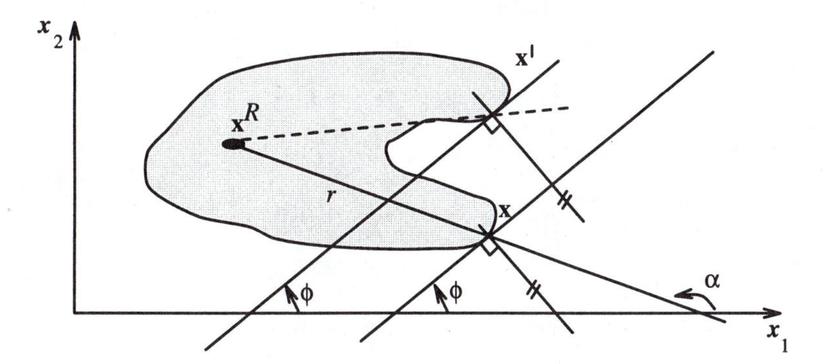

Figure 5.39: *Principles of the generalized Hough transform: geometry of R-table construction.*

An example of an R-table is given in Table 5.1. Assuming no rotation and known size, remaining description parameters required are the co-ordinates of the reference point (x_1^R, x_2^R). If size and rotation of the region may vary, the number of parameters increases to four. Each pixel $\mathbf{x}$ with a significant edge in the direction $\phi(\mathbf{x})$ has co-ordinates of potential reference points $\{x_1 + r(\phi)\cos[\alpha(\phi)], x_2 + r(\phi)\sin[\alpha(\phi)]\}$. These must be computed for all possible values of r and α according to the border direction $\phi(\mathbf{x})$ given in the R-table. The following algorithm presents the generalized Hough transform in the most general of cases in which rotation (τ) and size (S) may both change. If either there is no change in rotation ($\tau = 0$), or there is no size change ($S = 1$), the resulting accumulator data structure A is simpler.

ϕ_1	$(r_1^1, \alpha_1^1), (r_1^2, \alpha_1^2), \ldots, (r_1^{n_1}, \alpha_1^{n_1})$
ϕ_2	$(r_2^1, \alpha_2^1), (r_2^2, \alpha_2^2), \ldots, (r_2^{n_2}, \alpha_2^{n_2})$
ϕ_3	$(r_3^1, \alpha_3^1), (r_3^2, \alpha_3^2), \ldots, (r_3^{n_3}, \alpha_3^{n_3})$
...	...
ϕ_k	$(r_k^1, \alpha_k^1), (r_k^2, \alpha_k^2), \ldots, (r_k^{n_k}, \alpha_k^{n_k})$

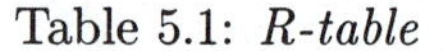

Table 5.1: *R-table*

Algorithm 5.16: Generalized Hough transform

1. Construct an R-table description of the desired object.
2. Form a data structure A that represents the potential reference points

$$A(x_1, x_2, S, \tau)$$

   Set all accumulator cell values $A(x_1, x_2, S, \tau)$ to zero.

3. For each pixel (x_1, x_2) in a thresholded gradient image, determine the edge direction $\Phi(\mathbf{x})$; find all potential reference points $\mathbf{x}^R$ and increase all $A(\mathbf{x}^R, S, \tau)$,

$$A(\mathbf{x}^R, S, \tau) = A(\mathbf{x}^R, S, \tau) + \Delta A$$

   for all possible values of rotation and size change,

$$x_1^R = x_1 + r(\phi) S \cos(\alpha(\phi) + \tau)$$

$$x_2^R = x_2 + r(\phi) S \sin(\alpha(\phi) + \tau)$$

4. The location of suitable regions is given by local maxima in the A data structure.

The Hough transform was initially developed to detect analytically defined shapes, such as lines, circles, or ellipses in general images, and the generalized Hough transform can be used to detect arbitrary shapes. However, even the generalized Hough transform requires the complete specification of the exact shape of the target object to achieve precise segmentation. Therefore, it allows detection of objects with complex but pre-determined, shapes. Other varieties exist that allow detection of objects whose exact shape is unknown, assuming a priori knowledge can be used to form an approximate model of the object [Philip 91].

The Hough transform has many desirable features [Illingworth and Kittler 88]. It recognizes partial or slightly deformed shapes, therefore behaving extremely well in recognition of occluded objects. It may be also used to measure similarity between a model and a detected object on the basis of size and spatial location of peaks in the parameter space. The Hough transform is very robust in the presence of additional structures in the image (other lines, curves, or objects) as well as being insensitive to image noise. Moreover, it may search for several occurrences of a shape during the same processing pass. Unfortunately, the conventional sequential approach requires a lot of storage and extensive computation. However, its inherent parallel character gives the potential for real-time implementations.

Many serious implementational problems were only touched upon here (shape parameterization, accumulation in parameter space, peak detection in parameter space, etc.). Details are discussed in surveys [Evans 85, Illingworth and Kittler 88, Princen et al. 94], where extensive lists of references may also be found. Because of the large time requirements in the sequential version, effort has been devoted to hierarchical approaches [Neveu 86, Princen et al. 89]; fast algorithms were developed [Guil et al. 95]; gray-scale Hough transforms working directly in image data were presented in [Lo and Tsai 95]; methods combining the Hough transform and automated line tracing were studied [Wang and Howarth 89, Lerner and Morelli 90], and many parallel implementations were tested [Oyster 87, Kannan and Chuang 88, Chandran and Davis 89, Hsu and Huang 90, Shankar and Asokan 90, Li et al. 93, Olariu et al. 93, Chung and Lin 95]. The unique properties of the Hough transform also provoke more and more applications [Kashyap and Koch 84, McDonnel et al. 87, Illingworth and Kittler 88, McKenzie and Protheroe 90, Fetterer et al. 90, Brummer 91]. Features of many existing varieties of the Hough transform, together with performance comparisons, are given in [Kalviainen et al. 95].

5.2.7 Border detection using border location information

If any information about boundary location or shape is known, it is of benefit to use it. The information may, for instance, be based on some higher-level knowledge, or can result from segmentation applied to a lower-resolution image.

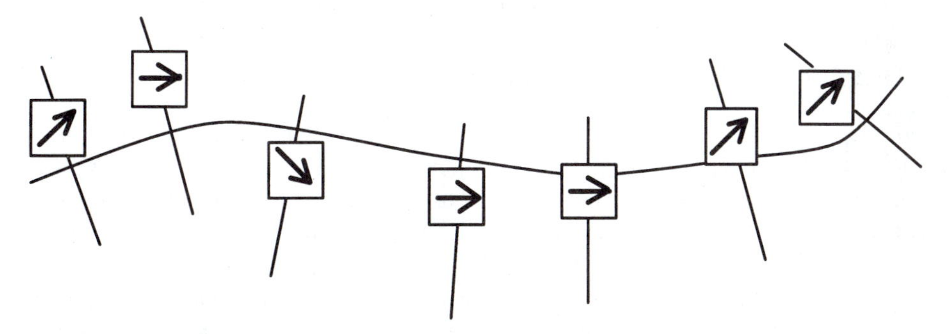

Figure 5.40: *A priori information about boundary location.*

One possibility is to determine a boundary in an image as the location of significant edges positioned close to an assumed border if the edge directions of these significant edges match the assumed boundary direction. The new border pixels are sought in directions perpendicular to the assumed border (see Figure 5.40). If a large number of border elements satisfying the given conditions is found, an approximate curve is computed based on these pixels, and a new, more accurate, border results.

Another possibility is based on prior knowledge of end points—this approach assumes low image noise and relatively straight boundaries. The process iteratively partitions the border and searches for the strongest edge located on perpendiculars to the line connecting end

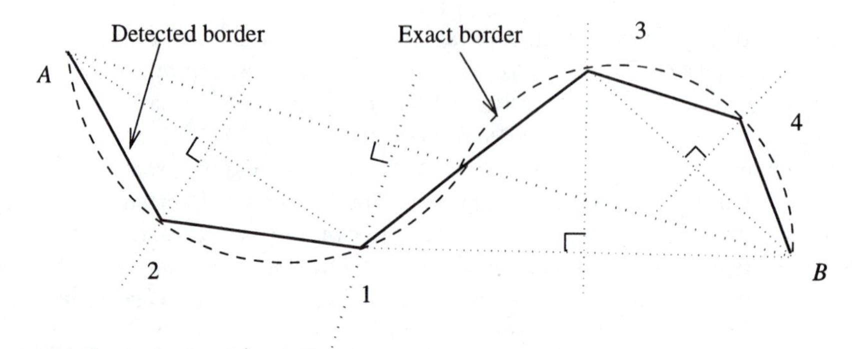

Figure 5.41: *Divide-and-conquer iterative border detection; numbers show the sequence of division steps.*

points of each partition; perpendiculars are located at the center of the connecting straight line—see Figure 5.41. The strongest significant edge located on the perpendicular that is close to the straight line connecting the end points of the current partition is accepted as a new border element. The iteration process is then repeated.

A recent approach to contour detection has been introduced [Kass et al. 87] in which active contour models (snakes) start their search for a contour taking advantage of user-provided knowledge about approximate position and shape of the required contour. An optimization method refines the starting contour estimate and matches the desired contour. This approach is discussed in Section 8.2.

5.2.8 Region construction from borders

All methods considered hitherto have focused on the detection of borders that partially or completely segmented the processed image. If a complete segmentation is achieved, the borders segment an image into regions; but if only a partial segmentation results, regions are not defined uniquely and region determination from borders may be a very complex task requiring cooperation with higher-level knowledge. However, methods exist that are able to construct regions from partial borders which do not form closed boundaries. These methods do not always find acceptable regions, but they are useful in many practical situations.

One of them is the **superslice** method [Milgram 79], which is applicable if regions have dominant gray-level properties. The approach assumes that some border-part locations are known in the image; the image data is then thresholded using different thresholds. Regions resulting from the thresholding for which the detected boundaries best coincide with assumed boundary segments are then accepted as correct.

Better results can be obtained by applying a method described in [Hong et al. 80] based on the existence of partial borders in the processed image. Region construction is based on probabilities that pixels are located inside a region closed by the partial borders. The border pixels are described by their positions and by pixel edge directions $\phi(\mathbf{x})$. The closest 'opposite' edge pixel is sought along a perpendicular to each significant image edge, and then closed borders are constructed from pairs of opposite edge pixels. A pixel is a potential region member if it is on a straight line connecting two opposite edge pixels. The final decision on which of the potential region pixels will form a region is probabilistic.

Algorithm 5.17: Region forming from partial borders

1. For each border pixel $\mathbf{x}$, search for an opposite edge pixel within a distance not exceeding a given maximum M. If an opposite edge pixel is not found, process the next border pixel in the image. If an opposite edge pixel is found, mark each pixel on the connecting straight line as a potential region member.

2. Compute the number of markers for each pixel in the image (the number of markers tells how often a pixel was on a connecting line between opposite edge pixels). Let $b(\mathbf{x})$ be the number of markers for the pixel $\mathbf{x}$.

3. The weighted number of markers $B(\mathbf{x})$ is then determined as follows:

$$\begin{aligned} B(\mathbf{x}) &= 0.0 \quad \text{for} \quad b(\mathbf{x}) = 0 \\ &= 0.1 \quad \text{for} \quad b(\mathbf{x}) = 1 \\ &= 0.2 \quad \text{for} \quad b(\mathbf{x}) = 2 \\ &= 0.5 \quad \text{for} \quad b(\mathbf{x}) = 3 \\ &= 1.0 \quad \text{for} \quad b(\mathbf{x}) > 3 \end{aligned} \tag{5.29}$$

   The confidence that a pixel $\mathbf{x}$ is a member of a region is given as the sum $\sum_i B(\mathbf{x_i})$ in a 3×3 neighborhood of the pixel $\mathbf{x}$. If the confidence that a pixel $\mathbf{x}$ is a region member is one or larger, then pixel $\mathbf{x}$ is marked as a region pixel, otherwise it is marked as a background pixel.

Note that this method allows the construction of bright regions on a dark background as well as dark regions on a bright background by taking either of the two options in the search for opposite edge pixels—step 1. Search orientation depends on whether relatively dark or bright regions are constructed. If $\phi(\mathbf{x})$ and $\phi(\mathbf{y})$ are directions of edges, the condition that must be satisfied for $\mathbf{x}$ and $\mathbf{y}$ to be opposite is

$$\frac{\pi}{2} < |(\phi(\mathbf{x}) - \phi(\mathbf{y})) \bmod (2\pi)| < \frac{3\pi}{2} \tag{5.30}$$

Note that it is possible to take advantage of prior knowledge of maximum region sizes—this information defines the value of M in step 1 of the algorithm, the maximum search length for the opposite edge pixel.

This method was applied to form texture primitives (Chapter 14 [Hong et al. 80, Sonka 86]) as shown in Figure 5.42. The differences between the results of this region detection method and those obtained by thresholding applied to the same data are clearly visible if Figures 5.42b and 5.42c are compared.

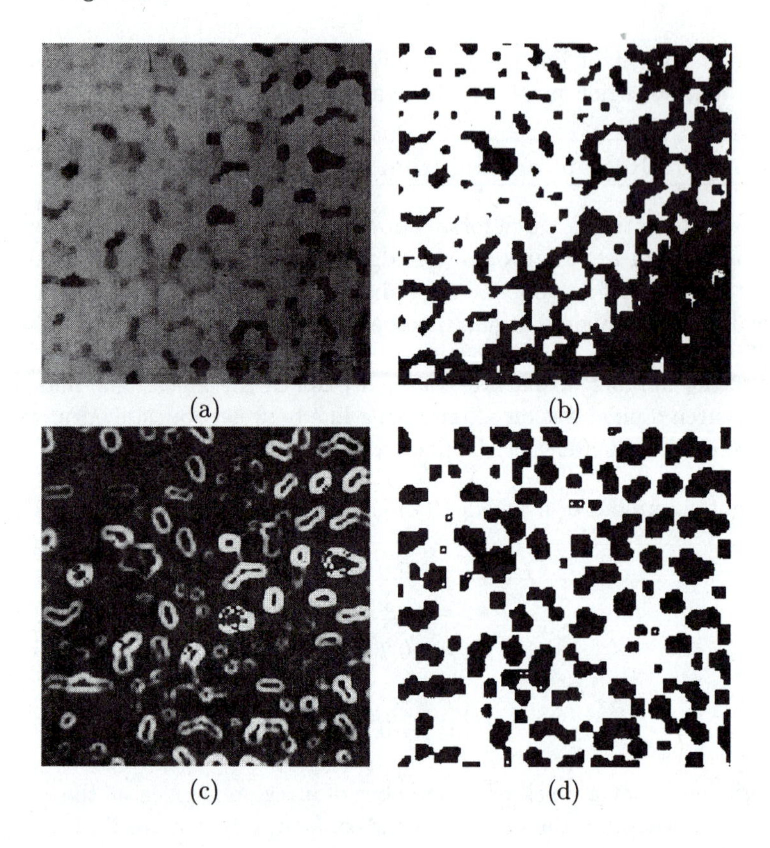

Figure 5.42: *Region forming from partial borders: (a) original image; (b) thresholding; (c) edge image; (d) regions formed from partial borders.*

5.3 Region-based segmentation

The aim of the segmentation methods described in the previous section was to find borders between regions; the following methods construct regions directly. It is easy to construct regions from their borders, and it is easy to detect borders of existing regions. However, segmentations resulting from edge-based methods and region-growing methods are not usually exactly the same, and a combination of results may often be a good idea. Region growing techniques are generally better in noisy images, where borders are extremely difficult to detect. Homogeneity is an important property of regions and is used as the main segmentation criterion in region growing, whose basic idea is to divide an image into zones of maximum homogeneity. The criteria for homogeneity can be based on gray-level, color, texture, shape, model (using semantic information), etc. [Haralick and Shapiro 85, Zamperoni 86, Grimson and Lozano-Perez 87, Pal and Pal 87, Adams and Bischof 94, Chang and Li 94, Chang and Li 95, Kurita 95, Baraldi and Parmiggiani 96]. Properties chosen to describe regions influence the form, complexity, and amount of prior information in the specific region-growing segmentation method. Methods that specifically address region-growing segmentation of color images are reported in [Schettini 93, Vlachos and Constantinides 93, Gauch and Hsia 92, Priese and

Rehrmann 93].

Regions have already been defined in Chapter 2 and discussed in Section 5.1, where equation (5.1) stated the basic requirements of segmentation into regions. Further assumptions needed in this section are that regions must satisfy the following conditions:

$$H(R_i) = \text{TRUE} \qquad i = 1, 2, \ldots, S \tag{5.31}$$

$$H(R_i \cup R_j) = \text{FALSE} \quad i \neq j, \quad R_i \text{ adjacent to } R_j \tag{5.32}$$

where S is the total number of regions in an image and $H(R_i)$ is a binary homogeneity evaluation of the region R_i. Resulting regions of the segmented image must be both homogeneous and maximal, where by 'maximal' we mean that the homogeneity criterion would not be true after merging a region with any adjacent region.

We will discuss simpler versions of region growing first, that is, the merging, splitting, and split-and-merge approaches, and will discuss the possible gains from using semantic information later, in Chapter 8. Of especial interest are the homogeneity criteria, whose choice is the most important factor affecting the methods mentioned; general and specific heuristics may also be incorporated. The simplest homogeneity criterion uses an average gray-level of the region, its color properties, simple texture properties, or an m-dimensional vector of average gray values for multi-spectral images. While the region growing methods discussed below deal with two-dimensional images, three-dimensional implementations are often possible. Considering three-dimensional connectivity constraints, homogeneous regions (volumes) of a three-dimensional image can be determined using three-dimensional region growing [Udupa 82]. Three-dimensional filling represents its simplest form and can be described as a three-dimensional connectivity-preserving variant of thresholding.

5.3.1 Region merging

The most natural method of region growing is to begin the growth in the raw image data, each pixel representing a single region. These regions almost certainly do not satisfy the condition of equation (5.32), and so regions will be merged as long as equation (5.31) remains satisfied.

Algorithm 5.18: Region merging (outline)

1. Define some starting method to segment the image into many small regions satisfying condition (5.31).
2. Define a criterion for merging two adjacent regions.
3. Merge all adjacent regions satisfying the merging criterion. If no two regions can be merged maintaining condition (5.31), stop.

This algorithm represents a general approach to region merging segmentation. Specific methods differ in the definition of the starting segmentation and in the criterion for merging. In the descriptions that follow, regions are those parts of the image that can be sequentially merged into larger regions satisfying equations (5.31) and (5.32). The result of region merging usually depends on the order in which regions are merged, meaning that segmentation results will probably differ if segmentation begins, for instance, in the upper left or lower right corner. This is because the merging order can cause two similar adjacent regions R_1 and R_2 not to be merged, since an earlier merge used R_1 and its new characteristics no longer allow it to be merged with region R_2. If the merging process used a different order, this merge may have been realized.

The simplest methods begin merging by starting the segmentation using regions of 2×2, 4×4, or 8×8 pixels. Region descriptions are then based on their statistical gray-level properties—a regional gray-level histogram is a good example. A region description is compared with the description of an adjacent region; if they match, they are merged into a larger region and a new region description is computed. Otherwise, regions are marked as non-matching. Merging of adjacent regions continues between all neighbors, including newly formed ones. If a region cannot be merged with any of its neighbors, it is marked 'final'; the merging process stops when all image regions are so marked.

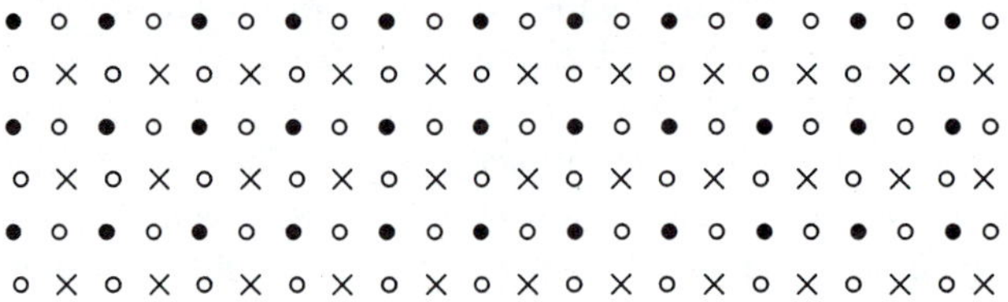

● ○ ● ○ ● ○ ● ○ ● ○ ● ○ ● ○ ● ○ ● ○ ● ○ ● ○
○ × ○ × ○ × ○ × ○ × ○ × ○ × ○ × ○ × ○ × ○ ×
● ○ ● ○ ● ○ ● ○ ● ○ ● ○ ● ○ ● ○ ● ○ ● ○ ● ○
○ × ○ × ○ × ○ × ○ × ○ × ○ × ○ × ○ × ○ × ○ ×
● ○ ● ○ ● ○ ● ○ ● ○ ● ○ ● ○ ● ○ ● ○ ● ○ ● ○
○ × ○ × ○ × ○ × ○ × ○ × ○ × ○ × ○ × ○ × ○ ×

Figure 5.43: *Supergrid data structure:* ×,*image data;* ○, *crack edges;* ●, *unused.*

State space search is one of the essential principles of problem solving in AI [Nilsson 82], whose application to image segmentation was first published in [Brice and Fennema 70]. According to this approach, pixels of the raw image are considered the starting state, each pixel being a separate region. A change of state can result from the merging of two regions or the splitting of a region into sub-regions. The problem can be described as looking for permissible changes of state while producing the best image segmentation. This state space approach brings two advantages; first, well-known methods of state space search can be applied which also include heuristic knowledge; second, higher-level data structures can be used which allow the possibility of working directly with regions and their borders, and no longer require the marking of each image element according to its region marking. Starting regions are formed by pixels of the same gray-level—these starting regions are small in real images. The first state changes are based on crack edge computations (Section 2.3.1), where local boundaries between regions are evaluated by the strength of crack edges along their common border. The data structure used in this approach (the so-called **supergrid**) carries all the necessary information (see Figure 5.43); this allows for easy region merging in 4-adjacency when crack edge values are stored in the '○' elements. Region merging uses the following two heuristics.

- Two adjacent regions are merged if a significant part of their common boundary consists of weak edges (*significance* can be based on the region with the shorter perimeter; the ratio of the number of *weak* common edges to the total length of the region perimeter may be used).

- Two adjacent regions are also merged if a significant part of their common boundary consists of weak edges, but in this case not considering the total length of the region borders.

Of the two given heuristics, the first is more general and the second cannot be used alone because it does not consider the influence of different region sizes.

Edge significance can be evaluated according to the formula

$$\begin{aligned} v_{ij} &= 0 \quad \text{if } \; s_{ij} < T_1 \\ &= 1 \quad \text{otherwise} \end{aligned} \tag{5.33}$$

where $v_{ij} = 1$ indicates a significant edge, $v_{ij} = 0$ a weak edge, T_1 is a preset threshold, and s_{ij} is the crack edge value $[s_{ij} = |f(\mathbf{x_i}) - f(\mathbf{x_j})|]$.

Algorithm 5.19: Region merging via boundary melting

1. Define a starting image segmentation into regions of constant gray-level. Construct a supergrid edge data structure in which to store the crack edge information.

2. Remove all weak crack edges from the edge data structure [using equation (5.33) and threshold T_1].

3. Recursively remove common boundaries of adjacent regions R_i, R_j, if

$$\frac{W}{\min(l_i, l_j)} \geq T_2$$

   where W is the number of weak edges on the common boundary, l_i, l_j are the perimeter lengths of regions R_i, R_j, and T_2 is another preset threshold.

4. Recursively remove common boundaries of adjacent regions R_i, R_j if

$$\frac{W}{l} \geq T_3 \tag{5.34}$$

   or, using a weaker criterion [Ballard and Brown 82],

$$W \geq T_3 \tag{5.35}$$

   where l is the length of the common boundary and T_3 is a third threshold.

Note that even if we have described a region growing method, the merging criterion is based on border properties and so the merging does not necessarily keep condition (5.31) true. The supergrid data structure allows precise work with edges and borders, but a big disadvantage of this data structure is that it is not suitable for the representation of regions—it is necessary to refer to each region as a part of the image, especially if semantic information about regions and neighboring regions is included. This problem can be solved by the construction and updating of a data structure describing region adjacencies and their boundaries, and for this purpose a good data structure to use can be a planar-region adjacency graph and a dual-region boundary graph [Pavlidis 77], (see Section 8.6).

Figure 5.44 gives a comparison of region merging methods. An original image and its pseudo-color representation (to see the small gray-level differences) are given in Figures 5.44a,b. The original image cannot be segmented by thresholding because of the significant and continuous gray-level gradient in all regions. Results of a recursive region merging method, which uses a simple merging criterion [equation (4.76)] are shown in Figure 5.44c; note the resulting horizontally elongated regions corresponding to vertical changes of image gray-levels. If region merging via boundary melting is applied, the segmentation results improve dramatically; see Figure 5.44d [Marik and Matas 89].

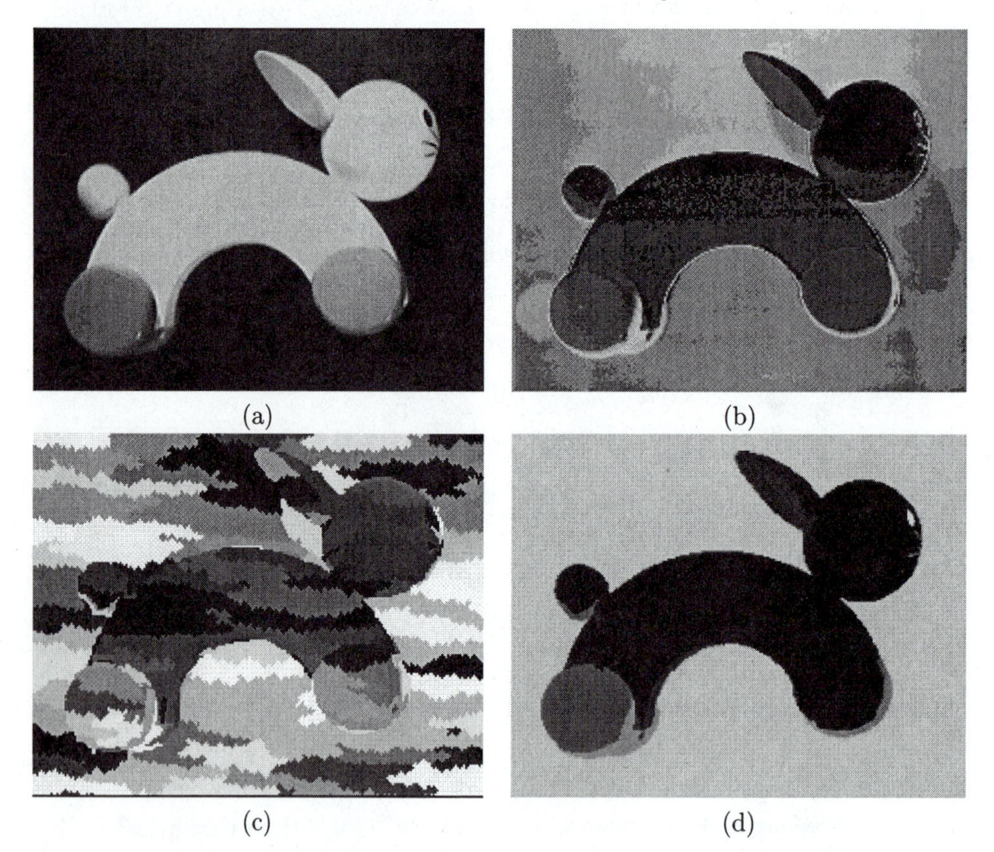

Figure 5.44: *Region merging segmentation: (a) original image; (b) pseudo-color representation of the original image; (c) recursive region merging; (d) region merging via boundary melting. Color versions of b, c, and d may be seen in the color inset. Courtesy R. Marik, Czech Technical University, Prague.*

5.3.2 Region splitting

Region splitting is the opposite of region merging, and begins with the whole image represented as a single region which does not usually satisfy condition (5.31). Therefore, the existing image regions are sequentially split to satisfy (5.1), (5.31) and (5.32). Even if this approach seems to be dual to region merging, region splitting does not result in the same segmentation even if the same homogeneity criteria are used. Some regions may be homogeneous during the splitting process and therefore are not split any more; considering the homogeneous regions created by region merging procedures, some may not be constructed because of the impossibility of merging smaller sub-regions earlier in the process. A fine black-and-white chessboard is an example: Let a homogeneity criterion be based on variance of average gray-levels in the quadrants of the evaluated region in the next lower pyramid level—if the segmentation process is based on region splitting, the image will not be split into sub-regions because its quadrants would have the same value of the measure as the starting region consisting of the whole image. The region merging approach, on the other hand, begins with merging single pixel regions into larger regions, and this process will stop when regions match the chessboard squares. Thus, if splitting is applied, the whole image will be considered one region; whereas if merging is applied, a chessboard will be segmented into squares as shown in Figure 5.45. In this particular case, considering gray-level variance within the entire region as a measure of region homogeneity, and not considering the variance of quadrants only, would also solve the problem. However, region merging and region splitting are not dual.

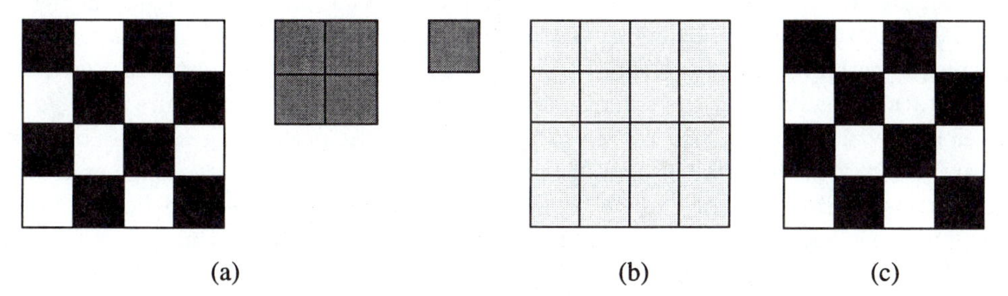

(a) (b) (c)

Figure 5.45: *Different segmentations may result from region splitting and region merging approaches: (a) chessboard image, corresponding pyramid; (b) region splitting segmentation (upper pyramid level is homogeneous, no splitting possible); (c) region merging segmentation (lowest pyramid level consists of regions that cannot be merged).*

Region splitting methods generally use similar criteria of homogeneity as region merging methods, and differ only in the direction of their application. The multi-spectral segmentation discussed in considering thresholding (Section 5.1.3) can be seen as an example of a region splitting method. As mentioned there, other criteria can be used to split regions (e.g., cluster analysis, pixel classification, etc.).

5.3.3 Splitting and merging

A combination of splitting and merging may result in a method with the advantages of both approaches [Horowitz and Pavlidis 74, Pavlidis 77]. Split-and-merge approaches work using pyramid image representations; regions are square shaped and correspond to elements of the appropriate pyramid level.

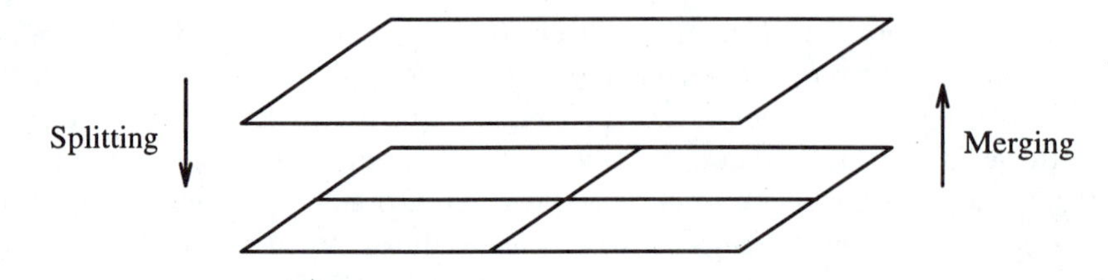

Figure 5.46: *Split-and-merge in a hierarchical data structure.*

If any region in any pyramid level is not homogeneous (excluding the lowest level), it is split into four sub-regions—these are elements of higher resolution at the level below. If four regions exist at any pyramid level with approximately the same value of homogeneity measure, they are merged into a single region in an upper pyramid level (see Figure 5.46). The segmentation process can be understood as the construction of a segmentation quadtree where each leaf node represents a homogeneous region—that is, an element of some pyramid level. Splitting and merging corresponds to removing or building parts of the segmentation quadtree—the number of leaf nodes of the tree corresponds to the number of segmented regions after the segmentation process is over. These approaches are sometimes called split-and-link methods if they use segmentation trees for storing information about adjacent regions. Split-and-merge methods usually store the adjacency information in region adjacency graphs (or similar data structures). Using segmentation trees, in which regions do not have to be contiguous, is both implementationally and computationally easier. An unpleasant drawback of segmentation quadtrees is the square-region shape assumption (see Figure 5.47), and it is therefore advantageous to add more processing steps that permit the merging of regions which are not part of the same branch of the segmentation tree. Starting image regions can either be chosen arbitrarily or can be based on prior knowledge. Because both split-and-merge processing options are available, the starting segmentation does not have to satisfy either condition (5.31) or (5.32).

The homogeneity criterion plays a major role in split-and-merge algorithms, just as it does in all other region growing methods. See [Chen et al. 91] for an adaptive split-and-merge algorithm and a review of region homogeneity analysis. If the image being processed is reasonably simple, a split-and-merge approach can be based on local image properties. If the image is very complex, even elaborate criteria including semantic information may not give acceptable results.

Algorithm 5.20: Split and merge

1. Define an initial segmentation into regions, a homogeneity criterion, and a pyramid data structure.

2. If any region R in the pyramid data structure is not homogeneous [$H(R) = \text{FALSE}$], split it into four child-regions; if any four regions with the same parent can be merged into a single homogeneous region, merge them. If no region can be split or merged, go to step 3.

3. If any two adjacent regions R_i, R_j (even if they are in different pyramid levels or do not have the same parent) can be merged into a homogeneous region, merge them.

4. Merge small regions with the most similar adjacent region if it is necessary to remove small-size regions.

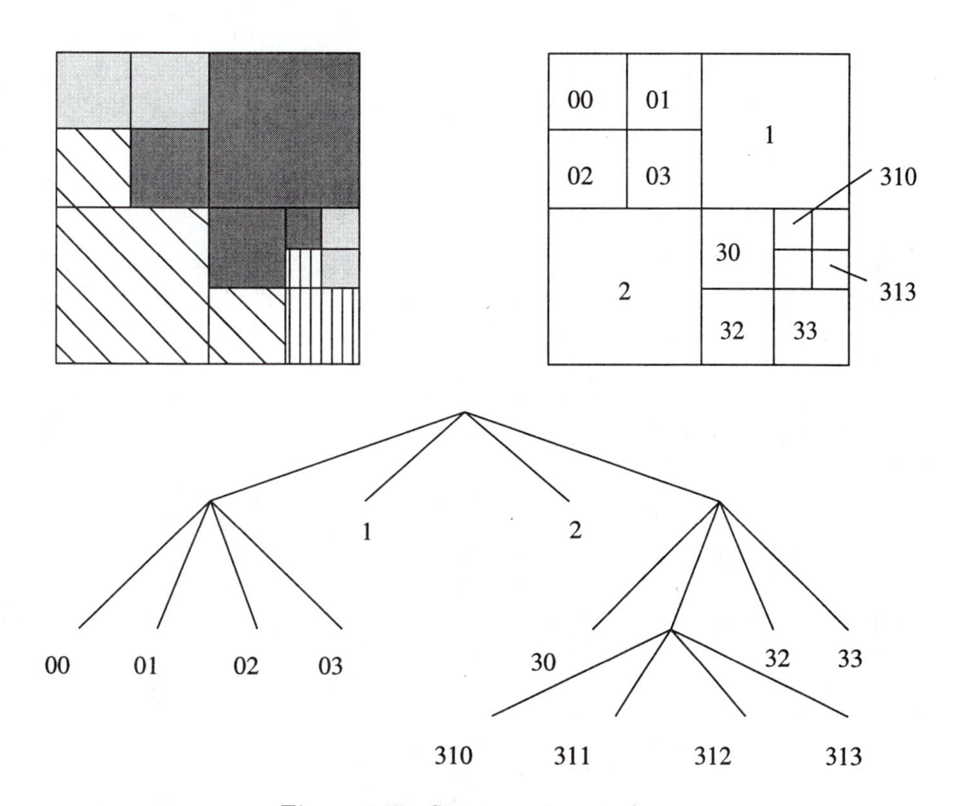

Figure 5.47: *Segmentation quadtree.*

A pyramid data structure with overlapping regions (Chapter 3) is an interesting modification of this method [Pietikainen and Rosenfeld 81, Hong 82, Pietikainen and Rosenfeld 82, Pietikainen et al. 82]. In this data structure, each region has four potential parent elements in the upper pyramid level and 16 possible child elements in the lower pyramid level. Segmentation tree generation begins in the lowest pyramid level. Properties of each region are compared with properties of each of its potential parents and the segmentation branch is linked to the most similar of them. After construction of the tree is complete, all the homogeneity values of all the elements in the pyramid data structure are recomputed to be based on child-region properties only. This recomputed pyramid data structure is used to generate a new segmentation tree, beginning again at the lowest level. The pyramid updating process and new segmentation tree generation is repeated until no significant segmentation changes can be detected between steps. Assume that the segmented image has a maximum of 2^n (non-contiguous) regions. Any of these regions must link to at least one element in the highest allowed pyramid level—let this pyramid level consist of 2^n elements. Each element of the highest pyramid level corresponds to one branch of the segmentation tree, and all the leaf nodes of this branch construct one region of the segmented image. The highest level of the segmentation tree must correspond to the expected number of image regions, and the pyramid height defines the maximum number of segmentation branches. If the number of regions in an image is less than 2^n, some regions can be represented by more than one element in the highest pyramid level. If this is the case, some specific processing steps can either allow merging of some elements in the highest pyramid level or can restrict some of these elements to be segmentation branch roots. If the number of image regions is larger than 2^n, the most similar regions will be merged into a single tree branch, and the method will not be able to give acceptable results.

Algorithm 5.21: Split and link to the segmentation tree

1. Define a pyramid data structure with overlapping regions. Evaluate the starting region description.

2. Build a segmentation tree starting with leaves. Link each node of the tree to that one of the four possible parents to which it has the most similar region properties. Build the whole segmentation tree. If there is no link to an element in the higher pyramid level, assign the value zero to this element.

3. Update the pyramid data structure; each element must be assigned the average of the values of all its existing children.

4. Repeat steps 2 and 3 until no significant segmentation changes appear between iterations (a small number of iterations is usually sufficient).

Considerably lower memory requirements can be found in a single-pass split-and-merge segmentation. A local 'splitting pattern' is detected in each 2×2 pixel image block and regions are merged in overlapping blocks of the same size [Suk and Chung 83]. In contrast to previous approaches, a single pass is sufficient here, although a second pass may be necessary for region identification (see Section 6.1). The computation is more efficient and the data structure implemented is very simple; the 12 possible splitting patterns for a 2×2 block are given in a list, starting with a homogeneous block up to a block consisting of four different pixels (see Figure 5.48). Pixel similarity can be evaluated adaptively according to the mean and variance of gray-levels of blocks throughout the image.

Algorithm 5.22: Single-pass split-and-merge

1. Search an entire image line by line except the last column and last line. Perform the following steps for each pixel.

2. Find a splitting pattern for a 2×2 pixel block.

3. If a mismatch between assigned labels and splitting patterns in overlapping blocks is found, try to change the assigned labels of these blocks to remove the mismatch (discussed below).

4. Assign labels to unassigned pixels to match a splitting pattern of the block.

5. Remove small regions if necessary.

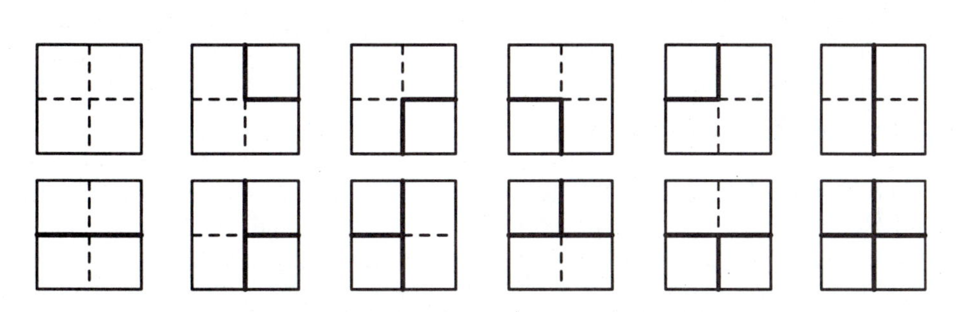

Figure 5.48: *Splitting of 2×2 image blocks, all 12 possible cases.*

The image blocks overlap during the image search. Except for locations at the image borders, three of the four pixels have been assigned a label in previous search locations, but these labels do not necessarily match the splitting pattern found in the processed block. If a mismatch is detected in step 3 of the algorithm, it is necessary to resolve possibilities of merging regions that were considered separate so far—to assign the same label to two regions previously labeled differently. Two regions R_1 and R_2 are merged into a region R_3 if

$$H(R_1 \cup R_2) = \text{TRUE} \tag{5.36}$$

$$|m_1 - m_2| < T \tag{5.37}$$

where m_1 and m_2 are the mean gray-level values in regions R_1 and R_2, and T is some appropriate threshold. If region merging is not allowed, regions keep their previous labels. To get a final segmentation, information about region merging must be stored and the merged-region characteristics must be updated after each merging operation. The assignment of labels to non-labeled pixels in the processed block is based on the block splitting pattern and on the labels of adjacent regions (step 4). If a match between a splitting pattern and the assigned labels was found in step 3, then it is easy to assign a label to the remaining pixel(s) to keep the label assignment and splitting pattern matched. Conversely, if a match was not found in step 3, an unassigned pixel is either merged with an adjacent region (the same label is assigned) or a new region is started. If a 2×2 block size is used, the only applicable pixel property is gray-level. If larger blocks are used, more complex image properties can be included in the homogeneity criteria (even if these larger blocks are divided into 2×2 sub-blocks to determine the splitting pattern).

Many other modifications exist, most of them trying to overcome the segmentation sensitivity to the order in which portions of the image are processed. The ideal solution would be to merge only the single most similar pair of adjacent regions in each iteration, which would result in very slow processing. A method performing the best merge within each of sets of local subimages (possibly overlapping) is described in [Tilton 89]. Another approach insensitive to scanning order is suggested in [Pramotepipop and Cheevasuvit 88].

Hierarchical merging where different criteria are employed at different stages of the segmentation process is discussed in [Goldberg and Zhang 87]. More and more information is incorporated into the merging criteria in later segmentation phases. A modified split-and-merge algorithm where splitting steps are performed with respect to the edge information and merging is based on gray-value statistics of merged regions is introduced in [Cornelis et al. 92, De Becker et al. 92, Deklerck et al. 93]. As splitting is not required to follow a quadtree segmentation pattern, segmentation borders are more natural than borders after the application of standard split-and-merge techniques.

Parallel implementations become more and more affordable, and parallel region growing algorithms may be found in [Mukund and Gonzalez 89, Celenk and Lakshman 89, Willebeek-Lemair and Reeves 90, Chang and Li 95]. Examples of region growing segmentations are given in [Cheevasuvit et al. 86, Cross 88, Laprade 88]. Additional sections describing more sophisticated methods of semantic region growing segmentation can be found in Chapter 8.

5.3.4 Watershed segmentation

The concepts of **watersheds** and **catchment basins** are well known in topography. Watershed lines divide individual catchment basins. The North American Continental Divide is a textbook example of a watershed line with catchment basins formed by the Atlantic and Pacific Oceans. Working with gradient images and following the concept introduced in Chapter 1, Figures 1.3 and 1.4, image data may be interpreted as a topographic surface where the gradient image gray-levels represent altitudes. Thus, region edges correspond to high watersheds and low-gradient region interiors correspond to catchment basins. According to equation (5.31), the goal of region growing segmentation is to create homogeneous regions; in watershed segmentation, catchment basins of the topographic surface are homogeneous in the sense that all pixels belonging to the same catchment basin are connected with the basin's region of minimum altitude (gray-level) by a *simple path* of pixels (Section 2.3.1) that have monotonically decreasing altitude (gray-level) along the path. Such catchment basins then represent the regions of the segmented image (Figure 5.49). While the concept of watersheds and catchment basins is quite straightforward, development of algorithms for watershed segmentation is a complex task, with many of the early methods resulting in either slow or inaccurate execution.

The first algorithms for watershed segmentation were developed for topographic digital elevation models [Collins 75, Puecker and Douglas 75, Marks et al. 84, Band 86, Soille and Ansoult 90]. Most of the existing algorithms start with extraction of potential watershed line pixels using a local 3×3 operation, which are then connected into geomorphological networks in subsequent steps. Due to the local character of the first step, these approaches are often inaccurate [Soille and Ansoult 90].

Somewhat independently, watersheds were investigated in digital image processing. In [Digabel and Lantuejoul 78, Beucher 82, Meyer and Beucher 90], a watershed transformation was introduced in the context of mathematical morphology; details can be found in

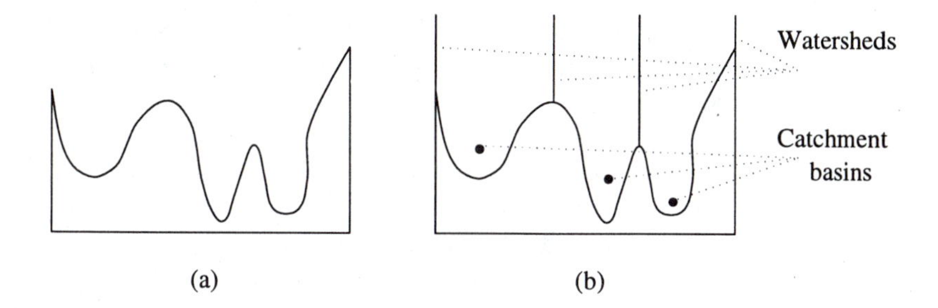

Figure 5.49: *One-dimensional example of watershed segmentation: (a) gray-level profile of image data; (b) watershed segmentation—local minima of gray-level (altitude) yield catchment basins, local maxima define the watershed lines.*

Chapter 11. Unfortunately, without special hardware, watershed transformations based on mathematical morphology are computationally demanding and therefore time consuming.

There are two basic approaches to watershed image segmentation. The first one starts with finding a *downstream* path from each pixel of the image to a local minimum of image surface altitude. A catchment basin is then defined as the set of pixels for which their respective downstream paths all end up in the same altitude minimum. While the downstream paths are easy to determine for continuous altitude surfaces by calculating the local gradients, no rules exist to define the downstream paths uniquely for digital surfaces.

While the given approaches were not efficient because of their extreme computational demands and inaccuracy, the second watershed segmentation approach represented by a seminal paper [Vincent and Soille 91] makes the idea practical. This approach is essentially dual to the first one; instead of identifying the downstream paths, the catchment basins fill from the bottom. As was explained earlier, each minimum represents one catchment basin, and the strategy is to start at the altitude minima. Imagine that there is a hole in each local minimum, and that the topographic surface is immersed in water. As a result, the water starts filling all catchment basins, minima of which are under the water level. If two catchment basins would merge as a result of further immersion, a dam is built all the way to the highest surface altitude and the dam represents the watershed line. An efficient algorithm for such watershed segmentation was presented in [Vincent and Soille 91]. The algorithm is based on *sorting* the pixels in increasing order of their gray values, followed by a *flooding* step consisting of a fast breadth-first scanning of all pixels in the order of their gray-levels.

During the sorting step, a brightness histogram is computed (Section 2.3.2). Simultaneously, a list of pointers to pixels of gray-level h is created and associated with each histogram gray-level to enable direct access to all pixels of any gray-level. Information about the image pixel sorting is used extensively in the flooding step. Suppose the flooding has been completed up to a level (gray-level, altitude) k. Then every pixel having gray-level less than or equal to k has already been assigned a unique catchment basin label. Next, pixels having gray-level $k+1$ must be processed; all such pixels can be found in the list that was prepared in the sorting step—consequently, all these pixels can be accessed directly. A pixel having gray-level $k+1$ may belong to a catchment basin labeled l if at least one of its neighbors already carries this label. Pixels that represent potential catchment basin members are put in a first-in first-out queue and await further processing. Geodesic influence zones are computed for all hitherto determined catchment basins. A geodesic influence zone of a catchment basin l_i is the locus of non-labeled image pixels of gray-level $k+1$ that are contiguous with the catchment basin l_i (contiguous within the region of pixels of gray-level $k+1$) for which their distance to l_i is smaller than their distance to any other catchment basin l_j (Figure 5.50). All pixels with gray-level $k+1$ that belong to the influence zone of a catchment basin labeled l are also labeled with the label l, thus causing the catchment basin to grow. The pixels from the queue are processed sequentially, and all pixels from the queue that cannot be assigned an existing label represent newly discovered catchment basins and are marked with new and unique labels.

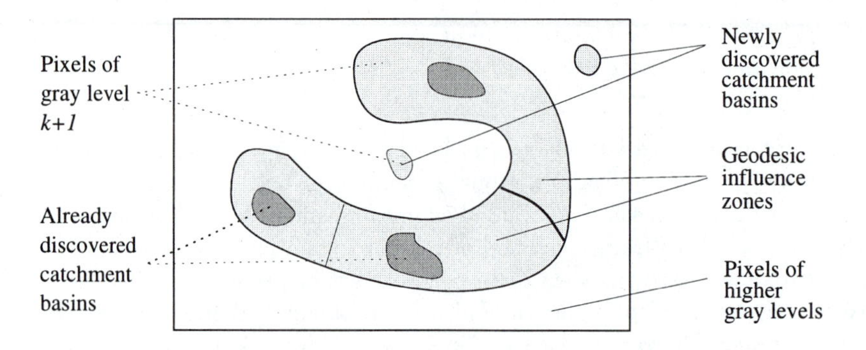

Figure 5.50: *Geodesic influence zones of catchment basins.*

Figure 5.51 shows an example of watershed segmentation. Note that the raw watershed segmentation produces a severely oversegmented image with hundreds or thousands of catchment basins (Figure 5.51c). To overcome this problem, region markers and other approaches have been suggested to generate good segmentation (Figure 5.51d) [Meyer and Beucher 90, Vincent and Soille 91, Higgins and Ojard 93].

While this method would work well in the continuous space with the watershed lines accurately dividing the adjacent catchment basins, the watersheds in images with large plateaus may be quite thick in discrete spaces. Figure 5.52 illustrates such a situation, consisting of pixels equidistant to two catchment basins in 4-connectivity. To avoid such behavior, detailed rules using successively ordered distances stored during the breadth-search process were developed that yield exact watershed lines. Full details, and pseudo-code for a fast watershed algorithm, are in found in [Vincent and Soille 91]; the method was found to be hundreds of times faster than several classical algorithms when using a conventional serial computer, is easily extensible to higher-dimensional images [Higgins and Ojard 93], and is applicable to square or hexagonal grids. Further improvements of the watershed segmentation based on immersion simulations are given in [Dobrin et al. 94].

5.3.5 Region growing post-processing

Images segmented by region growing methods often contain either too many regions (under-growing) or too few regions (over-growing) as a result of non-optimal parameter setting. To improve classification results, a variety of post-processors has been developed. Some of them combine segmentation information obtained from region growing and edge-based segmentation. An approach introduced in [Liow and Pavlidis 88, Pavlidis and Liow 90] solves several quadtree-related region growing problems and incorporates two post-processing steps. First, boundary elimination removes some borders between adjacent regions according to their contrast properties and direction changes along the border, taking resulting topology into consideration. Second, contours from the previous step are modified to be located precisely on appropriate image edges. Post-processing contour relaxation is suggested in [Aach et al. 89]. A combination of independent region growing and edge-based detected borders is described in [Koivunen and Pietikainen 90]. Other approaches combining region growing and edge detection can be found in [Venturi et al. 92, Manos et al. 93, Gambotto 93, Wu 93, Chu and Aggarwal 93, Vlachos and Constantinides 93, Lerner et al. 94, Falah et al. 94, Gevers and Smeulders 97].

Figure 5.51: *Watershed segmentation: (a) original; (b) gradient image, 3×3 Sobel edge detection, histogram equalized; (c) raw watershed segmentation; (d) watershed segmentation using region markers to control oversegmentation. Courtesy W. Higgins, Penn State University.*

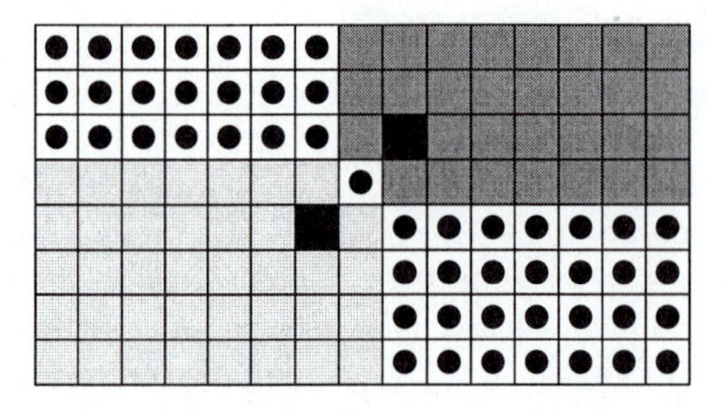

Figure 5.52: *Thick watershed lines may result in gray-level plateaus. Earlier identified catchment basins are marked as black pixels, and new catchment basin additions resulting from this processing step are shown in the two levels of gray. The thick watersheds are marked with •. To avoid thick watersheds, specialized rules must be developed.*

Simpler post-processors are based on general heuristics and decrease the number of small regions in the segmented image that cannot be merged with any adjacent region according to the originally applied homogeneity criteria. These small regions are usually not significant in further processing and can be considered as segmentation noise. It is possible to remove them from the image as follows.

Algorithm 5.23: Removal of small image regions

1. Search for the smallest image region $R_{\rm min}$.
2. Find the adjacent region R most similar to $R_{\rm min}$, according to the homogeneity criteria used. Merge R and $R_{\rm min}$.
3. Repeat steps 1 and 2 until all regions smaller than a pre-selected size are removed from the image.

This algorithm will execute much faster if all regions smaller than a pre-selected size are merged with their neighbors without having to order them by size.

5.4 Matching

Matching is another basic approach to segmentation that can be used to locate known objects in an image, to search for specific patterns, etc. Figure 5.53 shows an example of a desired pattern and its location found in the image. Matching is widely applicable; it can be used to determine stereoscopic scene properties if more than one image of the same scene taken from different locations is available. Matching in dynamic images (e.g., moving cars, clouds, etc.) is another application area. Generally speaking, one image can be used to extract objects or patterns, and directed search is used to look for the same (or similar) patterns in the remaining images. The best match is based on some criterion of optimality which depends on object properties and object relations.

Figure 5.53: *Segmentation by matching; matched pattern and location of the best match.*

Matched patterns can be very small, or they can represent whole objects of interest. While matching is often based on directly comparing gray-level properties of image sub-regions, it can be equally well performed using image-derived features or higher-level image descriptors. In such cases, the matching may become invariant to image transforms. Criteria of optimality can compute anything from simple correlations up to complex approaches of graph matching [Rosenfeld and Kak 82, Ballard and Brown 82].

5.4.1 Matching criteria

Match-based segmentation would be extremely easy if an exact copy of the pattern of interest could be expected in the processed image; however, some part of the pattern is usually corrupted in real images by noise, geometric distortion, occlusion, etc. Therefore, it is not possible to look for an absolute match, and a search for locations of maximum match is more appropriate.

Algorithm 5.24: Match-based segmentation

1. Evaluate a match criterion for each location and rotation of the pattern in the image.
2. Local maxima of this criterion exceeding a preset threshold represent pattern locations in the image.

Matching criteria can be defined in many ways; in particular, correlation between a pattern and the searched image data is a general matching criterion (see Section 2.1.2). Let f be an image to be processed, h be a pattern for which to search, and V be the set of all image pixels in the processed image. The following formulae represent good matching optimality criteria describing a match between f and h located at a position (u, v).

$$C_1(u, v) = \frac{1}{\max_{(i,j)\in V} |f(i + u, j + v) - h(i, j)| + 1} \tag{5.38}$$

$$C_2(u,v) = \frac{1}{\left(\sum_{(i,j)\in V} |f(i+u,j+v) - h(i,j)|\right) + 1} \tag{5.39}$$

$$C_3(u,v) = \frac{1}{\left(\sum_{(i,j)\in V} [f(i+u,j+v) - h(i,j)]^2\right) + 1} \tag{5.40}$$

Whether only those pattern positions entirely within the image are considered, or if partial pattern positions, crossing the image borders, are considered as well, depends on the implementation. A simple example of the C_3 optimality criterion values is given in Figure 5.54 for varying pattern locations—the best matched position is in in the upper left corner. An X-shaped correlation mask was used to detect positions of magnetic resonance markers in [Fisher et al. 91]; the original image and the correlation image are shown in Figure 5.55. The detected markers are further used in heart motion analysis (see Section 15.3.2).

$$\begin{vmatrix} 1 & 1 & 0 & 0 & 0 \\ 1 & 1 & 1 & 0 & 0 \\ 1 & 0 & 1 & 0 & 0 \\ 0 & 0 & 0 & 0 & 0 \\ 0 & 0 & 0 & 0 & 8 \end{vmatrix} \qquad \begin{vmatrix} 1 & 1 & 1 \\ 1 & 1 & 1 \\ 1 & 1 & 1 \end{vmatrix} \qquad \begin{vmatrix} \underline{\mathbf{1/3}} & 1/6 & 1/8 & \times & \times \\ 1/5 & 1/7 & 1/8 & \times & \times \\ 1/8 & 1/9 & 1/57 & \times & \times \\ \times & \times & \times & \times & \times \\ \times & \times & \times & \times & \times \end{vmatrix}$$

(a) (b) (c)

Figure 5.54: *Optimality matching criterion evaluation: (a) image data; (b) matched pattern; (c) values of the optimality criterion C_3 (the best match underlined).*

If a fast, effective Fourier transform algorithm is available, the convolution theorem can be used to evaluate matching. The correlation between a pattern h and image f can be determined by first taking the product of the Fourier transform F of the image f and the complex conjugate of the Fourier transform $H^{\#}$ of the pattern h and then applying the inverse transform (Chapter 12). Note that this approach considers an image to be periodic and therefore a target pattern is allowed to be positioned partially outside an image. To compute the product of Fourier transforms, F and $H^{\#}$ must be of the same size; if a pattern size is smaller, zero-valued lines and columns can be added to inflate it to the appropriate size. Sometimes, it may be better to add non-zero numbers, for example, the average gray-level of processed images can serve the purpose well.

A matching algorithm based on chamfering (see Section 2.3.1) can also be defined to locate features such as known boundaries in edge maps. This is based on the observation that matching features such as lines will produce a very good response in correct, or near correct, positions, but very poor elsewhere, meaning that matching may be very hard to optimize. To see this, consider two straight lines rotating about one another—they have exactly two matching positions (which are both perfect), but the crude match strength of all other positions is negligible. A more robust approach is to consider the total *distance* between the image feature and that for which a match is sought.

Recalling that the chamfer image (see Algorithm 2.1) computes distances from image subsets, we might construct such an image from an edge detection of the image under inspection.

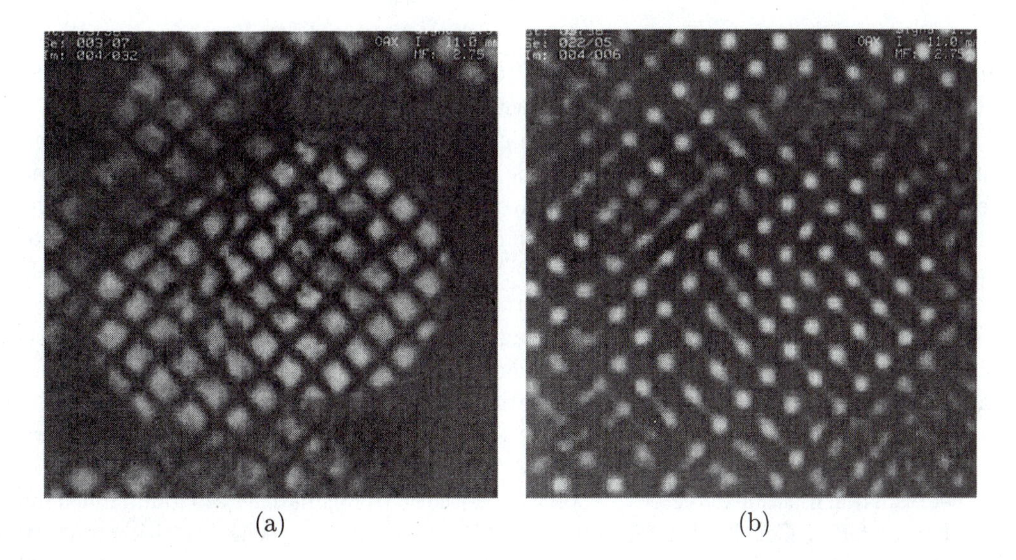

(a) (b)

Figure 5.55: *X-shaped mask matching: (a) original image (see also Figure 15.14); (b) correlation image; the better the local correlation with the X-shaped mask, the brighter the correlation image. Courtesy D. Fisher, S. Collins, The University of Iowa.*

Then, any position of a required boundary can be judged for fit by summing the corresponding pixel values under each of its component edges in a positioning over the image—low values will be good and high poor. Since the chamfering will permit gradual changes in this measure with changes in position, standard optimization techniques (see Section 7.6) can be applied to its movement in search of a best match. Examples of the use of chamfering can be found in [Barrow et al. 77, Gavrila and Davis 95].

5.4.2 Control strategies of matching

Match-based segmentation localizes all image positions at which close copies of the searched pattern are located. These copies must match the pattern in size and orientation, and the geometric distortion must be small. To adapt match-based methods to detect patterns that are rotated, and enlarged or reduced, it would be necessary to consider patterns of all possible sizes and rotations. Another option is to use just one pattern and match an image with all possible geometric transforms of this pattern, and this may work well if some information about the probable geometric distortion is available. Note that there is no difference in principle between these approaches.

However, matching can be used even if an infinite number of transformations are allowed. Let us suppose a pattern consists of parts, these parts being connected by rubber links. Even if a complete match of the whole pattern within an image may be impossible, good matches can often be found between pattern parts and image parts. Good matching locations may not be found in the correct relative positions, and to achieve a better match, the rubber connections between pattern parts must be either pushed or pulled. The final goal can be described as the search for good partial matches of pattern parts in locations that cause minimum force in rubber link connections between these parts. A good strategy is to look for the best partial matches first, followed by a heuristic graph construction of the best combination of these partial matches in which graph nodes represent pattern parts.

Match-based segmentation is time consuming even in the simplest cases with no geometric transformations, but the process can be made faster if a good operation sequence is found. The sequence of match tests must be data driven. Fast testing of image locations with a high probability of match may be the first step; then it is not necessary to test all possible pattern locations. Another speed improvement can be realized if a mismatch can be detected before all the corresponding pixels have been tested.

If a pattern is highly correlated with image data in some specific image location, then typically the correlation of the pattern with image data in some neighborhood of this specific location is good. In other words, the correlation changes slowly around the best matching location. If this is the case, matching can be tested at lower resolution first, looking for an exact match in the neighborhood of good low-resolution matches only.

The mismatch must be detected as soon as possible since mismatches are found much more often than matches. Considering formulae 5.38–5.40, testing in a specified position must stop when the value in the denominator (measure of mismatch) exceeds some preset threshold. This implies that it is better to begin the correlation test in pixels with a high probability of mismatch in order to get a steep growth in the mismatch criterion. This criterion growth will be faster than that produced by an arbitrary pixel order computation.

5.5 Advanced optimal border and surface detection approaches

Several fundamental approaches to edge-based segmentation were presented in Section 5.2. Of them, the concept of optimal border detection (Sections 5.2.4, 5.2.5) is extremely powerful and deserves more attention. In the following sections, two advanced graph-based border detection approaches are introduced. The first of them, the **simultaneous border detection** method, facilitates optimal identification of border pairs by finding a path in a three-dimensional graph. The second, the **optimal surface detection** method, uses multi-dimensional graph search for highly efficient determination of optimal surfaces in three- or higher-dimensional image data.

5.5.1 Simultaneous detection of border pairs

Border detection approaches discussed in Sections 5.2.4 and 5.2.5 identified individual region borders. If the goal is to determine borders of elongated objects, it may be advantageous to search for the pair of left and right borders **simultaneously** [Sonka et al. 93, Sonka et al. 95]. Such an approach facilitates more robust performance if the borders forming the border pair are interrelated, allowing information about one border to help identify the second. Examples include situations in which one border is locally noisy, ambiguous, or uncertain, where identifying borders individually may fail. Following a border of a road or river in a satellite image is an example. As seen in Figure 5.56a, the left and right borders, if considered individually, seem to be reasonable. However, if taken as a pair, it is unlikely that they represent left and right borders of, say, a river. Obviously, there is information contained in the position of one border that might be useful in identifying the position of the other border, and more probable borders may be detected if this is considered (Figure 5.56b).

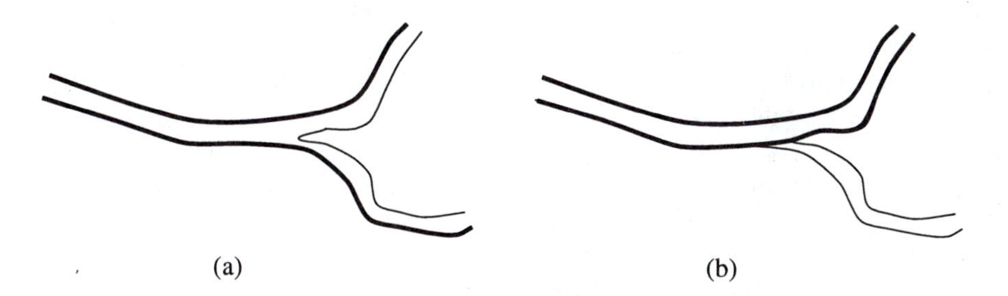

Figure 5.56: *Individual and simultaneous border detection: (a) individually identified borders may not be reasonable as a pair; (b) simultaneously identified borders satisfy border-pair properties.*

To search for an optimal border pair, the graph must be three-dimensional. Shown in Figure 5.57a are two adjacent but independent two-dimensional graphs, nodes in which correspond to pixels in the straightened edge image (Section 5.2.4). The column of nodes separating the left graph and the right graph corresponds to the pixels on the approximate region centerline. A row of nodes in the left graph corresponds to the resampled pixels along a line perpendicular to and left of the region centerline. If we connect nodes in the left graph as shown in Figure 5.57a, the resulting path corresponds to a possible position for the left border of the elongated region. Similarly, linking nodes together in the right graph produces a path corresponding to a possible position of the right region border. If the conventional border detection methods that were described earlier are applied, the 2D graphs would be searched independently to identify optimal left and right region borders.

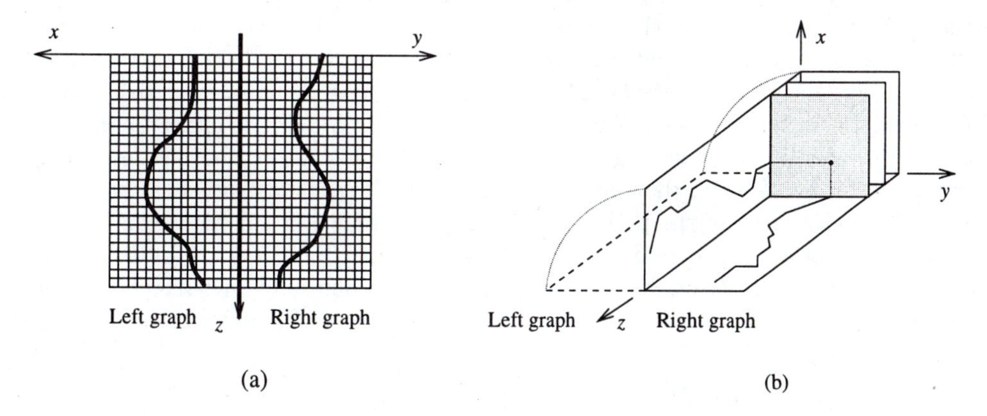

Figure 5.57: *Three-dimensional graph construction: (a) separate identification of the left and right borders by linking nodes in individual two-dimensional graphs corresponding to the left and right halves of the region segment of interest. (b) By rotating up the left graph, a three-dimensional graph results in which paths correspond to pairs of region borders.*

The process of constructing the three-dimensional graph can be visualized as one of rotating up the 2D graph corresponding to the pixels left of the approximate region centerline (Figure 5.57b). The result is a three-dimensional array of nodes in which each node corresponds to possible positions of the left and right region borders for a given point along the length of the elongated region, and a path through the graph corresponds to a possible pair of left and right region borders. Nodes in the 3D graph are referenced by their (x, y, z) co-ordinates; for a point along the region centerline defined by the co-ordinate z, a node with co-ordinates (x_1, y_1, z) corresponds to a left border that is x_1 pixels to the left of the centerline and a right border that is y_1 pixels to the right of the centerline.

As in the 2D case, it is necessary to specify a node successor rule, that is, the rule for linking nodes into complete paths. Since the left border must be continuous, each parent node in the 2D graph corresponding to the left border has three successors as discussed earlier, corresponding to a left border whose distance from the centerline decreases [successor co-ordinate of $(x-1, z+1)$], increases [successor co-ordinate of $(x+1, z+1)$], or stays the same [successor co-ordinate of $(x, z+1)$] as a function of position along the centerline. A similar statement holds for the right border. In the 3D graph, each parent node has nine successors corresponding to the possible combinations of change of positions of the left and right borders with respect to the centerline, thus forming a 3×3 successor window. With this successor rule, all paths through the 3D graph contain one and only one node from each **profile plane** in the 3D graph; that is, every path contains a single node derived from each of the left and right profile lines. This link definition ensures that region borders are continuous in the straightened image space.

Key aspects of the simultaneous approach for accurately identifying region borders are the assignment of costs to pairs of candidate borders and the identification of the optimal pair of region borders or lowest-cost path in the 3D graph. The cost function for a node in the 3D graph is derived by combining the edge costs associated with the corresponding pixels on the left and right profiles in a way that allows the position of the left border to influence the position of the right border and vice versa. This strategy resembles that employed by a human observer in situations where border positions are ambiguous. In designing the cost function, the aim is to discriminate against border pairs that are unlikely to correspond to the true region borders and to identify the border pairs that have the greatest overall probability of matching the actual borders. After the cost function is defined, either heuristic graph searching or dynamic programming methods can be used for optimal border detection.

Similarly to the 2D case, the cost of a path in the 3D graph is defined as the sum of the costs of the nodes forming the path. While many different cost functions can be designed corresponding to the general recommendations given in Section 5.2.4, the following one was found appropriate for description of border properties of a mutually inter-related border pair. Considering the cost minimization scheme, costs are assigned to nodes using the following function:

$$C_{\text{total}}(x, y, z) = [C_s(x, y, z) + C_{pp}(x, y, z)]w(x, y, z) - [P_L(z) + P_R(z)] \tag{5.41}$$

Each of the components of the cost function depends on the edge costs associated with image pixels. The edge costs of the left and right edge candidates located at positions x and y on profile z are inversely related to effective edge strength or other appropriate local border property descriptor $E_L(x, z)$, $E_R(y, z)$ and are given by

$$C_L(x, z) = \max_{x \in X, z \in Z} \{E_L(x, z)\} - E_L(x, z)$$

$$C_R(y,z) = \max_{y\in Y, z\in Z}\{E_R(y,z)\} - E_R(y,z) \tag{5.42}$$

X and Y are sets of integers ranging from 1 to the length of the left and right halves of the region profiles, and Z is the set of integers ranging from 1 to the length of the region centerline. To help avoid detection of regions adjacent to the region of interest, knowledge about the probable direction of the actual border may be incorporated into the local edge property descriptors $E_L(x,z)$, $E_R(y,z)$.

Considering the individual terms of the cost function [equation (5.41)], the term C_s is the sum of the costs for the left and right border candidates and causes the detected borders to *follow* image positions with low cost values. It is given by

$$C_s(x,y,z) = C_L(x,z) + C_R(y,z) \tag{5.43}$$

The C_{pp} term is useful in cases where one border has higher contrast (or other stronger border evidence) than the opposite border and causes the position of the low contrast border to be influenced by the position of the high-contrast border. It is given by

$$C_{pp}(x,y,z) = [C_L(x,z) - P_L(z)][C_R(y,z) - P_R(z)] \tag{5.44}$$

where

$$\begin{aligned} P_L(z) &= \max_{x\in X, z\in Z}\{E_L(x,z)\} - \max_{x\in X}\{E_L(x,z)\} \\ P_R(z) &= \max_{y\in Y, z\in Z}\{E_R(y,z)\} - \max_{y\in Y}\{E_R(y,z)\} \end{aligned} \tag{5.45}$$

Combining equations (5.42), (5.44), and (5.45), the C_{pp} term can also be expressed as

$$C_{pp}(x,y,z) = [\max_{x\in X}\{E_L(x,z)\} - E_L(x,z)][\max_{y\in Y}\{E_R(y,z)\} - E_R(y,z)] \tag{5.46}$$

The $w(x,y,z)$ component of the cost function incorporates a model of the region boundary in a way that causes the positions of the left and right borders to follow certain preferred directions relative to the model. This component has the effect of discriminating against borders that are unlikely to correspond to the actual region borders when considered as a pair. This is accomplished by including a weighting factor that depends on the direction by which a node is reached from its predecessor. For example, if the region is known to be approximately symmetric and its approximate centerline is known, the weighting factor may be given by (Figure 5.58)

$$\begin{aligned} & w(x,y,z) = 1 \quad \text{for} \\ & (x,y) \in \{(\hat{x}-1,\hat{y}-1),(\hat{x},\hat{y}),(\hat{x}+1,\hat{y}+1)\} \\ & w(x,y,z) = \alpha \quad \text{for} \\ & (x,y) \in \{(\hat{x}-1,\hat{y}),(\hat{x}+1,\hat{y}),(\hat{x},\hat{y}-1),(\hat{x},\hat{y}+1)\} \\ & w(x,y,z) = \beta \quad \text{for} \\ & (x,y) \in \{(\hat{x}-1,\hat{y}+1),(\hat{x}+1,\hat{y}-1)\} \end{aligned} \tag{5.47}$$

where the node at co-ordinates (x,y,z) is the successor of the node at $(\hat{x},\hat{y},z-1)$. In this case, the influence of the region model is determined by the values of α and β, typically

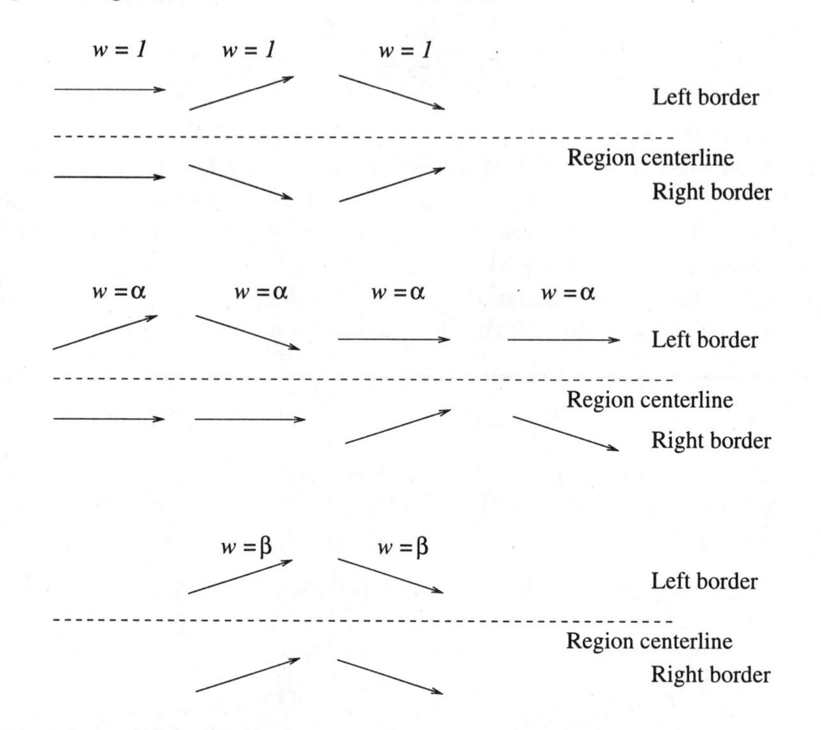

Figure 5.58: *The weighting factors $w(x, y, z)$ associated with local directions of the potential border elements for a symmetric region model.*

$\alpha > \beta$. In coronary border detection applications (Section 16.2.1), the values of α ranged from 1.2 to 1.8 and β from 1.4 to 2.2 [Sonka et al. 95]. The larger the values of α and β, the stronger is the model's influence on the detected borders.

As the number of possible paths in a 3D graph is very large, the identification of the optimal path can be computationally very demanding. For example, for a 3D graph with xyz nodes, where z is the length in pixels of the region centerline, the number of possible paths is approximately 9^z. With conventional border detection, described in Sections 5.2.4 and 5.2.5, the number of possible paths in the two two-dimensional graphs of the same size is about 3^z. Thus, the improvement in border detection accuracy achieved with simultaneous border detection is accomplished at the expense of a substantial increase in computational complexity.

Improving the graph search performance is of great importance, and the $P_L(z) + P_R(z)$ term in the cost function represents the lower-bound heuristic introduced in Section 5.2.4, and does not influence the detected border; it does, however, substantially improve search efficiency if a heuristic graph searching approach is used [Sonka et al. 93].

A second way to increase search efficiency is to use a multi-resolution approach (Section 8.1.5). First, the approximate positions of the region borders are identified in a low-resolution image; these approximate borders are used to guide the full-resolution search by limiting the portion of the full-resolution three-dimensional graph that is searched to find the precise region border positions.

To enhance border detection accuracy, a multi-stage border identification process may also be included. The goal of the first stage is to identify reliably the approximate borders of the region segment of interest while avoiding detection of other structures. Having identified the approximate border positions, the second stage is designed to localize the actual region borders accurately. In the first stage, the 3D simultaneous border detection algorithm is used to identify approximate region borders in a half-resolution image. Since this first stage is designed in part to avoid detection of structures other than the region of interest, a relatively strong region model is used. Region boundaries identified in the low-resolution image are used in the second stage to guide the search for the optimal borders in the full-resolution cost image, as described in the previous paragraph. A somewhat weaker region model may be used in the second stage to allow more influence from the image data (Section 8.1.5). An example of the method's application to simultaneous coronary border detection is given in Section 16.2. Further details about the cost function design can be found in [Sonka et al. 93, Sonka et al. 95].

5.5.2 Surface detection

If three-dimensional volumetric data are available, the task may be to identify three-dimensional surfaces representing object boundaries in the three-dimensional space. This task is common in segmentation of volumetric medical image data sets from magnetic resonance, X-ray, ultrasound, or other tomographic scanners, which produce 3D volumes consisting of stacked 2D image slices. Usually, the 2D images are more or less independently analyzed and the 2D results stacked to form the final 3D segmentation (see Section 16.3 for an example of such an analysis strategy). It is intuitively obvious that a set of 2D borders that were detected in individual slices may be far from optimal if the entire 3D volume is considered, and concurrent analysis of the entire 3D volume may give better results if a globally optimal surface is determined.

Consider an example of brain cortex visualization from three-dimensional magnetic resonance (MR) data sets of a human brain (Figure 5.59). Note that the internal cortex surfaces are not directly visible unless the brain is segmented into the right and left hemispheres. An example of such brain segmentation applied to an individual MR slice was given earlier in Figure 5.37. If the 3D case is considered, the goal is to identify the 3D surface that optimally divides the brain (Figure 5.60).

It is necessary to define a criterion of optimality for the surface. Since it must be contiguous in 3D space, it will consist of a mesh of 3D connected voxels. Consider a 3D graph that corresponds in size with the 3D image data volume; the graph nodes correspond to image voxels. If a cost is associated with each graph node, the optimal surface can be defined as that with the minimum total cost of all *legal* surfaces that can be defined in the 3D volume. The legality of the surface is defined by the 3D surface connectivity requirements that depend on the application at hand, and the total cost associated with a surface can be calculated as the sum of individual costs of all nodes forming the surface. Therefore, it should be possible to determine the optimal surface by application of optimal graph searching principles similar to those presented in Sections 5.2.4 and 5.2.5. Unfortunately, standard graph searching approaches cannot be directly extended from a search for a path to a search for a **surface** [Thedens et al. 95]. Generally, two distinct approaches can be developed to overcome this problem. New graph searching algorithms may be designed to search directly for a surface,

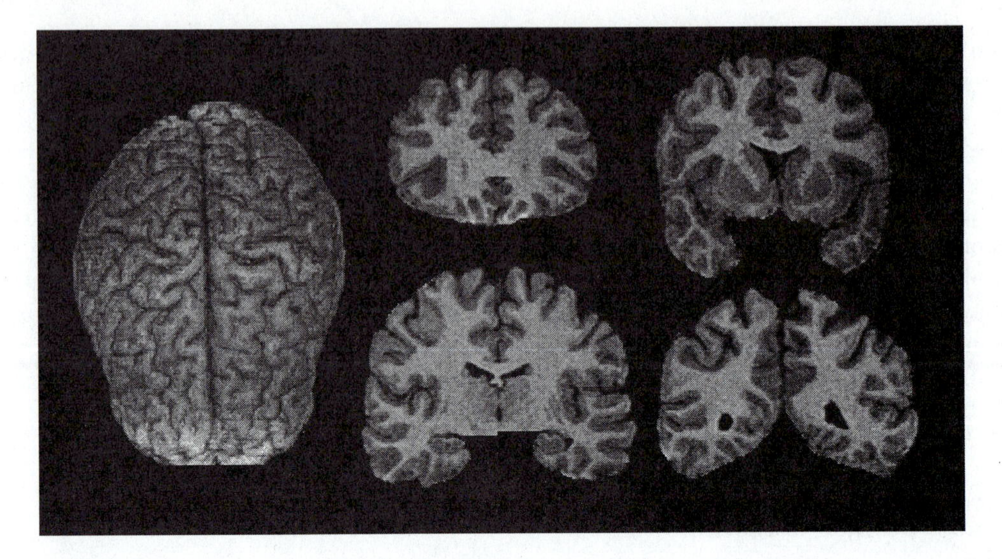

Figure 5.59: *Magnetic resonance images of human brain. Left: Three-dimensional surface rendering of original MR image data after segmentation of the brain from the skull. Right: Four of 120 two-dimensional slices that form the three-dimensional image volume. Courtesy R.J. Frank and H. Damasio, Human Neuroanatomy and Neuroimaging Laboratory, Department of Neurology, The University of Iowa.*

or a surface detection task may be represented in a way that permits conventional graph searching algorithms to be used.

Compared to the search for an optimal path through a graph (even through a 3D graph as shown in Section 5.5.1), the search for an optimal surface results in combinatorial explosion of the task's complexity, and the absence of an efficient searching algorithm has represented a limiting factor on 3D surface detection. One approach to optimal surface detection based on cost minimization in a graph was given in [Thedens et al. 90, Thedens et al. 95]. The method used standard graph searching principles applied to a transformed graph in which standard graph searching for a *path* was used to define a *surface*. While the method guaranteed surface optimality, it was impractical due to its enormous computational requirements. The same authors developed a heuristic approach to surface detection that was computationally feasible [Thedens et al. 95].

Using several ideas from [Thedens et al. 95], a sub-optimal approach to direct detection of surfaces was introduced in [Frank 96]. This approach is based on dynamic programming and avoids the problem of combinatorial explosion by introducing local conditions that must be satisfied by all legal surfaces. The paradigm is called **surface growing**. The graph size corresponds directly to the image size, and due to the local character of surface growing, the graph construction is straightforward and orderly. The entire approach is simple, elegant, computationally efficient, and fast. Additionally, it can be generalized to searching higher-dimensional spaces, e.g., time-variant three-dimensional surfaces. While the resulting surfaces typically represent good solutions, surface optimality is not guaranteed.

Consider a three-dimensional graph search problem with the graph consisting of $X \times Y \times Z$

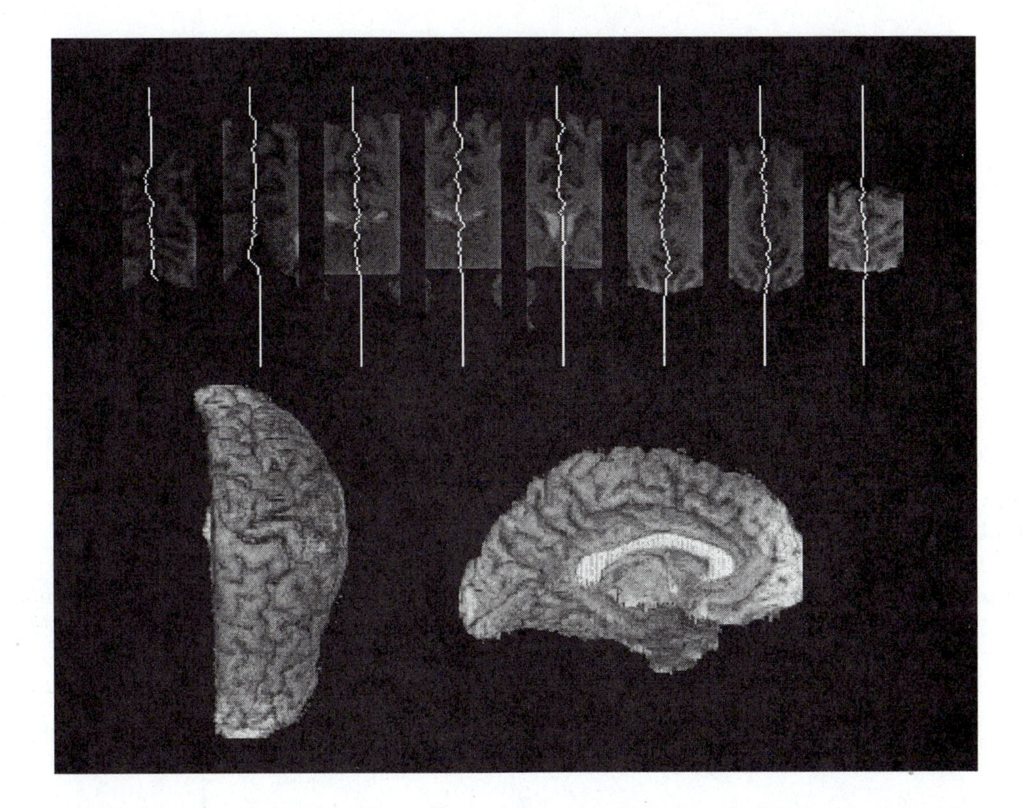

Figure 5.60: *Surface detection. Top: Borders between the left and right hemispheres forming the 3D surface are shown in eight of 120 individual slices. Bottom: After the segmentation in the left and right hemispheres, the internal cortex surfaces may be visualized. Courtesy R.J. Frank and H. Damasio, Human Neuroanatomy and Neuroimaging Laboratory, Department of Neurology, The University of Iowa.*

nodes as shown in Figure 5.61. Let the graph nodes be denoted by co-ordinate triplets (x, y, z) and the three-dimensional surface divide the volume in two parts as indicated in Figure 5.61. Then, the cost S of a surface is defined as

$$S = \sum_{x=1}^{X} \sum_{y=1}^{Y} C[x, y, z(x, y)] \tag{5.48}$$

where $C(x, y, z)$ represents the cost associated with the node (x, y, z) and $z(x, y)$ is a function of the two variables x, y that represents the legal z co-ordinate according to the following 3D surface connectivity constraints

$$\begin{aligned} &\forall (x, y) \in [1, X] \times [1, Y] : \\ &z(x, y) - z(x-1, y)| \leq N \quad \text{AND} \quad |z(x, y) - z(x, y-1)| \leq N \end{aligned} \tag{5.49}$$

The connectivity constraint guarantees surface continuity in 3D. More accurately, the parameter N represents the maximum allowed change in the z co-ordinate of the surface along the

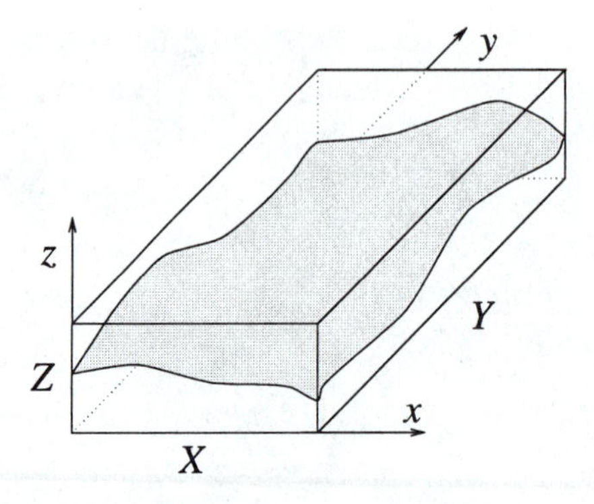

Figure 5.61: *The three-dimensional graph XYZ and the 3D surface dividing the graph into upper and lower parts.*

unit distance in the x and y directions. If N is small, the legal surface is stiff and the stiffness decreases with larger values of N. While N may differ for x and y directions or for different graph parts, a constant value of N will be considered for simplicity. Thus, each internal node of the graph may have $4(2N+1)$ legal neighbors that have to be examined when constructing the 3D graph. The legal neighbors of an internal node are shown in Figure 5.62a.

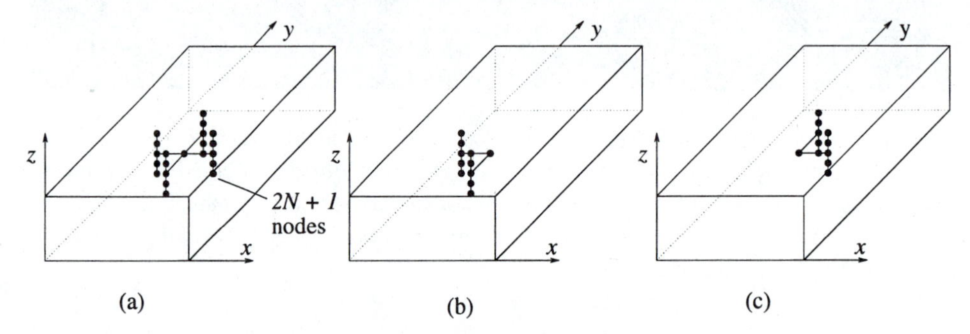

Figure 5.62: *Neighbors of an internal graph node. (a) Each internal node has* $4(2N+1)$ *legal neighbors. (b) Immediate predecessors. (c) Immediate successors.*

The main idea of the 3D graph search is based on the implicitly defined legitimacy of the explored surfaces. For a final surface to be legal, it is necessary to guarantee that all local neighbors are legal. Let the graph be searched starting from the vertical column with co-ordinates $(1,1,z)$ in z–x–y co-ordinate order toward the column (X,Y,z). Remember that the graph is constructed in agreement with the principles of dynamic programming (Section 5.2.5), meaning that a cost $C_{\text{cumulative}}$ is associated with each graph node (x,y,z) representing the minimum cumulative surface cost from the start column $(1,1,z)$ to the examined node (x,y,z). The cumulative surface cost is defined as the sum of the local cost associated with the node (x,y,z) (e.g., the inverted edge strength as discussed in Section 5.2.4) and the sum of the two cost minima identified in the two columns constructed in the 3D graph that represent the immediate predecessors. Considering the internal graph nodes, of the $4(2N+1)$ legal neighbors, one half represents the immediate predecessors and the other half the immediate successors as shown in Figure 5.62b,c.

$$\begin{aligned} C_{\text{cumulative}}(x,y,z) &= C(x,y,z) \\ &+ \min_{k\in[z-N,z+N]}\{C_{\text{cumulative}}(x-1,y,k)\} \\ &+ \min_{k\in[z-N,z+N]}\{C_{\text{cumulative}}(x,y-1,k)\} \end{aligned} \tag{5.50}$$

For any node, non-existent predecessors are not considered. Thus, the cumulative cost of any node reflects the cost of all its predecessors that formed the partial surface from the starting column $(1,1,z)$ to this node.

Recall that the graph is constructed in z–x–y co-ordinate order; therefore, the column of nodes constructed last is the column with co-ordinates (X,Y,z). Consequently, after the entire graph is constructed, the minimum of cumulative costs in the nodes of the column (X,Y,z) represents the globally minimum cost of all legal surfaces connecting the first column $(1,1,z)$ and the last column (X,Y,z) and passing through all the other columns. After determining the minimum cost surface node in the last graph column, the entire surface is specified by determining the back-tracking path by repeated identification of the minimum cost pairs of predecessors in the graph, starting from the last column (X,Y,z) and working backward toward the start column $(1,1,z)$ considering the connectivity constraints. The surface construction proceeds in reverse $z-y-x$ order. Thus, with the exception of the graph borders, for a column (x,y) there are always two nodes located in the columns $(x+1,y)$ and $(x,y+1)$ belonging to the surface already determined. The new node of the resulting surface located in column (x,y) must have a legal connectivity relationship with the surface nodes in columns $(x+1,y)$ and $(x,y+1)$. Propagation of this constraint guarantees the legality of the resulting surface. Therefore, the z co-ordinate of the surface node in the (x,y) column, denoted by $D(x,y)$, is defined as

$$D(x,y) = z \text{ for which } C_{\text{cumulative}}(x,y,z) = \min_{k\in[zmin,zmax]} (C_{\text{cumulative}}(x,y,k)) \tag{5.51}$$

where

$$\begin{aligned} zmax &= \min(Z, D(x+1,y)+N, D(x,y+1)+N) \\ zmin &= \max(1, D(x+1,y)-N, D(x,y+1)-N) \end{aligned} \tag{5.52}$$

The back-tracking process continues until the surface node in the column $(1,1,z)$ is identified.

Algorithm 5.25: Three-dimensional graph searching

1. Create a three-dimensional matrix $(X \times Y \times Z)$ corresponding in size to the 3D image volume.

2. *3D graph construction*: Starting from the column $(1,1,z)$ and proceeding in z–x–y co-ordinate order until the last matrix column (X,Y,z) is reached, calculate the costs of all graph nodes as specified by equation (5.50).

```
for y:=1 to Y do
    for x:=1 to X do
        for z:=1 to Z do
```

$$\begin{aligned} C_{\text{cumulative}}(x,y,z) \;&:= C(x,y,z) \\ &+ \min_{k\in[z-N,z+N]}\{C_{\text{cumulative}}(x-1,y,k)\} \\ &+ \min_{k\in[z-N,z+N]}\{C_{\text{cumulative}}(x,y-1,k)\} \end{aligned}$$

```
        end do
    end do
end do
```

3. *Surface construction*: Starting from the column (X, Y, z) and proceeding in the reverse z–y–x co-ordinate order until the first matrix column $(1, 1, z)$ is reached and considering the connectivity constraints, determine the minimum cumulative cost nodes defining the surface as specified by equations (5.51) and (5.52).

```
for x:=X downto 1 do
    for y:= Y downto 1 do
```

$$\begin{aligned} zmax &:= \min[Z, D(x+1,y)+N, D(x,y+1)+N] \\ zmin &:= \max[1, D(x+1,y)-N, D(x,y+1)-N] \end{aligned}$$

$$D(x,y) := z \text{ for which } C_{\text{cumulative}}(x,y,z) = \min_{k\in[zmin,zmax]} (C_{\text{cumulative}}(x,y,k))$$

```
    end do
end do
```

As can be seen, in the three-dimensional case, the algorithm essentially consists of two nested triple loops. If the dimension of the graph increases to M, the algorithm will consist of two M-deep nested loops. The complexity of the algorithm can be estimated as $(X)(Y)(M)(2N+1)$ additions and comparisons, including the back-tracking stage. Importantly, the complexity increases linearly with the increase of the graph dimension.

The graph searching approach presented above only facilitates plane-like surface detection; however, the 3D surface to be searched often has a cylindrical shape. The method can detect circular surfaces after a straightforward extension. Assuming a circular surface (shown in Figure 5.63), the 3D graph can be constructed by unfolding the circular structure, considering the additional set of links from the set of graph columns (X, y, z) to the columns $(1, y, z)$ to close the surface, and applying the graph searching algorithm 5.25. However, after the surface is identified by back-tracking through the graph, the first and last row of nodes may not satisfy the surface connectivity constraints. While it is possible to consider this constraint in the graph construction, a substantial increase in computational complexity results. There is a practical solution to the problem based on selecting a new cutset of nodes along the y co-ordinate to unfold the circular graph. Although not theoretically guaranteeing success under all circumstances, this approach has demonstrated practical applicability in intravascular ultrasound image sequences [Frank et al. 96].

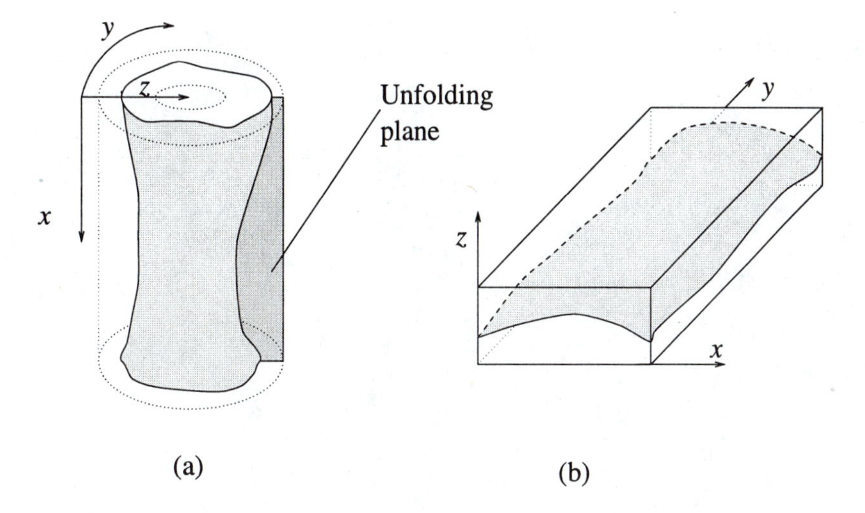

Figure 5.63: *For detection of cylindrical surfaces, the circular graph (image volume) must be unfolded to create a three-dimensional matrix of nodes.*

The three-dimensional graph searching method described was applied to brain cortex segmentation shown in Figures (5.59) and (5.60). The cost function was based on inverted gray-level values of the image voxels after the ventricles were three-dimensionally filled not to represent a large low-cost region. Figures 5.64 and 5.65 show the segmentation of an intravascular ultrasound pull-back sequence. The cost function was based on inverted edge strength and edge direction. The size of the graphs ranged from $100 \times 25 \times 10$ for the intravascular ultrasound image sequences to $256 \times 120 \times 40$ for the MR brain images.

5.6 Summary

- **Image segmentation**
  - The main goal of image segmentation is to divide an image into parts that have a strong correlation with objects or areas of the real world depicted in the image.
  - Segmentation methods can be divided into three groups: **thresholding**, **edge-based** segmentation and **region-based** segmentation.
  - Each region can be represented by its closed boundary, and each closed boundary describes a region.
  - Image data ambiguity is one of the main segmentation problems, often accompanied by information noise.
  - The more a priori information is available to the segmentation process, the better the segmentation results that can be obtained.

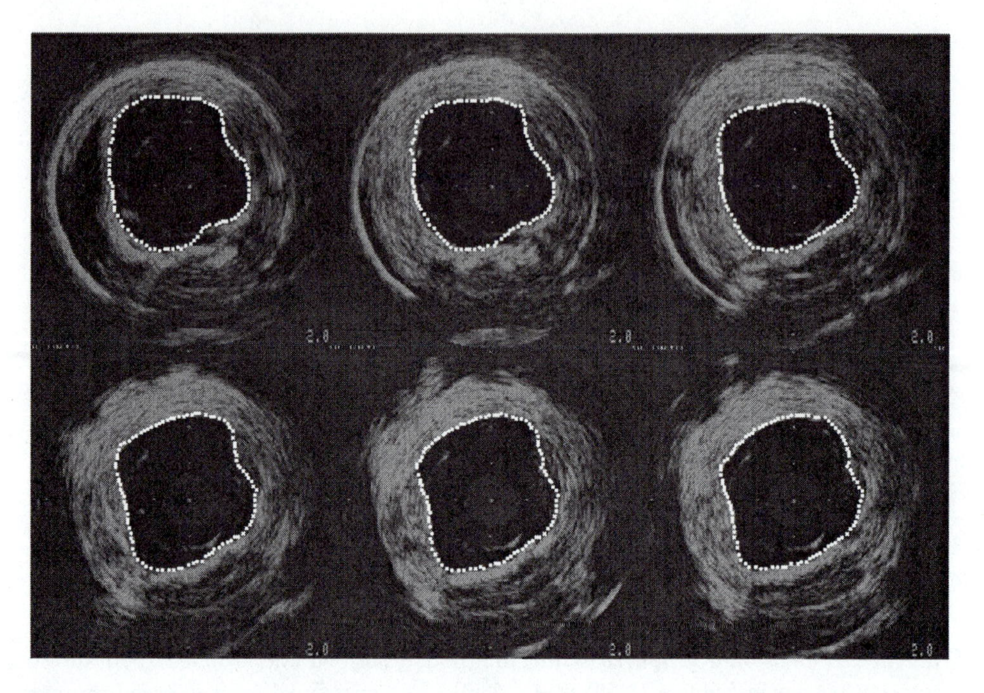

Figure 5.64: *Detection of a cylindrical surface: cross-sectional intravascular ultrasound images of an iliac artery. Overlaid borders show the detected cylindrical surface delineating the vessel lumen. Courtesy R.J. Frank, D.D. McPherson, K.B. Chandran, and E.L. Dove, Department of Neurology, The University of Iowa.*

- **Thresholding**
  - Thresholding represents the simplest image segmentation process, and it is computationally inexpensive and fast. A brightness constant called a **threshold** is used to segment objects and background.
  - Single thresholds can either be applied to the entire image (**global threshold**) or can vary in image parts (**local threshold**). Only under very unusual circumstances can thresholding be successful using a single threshold for the whole image.
  - Many modifications of thresholding exist: **local thresholding**, **band thresholding**, **semi-thresholding**, **multi-thresholding**, etc.
  - **Threshold detection** methods are used to determine the threshold automatically. If some property of an image after segmentation is known a priori, the task of threshold detection is simplified, since the threshold can be selected to ensure that this property is satisfied. Threshold detection can use p-**tile thresholding**, **histogram shape analysis**, **optimal thresholding**, etc.
  - In **bi-modal histograms**, the threshold can be determined as a minimum between the two highest local maxima.
  - **Optimal thresholding** determines the threshold as the closest gray-level corresponding to the minimum probability between the maxima of two or more normal

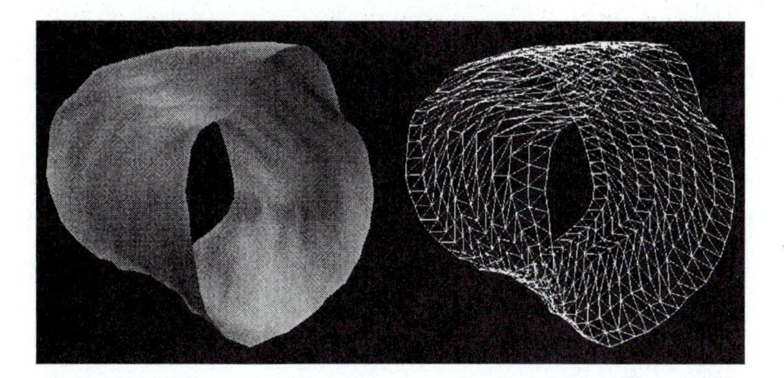

Figure 5.65: *Three-dimensional rendering of the arterial lumen segment, several cross sections of which were shown in Figure 5.64. Shaded surface display of the lumen surface is on the left, the wire-frame representation of the surface is shown on the right. Courtesy R.J. Frank, D.D. McPherson, K.B. Chandran, and E.L. Dove, Department of Neurology, The University of Iowa.*

distributions. Such thresholding results in minimum error segmentation.

- **Multi-spectral thresholding** is appropriate for color or multi-band images.
- Thresholding can be also performed in hierarchical data structures, the aim being to detect the presence of a region in a low-resolution image, and to give the region more precision in images of higher to full resolution.

- **Edge-based image segmentation**
  - Edge-based segmentation relies on edges found in an image by edge detecting operators—these edges mark image locations of discontinuities in gray-level, color, texture, etc.
  - The most common problems of edge-based segmentation, caused by image noise or unsuitable information in an image, are an edge presence in locations where there is no border, and no edge presence where a real border exists.
  - **Edge image thresholding** is based on construction of an edge image that is processed by an appropriate threshold.
  - In **edge relaxation**, edge properties are considered in the context of neighboring edges. If sufficient evidence of the border presence exists, local edge strength increases and vice versa. Using a global relaxation (optimization) process, continuous borders are constructed.
  - Three types of region borders may be formed: **inner**, **outer**, and **extended**. The inner border is always part of a region, but the outer border never is. Therefore, using inner or outer border definition, two adjacent regions never have a common border. Extended borders are defined as single common borders between adjacent regions still being specified by standard pixel co-ordinates.
  - If the criterion of optimality is defined, globally optimal borders can be determined using **(heuristic) graph searching** or **dynamic programming**. Graph-search-based border detection represents an extremely powerful segmentation approach.

The border detection process is transformed into a search for the optimal path in the weighted graph. Costs are associated with each graph node that reflect the likelihood that the border passes through the particular node (pixel). The aim is to find the optimal path (optimal border, with respect to some objective function) that connects two specified nodes or sets of nodes that represent the border's beginning and end.

- **Cost definition** (evaluation functions) is the key to successful border detection. Cost calculation complexity may range from simple inverted edge strength to complex representation of a priori knowledge about the sought borders, segmentation task, image data, etc.
- Graph searching uses Nilsson's **A-algorithm** and guarantees optimality. **Heuristic graph search** may substantially increase search speed, although the heuristics must satisfy additional constraints to guarantee optimality.
- **Dynamic programming** is based on the principle of optimality and presents an efficient way of simultaneously searching for optimal paths from multiple starting and ending points.
- Using the A-algorithm to search a graph, it is not necessary to construct the entire graph since the costs associated with expanded nodes are calculated only if needed. In dynamic programming, a complete graph must be constructed.
- If calculation of the local cost functions is computationally inexpensive, dynamic programming may represent a computationally less demanding choice. However, which of the two graph searching approaches (A-algorithm, dynamic programming) is more efficient for a particular problem depends on the evaluation functions and on the quality of heuristics for an A-algorithm.
- **Hough transform** segmentation is applicable if objects of known shape are to be detected within an image. The Hough transform can detect straight lines and curves (object borders) if their analytic equations are known. It is robust in recognition of occluded and noisy objects.
- The generalized Hough transform can be used if the analytic equations of the searched shapes are not available; the parametric curve (region border) description is based on sample situations and is determined in the learning stage.
- While forming the regions from complete borders is trivial, **region determination from partial borders** may be a very complex task. Region construction may be based on probabilities that pixels are located inside a region closed by the partial borders. Such methods do not always find acceptable regions but they are useful in many practical situations.

- **Region-based image segmentation**
  - **Region growing** segmentation should satisfy the following condition of complete segmentation [equation (5.1)]:

$$R = \bigcup_{i=1}^{S} R_i \qquad R_i \cap R_j = \emptyset \qquad i \neq j$$

and the maximum region homogeneity conditions [equations (5.31), (5.32)]:

$$H(R_i) = \text{TRUE} \qquad i = 1, 2, \dots, S$$

$$H(R_i \cup R_j) = \text{FALSE} \quad i \neq j, \quad R_i \text{ adjacent to } R_j$$

- Three basic approaches to region growing exist: **region merging, region splitting**, and **split-and-merge** region growing.
- **Region merging** starts with an oversegmented image in which regions satisfy equation (5.31). Regions are merged to satisfy condition (5.32) as long as equation (5.31) remains satisfied.
- **Region splitting** is the opposite of region merging. Region splitting begins with an undersegmented image which does not satisfy condition (5.31). Therefore, the existing image regions are sequentially split to satisfy conditions (5.1), (5.31), and (5.32).
- A combination of **splitting and merging** may result in a method with the advantages of both other approaches. Split-and-merge approaches typically use pyramid image representations. Because both split-and-merge processing options are available, the starting segmentation does not have to satisfy either condition (5.31) or (5.32).
- In **watershed** segmentation, catchment basins represent the regions of the segmented image. The first watershed segmentation approach starts with finding a downstream path from each pixel of the image to local minima of image surface altitude. A catchment basin is then defined as the set of pixels for which their respective downstream paths all end up in the same altitude minimum. In the second approach, each gray-level minimum represents one catchment basin and the strategy is to start filling the catchment basins from the bottom.
- Images segmented by region growing methods often contain either too many regions (under-growing) or too few regions (over-growing) as a result of non-optimal parameter setting. To improve classification results, a variety of **post-processors** has been developed. Simpler post-processors decrease the number of small regions in the segmented image. More complex post-processing may combine segmentation information obtained from region growing and edge-based segmentation.

- **Matching**
  - Matching can be used to locate objects of known appearance in an image, to search for specific patterns, etc. The best match is based on some criterion of optimality which depends on object properties and object relations.
  - Matching criteria can be defined in many ways; in particular, correlation between a pattern and the searched image data is often used as a general matching criterion.
  - Chamfer matching may be used to locate one-dimensional features that might otherwise defeat cost-based optimality approaches.

- **Advanced optimal border and surface detection**
  - **Simultaneous border detection** facilitates optimal identification of border pairs by finding an optimal path in a three-dimensional graph. It is based on the observation that there is information contained in the position of one border that might be useful in identifying the position of the other border. After a cost function that combines edge information from the left and right borders has been defined, either heuristic graph searching or dynamic programming methods can be used for optimal border detection.
  - **Optimal surface detection** uses multi-dimensional graph search for highly efficient determination of optimal surfaces in three- or higher-dimensional image data. The method called **surface growing** is based on dynamic programming and avoids the problem of combinatorial explosion by introducing local conditions that must be satisfied by all legal surfaces.

5.7 Exercises

Short-answer questions

1. Consider two gray-level images with a bi-modal histogram depicting several objects on background. The histogram of the first image is wide and shallow, the histogram of the second image is narrow and deep. How is the error in the area of segmented objects affected by the choice of threshold? In which image is the area of segmented objects more sensitive to threshold selection?
2. Does the bi-modality of the image histogram guarantee successful image segmentation if threshold-based image segmentation is applied and the threshold is selected in the minimum between the two histogram maxima?
3. What is the most typical problem of edge-based segmentation?
4. Explain the main concept of border detection using edge relaxation.
5. List the main advantages and disadvantages of inner, outer, and extended border definitions.
6. Consider a heuristic graph searching algorithm. Explain why the accuracy of estimate $\tilde{h}(\mathbf{x}_i)$ of the path cost from the current node $\mathbf{x}_i$ to the end node influences the search behavior. Specify under which conditions the optimality of the search is guaranteed. Describe how overestimation, underestimation, and exact estimation influence the path detection speed.
7. Explain why the lower-bound method for modification of node costs increases the A-algorithm graph search speed.
8. What is the main idea of the principle of optimality? How is it used in dynamic programming?
9. Which one of the live wire and live lane border detection methods relies more on operator interaction?
10. Using a polar representation, explain the main concept of the Hough transform for line detection. Draw several lines in the image space and sketch the corresponding Hough transforms in the parameter space. Label all important points, lines, and axes.

11. Explain, why the polar co-ordinate representation of lines

$$s = x \cos \theta + y \sin \theta$$

is more suitable for line detection using the Hough transform than the standard line representation

$$y = kx + q$$

12. Explain why a priori information about edge directions may increase the speed of Hough transform-based image segmentation.
13. Explain the main conceptual differences in edge-based and region-based approaches to image segmentation. Are these two approaches dual? If an edge-based and a region-based segmentation method are applied to the same image data, will the resulting segmentation be identical? Consider ideal noise-free image data and real-world digital images.
14. Specify mathematically the goal of region-based segmentation using the criterion of region homogeneity.
15. Explain the principles of and differences among the three basic approaches to region growing—merging, splitting, and split-and-merge.
16. Explain the main principles of watershed segmentation. Discuss why filling the catchment basins from the bottom is several orders of magnitude faster than the approaches using mathematical morphology.
17. Explain why watershed segmentation tends to oversegment images.
18. Explain why a strategy of fast determination of a mismatch can typically speed up the process of image matching more than a strategy of increasing the efficiency of proving a match.
19. Explain how an optimal pair of borders can be simultaneously detected using graph searching. How is the cost of a pair of border points determined from the underlying image data?
20. Explain why optimal surface detection results in combinatorial explosion of the search space.

Problems

1. Testing the functionality of image segmentation techniques and comparing their accuracy requires us to work with images for which the correct segmentation is known. Develop some test images by:

   - Create an image SEG1 containing artificial objects on a background of constant gray-level. Generate simple geometric objects such as squares, rectangles, diamonds, stars, circles, etc. each having a constant gray-level different from that of the background, some of them darker and some brighter than the background. Determine the area for each object and store it in an appropriate form.
   - Superimpose additive Gaussian noise with a given standard deviation, thus creating an image SEG2.
   - Superimpose random impulse noise of a given severity over the image SEG2, thus creating an image SEG3.

   By varying the shapes of objects, standard deviation of the Gaussian additive noise, and severity of the impulse noise, sets of images of controlled properties can be generated. To create a simple set of images for segmentation experiments, make a single image SEG1, apply three levels of Gaussian additive noise, and three levels of impulse noise. You will obtain a set of ten images that will be used in the segmentation problems below.

2. To assess the correctness of a segmentation, a set of measures must be developed to allow quantitative comparisons among methods. Develop a program for calculating the following two segmentation accuracy indices:

   - *Relative signed area error* A_{error} is expressed in percent and computed as

   $$A_{\text{error}} = \frac{\sum_{i=1}^{N} T_i - \sum_{j=1}^{M} A_j}{\sum_{i=1}^{N} T_i} \times 100$$

   where T_i is the true area of the ith object and A_i is the measured area of the jth object, N is the number of objects in the image, M is the number of objects after segmentation. Areas may be expressed in pixels.

   - *Labeling error* L_{error} is defined as a ratio of the number of incorrectly labeled pixels (object pixels labeled as background and vice versa) and the number of pixels of true objects $\sum_{i=1}^{N} T_i$ according to a priori knowledge, and is expressed in percent.

3. Implement the following methods for threshold-based segmentation and apply to the test images created in Problem 5.1. Determine the thresholds manually by trial and error. For each method and each image, quantitatively assess the segmentation accuracy using the indices developed in Problem 5.2. Compare the segmentation accuracy for individual methods.

   (a) Basic thresholding

   (b) Adaptive thresholding

   (c) Band thresholding

   (d) Multiple thresholding

   (e) Semithresholding

   (f) Adaptive thresholding

4. Implement the following threshold selection methods and apply the thresholds so determined to the test images created in Problem 5.1. Use basic thresholding. For each method and each image, quantitatively assess the segmentation accuracy using the indices developed in Problem 5.2. Compare the segmentation accuracy for individual methods. Use these approaches:

   (a) Mode method

   (b) p-tile method

   (c) Iterative optimal threshold selection method

5. Repeat Problem 5.4 using images of handwritten or printed text.

6. Implement adaptive thresholding and apply it to the test images created in Problem 5.1. Use the mode method for threshold selection. For each method and each image, quantitatively assess the segmentation accuracy using the indices developed in Problem 5.2.

7. Let an image be known to consist of objects and background. Let the mean gray-level and variance of the background be $\mu_1 = 30$, $\sigma_1^2 = 900$, and the mean gray-level and variance of the objects be $\mu_2 = 90$, $\sigma_2^2 = 400$. For both cases given below, sketch the gray-level histogram and determine the optimal threshold for segmenting the objects from background. Assume that the objects occupy the following percentage of the image area:

   (a) 30%

   (b) 60%

8. Implement recursive multi-spectral thresholding and apply it to color images represented by three RGB bands.

9. Repeat Problem 5.4 using edge images as input, the aim being to form the thresholded images of significant edges.

10. Implement Algorithms 5.5 and 5.6, and run them on a test set of images created in Problem 5.1 for a range of values of hysteresis thresholds t_0 and t_1. Make a subjective judgment about the best values, then try to relate these to properties of the distribution of edge magnitudes in the image (for example, does it help to select either threshold such that a given percentage of edges have higher or lower magnitudes?).

11. Develop a program for border detection using edge relaxation. Implement two modifications of the method—the first using edge confidence computation as described by equations (5.14) and (5.15), the second using equations (5.14), (5.15), and (5.16). By trial and error, find the best parameters of the method. Compare the resulting edge images with each other and with the edge images obtained in Problem 5.9.

12. Mark all pixels forming the

    (a) Outer

    (b) Extended

    border for the object given in Figure 5.66. If any pixel occurs more than once in the border, mark it more than once.

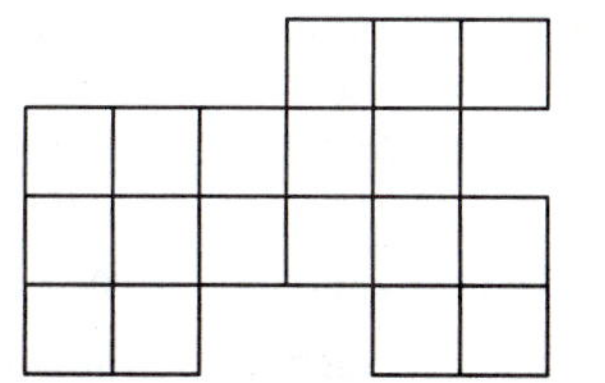

Figure 5.66: *Problems 5.12 and 5.27.*

13. Using 4- and/or 8-connectivity, develop a program for inner border tracing that finds borders of objects as well as borders of object holes. Test the program on a variety of binary images and on the segmentation results obtained in the previous problems. Make sure the program works well for single pixel objects and single-pixel-wide objects.

14. Modify the program developed in Problem 5.13 to determine outer object borders.

15. Develop a program for extended border tracing. Test in the same images as were used in Problem 5.13.

16. Using A-algorithm graph search, find the optimal path through the graph given in Figure 5.67, numbers within the circles specify node costs; consider three possible node successors. Show all steps of your graph searching, including the status of all associated data structures during each step.

17. Assume that an optimal border is detected using A-algorithm graph searching in a rectangular graph, with costs associated with each node. Let the optimal border be specified as the minimum-cost path through the graph. Discuss how changes in node costs specified below influence the resulting border (whether the same optimal border results or not) and what happens with the number of graph nodes expanded during the graph search (whether it remains

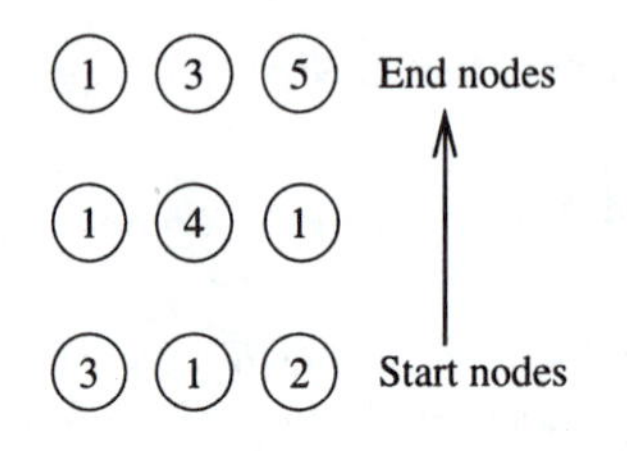

Figure 5.67: *Problem 5.16.*

the same, or becomes smaller or greater). Consider the following modifications of the graph node costs:

(a) The costs of all nodes are increased by a constant.

(b) The costs of all nodes are decreased by a constant but no cost becomes negative.

(c) The costs of all nodes are decreased by a constant and some costs become negative.

(d) The costs of one or more (but not all) graph profiles are increased by a constant.

(e) The costs of one or more (but not all) graph profiles are decreased by a constant but no cost becomes negative.

(f) The costs of one or more (but not all) graph profiles are decreased by a constant and some costs become negative.

18. Using dynamic programming, find the optimal path through the graph given in Figure 5.67 (numbers within the circles specify node costs); consider three possible node successors. Show all steps of your graph searching, including the status of all associated data structures during each step.
19. Repeat Problem 5.17 considering graph searching based on dynamic programming.
20. Consider the main differences between A-algorithm graph searching and dynamic programming. Explain why dynamic programming is often faster than the A-algorithm.
21. Develop programs for border detection using the following graph searching methods. Use the programs to determine low-quality borders in noisy images. If you want to avoid edge image warping to obtain the straightened edge image, you may consider detection of (approximately) straight borders of objects in the images. Implement:

    (a) The A-algorithm

    (b) A heuristic graph searching method [non-zero $\tilde{h}(\mathbf{x}_i)$]

    (c) Dynamic programming

22. When using the Hough transform for line detection, it is advantageous to represent the lines in polar co-ordinates. Develop a method for conversion from slope–intercept line description to polar co-ordinates description.
23. Consider a straight line formed by the following noisy line points in the image space: (0,0), (1,1), (2,2), (3,3), and (2,4). Find the straight line using the Hough transform. Show all your work.
24. Using the polar co-ordinate representation of lines, find the Hough transform of line objects shown in Figure 5.68 (show the Hough space separately for each object).
25. What is the dimensionality of a Hough accumulator array if circles of all sizes are to be determined in a two-dimensional image? List all necessary accumulator array variables.

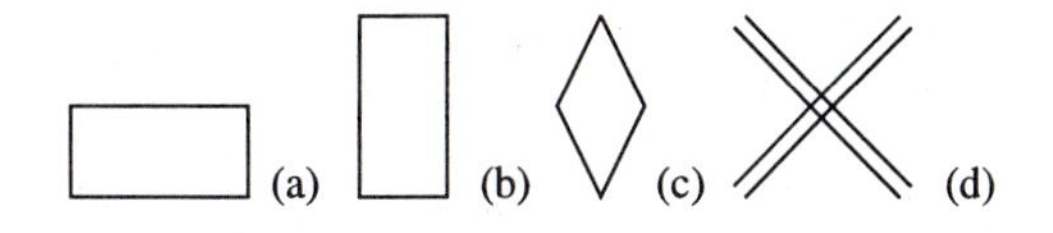

Figure 5.68: *Problem 5.24.*

26. Develop a program for Hough transform-based segmentation that finds circles of all sizes in the images created in Problem 5.1. To simplify the problem, you can use a priori knowledge about the number of circles to be identified in the image.

27. Construct an R-table for the object given in Figure 5.66. Let the third pixel from the left in the second row from the top represent the reference point.

28. In Problems 5.9 and 5.11, edge-based segmentation may not have produced closed borders. To achieve complete segmentation, implement a method for region forming from partial borders. For each resulting image segmentation, quantitatively assess the segmentation accuracy using the indices developed in Problem 5.2.

29. Design three criteria of homogeneity suitable for region growing-based segmentation of the images created in Problem 5.1.

30. Implement the following approaches to region growing and apply to the test images created in Problem 5.1. For each method and each image, quantitatively assess the segmentation accuracy using the indices developed in Problem 5.2. Compare the segmentation accuracy for individual methods.

    (a) Region merging
    (b) Region splitting
    (c) Split-and-merge
    (d) Region merging via boundary melting
    (e) Split and link to the segmentation tree
    (f) Single pass split-and-merge

31. Implement watershed segmentation and apply it to the test images created in Problem 5.1. Design a strategy to avoid severe oversegmentation of noisy images. For each image, quantitatively assess the segmentation accuracy using the indices developed in Problem 5.2.

32. Implement a method for removal of small regions in a region-growing post-processing step. Apply it to the images resulting from application of the methods developed in Problems 5.28 and 5.30.

33. In the previous examples, you have developed a variety of image segmentation methods. You have quantitatively evaluated their segmentation accuracy in the same set of test images. Using the previously achieved results, rank the implemented image segmentation methods. (If you did not implement all of the methods, rank the subset available).

34. In the previous problems, objects were formed from regions of a constant gray-level. Now consider a situation when the objects are acquired under non-homogeneous light conditions. You can model this sort of lighting by superimposing a gray-scale wedge over the images created in Problem 5.1. Select one of the previously developed approaches or design a new segmentation approach that would successfully segment objects from such images. For each image, quantitatively assess the segmentation accuracy using the indices developed in Problem 5.2.

35. Design three image matching criteria that are different from those given in the text and are computationally efficient.

36. Develop a simple artificial example demonstrating that the surface detection approach described in Section 5.5.2 may determine a non-optimal surface.

37. Implement the surface detection Algorithm 5.25 and show its applicability to 3D surface detection despite its lack of optimality.

5.8 References

[Aach et al. 89] T Aach, U Franke, and R Mester. Top-down image segmentation using object detection and contour relaxation. In *Proceedings—ICASSP, IEEE International Conference on Acoustics, Speech and Signal Processing,* Glasgow, Scotland, volume III, pages 1703–1706, IEEE, Piscataway, NJ, 1989.

[Adams and Bischof 94] R Adams and L Bischof. Seeded region growing. *IEEE Transactions on Pattern Analysis and Machine Intelligence*, 16:641–647, 1994.

[Ballard 81] D H Ballard. Generalizing the Hough transform to detect arbitrary shapes. *Pattern Recognition*, 13:111–122, 1981.

[Ballard and Brown 82] D H Ballard and C M Brown. *Computer Vision*. Prentice-Hall, Englewood Cliffs, NJ, 1982.

[Band 86] L E Band. Topographical partition of watersheds with digital elevation models. *Water Resources Research*, 22(1):15–24, 1986.

[Baraldi and Parmiggiani 96] A Baraldi and F Parmiggiani. Single linkage region growing algorithms based on the vector degree of match. *IEEE Transactions on Geoscience and Remote Sensing*, 34:137–148, 1996.

[Barrett and Mortensen 96] W A Barrett and E N Mortensen. Fast, accurate, and reproducible live-wire boundary extraction. In *Visualization in Biomedical Computing*, pages 183–192, Springer Verlag, Berlin, Heidelberg, 1996.

[Barrow et al. 77] H G Barrow, J M Tenebaum, R C Bolles, and H C Wolf. Parametric correspondence and chamfer matching: Two new techniques for image matching. In *5th International Joint Conference on Artificial Intelligence,* Cambridge, CA, pages 659–663. Carnegie-Mellon University, 1977.

[Baugher and Rosenfeld 86] E S Baugher and A Rosenfeld. Boundary localization in an image pyramid. *Pattern Recognition*, 19(5):373–396, 1986.

[Beghdadi et al. 95] A Beghdadi, A L Negrate, and P V de Lesegno. Entropic thresholding using a block source model. *Graphical Models and Image Processing*, 57:197–205, 1995.

[Bellmann 57] R Bellmann. *Dynamic Programming*. Princeton University Press, Princeton, NJ, 1957.

[Beucher 82] S Beucher. Watersheds of functions and picture segmentation. In *Proceedings IEEE International Conference Accoustics, Speech, and Signal Processing,* Paris, France, pages 1928–1931, IEEE, Los Alamitos, CA, 1982.

[Brice and Fennema 70] C R Brice and C L Fennema. Scene analysis using regions. *Artificial Intelligence*, 1:205–226, 1970.

[Brink 95] A D Brink. Minimum spatial entropy threshold selection. *IEE Proceedings Vision, Image and Signal Processing*, 142:128–132, 1995.

[Bruel 88] E Bruel. Precision of line following in digital images. PhD thesis, ETN-89-93329, Technische University, Delft, The Netherlands, 1988.

[Brummer 91] M E Brummer. Hough transform detection of the longitudinal fissure in tomographic head images. *IEEE Transactions on Medical Imaging*, 10(1):74–81, 1991.

[Canny 86] J F Canny. A computational approach to edge detection. *IEEE Transactions on Pattern Analysis and Machine Intelligence*, 8(6):679–698, 1986.

[Celenk and Lakshman 89] M Celenk and P Lakshman. Parallel implementation of the split and merge algorithm on hypercube processors for object detection and recognition. In *Applications of Artificial Intelligence VII; Proceedings of the Meeting,* Orlando, FL, pages 251–262, Society of Photo-Optical Instrumentation Engineers, Bellingham, WA, 1989.

[Chandran and Davis 89] S Chandran and L S Davis. Parallel vision algorithms—an approach. In *Parallel Processing for Scientific Computing; Proceedings of the Third SIAM Conference,* Los Angeles, CA, pages 235–249, Society for Industrial and Applied Mathematics, Philadelphia, PA, 1989.

[Chang and Li 94] Y L Chang and X Li. Adaptive image region-growing. *IEEE Transactions on Image Processing*, 3:868–872, 1994.

[Chang and Li 95] Y L Chang and X Li. Fast image region growing. *Image and Vision Computing*, 13:559–571, 1995.

[Cheevasuvit et al. 86] F Cheevasuvit, H Maitre, and D Vidal-Madjar. A robust method for picture segmentation based on a split-and-merge procedure. *Computer Vision, Graphics, and Image Processing*, 34:268–281, 1986.

[Chen et al. 91] S Y Chen, W C Lin, and C T Chen. Split-and-merge image segmentation based on localized feature analysis and statistical tests. *CVGIP - Graphical Models and Image Processing*, 53(5):457–475, 1991.

[Chien and Fu 74] Y P Chien and K S Fu. A decision function method for boundary detection. *Computer Graphics and Image Processing*, 2:125–140, 1974.

[Cho et al. 89] S Cho, R Haralick, and S Yi. Improvement of Kittler and Illingworth's minimum error thresholding. *Pattern Recognition*, 22(5):609–617, 1989.

[Chow and Kaneko 72] C K Chow and T Kaneko. Automatic boundary detection of the left ventricle from cineangiograms. *Computers in Biomedical Research*, 5:388–410, 1972.

[Chu and Aggarwal 93] C C Chu and J K Aggarwal. The integration of image segmentation maps using region and edge information. *IEEE Transactions on Pattern Analysis and Machine Intelligence*, 15:1241–1252, 1993.

[Chung and Lin 95] K L Chung and H Y Lin. Hough transform on reconfigurable meshes. *Computer Vision and Image Understanding*, 61:278–284, 1995.

[Clark 91] D Clark. Image edge relaxation on a hypercube. Technical Report Project 55:295, University of Iowa, 1991.

[Collins 75] S H Collins. Terrain parameters directly from a digital terrain model. *Canadian Surveyor*, 29(5):507–518, 1975.

[Collins and Skorton 86] S M Collins and D J Skorton. *Cardiac Imaging and Image Processing.* McGraw-Hill, New York, 1986.

[Collins et al. 91] S M Collins, C J Wilbricht, S R Fleagle, S Tadikonda, and M D Winniford. An automated method for simultaneous detection of left and right coronary borders. In *Computers in Cardiology 1990,* Chicago, IL, page 7, IEEE, Los Alamitos, CA, 1991.

[Cornelis et al. 92] J Cornelis, J De Becker, M Bister, C Vanhove, G Demonceau, and A Cornelis. Techniques for cardiac image segmentation. In *Proceedings of the 14th IEEE EMBS Conference, Vol. 14,* Paris, France, pages 1906–1908, IEEE, Piscataway, NJ, 1992.

[Cristi 88] R Cristi. Application of Markov random fields to smoothing and segmentation of noisy pictures. In *Proceedings—ICASSP, IEEE International Conference on Acoustics, Speech and Signal Processing 1988, New York, NY,* pages 1144–1147, IEEE, New York, 1988.

[Cross 88] A M Cross. Segmentation of remotely-sensed images by a split-and-merge process. *International Journal of Remote Sensing,* 9:1329–1345, 1988.

[Davis 82] L S Davis. Hierarchical generalized Hough transforms and line segment based generalized Hough transforms. *Pattern Recognition,* 15(4):277–285, 1982.

[De Becker et al. 92] J De Becker, M Bister, N Langloh, C Vanhove, G Demonceau, and J Cornelis. A split-and-merge algorithm for the segmentation of 2-D, 3-D, 4-D cardiac images. In *Proceedings of the IEEE Satellite Symposium on 3D Advanced Image Processing in Medicine,* Rennes, France, pages 185–189, IEEE, Piscataway, NJ, 1992.

[Degunst 90] M E Degunst. Automatic extraction of roads from SPOT images. PhD thesis, ETN-91-99417, Technische University, Delft, The Netherlands, 1990.

[Deklerck et al. 93] R Deklerck, J Cornelis, and M Bister. Segmentation of medical images. *Image and Vision Computing,* 11:486–503, 1993.

[Demi 96] M Demi. Contour tracking by enhancing corners and junctions. *Computer Vision and Image Understanding,* 63:118–134, 1996.

[Derin and Elliot 87] H Derin and H Elliot. Modelling and segmentation of noisy and textured images using Gibbs random fields. *IEEE Transactions on Pattern Analysis and Machine Intelligence,* 9(1):39–55, 1987.

[Digabel and Lantuejoul 78] H Digabel and C Lantuejoul. Iterative algorithms. In J L Chermant, editor, *Proceedings of 2nd European Symposium Quantitative Analysis of Microstructures in Material Science, Biology, and Medicine,* Caen, France, 1977, pages 85–99, Riederer Verlag, Stuttgart, 1978.

[Dobrin et al. 94] B P Dobrin, T Viero, and M Gabbouj. Fast watershed algorithms: Analysis and extensions. In *Proceedings of the SPIE Vol. 2180,* pages 209–220, SPIE, Bellingham, WA, 1994.

[Ducksbury 90] P G Ducksbury. Parallelisation of a dynamic programming algorithm suitable for feature detection. Technical report, RSRE-MEMO-4349; BR113300; ETN-90-97521, Royal Signals and Radar Establishment, Malvern, England, 1990.

[Duda and Hart 72] R O Duda and P E Hart. Using the Hough transforms to detect lines and curves in pictures. *Communications of the ACM,* 15(1):11–15, 1972.

[Duda and Hart 73] R O Duda and P E Hart. *Pattern Classification and Scene Analysis.* Wiley, New York, 1973.

[Dudani 76] S A Dudani. Region extraction using boundary following. In C H Chen, editor, *Pattern Recognition and Artificial Intelligence,* pages 216–232. Academic Press, New York, 1976.

[Evans 85] F Evans. Survey and comparison of the Hough transform. In *IEEE Computer Society Workshop on Computer Architecture for Pattern Analysis and Image Database Management 1985,* Miami Beach, FL, pages 378–380, IEEE, New York, 1985.

[Falah et al. 94] R K Falah, P Bolon, and J P Cocquerez. A region-region and region-edge cooperative approach of image segmentation. In *Proceedings of the IEEE International Conference on Image Processing,* Austin, TX, pages 470–474, IEEE, Los Alamitos, CA, 1994.

[Falcao et al. 95] A X Falcao, J K Udupa, S Samarasekera, S Sharma, B E Hirsch, and R A Lotufo. User-steered image segmentation paradigms: Live wire and live lane. Technical Report MIPG213, Deptartment of Radiology, University of Pennsylvania., 1995.

[Fetterer et al. 90] F M Fetterer, A E Pressman, and R L Crout. Sea ice lead statistics from satellite imagery of the Lincoln Sea during the iceshelf acoustic exercise. Technical report, AD-A228735; NOARL-TN-50, Naval Oceanographic and Atmospheric Research Laboratory, Bay Saint Louis, MS, Spring 1990.

[Fisher et al. 91] D J Fisher, J C Ehrhardt, and S M Collins. Automated detection of noninvasive magnetic resonance markers. In *Computers in Cardiology,* Chicago, IL, pages 493–496, IEEE, Los Alamitos, CA, 1991.

[Fleagle et al. 89] S R Fleagle, M R Johnson, C J Wilbricht, D J Skorton, R F Wilson, C W White, M L Marcus, and S M Collins. Automated analysis of coronary arterial morphology in cineangiograms: Geometric and physiologic validation in humans. *IEEE Transactions on Medical Imaging*, 8(4):387–400, 1989.

[Flynn 72] M J Flynn. Some computer organizations and their effectivness. *IEEE Transactions on Computers*, 21(9):948–960, 1972.

[Frank 96] R J Frank. Optimal surface detection using multi-dimensional graph search: Applications to intravascular ultrasound. Master's thesis, University of Iowa, 1996.

[Frank et al. 95] R J Frank, T J Grabowski, and H Damasio. Voxelvise percentage tissue segmentation of human brain magnetic resonance images (abstract). In *Abstracts, 25th Annual Meeting, Society for Neuroscience*, page 694, Society for Neuroscience, Washington, DC, 1995.

[Frank et al. 96] R J Frank, D D McPherson, K B Chandran, and E L Dove. Optimal surface detection in intravascular ultrasound using multi-dimensional graph search. In *Computers in Cardiology*, pages 45–48, IEEE, Los Alamitos, CA, 1996.

[Furst 86] M A Furst. Edge detection with image enhancement via dynamic programming. *Computer Vision, Graphics, and Image Processing*, 33:263–279, 1986.

[Gambotto 93] J P Gambotto. A new approach to combining region growing and edge detection. *Pattern Recognition Letters*, 14:869–875, 1993.

[Gauch and Hsia 92] J Gauch and C W Hsia. A comparison of three color image segmentation algorithms in four color spaces. In *Proceedings of the SPIE Vol. 1818*, pages 1168–1181, SPIE, Bellingham, WA, 1992.

[Gavrila and Davis 95] D M Gavrila and L S Davis. Towards 3D model-based tracking and recognition of human movement: A multi-view approach. In *Proceedings of the International Workshop on Face and Gesture Recognition,* Zurich, 1995.

[Gerbrands 88] J J Gerbrands. Segmentation of noisy images. PhD thesis, ETN-89-95461, Technische University, Delft, The Netherlands, 1988.

[Gevers and Smeulders 97] T Gevers and A W M Smeulders. Combining region splitting and edge detection through guided Delaunay image subdivision. In *Computer Vision and Pattern Recognition*, pages 1021–1026, IEEE Computer Society, Los Alamitos, CA, 1997.

[Glasbey 93] C A Glasbey. An analysis of histogram-based thresholding algorithms. *CVGIP - Graphical Models and Image Processing*, 55:532–537, 1993.

[Goldberg and Zhang 87] M Goldberg and J Zhang. Hierarchical segmentation using a composite criterion for remotely sensed imagery. *Photogrammetria*, 42:87–96, 1987.

[Gonzalez and Wintz 87] R C Gonzalez and P Wintz. *Digital Image Processing.* Addison-Wesley, Reading, MA, 2nd edition, 1987.

[Grimson and Lozano-Perez 87] W E L Grimson and T Lozano-Perez. Localizing overlapping parts by searching the interpretation tree. *IEEE Transactions on Pattern Analysis and Machine Intelligence*, 9(4):469–482, 1987.

[Gross and Rosenfeld 87] A D Gross and A Rosenfeld. Multiresolution object detection and delineation. *Computer Vision, Graphics, and Image Processing*, 39:102–115, 1987.

[Guil et al. 95] N Guil, J Villalba, and E L Zapata. A fast Hough transform for segment detection. *IEEE Transactions on Image Processing*, 4:1541–1548, 1995.

[Hancock and Kittler 90] E R Hancock and J Kittler. Edge-labeling using dictionary-based relaxation. *IEEE Transactions on Pattern Analysis and Machine Intelligence*, 12(2):165–181, 1990.

[Hanson and Riseman 78a] A R Hanson and E M Riseman, editors. *Computer Vision Systems.* Academic Press, New York, 1978.

[Hanson and Riseman 78b] A R Hanson and E M Riseman. Segmentation of natural scenes. In A R Hanson and E M Riseman, editors, *Computer Vision Systems*, pages 129–164. Academic Press, New York, 1978.

[Haralick and Shapiro 85] R M Haralick and L G Shapiro. Image segmentation techniques. *Computer Vision, Graphics, and Image Processing*, 29:100–132, 1985.

[Hartley 82] R L Hartley. Segmentation of images FLIR—a comparative study. *IEEE Transactions on Systems, Man and Cybernetics*, 12(4):553–566, 1982.

[Hassan 89] M H Hassan. A class of iterative thresholding algorithms for real-time image segmentation. In *Intelligent Robots and Computer Vision; Proceedings of the Seventh Meeting,* Cambridge, MA, pages 182–193, Society of Photo-Optical Instrumentation Engineers, Bellingham, WA, 1989.

[Herman and Liu 78] G T Herman and H K Liu. Dynamic boundary surface detection. *Computer Graphics and Image Processing*, 7:130–138, 1978.

[Higgins and Ojard 93] W E Higgins and E J Ojard. 3D images and use of markers and other topological information to reduce oversegmentations. *Computers in Medical Imaging and Graphics*, 17:387–395, 1993.

[Hong 82] T H Hong. Image smoothing and segmentation by multiresolution pixel linking further experiments. *IEEE Transactions on Systems, Man and Cybernetics*, 12(5):611–622, 1982.

[Hong et al. 80] T H Hong, C R Dyer, and A Rosenfeld. Texture primitive extraction using an edge-based approach. *IEEE Transactions on Systems, Man and Cybernetics*, 10(10):659–675, 1980.

[Horowitz and Pavlidis 74] S L Horowitz and T Pavlidis. Picture segmentation by a directed split-and-merge procedure. In *Proceedings of the 2nd International Joint Conference on Pattern Recognition*, pages 424–433, Copenhagen, Denmark, 1974.

[Hough 62] P V C Hough. *A Method and Means for Recognizing Complex Patterns.* US Patent 3,069,654, 1962.

[Hsu and Huang 90] C C Hsu and J S Huang. Partitioned Hough transform for ellipsoid detection. *Pattern Recognition*, 23(3–4):275–282, 1990.

[Illingworth and Kittler 87] J Illingworth and J Kittler. The adaptive Hough transform. *IEEE Transactions on Pattern Analysis and Machine Intelligence*, 9(5):690–698, 1987.

[Illingworth and Kittler 88] J Illingworth and J Kittler. Survey of the Hough transform. *Computer Vision, Graphics, and Image Processing*, 44(1):87–116, 1988.

[Kalviainen et al. 95] H Kalviainen, P Hirvonen, L Xu, and E Oja. Probabilistic and non-probabilistic Hough transforms: Overview and comparisons. *Image and Vision Computing*, 13:239–252, 1995.

[Kannan and Chuang 88] C S Kannan and Y H Chuang. Fast Hough transform on a mesh connected processor array. In *Intelligent Robots and Computer Vision; Proceedings of the Meeting,* Cambridge, MA, pages 581–585, Society of Photo-Optical Instrumentation Engineers, Bellingham, WA, 1988.

[Kashyap and Koch 84] R L Kashyap and Mark W Koch. Computer vision algorithms used in recognition of occluded objects. In *First Conference on Artificial Intelligence Applications,* Denver, CO, pages 150–155, IEEE, New York, 1984.

[Kass et al. 87] M Kass, A Witkin, and D Terzopoulos. Snakes: Active contour models. In *1st International Conference on Computer Vision,* London, England, pages 259–268, IEEE, Piscataway, NJ, 1987.

[Kim and Cho 94] J S Kim and H S Cho. A fuzzy logic and neural network approach to boundary detection for noisy imagery. *Fuzzy Sets and Systems*, 65:141–159, 1994.

[Kittler and Illingworth 85] J Kittler and J Illingworth. On threshold selection using clustering criteria. *IEEE Transactions on Systems, Man and Cybernetics*, 15(5):652–655, 1985.

[Kittler and Illingworth 86] J Kittler and J Illingworth. Minimum error thresholding. *Pattern Recognition*, 19:41–47, 1986.

[Koivunen and Pietikainen 90] V Koivunen and M Pietikainen. Combined edge and region-based method for range image segmentation. In *Proceedings of SPIE—The International Society for Optical Engineering*, volume 1381, pages 501–512, Society for Optical Engineering, Bellingham, WA, 1990.

[Kovalevsky 90] V A Kovalevsky. New definition and fast recognition of digital straight segments and arcs. In *International Conference on Pattern Recognition*, pages 31–34, IEEE, Los Alamitos, CA, 1990.

[Kovalevsky 92] V A Kovalevsky. Finite topology and image analysis. In P Hawkes, editor, *Image Mathematics and Image Processing, Series Advances in Electronics and Electron Physics, Vol. 84*, pages 197–259. Academic Press, New York, 1992.

[Kovalevsky 94] V A Kovalevsky. Topological foundations of shape analysis. In *Shape in Picture*, pages 21–36. Springer Verlag, Berlin, 1994.

[Kundu and Mitra 87] A Kundu and S K Mitra. A new algorithm for image edge extraction using a statistical classifier approach. *IEEE Transactions on Pattern Analysis and Machine Intelligence*, 9(4):569–577, 1987.

[Kurita 95] T Kurita. An efficient clustering algorithm for region merging. *IEICE Transactions on Information and Systems*, E78-D:1546–1551, 1995.

[Laprade 88] R H Laprade. Split-and-merge segmentation of aerial photographs. *Computer Vision, Graphics, and Image Processing*, 44(1):77–86, 1988.

[Lavagetto 90] F Lavagetto. Infrared image segmentation through iterative thresholding. In *Real-Time Image Processing II,* Orlando, FL, volume 1295, pages 29–38, The International Society for Optical Engineering, Bellingham, WA, 1990.

[Law et al. 96] T Law, H Itoh, and H Seki. Image filtering, edge detection, and edge tracing using fuzzy reasoning. *IEEE Transactions on Pattern Analysis and Machine Intelligence*, 18:481–491, 1996.

[Lerner and Morelli 90] B T Lerner and M V Morelli. Extensions of algebraic image operators: An approach to model-based vision. In *Third Annual Workshop on Space Operations Automation and Robotics (SOAR 1989)*, pages 687–695, NASA, Lyndon B Johnson Space Center, 1990.

[Lerner et al. 94] B T Lerner, W J Campbell, and J LeMoigne. Image segmentation by integration of edge and region data: The influence of edge detection algorithms. In *ARPA Image Understanding Workshop,* Monterey, CA, pages 1541–1545, ARPA, Los Altos, CA, 1994.

[Lester 78] J M Lester. Two graph searching techniques for boundary finding in white blood cell images. *Computers in Biology and Medicine*, 8:193–308, 1978.

[Levy 88] M Levy. New theoretical approach to relaxation, application to edge detection. In *9th International Conference on Pattern Recognition,* Rome, Italy, pages 208–212, IEEE, New York, 1988.

[Li et al. 93] Z N Li, B Yao, and F Tong. Linear generalized Hough transform and its parallelization. *Image and Vision Computing*, 11:11–24, 1993.

[Liow 91] Y T Liow. A contour tracing algorithm that preserves common boundaries between regions. *CVGIP - Image Understanding*, 53(3):313–321, 1991.

[Liow and Pavlidis 88] Y Liow and T Pavlidis. Enhancements of the split-and-merge algorithm for image segmentation. In *1988 IEEE International Conference on Robotics and Automation,* Philadelphia, PA, pages 1567–1572, Computer Society Press, Washington, DC, 1988.

[Liu 77] H K Liu. Two- and three-dimensional boundary detection. *Computer Graphics and Image Processing*, 6:123–134, 1977.

[Lo and Tsai 95] R C Lo and W H Tsai. Gray-scale Hough transform for thick line detection in gray-scale images. *Pattern Recognition*, 28:647–661, 1995.

[Manos et al. 93] G Manos, A Y Cairns, I W Ricketts, and D Sinclair. Automatic segmentation of hand-wrist radiographs. *Image and Vision Computing*, 11:100–111, 1993.

[Mardia and Hainsworth 88] K V Mardia and T J Hainsworth. A spatial thresholding method for image segmentation. *IEEE Transactions on Pattern Analysis and Machine Intelligence*, 10:919–927, 1988.

[Marik and Matas 89] R Marik and J Matas. Membrane method for graph construction. In *Computer Analysis of Images and Patterns. Third International Conference on Automatic Image Processing,* Leipzig, Germany. Scientific-Technological Society for Measurement and Automatic Control, September 1989.

[Marks et al. 84] D Marks, J Dozier, and J Frew. Automated basin deliniation from digital elevation data. *Geoprocessing*, 2:299–311, 1984.

[Marquardt 63] D W Marquardt. An algorithm for least squares estimation of non-linear parameters. *Journal of the Society for Industrial and Applied Mathematics*, 11:431–444, 1963.

[Martelli 72] A Martelli. Edge detection using heuristic search methods. *Computer Graphics and Image Processing*, 1:169–182, 1972.

[Martelli 76] A Martelli. An application of heuristic search methods to edge and contour detection. *Communications of the ACM*, 19(2):73–83, 1976.

[McDonnel et al. 87] M M McDonnel, M Lew, and T S Huang. Finding wheels of vehicles in stereo images. Technical report, AD-A194372; ETL-R-141, Army Engineer Topographic Labotatories, Fort Belvoir, VA, 1987.

[McKenzie and Protheroe 90] D S McKenzie and S R Protheroe. Curve description using the inverse Hough transform. *Pattern Recognition*, 23(3–4):283–290, 1990.

[Meyer and Beucher 90] F Meyer and S Beucher. Morphological segmentation. *Journal of Visual Communication and Image Representation*, 1:21–46, 1990.

[Milgram 79] D L Milgram. Region extraction using convergent evidence. *Computer Graphics and Image Processing*, 11:1–12, 1979.

[Mortensen et al. 92] E Mortensen, B Morse, W Barrett, and J Udupa. Adaptive boundary detection using 'live-wire' two-dimensional dynamic programming. In *Computers in Cardiology*, pages 635–638, IEEE Computer Society Press, Los Alamitos, CA, 1992.

[Mukund and Gonzalez 89] P R Mukund and R C Gonzalez. Generalized approach to split and merge segmentation on parallel architectures. In *Proceedings of SPIE—The International Society for Optical Engineering*, volume 1197, pages 254–264, Society for Optical Engineering, Bellingham, WA, 1989.

[Nagao and Matsuyama 80] M Nagao and T Matsuyama. *A Structural Analysis of Complex Aerial Photographs.* Plenum Press, New York, 1980.

[Narendra and Goldberg 77] P M Narendra and M Goldberg. A non-parametric clustering scheme for Landsat. *Pattern Recognition*, 9:207–215, 1977.

[Neveu 86] C F Neveu. Two-dimensional object recognition using multiresolution models. *Computer Vision, Graphics, and Image Processing*, 34(1):52–65, 1986.

[Ney 92] H Ney. A comparative study of two search strategies for connected word recognition: Dynamic programming and heuristic search. *IEEE Transactions on Pattern Analysis and Machine Intelligence*, 14(5):586–595, 1992.

[Nilsson 82] N J Nilsson. *Principles of Artificial Intelligence.* Springer Verlag, Berlin, 1982.

[Ohta et al. 80] Y I Ohta, T Kanade, and T Sakai. Color information for region segmentation. *Computer Graphics and Image Processing*, 13:222–241, 1980.

[Olariu et al. 93] S Olariu, J L Schwing, and J Zhang. Computing the Hough transform on reconfigurable meshes. *Image and Vision Computing*, 11:623–628, 1993.

[Otsu 79] N Otsu. A threshold selection method from gray–level histograms. *IEEE Transactions on Systems, Man and Cybernetics*, 9(1):62–66, 1979.

[Oyster 87] J M Oyster. Associative network applications to low-level machine vision. *Applied Optics*, 26:1919–1926, 1987.

[Pal and Pal 87] N R Pal and S K Pal. Segmentation based on contrast homogeneity measure and region size. *IEEE Transactions on Systems, Man and Cybernetics*, 17(5):857–868, 1987.

[Pavlidis 77] T Pavlidis. *Structural Pattern Recognition.* Springer Verlag, Berlin, 1977.

[Pavlidis and Liow 90] T Pavlidis and Y Liow. Integrating region growing and edge detection. *IEEE Transactions on Pattern Analysis and Machine Intelligence*, 12(3):225–233, 1990.

[Philip 91] K P Philip. Automatic detection of myocardial contours in cine computed tomographic images. PhD thesis, University of Iowa, 1991.

[Philip et al. 90] K P Philip, E L Dove, and K B Chandran. A graph search based algorithm for detection of closed contours in images. In *Proceedings: Annual International Conference IEEE—Engineering in Medicine and Biology Society*, IEEE, Philadelphia, 1990.

[Pietikainen and Rosenfeld 81] M Pietikainen and A Rosenfeld. Image segmentation by texture using pyramid node linking. *IEEE Transactions on Systems, Man and Cybernetics*, 11(12):822–825, 1981.

[Pietikainen and Rosenfeld 82] M Pietikainen and A Rosenfeld. Gray level pyramid linking as an aid in texture analysis. *IEEE Transactions on Systems, Man and Cybernetics*, 12(3):422–429, 1982.

[Pietikainen et al. 82] M Pietikainen, A Rosenfeld, and I Walter. Split–and–link algorithms for image segmentation. *Pattern Recognition*, 15(4):287–298, 1982.

[Pontriagin 62] L S Pontriagin. *The Mathematical Theory of Optimal Processes.* Interscience, New York, 1962.

[Pontriagin 90] L S Pontriagin. *Optimal Control and Differential Games: Collection of Papers.* American Mathematical Society, Providence, RI, 1990.

[Pope et al. 85] D L Pope, D L Parker, P D Clayton, and D E Gustafson. Left ventricular border detection using a dynamic search algorithm. *Radiology*, 155:513–518, 1985.

[Prager 80] J M Prager. Extracting and labeling boundary segments in natural scenes. *IEEE Transactions on Pattern Analysis and Machine Intelligence*, 2(1):16–27, 1980.

[Pramotepipop and Cheevasuvit 88] Y Pramotepipop and F Cheevasuvit. Modification of split-and-merge algorithm for image segmentation. In *Asian Conference on Remote Sensing, 9th,* Bangkok, Thailand, pages Q–26–1–Q–26–6, Asian Association on Remote Sensing, Tokyo, 1988.

[Press et al. 92] W H Press, S A Teukolsky, W T Vetterling, and B P Flannery. *Numerical Recipes in C: The Art of Scientific Computing.* Cambridge University Press, Cambridge, 2nd edition, 1992.

[Priese and Rehrmann 93] L Priese and V Rehrmann. On hierarchical color segmentation and applications. In *Computer Vision and Pattern Recognition (Proceedings)*, pages 633–634, IEEE, Los Alamitos, CA, 1993.

[Princen et al. 89] J Princen, J Illingworth, and J Kittler. Hierarchical approach to line extraction. In *Proceedings: IEEE Computer Society Conference on Computer Vision and Pattern Recognition,* Rosemont, IL, pages 92–97, IEEE, Piscataway, NJ, 1989.

[Princen et al. 94] J Princen, J Illingworth, and J Kittler. Hypothesis testing: A framework for analyzing and optimizing Hough transform performance. *IEEE Transactions on Pattern Analysis and Machine Intelligence*, 16:329–341, 1994.

[Puecker and Douglas 75] T K Puecker and D H Douglas. Detection of surface-specific points by local parallel processing of discrete terrain elevation data. *Computer Vision, Graphics, and Image Processing*, 4:375–387, 1975.

[Ramer 75] U Ramer. Extraction of line structures from photographs of curved objects. *Computer Graphics and Image Processing*, 4:425–446, 1975.

[Ramesh et al. 95] N Ramesh, J H Yoo, and I K Sethi. Thresholding based on histogram approximations. *IEE Proceedings Vision, Image and Signal Processing*, 142:271–279, 1995.

[Reddi et al. 84] S S Reddi, S F Rudin, and H R Keshavan. An optimal multiple threshold scheme for image segmentation. *IEEE Transactions on Systems, Man and Cybernetics*, 14:661–665, 1984.

[Ridler and Calvard 78] T W Ridler and S Calvard. Picture thresholding using an iterative selection method. *IEEE Transactions on Systems, Man and Cybernetics*, 8(8):630–632, 1978.

[Riseman and Arbib 77] E M Riseman and M A Arbib. Computational techniques in the visual segmentation of static scenes. *Computer Graphics and Image Processing*, 6:221–276, 1977.

[Rosenfeld 84] A Rosenfeld, editor. *Multiresolution Image Processing and Analysis.* Springer Verlag, Berlin, 1984.

[Rosenfeld and de la Torre 83] A Rosenfeld and P de la Torre. Histogram concavity analysis as an aid in threshold selection. *IEEE Transactions on Systems, Man and Cybernetics*, 13(3):231–235, 1983.

[Rosenfeld and Kak 82] A Rosenfeld and A C Kak. *Digital Picture Processing.* Academic Press, New York, 2nd edition, 1982.

[Rosenfeld et al. 76] A Rosenfeld, R A Hummel, and S W Zucker. Scene labelling by relaxation operations. *IEEE Transactions on Systems, Man and Cybernetics*, 6:420–433, 1976.

[Sahoo et al. 88] P K Sahoo, S Soltani, A K C Wong, and Y C Chen. Survey of thresholding techniques. *Computer Vision, Graphics, and Image Processing*, 41(2):233–260, 1988.

[Santago and Gage 93] P Santago and H D Gage. Quantification of MR brain images by mixture density and partial volume modeling. *IEEE Transactions on Medical Imaging*, 12:566–574, 1993.

[Schettini 93] R Schettini. A segmentation algorithm for color images. *Pattern Recognition Letters*, 14:499–506, 1993.

[Shankar and Asokan 90] R V Shankar and N Asokan. A parallel implementation of the Hough transform method to detect lines and curves in pictures. In *Proceedings of the 32nd Midwest Symposium on Circuits and Systems,* Champaign, IL, pages 321–324, IEEE, Piscataway, NJ, 1990.

[Sher 92] D B Sher. A technique for deriving the distribution of edge neighborhoods from a library of occluding objects. In *Proceedings of the 6th International Conference on Image Analysis and Processing. Progress in Image Analysis and Processing II*, pages 422–429, World Scientific, Singapore, 1992.

[Soille and Ansoult 90] P Soille and M Ansoult. Automated basin delineation from DEMs using mathematical morphology. *Signal Processing*, 20:171–182, 1990.

[Song et al. 90] S Song, M Liao, and J Qin. Multiresolution image dynamic thresholding. *Machine Vision and Applications*, 3(1):13–16, 1990.

[Sonka 86] M Sonka. A new texture recognition method. *Computers and Artificial Intelligence*, 5(4):357–364, 1986.

[Sonka et al. 93] M Sonka, C J Wilbricht, S R Fleagle, S K Tadikonda, M D Winniford, and S M Collins. Simultaneous detection of both coronary borders. *IEEE Transactions on Medical Imaging*, 12(3):588–599, 1993.

[Sonka et al. 94] M Sonka, M D Winniford, X Zhang, and S M Collins. Lumen centerline detection in complex coronary angiograms. *IEEE Transactions on Biomedical Engineering*, 41:520–528, 1994.

[Sonka et al. 95] M Sonka, M D Winniford, and S M Collins. Robust simultaneous detection of coronary borders in complex images. *IEEE Transactions on Medical Imaging*, 14(1):151–161, 1995.

[Suk and Chung 83] M Suk and S M Chung. A new image segmentation technique based on partition mode test. *Pattern Recognition*, 16(5):469–480, 1983.

[Tadikonda et al. 92] S K Tadikonda, M Sonka, and S M Collins. Efficient coronary border detection using heuristic graph searching. In *Proceedings of the Annual International Conference of the IEEE EMBS,* Paris, France, volume 14, pages 1897–1899, IEEE, Piscataway, NJ, 1992.

[Tanimoto 78] S Tanimoto. Regular hierarchical image and processing structures in machine vision. In A R Hanson and E M Riseman, editors, *Computer Vision Systems*, pages 165–174. Academic Press, New York, 1978.

[Tanimoto and Pavlidis 75] S Tanimoto and T Pavlidis. A hierarchical data structure for picture processing. *Computer Graphics and Image Processing*, 4:104–119, 1975.

[Thedens et al. 90] D R Thedens, D J Skorton, and S R Fleagle. A three-dimensional graph searching technique for cardiac border detection in sequential images and its application to magnetic resonance image data. In *Computers in Cardiology*, pages 57–60, IEEE, Los Alamitos, CA, 1990.

[Thedens et al. 95] D R Thedens, D J Skorton, and S R Fleagle. Methods of graph searching for border detection in image sequences with application to cardiac magnetic resonance imaging. *IEEE Transactions on Medical Imaging*, 14:42–55, 1995.

[Tilton 89] J C Tilton. Image segmentation by iterative parallel region growing and splitting. In *Proceedings of IGARSS '89 and Canadian Symposium on Remote Sensing,* Vancouver, Canada, pages 2420–2423, IEEE, New York, 1989.

[Udupa 82] J K Udupa. Interactive segmentation and boundary surface formation for 3-D digital images. *Computer Graphics and Image Processing*, 18:213–235, 1982.

[Udupa et al. 92] J K Udupa, S Samarasekera, and W A Barrett. Boundary detection via dynamic programming. In *Visualization in Biomedical Computing, Proc. SPIE Vol. 1808*, pages 33–39, SPIE, Bellingham, WA, 1992.

[van der Heijden 95] F van der Heijden. Edge and line feature extraction based on covariance models. *IEEE Transactions on Pattern Analysis and Machine Intelligence*, 17:69–77, 1995.

[van der Zwet and Reiber 92] P M J van der Zwet and J H C Reiber. A new algorithm to detect irregular coronary boundaries: The gradient field transform. In *Computers in Cardiology,* pages 107–110, IEEE, Los Alamitos, CA, 1992.

[van der Zwet and Reiber 94] P M J van der Zwet and J H C Reiber. A new approach for the quantification of complex lesion morphology: The gradient field transform; basic principles and validation results. *Journal of the Amercian College of Cardiologists*, 82:216–224, 1994.

[Venkateswarlu and Boyle 95] N B Venkateswarlu and R D Boyle. New segmentation techniques for document image analysis. *Image and Vision Computing*, 13:573–583, 1995.

[Venturi et al. 92] G Venturi, A Di Giuliani, and G Vernazza. Image segmentation using edge and region information. In *Proceedings of the 6th International Conference on Image Analysis and Processing*, pages 78–82, World Scientific, Singapore, 1992.

[Vincent and Soille 91] L Vincent and P Soille. Watersheds in digital spaces: An efficient algorithm based on immersion simulations. *IEEE Transactions on Pattern Analysis and Machine Intelligence*, 13(6):583–598, 1991.

[Vlachos and Constantinides 93] T Vlachos and A G Constantinides. Graph-theoretical approach to colour picture segmentation and contour classification. *IEE Proceedings Communication, Speech and Vision*, 140:36–45, 1993.

[Wang and Howarth 87] J F Wang and P J Howarth. Automated road network extraction from Landsat TM imagery. In *American Society for Photogrammetry and Remote Sensing and ACSM, Annual Convention,* Baltimore, MD, pages 429–438, American Society for Photogrammetry and Remote Sensing and ACSM, Falls Church, VA, 1987.

[Wang and Howarth 89] J F Wang and P J Howarth. Edge following as graph searching and Hough transform algorithms for lineament detection. In *Proceedings of IGARSS '89 and Canadian Symposium on Remote Sensing,* Vancouver, Canada, pages 93–96, IEEE, New York, 1989.

[Weszka and Rosenfeld 79] J S Weszka and A Rosenfeld. Histogram modification for threshold selection. *IEEE Transactions on Systems, Man and Cybernetics*, 9(1):38–52, 1979.

[Weszka et al. 76] J S Weszka, C Dyer, and A Rosenfeld. A comparative study of texture measures for terrain classification. *IEEE Transactions on Systems, Man and Cybernetics*, 6(4):269–285, 1976.

[Willebeek-Lemair and Reeves 90] M Willebeek-Lemair and A Reeves. Solving nonuniform problems on SIMD computers—case study on region growing. *Journal of Parallel and Distributed Computing*, 8:135–149, 1990.

[Wood 85] J W Wood. Line finding algorithms for SAR. Technical report, AD-A162024; RSRE-MEMO-3841; BR97301, Royal Signals and Radar Establishment, Malvern, England, 1985.

[Wu 93] X Wu. Adaptive split-and-merge segmentation based on piecewise least-square approximation. *IEEE Transactions on Pattern Analysis and Machine Intelligence*, 15:808–815, 1993.

[Wu and Leahy 93] Z Wu and R Leahy. An optimal graph theoretic approach to data clustering: Theory and its application to image segmentation. *IEEE Transactions on Pattern Analysis and Machine Intelligence*, 15:1101–1113, 1993.

[Xu and Oja 93] L Xu and E Oja. Randomized Hough transform (RHT): Basic mechanisms, algorithms, and computational complexities. *CVGIP - Image Understanding*, 57:131–154, 1993.

[Yuen and Hlavac 91] S Y K Yuen and V Hlavac. An approach to quantization of the Hough space. In *Proceedings of the 7th Scandinavian Conference on Image Analysis,* Aalborg, Denmark, pages 733–740, Pattern Recognition Society of Denmark, Copenhagen, Copenhagen, Denmark, August 1991.

[Zamperoni 86] P Zamperoni. Analysis of some region growing operators for image segmentation. In V Cappelini and R Marconi, editors, *Advances in Image Processing and Pattern Recognition*, pages 204–208. North Holland, Amsterdam, 1986.

[Zucker 76] S W Zucker. Relaxation labelling, local ambiguity, and low-level vision. In C H Chen, editor, *Pattern Recognition and Artificial Intelligence*, pages 593–616, Academic Press, New York, 1976.

Chapter 6

Shape representation and description

The last chapter was devoted to image segmentation methods and showed how to construct homogeneous regions of images and/or their boundaries. Recognition of image regions is an important step on the way to understanding image data, and requires an exact region description in a form suitable for a classifier (Chapter 7). This description should generate a numeric feature vector, or a non-numeric syntactic description word, which characterizes properties (for example, shape) of the region. Region description is the third of the four levels given in Chapter 3, implying that the description already comprises some abstraction—for example, 3D objects can be represented in a 2D plane and shape properties that are used for description are usually computed in two dimensions. If we are interested in a 3D object description, we have to process at least two images of the same object taken from different viewpoints (stereo vision), or derive the 3D shape from a sequence of images if the object is in motion. A 2D shape representation is sufficient in the majority of practical applications, but if 3D information is necessary—if, say, 3D object reconstruction is the processing goal, or the 3D characteristics bear the important information—the object description task is much more difficult; these topics are introduced in Chapter 9. In the following sections, we will limit our discussion to 2D shape features and proceed under the assumption that object descriptions result from the image segmentation process.

Defining the shape of an object can prove to be very difficult. Shape is usually represented verbally or in figures, and people use terms such as *elongated*, *rounded*, *with sharp edges*, etc. The computer era has introduced the necessity to describe even very complicated shapes precisely, and while many practical shape description methods exist, there is no generally accepted methodology of shape description. Further, it is not known what is important in shape. Current approaches have both positive and negative attributes; computer graphics [Woodwark 86] or mathematics [Lord and Wilson 84] use effective shape representations which are unusable in shape recognition [Juday 88] and vice versa. In spite of this, it is possible to find features common to most shape description approaches. Location and description of substantial variations in the first derivative of object boundaries often yield suitable information. Examples include alphanumeric optical character recognition (OCR), technical drawings, electro-cardiogram (ECG) curve characterization, etc.

Shape is an object property which has been carefully investigated in recent years and many papers may be found dealing with numerous applications—OCR, ECG analysis, electroencephalogram (EEG) analysis, cell classification, chromosome recognition, automatic inspection, technical diagnostics, etc. Despite this variety, differences among many approaches are limited mostly to terminology. These common methods can be characterized from different points of view [Pavlidis 78, Pavlidis 80, Ballard and Brown 82, Brady 84, Besl 88, Marshall 89b, Koenderink 90, Watt 93, Hogg 93].

- Input representation form: Object description can be based on boundaries (contour-based, external) or on more complex knowledge of whole regions (region-based, internal).
- Object reconstruction ability: That is, whether an object's shape can or cannot be reconstructed from the description. Many varieties of shape-preserving methods exist. They differ in the degree of precision with respect to object reconstruction.
- Incomplete shape recognition ability: That is, to what extent an object's shape can be recognized from the description if objects are occluded and only partial shape information is available.
- Local/global description character: Global descriptors can only be used if complete object data are available for analysis. Local descriptors describe local object properties using partial information about the objects. Thus, local descriptors can be used for description of occluded objects.
- Mathematical and heuristic techniques: A typical mathematical technique is shape description based on the Fourier transform. A representative heuristic method may be elongatedness.
- Statistical or syntactic object description (Chapter 7).
- A robustness of description to translation, rotation, and scale transformations: Shape description properties in different resolutions.

The role of different description methods in image analysis and image understanding is illustrated by the flowchart shown in Figure 6.1.

Problems of scale (resolution) are common in digital images. Sensitivity to scale is even more serious if a shape description is derived, because shape may change substantially with image resolution. Contour detection may be affected by noise in high resolution, and small details may disappear in low resolution (see Figure 6.2). Therefore, shape has been studied in multiple resolutions which again causes difficulties with matching corresponding shape representations from different resolutions. Moreover, the conventional shape descriptions change discontinuously. A **scale-space** approach has been presented in [Babaud et al. 86, Witkin 86, Yuille and Poggio 86, Maragos 89] that aims to obtain continuous shape descriptions if the resolution changes continuously. This approach is not a new technique itself, but is an extension of existing techniques, and more robust shape methods may result from developing and retaining their parameters over a range of scales [Marshall 89b]. This approach will be mentioned in more detail in Section 6.2.4.

In many tasks, it is important to represent classes of shapes properly, e.g., shape classes of apples, oranges, pears, bananas, etc. The **shape classes** should represent the generic shapes

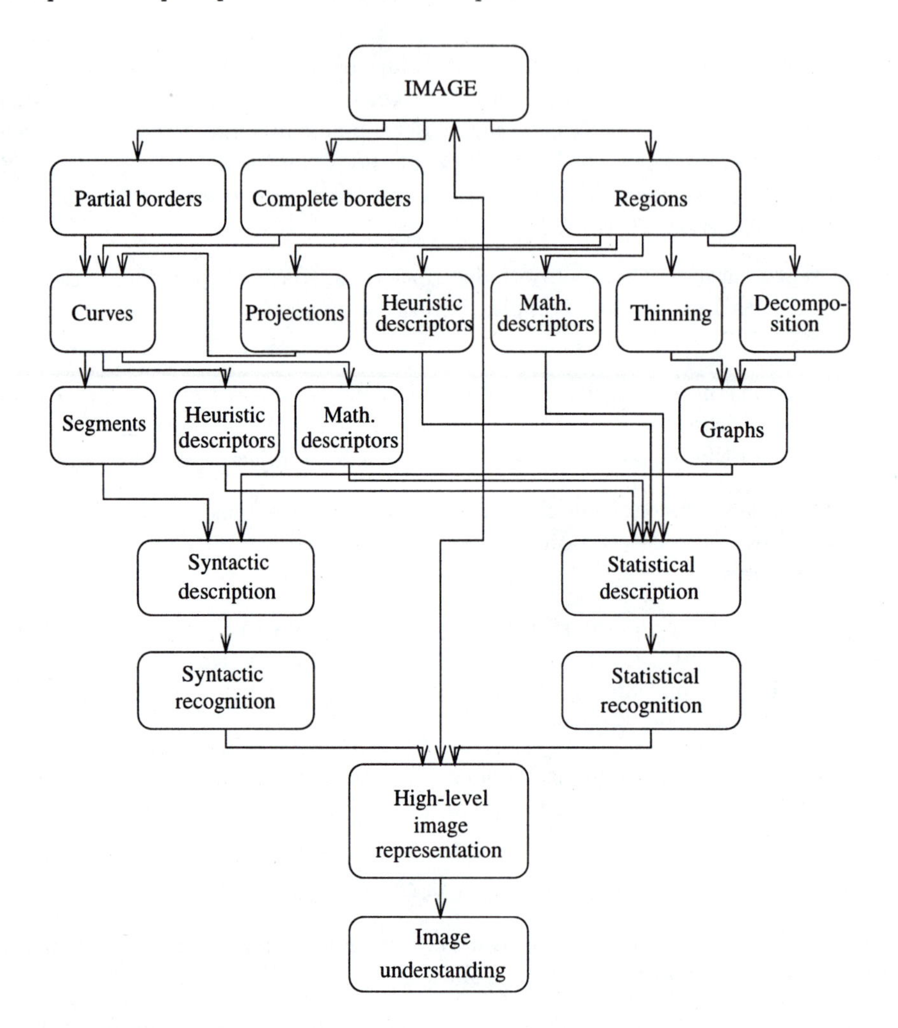

Figure 6.1: *Image analysis and understanding methods.*

of the objects belonging to the same classes well. Obviously, shape classes should emphasize shape differences among classes, while the influence of shape variations within classes should not be reflected in the class description. Current research challenges includes development of approaches to automated learning about shape and reliable definition of shape classes (Section 6.4).

Object representation and shape description methods discussed in the following sections are not an exhaustive list—we will try to introduce generally applicable methods. It is necessary to apply a problem-oriented approach to the solution of specific problems of description and recognition. This means that the following methods are appropriate for a large variety of descriptive tasks and the following ideas may be used to build a specialized, highly efficient method suitable for a particular problem description. Such a method will no longer be general since it will take advantage of a priori knowledge about the problem. This is the way human

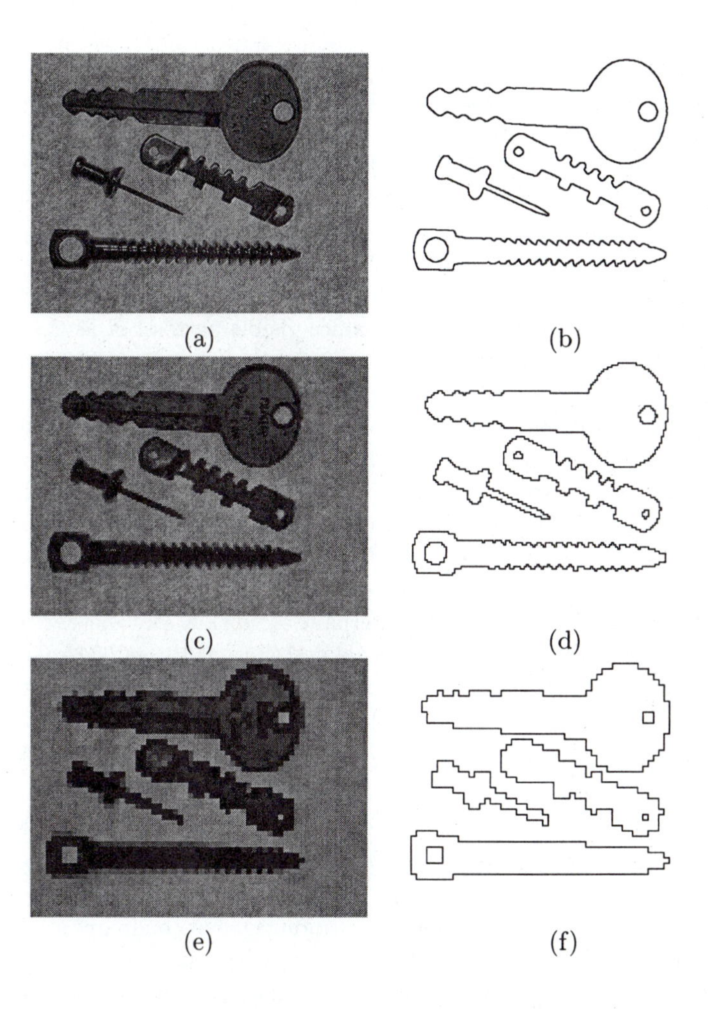

Figure 6.2: *(a) Original image 640 × 480. (b) Contours of a. (c) Original image 160 × 120. (d) Contours of c. (e) Original image 64 × 48. (f) Contours of e.*

beings can solve their vision and recognition problems, by using highly specialized knowledge.

It should be understood that despite the fact that we are dealing with two-dimensional shape and its description, our world is three-dimensional and the same objects, if seen from different angles (or changing position/orientation in space), may form very different 2D projections (see Chapter 9). The ideal case would be to have a universal shape descriptor capable of overcoming these changes—to design projection-invariant descriptors. Consider an object with planar faces and imagine how many very different 2D shapes may result from a given face if the position and 3D orientation of this simple object changes with respect to an observer. In some special cases, such as circles which transform to ellipses, or planar polygons, projectively invariant features (called **invariants**) can be found. Unfortunately, no existing shape descriptor is perfect; in fact, they are all far from being perfect. Therefore, a very careful choice of descriptors resulting from detailed analysis of the shape recognition problem must precede any implementation, and whether or not a 2D representation is capable of describing a 3D shape must also be considered. For some 3D shapes, their 2D projection may bear enough information for recognition—aircraft contours are a good example; successful recognition of airplanes from projections are known even if they change their position and orientation in space. In many other cases, objects must be seen from a specific direction to get enough descriptive information—human faces are such a case.

Object occlusion is another hard problem in shape recognition. However, the situation is easier here (if pure occlusion is considered, not combined with orientation variations yielding changes in 2D projections as discussed above), since visible parts of objects may be used for description. Here, the shape descriptor choice must be based on its ability to describe local object properties—if the descriptor gives only a global object description (e.g., object size, average boundary curvature, perimeter), such a description is useless if only a part of an object is visible. If a local descriptor is applied (e.g., description of local boundary changes), this information may be used to compare the visible part of the object to all objects which may appear in the image. Clearly, if object occlusion occurs, the local or global character of the shape descriptor must be considered first.

In Sections 6.2 and 6.3, descriptors are sorted according to whether they are based on object boundary information (contour-based, external description) or whether the information from object regions is used (region-based, internal description) [Li and MA 94]. This classification of shape description methods corresponds to previously described boundary-based and region-based segmentation methods. However, both contour-based and region-based shape descriptors may be local or global and differ in sensitivity to translation, rotation, scaling, etc.

6.1 Region identification

Region identification is necessary for region description. One of the many methods for region identification is to label each region (or each boundary) with a unique (integer) number; such identification is called **labeling** or **coloring** (also connected component labeling), and the largest integer label usually gives the number of regions in the image. Another method is to use a smaller number of labels (four is theoretically sufficient [Appel and Haken 77, Saaty and Kainen 77, Nishizeki and Chiba 88, Wilson and Nelson 90]), and ensure that no two neighboring regions have the same label; then information about some region pixel must be added to the description to provide full region reference. This information is usually stored in a separate data structure. Alternatively, mathematical morphology approaches (Chapter 11) may be used for region identification.

Assume that the segmented image R consists of m disjoint regions R_i [as in equation (5.1)]. The image R often consists of objects and a background

$$R_b^C = \bigcup_{i=1, i \neq b}^{m} R_i$$

where R^C is the set complement, R_b is considered background, and other regions are considered objects. Input to a labeling algorithm is usually either a binary or multi-level image, where background may be represented by zero pixels, and objects by non-zero values. A multi-level image is often used to represent the labeling result, background being represented by zero values, and regions represented by their non-zero labels. Algorithm 6.1 presents a sequential approach to labeling a segmented image.

Algorithm 6.1: 4-neighborhood and 8-neighborhood region identification

1. First pass: Search the entire image R row by row and assign a non-zero value v to each non-zero pixel $R(i,j)$. The value v is chosen according to the labels of the pixel's neighbors, where the property *neighboring* is defined by Figure 6.3. ('neighbors' outside the image R are not considered),

    - If all the neighbors are background pixels (with pixel value zero), $R(i,j)$ is assigned a new (and as yet) unused label.
    - If there is just one neighboring pixel with a non-zero label, assign this label to the pixel $R(i,j)$.
    - If there is more than one non-zero pixel among the neighbors, assign the label of any one to the labeled pixel. If the labels of any of the neighbors differ (*label collision*), store the label pair as being equivalent. Equivalence pairs are stored in a separate data structure—an equivalence table.

2. Second pass: All of the region pixels were labeled during the first pass, but some regions have pixels with different labels (due to label collisions). The whole image is scanned again, and pixels are re-labeled using the equivalence table information (for example, with the lowest value in an equivalence class).

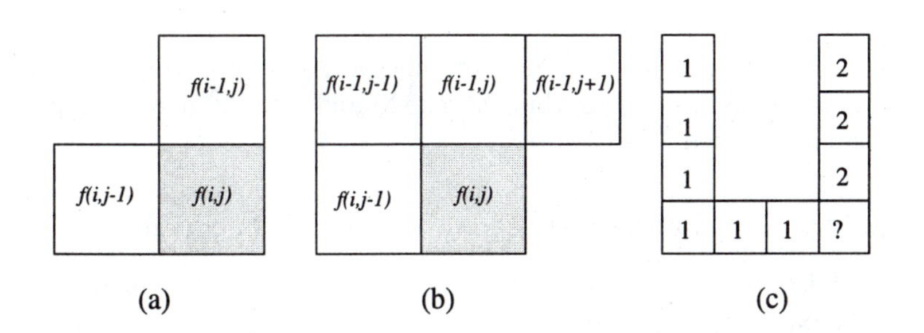

Figure 6.3: *Masks for region identification: (a) in 4-connectivity; (b) in 8-connectivity; (c) label collision.*

Label collision is a very common occurrence—examples of image shapes experiencing this are U-shaped objects, mirrored E (∃) objects, etc. (see Figure 6.3c). The equivalence table is a list of all label pairs present in an image; all equivalent labels are replaced by a unique label in the second step. Since the number of label collisions is usually not known beforehand, it is necessary to allocate sufficient memory to store the equivalence table in an array. A dynamically allocated data structure is recommended. Further, if pointers are used for label specification, scanning the image for the second time is not necessary (the second pass of the algorithm) and only rewriting labels to which these pointers are pointing is much faster.

The algorithm is basically the same in 4-connectivity and 8-connectivity, the only difference being in the neighborhood mask shape (Figure 6.3b). It is useful to assign the region labels incrementally to permit the regions to be counted easily in the second pass. An example of partial results is given in Figure 6.4.

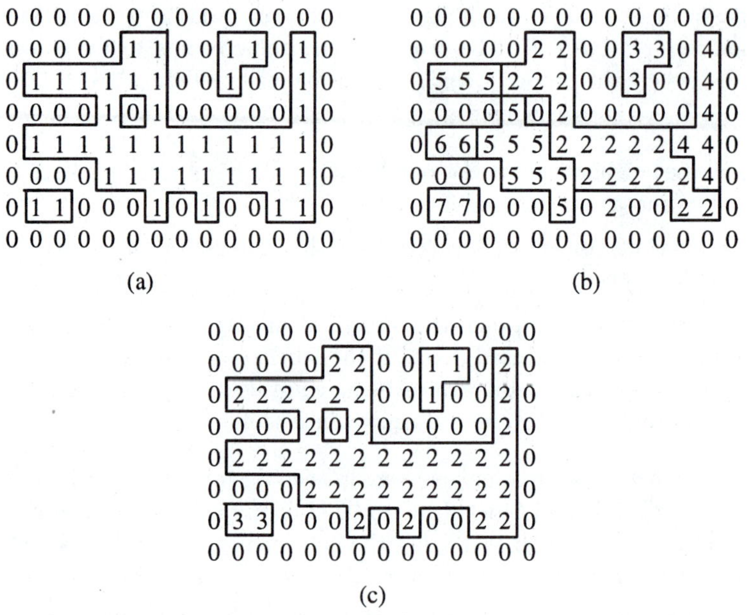

```
0 0 0 0 0 0 0 0 0 0 0 0 0 0
0 0 0 0 0 1 1 0 0 1 1 0 1 0
0 1 1 1 1 1 1 0 0 1 0 0 1 0
0 0 0 0 1 0 1 0 0 0 0 0 1 0
0 1 1 1 1 1 1 1 1 1 1 1 1 0
0 0 0 0 1 1 1 1 1 1 1 1 1 0
0 1 1 0 0 0 1 0 1 0 0 1 1 0
0 0 0 0 0 0 0 0 0 0 0 0 0 0
```

(a)

```
0 0 0 0 0 0 0 0 0 0 0 0 0 0
0 0 0 0 0 2 2 0 0 3 3 0 4 0
0 5 5 5 2 2 2 0 0 3 0 0 4 0
0 0 0 0 5 0 2 0 0 0 0 0 4 0
0 6 6 5 5 5 2 2 2 2 2 4 4 0
0 0 0 0 5 5 5 2 2 2 2 2 4 0
0 7 7 0 0 0 5 0 2 0 0 2 2 0
0 0 0 0 0 0 0 0 0 0 0 0 0 0
```

(b)

```
0 0 0 0 0 0 0 0 0 0 0 0 0 0
0 0 0 0 0 2 2 0 0 1 1 0 2 0
0 2 2 2 2 2 2 0 0 1 0 0 2 0
0 0 0 0 2 0 2 0 0 0 0 0 2 0
0 2 2 2 2 2 2 2 2 2 2 2 2 0
0 0 0 0 2 2 2 2 2 2 2 2 2 0
0 3 3 0 0 0 2 0 2 0 0 2 2 0
0 0 0 0 0 0 0 0 0 0 0 0 0 0
```

(c)

Figure 6.4: *Object identification in 8-connectivity: (a),(b),(c) algorithm steps. Equivalence table after step (b): 2-5, 5-6, 2-4.*

Region identification can be performed on images that are not represented as straightforward matrices; the following algorithm [Rosenfeld and Kak 82] may be applied to images that are run length encoded (see Chapter 3).

Algorithm 6.2: Region identification in run length encoded data

1. First pass: Use a new label for each continuous run in the first image row that is not part of the background.

2. For the second and subsequent rows, compare positions of runs.

   - If a run in a row does not neighbor (in the 4- or 8-sense) any run in the previous row, assign a new label.
   - If a run neighbors precisely one run in the previous row, assign its label to the new run.
   - If the new run neighbors more than one run in the previous row, a label collision has occurred.

Collision information is stored in an equivalence table, and the new run is labeled using the label of any one of its neighbors.

3. Second pass: Search the image row by row and re-label the image according to the equivalence table information.

If the segmented image is represented by a quadtree data structure, the following algorithm may be applied.

Algorithm 6.3: Quadtree region identification

1. First pass: Search quadtree nodes in a given order—e.g., beginning from the root and in the NW, NE, SW, SE directions. Whenever an unlabeled non-zero leaf node is entered, a new label is assigned to it. Then search for neighboring leaf nodes in the E and S directions (plus SE in 8-connectivity). If those leaves are non-zero and have not yet been labeled, assign the label of the node from which the search started. If the neighboring leaf node has already been labeled, store the collision information in an equivalence table.

2. Repeat step 1 until the whole tree has been searched.

3. Second pass: Re-label the leaf nodes of the quadtree according to the equivalence table.

Algorithmic details and the procedure for looking for neighboring leaf nodes can be found in [Rosenfeld and Kak 82, Samet 84].

The **region counting** task is closely related to the region identification problem. As we have seen, object counting can be an intermediate result of region identification. If it is only necessary to count regions with no need to identify them, a one-pass algorithm is sufficient [Rosenfeld and Kak 82, Atkinson et al. 85].

6.2 Contour-based shape representation and description

Region borders must be expressed in some mathematical form. The **rectangular** representation of $\mathbf{x}_n$ pixel co-ordinates as a function of the path length n is most common. Other useful representations are (see Figure 6.5)

- **Polar** co-ordinates, in which border elements are represented as pairs of angle ϕ and distance r;

- **Tangential** co-ordinates, which codes the tangential directions $\theta(\mathbf{x}_n)$ of curve points as a function of path length n.

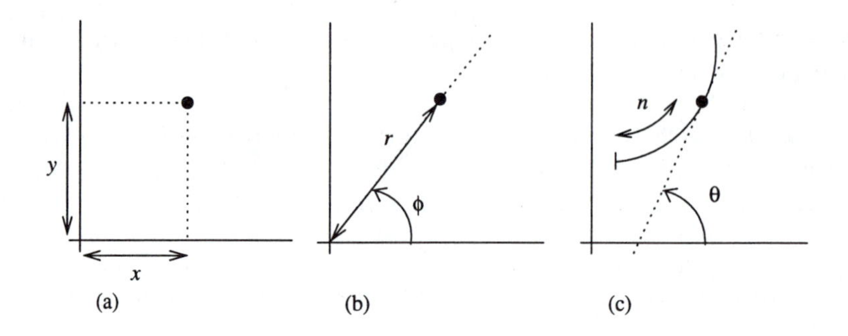

Figure 6.5: *Co-ordinate systems: (a) rectangular (Cartesian); (b) polar; (c) tangential.*

6.2.1 Chain codes

Chain codes describe an object by a sequence of unit-size line segments with a given orientation (see Section 3.2.2). The first element of such a sequence must bear information about its position to permit the region to be reconstructed. The process results in a sequence of numbers (see Figure 6.6); to exploit the position invariance of chain codes the first element, which contains the position information, is omitted. This definition of the chain code is known as **Freeman's** code [Freeman 61]. Note that a chain code object description may easily be obtained as a by-product of border detection; see Section 5.2.3 for a description of border detection algorithms.

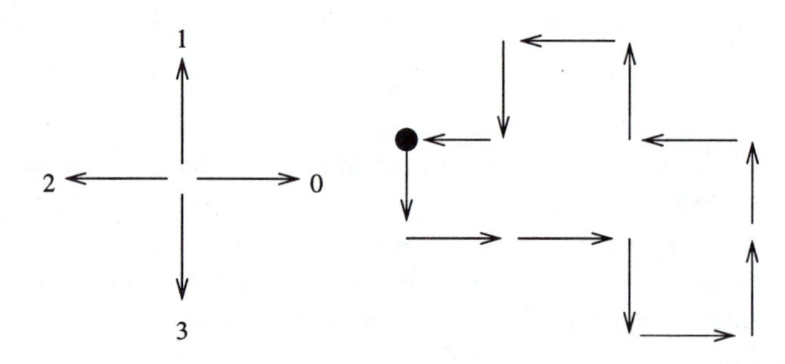

Figure 6.6: *Chain code in 4-connectivity, and its derivative. Code: 3, 0, 0, 3, 0, 1, 1, 2, 1, 2, 3, 2; derivative: 1, 0, 3, 1, 1, 0, 1, 3, 1, 1, 3, 1.*

If the chain code is used for matching, it must be independent of the choice of the first border pixel in the sequence. One possibility for normalizing the chain code is to find the pixel in the border sequence which results in the minimum integer number if the description chain is interpreted as a base 4 number—that pixel is then used as the starting pixel [Tsai and Yu 85]. A *mod 4* or *mod 8* difference code, called a chain code **derivative**, is another numbered sequence that represents relative directions of region boundary elements, measured as multiples of counter-clockwise 90° or 45° direction changes (Figure 6.6). A chain code is very sensitive to noise, and arbitrary changes in scale and rotation may cause problems if used for recognition. The smoothed version of the chain code (averaged directions along a specified path length) is less noise sensitive [Li and Zhiying 88].

6.2.2 Simple geometric border representation

The following descriptors are based mostly on geometric properties of described regions. Because of the discrete character of digital images, all of them are sensitive to image resolution.

Boundary length

Boundary length is an elementary region property, that is simply derived from the chain code representation. Vertical and horizontal steps have unit length, and the length of diagonal steps in 8-connectivity is $\sqrt{2}$. It can be shown that the boundary is longer in 4-connectivity, where a diagonal step consists of two rectangular steps with a total length of 2. A closed-boundary length (**perimeter**) can also be easily evaluated from run length [Rosenfeld and Kak 82] or quadtree representations [Samet 81, Crowley 84]. Boundary length increases as the image raster resolution increases; on the other hand, region area is not affected by higher resolution and converges to some limit (see also the description of fractal dimension in Section 14.1.6). To provide continuous-space perimeter properties (area computation from the boundary length, shape features, etc.), it is better to define the region border as being the outer or extended border (see Section 5.2.3). If inner borders are used, some properties are not satisfied—e.g., the perimeter of a 1-pixel region is 4 if the outer boundary is used, and 1 if the inner is used.

Curvature

In the continuous case, curvature is defined as the rate of change of slope. In discrete space, the curvature description must be slightly modified to overcome difficulties resulting from violation of curve smoothness. The curvature scalar descriptor (also called boundary straightness) finds the ratio between the total number of boundary pixels (length) and the number of boundary pixels where the boundary direction changes significantly. The smaller the number of direction changes, the straighter the boundary. The evaluation algorithm is based on the detection of angles between line segments positioned b boundary pixels from the evaluated boundary pixel in both directions. The angle need not be represented numerically; rather, relative position of line segments can be used as a property. The parameter b determines sensitivity to local changes of the boundary direction (Figure 6.7). Curvature computed from the chain code can be found in [Rosenfeld 74], and the tangential border representation is also suitable for curvature computation. Values of the curvature at all boundary pixels can be represented by a histogram; relative numbers then provide information on how common specific boundary direction changes are. Histograms of boundary angles, such as the β angle in Figure 6.7, can be built in a similar way—such histograms can be used for region description. Another approach to calculating curvature from digital curves is based on convolution with the truncated Gaussian kernel [Lowe 89], and an improved version not suffering from systematic bias caused by curvature smoothing effect is given in [Hlavac et al. 94].

Bending energy

The bending energy (BE) of a border (curve) may be understood as the energy necessary to bend a rod to the desired shape, and can be computed as a sum of squares of the border

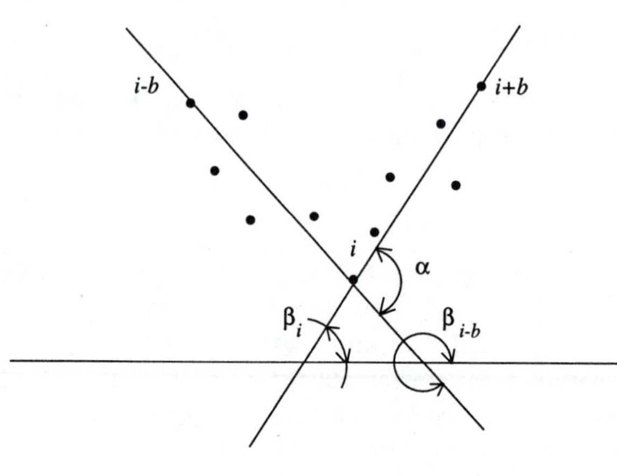

Figure 6.7: *Curvature.*

curvature $c(k)$ over the border length L.

$$\text{BE} = \frac{1}{L} \sum_{k=1}^{L} c^2(k) \tag{6.1}$$

Bending energy can easily be computed from Fourier descriptors using Parseval's theorem [Oppenheim et al. 83, Papoulis 91]. To represent the border, Freeman's chain code or its smoothed version may be used [Smeulders et al. 80]; see Figure 6.8. Bending energy does not permit shape reconstruction.

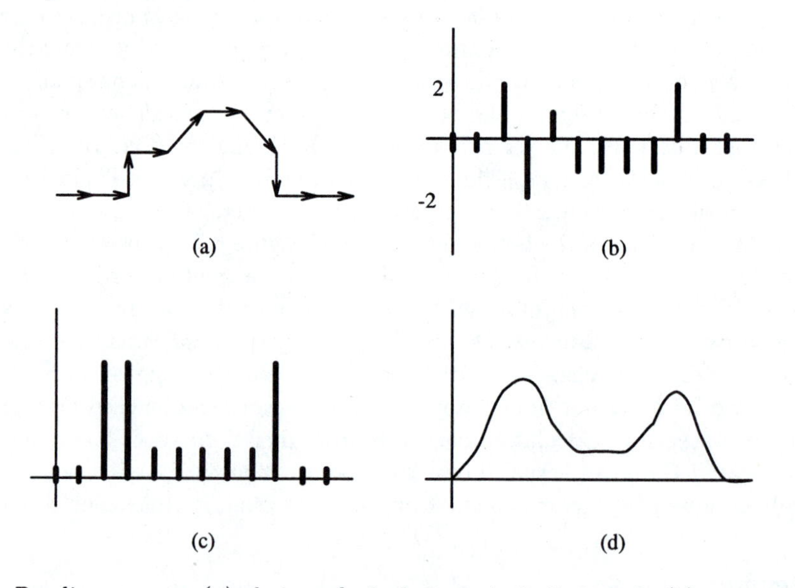

Figure 6.8: *Bending energy: (a) chain code 0, 0, 2, 0, 1, 0, 7, 6, 0, 0; (b) curvature 0, 2, -2, 1, -1, -1, -1, 2, 0; (c) sum of squares gives the bending energy; (d) smoothed version.*

Signature

The signature of a region may be obtained as a sequence of normal contour distances. The normal contour distance is calculated for each boundary element as a function of the path length. For each border point A, the shortest distance to an opposite border point B is sought in a direction perpendicular to the border tangent at point A; see Figure 6.9. Note that *being opposite* is not a symmetric relation (compare Algorithm 5.17). Signatures are noise sensitive, and using smoothed signatures or signatures of smoothed contours reduces noise sensitivity. Signatures may be applied to the recognition of overlapping objects or whenever only partial contours are available [Vernon 87]. Position, rotation, and scale-invariant modifications based on gradient-perimeter and angle-perimeter plots are discussed in [Safaee-Rad et al. 89].

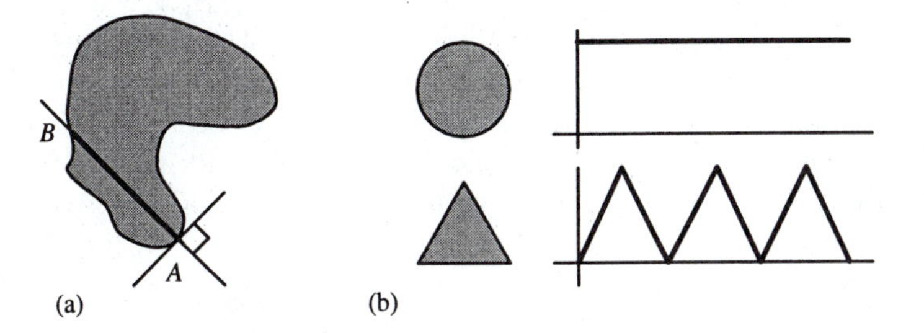

Figure 6.9: *Signature: (a) construction; (b) signatures for a circle and a triangle.*

Chord distribution

A line joining any two points of the region boundary is a chord, and the distribution of lengths and angles of all chords on a contour may be used for shape description. Let $b(x,y) = 1$ represent the contour points, and $b(x,y) = 0$ represent all other points. The chord distribution can be computed (see Figure 6.10a) as

$$h(\Delta x, \Delta y) = \int \int b(x,y) b(x + \Delta x, y + \Delta y)\, dx\, dy \tag{6.2}$$

or in digital images as

$$h(\Delta x, \Delta y) = \sum_i \sum_j b(i,j) b(i + \Delta x, j + \Delta y) \tag{6.3}$$

To obtain the rotation-independent radial distribution $h_r(r)$, the integral over all angles is computed (Figure 6.10b).

$$h_r(r) = \int_{-\pi/2}^{\pi/2} h(\Delta x, \Delta y) r\, d\theta \tag{6.4}$$

where $r = \sqrt{\Delta x^2 + \Delta y^2}$, $\theta = \sin^{-1}(\Delta y / r)$. The distribution $h_r(r)$ varies linearly with scale. The angular distribution $h_a(\theta)$ is independent of scale, while rotation causes a proportional offset.

$$h_a(\theta) = \int_0^{\max(r)} h(\Delta x, \Delta y)\, dr \tag{6.5}$$

Combination of both distributions gives a robust shape descriptor [Smith and Jain 82, Cootes et al. 92].

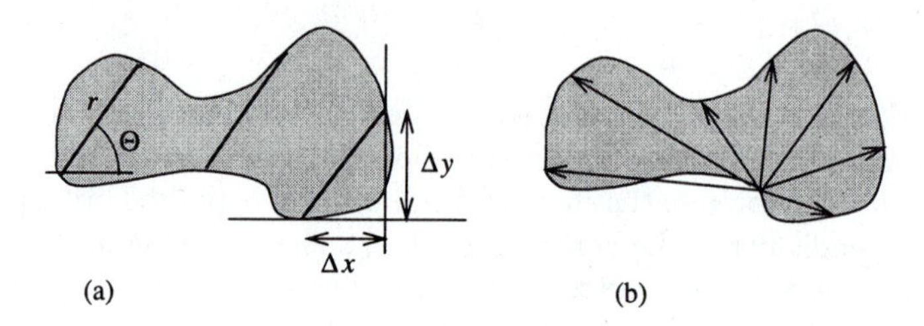

Figure 6.10: *Chord distribution.*

6.2.3 Fourier transforms of boundaries

Suppose C is a closed curve (boundary) in the complex plane (Figure 6.11a). Traveling anti-clockwise along this curve keeping constant speed, a complex function $z(t)$ is obtained, where t is a time variable. The speed should be chosen such that one circumnavigation of the boundary takes time 2π; then a periodic function with period 2π is obtained after multiple passes around the curve. This permits a Fourier representation of $z(t)$ (see Section 2.1.3),

$$z(t) = \sum_n T_n e^{\mathrm{int}} \tag{6.6}$$

The coefficients T_n of the series are called the **Fourier descriptors** of the curve C. It is more useful to consider the curve distance s in comparison to time

$$t = 2\pi s / L \tag{6.7}$$

where L is the curve length. The Fourier descriptors T_n are given by

$$T_n = \frac{1}{L} \int_0^L z(s) e^{-i(2\pi/L)ns} \, ds \tag{6.8}$$

The descriptors are influenced by the curve shape and by the initial point of the curve. Working with digital image data, boundary co-ordinates are discrete and the function $z(s)$ is not continuous. Assume that $z(k)$ is a discrete version of $z(s)$, where 4-connectivity is used to get a constant sampling interval; the descriptors T_n can be computed from the discrete Fourier transform (DFT, Section 12.2) of $z(k)$:

$$z(k) \longleftarrow \text{DFT} \longrightarrow T_n \tag{6.9}$$

The Fourier descriptors can be invariant to translation and rotation if the co-ordinate system is appropriately chosen [Pavlidis 77, Persoon and Fu 77, Wallace and Wintz 80, Grimmins 82, Lin and Chellappa 87]. They have been used for handwritten alphanumeric character description in [Shridhar and Badreldin 84]; the character boundary in this description was represented by co-ordinate pairs (x_m, y_m) in 4-connectivity, $(x_1, y_1) = (x_L, y_L)$. Then

$$a_n = \frac{1}{L-1} \sum_{m=1}^{L-1} x_m e^{-i[2\pi/(L-1)]nm} \tag{6.10}$$

$$b_n = \frac{1}{L-1} \sum_{m=1}^{L-1} y_m e^{-i[2\pi/(L-1)]nm} \tag{6.11}$$

The coefficients a_n, b_n are not invariant but after the transform,

$$r_n = (|\, a_n \,|^2 + |\, b_n \,|^2)^{1/2} \tag{6.12}$$

r_n are translation and rotation invariant. To achieve a magnification invariance the descriptors w_n are used:

$$w_n = r_n / r_1 \tag{6.13}$$

The first 10–15 descriptors w_n are found to be sufficient for character description.

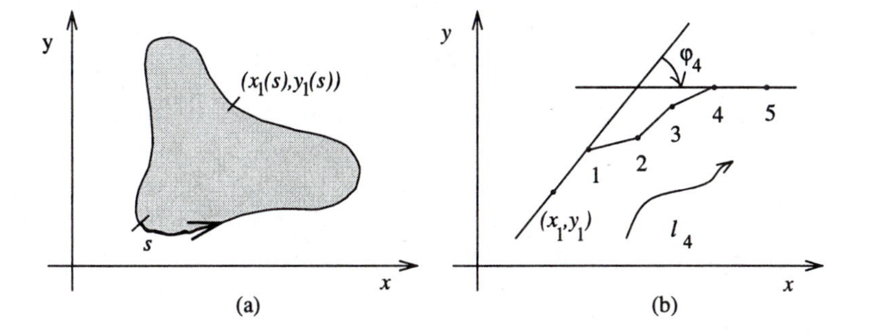

Figure 6.11: *Fourier description of boundaries: (a) Descriptors T_n, (b) descriptors S_n.*

A closed boundary can be represented as a function of angle tangents versus the distance between the boundary points from which the angles were determined (Figure 6.11b). Let φ_k be the angle measured at the kth boundary point, and let l_k be the distance between the boundary starting point and the kth boundary point. A periodic function can be defined

$$a(l_k) = \varphi_k + u_k \tag{6.14}$$

$$u_k = 2\pi l_k / L \tag{6.15}$$

The descriptor set is then

$$S_n = \frac{1}{2\pi} \int_0^{2\pi} a(u) e^{-inu}\, du \tag{6.16}$$

The discrete Fourier transform is used in all practical applications [Pavlidis 77].

The high-quality boundary shape representation obtained using only a few lower-order coefficients is a favorable property common to Fourier descriptors. We can compare the results of using the S_n and T_n descriptors: The S_n descriptors have more high-frequency components present in the boundary function due to more significant changes of tangent angles, and as a result, they do not decrease as fast as the T_n descriptors. In addition, the S_n descriptors are not suitable for boundary reconstruction since they often result in a non-closed boundary. A method for obtaining a closed boundary using S_n descriptors is given in [Strackee and Nagelkerke 83]. The T_n descriptor values decrease quickly for higher frequencies, and their reconstruction always results in a closed boundary. Moreover, the S_n descriptors cannot be applied for squares, equilateral triangles, etc. [Wallace 81] unless the solution methods introduced in [Wallace and Wintz 80] are applied.

Fourier descriptors can also be used for calculation of region area, location of centroid, and computation of second-order moments [Kiryati and Maydan 89]. Fourier descriptors are a general technique, but problems with describing local information exist. A modified technique using a combined frequency-position space that deals better with local curve properties is described in [Eichmann et al. 90], and another modification that is invariant under rotation, translation, scale, mirror reflection, and shifts in starting points is discussed in [Krzyzak et al. 89]. Conventional Fourier descriptors cannot be used for recognition of occluded objects. Nevertheless, classification of partial shapes using Fourier descriptors is introduced in [Lin and Chellappa 87]. Boundary detection and description using elliptic Fourier decomposition of the boundary is described in [Staib and Duncan 92].

6.2.4 Boundary description using segment sequences

Representation of a boundary using **segments** with specified properties is another option for boundary (and curve) description. If the segment type is known for all segments, the boundary can be described as a chain of segment types, a code word consisting of representatives of a type alphabet. An example is given in Figure 6.14 which will be discussed later in more detail. This sort of description is suitable for syntactic recognition (see Section 7.4). A trivial segment chain is used to obtain the Freeman code description discussed in Section 6.2.1.

A **polygonal representation** approximates a region by a polygon, the region being represented using its vertices. Polygonal representations are obtained as a result of a simple boundary segmentation. The boundary can be approximated with varying precision; if a more precise description is necessary, a larger number of line segments may be employed. Any two boundary points $\mathbf{x}_1$, $\mathbf{x}_2$ define a line segment, and a sequence of points $\mathbf{x}_1$, $\mathbf{x}_2$, $\mathbf{x}_3$ represents a chain of line segments—from the point $\mathbf{x}_1$ to the point $\mathbf{x}_2$, and from $\mathbf{x}_2$ to $\mathbf{x}_3$. If $\mathbf{x}_1=\mathbf{x}_3$, a closed boundary results. There are many types of straight-segment boundary representations [Pavlidis 77, Koch and Kashyap 87, Matas and Kittler 93, Lindenbaum and Bruckstein 93, Ji and Haralick 97]; the problem lies in determining the location of boundary vertices, one solution to which is to apply a split-and-merge algorithm. The merging step consists of going through a set of boundary points and adding them to a straight segment as long as a segment straightness criterion is satisfied. If the straightness characteristic of the segment is lost, the last connected point is marked as a vertex and construction of a new straight segment begins. This general approach has many variations, some of which are described in [Pavlidis 77].

Boundary vertices can be detected as boundary points with a significant change of boundary direction using the curvature (boundary straightness) criterion (see Section 6.2.2). This approach works well for boundaries with rectilinear boundary segments.

Another method for determining the boundary vertices is a **tolerance interval approach** based on setting a maximum allowed difference e. Assume that point $\mathbf{x}_1$ is the end point of a previous segment and so by definition the first point of a new segment. Define points $\mathbf{x}_2$, $\mathbf{x}_3$ positioned a distance e from the point $\mathbf{x}_1$ to be rectilinear—$\mathbf{x}_1$, $\mathbf{x}_2$, $\mathbf{x}_3$ are positioned on a straight line—see Figure 6.12. The next step is to locate a segment which can fit between parallels directed from points $\mathbf{x}_2$ and $\mathbf{x}_3$. Resulting segments are sub-optimal, although optimality can be achieved with a substantial increase in computational effort [Tomek 74].

The methods introduced above represent single-pass algorithms of boundary segmentation using a segment-growing approach. Often they do not result in the best possible boundary segmentation because the vertex which is located often indicates that the real vertex should

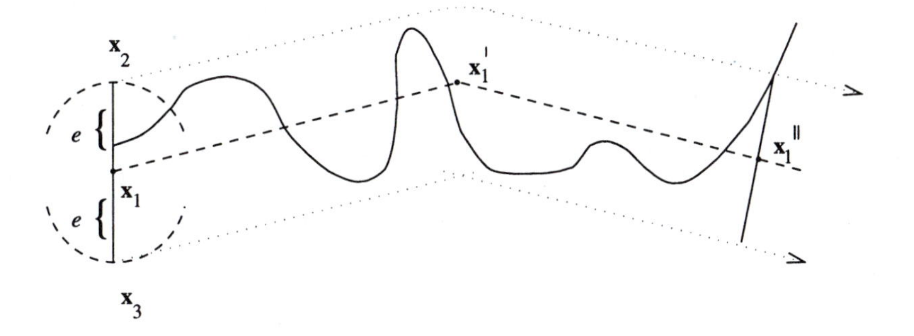

Figure 6.12: *Tolerance interval.*

have been located a few steps back. The splitting approach of segmenting boundaries into smaller segments can sometimes help, and the best results can be anticipated using a combination of both methods. If the splitting approach is used, segments are usually divided into two new, smaller segments until the new segments meet the final requirements [Duda and Hart 73, Pavlidis 77]. A simple procedure for splitting begins from end points $\mathbf{x}_1$ and $\mathbf{x}_2$ of a curve; these end points are connected by a line segment. The next step searches all the curve points for the curve point $\mathbf{x}_3$ with the largest distance from the line segment. If the point located is within a preset distance between itself and the line segment, the segment $\mathbf{x}_1$–$\mathbf{x}_2$ is an end segment and all curve vertices are found, the curve being represented polygonally by vertices $\mathbf{x}_1$ and $\mathbf{x}_2$. Otherwise the point $\mathbf{x}_3$ is set as a new vertex and the process is applied recursively to both resulting segments $\mathbf{x}_1$–$\mathbf{x}_3$ and $\mathbf{x}_3$–$\mathbf{x}_2$ (see Figure 6.13 and Section 5.2.7).

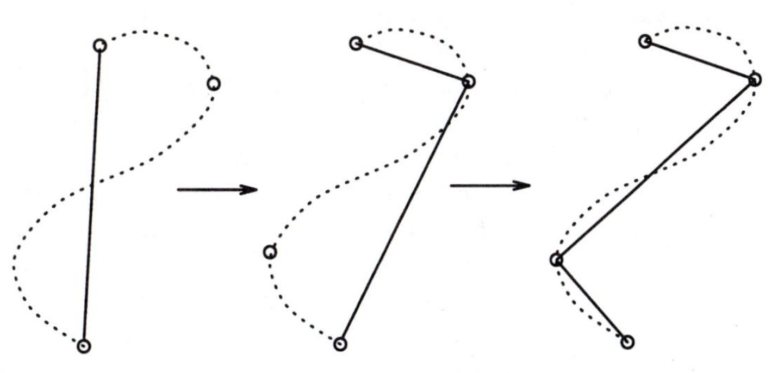

Figure 6.13: *Recursive boundary splitting.*

Boundary segmentation into segments of **constant curvature** is another possibility for boundary representation. The boundary may also be split into segments which can be represented by polynomials, usually of second order, such as circular, elliptic, or parabolic segments [Costabile et al. 85, Wuescher and Boyer 91]. Curve segmentation into circular arcs and straight lines is presented in [Rosin and West 89]. Segments are considered as primitives for syntactic shape recognition procedures—a typical example is the syntactic description and recognition of chromosomes [Fu 74], where boundary segments are classified as convex segments of large curvature, concave segments of large curvature, straight segments, etc., as illustrated in Figure 6.14.

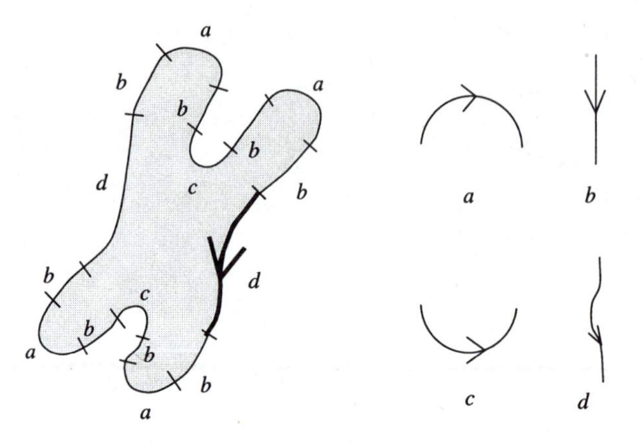

Figure 6.14: *Structural description of chromosomes by a chain of boundary segments, code word: d, b, a, b, c, b, a, b, d, b, a, b, c, b, a, b (adapted from [Fu 74]).*

Other syntactic object recognition methods based on a contour partitioning into primitives from a specified set are described in [Jakubowski 85, Jakubowski 90, Tampi and Sridhar 90]. Partitioning of the contour using location of points with high positive curvatures (corners) is described in [Chien and Aggarwal 89], together with applications to occluded contours. A discrete curvature function based on a chain code representation of a boundary is used with a morphological approach to obtain segments of constant curvature in [Leymarie and Levine 89]. Contour partitioning using segments of constant intensity is suggested in [Marshall 89a], and polygonal representation used in a *hypothesize and verify* approach to recognition of occluded objects may be found in [Koch and Kashyap 87].

Sensitivity of shape descriptors to scale (image resolution) has already been mentioned as an undesirable feature of a majority of descriptors. In other words, shape description varies with scale, and different results are achieved at different resolutions. This problem is no less important if a curve is to be divided into segments; some curve segmentation points exist in one resolution and disappear in others without any direct correspondence. Considering this, a **scale-space** approach to curve segmentation that guarantees a continuously changing position of segmentation points is a significant achievement [Babaud et al. 86, Witkin 86, Yuille and Poggio 86, Maragos 89, Florack et al. 92, Griffin et al. 92]. In this approach, only new segmentation points can appear at higher resolutions, and no existing segmentation points can disappear. This is in agreement with our understanding of varying resolutions; finer details can be detected in higher resolution, but significant details should not disappear if the resolution increases. This technique is based on application of a unique Gaussian smoothing kernel to a one-dimensional signal (e.g., a curvature function) over a range of sizes and the result is differentiated twice. To determine the peaks of curvature, the zero-crossing of the second derivative is detected; the positions of zero-crossings give the positions of curve segmentation points. Different locations of segmentation points are obtained at varying resolution (different Gaussian kernel size). An important property of the Gaussian kernel is that the location of segmentation points changes continuously with resolution which can be seen in the **scale-space image** of the curve, Figure 6.15a. Fine details of the curve disappear in pairs with increasing size of the Gaussian smoothing kernel, and two segmentation points always merge to form a closed contour, showing that any segmentation point existing in coarse resolution must also exist in finer resolution. Moreover, the position of a segmentation point is most accurate in finest resolution, and this position can be traced from coarse to fine resolution using the scale-space image. A multi-scale curve description can be represented by an **interval tree**, Figure 6.15b. Each pair of zero-crossings is represented by a rectangle, its position corresponding with segmentation point locations on the curve, its height showing the lowest resolution at which the segmentation point can be detected. Interval trees can be used for curve decomposition in different scales, keeping the possibility of segment description using higher-resolution features.

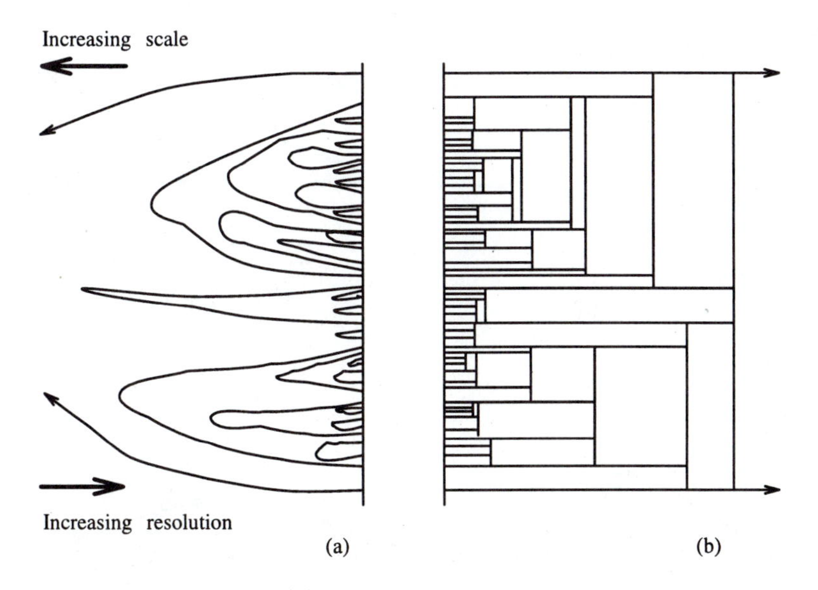

Figure 6.15: *Scale-space image: (a) varying number and locations of curve segmentation points as a function of scale; (b) curve representation by an interval tree.*

Another scale-space approach to curve decomposition is the **curvature primal sketch** [Asada and Brady 86] (compare Section 9.1.1). A set of primitive curvature discontinuities is defined and convolved with first and second derivatives of a Gaussian in multiple resolutions. The curvature primal sketch is computed by matching the multi-scale convolutions of a shape. The curvature primal sketch then serves as a shape representation; shape reconstruction may be based on polygons or splines. Another multi-scale border-primitive detection technique that aggregates curve primitives at one scale into curve primitives at a coarser scale is described in [Saund 90]. A robust approach to multi-scale curve corner detection that uses additional information extracted from corner behavior in the whole multi-resolution pyramid is given in [Fermuller and Kropatsch 92].

6.2.5 B-spline representation

Representation of curves using piecewise polynomial interpolation to obtain smooth curves is widely used in computer graphics. B-splines are piecewise polynomial curves whose shape is closely related to their control polygon—a chain of vertices giving a polygonal representation of a curve. B-splines of the third order are most common because this is the lowest order which includes the change of curvature. Splines have very good representation properties and are easy to compute: First, they change their shape less then their control polygon, and they do not oscillate between sampling points as many other representations do. Furthermore, a spline curve is always positioned inside a convex $n+1$-polygon for a B-spline of the nth order—Figure 6.16. Second, the interpolation is local in character. If a control polygon vertex changes its position, a resulting change of the spline curve will occur in only a small neighborhood of that vertex. Third, methods of matching region boundaries represented by splines to image data are based on a direct search of original image data. These methods are similar to the segmentation methods described in Section 5.2.6. A spline direction can be derived directly from its parameters.

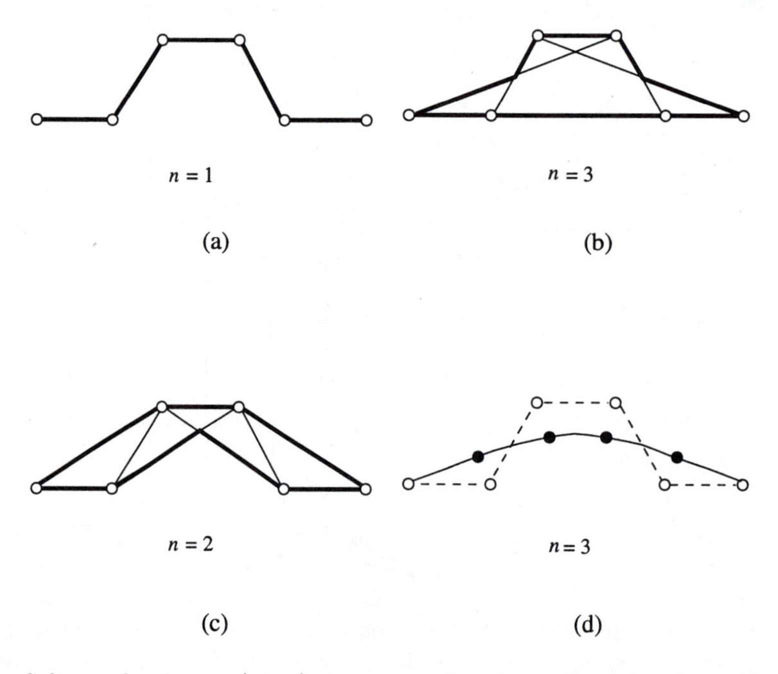

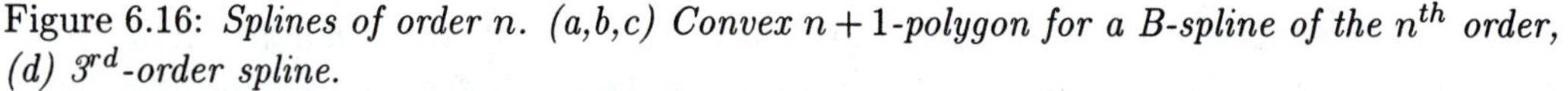

Figure 6.16: *Splines of order n. (a,b,c) Convex $n+1$-polygon for a B-spline of the nth order, (d) 3rd-order spline.*

Let $\mathbf{x}_i$, $i = 1, \ldots, n$ be points of a B-spline interpolation curve; call this interpolation curve $\mathbf{x}(s)$. The s parameter changes linearly between points $\mathbf{x}_i$—that is, $\mathbf{x}_i = \mathbf{x}(i)$. Each part of a cubic B-spline curve is a third-order polynomial, meaning that it and its first and second derivatives are continuous. B-splines are given by

$$\mathbf{x}(s) = \sum_{i=0}^{n+1} \mathbf{v}_i B_i(s) \tag{6.17}$$

where $\mathbf{v}_i$ are coefficients representing a spline curve, and $B_i(s)$ are base functions whose shape is given by the spline order. The coefficients $\mathbf{v}_i$ bear information dual to information about the spline curve points $\mathbf{x}_i$—the values $\mathbf{v}_i$ can be derived from $\mathbf{x}_i$ values and vice versa. The coefficients $\mathbf{v}_i$ represent vertices of the control polygon, and if there are n points $\mathbf{x}_i$, there must be $n+2$ points $\mathbf{v}_i$. The two end points $\mathbf{v}_0$, $\mathbf{v}_{n+1}$ are specified by binding conditions. If the curvature of a B-spline curvature is to be zero at the curve beginning and end, then

$$\mathbf{v}_0 = 2\mathbf{v}_1 - \mathbf{v}_2$$

$$\mathbf{v}_{n+1} = 2\mathbf{v}_n - \mathbf{v}_{n-1} \tag{6.18}$$

If the curve is closed, then $\mathbf{v}_0 = \mathbf{v}_n$ and $\mathbf{v}_{n+1} = \mathbf{v}_1$.

The base functions are non-negative and are of local importance only. Each base function $B_i(s)$ is non-zero only for $s \in (i-2, i+2)$, meaning that for any $s \in (i, i+1)$, there are only four non-zero base functions for any i: $B_{i-1}(s)$, $B_i(s)$, $B_{i+1}(s)$, and $B_{i+2}(s)$. If the distance between the $\mathbf{x}_i$ points is constant (e.g., unit distances), all the base functions are of the same form and consist of four parts $C_j(t)$, $j = 0, \dots, 3$.

$$\begin{aligned}
C_0(t) &= \frac{t^3}{6} \\
C_1(t) &= \frac{-3t^3 + 3t^2 + 3t + 1}{6} \\
C_2(t) &= \frac{3t^3 - 6t^2 + 4}{6} \\
C_3(t) &= \frac{-t^3 + 3t^2 - 3t + 1}{6}
\end{aligned} \tag{6.19}$$

Because of equation (6.17) and zero-equal base functions for $s \notin (i-2, i+2)$, $\mathbf{x}(s)$ can be computed from the addition of only four terms for any s.

$$\mathbf{x}(s) = C_{i-1,3}(s)\mathbf{v}_{i-1} + C_{i,2}(s)\mathbf{v}_i + C_{i+1,1}(s)\mathbf{v}_{i+1} + C_{i+2,0}(s)\mathbf{v}_{i+2} \tag{6.20}$$

Here, $C_{i,j}(s)$ means that we use the jth part of the base function B_i (see Figure 6.17). Note that

$$C_{i,j}(s) = C_j(s-i) \tag{6.21}$$

$$i = 0, \dots, n+1 \qquad j = 0, 1, 2, 3$$

To work with values inside the interval $[i, i+1)$, the interpolation curve $\mathbf{x}(s)$ can be computed as

$$\mathbf{x}(s) = C_3(s-i)\mathbf{v}_{i-1} + C_2(s-i)\mathbf{v}_i + C_1(s-i)\mathbf{v}_{i+1} + C_0\mathbf{v}_{i+2} \tag{6.22}$$

Specifically, if $s = 5$, s is positioned at the beginning of the interval $[i, i+1)$, therefore $i = 5$ and

$$\mathbf{x}(5) = C_3(0)\mathbf{v}_4 + C_2(0)\mathbf{v}_5 + C_1(0)\mathbf{v}_6 = \frac{1}{6}\mathbf{v}_4 + \frac{4}{6}\mathbf{v}_5 + \frac{1}{6}\mathbf{v}_6 \tag{6.23}$$

or, if $s = 7.7$, then $i = 7$ and

$$\mathbf{x}(5) = C_3(0.7)\mathbf{v}_6 + C_2(0.7)\mathbf{v}_7 + C_1(0.7)\mathbf{v}_8 + C_0(0.7)\mathbf{v}_9 \tag{6.24}$$

Other useful formulae can be found in [DeBoor 78, Ballard and Brown 82, Ikebe and Miyamoto 82].

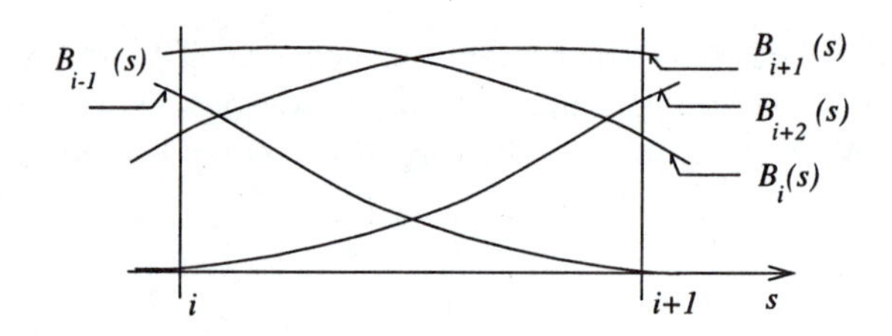

Figure 6.17: *The only four non-zero base functions for* $s \in (i, i+1)$.

Splines generate curves which are usually considered pleasing. They allow a good curve approximation, and can easily be used for image analysis curve representation problems. A technique transforming curve samples to B-spline control polygon vertices is described in [Paglieroni and Jain 88] together with a method of efficient computation of boundary curvature, shape moments, and projections from control polygon vertices. Splines differ in their complexity; one of the simplest applies the B-spline formula for curve modeling as well as for curve extraction from image data [DeBoor 78]. Splines are used in computer vision to form exact and flexible inner model representations of complex shapes which are necessary in model-driven segmentation and in complex image understanding tasks. On the other hand, splines are highly sensitive to change in scale.

6.2.6 Other contour-based shape description approaches

Many other methods and approaches can be used to describe two-dimensional curves and contours.

The **Hough transform** has excellent shape description abilities and is discussed in detail in the image segmentation context in Section 5.2.6 (see also [McKenzie and Protheroe 90]). Region-based shape description using **statistical moments** is covered in Section 6.3.2 where a technique of contour-based moments computation from region borders is also included. Further, it is necessary to mention the **fractal** approach to shape [Mandelbrot 82, Barnsley 88, Falconer 90] which is gaining growing attention in image shape description [Frisch et al. 87, Chang and Chatterjee 89, Vemuri and Radisavljevic 93, Taylor and Lewis 94].

Mathematical morphology can be used for shape description, typically in connection with region skeleton construction (see Section 6.3.4) [Reinhardt and Higgins 96]. A different approach is introduced in [Loui et al. 90], where a **geometrical correlation function** represents two-dimensional continuous or discrete curves. This function is translation, rotation, and scale invariant and may be used to compute basic geometrical properties.

Neural networks (Section 7.3) can be used to recognize shapes in raw boundary representations directly. Contour sequences of noiseless reference shapes are used for training, and noisy data are used in later training stages to increase robustness; effective representations of closed planar shapes result [Gupta et al. 90]. Another neural network shape representation system uses a modified Walsh-Hadamard transform (Chapter 12) to achieve position-invariant shape representation [Minnix et al. 90].

6.2.7 Shape invariants

Shape invariants represent a very active current research area in machine vision. Although the importance of shape invariance has been known for a long time, the first machine vision-related paper about shape invariants [Weiss 88] appeared in 1988, followed by a book [Kanatani 90] in 1990. The following section gives a brief overview of this topic and is based mostly on a paper [Forsyth et al. 91] and on a book [Mundy and Zisserman 92] in which additional details can be found. The book [Mundy and Zisserman 92] gives an overview of this topic in its Introduction, and its Appendix presents an excellent and detailed survey of projective geometry for machine vision. Even if shape invariance is a novel approach in machine vision, invariant theory is not new, and many of its principles were introduced in the nineteenth century.

As has been mentioned many times, object description is necessary for object recognition. Unfortunately, all the shape descriptors discussed so far depend on viewpoint, meaning that object recognition may often be impossible as a result of changed object or observer position, as illustrated in Figure 6.18. The role of shape description invariance is obvious—shape invariants represent properties of such geometric configurations which remain unchanged under an appropriate class of transforms [Mundy and Zisserman 92, Reiss 93]. Machine vision is especially concerned with the class of projective transforms.

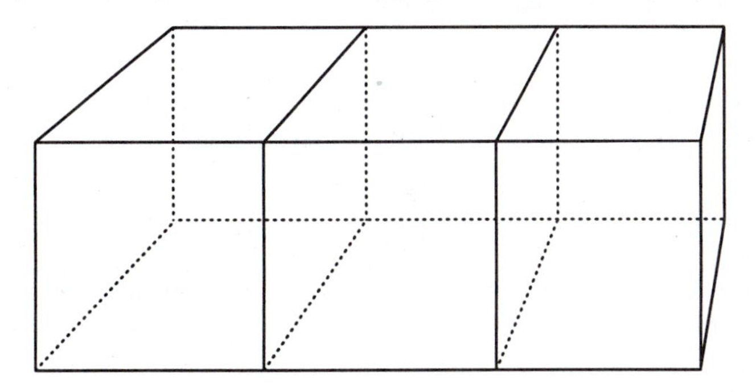

Figure 6.18: *Change of shape caused by a projective transform. The same rectangular cross section is represented by different polygons in the image plane.*

Collinearity is the simplest example of a projectively invariant image feature. Any straight line is projected as a straight line under any projective transform. Similarly, the basic idea of the projection-invariant shape description is to find such shape features that are unaffected by the transform between the object and the image plane.

A standard technique of projection-invariant description is to hypothesize the pose (position and orientation) of an object and transform this object into a specific co-ordinate system; then shape characteristics measured in this co-ordinate system yield an invariant description. However, the pose must be hypothesized for each object and each image, which makes this approach difficult and unreliable.

Application of **invariant theory**, where invariant descriptors can be computed directly from image data without the need for a particular co-ordinate system, represents another approach. In addition, invariant theory can determine the total number of functionally independent invariants for a given situation, therefore showing completeness of the description invariant set. Invariant theory is based on a collection of transforms that can be composed and inverted. In vision, the **plane-projective group** of transforms is considered which contains all the perspectives as a subset. The **group approach** provides a mathematical tool for generating invariants; if the transform does not satisfy the properties of a group, this machinery is not available [Mundy and Zisserman 92]. Therefore, the change of co-ordinates due to the plane-projective transform is generalized as a **group action. Lie group** theory is especially useful in designing new invariants.

Let corresponding entities in two different co-ordinate systems be distinguished by capital and lowercase letters. An invariant of a linear transformation is defined in [Mundy and Zisserman 92] as follows:

> An invariant, $I(\mathbf{P})$, of a geometric structure described by a parameter vector $\mathbf{P}$, subject to a linear transformation $\mathbf{T}$ of the co-ordinates $\mathbf{x} = \mathbf{TX}$, is transformed according to $I(\mathbf{p}) = I(\mathbf{P})|\mathbf{T}|^w$. Here $I(\mathbf{p})$ is the function of the parameters after the linear transformation, and $|\mathbf{T}|$ is the determinant of the matrix $\mathbf{T}$.

In this definition, w is referred to as the weight of the invariant. If $w = 0$, the invariants are called **scalar invariants**, which are considered below. Invariant descriptors are unaffected by object pose, by perspective projection, and by the intrinsic parameters of the camera.

Several examples of invariants are now given.

1. **Cross ratio**: The cross ratio represents a classic invariant of a projective line. As mentioned earlier, a straight line is always projected as a straight line. Any four collinear points A, B, C, D may be described by the cross-ratio invariant

$$I = \frac{(A-C)(B-D)}{(A-D)(B-C)} \tag{6.25}$$

where $(A - C)$ represents the distance between points A and C (see Figure 6.19). Note that the cross ratio depends on the order in which the four collinear points are labeled.

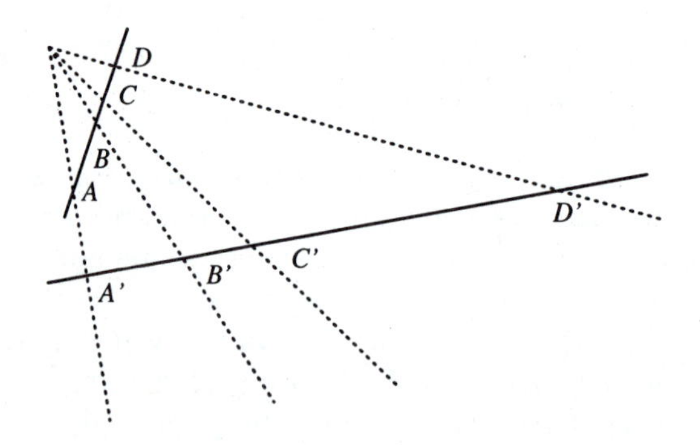

Figure 6.19: *Cross ratio; four collinear points form a projective invariant.*

2. **Systems of lines or points**: A system of four co-planar concurrent lines (meeting at the same point) is dual to a system of four collinear points and the cross ratio is its invariant; see Figure 6.19.

A system of five general co-planar lines forms two invariants,

$$I_1 = \frac{|\mathbf{M}_{431}||\mathbf{M}_{521}|}{|\mathbf{M}_{421}||\mathbf{M}_{531}|} \qquad I_2 = \frac{|\mathbf{M}_{421}||\mathbf{M}_{532}|}{|\mathbf{M}_{432}||\mathbf{M}_{521}|} \tag{6.26}$$

where $\mathbf{M}_{ijk} = (\mathbf{l}_i, \mathbf{l}_j, \mathbf{l}_k)$. $\mathbf{l}_i = (l_i^1, l_i^2, l_i^3)^T$ is a representation of a line $l_i^1 x + l_i^2 y + l_i^3 = 0$, where $i \in [1,5]$, and $|\mathbf{M}|$ is the determinant of $\mathbf{M}$. If the three lines forming the matrix $\mathbf{M}_{ijk}$ are concurrent, the matrix becomes singular and the invariant is undefined.

A system of five co-planar points is dual to a system of five lines and the same two invariants are formed. These two functional invariants can also be formed as two cross ratios of two co-planar concurrent line quadruples; see Figure 6.20. Note that even though combinations other than those given in Figure 6.20 may be formed, only the two presented functionally independent invariants exist.

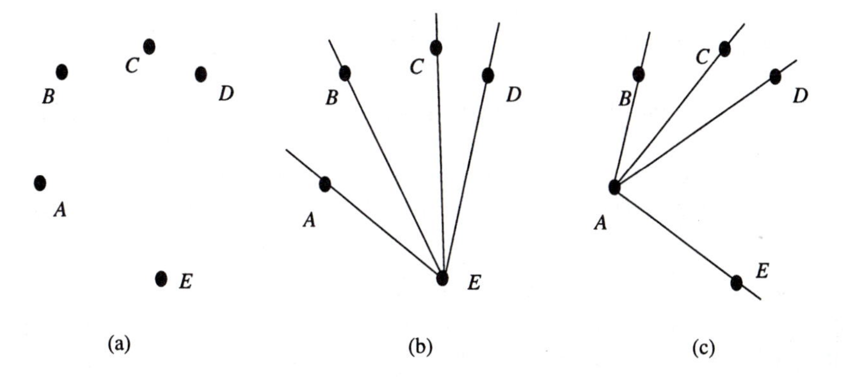

Figure 6.20: *Five co-planar points form two cross-ratio invariants: (a) co-planar points; (b) five points form a system of four concurrent lines; (c) the same five points form another system of four co-planar lines.*

3. **Plane conics**: A plane conic may be represented by an equation

$$ax^2 + bxy + cy^2 + dx + ey + f = 0 \tag{6.27}$$

for $\mathbf{x} = (x, y, 1)^T$. Then the conic may also be defined by a matrix $\mathbf{C}$,

$$\mathbf{C} = \begin{vmatrix} a & b/2 & d/2 \\ b/2 & c & e/2 \\ d/2 & e/2 & f \end{vmatrix}$$

and

$$\mathbf{x}^T \mathbf{C} \mathbf{x} = 0 \tag{6.28}$$

For any conic represented by a matrix $\mathbf{C}$, and any two co-planar lines not tangent to the conic, one invariant may be defined:

$$I = \frac{(\mathbf{l}_1^T \mathbf{C}^{-1} \mathbf{l}_2)^2}{(\mathbf{l}_1^T \mathbf{C}^{-1} \mathbf{l}_1)(\mathbf{l}_2^T \mathbf{C}^{-1} \mathbf{l}_2)} \tag{6.29}$$

The same invariant can be formed for a conic and two co-planar points.

Two invariants can be determined for a pair of conics represented by their respective matrices $\mathbf{C}_1, \mathbf{C}_2$ normalized so that $|\mathbf{C}_i| = 1$.

$$I_1 = \text{Trace}[\mathbf{C}_1^{-1}\mathbf{C}_2] \qquad I_2 = \text{Trace}[\mathbf{C}_2^{-1}\mathbf{C}_1] \tag{6.30}$$

(The trace of a matrix is calculated as the sum of elements on the main diagonal.) For non-normalized conics, the invariants of associated quadratic forms are

$$I_1 = \text{Trace}[\mathbf{C}_1^{-1}\mathbf{C}_2]\left(\frac{|\mathbf{C}_1|}{|\mathbf{C}_2|}\right)^{\frac{1}{3}} \qquad I_2 = \text{Trace}[\mathbf{C}_2^{-1}\mathbf{C}_1]\left(\frac{|\mathbf{C}_2|}{|\mathbf{C}_1|}\right)^{\frac{1}{3}} \tag{6.31}$$

and two true invariants of the conics are [Quan et al. 92]

$$I_1 = \frac{\text{Trace}[\mathbf{C}_1^{-1}\mathbf{C}_2]}{(\text{Trace}[\mathbf{C}_2^{-1}\mathbf{C}_1])^2}\frac{|\mathbf{C}_1|}{|\mathbf{C}_2|} \qquad I_2 = \frac{\text{Trace}[\mathbf{C}_2^{-1}\mathbf{C}_1]}{(\text{Trace}[\mathbf{C}_1^{-1}\mathbf{C}_2])^2}\frac{|\mathbf{C}_2|}{|\mathbf{C}_1|} \tag{6.32}$$

An interpretation of these invariants is given in [Maybank 92]. Two plane conics uniquely determine four points of intersection, and any point that is not an intersection point may be chosen to form a five-point system together with the four intersection points. Therefore, two invariants exist for the pair of conics, as for the five-point system.

Many man-made objects consist of a combination of straight lines and conics, and these invariants may be used for their description. However, if the object has a contour which cannot be represented by an algebraic curve, the situation is much more difficult. **Differential invariants** can be formed (e.g., curvature, torsion, Gaussian curvature) which are not affected by projective transforms. These invariants are local—that is, the invariants are found for each point on the curve, which may be quite general. Unfortunately, these invariants are extremely large and complex polynomials, requiring up to seventh derivatives of the curve, which makes them practically unusable due to image noise and acquisition errors, although noise-resistant local invariants are beginning to appear [Weiss 92]. However, if additional information is available, higher derivatives may be avoided. In [Brill et al. 92, Van Gool et al. 92], higher derivatives are traded for extra reference points which can be detected on curves in different projections, although the necessity of matching reference points in different projections brings other difficulties.

Designing new invariants is an important part of invariant theory in its application to machine vision. The easiest way is to combine primitive invariants, forming new ones from these combinations. Nevertheless, no new information is obtained from these combinations. Further, complete tables of invariants for systems of vectors under the action of the rotation group, the affine transform group, and the general linear transform group may be found in [Weyl 46]. To obtain new sets of functional invariants, several methods (eliminating transform parameters, the infinitesimal method, the symbolic method) can be found in [Forsyth et al. 91, Mundy and Zisserman 92].

Stability of invariants is another crucial property which affects their applicability. The robustness of invariants to image noise and errors introduced by image sensors is of prime importance, although not much is known about this. Results of plane-projective invariant stability testing (cross ratio, five co-planar points, two co-planar conics) can be found in

[Forsyth et al. 91, Hopcroft et al. 92]. Further, different invariants have different stabilities and distinguishing powers. It was found, for example [Rothwell et al. 92a], that measuring a single conic and two lines in a scene is too computationally expensive to be worthwhile. It is recommended to combine different invariants to enable fast object recognition.

An example of recognition of man-made objects using invariant description of four co-planar lines, a conic and two lines, and a pair of co-planar conics is given in [Rothwell et al. 92a]. The recognition system is based on a model library containing over 30 object models—significantly more than are reported for other recognition systems. Moreover, the construction of the model library is extremely easy; no special measurements are needed, the object is digitized in a standard way, and the projectively invariant description is stored as a model [Rothwell et al. 92b]. Further, there is no need for camera calibration. The recognition accuracy is 100% for occluded objects viewed from different viewpoints if the objects are not severely disrupted by shadows and specularities. An example of such object recognition is given in Figure 6.21.

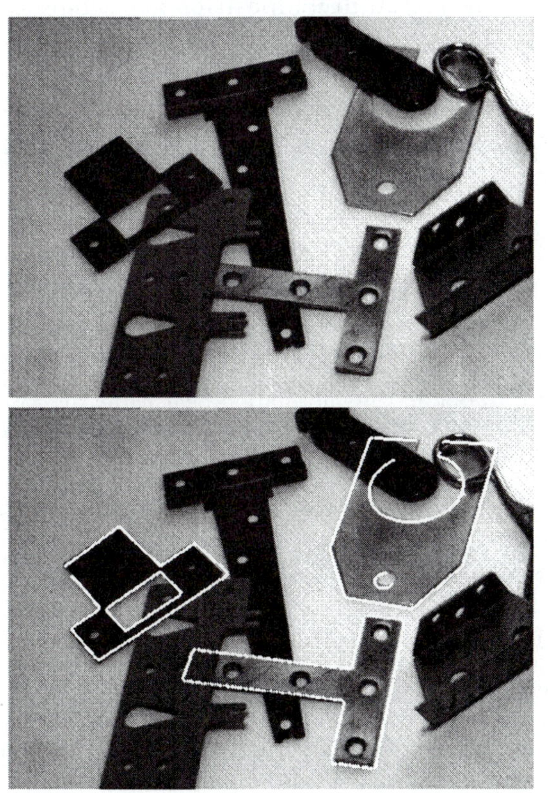

Figure 6.21: *Object recognition based on shape invariants: (a) original image of overlapping objects taken from an arbitrary viewpoint; (b) object recognition based on line and conic invariants. Courtesy D. Forsyth, The University of Iowa; C. Rothwell, A. Zisserman, University of Oxford; J. Mundy, General Electric Corporate Research and Development, Schenectady, NY.*

6.3 Region-based shape representation and description

We can use boundary information to describe a region, and shape can be described from the region itself. A large group of shape description techniques is represented by heuristic approaches which yield acceptable results in description of simple shapes. Region area, rectangularity, elongatedness, direction, compactness, etc., are examples of these methods. Unfortunately, they cannot be used for region reconstruction and do not work for more complex shapes. Other procedures based on region decomposition into smaller and simpler sub-regions must be applied to describe more complicated regions, then sub-regions can be described separately using heuristic approaches. Objects are represented by a planar graph with nodes representing sub-regions resulting from region decomposition, and region shape is then described by the graph properties [Rosenfeld 79, Bhanu and Faugeras 84, Turney et al. 85]. There are two general approaches to acquiring a graph of sub-regions: The first one is region thinning leading to the **region skeleton**, which can be described by a graph. The second option starts with the **region decomposition** into sub-regions, which are then represented by nodes, while arcs represent neighborhood relations of sub-regions. It is common to stipulate that sub-regions be convex.

Graphical representation of regions has many advantages; the resulting graphs

- Are translation and rotation invariant; position and rotation can be included in the graph definition
- Are insensitive to small changes in shape
- Are highly invariant with respect to region magnitude
- Generate a representation which is understandable
- Can easily be used to obtain the information-bearing features of the graph
- Are suitable for syntactic recognition

On the other hand, the shape representation can be difficult to obtain and the classifier-learning stage is not easy either (see Chapter 7). Nevertheless, if we are to get closer to the reality of computer vision, and to understand complex images, there is no alternative.

6.3.1 Simple scalar region descriptors

A number of simple heuristic shape descriptors exist which relate to statistical feature description. These methods are basic and are used for description of sub-regions in complex regions, and may then be used to define graph node classification [Bribiesca and Guzman 80].

Area

The simplest and most natural property of a region is its area, given by the number of pixels of which the region consists. The *real* area of each pixel may be taken into consideration to get the *real size* of a region, noting that in many cases, especially in satellite imagery, pixels in different positions correspond to different areas in the real world. If an image is represented as a rectangular raster, simple counting of region pixels will provide its area. If the image is represented by a quadtree, however, it may be more difficult to find the region area. Assuming that regions have been identified by labeling, the following algorithm may be used.

Algorithm 6.4: Calculating area in quadtrees

1. Set all region area variables to zero, and determine the global quadtree depth H; for example, the global quadtree depth is $H = 8$ for a 256×256 image.

2. Search the tree in a systematic way. If a leaf node at a depth h has a non-zero label, proceed to step 3.

3. Compute:
$$area[region\_label] = area[region\_label] + 4^{(H-h)}$$

4. The region areas are stored in variables $area[region\_label]$.

The region can be represented by n polygon vertices (i_k, j_k), and $(i_0, j_0) = (i_n, j_n)$. The area is given by

$$area = \frac{1}{2}\left|\sum_{k=0}^{n-1}(i_k j_{k+1} - i_{k+1} j_k)\right| \tag{6.33}$$

—the sign of the sum represents the polygon orientation. If a smoothed boundary is used to overcome noise sensitivity problems, the region area value resulting from equation (6.33) is usually somewhat reduced. Various smoothing methods and accurate area-recovering techniques are given in [Koenderink and van Doorn 86].

If the region is represented by the (anti-clockwise) Freeman chain code, the following algorithm provides the area.

Algorithm 6.5: Region area calculation from Freeman 4-connectivity chain code representation

1. Set the region *area* to zero. Assign the value of the starting point i co-ordinate to the variable *vertical_position*.

2. For each element of the chain code (values 0, 1, 2, 3) do

```
switch(code) {
    case 0:
        area := area - vertical_position;
        break;
    case 1:
        vertical_position := vertical_position + 1;
        break;
    case 2:
        area := area + vertical_position;
```

```
            break;
        case 3:
            vertical_position := vertical_position - 1;
            break;
}
```

3. If all boundary chain elements have been processed, the region area is stored in the variable *area*.

Euler's number

Euler's number ϑ (sometimes called **genus** or the **Euler-Poincaré characteristic**) describes a simple, topologically invariant property of the object. It is based on S, the number of contiguous parts of an object, and N, the number of holes in the object (an object can consist of more than one region, otherwise the number of contiguous parts is equal to one; see Section 2.3.1). Then

$$\vartheta = S - N \tag{6.34}$$

Special procedures to compute Euler's number can be found in [Dyer 80, Rosenfeld and Kak 82, Pratt 91], and in Chapter 11.

Projections

Horizontal and vertical region projections $p_h(i)$ and $p_v(j)$ are defined as

$$p_h(i) = \sum_j f(i,j) \qquad\qquad p_v(j) = \sum_i f(i,j) \tag{6.35}$$

Region description by projections is usually connected to binary image processing. Projections can serve as a basis for definition of related region descriptors; for example, the width (height) of a region with no holes is defined as the maximum value of the horizontal (vertical) projection of a binary image of the region. These definitions are illustrated in Figure 6.22. Note that projections can be defined in any direction. A practical example exploiting the use of projections is described in Section 16.1.

Eccentricity

The simplest eccentricity characteristic is the ratio of the length of the maximum chord A to the maximum chord B which is perpendicular to A (the ratio of major and minor axes of an object)—see Section 6.2.2, Figure 6.23. Another approximate eccentricity measure is based on a ratio of main region axes of inertia [Ballard and Brown 82, Jain 89].

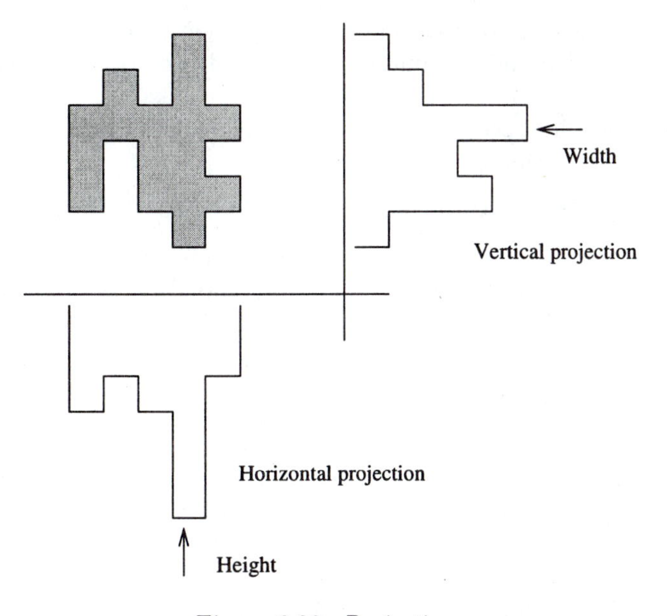

Figure 6.22: *Projections.*

Elongatedness

Elongatedness is a ratio between the length and width of the region bounding rectangle. This is the rectangle of minimum area that bounds the shape, which is located by turning in discrete steps until a minimum is located (see Figure 6.24a). This criterion cannot succeed in curved regions (see Figure 6.24b), for which the evaluation of elongatedness must be based on maximum region thickness. Elongatedness can be evaluated as a ratio of the region area and the square of its thickness. The maximum region thickness (holes must be filled if present) can be determined as the number of erosion steps (see Chapter 11) that may be applied before the region totally disappears. If the number of erosion steps is d, elongatedness is then

$$elongatedness = \frac{area}{(2d)^2} \tag{6.36}$$

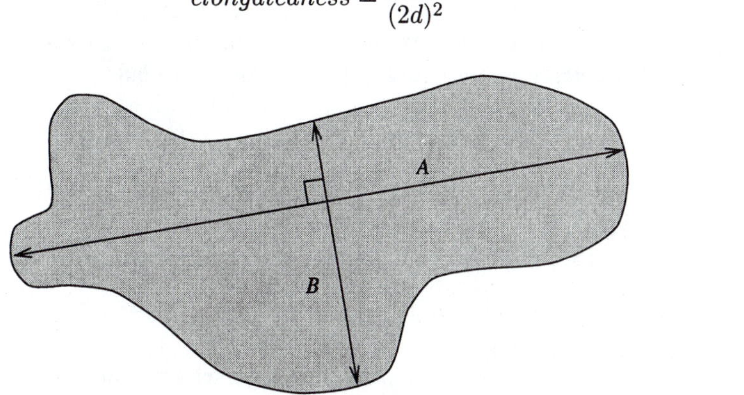

Figure 6.23: *Eccentricity.*

Another method based on longest central line detection is described in [Nagao and Matsuyama 80]; representation and recognition of elongated regions is also discussed in [Lipari and Harlow 88].

Note that the bounding rectangle can be computed efficiently from boundary points, if its direction θ is known. Defining

$$\alpha(x, y) = x \cos\theta + y \sin\theta \qquad \beta(x, y) = -x \sin\theta + y \cos\theta \tag{6.37}$$

search for the minimum and maximum of α and β over all boundary points (x, y). The values of $\alpha_{\min}, \alpha_{\max}, \beta_{\min}, \beta_{\max}$ then define the bounding rectangle, and $l_1 = (\alpha_{\max} - \alpha_{\min})$ and $l_2 = (\beta_{\max} - \beta_{\min})$ are its length and width.

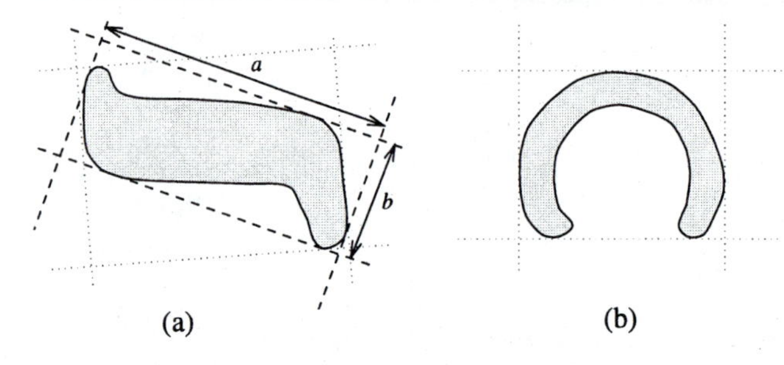

Figure 6.24: *Elongatedness: (a) bounding rectangle gives acceptable results; (b) bounding rectangle cannot represent elongatedness.*

Rectangularity

Let F_k be the ratio of region area and the area of a bounding rectangle, the rectangle having the direction k. The rectangle direction is turned in discrete steps as before, and **rectangularity** measured as a maximum of this ratio F_k:

$$rectangularity = \max_k(F_k) \tag{6.38}$$

The direction need only be turned through one quadrant. Rectangularity assumes values from the interval $(0, 1]$, with 1 representing a perfectly rectangular region. Sometimes, it may be more natural to draw a bounding triangle; a method for similarity evaluation between two triangles called **sphericity** is presented in [Ansari and Delp 90].

Direction

Direction is a property which makes sense in elongated regions only. If the region is elongated, **direction** is the direction of the longer side of a minimum bounding rectangle. If the shape moments are known (Section 6.3.2), the direction θ can be computed as

$$\theta = \frac{1}{2} \tan^{-1}\left(\frac{2\mu_{11}}{\mu_{20} - \mu_{02}}\right) \tag{6.39}$$

It should be noted that elongatedness and rectangularity are independent of linear transformations-translation, rotation, and scaling. Direction is independent on all linear transformations which do not include rotation. Mutual direction of two rotating objects is rotation invariant.

Compactness

Compactness is a popular shape description characteristic independent of linear transformations given by

$$compactness = \frac{(region\_border\_length)^2}{area} \tag{6.40}$$

The most compact region in a Euclidean space is a circle. Compactness assumes values in the interval $[1, \infty)$ in digital images if the boundary is defined as an inner boundary (see Section 5.2.3); using the outer boundary, compactness assumes values in the interval $[16, \infty)$. Independence from linear transformations is gained only if an outer boundary representation is used. Examples of a compact and a non-compact region are shown in Figure 6.25.

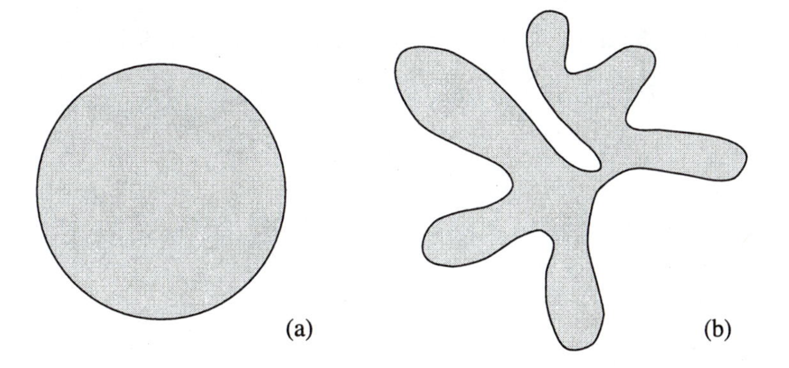

Figure 6.25: *Compactness: (a) compact; (b) non-compact.*

6.3.2 Moments

Region moment representations interpret a normalized gray-level image function as a probability density of a 2D random variable. Properties of this random variable can be described using statistical characteristics—**moments** [Papoulis 91]. Assuming that non-zero pixel values represent regions, moments can be used for binary or gray-level region description. A moment of order $(p+q)$ is dependent on scaling, translation, rotation, and even on gray-level transformations and is given by

$$m_{pq} = \int_{-\infty}^{\infty} \int_{-\infty}^{\infty} x^p y^q f(x, y)\, dx\, dy \tag{6.41}$$

In digitized images we evaluate sums:

$$m_{pq} = \sum_{i=-\infty}^{\infty} \sum_{j=-\infty}^{\infty} i^p j^q f(i, j) \tag{6.42}$$

where x, y, i, j are the region point co-ordinates (pixel co-ordinates in digitized images). Translation invariance can be achieved if we use the central moments,

$$\mu_{pq} = \int_{-\infty}^{\infty} \int_{-\infty}^{\infty} (x - x_c)^p (y - y_c)^q f(x, y)\, dx\, dy \tag{6.43}$$

or in digitized images,

$$\mu_{pq} = \sum_{i=-\infty}^{\infty} \sum_{j=-\infty}^{\infty} (i - x_c)^p (j - y_c)^q f(i,j) \tag{6.44}$$

where x_c, y_c are the co-ordinates of the region's center of gravity (centroid), which can be obtained using the following relationships:

$$x_c = \frac{m_{10}}{m_{00}} \tag{6.45}$$

$$y_c = \frac{m_{01}}{m_{00}}$$

In the binary case, m_{00} represents the region area [see equations (6.41) and (6.42)]. Scale-invariant features can also be found in scaled central moments η_{pq} (scale change $x' = \alpha x, y' = \alpha y$),

$$\eta_{pq} = \frac{\mu'_{pq}}{(\mu'_{00})^\gamma} \tag{6.46}$$

$$\gamma = \frac{p+q}{2} + 1$$

$$\mu'_{pq} = \frac{\mu_{pq}}{\alpha^{(p+q+2)}}$$

and normalized un-scaled central moments ϑ_{pq},

$$\vartheta_{pq} = \frac{\mu_{pq}}{(\mu_{00})^\gamma} \tag{6.47}$$

Rotation invariance can be achieved if the co-ordinate system is chosen such that $\mu_{11} = 0$ [Cash and Hatamian 87]. Many aspects of moment properties, normalization, descriptive power, sensitivity to noise, and computational cost are discussed in [Savini 88]. A less general form of invariance was given in [Hu 62] and is discussed in [Maitra 79, Jain 89, Pratt 91], in which seven rotation-, translation-, and scale-invariant moment characteristics are used.

$$\varphi_1 = \vartheta_{20} + \vartheta_{02} \tag{6.48}$$

$$\varphi_2 = (\vartheta_{20} - \vartheta_{02})^2 + 4\vartheta_{11}^2 \tag{6.49}$$

$$\varphi_3 = (\vartheta_{30} - 3\vartheta_{12})^2 + (3\vartheta_{21} - \vartheta_{03})^2 \tag{6.50}$$

$$\varphi_4 = (\vartheta_{30} + \vartheta_{12})^2 + (\vartheta_{21} + \vartheta_{03})^2 \tag{6.51}$$

$$\begin{aligned}\varphi_5 = \; & (\vartheta_{30} - 3\vartheta_{12})(\vartheta_{30} + \vartheta_{12})[(\vartheta_{30} + \vartheta_{12})^2 - 3(\vartheta_{21} + \vartheta_{03})^2] \\ & + (3\vartheta_{21} - \vartheta_{03})(\vartheta_{21} + \vartheta_{03})[3(\vartheta_{30} + \vartheta_{12})^2 - (\vartheta_{21} + \vartheta_{03})^2]\end{aligned} \tag{6.52}$$

$$\varphi_6 = (\vartheta_{20} - \vartheta_{02})[(\vartheta_{30} + \vartheta_{12})^2 - (\vartheta_{21} + \vartheta_{03})^2] + 4\vartheta_{11}(\vartheta_{30} + \vartheta_{12})(\vartheta_{21} + \vartheta_{03}) \tag{6.53}$$

$$\begin{aligned}\varphi_7 = \; & (3\vartheta_{21} - \vartheta_{03})(\vartheta_{30} + \vartheta_{12})[(\vartheta_{30} + \vartheta_{12})^2 - 3(\vartheta_{21} + \vartheta_{03})^2] \\ & - (\vartheta_{30} - 3\vartheta_{12})(\vartheta_{21} + \vartheta_{03})[3(\vartheta_{30} + \vartheta_{12})^2 - (\vartheta_{21} + \vartheta_{03})^2]\end{aligned} \tag{6.54}$$

where the ϑ_{pq} values can be computed from equation (6.47).

While the seven moment characteristics presented above were shown to be useful, they are invariant only to translation, rotation, and scaling. Recent algorithms for fast computation of translation-, rotation-, and scale-invariant moments were given in [Li and Shen 91, Jiang and Bunke 91]. However, these approaches do not yield descriptors that are invariant under general affine transforms. A complete set of four affine moment invariants derived from second- and third-order moments is presented in [Flusser and Suk 93].

$$I_1 = \frac{\mu_{20}\mu_{02} - \mu_{11}^2}{\mu_{00}^4} \tag{6.55}$$

$$I_2 = \frac{\mu_{30}^2\mu_{03}^2 - 6\mu_{30}\mu_{21}\mu_{12}\mu_{03} + 4\mu_{30}\mu_{12}^3 + 4\mu_{21}^3\mu_{03} - 3\mu_{21}^2\mu_{12}^2}{\mu_{00}^{10}} \tag{6.56}$$

$$I_3 = \frac{\mu_{20}(\mu_{21}\mu_{03} - \mu_{12}^2) - \mu_{11}(\mu_{30}\mu_{03} - \mu_{21}\mu_{12}) + \mu_{02}(\mu_{30}\mu_{12} - \mu_{21}^2)}{\mu_{00}^7} \tag{6.57}$$

$$\begin{aligned} I_4 = \quad & (\mu_{20}^3\mu_{03}^2 - 6\mu_{20}^2\mu_{11}\mu_{12}\mu_{03} - 6\mu_{20}^2\mu_{02}\mu_{21}\mu_{03} + 9\mu_{20}^2\mu_{02}\mu_{12}^2 \\ & +12\mu_{20}\mu_{11}^2\mu_{21}\mu_{03} + 6\mu_{20}\mu_{11}\mu_{02}\mu_{30}\mu_{03} - 18\mu_{20}\mu_{11}\mu_{02}\mu_{21}\mu_{12} \\ & -8\mu_{11}^3\mu_{30}\mu_{03} - 6\mu_{20}\mu_{02}^2\mu_{30}\mu_{12} + 9\mu_{20}\mu_{02}^2\mu_{21}^2 \\ & +12\mu_{11}^2\mu_{02}\mu_{30}\mu_{12} - 6\mu_{11}\mu_{02}^2\mu_{30}\mu_{21} + \mu_{02}^3\mu_{30}^2)/\mu_{00}^{11} \end{aligned} \tag{6.58}$$

Details of the process for the derivation of invariants and examples of invariant moment object descriptions can be found in [Flusser and Suk 93], and a complete proof and detailed discussion of the properties of them are given in [Flusser and Suk 91].

All moment characteristics are dependent on the linear gray-level transformations of regions; to describe region shape properties, we work with binary image data [$f(i,j) = 1$ in region pixels] and dependence on the linear gray-level transform disappears.

Moment characteristics can be used in shape description even if the region is represented by its boundary. A closed boundary is characterized by an ordered sequence $z(i)$ that represents the Euclidean distance between the centroid and all N boundary pixels of the digitized shape. No extra processing is required for shapes having spiral or concave contours. Translation-, rotation-, and scale-invariant one-dimensional normalized contour sequence moments $\overline{m}_r, \overline{\mu}_r$ are defined in [Gupta and Srinath 87]. The rth contour sequence moment m_r and the rth central moment μ_r can be estimated as

$$m_r = \frac{1}{N}\sum_{i=1}^{N}[z(i)]^r \tag{6.59}$$

$$\mu_r = \frac{1}{N}\sum_{i=1}^{N}[z(i) - m_1]^r \tag{6.60}$$

The rth normalized contour sequence moment $\overline{m}_r$ and normalized central contour sequence moment $\overline{\mu}_r$ are defined as

$$\overline{m}_r = \frac{m_r}{(\mu_2)^{r/2}} = \frac{\frac{1}{N}\sum_{i=1}^{N}[z(i)]^r}{[\frac{1}{N}\sum_{i=1}^{N}[z(i) - m_1]^2]^{r/2}} \tag{6.61}$$

$$\overline{\mu}_r = \frac{\mu_r}{(\mu_2)^{r/2}} = \frac{\frac{1}{N}\sum_{i=1}^{N}[z(i) - m_1]^r}{[\frac{1}{N}\sum_{i=1}^{N}[z(i) - m_1]^2]^{r/2}} \tag{6.62}$$

While the set of invariant moments $\overline{m}_r, \overline{\mu}_r$ can be used directly for shape representation, less noise-sensitive results can be obtained from the following shape descriptors [Gupta and Srinath 87]

$$F_1 = \frac{(\mu_2)^{1/2}}{m_1} = \frac{[\frac{1}{N}\sum_{i=1}^{N}[z(i) - m_1]^2]^{1/2}}{\frac{1}{N}\sum_{i=1}^{N} z(i)} \tag{6.63}$$

$$F_2 = \frac{\mu_3}{(\mu_2)^{3/2}} = \frac{\frac{1}{N}\sum_{i=1}^{N}[z(i) - m_1]^3}{[\frac{1}{N}\sum_{i=1}^{N}[z(i) - m_1]^2]^{3/2}} \tag{6.64}$$

$$F_3 = \frac{\mu_4}{(\mu_2)^2} = \frac{\frac{1}{N}\sum_{i=1}^{N}[z(i) - m_1]^4}{[\frac{1}{N}\sum_{i=1}^{N}[z(i) - m_1]^2]^2} \tag{6.65}$$

$$F_4 = \overline{\mu}_5 \tag{6.66}$$

Lower probabilities of error classification were obtained using contour sequence moments than area-based moments [equations (6.48)–(6.54)] in a shape recognition test; also, contour sequence moments are less computationally demanding.

6.3.3 Convex hull

A region R is convex if and only if for any two points $\mathbf{x}_1, \mathbf{x}_2 \in R$, the whole line segment $\mathbf{x}_1\mathbf{x}_2$ defined by its end points $\mathbf{x}_1, \mathbf{x}_2$ is inside the region R. The convex hull of a region is the smallest convex region H which satisfies the condition $R \subseteq H$—see Figure 6.26. The convex hull has some special properties in digital data which do not exist in the continuous case. For instance, concave parts can appear and disappear in digital data due to rotation, and therefore the convex hull is not rotation invariant in digital space [Gross and Latecki 95]. The convex hull can be used to describe region shape properties and can be used to build a tree structure of region concavity.

A discrete convex hull can be defined by the following algorithm which may also be used for convex hull construction. This algorithm has complexity $\mathcal{O}(n^2)$ and is presented here as an intuitive way of detecting the convex hull. Algorithm 6.7 describes a more efficient approach.

Algorithm 6.6: Region convex hull construction

1. Find all pixels of a region R with the minimum row co-ordinate; among them, find the pixel P_1 with the minimum column co-ordinate. Assign $\mathbf{P}_k = \mathbf{P}_1$, $\mathbf{v} = (0, -1)$; the vector $\mathbf{v}$ represents the direction of the previous line segment of the convex hull.

2. Search the region boundary in an anti-clockwise direction (Algorithm 5.8) and compute the angle orientation φ_n for every boundary point $\mathbf{P}_n$ which lies after the point $\mathbf{P}_1$ (in the direction of boundary search—see Figure 6.26). The angle orientation φ_n is the angle of vector $\mathbf{P}_k\mathbf{P}_n$. The point $\mathbf{P}_q$ satisfying the condition $\varphi_q = \min_n \varphi_n$ is an element (vertex) of the region convex hull.

3. Assign $\mathbf{v} = \mathbf{P}_k\mathbf{P}_q$, $\mathbf{P}_k = \mathbf{P}_q$.

4. Repeat steps 2 and 3 until $\mathbf{P}_k = \mathbf{P}_1$.

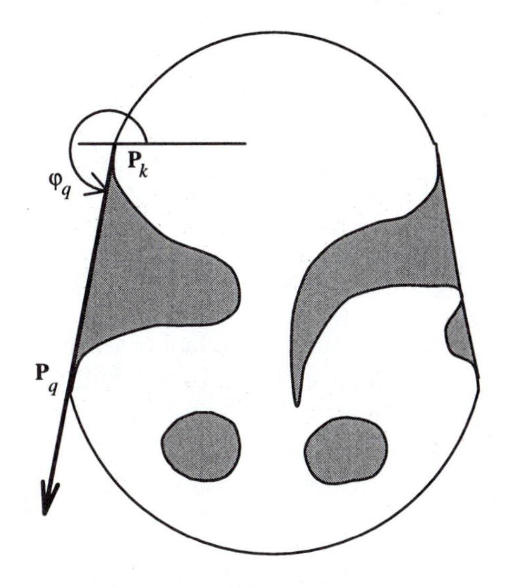

Figure 6.26: *Convex hull.*

The first point $\mathbf{P}_1$ need not be chosen as described in the given algorithm, but it must be an element of a convex segment of the inner region boundary.

As has been mentioned, more efficient algorithms exist, especially if the object is defined by an ordered sequence $P = \{\mathbf{v}_1, \mathbf{v}_2, \ldots, \mathbf{v}_n\}$ of n vertices, $\mathbf{v}_i$ representing a polygonal boundary of the object. Many algorithms [Toussaint 85] exist for detection of the convex hull with computational complexity $\mathcal{O}(n \log n)$ in the worst case; these algorithms and their implementations vary in speed and memory requirements. As discussed in [Toussaint 91], the code of [Bhattacharya and Toussaint 83] (in which a Fortran listing appears) seems to be the fastest to date, using only $5n$ storage space.

If the polygon P is a *simple* polygon (self-non-intersecting polygon) which is always the case in a polygonal representation of object borders, the convex hull may be found in linear time $\mathcal{O}(n)$. In the past two decades, many linear-time convex hull detection algorithms have been published; however more than half of them were later discovered to be incorrect [Toussaint 85, Toussaint 91], with counter-examples published. The algorithm of [McCallum and Avis 79] was the first correct linear-time one. The simplest correct convex hull algorithm was given in [Melkman 87] and was based on previous work [Lee 83, Bhattacharya and Gindy 84, Graham and Yao 84]. Melkman's convex hull detection algorithm is now discussed further.

Let the polygon for which the convex hull is to be determined be a simple polygon $P = \{\mathbf{v}_1, \mathbf{v}_2, \ldots, \mathbf{v}_n\}$ and let the vertices be processed in this order. For any three vertices $\mathbf{x}$,$\mathbf{y}$,$\mathbf{z}$

in an ordered sequence, a directional function δ may be evaluated (Figure 6.27).

$$\begin{aligned}\delta(\mathbf{x},\mathbf{y},\mathbf{z}) &= 1 \quad \text{if } \mathbf{z} \text{ is to the right of the directed line } \mathbf{xy}\\ &= 0 \quad \text{if } \mathbf{z} \text{ is collinear with the directed line } \mathbf{xy}\\ &= -1 \quad \text{if } \mathbf{z} \text{ is to the left of the directed line } \mathbf{xy}\end{aligned}$$

The main data structure H is a list of vertices (deque) of polygonal vertices already processed.

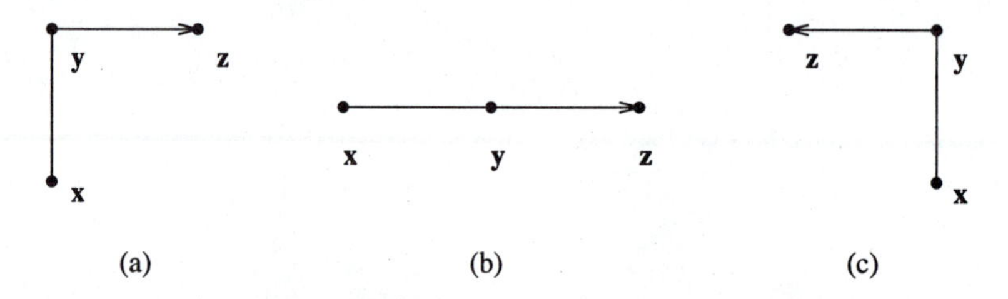

Figure 6.27: *Directional function δ: (a) $\delta(\mathbf{x},\mathbf{y},\mathbf{z}) = 1$; (b) $\delta(\mathbf{x},\mathbf{y},\mathbf{z}) = 0$; (c) $\delta(\mathbf{x},\mathbf{y},\mathbf{z}) = -1$.*

The current contents of H represents the convex hull of the currently processed part of the polygon, and after the detection is completed, the convex hull is stored in this data structure. Therefore, H always represents a closed polygonal curve, $H = \{d_b, \ldots, d_t\}$ where d_b points to the bottom of the list and d_t points to its top. Note that d_b and d_t always refer to the same vertex simultaneously representing the first and the last vertex of the closed polygon.

Here are the main ideas of the algorithm. The first three vertices A, B, C from the sequence P form a triangle (if not collinear) and this triangle represents a convex hull of the first three vertices—Figure 6.28a. The next vertex D in the sequence is then tested for being located inside or outside the current convex hull. If D is located inside, the current convex hull does not change—Figure 6.28b. If D is outside of the current convex hull, it must become a new convex hull vertex (Figure 6.28c) and, based on the current convex hull shape, either none, one, or several vertices must be removed from the current convex hull—Figure 6.28c,d. This process is repeated for all remaining vertices in the sequence P.

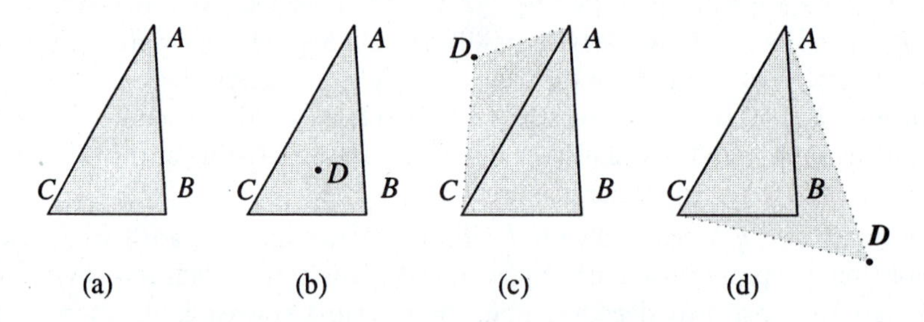

Figure 6.28: *Convex hull detection. (a) First three vertices A, B, C form a triangle. (b) If the next vertex D is positioned inside the current convex hull ABC, current convex hull does not change. (c) If the next vertex D is outside of the current convex hull, it becomes a new vertex of the new current convex hull $ABCDA$. (d) In this case, vertex B must be removed from the current convex hull and the new current convex hull is $ADCA$.*

Following the terminology used in [Melkman 87], the variable $\mathbf{v}$ refers to the input vertex under consideration, and the following operations are defined:

`push v`	:	$t := t+1;\quad d_t \rightarrow \mathbf{v}$
`pop` d_t	:	$t := t-1$
`insert v`	:	$b := b-1;\quad d_b \rightarrow \mathbf{v}$
`remove` d_b	:	$b := b+1$
`input v`	:	next vertex is entered from sequence P, if P is empty, stop.

where $\rightarrow$ means 'points to'. The algorithm is then as follows.

Algorithm 6.7: Simple polygon convex hull detection

1. Initialize.

```
t := -1;
b := 0;
input v1; input v2; input v3;
if ( δ(v1,v2,v3) > 0 )
      { push v1;
        push v2; }
   else
      { push v2;
        push v1; }
push v3;
insert v3;
```

2. If the next vertex $\mathbf{v}$ is inside the current convex hull H, enter and check a new vertex; otherwise process steps 3 and 4;

```
input v;
while ( δ(v,d_b,d_{b+1}) ≥ 0     AND     δ(d_{t-1},d_t,v) ≥ 0 )
   input v;
```

3. Rearrange vertices in H, top of the list.

```
while ( δ(d_{t-1},d_t,v) ≤ 0 )
   pop d_t;
push v;
```

4. Rearrange vertices in H, bottom of the list.

```
while ( δ(v,d_b,d_{b+1}) ≤ 0 )
   remove d_b;
insert v;
go to step 2;
```

The algorithm as presented may be difficult to follow, but a less formal version would be impossible to implement; a formal proof is given in [Melkman 87]. The following example makes the algorithm more understandable.

Let $P = \{A, B, C, D, E\}$ as shown in Figure 6.29a. The data structure H is created in the first step:

$$\begin{array}{lcccccc} t, b \ldots & & -1 & 0 & 1 & 2 \\ H & = & C & A & B & C \\ & & d_b & & & d_t \end{array}$$

In the second step, vertex D is entered (Figure 6.29b):

$$\begin{array}{rclcrcl} \delta(D, d_b, d_{b+1}) & = & \delta(D, C, A) & = & 1 & > & 0 \\ \delta(d_{t-1}, d_t, D) & = & \delta(B, C, D) & = & -1 & < & 0 \end{array}$$

Based on the values of the directional function δ, in this case, no other vertex is entered during this step. Step 3 results in the following current convex hull H;

$$\delta(B, C, D) = -1 \longrightarrow \texttt{pop } d_t \longrightarrow$$

$$\begin{array}{lcccccc} t, b \ldots & & -1 & 0 & 1 & 2 \\ H & = & C & A & B & C \\ & & d_b & & d_t & \end{array}$$

$$\delta(A, B, D) = -1 \longrightarrow \texttt{pop } d_t \longrightarrow$$

$$\begin{array}{lcccccc} t, b \ldots & & -1 & 0 & 1 & 2 \\ H & = & C & A & B & C \\ & & d_b & d_t & & \end{array}$$

$$\delta(C, A, D) = 1 \longrightarrow \texttt{push } D \longrightarrow$$

$$\begin{array}{lcccccc} t, b \ldots & & -1 & 0 & 1 & 2 \\ H & = & C & A & D & C \\ & & d_b & & d_t & \end{array}$$

In step 4—Figure 6.29c:

$$\delta(D, C, A) = 1 \longrightarrow \texttt{insert } D \longrightarrow$$

$$\begin{array}{lccccccc} t, b \ldots & & -2 & -1 & 0 & 1 & 2 \\ H & = & D & C & A & D & C \\ & & d_b & & & d_t & \end{array}$$

Go to step 2; vertex E is entered—Figure 6.29d:

$$\begin{array}{rclcl} \delta(E, D, C) & = & 1 & > & 0 \\ \delta(A, D, E) & = & 1 & > & 0 \end{array}$$

A new vertex should be entered from P, but there is no unprocessed vertex in the sequence P and the convex hull generating process stops. The resulting convex hull is defined by the sequence $H = \{d_b, \ldots, d_t\} = \{D, C, A, D\}$, which represents a polygon $DCAD$, always in the clockwise direction—Figure 6.29e.

A **region concavity tree** is another shape representation option [Sklansky 72]. A tree is generated recursively during the construction of a convex hull. A convex hull of the whole region is constructed first, and convex hulls of concave residua are found next. The resulting convex hulls of concave residua of the regions from previous steps are searched until no concave residuum exists. The resulting tree is a shape representation of the region. Concavity tree construction can be seen in Figure 6.30.

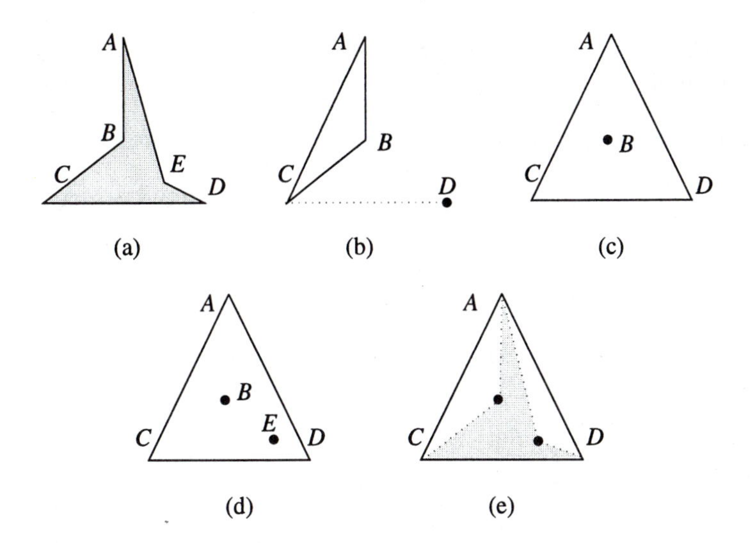

Figure 6.29: *Example of convex hull detection: (a) the processed region—polygon ABCDEA; (b) vertex D is entered and processed; (c) vertex D becomes a new vertex of the current convex hull ADC; (d) vertex E is entered and processed, E does not become a new vertex of the current convex hull; (e) the resulting convex hull DCAD.*

6.3.4 Graph representation based on region skeleton

This method corresponds significantly curving points of a region boundary to graph nodes. The main disadvantage of boundary-based description methods is that geometrically close points can be far away from one another when the boundary is described—graphical representation methods overcome this disadvantage. Shape properties are then derived from the graph properties.

The region graph is based on the region skeleton, and the first step is the skeleton construction. There are four basic approaches to skeleton construction:

- Thinning—iterative removal of region boundary pixels
- Wave propagation from the boundary
- Detection of local maxima in the distance-transformed image of the region
- Analytical methods

Most thinning procedures repeatedly remove boundary elements until a pixel set with maximum thickness of 1 or 2 is found. The following algorithm constructs a skeleton of maximum thickness 2.

Algorithm 6.8: Skeleton by thinning

1. Let R be the set of region pixels, $H_i(R)$ its inner boundary, and $H_o(R)$ its outer boundary. Let $S(R)$ be a set of pixels from the region R which have all their neighbors in

8-connectivity either from the inner boundary $H_i(R)$ or from the background—from the residuum of R. Assign $R_{\rm old} = R$.

2. Construct a region $R_{\rm new}$ which is a result of one-step thinning as follows:

$$R_{\rm new} = S(R_{\rm old}) \cup [R_{\rm old} - H_i(R_{\rm old})] \cup [H_o(S(R_{\rm old})) \cap R_{\rm old}]$$

3. If $R_{\rm new} = R_{\rm old}$, terminate the iteration and proceed to step 4. Otherwise assign $R_{\rm old} = R_{\rm new}$ and repeat step 2.

4. $R_{\rm new}$ is a set of skeleton pixels, the skeleton of the region R.

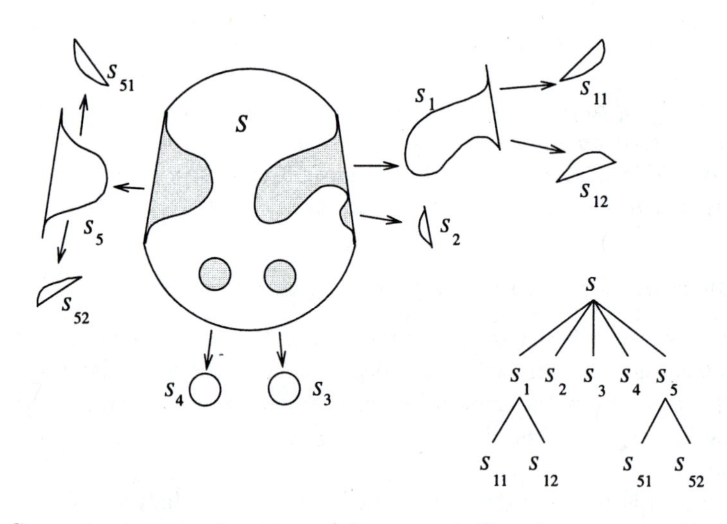

Figure 6.30: *Concavity tree construction: (a) convex hull and concave residua; (b) concavity tree.*

Steps of this algorithm are illustrated in Figure 6.31. If there are skeleton segments which have a thickness of 2 in the skeleton, one extra step can be added to reduce those to a thickness of 1, although care must be taken not to break the skeleton connectivity.

A large number of thinning algorithms can be found in the image processing literature [Hildich 69, Pavlidis 78]. If special prior conditions apply, these algorithms can be much simpler. Thinning is generally a time-consuming process, although sometimes it is not necessary to look for a skeleton, and one side of a parallel boundary can be used for skeleton-like region representation. Mathematical morphology is a powerful tool used to find the region skeleton, and thinning algorithms which use mathematical morphology are given in Chapter 11; see also [Maragos and Schafer 86], where the morphological approach is shown to unify many other approaches to skeletonization.

Thinning procedures often use a medial axis transform (also symmetric axis transform) to construct a region skeleton [Blum 73, Pavlidis 77, Samet 85, Arcelli and Sanniti di Baja 86, Pizer et al. 87, Lam et al. 92, Wright and Fallside 93]. Under the medial axis definition, the

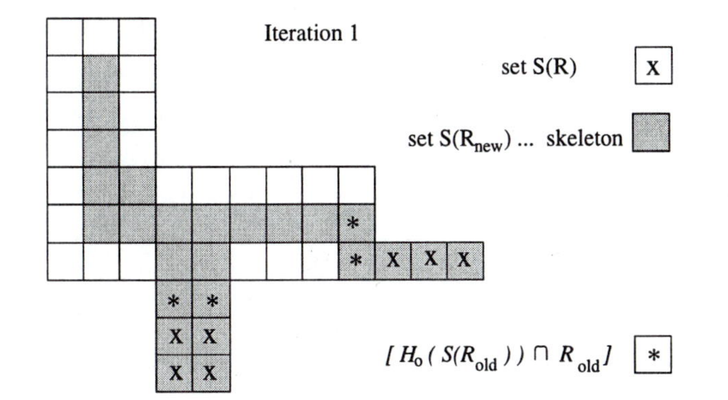

Figure 6.31: *Skeleton by thinning (Algorithm 6.8).*

skeleton is the set of all region points which have the same minimum distance from the region boundary for at least two separate boundary points. Examples of skeletons resulting from this condition are shown in Figures 6.32 and 6.33. Such a skeleton can be constructed using a distance transform which assigns a value to each region pixel representing its (minimum) distance from the region's boundary. The skeleton can be determined as a set of pixels whose distance from the region's border is locally maximal. As a post-processing step, local maxima can be detected using operators that detect linear features and roof profiles [Canny 83, Petrou 90, Wright and Fallside 93]. Every skeleton element can be accompanied by information about its distance from the boundary—this gives the potential to reconstruct a region as an envelope curve of circles with center points at skeleton elements and radii corresponding to the stored distance values. Shape descriptions, as discussed in Section 6.3.1 can be derived from this skeleton but, with the exception of elongatedness, the evaluation can be difficult. In addition, this skeleton construction is time-consuming, and a resulting skeleton is highly sensitive to boundary noise and errors. Small changes in the boundary may cause serious changes in the skeleton—see Figure 6.32. This sensitivity can be removed by first representing the region as a polygon, then constructing the skeleton. Boundary noise removal can be absorbed into the polygon construction. A multi-resolution (scale-space) approach to skeleton construction may also result in decreased sensitivity to boundary noise [Pizer et al. 87, Maragos 89]. Similarly, the approach using the Marr-Hildreth edge detector with varying smoothing parameter facilitates scale-based representation of the region's skeleton [Wright and Fallside 93].

A method of skeleton construction based on the Fourier coefficients of a boundary T_n and S_n (see Section 6.2.3) is given in [Persoon and Fu 77]. Neural networks [Krishnapuram and Chen 91] and a Voronoi diagram approach [Brandt and Algazi 92, Ogniewicz and Ilg 92, Mayya and Rajan 95] can also be applied to find the skeleton. Fast parallel algorithms for thinning are given in [Guo and Hall 92]. Use of the intensity axis of symmetry represents an unconventional approach to skeletonization that does not require explicit region segmentation [Gauch and Pizer 93]. If derived from boundary data considering the scale, the intensity axes of symmetry are often called **cores** [Morse et al. 93, Fritsch et al. 97]. The cores are invariant to translation, rotation, linear variation of intensity, and scale, and are insensitive to small-scale noise (spatially uncorrelated), small-scale blurring (compared to the object's width),

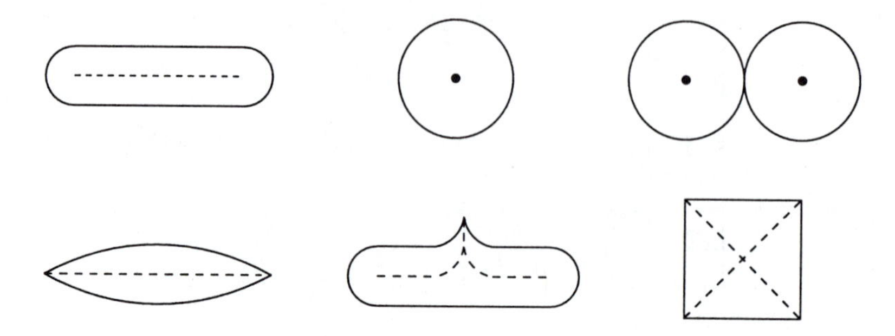

Figure 6.32: *Region skeletons; small changes in border can have a significant effect on the skeleton.*

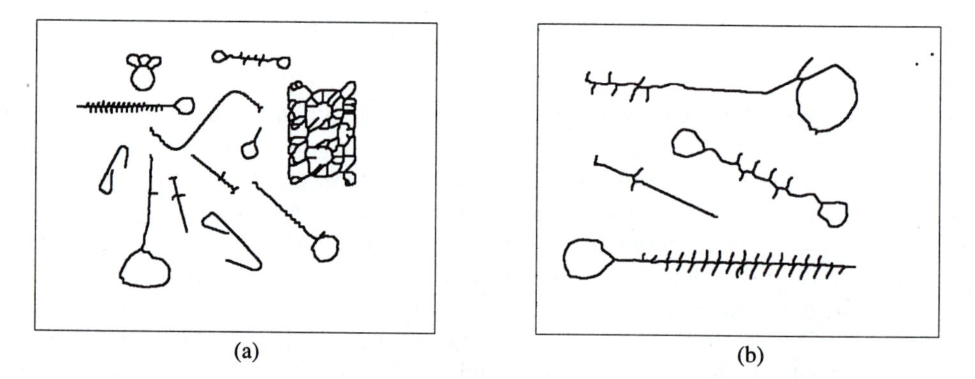

Figure 6.33: *Region skeletons, see Figures 5.1a and 6.2a for original images; thickened for visibility.*

and small-scale local deformation.

Skeleton construction algorithms do not result in graphs, but the transformation from skeletons to graphs is relatively straightforward. Consider first the medial axis skeleton, and assume that a minimum radius circle has been drawn from each point of the skeleton which has at least one point common with a region boundary. Let *contact* be each contiguous subset of the circle which is common to the circle and to the boundary. If a circle drawn from its center A has one contact only, A is a skeleton end point. If the point A has two contacts, it is a normal skeleton point. If A has three or more contacts, the point A is a skeleton node point.

Algorithm 6.9: Region graph construction from skeleton

1. Assign a point description to all skeleton points—end point, node point, normal point.

2. Let graph node points be all end points and node points. Connect any two graph nodes by a graph edge if they are connected by a sequence of normal points in the region skeleton.

It can be seen that boundary points of high curvature have the main influence on the graph. They are represented by graph nodes, and therefore influence the graph structure.

If other than medial axis skeletons are used for graph construction, end points can be defined as skeleton points having just one skeleton neighbor, normal points as having two skeleton neighbors, and node points as having at least three skeleton neighbors. It is no longer true that node points are never neighbors and additional conditions must be used to decide when node points should be represented as nodes in a graph and when they should not.

6.3.5 Region decomposition

The decomposition approach is based on the idea that shape recognition is a hierarchical process. Shape **primitives** are defined at the lower level, primitives being the simplest elements which form the region. A graph is constructed at the higher level—nodes result from primitives, arcs describe the mutual primitive relations. Convex sets of pixels are one example of simple shape primitives.

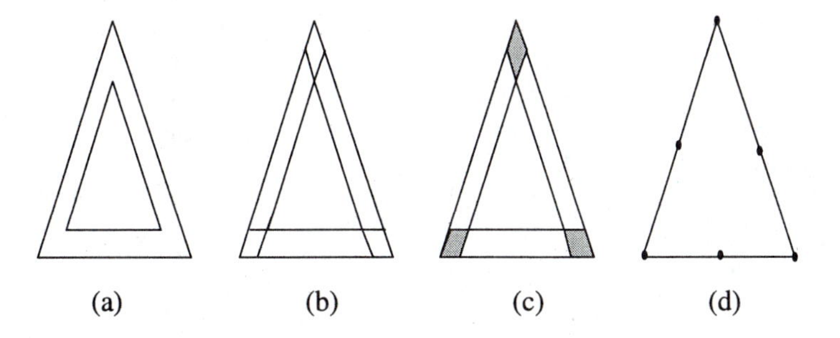

Figure 6.34: *Region decomposition: (a) region; (b) primary regions; (c) primary sub-regions and kernels; (d) decomposition graph.*

The solution to the decomposition problem consists of two main steps: The first step is to segment a region into simpler sub-regions (primitives), and the second is the analysis of primitives. Primitives are simple enough to be described successfully using simple scalar shape properties (see Section 6.3.1). A detailed description of how to segment a region into primary convex sub-regions, methods of decomposition to concave vertices, and graph construction resulting from a polygonal description of sub-regions are given in [Pavlidis 77]. The general idea of decomposition is shown in Figure 6.34, where the original region, one possible decomposition, and the resulting graph are presented. Primary convex sub-regions are labeled as primary sub-regions or kernels. Kernels (shown shaded in Figure 6.34c) are sub-regions which belong to several primary convex sub-regions. If sub-regions are represented by polygons, graph nodes bear the following information;

1. Node type representing primary sub-region or kernel

2. Number of vertices of the sub-region represented by the node

3. Area of the sub-region represented by the node

4. Main axis direction of the sub-region represented by the node

5. Center of gravity of the sub-region represented by the node

If a graph is derived using attributes 1–4, the final description is translation invariant. A graph derived from attributes 1–3 is translation and rotation invariant. Derivation using the first two attributes results in a description which is size invariant in addition to possessing translation and rotation invariance.

A decomposition of a region uses its structural properties, and a syntactic graph description is the result. Problems of how to decompose a region and how to construct the description graph are still open; an overview of some techniques that have been investigated can be found in [Feng and Pavlidis 75, Moayer and Fu 75, Pavlidis 77, Stallings 76, Shapiro 80, di Baja and Thiel 94, Held and Abe 94]. Shape decomposition into a complete set of convex parts ordered by size is described in [Cortopassi and Rearick 88], and a morphological approach to skeleton decomposition is used to decompose complex shapes into simple components in [Zhou and Venetsanopoulos 89, Pitas and Venetsanopoulos 90, Kanungo and Haralick 92, Loncaric and Dhawan 95, Xiaoqi and Baozong 95, Wang et al. 95, Reinhardt and Higgins 96]; the decomposition is shown to be invariant to translation, rotation, and scaling. Recursive sub-division of shape based on second central moments is another translation-, rotation-, scaling-, and intensity shift-invariant decomposition technique [Zhu and Poh 88]. Hierarchical decomposition and shape description that uses region and contour information, addresses issues of local versus global information, scale, shape parts, and axial symmetry is given in [Rom and Medioni 92, Rom and Medioni 93]. Multi-resolution approaches to decomposition are reported in [Loncaric and Dhawan 93, Cinque and Lombardi 95].

6.3.6 Region neighborhood graphs

Any time a region decomposition into sub-regions or an image decomposition into regions is available, the region or image can be represented by a region neighborhood graph (the region adjacency graph described in Section 3.2.3 being a special case). This graph represents every region as a graph node, and nodes of neighboring regions are connected by edges. A region neighborhood graph can be constructed from a quadtree image representation, from run length encoded image data, etc. Binary tree shape representation is described in [Leu 89], where merging of boundary segments results in shape decomposition into triangles, their relations being represented by the binary tree.

Very often, the relative position of two regions can be used in the description process—for example, a region A may be positioned to the *left of* a region B, or *above* B, or *close to* B, or a region C may lie *between* regions A and B, etc. We know the meaning of all of the given relations if A, B, C are points, but, with the exception of the relation *to be close*, they can become ambiguous if A, B, C are regions. For instance (see Figure 6.35), the relation *to be left of* can be defined in many different ways:

- All pixels of A must be positioned to the left of all pixels of B.
- At least one pixel of A must be positioned to the left of some pixel of B.
- The center of gravity of A must be to the left of the center of gravity of B.

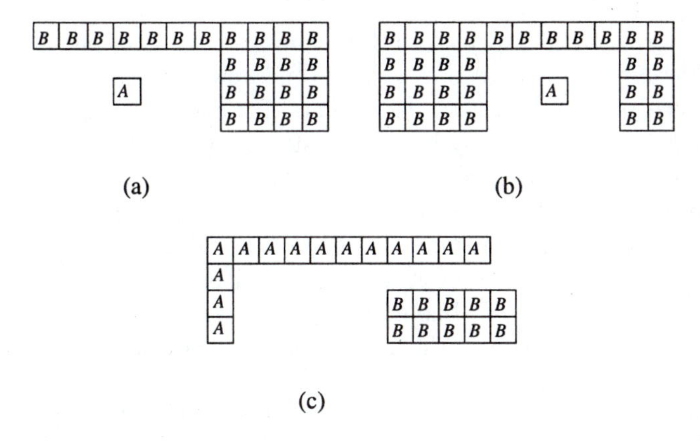

Figure 6.35: *Binary relation* to be left of; *see text.*

All of these definitions seem to be satisfactory in many cases, but they can sometimes be unacceptable because they do not meet the usual meaning of *being left of.* Human observers are generally satisfied with the definition:

- The center of gravity of A must be positioned to the left of the leftmost point of B and (logical AND) the rightmost pixel of A must be left of the rightmost pixel of B [Winston 75].

Many other inter-regional relations are defined in [Winston 75], where relational descriptions are studied in detail.

An example of applying geometrical relations between simply shaped primitives to shape representation and recognition may be found in [Shariat 90], where recognition is based on a **hypothesize and verify** control strategy. Shapes are represented by region neighborhood graphs that describe geometrical relations among primitive shapes. The model-based approach increases the shape recognition accuracy and makes partially occluded object recognition possible. Recognition of any new object is based on a definition of a new shape model.

6.4 Shape classes

Representation of **shape classes** is considered a challenging problem of shape description [Hogg 93]. The shape classes are expected to represent the generic shapes of the objects belonging to the class well and emphasize shape differences between classes, while the shape variations allowed within classes should not influence the description.

There are many ways to deal with such requirements. A widely used representation of in-class shape variations is determination of class-specific regions in the feature space. The feature space can be defined using a selection of shape features described earlier in this chapter (for more information about feature spaces, see Chapter 7). Another approach to shape class definition is to use a single prototype shape and determine a planar warping transform that if applied to the prototype produces shapes from the particular class. The prototype shape may be derived from examples.

If a set of landmarks can be identified on the regions belonging to specific shape classes, the landmarks can characterize the classes in a simple and powerful way. Landmarks are usually selected as easily recognizable border or region points. For planar shapes, a co-ordinate system can be defined that is invariant to similarity transforms of the plane (rotation, translation, scaling) [Bookstein 91, Rangarajan et al. 97]. If such a landmark model uses n points per 2D object, the dimensionality of the shape space is $2n$. Clearly, only a subset of the entire shape space corresponds to each shape class and the shape class definition reduces to the definition of the shape space subsets. In [Cootes et al. 92], principal components in the shape space are determined from training sets of shapes after the shapes are iteratively aligned. The first few principal components characterize the most significant variations in shape. Thus, a small number of shape parameters represent the major shape variation characteristics associated with the shape class. Such a shape class representation is referred to as **point distribution models** and is discussed in detail in Section 8.3 in the context of image interpretation.

6.5 Summary

- **Shape representation and description**
  - Region description generates a numeric feature vector or a non-numeric syntactic description word, which characterize properties (for example, shape) of the described region.
  - While many practical shape description methods exist, there is no generally accepted methodology of shape description. Further, it is not known what is important in shape.
  - Shape may change substantially with image resolution. Conventional shape descriptions change discontinuously with changes in resolution. A **scale-space** approach aims to obtain continuous shape descriptions for continuous resolution changes.
  - The **shape classes** represent the generic shapes of the objects belonging to the same classes. Shape classes should emphasize shape differences among classes, while the shape variations within classes should not be reflected in the shape class description.
- **Region identification**
  - Region identification assigns unique **labels** to image regions.
  - If nonrepeating ordered numerical labels are used, the largest integer label gives the number of regions in the image.
- **Contour-based shape descriptors**
  - **Chain codes** describe an object by a sequence of unit-size line segments with a given orientation, called **Freeman's** code.
  - **Simple geometric border representations** are based on geometric properties of described regions, e.g.:

* Boundary length
* Curvature
* Bending energy
* Signature
* Chord distribution

- **Fourier shape descriptors** can be applied to closed curves, co-ordinates of which can be treated as periodic signals.
- Shape can be represented as a sequence of **segments** with specified properties. If the segment type is known for all segments, the boundary can be described as a chain of segment types, a code word consisting of representatives of a type alphabet.
- **B-splines** are piecewise polynomial curves whose shape is closely related to their control polygon—a chain of vertices giving a polygonal representation of a curve. B-splines of third order are most common, representing the lowest order which includes the change of curvature.
- **Shape invariants** represent properties of geometric configurations that remain unchanged under an appropriate class of transforms; machine vision is especially concerned with the class of projective transforms.

• **Region-based shape descriptors**

- Simple geometric region descriptors use geometric properties of described regions:
  * Area
  * Euler's number
  * Projections
  * Height, width
  * Eccentricity
  * Elongatedness
  * Rectangularity
  * Direction
  * Compactness
- **Statistical moments** interpret a normalized gray-level image function as a probability density of a 2D random variable. Properties of this random variable can be described using statistical characteristics—**moments**. Moment-based descriptors can be defined to be independent of scaling, translation, and rotation.
- The **convex hull** of a region is the smallest convex region H which satisfies the condition $R \subset H$.
- More complicated shapes can be described using region decomposition into smaller and simpler sub-regions. Objects can be represented by a planar graph with nodes representing sub-regions resulting from region decomposition. Region shape can then be described by the graph properties. There are two general approaches to acquiring a graph of sub-regions:

    * Region thinning
    * Region decomposition

- **Region thinning** leads to the region **skeleton** that can be described by a graph. Thinning procedures often use a medial axis transform to construct a region skeleton. Under the medial axis definition, the skeleton is the set of all region points which have the same minimum distance from the region boundary for at least two separate boundary points.
- **Region decomposition** considers shape recognition to be a hierarchical process. Shape **primitives** are defined at the lower level, primitives being the simplest elements which form the region. A graph is constructed at the higher level—nodes result from primitives, arcs describe the mutual primitive relations.
- **Region neighborhood graphs** represents every region as a graph node, and nodes of neighboring regions are connected by edges. The **region adjacency graph** is a special case of the region neighborhood graph.

• **Shape classes**

- Shape classes represent the generic shapes of the objects belonging to the class and emphasize shape differences among classes.
- A widely used representation of in-class shape variations is determination of class-specific regions in the feature space.

6.6 Exercises

Short-answer questions

1. What are the prerequisites of shape description?
2. What are the main distinguishing aspects among various shape representation and shape description methods?
3. Explain how the problem of scale affects shape description.
4. Explain what shape classes are and why are they important.
5. Explain the rationale behind projection-invariant shape descriptors.
6. Define the three most common representations of region borders.
7. Define the boundary chain code in 4- and 8-connectivity.
8. Define the chain code derivative in 4- and 8-connectivity.
9. Define the following border-based region descriptors:
    (a) Boundary length
    (b) Curvature
    (c) Bending energy
    (d) Signature
    (e) Chord distribution
    (f) Fourier transform of boundaries using T_n descriptors

(g) Fourier transform of boundaries using S_n descriptors
(h) Polygonal segment representation
(i) Constant curvature representation
(j) Tolerance interval representation

10. Explain the concept of multi-scale curve description using interval trees in the scale space.
11. Describe the concept of B-spline curve interpolation.
12. Explain the role of the spline order.
13. Define the following projection-invariant shape descriptors:
    (a) Cross ratio
    (b) System of four co-planar concurrent lines
    (c) System of five co-planar concurrent lines
    (d) System of five co-planar points
    (e) Plane conics
14. Describe a subclass of boundary shapes to which the invariants listed in Question 6.13 can be applied.
15. Describe how differential invariants can help with invariant description of general shapes.
16. Explain the difference between local and global invariants.
17. Define the following region shape descriptors:
    (a) Area
    (b) Euler's number
    (c) Horizontal and vertical projections
    (d) Eccentricity
    (e) Elongatedness
    (f) Rectangularity
    (g) Direction
    (h) Compactness
    (i) Statistical moments
    (j) Convex hull
    (k) Region concavity tree
18. List the advantages of graph-based region shape descriptors.
19. Describe the principles of region skeletonization by thinning.
20. Describe the medial axis transform.
21. Describe the principles of shape description using graph decomposition.

Problems

1. Write a function (subroutine) for region identification in 4-neighborhood connectivity.
2. Write a function (subroutine) for region identification in 8-neighborhood connectivity.
3. Develop a program for region identification and region counting that will use function(s) developed in Problems 6.1 and 6.2. Test on binary segmented images.
4. Modify the program developed in Problem 6.3 to accept multi-level segmented images, assuming that the background gray-level is known.
5. Develop a program for region identification and region counting in run length encoded image data. Use the program developed in Problem 3.2 to generate run length encoded image data.
6. Develop a program for region identification and region counting in quadtrees. Use the program developed in Problem 3.5 to generate quadtree image data.
7. Write a function (subroutine) for chain code generation in 4-connectivity. Test on images in which the regions have been identified using one of the programs developed in Problems 6.3–6.6.
8. Write a function (subroutine) for chain code generation in 8-connectivity. Test on images in which the regions have been identified using one of the programs developed in Problems 6.3–6.6.
9. An object is described by the following chain code in 4-connectivity: 10123230.
   (a) Determine the normalized version of the chain code.
   (b) Determine the derivative of the original chain code.
10. Determine the Euler number of the following characters: 0, 4, 8, A, B, C, D.
11. Prove the statement that the most compact region in a Euclidean space is a circle. Compare the compactness values of a square and a rectangle of any aspect ratio—what can you conclude?
12. Write functions (subroutines) determining the following border descriptors:
    (a) Boundary length
    (b) Curvature
    (c) Bending energy
    (d) Signature
    (e) Chord distribution
    (f) Fourier transform of boundaries using T_n descriptors
    (g) Fourier transform of boundaries using S_n descriptors

    Use the functions in a program to determine shape features of binary objects.
13. Write functions (subroutines) determining the following region shape descriptors:
    (a) Area
    (b) Area from chain code border representation
    (c) Area from quadtree region representation
    (d) Euler's number
    (e) Horizontal and vertical projections
    (f) Eccentricity
    (g) Elongatedness

(h) Rectangularity

(i) Direction

(j) Compactness

(k) Affine transform invariant statistical moments

Use the functions in a program to determine shape features of binary objects.

14. Develop a program to determine the shape features listed in Problems 6.12–6.13 in images containing several objects simultaneously. The program should report the features in a table, and the objects should be identified by their centroid co-ordinates.

15. Develop a program to determine the shape features listed in Problems 6.12–6.13 from shapes encoded using run length code and/or quadtree image data.

16. Develop a program to generate digital images of simple shapes (rectangle, diamond, circle, etc.) in various sizes and rotations. Using the functions prepared in Problems 6.12–6.13, compare the shape features determined by individual shape descriptors as a function of size and a function of rotation.

17. Develop a program for simple polygon convex hull detection.

18. Develop a program for region concavity tree construction.

19. Determine a medial axis skeleton of a circle, square, rectangle, and triangle.

20. Develop a program constructing a medial axis of a binary region.

    (a) Apply the program to computer-generated letters and numerals.

    (b) Apply the program to printed letters and numerals after their digitization using a video camera or a scanner.

    (c) Explain the differences in performance of your algorithm.

    (d) Develop a practically applicable thinning algorithm that constructs line shapes from scanned characters.

6.7 References

[Ansari and Delp 90] N Ansari and E J Delp. Distribution of a deforming triangle. *Pattern Recognition*, 23(12):1333–1341, 1990.

[Appel and Haken 77] K Appel and W Haken. Every planar map is four colourable: Part I: discharging. *Illinois Journal of Mathematics*, 21:429–490, 1977.

[Arcelli and Sanniti di Baja 86] C Arcelli and G Sanniti di Baja. Endoskeleton and exoskeleton of digital figures, an effective procedure. In V Cappellini and R Marconi, editors, *Advances in Image Processing and Pattern Recognition*, pages 224–228, North Holland, Amsterdam, 1986.

[Asada and Brady 86] H Asada and M Brady. The curvature primal sketch. *IEEE Transactions on Pattern Analysis and Machine Intelligence*, 8(1):2–14, 1986.

[Atkinson et al. 85] H H Atkinson, Gargantini, I, and T R S Walsh. Counting regions, holes and their nesting level in time proportional to the border. *Computer Vision, Graphics, and Image Processing*, 29:196–215, 1985.

[Babaud et al. 86] J Babaud, A P Witkin, M Baudin, and R O Duda. Uniqueness of the Gaussian kernel for scale-space filtering. *IEEE Transactions on Pattern Analysis and Machine Intelligence*, 8(1):26–33, 1986.

[Ballard and Brown 82] D H Ballard and C M Brown. *Computer Vision.* Prentice-Hall, Englewood Cliffs, NJ, 1982.

[Barnsley 88] M F Barnsley. *Fractals Everywhere.* Academic Press, Boston, 1988.

[Besl 88] P J Besl. Geometric modelling and computer vision. *Proceedings of the IEEE*, 76:936–958, 1988.

[Bhanu and Faugeras 84] B Bhanu and O D Faugeras. Shape matching of two–dimensional objects. *IEEE Transactions on Pattern Analysis and Machine Intelligence*, 6(2):137–155, 1984.

[Bhattacharya and Gindy 84] B K Bhattacharya and H E Gindy. A new linear convex hull algorithm for simple polygons. *IEEE Transactions on Information Theory*, 30:85–88, 1984.

[Bhattacharya and Toussaint 83] B K Bhattacharya and G T Toussaint. Time-and-storage-efficient implementation of an optimal planar convex hull algorithm. *Image and Vision Computing*, 1(3):140–144, 1983.

[Blum 73] H Blum. Biological shape and visual science (part 1). *Journal of Theoretical Biology*, 38:205–287, 1973.

[Bookstein 91] F L Bookstein. *Morphometric Tools for Landmark Data.* Cambridge University Press, Cambridge, 1991.

[Brady 84] M Brady. Representing shape. In M Brady, L A Gerhardt, and H F Davidson, editors, *Robotics and Artificial Intelligence*, pages 279–300. Springer + NATO, Berlin, 1984.

[Brandt and Algazi 92] J W Brandt and V R Algazi. Continuous skeleton computation by Voronoi diagram. *CVGIP - Image Understanding*, 55(3), 1992.

[Bribiesca and Guzman 80] E Bribiesca and A Guzman. How to describe pure form and how to measure differences in shapes using shape numbers. *Pattern Recognition*, 12(2):101–112, 1980.

[Brill et al. 92] M H Brill, E B Barrett, and P M Payton. Projective invariants for curves in two and three dimensions. In J L Mundy and A Zisserman, editors, *Geometric Invariance in Computer Vision.* MIT Press, Cambridge, MA; London, 1992.

[Canny 83] J F Canny. Finding edges and lines in images. Technical Report AI-TR-720, MIT, Artificial Intelligence Laboratory, Cambridge, MA, 1983.

[Cash and Hatamian 87] G L Cash and M Hatamian. Optical character recognition by the method of moments. *Computer Vision, Graphics, and Image Processing*, 39:291–310, 1987.

[Chang and Chatterjee 89] C Chang and S Chatterjee. Fractal based approach to shape description, reconstruction and classification. In *Twenty-Third Annual Asilomar Conference on Signals, Systems and Computers,* Pacific Grove, CA, pages 172–176, IEEE, Los Alamitos, CA, 1989.

[Chien and Aggarwal 89] C H Chien and J K Aggarwal. Model construction and shape recognition from occluding contours. *IEEE Transactions on Pattern Analysis and Machine Intelligence*, 11(4):372–389, 1989.

[Cinque and Lombardi 95] L Cinque and L Lombardi. Shape description and recognition by a multiresolution approach. *Image and Vision Computing*, 13:599–607, 1995.

[Cootes et al. 92] T F Cootes, D H Cooper, C J Taylor, and J Graham. Trainable method of parametric shape description. *Image and Vision Computing*, 10(5), 1992.

[Cortopassi and Rearick 88] P P Cortopassi and T C Rearick. Computationally efficient algorithm for shape decomposition. In *CVPR '88: Computer Society Conference on Computer Vision and Pattern Recognition,* Ann Arbor, MI, pages 597–601, IEEE, Los Alamitos, CA, 1988.

[Costabile et al. 85] M F Costabile, C Guerra, and G G Pieroni. Matching shapes: A case study in time-varying images. *Computer Vision, Graphics, and Image Processing*, 29:296–310, 1985.

[Crowley 84] J L Crowley. A multiresolution representation for shape. In A Rosenfeld, editor, *Multiresolution Image Processing and Analysis*, pages 169–189. Springer Verlag, Berlin, 1984.

[DeBoor 78] C A DeBoor. *A Practical Guide to Splines.* Springer Verlag, New York, 1978.

[di Baja and Thiel 94] G Sanniti di Baja and E Thiel. Shape description via weighted skeleton partition. In *Proceedings of the 7th International Conference on Image Analysis and Processing. Progress in Image Analysis and Processing III*, pages 87–94, 1994.

[Duda and Hart 73] R O Duda and P E Hart. *Pattern Classification and Scene Analysis.* Wiley, New York, 1973.

[Dyer 80] C R Dyer. Computing the Euler number of an image from its quadtree. *Computer Graphics and Image Processing*, 13:270–276, 1980.

[Eichmann et al. 90] G Eichmann, C Lu, M Jankowski, and R Tolimeiri. Shape representation by Gabor expansion. In *Hybrid Image and Signal Processing II,* Orlando, FL, pages 86–94, Society for Optical Engineering, Bellingham, WA, 1990.

[Falconer 90] K Falconer. *Fractal Geometry: Mathematical Foundations and Applications.* Wiley, Chichester, New York, 1990.

[Feng and Pavlidis 75] H Y Feng and T Pavlidis. Decomposition of polygons into simpler components. *IEEE Transactions on Computers*, 24:636–650, 1975.

[Fermuller and Kropatsch 92] C Fermuller and W Kropatsch. Multi-resolution shape description by corners. In *Proceedings, 1992 Computer Vision and Pattern Recognition,* Champaign, IL, pages 271–276, IEEE, Los Alamitos, CA, 1992.

[Florack et al. 92] L M J Florack, B M Haar-Romeny, J J Koenderink, and M A Viergever. Scale and the differential structure of images. *Image and Vision Computing*, 10(6):376–388, 1992.

[Flusser and Suk 91] J Flusser and T Suk. Classification of objects by affine moment invariants. Technical Report UTIA-1736, Czechoslovak Academy of Sciences, Prague, 1991.

[Flusser and Suk 93] J Flusser and T Suk. Pattern recognition by affine moment invariants. *Pattern Recognition*, 26:167–174, 1993.

[Forsyth et al. 91] D Forsyth, J L Mundy, A Zisserman, C Coelho, A Heller, and C Rothwell. Invariant descriptors for 3D object recognition and pose. *IEEE Transactions on Pattern Analysis and Machine Intelligence*, 13(10):971–991, 1991.

[Freeman 61] H Freeman. On the encoding of arbitrary geometric configuration. *IRE Transactions on Electronic Computers*, EC–10(2):260–268, 1961.

[Frisch et al. 87] A A Frisch, D A Evans, J P Hudson, and J Boon. Shape discrimination of sand samples using the fractal dimension. In *Coastal Sediments '87, Proceedings of a Specialty Conference on Advances in Understanding of Coastal Sediment Processes,* New Orleans, LA, pages 138–153, ASCE, Dallas, TX, 1987.

[Fritsch et al. 97] D Fritsch, S Pizer, L Yu, V Johnson, and E Chaney. Segmentation of medical image objects using deformable shape loci. In J Duncan and G Gindi, editors, *Information Processing in Medical Imaging*, pages 127–140, Springer Verlag, Berlin; New York, 1997.

[Fu 74] K S Fu. *Syntactic Methods in Pattern Recognition.* Academic Press, New York, 1974.

[Gauch and Pizer 93] J M Gauch and S M Pizer. The intensity axis of symmetry and its application to image segmentation. *IEEE Transactions on Pattern Analysis and Machine Intelligence*, 15:753–770, 1993.

[Graham and Yao 84] R L Graham and F F Yao. Finding the convex hull of a simple polygon. *Journal of Algorithms*, 4:324–331, 1984.

[Griffin et al. 92] L D Griffin, A C F Colchester, and G P Robinson. Scale and segmentation of grey-level images using maximum gradient paths. *Image and Vision Computing*, 10(6):389–402, 1992.

[Grimmins 82] T R Grimmins. A complete set of Fourier descriptors for two–dimensional shapes. *IEEE Transactions on Systems, Man and Cybernetics*, 12(6):923–927, 1982.

[Gross and Latecki 95] A Gross and L Latecki. Digital geometric invariance and shape representation. In *Proceedings of the International Symposium on Computer Vision*, pages 121–126, IEEE, Los Alamitos, CA, 1995.

[Guo and Hall 92] Z Guo and R W Hall. Fast fully parallel thinning algorithms. *CVGIP – Image Understanding*, 55(3):317–328, 1992.

[Gupta and Srinath 87] L Gupta and M D Srinath. Contour sequence moments for the classification of closed planar shapes. *Pattern Recognition*, 20(3):267–272, 1987.

[Gupta et al. 90] L Gupta, M R Sayeh, and R Tammana. Neural network approach to robust shape classification. *Pattern Recognition*, 23(6):563–568, 1990.

[Held and Abe 94] A Held and K Abe. On the decomposition of binary shapes into meaningful parts. *Pattern Recognition*, 27:637–647, 1994.

[Hildich 69] C J Hildich. Linear skeletons from square cupboards. In B Meltzer and D Michie, editors, *Machine Intelligence IV*, pages 403–420. Elsevier, New York, 1969.

[Hlavac et al. 94] V Hlavac, T Pajdla, and M Sommer. Improvements of the curvature computation. In *International Conference on Pattern Recognition*, pages 536–538, IEEE, Los Alamitos, CA, 1994.

[Hogg 93] D C Hogg. Shape in machine vision. *Image and Vision Computing*, 11:309–316, 1993.

[Hopcroft et al. 92] J E Hopcroft, D P Huttenlocher, and P C Wayner. Affine invariants for model-based recognition. In J L Mundy and A Zisserman, editors, *Geometric Invariance in Computer Vision*. MIT Press, Cambridge, MA; London, 1992.

[Hu 62] M K Hu. Visual pattern recognition by moment invariants. *IRE Transactions Information Theory*, 8(2):179–187, 1962.

[Ikebe and Miyamoto 82] Y Ikebe and S Miyamoto. Shape design, representation, and restoration with splines. In K S Fu and H Kunii, editors, *Picture Engineering*. Springer Verlag, Berlin, 1982.

[Jain 89] A K Jain. *Fundamentals of Digital Image Processing*. Prentice-Hall, Englewood Cliffs, NJ, 1989.

[Jakubowski 85] R Jakubowski. Extraction of shape features for syntactic recognition of mechanical parts. *IEEE Transactions on Systems, Man and Cybernetics*, 15(5):642–651, 1985.

[Jakubowski 90] R Jakubowski. Decomposition of complex shapes for their structural recognition. *Information Sciences*, 50(1):35–71, 1990.

[Ji and Haralick 97] Q Ji and R M Haralick. Corner detection with covariance propagation. In *Computer Vision and Pattern Recognition*, pages 362–367, IEEE Computer Society, Los Alamitos, CA, 1997.

[Jiang and Bunke 91] X Y Jiang and H Bunke. Simple and fast computation of moments. *Pattern Recognition*, 24:801–806, 1991.

[Juday 88] R D Juday, editor. *Digital and Optical Shape Representation and Pattern Recognition,* Orlando, FL, Bellingham, WA, 1988. SPIE.

[Kanatani 90] K Kanatani. *Group-Theoretical Methods in Image Understanding.* Springer Verlag, Berlin, 1990.

[Kanungo and Haralick 92] T Kanungo and R M Haralick. Vector-space solution for a morphological shape-decomposition problem. *Journal of Mathematical Imaging and Vision*, 2:51–82, 1992.

[Kiryati and Maydan 89] N Kiryati and D Maydan. Calculating geometric properties from Foutier representation. *Pattern Recognition*, 22(5):469–475, 1989.

[Koch and Kashyap 87] M W Koch and R L Kashyap. Using polygons to recognize and locate partially occluded objects. *IEEE Transactions on Pattern Analysis and Machine Intelligence*, 9(4):483–494, 1987.

[Koenderink 90] J J Koenderink. *Solid Shape.* MIT Press, Cambridge, MA, 1990.

[Koenderink and van Doorn 86] J J Koenderink and A J van Doorn. Dynamic shape. Technical report, Department of Medical and Physiological Physics, State University, Utrecht, The Netherlands, 1986.

[Krishnapuram and Chen 91] R Krishnapuram and L F Chen. Iterative neural networks for skeletonization and thinning. In *Intelligent Robots and Computer Vision IX: Neural, Biological, and 3D Methods,* Boston, pages 271–281, Society for Optical Engineering, Bellingham, WA, 1991.

[Krzyzak et al. 89] A Krzyzak, S Y Leung, and C Y Suen. Reconstruction of two-dimensional patterns from Fourier descriptors. *Machine Vision and Applications*, 2(3):123–140, 1989.

[Lam et al. 92] L Lam, S W Lee, and C Y Suen. Thinning methodologies—a comprehensive survey. *IEEE Transactions on Pattern Analysis and Machine Intelligence*, 14(9):869–885, 1992.

[Lee 83] D T Lee. On finding the convex hull of a simple polygon. *International Journal of Computer and Information Sciences*, 12:87–98, 1983.

[Leu 89] J G Leu. View-independent shape representation and matching. In *IEEE International Conference on Systems Engineering,* Fairborn, OH, pages 601–604, IEEE, Piscataway, NJ, 1989.

[Leymarie and Levine 89] F Leymarie and M D Levine. Shape features using curvature morphology. In *Visual Communications and Image Processing IV,* Philadelphia, PA, pages 390–401, SPIE, Bellingham, WA, 1989.

[Li and MA 94] B Li and S D MA. On the relation between region and contour representation. In *International Conference on Pattern Recognition*, pages 352–355, IEEE, Los Alamitos, CA, 1994.

[Li and Shen 91] B C Li and J Shen. Fast computation of moment invariants. *Pattern Recognition*, 24:807–813, 1991.

[Li and Zhiying 88] X Li and Z Zhiying. Group direction difference chain codes for the representation of the border. In *Digital and Optical Shape Representation and Pattern Recognition,* Orlando, FL, pages 372–376, SPIE, Bellingham, WA, 1988.

[Lin and Chellappa 87] C C Lin and R Chellappa. Classification of partial 2D shapes using Fourier descriptors. *IEEE Transactions on Pattern Analysis and Machine Intelligence*, 9(5):686–690, 1987.

[Lindenbaum and Bruckstein 93] M Lindenbaum and A Bruckstein. On recursive, o(n) partitioning of a digitized curve into digital straight segments. *IEEE Transactions on Pattern Analysis and Machine Intelligence*, 15:949–953, 1993.

[Lipari and Harlow 88] C Lipari and C A Harlow. Representation and recognition of elongated regions in aerial images. In *Applications of Artificial Intelligence VI,* Orlando, FL, pages 557–567, SPIE, Bellingham, WA, 1988.

[Loncaric and Dhawan 93] S Loncaric and A P Dhawan. A morphological signature transform for shape description. *Pattern Recognition*, 26:1029–1037, 1993.

[Loncaric and Dhawan 95] S Loncaric and A P Dhawan. Near-optimal MST-based shape description using genetic algorithm. *Pattern Recognition*, 28:571–579, 1995.

[Lord and Wilson 84] E A Lord and C B Wilson. *The Mathematical Description of Shape and Form.* Halsted Press, Chichester, England, 1984.

[Loui et al. 90] A C P Loui, A N Venetsanopoulos, and K C Smith. Two-dimensional shape representation using morphological correlation functions. In *Proceedings of the 1990 International Conference on Acoustics, Speech, and Signal Processing—ICASSP 90,* Albuquerque, NM, pages 2165–2168, IEEE, Piscataway, NJ, 1990.

[Lowe 89] D G Lowe. Organization of smooth image curves at multiple scales. *International Journal of Computer Vision*, 1:119–130, 1989.

[Maitra 79] S Maitra. Moment invariants. *Proceedings IEEE*, 67(4):697–699, 1979.

[Mandelbrot 82] B B Mandelbrot. *The Fractal Geometry of Nature.* Freeman, New York, 1982.

[Maragos 89] P Maragos. Pattern spectrum and multiscale shape representation. *IEEE Transactions on Pattern Analysis and Machine Intelligence*, 11:701–716, 1989.

[Maragos and Schafer 86] P A Maragos and R W Schafer. Morphological skeleton representation and coding of binary images. *IEEE Transactions on Acoustics, Speech and Signal Processing*, 34(5):1228–1244, 1986.

[Marshall 89a] S Marshall. Application of image contours to three aspects of image processing; compression, shape recognition and stereopsis. In *Third International Conference on Image Processing and its Applications,* Coventry, England, pages 604–608, IEE, Michael Faraday House, Stevenage, England, 1989.

[Marshall 89b] S Marshall. Review of shape coding techniques. *Image and Vision Computing*, 7(4):281–194, 1989.

[Matas and Kittler 93] J Matas and J Kittler. Junction detection using probabilistic relaxation. *Image and Vision Computing*, 11:197–202, 1993.

[Maybank 92] S J Maybank. The projection of two non-coplanar conics. In J L Mundy and A Zisserman, editors, *Geometric Invariance in Computer Vision.* MIT Press, Cambridge, MA; London, 1992.

[Mayya and Rajan 95] N Mayya and V T Rajan. An efficient shape representation scheme using voronoi skeletons. *Pattern Recognition Letters*, 16:147–160, 1995.

[McCallum and Avis 79] D McCallum and D Avis. A linear algorithm for finding the convex hull of a simple polygon. *Information Processing Letters*, 9:201–206, 1979.

[McKenzie and Protheroe 90] D S McKenzie and S R Protheroe. Curve description using the inverse Hough transform. *Pattern Recognition*, 23(3–4):283–290, 1990.

[Melkman 87] A V Melkman. On-line construction of the convex hull of a simple polyline. *Information Processing Letters*, 25(1):11–12, 1987.

[Minnix et al. 90] J I Minnix, E S McVey, and R M Inigo. Multistaged neural network architecture for position invariant shape recognition. In *Visual Communications and Image Processing '90,* Lausanne, Switzerland, pages 58–68, SPIE, Bellingham, WA, 1990.

[Moayer and Fu 75] B Moayer and K S Fu. A tree system approach for fingerprint pattern recognition. *IEEE Transactions on Computers*, 24(4):436–450, 1975.

[Morse et al. 93] B S Morse, S M Pizer, and A Liu. Multiscale medial analysis of medical images. In Barrett and Gmitro, editors, *Information Processing in Medical Imaging*, pages 112–131. Springer Verlag, Berlin, 1993.

[Mundy and Zisserman 92] J L Mundy and A Zisserman. *Geometric Invariance in Computer Vision.* MIT Press, Cambridge, MA; London, 1992.

[Nagao and Matsuyama 80] M Nagao and T Matsuyama. *A Structural Analysis of Complex Aerial Photographs.* Plenum Press, New York, 1980.

[Nishizeki and Chiba 88] T Nishizeki and N Chiba. *Planar Graphs: Theory and Algorithms.* North Holland, Amsterdam–New York–Tokyo, 1988.

[Ogniewicz and Ilg 92] R Ogniewicz and M Ilg. Voronoi skeletons: Theory and applications. In *Proceedings, 1992 Computer Vision and Pattern Recognition,* Champaign, IL, pages 63–69, IEEE, Los Alamitos, CA, 1992.

[Oppenheim et al. 83] A V Oppenheim, A S Willsky, and I T Young. *Signals and Systems.* Prentice-Hall, Englewood Cliffs, NJ, 1983.

[Paglieroni and Jain 88] D W Paglieroni and A K Jain. Control point transforms for shape representation and measurement. *Computer Vision, Graphics, and Image Processing*, 42(1):87–111, 1988.

[Papoulis 91] A Papoulis. *Probability, Random Variables, and Stochastic Processes.* McGraw-Hill, New York, 3rd edition, 1991.

[Pavlidis 77] T Pavlidis. *Structural Pattern Recognition.* Springer Verlag, Berlin, 1977.

[Pavlidis 78] T Pavlidis. A review of algorithms for shape analysis. *Computer Graphics and Image Processing*, 7:243–258, 1978.

[Pavlidis 80] T Pavlidis. Algorithms for shape analysis of contours and waveforms. *IEEE Transactions on Pattern Analysis and Machine Intelligence*, 2(4):301–312, 1980.

[Persoon and Fu 77] E Persoon and K S Fu. Shape discrimination using Fourier descriptors. *IEEE Transactions on Systems, Man and Cybernetics*, 7:170–179, 1977.

[Petrou 90] M Petrou. Optimal convolution filters and an algorithm for the detection of linear features. Technical Report Dept. of Electronic and Electrical Eng., University of Surrey, United Kingdom, 1990.

[Pitas and Venetsanopoulos 90] I Pitas and A N Venetsanopoulos. Morphological shape decomposition. *IEEE Transactions on Pattern Analysis and Machine Intelligence*, 12(1):38–45, 1990.

[Pizer et al. 87] S M Pizer, W R Oliver, and S H Bloomberg. Hierarchical shape description via the multiresolution symmetric axis transform. *IEEE Transactions on Pattern Analysis and Machine Intelligence*, 9(4):505–511, 1987.

[Pratt 91] W K Pratt. *Digital Image Processing.* Wiley, New York, 2nd edition, 1991.

[Quan et al. 92] L Quan, P Gros, and R Mohr. Invariants of a pair of conics revisited. *Image and Vision Computing*, 10(5):319–323, 1992.

[Rangarajan et al. 97] A Rangarajan, H Chui, and F L Bookstein. The Softassign Procrustes matching algorithm. In J Duncan and G Gindi, editors, *Information Processing in Medical Imaging*, pages 29–42, Springer Verlag, Berlin; New York, 1997.

[Reinhardt and Higgins 96] J M Reinhardt and W E Higgins. Efficient morphological shape representation. *IEEE Transactions on Image Processing*, 5:89–101, 1996.

[Reiss 93] T H Reiss. *Recognizing Planar Objects using Invariant Image Features.* Springer Verlag, Berlin; New York, 1993.

[Rom and Medioni 92] H Rom and G Medioni. Hierarchical decomposition and axial shape description. In *Proceedings, 1992 Computer Vision and Pattern Recognition,* Champaign, IL, pages 49–55, IEEE, Los Alamitos, CA, 1992.

[Rom and Medioni 93] H Rom and G Medioni. Hierarchical decomposition and axial shape description. *IEEE Transactions on Pattern Analysis and Machine Intelligence*, 15:973–981, 1993.

[Rosenfeld 74] A Rosenfeld. Digital straight line segments. *IEEE Transactions on Computers*, 23:1264–1269, 1974.

[Rosenfeld 79] A Rosenfeld. *Picture Languages—Formal Models for Picture Recognition.* Academic Press, New York, 1979.

[Rosenfeld and Kak 82] A Rosenfeld and A C Kak. *Digital Picture Processing.* Academic Press, New York, 2nd edition, 1982.

[Rosin and West 89] P L Rosin and G A W West. Segmentation of edges into lines and arcs. *Image and Vision Computing*, 7(2):109–114, 1989.

[Rothwell et al. 92a] C A Rothwell, A Zisserman, D A Forsyth, and J L Mundy. Fast recognition using algebraic invariants. In J L Mundy and A Zisserman, editors, *Geometric Invariance in Computer Vision.* MIT Press, Cambridge, MA; London, 1992.

[Rothwell et al. 92b] C A Rothwell, A Zisserman, J L Mundy, and D A Forsyth. Efficient model library access by projectively invariant indexing functions. In *Proceedings, 1992 Computer Vision and Pattern Recognition,* Champaign, IL, pages 109–114, IEEE, Los Alamitos, CA, 1992.

[Saaty and Kainen 77] T L Saaty and P C Kainen. *The Four Colour Problem.* McGraw-Hill, New York, 1977.

[Safaee-Rad et al. 89] R Safaee-Rad, B Benhabib, K C Smith, and K M Ty. Position, rotation, and scale-invariant recognition of 2 dimensional objects using a gradient coding scheme. In *IEEE Pacific RIM Conference on Communications, Computers and Signal Processing,* Victoria, BC, Canada, pages 306–311, IEEE, Piscataway, NJ, 1989.

[Samet 81] H Samet. Computing perimeters of images represented by quadtrees. *IEEE Transactions on Pattern Analysis and Machine Intelligence*, 3:683–687, 1981.

[Samet 84] H Samet. A tutorial on quadtree research. In A Rosenfeld, editor, *Multiresolution Image Processing and Analysis*, pages 212–223. Springer Verlag, Berlin, 1984.

[Samet 85] H Samet. Reconstruction of quadtree medial axis transforms. *Computer Vision, Graphics, and Image Processing*, 29:311–328, 1985.

[Saund 90] E Saund. Symbolic construction of a 2D scale-space image. *IEEE Transactions on Pattern Analysis and Machine Intelligence*, 12:817–830, 1990.

[Savini 88] M Savini. Moments in image analysis. *Alta Frequenza*, 57(2):145–152, 1988.

[Shapiro 80] L Shapiro. A structural model of shape. *IEEE Transactions on Pattern Analysis and Machine Intelligence*, 2(2):111–126, 1980.

[Shariat 90] H Shariat. A model-based method for object recognition. In *IEEE International Conference on Robotics and Automation,* Cincinnati, OH, pages 1846–1851, IEEE, Los Alamitos, CA, 1990.

[Shridhar and Badreldin 84] M Shridhar and A Badreldin. High accuracy character recognition algorithms using Fourier and topological descriptors. *Pattern Recognition*, 17(5):515–524, 1984.

[Sklansky 72] J Sklansky. Measuring concavity on a rectangular mosaic. *IEEE Transactions on Computers*, 21(12):1355–1364, 1972.

[Smeulders et al. 80] A W M Smeulders, A M Vossepoel, J Vrolijk, J S Ploem, and C J Cornelisse. Some shape parameters for cell recognition. In *Proceedings of Pattern Recognition in Practice*, pages 131–142, North Holland, Amsterdam, 1980.

[Smith and Jain 82] S Smith and A Jain. Cord distribution for shape matching. *Computer Graphics and Image Processing*, 20:259–265, 1982.

[Staib and Duncan 92] L H Staib and J S Duncan. Boundary finding with parametrically deformable models. *IEEE Transactions on Pattern Analysis and Machine Intelligence*, 14(11):1061–1075, 1992.

[Stallings 76] W Stallings. Approaches to Chinese character recognition. *Pattern Recognition*, 8(1):87–98, 1976.

[Strackee and Nagelkerke 83] J Strackee and N J D Nagelkerke. On closing the Fourier descriptor presentation. *IEEE Transactions on Pattern Analysis and Machine Intelligence*, 5(6):660–661, 1983.

[Tampi and Sridhar 90] K R Tampi and C S Sridhar. Shape detection using word like image description. In *Proceedings of the 1990 International Conference on Acoustics, Speech, and Signal Processing—ICASSP 90* Albuquerque, NM, pages 2041–2043, IEEE, Piscataway, NJ, 1990.

[Taylor and Lewis 94] R I Taylor and P H Lewis. 2D shape signature based on fractal measurements. *IEE Proceedings—Vision, Image and Signal Processing*, 141:422–430, 1994.

[Tomek 74] I Tomek. Two algorithms for piecewise linear continuous approximation of functions of one variable. *IEEE Transactions on Computers*, 23(4):445–448, 1974.

[Toussaint 85] G Toussaint. A historical note on convex hull finding algorithms. *Pattern Recognition Letters*, 3(1):21–28, 1985.

[Toussaint 91] G Toussaint. A counter-example to a convex hull algorithm for polygons. *Pattern Recognition*, 24(2):183–184, 1991.

[Tsai and Yu 85] W H Tsai and S S Yu. Attributed string matching with merging for shape recognition. *IEEE Transactions on Pattern Analysis and Machine Intelligence*, 7(4):453–462, 1985.

[Turney et al. 85] J L Turney, T N Mudge, and R A Volz. Recognizing partially occluded parts. *IEEE Transactions on Pattern Analysis and Machine Intelligence*, 7(4):410–421, 1985.

[Van Gool et al. 92] L J Van Gool, T Moons, E Pauwels, and A Oosterlinck. Semi-differential invariants. In J L Mundy and A Zisserman, editors, *Geometric Invariance in Computer Vision.* MIT Press, Cambridge, MA; London, 1992.

[Vemuri and Radisavljevic 93] B C Vemuri and A Radisavljevic. From global to local, a continuum of shape models with fractal priors. In *Computer Vision and Pattern Recognition*, pages 307–313, IEEE, Los Alamitos, CA, 1993.

[Vernon 87] D Vernon. Two-dimensional object recognition using partial contours. *Image and Vision Computing*, 5(1):21–27, 1987.

[Wallace 81] T P Wallace. Comments on algorithms for shape analysis of contours and waveforms. *IEEE Transactions on Pattern Analysis and Machine Intelligence*, 3(5), 1981.

[Wallace and Wintz 80] T P Wallace and P A Wintz. An efficient three-dimensional aircraft recognition algorithm using normalized Fourier descriptors. *Computer Graphics and Image Processing*, 13:99–126, 1980.

[Wang et al. 95] D Wang, V Haese-Coat, and J Ronsin. Shape decomposition and representation using a recursive morphological operation. *Pattern Recognition*, 28:1783–1792, 1995.

[Watt 93] R J Watt. Issues in shape perception. *Image and Vision Computing*, 11:389–394, 1993.

[Weiss 88] I Weiss. Projective invariants of shapes. In *Proceedings of the DARPA Image Understanding Workshop,* Cambridge, MA, volume 2, pages 1125–1134. DARPA, 1988.

[Weiss 92] I Weiss. Noise resistant projective and affine invariants. In *Proceedings, 1992 Computer Vision and Pattern Recognition,* Champaign, IL, pages 115–121, IEEE, Los Alamitos, CA, 1992.

[Weyl 46] H Weyl. *The Classical Groups and Their Invariants.* Princeton University Press, Princeton, NJ, 1946.

[Wilson and Nelson 90] R Wilson and R Nelson. *Graph Colourings.* Longman Scientific and Technical; Wiley, Essex, England, and New York, 1990.

[Winston 75] P H Winston, editor. *The Psychology of Computer Vision.* McGraw-Hill, New York, 1975.

[Witkin 86] A P Witkin. Scale-space filtering. In A P Pentland, editor, *From Pixels to Predicates,* pages 5–19. Ablex, Norwood, NJ, 1986.

[Woodwark 86] J Woodwark. *Computing Shape: An Introduction to the Representation of Component and Assembly Geometry for Computer-Aided Engineering.* Butterworths, London–Boston, 1986.

[Wright and Fallside 93] M W Wright and F Fallside. Skeletonisation as model-based feature detection. *IEE Proceedings Communication, Speech and Vision*, 140:7–11, 1993.

[Wuescher and Boyer 91] D M Wuescher and K L Boyer. Robust contour decomposition using a constant curvature criterion. *IEEE Transactions on Pattern Analysis and Machine Intelligence*, 13(10):41–51, 1991.

[Xiaoqi and Baozong 95] Z Xiaoqi and Y Baozong. Shape description and recognition using the high order morphological pattern spectrum. *Pattern Recognition*, 28:1333–1340, 1995.

[Yuille and Poggio 86] A L Yuille and T A Poggio. Scaling theorems for zero-crossings. *IEEE Transactions on Pattern Analysis and Machine Intelligence*, 8(1):15–25, 1986.

[Zhou and Venetsanopoulos 89] Z Zhou and A N Venetsanopoulos. Generic ribbons: A morphological approach towards natural shape decomposition. In *Visual Communications and Image Processing IV, Philadelphia, PA*, pages 170–180, Society for Optical Engineering, Bellingham, WA, 1989.

[Zhu and Poh 88] Q Zhu and L Poh. Transformation-invariant recursive subdivision method for shape analysis. In *9th International Conference on Pattern Recognition,* Rome, Italy, pages 833–835, IEEE, New York, 1988.

Chapter 7

Object recognition

Not even the simplest machine vision tasks can be solved without the help of recognition. Pattern recognition is used for region and object classification, and basic methods of pattern recognition must be understood in order to study more complex machine vision processes.

Classification of objects or regions has been mentioned several times; recognition is then the last step of the bottom-up image processing approach. It is also often used in other control strategies for image understanding. Almost always, when information about an object or region class is available, some pattern recognition method is used.

Consider a simple recognition problem. Two different parties take place at the same hotel at the same time—the first is a celebration of a successful basketball season, and the second a yearly meeting of jockeys. The doorman is giving directions to guests, asking which party they are to attend. After a while the doorman discovers that no questions are necessary and he directs the guests to the right places, noticing that instead of questions, he can just use the obvious physical features of basketball players and jockeys. Maybe he uses two features to make a decision, the weight and the height of the guests. All small and light men are directed to the jockey party, all tall and heavier guests are sent to the basketball party. Representing this example in terms of recognition theory, the early guests answered the doorman's question as to which party they are going to visit. This information, together with characteristic features of these guests, resulted in the ability of the doorman to classify them based only on their features. Plotting the guests' height and weight in a two-dimensional space (see Figure 7.1), it is clear that jockeys and basketball players form two easily separable classes and that this recognition task is extremely simple. Although real object recognition problems are often more difficult, and the classes do not differ so substantially, the main principles remain the same.

The theory of pattern recognition is thoroughly discussed in several references [Fu 82, Devijver and Kittler 82, Oja 83, Devijver and Kittler 86, Patrick and Fattu 86, Fukunaga 90, Dasarathy 91, Sethi and Jain 91, Schalkoff 92, Chen et al. 93, Pavel 93, Nigrin 93, Cherkassky et al. 94, Hlavac and Sara 95], and here only a brief introduction will be given. In addition, we will introduce some other related techniques: graph matching, neural nets, genetic algorithms, simulated annealing, and fuzzy logic.

No recognition is possible without knowledge. Decisions about classes or groups into which recognized objects are classified are based on such knowledge—knowledge about objects and their classes gives the necessary information for object classification. Both specific knowledge

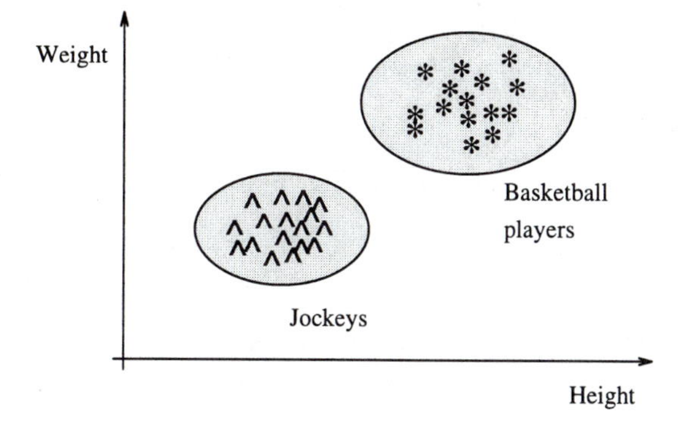

Figure 7.1: *Recognition of basketball players and jockeys; features are weight and height.*

about the objects being processed and hierarchically higher and more general knowledge about object classes is required. First, common knowledge representation techniques will be introduced, because the concept of knowledge representation in suitable form for a computer may not be straightforward.

7.1 Knowledge representation

Knowledge as well as knowledge representation problems are studied in artificial intelligence (AI), and computer vision takes advantage of these results. Use of AI methods is very common in higher processing levels, and a study of AI is necessary for a full appreciation of computer vision and image understanding. Here we present a short outline of common techniques as they are used in AI, and an overview of some basic knowledge representations. More detailed coverage of knowledge representation can be found in [Michalski et al. 83, Winston 84, Simons 84, Devijver and Kittler 86, Wechsler 90, Reichgelt 91, Lakemeyer and Nebel 94, Masuch and Polos 94].

Experience shows that a good knowledge representation design is the most important part of solving the understanding problem. Moreover, a small number of relatively simple control strategies is often sufficient for AI systems to show complex behavior, assuming an appropriately complex knowledge base is available. In other words, a high degree of control sophistication is not required for intelligent behavior, but a rich, well structured representation of a large set of a priori data and hypotheses is needed [Schutzer 87].

Other terms of which regular use will be made are **syntax** and **semantics** [Winston 84].

- The **syntax** of a representation specifies the symbols that may be used and the ways that they may be arranged.
- The **semantics** of a representation specifies how meaning is embodied in the symbols and the symbol arrangement allowed by the syntax.
- A **representation** is a set of syntactic and semantic conventions that make it possible to describe things.

The main knowledge representation techniques used in AI are formal grammars and languages, predicate logic, production rules, semantic nets, and frames. Even if features and descriptions are not usually considered knowledge representations, they are added for practical reasons; these low-level forms of knowledge representation will be mentioned many times throughout the coming sections.

Note that knowledge representation data structures are mostly extensions of conventional data structures such as lists, trees, graphs, tables, hierarchies, sets, rings, nets, and matrices.

Descriptions, features

Descriptions and features cannot be considered pure knowledge representations. Nevertheless, they can be used for representing knowledge as a part of a more complex representation structure.

Descriptions usually represent some scalar properties of objects, and are called **features**. Typically, a single description is insufficient for object representation, therefore the descriptions are combined into **feature vectors**. Numerical feature vectors are inputs for statistical pattern recognition techniques (see Section 7.2).

Figure 7.2: *Feature description of simple objects.*

A simple example of feature description of objects is shown in Figure 7.2. The *size* feature can be used to represent an area property, and the *compactness* feature describes circularity (see Section 6.3.1). Then the feature vector $\mathbf{x} = (size, compactness)$ can be used for object classification into the following classes of objects: small, large, circular, noncircular, small and circular, small and noncircular, etc., assuming information about what is considered small/large and circular/noncircular is available.

Grammars, languages

If an object's structure needs to be described, feature description is not appropriate. A structural description is formed from existing primitives and the relations between them.

Primitives are represented by information about their types. The simplest form of structure representations are chains, trees, and general graphs. Structural description of chromosomes using border segments as primitives is a classic example of structural object description [Fu 82] (see Figure 6.14), where borders are represented by a chain of symbols, the symbols representing specific types of border primitives. Hierarchical structures can be represented by trees—the concavity tree of Figure 6.30 serves as an example. A more general graph representation is used in Chapter 14, where a graph grammar (Figure 14.6) is used for texture description. Many examples of syntactic object description may be found in [Fu 74, Fu 77, Fu 80, Fu 82].

One object can be described by a chain, a tree, a graph, etc., of symbols. Nevertheless, the whole class of objects cannot be described by a single chain, a single tree, etc., but a class of structurally described objects can be represented by **grammars** and **languages**. Grammars and languages (similar to natural languages) provide rules defining how the chains, trees, or graphs can be constructed from a set of symbols (primitives). A more specific description of grammars and languages is given in Section 7.4.

Predicate logic

Predicate logic plays a very important role in knowledge representation—it introduces a mathematical formalism to derive new knowledge from old knowledge by applying mathematical deduction. Predicate logic works with combinations of logic variables, quantifiers ($\exists, \forall$), and logic operators (*and*, *or*, *not*, *implies*, *equivalent*). The logic variables are binary (*true*, *false*). The idea of proof and rules of inference such as **modus ponens** and **resolution** are the main building blocks of predicate logic [Pospesel 76].

Predicate logic forms the essence of the programming language PROLOG, which is widely used if objects are described by logic variables. Requirements of 'pure truth' represent the main weakness of predicate logic in knowledge representation, since it does not allow work with uncertain or incomplete information. Predicate logic incorporates logic conditions and constraints into knowledge processing (see Section 8.5) [Hayes 77, Kowalski 79, Clocksin and Mellish 81].

Production rules

Production rules represent a wide variety of knowledge representations that are based on **condition action** pairs. The essential model of behavior of a system based on production rules (a production system) can be described as follows:

if condition X holds *then* action Y is appropriate

Information about what action is appropriate at what time represents knowledge. The procedural character of knowledge represented by production rules is another important property—not all the information about objects must be listed as an object property. Consider a simple knowledge base where the following knowledge is present:

$$if \quad \text{ball} \quad then \quad \text{circular} \tag{7.1}$$

Let the knowledge base also include the statements

$$\begin{array}{lll} \text{object A} & is\_a & \text{ball} \\ \text{object B} & is\_a & \text{ball} \\ \text{object C} & is\_a & \text{shoe} \\ & \text{etc.} & \end{array} \tag{7.2}$$

To answer the question *how many objects are circular?*, if enumerative knowledge representation is used, the knowledge must be listed as

object A is_a (ball, circular)

object B *is_a* (ball, circular)

etc.

If procedural knowledge is used, the knowledge base (7.2) together with the knowledge (7.1) gives the same information in a significantly more efficient manner.

Both production rule knowledge representation and production systems appear frequently in computer vision and image understanding problems. Furthermore, production systems, together with a mechanism for handling uncertainty information, form a basis of expert systems.

Fuzzy logic

Fuzzy logic has been developed [Zadeh 65, Zimmermann et al. 84] to overcome the obvious limitations of numerical or crisp representation of information. Consider the use of knowledge represented by equation (7.1) for recognition of balls; using the production rule, the knowledge about balls may be represented as

$$if \quad \text{circular} \quad then \quad \text{ball} \tag{7.3}$$

If the object in a two-dimensional image is considered circular then it may represent a ball. Our experience with balls, however, says that they are usually close to, but not perfectly, circular. Thus, it is necessary to define some circularity threshold so that all *reasonably* circular objects from our set of objects are labeled as balls. Here is the fundamental problem of crisp descriptions: how circular must an object be to be considered circular?

If humans represent such knowledge, the rule for ball circularity may look like

$$if \text{ circularity is HIGH} \quad then \quad \text{object is a ball with HIGH confidence} \tag{7.4}$$

Clearly, high circularity is a preferred property of balls. Such knowledge representation is very close to common sense representation of knowledge, with no need for exact specification of the circularity/non-circularity threshold. **Fuzzy rules** are of the form

$$if \quad X \text{ is } A \quad then \quad Y \text{ is } B \tag{7.5}$$

where X and Y represent some properties and A and B are **linguistic variables**. Fuzzy logic can be used to solve object recognition and other decision-making tasks, among others; this is discussed further in Section 7.7.

Semantic nets

Semantic nets are a special variation of relational data structures (see Chapter 3). The semantics distinguish them from general nets—semantic nets consist of objects, their description, and a description of relations between objects (often just relations between neighbors). Logical forms of knowledge can be included in semantic nets, and predicate logic can be used to represent and/or evaluate the local information and local knowledge. Semantic nets can also represent common sense knowledge that is often imprecise and needs to be treated in a probabilistic way. Semantic nets have a hierarchical structure; complex representations consist of less complex representations, which can in turn be divided into simpler ones, etc. Relations between partial representations are described at all appropriate hierarchical levels.

Evaluated graphs are used as a semantic net data structure; nodes represent objects and arcs represent relations between objects. The following definition of a human face is an example of a simple semantic net:

- A *face* is a circular part of the human body that consists of two eyes, one nose, and one mouth.
- One eye is positioned left of the other eye.
- The nose is between and below the eyes.
- The mouth is below the nose.
- An eye is approximately circular.
- The nose is vertically elongated.
- The mouth is horizontally elongated.

The semantic net representing this knowledge is shown in Figure 7.3.

It is clear that the descriptive structures found in real images match the knowledge represented by a semantic net with varying degrees of closeness. The question of whether the described structure is similar to that represented by the semantic net is discussed in Section 7.5 and in Chapter 8.

A detailed discussion of semantic nets related to image information can be found in [Niemann 90], and more general properties of semantic nets are described in [Michalski et al. 83, Sharples et al. 89].

Frames, scripts

Frames provide a very general method for knowledge representation which may contain all the knowledge representation principles discussed so far. They are sometimes called **scripts** because of their similarity to film scripts. Frames are suitable for representing common sense knowledge under specific circumstances. Consider a frame called *plane_start*; this frame may consist of the following sequence of actions:

1. Start the engines.
2. Taxi to the runway.
3. Increase RPMs of engines to maximum.
4. Travel along runway increasing speed.
5. Fly.

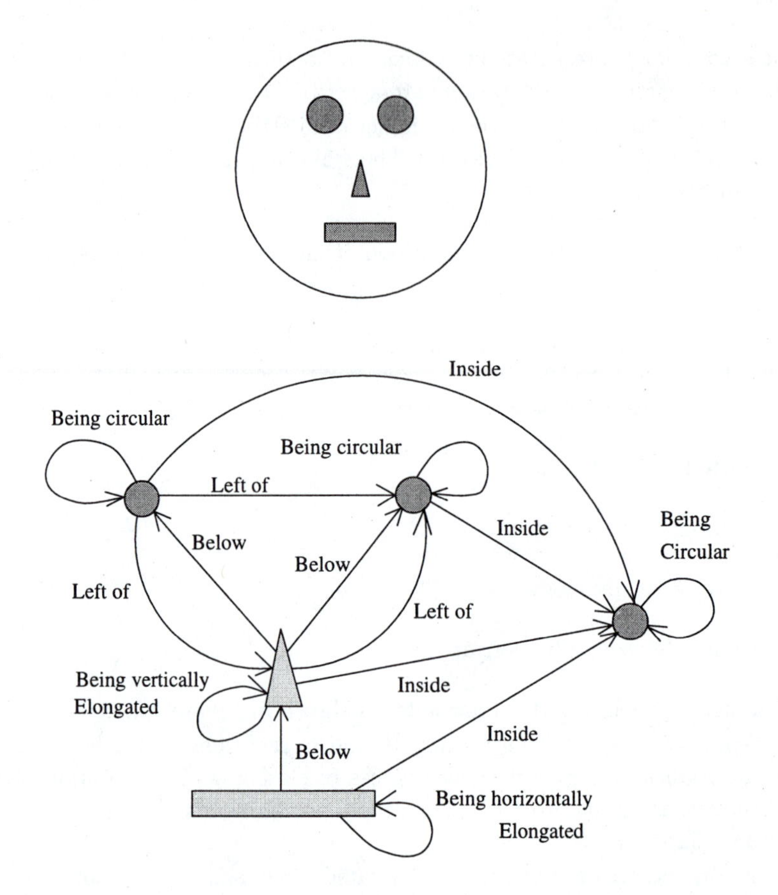

Figure 7.3: *Semantic nets: human face model and its net.*

Assuming this frame represents knowledge of how planes *usually* start, the situation of a plane standing on a runway with engines running causes the prediction that the plane will start in a short time. The frame can be used as a substitute for missing information which may be extremely important in vision-related problems.

Assuming that one part of the runway is not visible from the observation point, using the *plane_start* frame, a computer vision system can overcome the lack of continuous information between the plane moving at the beginning of the runway and flying when it next appears. If it is a passenger plane, the frame may have additional items such as *time of departure, time of arrival, departure city, arrival city, airline, flight number*, etc., because in a majority of cases it makes sense to be interested in this information if we identify a passenger plane.

From a formal point of view, a frame is represented by a general semantic net accompanied by a list of relevant variables, concepts, and concatenation of situations. No standard form of frame exists. Frames represent a tool for organizing knowledge in prototypical objects, and for description of mutual influences of objects using stereotypes of behavior in specific situations. Examples of frames can be found elsewhere [Michalski et al. 83, Winston 84, Schutzer 87, Sharples et al. 89]. Frames are considered high-level knowledge representations.

7.2 Statistical pattern recognition

An object is a physical unit, usually represented in image analysis and computer vision by a region in a segmented image. The set of objects can be divided into disjoint subsets, that, from the classification point of view, have some common features and are called **classes**. The definition of how the objects are divided into classes is ambiguous and depends on the classification goal.

Object recognition is based on assigning classes to objects, and the device that does these assignments is called the **classifier**. The number of classes is usually known beforehand, and typically can be derived from the problem specification. Nevertheless, there are approaches in which the number of classes may not be known (see Section 7.2.4).

The classifier (similarly to a human) does not decide about the class from the object itself—rather, sensed object properties serve this purpose. For example, to distinguish steel from sandstone, we do not have to determine their molecular structures, although this would describe these materials well. Properties such as texture, specific weight, hardness, etc., are used instead. This sensed object is called the **pattern**, and the classifier does not actually recognize objects, but recognizes their patterns. Object recognition and pattern recognition are considered synonymous.

The main pattern recognition steps are shown in Figure 7.4. The block 'Construction of formal description' is based on the experience and intuition of the designer. A set of elementary properties is chosen which describe some characteristics of the object; these properties are measured in an appropriate way and form the description pattern of the object. These properties can be either quantitative or qualitative in character and their form can vary (numerical vectors, chains, etc.). The theory of recognition deals with the problem of designing the classifier for the specific (chosen) set of elementary object descriptions.

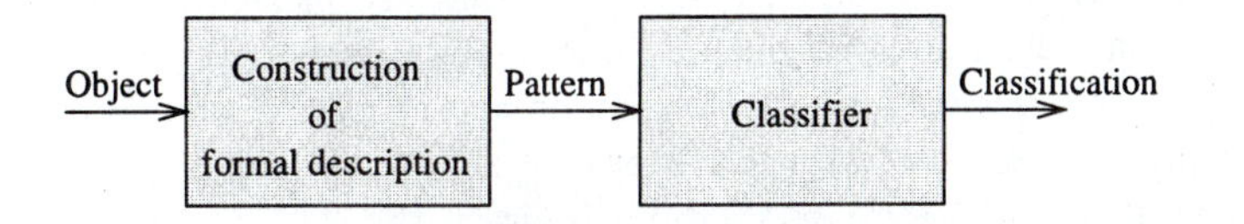

Figure 7.4: *Main pattern recognition steps.*

Statistical object description uses elementary numerical descriptions called **features**, $x_1, x_2, \ldots, x_n$; in image analysis, the features result from object description as discussed in Chapter 6. The pattern (also referred to as pattern vector, or feature vector) $\mathbf{x} = (x_1, x_2, \ldots, x_n)$ that describes an object is a vector of elementary descriptions, and the set of all possible patterns forms the **pattern space** X (also called **feature space**). If the elementary descriptions were appropriately chosen, similarity of objects in each class results in the proximity of their patterns in pattern space. The classes form clusters in the feature space, which can be separated by a discrimination curve (or hyper-surface in a multi-dimensional feature space)—see Figure 7.5.

If a discrimination hyper-surface exists which separates the feature space such that only objects from one class are in each separated region, the problem is called a recognition task with **separable classes**. If the discrimination hyper-surfaces are hyper-planes, it is called a **linearly separable** task. If the task has separable classes, each pattern will represent only

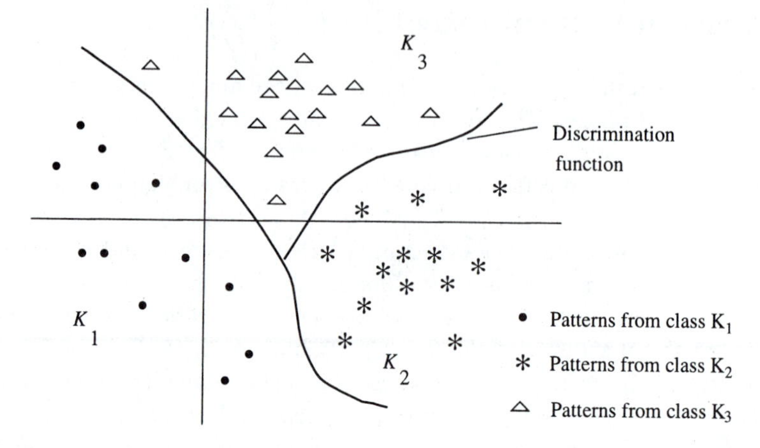

Figure 7.5: *General discrimination functions.*

objects from one class. Intuitively, we may expect that separable classes can be recognized without errors.

The majority of object recognition problems do not have separable classes, in which case the locations of the discrimination hyper-surfaces in the feature space can never separate the classes correctly and some objects will always be misclassified.

7.2.1 Classification principles

A statistical classifier is a device with n inputs and 1 output. Each input is used to enter the information about one of n features $x_1, x_2, \ldots, x_n$ that are measured from an object to be classified. An R-class classifier will generate one of R symbols $\omega_1, \omega_2, \ldots, \omega_R$ as an output, and the user interprets this output as a decision about the class of the processed object. The generated symbols ω_r are the **class identifiers**.

The function $d(\mathbf{x}) = \omega_r$ describes relations between the classifier inputs and the output; this function is called the **decision rule**. The decision rule divides the feature space into R disjoint subsets K_r, $r = 1, \ldots, R$, each of which includes all the feature representation vectors $\mathbf{x}'$ of objects for which $d(\mathbf{x}') = \omega_r$. The borders between subsets K_r, $r = 1, \ldots, R$ form the discrimination hyper-surfaces mentioned earlier. The determination of discrimination hyper-surfaces (or definition of the decision rule) is the goal of classifier design.

The discrimination hyper-surfaces can be defined by R scalar functions $g_1(\mathbf{x}), g_2(\mathbf{x}), \ldots, g_R(\mathbf{x})$ called **discrimination functions**. The design of discrimination functions must satisfy the following formula for all $\mathbf{x} \in K_r$ and for any $s \in \{1, \ldots, R\}$, $s \neq r$:

$$g_r(\mathbf{x}) \geq g_s(\mathbf{x}) \tag{7.6}$$

Therefore, the discrimination hyper-surface between class regions K_r and K_s is defined by

$$g_r(\mathbf{x}) - g_s(\mathbf{x}) = 0 \tag{7.7}$$

The decision rule results from this definition. The object pattern $\mathbf{x}$ will be classified into the class whose discrimination function gives a maximum of all the discrimination functions:

$$d(\mathbf{x}) = \omega_r \iff g_r(\mathbf{x}) = \max_{s=1,\ldots,R} g_s(\mathbf{x}) \tag{7.8}$$

Linear discrimination functions are the simplest and are widely used. Their general form is

$$g_r(\mathbf{x}) = q_{r0} + q_{r1}x_1 + \ldots + q_{rn}x_n \tag{7.9}$$

for all $r = 1, \ldots, R$. If all the discrimination functions of the classifier are linear, it is called a **linear classifier**.

Another possibility is to construct classifiers based on the **minimum distance** principle. The resulting classifier is just a special case of classifiers with discrimination functions, but they have computational advantages and may easily be implemented on digital computers. Assume that R points are defined in the feature space, $\mathbf{v}_1, \mathbf{v}_2, \ldots, \mathbf{v}_R$ that represent **exemplars** (sample patterns) of classes $\omega_1, \omega_2, \ldots, \omega_R$. A minimum distance classifier classifies a pattern $\mathbf{x}$ into the class to whose exemplar it is closest.

$$d(\mathbf{x}) = \omega_r \iff |\mathbf{v}_r - \mathbf{x}| = \min_{s=1,\ldots,R} |\mathbf{v}_s - \mathbf{x}| \tag{7.10}$$

Each discrimination hyper-plane is perpendicular to the line segment $\mathbf{v}_s\mathbf{v}_r$ and bisects it (Figure 7.6).

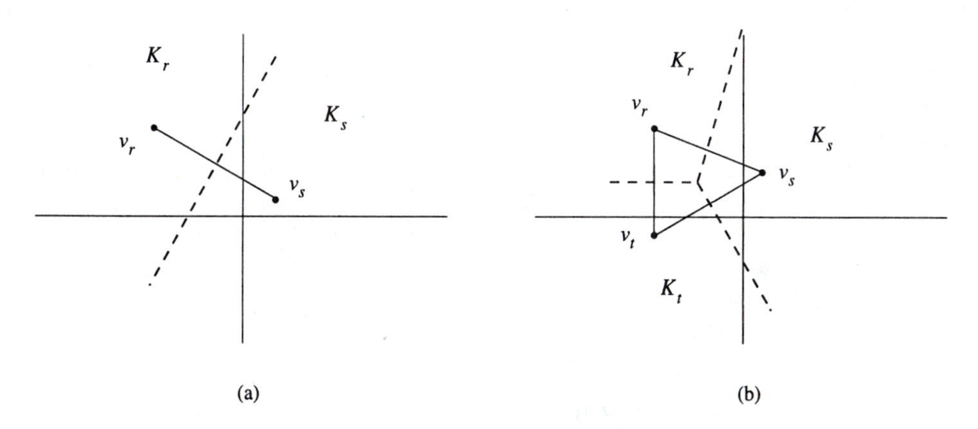

Figure 7.6: *Minimum distance discrimination functions: (a) two-class problem; (b) three-class problem.*

If each class is represented by just one exemplar, a linear classifier results. If more than one exemplar represents some class, the classifier results in piecewise linear discrimination hyper-planes. An algorithm for learning and classification using a minimum distance classifier can be found in Section 7.2.3, Algorithm 7.2.

Non-linear classifiers usually transform the original feature space X^n into a new feature space X^m applying some appropriate non-linear function $\mathbf{\Phi}$, where the superscripts n, m refer to the space dimensionality.

$$\mathbf{\Phi} = (\phi_1, \phi_2, \ldots, \phi_m) : X^n \to X^m \tag{7.11}$$

After the non-linear transformation, a linear classifier is applied in the new feature space—the role of the function $\mathbf{\Phi}$ is to 'straighten' the non-linear discrimination hyper-surfaces of the original feature space into hyper-planes in the transformed feature space. This approach to feature space transformation is called a **Φ-classifier**.

The discrimination functions of a Φ-classifier are

$$g_r(\mathbf{x}) = q_{r0} + q_{r1}\phi_1(\mathbf{x}) + \ldots + q_{rm}\phi_m(\mathbf{x}) \tag{7.12}$$

where $r = 1, \ldots, R$. We may re-write the formula in vector representation

$$g_r(\mathbf{x}) = \mathbf{q}_r \cdot \mathbf{\Phi}(\mathbf{x}) \tag{7.13}$$

where $\mathbf{q}_r$, $\mathbf{\Phi}(\mathbf{x})$ are vectors consisting of $q_{r0}, \ldots, q_{rm}$ and $\phi_0(\mathbf{x}), \ldots, \phi_m(\mathbf{x})$, respectively, $\phi_0(\mathbf{x}) \equiv 1$. Non-linear classifiers are described in detail in [Sklansky 81, Devijver and Kittler 82].

7.2.2 Classifier setting

A classifier based on discrimination functions is a deterministic machine—one pattern $\mathbf{x}$ will always be classified into the same class. Note that the pattern $\mathbf{x}$ may represent objects from different classes, meaning that the classifier decision may be correct for some objects and incorrect for others. Therefore, setting of the optimal classifier should be probabilistic. Incorrect classifier decisions cause some losses to the user, and according to the definition of loss, different criteria for optimal classifier settings will be obtained. Discussing these optimality criteria from the mathematical point of view, criteria represent the value of the mean loss caused by classification.

Let the classifier be considered a universal machine that can be set to represent any decision rule from the rule set D. The set D may be ordered by a parameter vector $\mathbf{q}$ that refers to particular discrimination rules. The value of the mean loss $J(\mathbf{q})$ depends on the decision rule that is applied: $\omega = d(\mathbf{x}, \mathbf{q})$. In comparison with the definition of decision rule used in the previous section, the parameter vector $\mathbf{q}$ has been added to represent the specific decision rule used by the classifier. The decision rule

$$\omega = d(\mathbf{x}, \mathbf{q}^*) \tag{7.14}$$

that gives the minimum mean loss $J(\mathbf{q})$ is called the optimum decision rule, and $\mathbf{q}^*$ is called the vector of optimal parameters.

$$J(\mathbf{q}^*) = \min_{\mathbf{q}} J(\mathbf{q}) \qquad d(\mathbf{x}, \mathbf{q}) \in D \tag{7.15}$$

The **minimum error criterion** (Bayes criterion, maximum likelihood) uses loss functions of the form $\lambda(\omega_r|\omega_s)$, where $\lambda(.)$ is the number that describes quantitatively the loss incurred if a pattern $\mathbf{x}$ which should be classified into the class ω_s is incorrectly classified into the class ω_r.

$$\omega_r = d(\mathbf{x}, \mathbf{q}) \tag{7.16}$$

The mean loss is

$$J(\mathbf{q}) = \int_X \sum_{s=1}^{R} \lambda[d(\mathbf{x}, \mathbf{q})|\omega_s] p(\mathbf{x}|\omega_s) P(\omega_s) \, d\mathbf{x} \tag{7.17}$$

where $P(\omega_s)$, $s = 1, \ldots, R$ are the a priori probabilities of classes, and $p(\mathbf{x}|\omega_s)$, $s = 1, \ldots, R$ are the conditional probability densities of objects $\mathbf{x}$ in the class ω_s.

A classifier that has been set according to the minimum-loss optimality criterion is easy to construct using discrimination functions; usually, unit loss functions are considered,

$$\begin{aligned} \lambda(\omega_r|\omega_s) &= 0 \text{ for } r = s \\ &= 1 \text{ for } r \neq s \end{aligned} \tag{7.18}$$

and the discrimination functions are

$$g_r(\mathbf{x}) = p(\mathbf{x}|\omega_r)P(\omega_r) \qquad r = 1, \ldots, R \tag{7.19}$$

where $g_r(\mathbf{x})$ corresponds (up to a multiplicative constant) to the value of the a posteriori probability $P(\omega_r|\mathbf{x})$.

This probability describes how often a pattern $\mathbf{x}$ is from the class ω_r. Clearly, the optimal decision is to classify a pattern $\mathbf{x}$ to a class ω_r if the a posteriori probability $P(\omega_r|\mathbf{x})$ is the highest of all possible a posteriori probabilities,

$$P(\omega_r|\mathbf{x}) = \max_{s=1,\ldots,R} P(\omega_s|\mathbf{x}) \tag{7.20}$$

A posteriori probability may be computed from a priori probabilities using the Bayes formula

$$P(\omega_s|\mathbf{x}) = \frac{p(\mathbf{x}|\omega_s)P(\omega_s)}{p(\mathbf{x})} \tag{7.21}$$

where $p(\mathbf{x})$ is the mixture density. The mean loss is equal to the probability of an incorrect decision and represents a theoretical optimum—no other classifier setting can give a lower probability of the decision loss. Plots of a posteriori probabilities are shown in Figure 7.7, and corresponding discrimination hyper-surfaces for a three-class classifier can be seen in Figure 7.8.

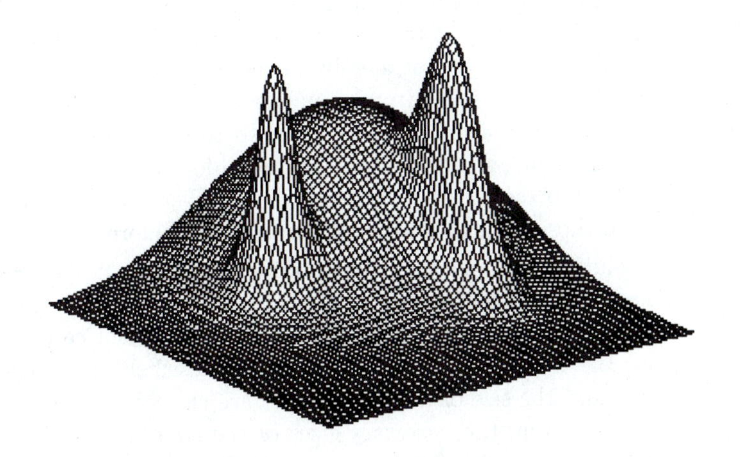

Figure 7.7: *Minimum error classifier: a posteriori probabilities.*

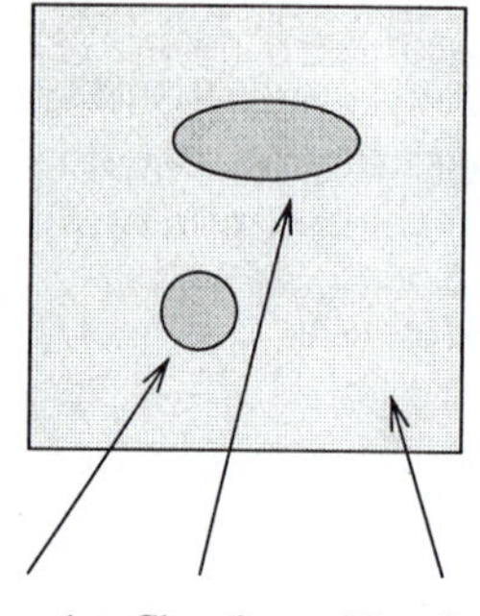

Figure 7.8: *Minimum error classifier: discrimination hyper-surfaces and resulting classes.*

Another criterion is the **best approximation criterion**, which is based on the best approximation of discrimination functions by linear combinations of pre-determined functions $\phi_i(\mathbf{x})$, $i = 1, \ldots, n$. The classifier is then constructed as a Φ-classifier.

Analytic minimization of the extrema problem (7.14) is in many practical cases impossible because the multi-dimensional probability densities are not available. Criteria for loss function evaluation can be found in [Sklansky 81, Devijver and Kittler 82]. The requirements for classification correctness, and the set of objects accompanied by information about their classes, are usually available in practical applications—very often this is all the information that can be used for the classifier design and setting.

The ability to set classification parameters from a set of examples is very important and is called **classifier learning**. The classifier setting is based on a set of objects (represented by their feature vectors), each object being accompanied by information about its proper classification—this set of patterns and their classes is called the **training set**. Clearly, the quality of the classifier setting depends on the quality and size of the training set, which is always finite. Therefore, to design and set a classifier, it is not possible to use all the objects which will later need classifying; that is, the patterns that were not used for classifier design and setting will also enter the classifier, not merely the patterns contained in the training set. The classifier setting methods must be **inductive** in the sense that the information obtained from the elements of the training set must be generalized to cover the whole feature space, implying that the classifier setting should be (near) optimal for all feasible patterns, not only for those patterns that were present in the training set. In other words, the classifier should be able to recognize even those objects that it has never 'seen' before.

It may be that a solution for a given problem does not exist. If the requirements for classification correctness together with the set of training examples are given, it may be impossible to give an immediate answer as to whether the assignment can be fulfilled. The larger the training set, the better the guarantee that the classifier may be set correctly—classification correctness and the size of the training set are closely related. If the statistical properties of patterns are known, the necessary sizes of the training sets can be estimated, but the problem is that in reality they are not usually known. The training set is actually supposed to substitute this missing statistical information. Only after processing of the training set can the designer know whether it was sufficient, and whether an increase in the training set size is necessary.

The training set size will typically be increased several times until the correct classification setting is achieved. The problem, which originally could not be solved, uses more and more information as the training set size increases, until the problem specifications can be met.

The general idea of sequential increase in training set size can be understood as presenting small portions of a large training set to the classifier, whose performance is checked after each portion. The smallest portion size is one element of the training set. Sequential processing of information (which cannot be avoided in principle) has some substantial consequences in the classifier setting process.

All the properties of the classifier setting methods given have analogies in the learning process of living organisms. The basic properties of learning can be listed as follows.

- **Learning** is the process of automated system optimization based on the sequential presentation of examples.
- The **goal of learning** is to minimize the optimality criterion. The criterion may be represented by the mean loss caused by incorrect decisions.
- The finite size of the training set requires the **inductive** character of learning. The goal of learning must be achieved by generalizing the information from examples, before all feasible examples have been presented. The examples may be chosen at random.
- The unavoidable requirements of sequential information presentation and the finite size of system memory necessitate the **sequential character of learning**. Therefore, learning is not a one-step process, but rather a step-by-step process of improvement.

The learning process searches out the optimal classifier setting from examples. The classifier system is constructed as a universal machine that becomes optimal after processing the training set examples (supervised learning), meaning that it is not necessary to repeat the difficult optimal system design if a new application appears. Learning methods do not depend on the application; the same learning algorithm can be applied if a medical diagnostics classifier is set just as when an object recognition classifier for a robot is set.

The quality of classifier decisions is closely related to the quality and amount of information that is available. From this point of view, the patterns should represent as complex a description as possible. On the other hand, a large number of description features would result. Therefore, the object description is always a trade-off between the permissible classification error, the complexity of the classifier construction, and the time required for classification. This results in a question of how to choose the best features from a set of available features, and how to detect the features with the highest contribution to the recognition success. Methods of determination of **informativity** and **discriminativity** of measured features can be found in [Fu 68, Young and Calvert 74, Devijver and Kittler 82, Pudil et al. 94a, Pudil et al. 94b].

7.2.3 Classifier learning

Two common learning strategies will be presented in this section:

- **Probability density estimation** estimates the probability densities $p(\mathbf{x}|\omega_r)$ and probabilities $P(\omega_r)$, $r = 1, \ldots, R$. The discrimination functions are computed according to the minimum error criterion [equation (7.19)].

- **Direct loss minimization** finds the decision rule $\omega = d(\mathbf{x}, \mathbf{q}^*)$ by direct minimization of losses $J(\mathbf{q})$ without estimation of probability densities and probabilities. The criterion of the best approximation is applied.

Probability density estimation methods differ in computational difficulty according to the amount of prior information available about them. If some prior information is available, it usually describes the shape of probability density functions $p(\mathbf{x}|\omega_r)$. The parameters describing the distribution are not usually known, and learning must find the estimate of these parameters. Therefore, this class of learning methods is sometimes called **parametric learning**.

Assume the patterns in the rth class can be described by a normal distribution. The probability density for the normal distribution $N(\boldsymbol{\mu}_r, \boldsymbol{\Psi}_r)$ can be computed for patterns from the class ω_r:

$$p(\mathbf{x}|\omega_r) = \frac{1}{(2\pi)^{n/2}\sqrt{(\det \boldsymbol{\Psi}_r)}} \exp\left[-\frac{1}{2}(\mathbf{x} - \boldsymbol{\mu}_r)^T \, \boldsymbol{\Psi}_r^{-1} \, (\mathbf{x} - \boldsymbol{\mu}_r)\right] \tag{7.22}$$

where $\boldsymbol{\Psi}_r$ is the dispersion matrix (and we recall that $\mathbf{x}$, $\boldsymbol{\mu}_i$ are column vectors). Details about multi-variate probability density function estimation may be found in [Rao 65, Johnson and Wichern 90]. The computation process depends on additional information about the vector of values $\boldsymbol{\mu}_r$ and $\boldsymbol{\Psi}_r$; three cases can be distinguished:

1. The dispersion matrix $\boldsymbol{\Psi}_r$ is known, but the mean value vector $\boldsymbol{\mu}_r$ is unknown. One of the feasible estimates of the mean value may be the average

$$\tilde{\boldsymbol{\mu}}_r = \overline{\mathbf{x}} \tag{7.23}$$

which can be computed recursively:

$$\overline{\mathbf{x}}(k+1) = \frac{1}{k+1}[k \; \overline{\mathbf{x}}(k) + \mathbf{x}_{k+1}] \tag{7.24}$$

where $\overline{\mathbf{x}}(k)$ is the average computed from k samples, and $\mathbf{x}_{k+1}$ is the $(k+1)^{st}$ pattern from the class r from the training set. This estimate is unbiased, consistent, efficient, and linear.

Alternatively, if the a priori estimate of the mean $\tilde{\boldsymbol{\mu}}_r(0)$ is available, the Bayes approach to estimation of the normal distribution parameters can be used. Then, the estimate can be computed recursively:

$$\tilde{\boldsymbol{\mu}}_r(k+1) = \frac{a+k}{a+k+1}\tilde{\boldsymbol{\mu}}_r(k) + \frac{1}{a+k+1}\mathbf{x}_{k+1} \tag{7.25}$$

The parameter a represents the confidence in the a priori estimate $\tilde{\boldsymbol{\mu}}_r(0)$. In training, a specifies the number of steps during which the designer believes more in the a priori estimate than in the mean value so far determined by training. Note that for $a = 0$, the Bayes estimate is identical to that given in equation (7.24).

2. The dispersion matrix $\mathbf{\Psi}_r$ is unknown, but the mean value vector $\boldsymbol{\mu}_r$ is known. The estimate of the dispersion matrix $\mathbf{\Psi}_r$ if the mean value $\boldsymbol{\mu}_r$ is known is usually taken as

$$\tilde{\mathbf{\Psi}}_r = \frac{1}{K} \sum_{k=1}^{K} (\mathbf{x}_k - \boldsymbol{\mu}_r) \ (\mathbf{x}_k - \boldsymbol{\mu}_r)^T \tag{7.26}$$

or, in recursive form,

$$\tilde{\mathbf{\Psi}}_r(k+1) = \frac{1}{k+1} [k \ \tilde{\mathbf{\Psi}}_r(k) + (\mathbf{x}_{k+1} - \boldsymbol{\mu}_r) \ (\mathbf{x}_{k+1} - \boldsymbol{\mu}_r)^T] \tag{7.27}$$

This estimate is unbiased and consistent.

As another option, if the a priori estimate $\tilde{\mathbf{\Phi}}_r(0)$ of the dispersion matrix $\mathbf{\Psi}_r$ is known, the Bayes estimation approach can be applied. Let K be the number of samples in the training set, and $\tilde{\mathbf{\Psi}}_r(K)$ be calculated as in equation (7.27). Then

$$\tilde{\mathbf{\Phi}}_r(K) = \frac{b \ \tilde{\mathbf{\Phi}}_r(0) + K \ \tilde{\mathbf{\Psi}}_r(K)}{b + K} \tag{7.28}$$

and $\tilde{\mathbf{\Phi}}_r(K)$ is considered the Bayes estimate of the dispersion matrix $\mathbf{\Psi}_r$. Parameter b represents the confidence in the a priori estimate $\tilde{\mathbf{\Phi}}_r(0)$.

3. Both the dispersion matrix $\mathbf{\Psi}_r$ and the mean value vector $\boldsymbol{\mu}_r$ are unknown. The following estimates can be used

$$\tilde{\boldsymbol{\mu}}_r = \overline{\mathbf{x}} \tag{7.29}$$

$$\tilde{\mathbf{\Psi}}_r = \mathbf{S} = \frac{1}{K-1} \sum_{k=1}^{K} (\mathbf{x}_k - \overline{\mathbf{x}}) \ (\mathbf{x}_k - \overline{\mathbf{x}})^T \tag{7.30}$$

or, in the recursive form,

$$\begin{aligned} \mathbf{S}(k+1) = \ & \tfrac{1}{k}\{(k-1)\mathbf{S}(k) \\ & +[\mathbf{x}_{k+1} - \overline{\mathbf{x}}(k+1)] \ [\mathbf{x}_{k+1} - \overline{\mathbf{x}}(k+1)]^T \\ & +k[\overline{\mathbf{x}}(k) - \overline{\mathbf{x}}(k+1)] \ [\overline{\mathbf{x}}(k) - \overline{\mathbf{x}}(k+1)]^T\} \end{aligned} \tag{7.31}$$

Alternatively, if the a priori estimate $\tilde{\mathbf{\Phi}}_r(0)$ of the dispersion matrix $\mathbf{\Psi}_r$ and the a priori estimate $\tilde{\boldsymbol{\nu}}_r(0)$ of the mean vector for class r are known, the Bayes estimates can be determined as follows:

$$\tilde{\boldsymbol{\nu}}_r(K) = \frac{a\tilde{\boldsymbol{\mu}}_r(0) + K\tilde{\boldsymbol{\mu}}_r(K)}{a + K} \tag{7.32}$$

where K is the number of samples in the training set and $\tilde{\boldsymbol{\mu}}_r(K)$ is determined using equations (7.23) and (7.24). The dispersion matrix estimate is calculated as

$$\begin{aligned} \tilde{\mathbf{\Phi}}_r(K) = \ & \tfrac{b}{b+K} \ \tilde{\mathbf{\Phi}}_r(0) + a\tilde{\boldsymbol{\nu}}_r(0)\tilde{\boldsymbol{\nu}}_r(0)^T + (K-1)\tilde{\mathbf{\Psi}}_r(K) \\ & +K\tilde{\boldsymbol{\mu}}_r(K)\tilde{\boldsymbol{\mu}}_r(K)^T - (a+K)\tilde{\boldsymbol{\nu}}_r(K)\tilde{\boldsymbol{\nu}}_r(K)^T \end{aligned} \tag{7.33}$$

where $\tilde{\mathbf{\Psi}}_r(K)$ is calculated as given in equation (7.27). Then, $\tilde{\boldsymbol{\nu}}_r(K)$ and $\tilde{\mathbf{\Phi}}_r(K)$ are considered the Bayes estimates of the mean vector and the dispersion matrix for class r, respectively. Again, parameters a, b represent the confidence in the a priori estimates of $\tilde{\mathbf{\Phi}}_r(0)$ and $\tilde{\boldsymbol{\nu}}_r(0)$.

The a priori probabilities of classes $P(\omega_r)$ are estimated as relative frequencies

$$P(\omega_r) = \frac{K_r}{K} \tag{7.34}$$

where K is the total number of objects in the training set; K_r is the number of objects from the class r in the training set.

Algorithm 7.1: Learning and classification based on estimates of probability densities assuming the normal distribution

1. Learning: Compute the estimates of the mean value vector $\boldsymbol{\mu}_r$ and the dispersion matrix $\boldsymbol{\Psi}_r$, equations (7.24) and/or (7.27), (7.31).

2. Compute the estimates of the a priori probability densities $p(\mathbf{x}|\omega_r)$, equation (7.22).

3. Compute the estimates of the a priori probabilities of classes, equation (7.34).

4. Classification: Classify all patterns into the class r if

$$\omega_r = \max_{i=1,\ldots,s} \left[p(\mathbf{x}|\omega_i)\, P(\omega_i) \right]$$

[equations (7.19) and (7.8)].

If no prior information is available (i.e., even the distribution type is not known), the computation is more complex. In such cases, if it is not necessary to use the minimum error criterion, it is advantageous to use a direct loss minimization method.

No probability densities or probabilities are estimated in the second group of methods based on direct minimization of losses. The minimization process can be compared to gradient optimization methods, but pure gradient methods cannot be used because of unknown probability densities, so the gradient cannot be evaluated. Nevertheless, the minimum can be found using methods of **stochastic approximations** that are discussed in [Sklansky 81].

The most important conclusion is that the learning algorithms can be represented by recursive formulae in both groups of learning methods and it is easy to implement them.

We have noted that the most common and easily implementable classifier is the minimum distance classifier. Its learning and classification algorithm as follows.

Algorithm 7.2: Minimum distance classifier learning and classification

1. Learning: For all classes, compute class exemplars $\mathbf{v}_i$ based on the training set

$$\mathbf{v}_i(k_i + 1) = \frac{1}{k_i + 1} [k_i \mathbf{v}_i(k_i) + \mathbf{x}_i(k_i + 1)] \tag{7.35}$$

where $\mathbf{x}_i(k_i + 1)$ are objects from the class i and k_i denotes the number of objects from class i used thus far for learning.

2. Classification: For an object description vector $\mathbf{x}$, determine the distance of $\mathbf{x}$ from the class exemplars $\mathbf{v}_i$. Classify the object into the class j if the distance of $\mathbf{x}$ from $\mathbf{v}_j$ is the minimum such [equation (7.10)].

7.2.4 Cluster analysis

We noted earlier that classification methods exist which do not need training sets for learning. In particular, they do not need information about the class of objects in the learning stage, but learn them without a teacher (unsupervised learning). One such group of classification methods is called **cluster analysis**. Cluster analysis can be applied in classification if for any reason the training set cannot be prepared, or if examples with known class evaluation are not available.

Cluster analysis methods divide the set of processed patterns into subsets (clusters) based on the mutual similarity of subset elements. Each cluster contains patterns representing objects that are similar according to the selected object description and similarity criteria. Objects that are not similar reside in different clusters.

There are two main groups of cluster analysis methods—the first is hierarchical and the second non-hierarchical. Hierarchical methods construct a clustering tree; the set of patterns is divided into the two most dissimilar subsets, and each subset is divided into other different subsets, etc. Non-hierarchical methods sequentially assign each pattern to one cluster. Methods and algorithms for cluster analysis can be found in [Duda and Hart 73, Dubes and Jain 80, Devijver and Kittler 82, Blashfield et al. 82, Romesburg 84, McQuitty 87, Kaufman and Rousseeuw 90, Schalkoff 92, Everitt and Brian 93, Arabie et al. 96].

Non-hierarchical cluster analysis methods are either parametric or non-parametric. Parametric approaches are based on known class-conditioned distributions and require distribution parameter estimation that is similar to that used in minimum error classification described in Section 7.2.3. Parametric clustering approaches used for threshold-based image segmentation were also described in Section 5.1.2.

Non-parametric cluster analysis is a popular, simple, and practically useful non-hierarchical approach to cluster analysis. The **MacQueen k-means** cluster analysis method is a well-known example of this approach [MacQueen 67]. We need to assume that the number of clusters k is known—if it is not, it can be determined as the number of classes that gives the maximum confidence in results, or some more complex clustering method can be applied that does not need this information. The starting cluster points are constructed in the first step, represented by k points in the n-dimensional feature space. These points can either be selected at random from the clustered set of patterns, or the first k patterns from the set can be chosen. If exemplars of clusters are available, even if these exemplars are unreliable, it is worthwhile using them as the starting cluster points. The method has two main stages; patterns are allocated to one of the existing clusters in the first stage according to their distance from the cluster exemplars, choosing the closest. Then the exemplar is re-computed as the center of gravity of all patterns in that cluster. If all the patterns from the set have been processed, the current exemplars of clusters are considered final; all the patterns are assigned to one of the clusters, represented by the exemplars determined in the first stage. Then the patterns are (re-)assigned to clusters according to their distance from the exemplars, patterns being assigned to the closest cluster. The exemplars are not recomputed in the second stage. It should be clear that elements that were used for the starting cluster point definitions need not be members of the same clusters at the end.

Algorithm 7.3: MacQueen k-means cluster analysis

1. Define the number of clusters.
2. Initialize the cluster starting points (exemplars, initial guesses) $\mathbf{v}_1, \mathbf{v}_2, \ldots, \mathbf{v}_k$. Usually some patterns are chosen to serve as cluster starting points, perhaps chosen at random.
3. First pass: Decide to which cluster each pattern belongs, choosing the closest (do not process those patterns that were used to initialize clusters). Re-compute the relevant exemplar after an object is added to a cluster, possibly using equation (7.35).
4. Second pass: Let the final exemplars be exemplars of resulting clusters. Classify all objects (including those used to form starting exemplars) using the final exemplars from the first pass. Use the same distance criterion as in the first pass.

Because of its simplicity, the MacQueen method has its limitations. There are many variations on this algorithm; one is to repeat the second stage until convergence. The ISODATA cluster analysis method [Dubes and Jain 76, Kaufman and Rousseeuw 90] may solve a complex clustering problem better. ISODATA uses two parameter sets, one which does not change during the clustering and another which can be interactively adjusted until an acceptable clustering result is obtained. ISODATA represents a set of non-hierarchical cluster analysis methods from which the best can be picked.

Determining the number of clusters has not been mentioned—for example, what metric is the most suitable in n-dimensional space, etc. Answers to these and many other questions can be found in [Romesburg 84, McQuitty 87, Kaufman and Rousseeuw 90].

Note that statistical pattern recognition and cluster analysis can be combined. For instance, the minimum distance classifier can be taught using cluster analysis methods, cluster exemplars can be considered class exemplars, these exemplars can be assigned appropriate names, and other patterns can be recognized using the resulting classifier [Sonka 86]. Additionally, fuzzy clustering approaches were reported with good results [Bezdek 81] (Section 7.7).

7.3 Neural nets

Neural nets have seen an explosion of interest since their re-discovery as a pattern recognition paradigm in the early 1980s. The value of some of the applications for which they are used may be arguable, but there is no doubt that they represent a tool of great value in various areas generally regarded as 'difficult', particularly speech and visual pattern recognition.

Most neural approaches are based on combinations of elementary processors (neurons), each of which takes a number of inputs and generates a single output. Associated with each input is a weight, and the output (in most cases) is then a function of the weighted sum

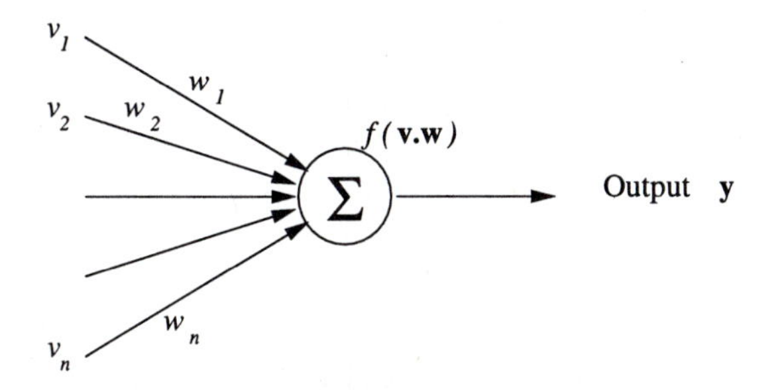

Figure 7.9: *A simple (McCulloch-Pitts) neuron.*

of inputs; this output function may be discrete or continuous, depending on the variety of network in use. A simple neuron is shown in Figure 7.9—this model is derived from pioneering work on neural simulation conducted over 50 years ago [McCulloch and Pitts 43]. The inputs are denoted by $v_1, v_2, \ldots$, and the weights by $w_1, w_2, \ldots$; the total input to the neuron is then

$$x = \sum_{i=1}^{n} v_i w_i \tag{7.36}$$

or, more generally,

$$x = \sum_{i=1}^{n} v_i w_i - \theta \tag{7.37}$$

where θ is a threshold associated with this neuron. Also associated with the neuron is a **transfer function** $f(x)$ which provides the output; common examples are

$$f(x) = \begin{cases} 0 & \text{if } x \leq 0 \\ 1 & \text{if } x > 0 \end{cases} \tag{7.38}$$

$$f(x) = \frac{1}{1+e^{-x}} \tag{7.39}$$

This model saw a lot of enthusiastic use during an early phase, culminating in Rosenblatt's **perceptron** [Rosenblatt 62].

The general idea of collections (networks) of these neurons is that they are interconnected (so the output of one becomes the input of another, or others)—this idea mimics the high level of interconnection of elementary neurons found in brains, which is thought to explain the damage resistance and recall capabilities of humans. Such an interconnection may then take some number of external inputs and deliver up some (possibly different) number of external outputs—see Figure 7.10. What lies between then specifies the network: This may mean a large number of heavily interconnected neurons, or some highly structured (e.g., layered) interconnection, or, pathologically, nothing (so that inputs are connected straight to outputs).

There are many uses to which such a structure may be put; the general task being performed is vector association. Examples may be

- Classification: If the output vector (m-dimensional) is binary and contains only a single one, the position of the one classifies the input pattern into one of m categories.

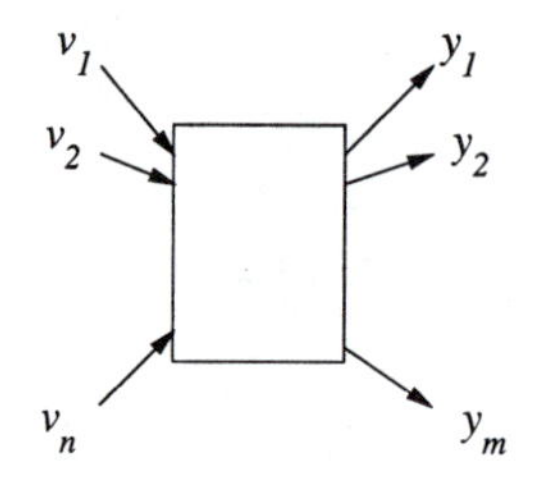

Figure 7.10: *A neural network as a vector associator.*

- Auto-association: Some uses of neural networks cause them to re-generate the input pattern at the outputs (so $m = n$ and $v_i = y_i$); the purpose of this may be to derive a more compact vector representation from within the network internals.
- General association: At their most interesting, the vectors $\mathbf{v}$ and $\mathbf{y}$ represent patterns in different domains, and the network is forming a correspondence between them. One of the most quoted examples of this is NetTalk [Sejnowski and Rosenberg 87], in which the inputs represent a stream of written text and the outputs are phonemes—thus the network is a speech generator.

7.3.1 Feed-forward networks

The first neural networks involved no 'internals' (so the box in Figure 7.10 was empty); these early perceptrons had a training algorithm developed which was shown to converge *if a solution to the problem at hands exists* [Minsky 88]; unfortunately, this caveat proved very restrictive, requiring that the classification being performed be linearly separable (vector clusters of interest lay in distinct half-spaces). This restriction was overcome by the now very popular **back-propagation** algorithm [Rumelhart and McClelland 86], which trains strictly layered networks in which it is assumed that at least one layer exists between input and output (it fact, it can be shown that two such 'hidden' layers always suffice [Kolmogorov 63, Hecht-Nielson 87]). Such a network is shown in Figure 7.11, and is an example of a **feed-forward** network, in which data are admitted at the inputs and travel in one direction toward the outputs, at which the 'answer' may be read.

The standard approach to use of such networks is to obtain a training set of data—a set of vectors for which the 'answer' is already known. This is used to teach a network with some training algorithm, such that the network can perform the association accurately. Then, in classification (or 'live') mode, unknown patterns are fed into the net and it produces answers based on generalizing what it has learned.

Back-propagation proceeds by comparing the output of the network to that expected, and computing an error measure based on sum of square differences. This is then minimized using gradient descent by altering the weights of the network. Denoting a member of the training set by $\mathbf{v}^i$, the actual outputs by $\mathbf{y}^i$, and the *desired* outputs by ω^i, the error is

$$E = \sum_i \sum_j (y_j^i - \omega_j^i)^2$$

(thus summing square difference over the entire training set) and the algorithm performs the updates,

$$w_{ij}(k+1) = w_{ij}(k) - \epsilon \frac{\partial E}{\partial w_{ij}} \tag{7.40}$$

iteratively until 'good' performance is seen (k here counts the iterations of the updates).

The literature on back-propagation is large and thorough, and we present here a summary of the algorithm only.

Algorithm 7.4: Back-propagation learning

1. Assign small random numbers to the weights w_{ij}, and set $k = 0$.
2. Input a pattern $\mathbf{v}$ from the training set and evaluate the neural net output $\mathbf{y}$.
3. If $\mathbf{y}$ does not match the required output vector ω, adjust the weights

$$w_{ij}(k+1) = w_{ij}(k) + \epsilon \delta_j z_i(k) \tag{7.41}$$

where ϵ is called the **learning constant** or **learning rate**, $z_i(k)$ is the output of the node i, k is the iteration number, δ_j is an error associated with the node j in the adjacent upper level

$$\delta_j = \begin{cases} y_j(1-y_j)(\omega_j - y_j) & \text{for output node } j \\ z_j(1-z_j)(\sum_l \delta_l w_{jl}) & \text{for hidden node } j \end{cases} \tag{7.42}$$

4. Go to step 2 and fetch the next input pattern.
5. Increment k, and repeat steps 2 to 4 until each training pattern outputs a suitably good approximation to that expected. Each circuit of this loop is termed an **epoch**.

The convergence process can be very slow, and there is an extensive literature on speeding the algorithm (see, for example, [Haykin 94]). The best known of these techniques is the introduction of **momentum**, which accelerates convergence across plateaux of the cost surface, and controls behavior in steep ravines. This approach rewrites equation (7.40) as

$$\Delta w_{ij} = \epsilon \frac{\partial E}{\partial w_{ij}}$$

and updates it to

$$\Delta w_{ij} := \epsilon \frac{\partial E}{\partial w_{ij}} + \epsilon \Delta w_{ij}$$

which updates equation (7.41) to

$$w_{ij}(k+1) = w_{ij}(k) + \epsilon \delta_j z_i(k) + \alpha[w_{ij}(k) - w_{ij}(k-1)] \tag{7.43}$$

α is called the **momentum constant**, and is chosen to be between 0 and 1, having the effect of contributing a proportion of the update of the previous iteration into the current one. Thus, in areas of very low gradient, some movement continues.

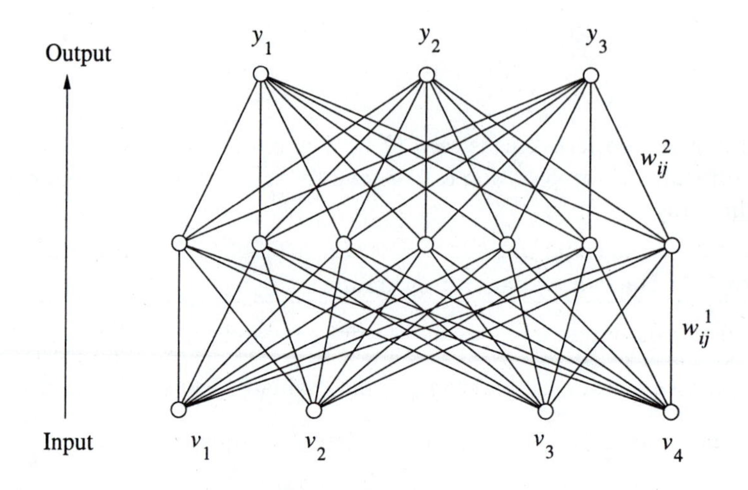

Figure 7.11: *A three-layered neural net structure.*

7.3.2 Unsupervised learning

A different class of networks are self-teaching—that is, they do not depend on the net being exposed to a a training set with known information about classes, but are able to self-organize themselves to recognize patterns automatically. Various types of networks exist under this general heading, of which the best known are Kohonen feature maps.

Kohonen maps take as input n-dimensional data vectors and generate an n-dimensional output that, within the domain of the problem at hand, 'best represents' the particular input given. More precisely, the network has a layer of neurons, each of which is connected to all n input vector components, each neuron calculates its input [equation (7.36)], and that with the largest input is regarded as the 'winner'; the n weights associated with the input arcs to this node then represent the output. Figure 7.12 illustrates this. The weights are updated using

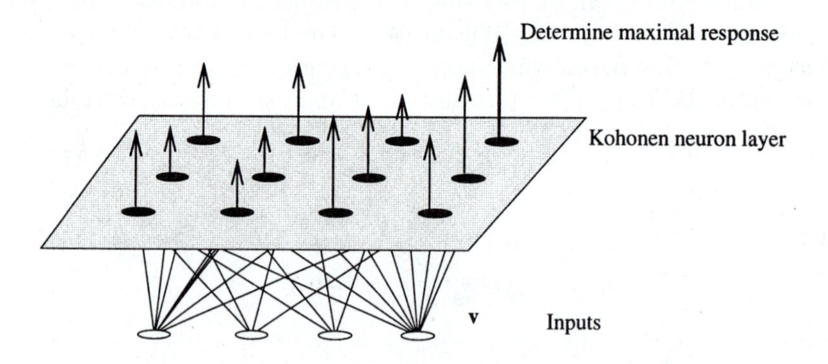

Figure 7.12: *Kohonen self-organizing neural net.*

a learning algorithm that finds the data structure for itself (that is, no prior classification is needed or indeed known). It may be clear that such a network is performing the role of clustering—similar inputs generate the same output.

The theory underlying Kohonen networks is derived from the operation of biological neurons, which are known to exist in locally 2D layers and in which neural responses are known to cluster. The derivation of the algorithm may be found in various standard texts [Kohonen 88, Kohonen 95] but may be summarized as follows.

Algorithm 7.5: Unsupervised learning of the Kohonen feature map

1. Assign random numbers with a small variance around the average values of the input feature vector elements to the weights w_{ij}.

2. Collect a sample of vectors $V = \{\mathbf{v}\}$ from the set to be analyzed.

3. Select a new vector $\mathbf{v} \in V$, and determine the neuron with the biggest input:

$$j^* = \text{argmax}_j \sum_i w_{ij} v_i$$

4. For all neurons n_j within a neighborhood of radius r of n_{j^*}, perform the weight update with step size $\alpha > 0$ (learning rate)

$$w_{ij} := w_{ij} + \alpha(v_i - w_{ij}) \tag{7.44}$$

5. Go to step 3

6. Reduce r and α, and go to step 3.

Kohonen networks enjoy considerable use, often as components of larger systems which may include other varieties of neural network.

Several other varieties of self-teaching net exist, of which the best known is perhaps ART (Adaptive Resonance Theory) [Carpenter and Grossberg 87a, Carpenter and Grossberg 87b]. More specialized texts provide ample detail.

7.3.3 Hopfield neural nets

Hopfield nets are used mostly in optimization problems [Hopfield and Tank 85, Hopfield and Tank 86]; however, it is possible to represent recognition as an optimization task—find the maximum similarity between a pattern $\mathbf{x}$ and one of the existing exemplars $\mathbf{v}$.

In the Hopfield neural model, the network does not have designated inputs and outputs, but rather the current configuration represents its state. The neurons, which are fully interconnected, have discrete (0/1 or -1/1) outputs, calculated from equation (7.38). Weights between neurons do not evolve (learn), but are computed from a set of known exemplars at initialization,

$$w_{ij} = \sum_r (v_i^r v_j^r) \qquad (i \neq j) \tag{7.45}$$

where w_{ij} is the interconnection weight between nodes i and j; and v_i^r is the ith element of the rth exemplar; $w_{ii} = 0$ for any i.

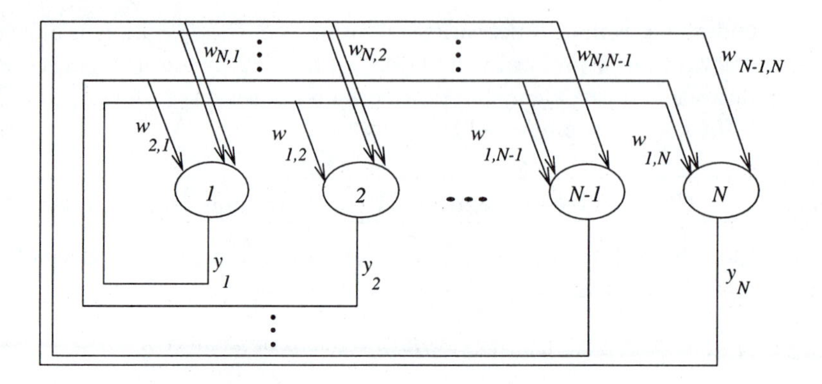

Figure 7.13: *Hopfield recurrent neural net.*

The Hopfield net acts as an associative memory where the exemplars are stored; its architecture is shown in Figure 7.13. When used for recognition, the feature vector to be classified enters the net in the form of initial values of node outputs. The Hopfield net then recurrently iterates using existing interconnections with fixed weights until a stable state is found—that such a state is reached can be proved under certain conditions [for which equation (7.45) is sufficient]. The resulting stable state should be equal to the values of the exemplar that is closest to the processed feature vector in the Hamming metric sense. Supposing these class exemplars $\mathbf{v}^r$ are known, the recognition algorithm is as follows.

Algorithm 7.6: Recognition using a Hopfield net

1. Based on existing exemplars $\mathbf{v}^i$ of r classes, compute the interconnection weights w_{ij} [equation (7.45)].

2. Apply an unknown feature vector $\mathbf{x}$ as initial outputs $\mathbf{y}(0)$ of the net.

3. Iterate until the net converges (output $\mathbf{y}$ does not change):

$$y_j(k+1) = f\left\{\sum_{i=1}^{N}[w_{ij}y_i(k)]\right\} \tag{7.46}$$

The final output vector $\mathbf{y}$ represents the exemplar of the class into which the processed feature vector $\mathbf{x}$ is classified. In other words, a Hopfield neural net transforms a non-ideal representation of an object (fuzzy, noisy, incomplete, etc.) to the ideal exemplar representation. The transformation of a noisy binary image of characters to a clear letter is one vision-related application; binary image recognition examples can be found in [Kosko 91, Rogers and Kabrisky 91].

The Hopfield neural net converges by seeking a minimum of a particular function—this is usually a *local* minimum, which may mean that the correct exemplar (the *global* minimum) is not found. Moreover, the number of local minima grows rapidly with the number of exemplars stored in the associative network. It can be shown that the minimum number of nodes N required is about seven times the number of memories M to be stored (this is known as the $0.15N \geq M$ rule) [McEliece et al. 87, Amit 89], causing a rapid increase in the number of necessary nodes.

This overview has shown only the main principles of neural nets and their connections to conventional statistical pattern recognition, and we have not discussed many alternative neural net techniques, methods, and implementations. The state of the art and many references may be found in a set of selected papers [Carpenter and Grossberg 91] and in [Amit 89, Hecht-Nielsen 90, Judd 90, Simpson 90, Mozer 91, Zhou 92, Masters 95], and in various useful introductory texts [Wasserman 89, Carling 92, Nigrin 93, Fausett 94, Braspenning et al. 95]. Some interesting applications, many in the visual domain, may be found in [Linggard et al. 92].

7.4 Syntactic pattern recognition

Quantitative description of objects using numeric parameters (the feature vector) is used in statistical pattern recognition, while **qualitative** description of an object is a characteristic of syntactic pattern recognition. The object structure is contained in the syntactic description. Syntactic object description should be used whenever feature description is not able to represent the complexity of the described object and/or when the object can be represented as a hierarchical structure consisting of simpler parts. The elementary properties of the syntactically described objects are called **primitives** (Section 6.2.4 covered the syntactic description of object borders using border primitives, these border primitives representing parts of borders with a specific shape). Graphical or relational descriptions of objects where primitives represent sub-regions of specific shape is another example (see Sections 6.3.3 to 6.3.5). After each primitive has been assigned a symbol, relations between primitives in the object are described, and a **relational structure** results (Chapters 3 and 6). As in statistical recognition, the design of description primitives and their relation is not algorithmic. The design is based on the analysis of the problem, designer experience, and abilities. However, there are some principles that are worth following:

1. The number of primitive types should be small.
2. The primitives chosen must be able to form an appropriate object representation.
3. Primitives should be easily segmentable from the image.
4. Primitives should be easily recognizable using some statistical pattern recognition method.
5. Primitives should correspond with significant natural elements of the object (image) structure being described.

For example, if technical drawings are described, primitives are line and curve segments, binary relations describe relations such as *to be adjacent*, *to be left of*, *to be above*, etc. This description structure can be compared with the structure of a natural language. The text consists of sentences, sentences consist of words, words are constructed by concatenation of letters. Letters are considered primitives in this example; the set of all letters is called the

alphabet. The set of all words in the alphabet that can be used to describe objects from one class (the set of all feasible descriptions) is named the **description language** and represents descriptions of all objects in the specific class. In addition, a **grammar** represents a set of rules that must be followed when words of the specific language are constructed from letters (of the alphabet). Grammars can describe infinite languages as well. These definitions will be considered in more detail in Section 7.4.1.

Assume that the object is appropriately described by some primitives and their relations. Moreover, assume that the grammar is known for each class that generates descriptions of all objects of the specified class. Syntactic recognition decides whether the description word is or is not syntactically correct according to the particular class grammars, meaning that each class consists only of objects whose syntactic description can be generated by the particular grammar. Syntactic recognition is a process that looks for the grammar that can generate the syntactic word that describes an object.

We mentioned relational structure in the correspondence with the syntactic description of objects. Each relational structure with multiple relations can be transformed to a relational structure with at most binary relations; the image object is then represented by a **graph** which is **planar** if relations with adjacent regions only are considered. A graphical description is very natural, especially in the description of segmented images—examples were given in Section 6.3. Each planar graph can be represented either by a graph grammar or by a sequence of symbols (chain, word, etc.) over an alphabet. Sequential representation is not always advantageous in image object recognition because the valuable correspondence between the syntactic description and the object may be lost. Nevertheless, work with chain grammars is more straightforward and understandable, and all the main features of more complex grammars are included in chain grammars. Therefore, we will discuss principally sequential syntactic descriptions and chain grammars. More precise and detailed discussion of grammars, languages, and syntactic recognition methods can be found in [Fu 74, Fu 77, Chen 76, Pavlidis 77, Rosenfeld 79b, Fu 80, Pavlidis 80].

The syntactic recognition process is described by the following algorithm.

Algorithm 7.7: Syntactic recognition

1. Learning: Based on the problem analysis, define the primitives and their possible relations.
2. Construct a description grammar for each class of objects using either hand analysis of syntactic descriptions or automated grammar inference (see Section 7.4.3).
3. Recognition: For each object, extract its primitives first; recognize the primitives' classes and describe the relations between them. Construct a description word representing an object.
4. Based on the results of the syntactic analysis of the description word, classify an object into that class for which its grammar (constructed in step 2) can generate the description word.

It can be seen that the main difference between statistical and syntactic recognition is in the learning process. Grammar construction can rarely be algorithmic using today's approaches, requiring significant human interaction. It is usually found that the more complex the primitives are, the simpler is the grammar, and the simpler and faster is the syntactic analysis. More complex description primitives on the other hand make step 3 of the algorithm more difficult and more time consuming; also, primitive extraction and evaluation of relations may not be simple.

7.4.1 Grammars and languages

Assuming that the primitives have been successfully extracted, all the inter-primitive relations can then be described syntactically as n-ary relations; these relations form structures (chains, trees, graphs) called **words** that represent the object or the pattern. Each pattern is therefore described by a word. Primitive classes can be understood as letters from the alphabet of symbols called **terminal symbols**. Let the alphabet of terminal symbols be V_t.

The set of patterns from a particular class corresponds to a set of words. This set of words is called the **formal language** and is described by a **grammar**. The grammar is a mathematical model of a generator of syntactically correct words (words from the particular language); it is a quadruple,

$$G = [V_n, V_t, P, S] \tag{7.47}$$

where V_n and V_t are disjoint alphabets, elements of V_n are called **non-terminal symbols**, and elements of V_t are terminal symbols. Define V^* to be the set of all empty or non-empty words built from the terminal and/or non-terminal symbols. The symbol S is the grammar axiom or the *start* symbol. The set P is a non-empty finite subset of the set $V^* \times V^*$; elements of P are called the substitution rules. The set of all words that can be generated by the grammar G is called the **language** $L(G)$. Grammars that generate the same language are called **equivalent**.

A simple example will illustrate this terminology. Let the words generated by the grammar be squares of arbitrary size with sides parallel to the co-ordinate axes, and let the squares be represented by the Freeman chain code of the border in 4-connectivity (see Section 6.2.1). There are four terminal symbols (primitives) of the grammar in this case, $V_t = \{0, 1, 2, 3\}$. Let the non-terminal symbols be $V_n = \{s, a, b, c, d\}$. Note that the terminal symbols correspond to natural primitives of the 4-connectivity Freeman code; the non-terminal symbols were chosen from an infinite set of feasible symbols. The set of substitution rules P demonstrates how the start symbol $S = s$ can be transformed to words corresponding to the Freeman chain code description of squares:

$$\begin{aligned} P: \quad (1) \quad & s && \rightarrow \quad abcd \\ (2) \quad & aAbBcCdD && \rightarrow \quad a1Ab2Bc3Cd0D \\ (3) \quad & aAbBcCdD && \rightarrow \quad ABCD \end{aligned} \tag{7.48}$$

where A (B, C, D, respectively) is a variable representing any chain (including an empty one) consisting only of terminal symbols 1 $(2, 3, 0)$. Rule 3 stops the word generating process. For example, a square with a side length $l = 2$ with the Freeman chain description 11223300 is generated by the following sequence of substitution rules (see Figure 7.14):

$$s \rightarrow^1 abcd \rightarrow^2 a1b2c3d0 \rightarrow^2 a11b22c33d00 \rightarrow^3 11223300$$

where the arrow superscript refers to the appropriate substitution rule. The simple analysis of generated words shows that the language generated consists only of Freeman chain code representations of squares with sides parallel to the plane co-ordinates.

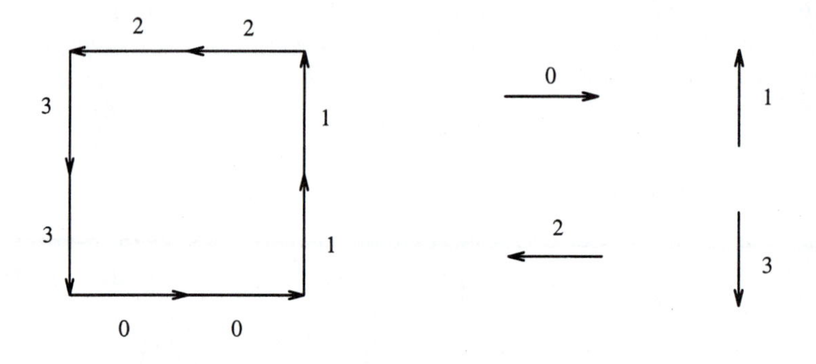

Figure 7.14: *Square shape description.*

Grammars can be divided into four main groups ordered from the general to the specific [Chomsky 66, Chomsky et al. 71]:

1. Type 0—**General Grammars**
   There are no limitations for the substitution rules.

2. Type 1—**Context-Sensitive Grammars**
   Substitution rules can be of the form
   $$W_1 \alpha W_2 \to W_1 U W_2 \tag{7.49}$$
   which can contain the substitution rule $S \to e$, where e is an empty word; words W_1, W_2, U consist of elements of V^*, $U \neq e$, $\alpha \in V_n$. This means that the non-terminal symbols can be substituted by the word U in the context of words W_1 and W_2.

3. Type 2—**Context-Free Grammars**
   Substitution rules have the form
   $$\alpha \to U \tag{7.50}$$
   where $U \in V^*$, $U \neq e$, $\alpha \in V_n$. Grammars can contain the rule $S \to e$. This means that the non-terminal symbol can be substituted by a word U independently of the context of α.

4. Type 3—**Regular Grammars**
   The substitution rules of regular grammars are of the form
   $$\alpha \to x\beta \qquad or \qquad \alpha \to x \tag{7.51}$$
   where $\alpha, \beta \in V_n$, $x \in V_t$. The substitution rule $S \to e$ may be included.

All the grammars discussed so far have been **non-deterministic**. The same left-hand side might appear in several substitution rules with different right-hand sides, and no rule exists that specifies which rule should be chosen. A non-deterministic grammar generates a language in which no words are 'preferred'. If it is advantageous to generate some words (those more probable) more often than others, substitution rules can be accompanied by numbers (for instance, by probabilities) that specify how often the substitution rule should be applied. If the substitution rules are accompanied by probabilities, the grammar is called **stochastic**. If the accompanying numbers do not satisfy the properties of probability (unit sum of probabilities for all rules with the same left-hand side), the grammar is called **fuzzy** [Zimmermann et al. 84].

Note that the evaluation of the frequency with which each substitution rule should be used can substantially increase the efficiency of syntactic analysis in the recognition stage [Fu 74].

7.4.2 Syntactic analysis, syntactic classifier

If appropriate grammars exist that can be used for representation of all patterns in their classes, the last step is to design a syntactic classifier which assigns the pattern (the word) to an appropriate class. It is obvious that the simplest way is to construct a separate grammar for each class; an unknown pattern x enters a parallel structure of blocks that can decide if $x \in L(G_j)$, where $j = 1, 2, \ldots R$ and R is the number of classes; $L(G_j)$ is the language generated by the jth grammar. If the jth block's decision is positive, the pattern is accepted as a pattern from the jth class and the classifier assigns the pattern to the class j. Note that generally more than one grammar can accept a pattern as belonging to its class.

The decision of whether or not the word can be generated by a particular grammar is made during **syntactic analysis**. Moreover, syntactic analysis can construct the pattern derivation tree which can represent the structural information about the pattern.

If a language is finite (and of a reasonable size), the syntactic classifier can search for a match between the word being analyzed and all the words of the language. Another simple syntactic classifier can be based on comparisons of the chain word descriptions with typical representatives of classes comparing primitive type presence only. This method is very fast and easily implemented, though it does not produce reliable results since the syntactic information is not used at all. However, impossible classes can be rejected in this step, which can speed up the syntactic analysis process.

Syntactic analysis is based on efforts to construct the tested pattern by the application of some appropriate sequence of substitution rules to the start symbol. If the substitution process is successful, the analysis process stops and the tested pattern can be generated by the grammar. The pattern can be classified into the class represented by the grammar. If the substitution process is unsuccessful, the pattern is not accepted as representing an object of this class.

If the class description grammar is regular (type 3), syntactic analysis is very simple. The grammar can be substituted with a finite non-deterministic automaton and it is easy to decide if the pattern word is accepted or rejected by the automaton [Fu 82]. If the grammar is context free (type 2), the syntactic analysis is more difficult. Nevertheless, it can be designed using stack automata.

Generally, which process of pattern word construction is chosen is not important; the transformation process can be done in top-down or bottom-up manner.

A top-down process begins with the start symbol and substitution rules are applied in the appropriate way to obtain the same pattern word as that under analysis. The final goal of syntactic analysis is to generate the same word as the analyzed word; every partial substitution creates a set of sub-goals, just as new branches are created in the generation tree. Effort is always devoted to fulfill the current sub-goal. If the analysis is not successful in fulfilling the sub-goal, it indicates an incorrect choice of the substitution rule somewhere in the previous substitutions, and back-tracking is invoked to get back to the nearest higher tree level (closer to the root), and to pick another applicable rule. The process of rule applications and back-tracking is repeated until the required pattern word results. If the whole generating process ends unsuccessfully, the grammar does not generate the word, and the analyzed pattern does not belong to the class.

This top-down process is a series of expansions starting with the start symbol S. A bottom-up process starts with the analyzed word, which is **reduced** by applying reverse substitutions, the final goal being to reduce the word to the start symbol S. The main principle of bottom-up analysis is to detect sub-words in the analyzed word that match the pattern on the right-hand side of some substitution rule, then the reduction process substitutes the former right-hand side with the left-hand side of the rule in the analyzed word. The bottom-up method follows no sub-goals; all the effort is devoted to obtaining a reduced and simplified word pattern until the start symbol is obtained. Again, if the process is not successful, the grammar does not generate the analyzed word.

The pure top-down approach is not very efficient, since too many incorrect paths are generated. The number of misleading paths can be decreased by application of consistency tests. For example, if the word starts with a non-terminal symbol I, only rules with the right-hand side starting with I should be considered. Many more consistency tests can be designed that take advantage of prior knowledge. This approach is also called **tree pruning** (See Figure 7.15) [Nilsson 71, Nilsson 82].

Tree pruning is often used if an exhaustive search cannot be completed because the search effort would exceed any reasonable bounds. Note that pruning can mean that the final solution is not optimal or may not be found at all (especially if tree search is used to find the best path through the graph, Section 5.2.4). This depends on the quality of the a priori information that is applied during the pruning process.

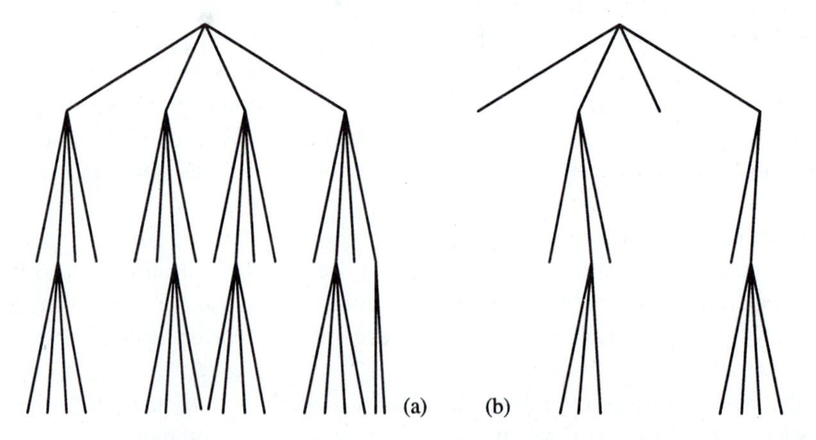

Figure 7.15: *Tree pruning: (a) original tree; (b) pruning decreases size of the searched tree.*

There are two main principles for recovery from following a wrong path. The first one is represented by the back-tracking mechanism already mentioned, meaning that the generation of words returns to the nearest point in the tree where another substitution rule can be applied which has not yet been applied. This approach requires the ability to re-construct the former appearances of generated sub-words and/or remove some branches of the derivation tree completely.

The second approach does not include back-tracking. All possible combinations of the substitution rules are applied in parallel and several generation trees are constructed simultaneously. If any tree succeeds in generating the analyzed word, the generation process ends. If any tree generation ends with a non-successful word, this tree is abandoned. The latter approach uses more brute force, but the algorithm is simplified by avoiding back-tracking.

It is difficult to compare the efficiency of these two and the choice depends on the application; Bottom-up analysis is more efficient for some grammars, and top-down is more efficient for others. As a general observation, the majority of syntactic analyzers which produce all generated words is based on the top-down principle. This approach is appropriate for most grammars but is usually less efficient.

Another approach to syntactic analysis uses example relational structures of classes. The syntactic analysis consists of matching the relational structure that represents the analyzed object with the example relational structure. The main goal is to find an **isomorphism** of both relational structures. These methods can be applied to n-ary relational structures as well. Relational structure matching is a perspective approach to syntactic recognition, a perspective way of image understanding (see Section 7.5). A simple example of relational structure matching is shown in Figure 7.16. A detailed description of relational structure matching approaches can be found in [Barrow and Popplestone 71, Pavlidis 77, Ballard and Brown 82, Baird 84].

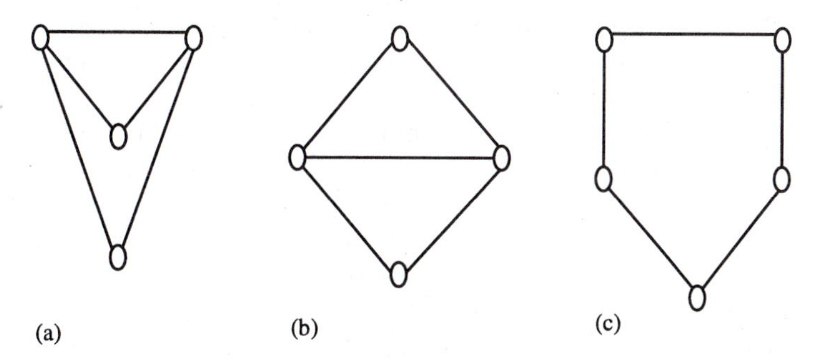

Figure 7.16: *Matching relational structures: (a) and (b) match assuming nodes and relations of the same type; (c) does not match either (a) or (b).*

7.4.3 Syntactic classifier learning, grammar inference

To model a language of any class of patterns as closely as possible, the grammar rules should be extracted from a training set of example words. This process of grammar construction from examples is known as **grammar inference**, the essence of which can be seen in Figure 7.17.

The source of words generates finite example words consisting of the terminal symbols. Assume that these examples include structural features that should be represented by a grammar G which will serve as a model of this source. All the words that can be generated by the source are included in the language $L(G)$, and the words that cannot be generated by the source represent a residuum of this set $L^C(G)$. This information enters the inference algorithm whose goal is to find and describe the grammar G. Words that are included in the language $L(G)$ can be acquired simply from the source of examples. However, the elements of $L^C(G)$ must be presented by a teacher that has additional information about the grammar properties [Gonzalez and Thomason 74, Barrero 91, Schalkoff 92].

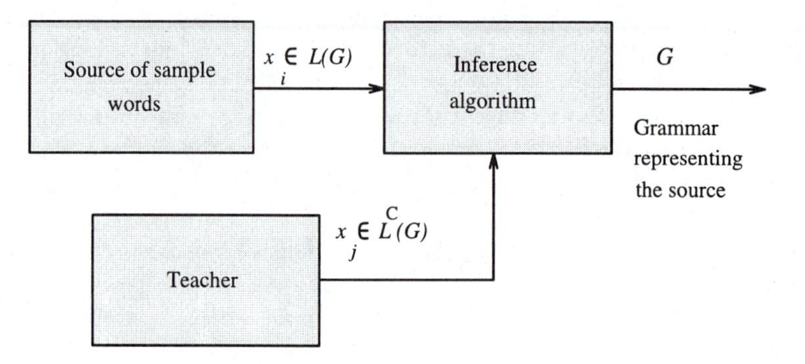

Figure 7.17: *Grammar inference.*

Note that the number of example words generated by the source is finite and it is therefore not sufficient to define the possibly infinite language $L(G)$ unambiguously. Any finite set of examples can be represented by an infinite set of languages, making it impossible to identify unambiguously the grammar that generated the examples. Grammar inference is expected to construct the grammar that describes the training set of examples, plus another set of words that in some sense have the same structure as the examples.

The inference methods can be divided into two groups, based on **enumeration** and **induction**. Enumeration detects the grammar G from the finite set M of grammars that can generate the entire training set of examples or its main part. The difficulty is in the definition of the set M of grammars and in the procedure to search for the grammar G. Induction-based methods start with the analysis of words from the training set; the substitution rules are derived from these examples using patterns of similar words.

There is no general method for grammar inference that constructs a grammar from a training set. Existing methods can be used to infer regular and context-free grammars, and may furthermore be successful in some other special cases. Even if simple grammars are considered, the inferred grammar usually generates a language that is much larger than the minimum language that can be used for appropriate representation of the class. This property of grammar inference is extremely unsuitable for syntactic analysis because of the computational complexity. Therefore, the main role in syntactic analyzer learning is still left to a human analyst, and the grammar construction is based on heuristics, intuition, experience, and prior information about the problem.

If the recognition is based on sample relational structures, the main problem is in its automated construction. The conventional method for the sample relational structure construction is described in [Winston 75], where the relational descriptions of objects from the training set are used. The training set consists of examples and counter-examples. The counter-examples should be chosen to have only one typical difference in comparison with a pattern that is a representative of the class.

7.5 Recognition as graph matching

The following section is devoted to recognition methods based on graph comparisons. Graphs with evaluated nodes and evaluated arcs will be considered as they appear in the image description using relational structures. The aim is to decide whether the reality represented by an image matches prior knowledge about the image incorporated into the graphical models. An example of a typical graph matching task is in Figure 7.18.

If this task is presented as an object recognition problem, the object graph must match the object model graph exactly. If the problem is to find an object (represented by a model graph) in the graphical representation of the image, the model must match a sub-graph in the image graph exactly. An exact match of graphs is called graph **isomorphism**—for example, the graphs in Figure 7.18 are isomorphic.

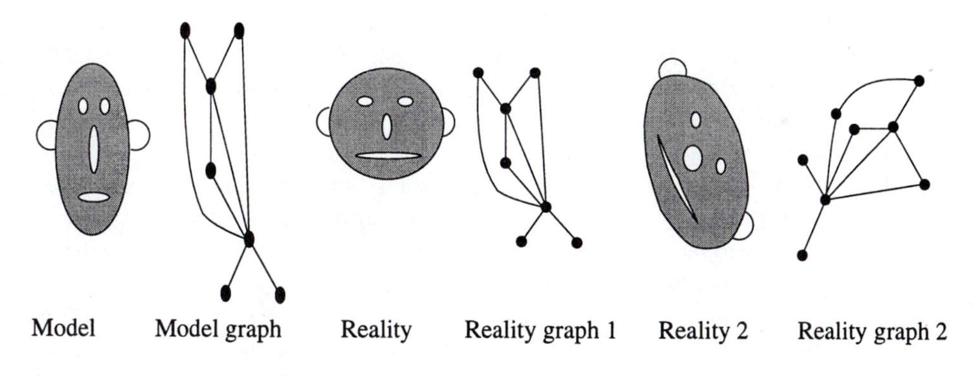

Figure 7.18: *Graph matching problem.*

Graph isomorphism and sub-graph isomorphism evaluation is a classical problem in graph theory and is important from both practical and theoretical points of view. Graph theory is covered in [Harary 69, Berge 76, Mohring 91, Nagl 90, Tucker 95], and graph theoretical algorithms can be studied in [Even 79, Lau 89, McHugh 90]. The problem is actually more complex in reality, since the requirement of an exact match is very often too strict in recognition problems.

Because of imprecise object descriptions, image noise, overlapping objects, lighting conditions, etc., the object graph usually does not match the model graph exactly. Graph matching is a difficult problem, and evaluation of graph **similarity** is not any easier. An important problem in evaluation of graph similarity is to design a metric which determines how similar two graphs are.

7.5.1 Isomorphism of graphs and sub-graphs

Regardless of whether graph or sub-graph isomorphism is required, the problems can be divided into three main classes [Harary 69, Berge 76, Ballard and Brown 82].

1. **Graph isomorphism.** Given two graphs $G_1 = (V_1, E_1)$ and $G_2 = (V_2, E_2)$, find a *one-to-one* and *onto* mapping (an isomorphism) f between V_1 and V_2 such that for each edge of E_1 connecting any pair of nodes $v, v' \in V_1$, there is an edge of E_2 connecting $f(v)$ and $f(v')$; further, if $f(v)$ and $f(v')$ are connected by an edge in G_2, v and v' are connected in G_1.

2. **Sub-graph isomorphism.** Find an isomorphism between a graph G_1 and sub-graphs of another graph G_2. This problem is more difficult than the previous one.

3. **Double sub-graph isomorphism.** Find all isomorphisms between sub-graphs of a graph G_1 and sub-graphs of another graph G_2. This problem is of the same order of difficulty as number 2.

The sub-graph isomorphism and the double sub-graph isomorphism problems are NP-complete, meaning that, using known algorithms, the solution can only be found in time proportional to an exponential function of the length of the input. It is still not known whether the graph isomorphism problem is NP-complete (see [Even 79, Sedgewick 84, Blum and Rivest 88] for details and examples). Despite extensive effort, there is neither an algorithm that can test for graph isomorphism in polynomial time, nor is there a proof that such an algorithm cannot exist. However, non-deterministic algorithms for graph isomorphism that use heuristics and look for sub-optimal solutions give a solution in polynomial time in both graph and sub-graph isomorphism testing.

Isomorphism testing is computationally expensive for both non-evaluated and evaluated graphs. Evaluated graphs are more common in recognition and image understanding, where nodes are evaluated by properties of regions they represent, and graph arcs are evaluated by relations between nodes they connect (see Section 7.1).

The evaluations can simplify the isomorphism testing. More precisely, the evaluation may make disproof of isomorphism easier. Isomorphic evaluated graphs have the same number of nodes with the same evaluation, and the same number of arcs with the same evaluation. An isomorphism test of two evaluated graphs $G_1 = (V_1, E_1)$ and $G_2 = (V_2, E_2)$ can be based on partitioning the node sets V_1 and V_2 in a consistent manner looking for inconsistencies in the resulting set partitions. The goal of the partitioning is to achieve a one-to-one correspondence between nodes from sets V_1 and V_2 for all nodes of the graphs G_1 and G_2. The algorithm consists of repeated node set partitioning steps, and the necessary conditions of isomorphism are tested after each step (the same number of nodes of equivalent properties in corresponding sets of both graphs). The node set partitioning may, for example, be based on the following properties:

- Node attributes (evaluations)
- The number of adjacent nodes (connectivity)
- The number of edges of a node (node degree)

- Types of edges of a node
- The number of edges leading from a node back to itself (node order)
- The attributes of adjacent nodes

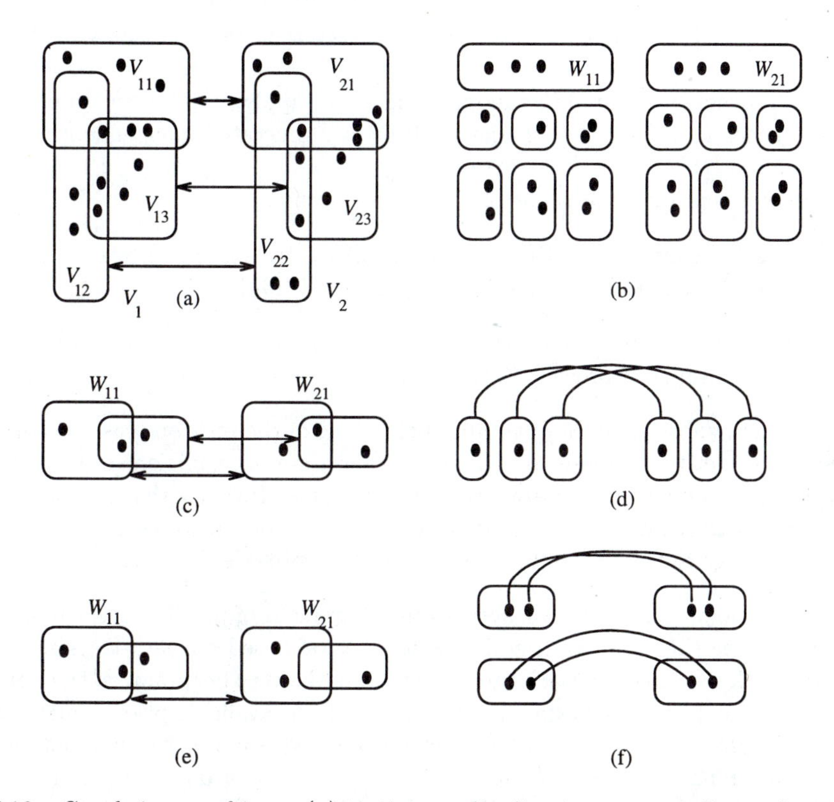

Figure 7.19: *Graph isomorphism: (a) testing cardinality in corresponding subsets; (b) partitioning node subsets; (c) generating new subsets; (d) subset isomorphism found; (e) graph isomorphism disproof; (f) situation when arbitrary search is necessary.*

After the new subsets are generated based on one of the listed criteria, the cardinality of corresponding subsets of nodes in graphs G_1 and G_2 are tested; see Figure 7.19a. Obviously, if v_{1i} is in several subsets V_{1j}, then the corresponding node v_{2i} must also be in the corresponding subsets V_{2j}, or the isomorphism is disproved.

$$v_{2i} \in \bigcap_{j | v_{1i} \in V_{1j}} V_{2j} \tag{7.52}$$

If all the generated subsets satisfy the necessary conditions of isomorphism in step i, the subsets are split into new sets of nodes W_{1n}, W_{2n} (Figure 7.19b):

$$\begin{aligned} W_{1i} \bigcap W_{1j} = \emptyset \quad & \text{for } i \neq j \\ W_{2i} \bigcap W_{2j} = \emptyset \quad & \text{for } i \neq j \end{aligned} \tag{7.53}$$

Clearly, if $V_{1j} = V_{2j}$ and if $v_{1i} \notin V_{1k}$, then $v_{2i} \in V_{2k}^C$, where V^C is the set complement. Therefore, by equation (7.52), corresponding elements v_{1i}, v_{2i} of W_{1n}, W_{2n} must satisfy [Niemann 90]

$$v_{2i} \in \{ \bigcap_{\{j|v_{1i} \in W_{1j}\}} W_{2j}\} \cap \{ \bigcap_{\{k|v_{1i} \notin W_{1k} \wedge W_{1k} = W_{2k}\}} W_{2k}^C\} \tag{7.54}$$

The cardinality of all the corresponding sets W_{1n}, W_{2n} is tested to disprove the graph isomorphism.

The same process is repeated in the following steps, applying different criteria for graph node subset generation. Note that the new subsets are generated independently in W_{1i}, W_{2i} (Figure 7.19c).

The process is repeated unless one of three cases occurs:

1. The set partitioning reaches the stage when all the corresponding sets W_{1i}, W_{2i} contain one node each. The isomorphism is found (Figure 7.19d).

2. The cardinality condition is not satisfied in at least one of the corresponding subsets. The isomorphism is disproved (Figure 7.19e).

3. No more new subsets can be generated before one of the previous cases occurs. In that situation, either the node set partitioning criteria are not sufficient to establish an isomorphism, or more than one isomorphism is possible. If this is the case, the systematic arbitrary assignment of nodes that have more than one possible corresponding node and cardinality testing after each assignment may provide the solution (Figure 7.19f).

The last part of the process, based on systematic assignment of possibly corresponding nodes and isomorphism testing after each assignment, may be based on back-tracking principles. Note that the back-tracking approach can be used from the very beginning of the isomorphism testing, but it is more efficient to start the test using all the available prior information about the matched graphs. The back-tracking process is applied if more than one potential correspondence between nodes is encountered. Back-tracking tests for directed graph isomorphism and a recursive algorithm are given in [Ballard and Brown 82] together with accompanying hints for improving the efficiency of back-track searches [Bittner and Reingold 75, Haralick and Elliott 79, Nilsson 82]. The process presented above, graph isomorphism testing, is summarized in the following algorithm.

Algorithm 7.8: Graph isomorphism

1. Take two graphs $G_1 = (V_1, E_1), G_2 = (V_2, E_2)$.

2. Use a node property criterion to generate subsets V_{1i}, V_{2i} of the node sets V_1 and V_2. Test whether the cardinality conditions hold for corresponding subsets. If not, the isomorphism is disproved.

3. Partition the subsets V_{1i}, V_{2i} into subsets W_{1j}, W_{2j} satisfying the conditions given in equation (7.53) (no two subsets W_{1j} or W_{2j} contain the same node). Test whether the cardinality conditions hold for all the corresponding subsets W_{1j}, W_{2j}. If not, the isomorphism is disproved.

4. Repeat steps 2 and 3 using another node property criterion in all subsets W_{1j}, W_{2j} generated so far. Stop if one of the three above-mentioned situations occurs.

5. Based on the situation that stopped the repetition process, the isomorphism either was proved, disproved, or some additional procedures (such as back-tracking) must be applied to complete the proof or disproof.

A classic approach to sub-graph isomorphism can be found in [Ullmann 76]. A brute force enumeration process is described as a depth-first tree search algorithm. As a way of improving the efficiency of the search, a refinement procedure is entered after each node is searched in the tree—the procedure reduces the number of node successors, which yields a shorter execution time. An alternative approach testing isomorphism of graphs and sub-graphs transforms the graph problem into a linear programming problem [Zdrahal 81].

The double sub-graph isomorphism problem can be translated into a sub-graph isomorphism problem using the **clique**—a complete (totally connected) sub-graph—approach. A clique is said to be maximal if no other clique properly includes it. Note that a graph may have more than one maximal clique; however, it is often important to find the largest maximal clique (that with the largest number of elements). (Other definitions consider a clique always to be maximal [Harary 69].)

The search for the maximal clique is a well-known problem in graph theory. An example algorithm for finding all cliques of an undirected graph can be found in [Bron and Kerbosch 73]. The maximal clique $G_{\text{clique}} = (V_{\text{clique}}, E_{\text{clique}})$ of the graph $G = (V, E)$ can be found as follows [Niemann 90].

Algorithm 7.9: Maximal clique location

1. Take an arbitrary node $v_j \in V$; construct a subset $V_{\text{clique}} = \{v_j\}$.

2. In the set V_{clique}^C search for a node v_k that is connected with all nodes in V_{clique}. Add the node v_k to a set V_{clique}.

3. Repeat step 2 as long as new nodes v_k can be found.

4. If no new node v_k can be found, V_{clique} represents the node set of the maximal clique sub-graph G_{clique} (the maximal clique that contains the node v_j).

To find the largest maximal clique, an additional maximizing search is necessary. Other clique-finding algorithms are discussed in [Ballard and Brown 82, Yang et al. 89].

The search for isomorphism of two sub-graphs (the double sub-graph isomorphism) is transformed to a clique search using the **assignment graph** [Ambler 75]. A pair (v_1, v_2), $v_1 \in V_1$, $v_2 \in V_2$ is called an **assignment** if the nodes v_1 and v_2 have the same node property descriptions, and two assignments (v_1, v_2) and (v_1', v_2') are **compatible** if (in addition) all relations between v_1 and v_1' also hold for v_2 and v_2' (graph arcs between v_1,v_1' and v_2,v_2' must have the same evaluation, including the no-edge case). The set of assignments defines the set of nodes V_a of the assignment graph G_a. Two nodes in V_a (two assignments) are connected by an arc in the assignment graph G_a if these two nodes are compatible. The search for the maximum matching sub-graphs of graphs G_1 and G_2 is a search for the maximum totally connected sub-graph in G_a (the maximum totally compatible subset of assignments).

The maximum totally connected sub-graph is a maximal clique, and the maximal clique-finding algorithm can be applied to solve this problem.

7.5.2 Similarity of graphs

All the approaches mentioned above tested for a perfect match between graphs and/or sub-graphs. This cannot be anticipated in real applications, and these algorithms are not able to distinguish between a small mismatch of two very similar graphs and the case when the graphs are not similar at all. Moreover, if graph similarity is tested, the main stress is given to the ability to quantify the similarity. Having three graphs G_1, G_2, G_3, the question as to which two are more similar is a natural one [Buckley 90].

The similarity of two strings (chains) can be based on the **Levenshtein distance**, which is defined as the smallest number of deletions, insertions, and substitutions necessary to convert one string into the other. Transformations of string elements can be assigned a specific transition cost to make the computed similarity (distance) more flexible and more sensitive. This principle can be applied to graph similarity as well. The set of feasible transformations of nodes and arcs (insertion, deletion, substitution, relabeling) is defined, and these transformations are accompanied by transition costs. Any sequence of transformations is assigned a combination of single step costs (like the sum of individual costs). The set of transformations that has the minimum cost and transforms one graph to another graph defines a distance between them [Niemann 90, Shapiro and Haralick 80].

Note that similarity can be searched for in hierarchical graph structures. The graphs consist of a number of sub-graphs in which isomorphism (or similarity) has already been proved. The next step is to detect, describe, and evaluate relations between these sub-graphs (Figure 7.20, cf. Figure 7.18).

To explain the principles, a physical analogy of templates and springs [Fischler and Elschlager 73] is usually considered. The templates (sub-graphs) are connected by springs (relations between sub-graphs). The quality of the match of two graphs relates to the quality of the local fit (in corresponding templates) and to the amount of energy used to stretch the springs to match one graph onto the second (reference) graph. To make the graph similarity measure more flexible, extra costs may be added for missing parts of the graph as well as for some extra ones. The spring energy penalty may be made highly non-linear, better to reflect the descriptive character in particular applications.

7.6 Optimization techniques in recognition

Optimization itself is much more flexible than is usually recognized. Considering image recognition and understanding, the best image representation is sought (the best matching between the image and the model is required, the best image understanding is the goal). Whenever 'the best' is considered, some objective function of *goodness* must be available, implying that

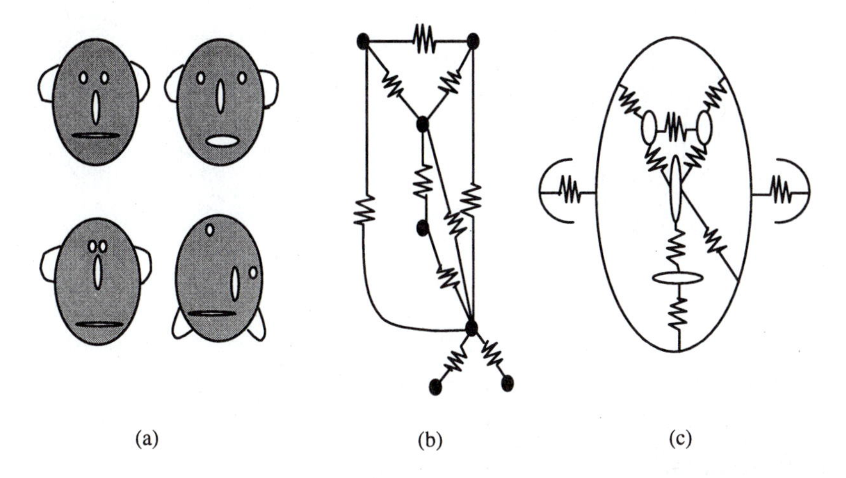

(a) (b) (c)

Figure 7.20: *Templates and springs principle: (a) different objects having the same description graphs; (b),(c) nodes (templates) connected by springs, graph nodes may represent other graphs in finer resolution.*

an optimization technique can be applied which looks for the evaluation function maximum ... for *the best*.

A function optimization problem may be defined as follows: Given some finite domain D and a function $f : D \to R$, R being the set of real numbers, find the *best* value in D under f. Finding the *best* value in D is understood as finding a value $\mathbf{x} \in D$ yielding either the minimum (function minimization) or the maximum (function maximization) of the function f:

$$f_{\min}(\mathbf{x}) = \min_{\mathbf{x} \in D} f(\mathbf{x}) \qquad\qquad f_{\max}(\mathbf{x}) = \max_{\mathbf{x} \in D} f(\mathbf{x}) \tag{7.55}$$

The function f is called the **objective** function. Maximization of the objective function will be considered here, as it is typical in image interpretation applications, discussed in Chapter 8. However, optimization methods for seeking maxima and minima are logically equivalent, and optimization techniques can be equally useful if either an objective function maximum or function minimum is required.

It should be noted that no optimization algorithm can guarantee finding a good solution to the problem if the objective function does not reflect the *goodness* of the solution. Therefore, the design of the objective function is a key factor in the performance of any optimization algorithm (similarly, appropriate feature selection is necessary for the success of a classifier).

Most of the conventional approaches to optimization use calculus-based methods which can be compared to climbing a hill (in the case of maximization)—the gradient of the objective function gives the steepest direction to climb. The main limitation of calculus-based methods is their local behavior; the search can easily end in a local maximum, and the global maximum can be missed (see Figure 7.21).

Several methods improve the probability of finding the global maximum; to start the hill climbing at several points in the search space, to apply enumerative searches such as dynamic programming, to apply random searches, etc. Among these possibilities are genetic algorithms and simulated annealing.

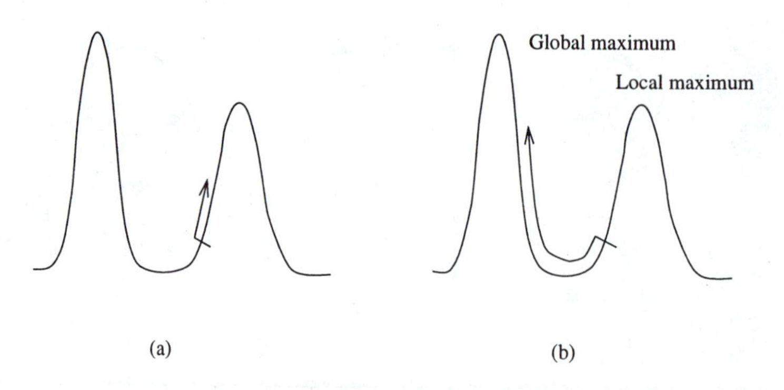

Figure 7.21: *Limitations of hill climbing methods.*

7.6.1 Genetic algorithms

Genetic algorithms (GA) use natural evolution mechanisms to search for the maximum of an objective function; as with any optimization technique, they can be used in recognition and machine learning.

Genetic algorithms do not guarantee that the global optimum will be found, but empirical results from many applications show that the final solution is usually very close to it. This is very important in image understanding applications, as will be seen in the next chapter. There are almost always several consistent (feasible) solutions that are locally optimal in image understanding, or matching, and only one of those possible solutions is the best one represented by the global maximum. The opportunity to find the (near) global optimum is very valuable in these tasks.

Genetic algorithms differ substantially from other optimization methods in the following ways [Goldberg 89].

1. GAs work with a coding of the parameter set, not the parameters themselves. Genetic algorithms require the natural parameter set of the optimization problem to be coded as a finite-length string over some finite alphabet. This implies that any optimization problem representation must be transformed to a string representation; binary strings are often used (the alphabet consists of the symbols 0 and 1 only). The design of the problem representation as a string is an important part of the GA method.

2. GAs search from a population of points, not a single point. The population of solutions that is processed in each step is large, meaning that the search for the optimum is driven from many places in the search space simultaneously. This gives a better chance of finding the global optimum.

3. GAs use the objective function directly, not derivatives or other auxiliary knowledge. The search for new, better solutions depends on the values of the evaluation function only. Note that, as in other recognition methods, the GAs find the (near) global optimum of the evaluation function but there is no guarantee at all that the evaluation function is relevant to the problem. The evaluation function describes the *goodness* of the particular string. The value of the evaluation function is called **fitness** in GAs.

4. GAs use probabilistic transition rules, not deterministic rules. Rules of transition from the current population of strings to a new and better population of strings are based on the natural idea of supporting good strings with higher fitness and removing poor strings with lower fitness. This is the key idea of genetic algorithms. The best strings representing the best solutions are allowed to survive the evolution process with a higher probability.

   The survival of the fittest and the death of the poor code strings is achieved by applying three basic operations: **reproduction, crossover**, and **mutation**.

The population of strings represents all the strings that are being processed in the current step of the GA. The sequence of reproduction, crossover, and mutation generates a new population of strings from the previous population.

Reproduction

The reproduction operator is responsible for the survival of the fittest and for the death of others based on a probabilistic treatment.

The reproduction mechanism copies strings with highest fitness into the next generation of strings. The selection process is usually probabilistic, the probability that a string is reproduced into the new population being given by its relative fitness in the current population—this is their mechanism of survival. The lower the fitness of the string, the lower the chances for survival. This process results in a set of strings where some strings of higher fitness may be copied more than once into the next population. The total number of strings in the population usually remains unchanged, and the average fitness of the new generation is higher than it was before.

Crossover

There are many variations on the crossover. The basic idea is to mate the newly reproduced strings at random, randomly choosing a position for the border of each pair of strings, and to produce new strings by swapping all characters between the beginning of the string pairs and the border position; see Figure 7.22.

Not all newly reproduced strings are subject to the crossover. There is a probability parameter representing the number of pairs which will be processed by crossover; also, it may be performed such that the best reproduced strings are kept in an unchanged form.

The crossover operation together with reproduction represent the main power of GAs. However, there is one more idea in the crossover operation: Blocks of characters can be detected in the strings that have locally correct structure even if the string as a whole does not represent a good solution. These blocks of characters in strings are called **schemata**. Schemata are sub-strings that can represent building blocks of the string, and can be understood as the local pattern of characters. Clearly, if schemata can be manipulated as locally correct blocks, the optimal solution can be located faster than if all the characters are handled independently. In every generation of n strings, about n^3 schemata are processed. This is called the **implicit parallelism** of genetic algorithms [Goldberg 89].

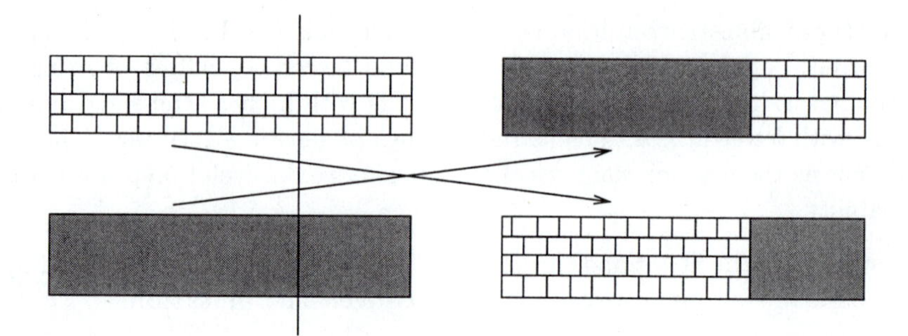

Figure 7.22: *Principle of crossover. Two strings before (left) and after the crossover (right).*

Mutation

The mutation operator plays only a secondary role in GAs. Its principle is randomly to change one character of some string of the population from time to time—it might, for example, take place approximately once per thousand bit transfers (i.e., one bit mutation per one thousand bits transferred from generation to generation). The main reason for mutation is the fact that some local configurations of characters in strings of the population can be totally lost as a result of reproduction and crossover operations. Mutation protects GAs against such irrecoverable loss of good solution features.

Population convergence in GAs is a serious question. For practical purposes this question becomes one of when to stop generating the new string populations. A common and practically proven criterion recommends that the population-generating process be stopped when the maximum achieved fitness in the population has not improved substantially through several previous generations.

We have not yet discussed how to create the starting population, which usually consists of a large number of strings, the number depending on the application. The starting population can be generated at random, assuming the alphabet of characters and the desired length of strings are known. Nevertheless, as always, if some prior knowledge about the solution is available (the probable local patterns of characters, the probable percentages of characters in strings, etc.), then it is advantageous to use this information to make the starting population as fit as possible. The better the starting population, the easier and faster the search for the optimum.

The simplified version of the genetic algorithm is as follows.

Algorithm 7.10: Genetic algorithm

1. Create a starting population of code strings, and find the value of their objective functions.

2. Probabilistically reproduce high fitness strings in the new population, remove poor fitness strings (reproduction).

3. Construct new strings combining reproduced code strings from the previous population (crossover).

4. From time to time, change one character of some string at random (mutation).
5. Order code strings of the current population according to the value of their objective function (fitness).
6. If the maximum achieved string fitness does not increase over several steps, stop. The desired optimum is represented by the current string of maximum fitness. Otherwise, repeat the sequence of steps starting at step 2.

See Section 8.6.2 for an example of the algorithm. A more detailed and precise description of genetic algorithms can be found in [Goldberg 89, Rawlins 91, Michalewicz 94, Adeli and Hung 95, Chambers 95, Mitchell 96]. Many examples and descriptions of related techniques are included there as well, such as knowledge implementation into mutation and crossover, GA learning systems, hybrid techniques that combine good properties of conventional hill climbing searches and GAs, etc.

7.6.2 Simulated annealing

Simulated annealing [Kirkpatrick et al. 83, Cerny 85] represents another group of robust optimization methods. Similarly to genetic algorithms, simulated annealing searches for a minimum of an objective function (cost function) that represents the goodness of some complex system. Searching for minima is considered in this section because it simplifies energy-related correspondences with the natural behavior of matter. Simulated annealing may be suitable for NP-complete optimization problems; simulated annealing does not guarantee that the global optimum is found, but the solution is usually near-optimal.

Cerny [Cerny 85] often uses the following example to explain the principle of simulated annealing optimization. Imagine a sugar bowl freshly filled with cube sugar. Usually, some cubes do not fit in the sugar bowl and the lid cannot be closed. From experience, everybody knows that shaking the sugar bowl will result in better placement of the cubes inside the bowl and the lid will close properly. In other words, considering the number of cubes that can be inside the bowl as an evaluation function, shaking the bowl results in a near-minimal solution (considering sugar space requirements). The degree of shaking is a parameter of this optimization process and corresponds to the heating and cooling process as described below.

Simulated annealing combines two basic optimization principles, **divide and conquer** and **iterative improvement** (hill climbing). This combination avoids getting stuck in local optima. A strong connection between statistical mechanics or thermodynamics, and multivariate or combinatorial optimization is the basis for annealing optimization.

In statistical mechanics, only the most probable change of state of a system in thermal equilibrium at a given temperature is observed in experiments; each configuration (state) defined by the set of atomic positions $\{x_i\}$ of the system is weighted by its Boltzmann constant probability factor,

$$\exp\left[\frac{-E(\{x_i\})}{k_B T}\right] \tag{7.56}$$

where $E(\{x_i\})$ is the energy of the state, k_B is the Boltzmann constant, and T is the temperature [Kirkpatrick et al. 83].

One of the main characteristics of the Boltzmann density is that at high temperature each state has an almost equal chance of becoming the new state, but at low temperature only states with low energies have a high probability of becoming current. The optimization can be compared with the ability of matter to form a crystalline structure that represents an energy minimum if the matter is melted and cooled down slowly. This minimum can be considered the optimization minimum for the energy function playing the role of the objective function. The crystalization process depends on the cooling speed of the molten liquid; if the cooling is too fast, the crystal includes many local defects and the global energy minimum is not reached.

Simulated annealing consists of downhill iteration steps combined with controlled uphill steps that make it possible to escape from local minima (see Figure 7.23).

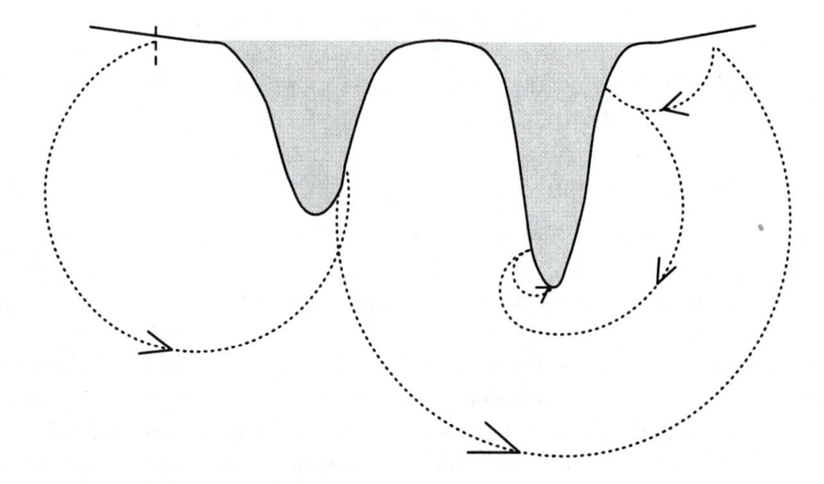

Figure 7.23: *Uphill steps make it possible to get out of local minima; the dotted line shows a possible convergence route.*

The physical model of the process starts with heating the matter until it melts; then the resulting liquid is cooled down slowly to keep the quasi-equilibrium. The cooling algorithm [Metropolis et al. 53] consists of repeated random displacements (state changes) of atoms in the matter, and the energy change ΔE is evaluated after each state change. If $\Delta E \leq 0$ (lower energy), the state change is accepted, and the new state is used as the starting state of the next step. If $\Delta E > 0$, the state is accepted with probability

$$P(\Delta E) = \exp(\frac{-\Delta E}{k_B T}) \tag{7.57}$$

To apply this physical model to an optimization problem, the temperature parameter T must be decreased in a controlled manner during optimization. The random part of the algorithm can be implemented by generating random numbers uniformly distributed in the interval (0,1); one such random number is selected and compared with $P(\Delta E)$.

Algorithm 7.11: Simulated annealing optimization

1. Let $\mathbf{x}$ be a vector of optimization parameters; compute the value of the objective function $J(\mathbf{x})$.

2. Repeat steps 3 and 4 $n(T)$ times.

3. Perturb the parameter vector $\mathbf{x}$ slightly, creating the vector $\mathbf{x}_{\text{new}}$, and compute the new value of the optimization function $J(\mathbf{x}_{\text{new}})$.

4. Generate a random number $r \in (0,1)$, from a uniform distribution in the interval (0,1). If
$$r < \exp\left\{\frac{-[J(\mathbf{x}_{\text{new}}) - J(\mathbf{x})]}{k_B T}\right\} \tag{7.58}$$
then assign $\mathbf{x} = \mathbf{x}_{\text{new}}$ and $J(\mathbf{x}) = J(\mathbf{x}_{\text{new}})$.

5. Repeat steps 2 to 4 until a convergence criterion is met.

6. The parameter vector $\mathbf{x}$ now represents the solution of the optimization problem.

Note that nothing is known beforehand about how many steps n, what perturbations to the parameter (state changes), what choice of temperatures T, and what speed of cooling down should be applied to achieve the best (or even good) results, although some general guidelines exist and appropriate parameters can be found for particular problems. It is only known that the annealing process must continue long enough to reach a steady state for each temperature. Remember the sugar bowl example: the shaking is much stronger at the beginning and gradually decreases for the best results.

The sequence of temperatures and the number n of steps necessary to achieve equilibrium in each temperature is called the **annealing schedule**. Large values of n and small decrements of T yield low final values of the optimization function (the solution is close to the global minimum) but require long computation time. A small number of repetitions n and large decrements in T proceed faster, but the results may not be close to the global minimum. The values T and n must be chosen to give a solution close to the minimum without wasting too much computation time. There is no known practically applicable way to design an optimal annealing schedule.

The annealing algorithm is easy to implement. Annealing has been applied to many optimization problems, including pattern recognition, graph partitioning, and many others, and has been demonstrated to be of great value (although examples of optimization problems exist in which it performs less well than standard algorithms and other heuristics). In the computer vision area, applications include stereo correspondence [Barnard 87], boundary detection [Geman et al. 90], texture segmentation [Bouman and Liu 91], and edge detection [Tan et al. 92]. Implementation details and annealing algorithm properties together with an extensive list of references can be found in [Aarts and van Laarhoven 86, van Laarhoven and Aarts 87, van Laarhoven 88, Otten and van Ginneken 89, Azencott 92].

7.7 Fuzzy systems

Fuzzy systems are capable of representing diverse, non-exact, uncertain, and inaccurate knowledge or information. They use qualifiers that are very close to the human way of expressing knowledge, such as bright, medium dark, dark, etc. Fuzzy systems can represent complex knowledge and even knowledge from contradictory sources. They are based on fuzzy logic, which represents a powerful approach to decision making [Zadeh 65, Kaufmann 75, Bezdek 81, Kandel 82, Pal and Majumder 86, Zimmermann 87, Pal 91, Zimmermann 91, Zadeh and Kacprzyk 92, Kosko 92, Cox 94, Furuhashi 95, Pedrycz 95, Adeli and Hung 95]. The fundamental principles of fuzzy logic were presented in Section 7.1; here, **fuzzy sets, fuzzy membership functions**, and **fuzzy systems** are introduced and fundamental fuzzy reasoning approaches are presented.

7.7.1 Fuzzy sets and fuzzy membership functions

When humans describe objects, they often use imprecise descriptors such as *bright, large, rounded, elongated,* etc. For instance, fair-weather clouds may be described as small, medium dark or bright, somewhat rounded regions; thunderstorm clouds may be described as dark or very dark, large regions—people are quite comfortable with such descriptions. However, if the task is to recognize clouds from photographs of the sky automatically by using pattern recognition approaches, it becomes obvious that crisp boundaries (discrimination functions) must be drawn that separate the cloud classes. It may be quite arbitrary to make a decision about the boundary location—a decision that a cloud region R_1 characterized by average gray-level g, roundness r and size s represents a thunderstorm cloud, while the region R_2 characterized by average gray-level $g+1$, the same roundness r, and size s does not. It may be more appropriate to consider a region R_1 as belonging to the set of fair-weather clouds with some degree of membership and belonging to the set of thunderstorm clouds with another degree of membership. Similarly, another region R_2 might belong to both cloud sets with some other degrees of membership. Fuzzy logic thus facilitates simultaneous membership of regions in different fuzzy sets. Figures 7.24a,b demonstrate the difference between the crisp and fuzzy sets representing the average gray-level of the cloud regions.

A **fuzzy set** S in a fuzzy space X is a set of ordered pairs

$$S = \{(x, \mu_S(x)) | x \in X\} \tag{7.59}$$

where $\mu_S(x)$ represents the grade of membership of x in S. The range of the membership function is a subset of non-negative real numbers whose supremum is finite. For convenience, a unit supremum is widely used

$$\sup_{x \in X} \mu_S(x) = 1 \tag{7.60}$$

The fuzzy sets are often denoted solely by its membership function.

The description of DARK regions presented in Figure 7.24b is a classical example of a fuzzy set and illustrates the properties of fuzzy spaces. The **domain** of the fuzzy set is depicted along the x axis and ranges from black to white (0–255). The degree of membership $\mu(x)$ can be seen along the vertical axis. The membership is between zero and one, zero representing no membership and one representing the complete membership. Thus, a white region with

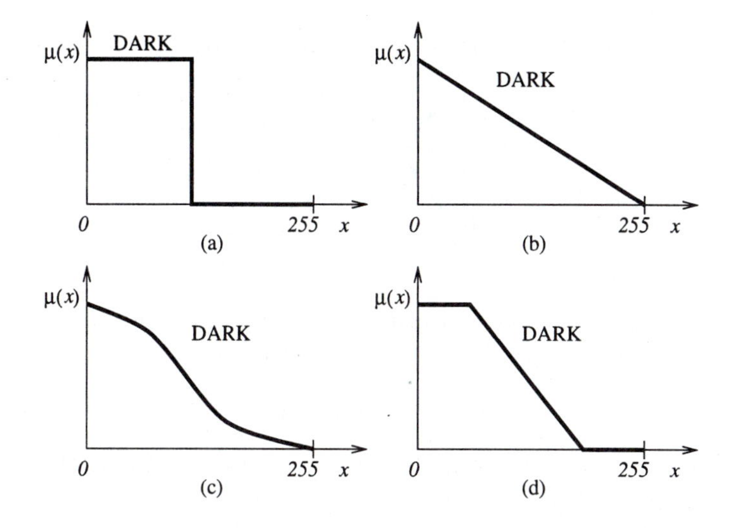

Figure 7.24: *Crisp and fuzzy sets representing cloud regions of the same size and roundness, varying average gray-level g: (a) crisp set showing the Boolean nature of the* DARK *set; (b) fuzzy set* DARK*; (c) another possible membership function associated with the fuzzy set* DARK*; (d) yet another possible membership function.*

an average gray-level of 255 has zero membership in the DARK fuzzy set, while the black region (average gray-level = 0) has complete membership in the DARK fuzzy set. As shown in Figure 7.24b, the membership function may be linear, but a variety of other curves may also be used (Figure 7.24c,d).

Consider average gray-levels of fair-weather and thunderstorm clouds; Figure 7.25 shows possible membership functions associated with the fuzzy sets DARK, MEDIUM_DARK, BRIGHT. As the figure shows, a region with a specific average gray-level g may belong to several fuzzy sets simultaneously. Thus, the memberships $\mu_{DARK}(g)$, $\mu_{MEDIUM\_DARK}(g)$, $\mu_{BRIGHT}(g)$ represent the fuzziness of the description since they assess the degree of certainty about the membership of the region in the particular fuzzy set. The maximum membership value associated with any fuzzy set is called the **height** of the fuzzy set.

In fuzzy system design, normalized versions of membership functions are used. The **minimum normal form** requires at least one element of the fuzzy set domain to have a membership value of one, and the **maximum normal form** is such minimum normal forms for which at least one element of the domain has a membership value of zero.

In fuzzy reasoning systems, fuzzy membership functions are usually generated in the minimum normal form; a long list of possible fuzzy membership functions (linear, sigmoid, beta curve, triangular curve, trapezoidal curve, shouldered curve, arbitrary curve, etc.), fuzzy numbers, fuzzy quantities, and fuzzy counts can be found together with their definitions in [Cox 94].

Shape of fuzzy membership functions can be modified using **fuzzy set hedges**. Hedges may intensify, dilute, form a complement, narrowly or broadly approximate, etc., the membership of the fuzzy set elements. Zero or more hedges and the associated fuzzy set constitute a single semantic entity called a **linguistic variable**. Suppose $\mu_{DARK}(x)$ represents the

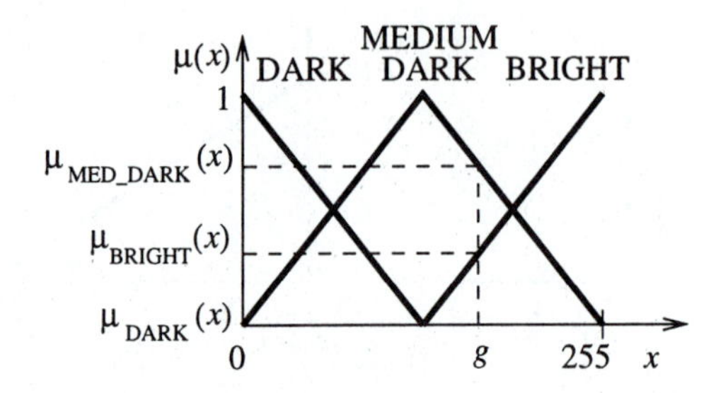

Figure 7.25: *Membership functions associated with fuzzy sets* DARK, MEDIUM_DARK, *and* BRIGHT. *Note that several membership values may be associated with a specific average gray-level g.*

membership function of the fuzzy set DARK; then the intensified fuzzy set VERY_DARK will have the membership function (Figure 7.26a)

$$\mu_{VERY\_DARK}(x) = \mu^2_{DARK}(x) \tag{7.61}$$

Similarly, a diluting hedge creating a fuzzy set SOMEWHAT_DARK will have a membership function (Figure 7.26b)

$$\mu_{SOMEWHAT\_DARK}(x) = \sqrt{\mu_{DARK}(x)} \tag{7.62}$$

Multiple hedges can be applied to a single fuzzy membership function and a fuzzy set VERY_VERY_DARK can be created as

$$\mu_{VERY\_VERY\_DARK}(x) = \mu^2_{DARK}(x) \cdot \mu^2_{DARK}(x) = \mu^4_{DARK}(x) \tag{7.63}$$

There are no theoretically solid reasons for these hedge formulae, but they have the merit of success in practice—they simply 'seem to work' [Cox 94].

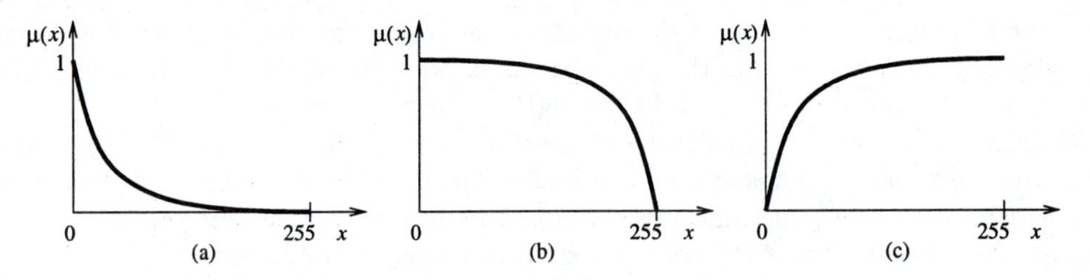

Figure 7.26: *Fuzzy set hedges. Fuzzy set* DARK *is shown in Figure 7.24b. (a) Fuzzy set* VERY_DARK. *(b) Fuzzy set* SOMEWHAT_DARK. *(c) Fuzzy set* NOT_VERY_DARK.

7.7.2 Fuzzy set operators

Rarely can a recognition problem be solved using a single fuzzy set and the associated single membership function. Therefore, tools must be made available that combine various fuzzy sets and allow one to determine membership functions of such combinations. In conventional logic, membership functions are either zero or one (Figure 7.24) and for any class set S, a rule of noncontradiction holds: An intersection of a set S with its complement S^c is an empty set.

$$S \cap S^c = \emptyset \tag{7.64}$$

Clearly, this rule does not hold in fuzzy logic, since domain elements may belong to fuzzy sets and their complements simultaneously. There are three basic **Zadeh operators** on fuzzy sets: **fuzzy intersection**, **fuzzy union**, and **fuzzy complement**. Let $\mu_A(x)$ and $\mu_B(y)$ be two membership functions associated with two fuzzy sets A and B with domains X and Y. Then the intersection, union, and complement are pointwise defined for all $x \in X, y \in Y$ (note that other definitions also exist) as

$$\begin{aligned} \text{Intersection} \quad A \cap B: \qquad & \mu_{A \cap B}(x,y) = && \min[\mu_A(x), \mu_B(y)] \\ \text{Union} \quad A \cup B: \qquad & \mu_{A \cup B}(x,y) = && \max[\mu_A(x), \mu_B(y)] \\ \text{Complement} \quad A^c: \qquad & \mu_{A^c}(x) = && 1 - \mu_A(x) \end{aligned} \tag{7.65}$$

Note that the fuzzy set operators may be combined with the hedges and new fuzzy sets may be constructed; e.g., a fuzzy set NOT_VERY_DARK would be constructed as NOT (VERY (DARK))

$$\mu_{NOT\_VERY\_DARK}(x) = 1 - \mu^2_{DARK}(x)$$

(see Figure 7.26).

7.7.3 Fuzzy reasoning

In fuzzy reasoning, information carried in individual fuzzy sets is combined to make a decision. The functional relationship determining the degree of membership in related fuzzy membership functions is called **method of composition** (method of implication) and results in the definition of a **fuzzy solution space**. To arrive at the decision, a **de-fuzzification** (decomposition) process determines a functional relationship between the fuzzy solution space and the decision. Processes of composition and de-fuzzification form the basis of fuzzy reasoning (Figure 7.27), which is performed in the context of a **fuzzy system model** that consists of control, solution, and working data variables; fuzzy sets; hedges; fuzzy rules; and a control mechanism. Fuzzy models use a series of unconditional and conditional propositions called

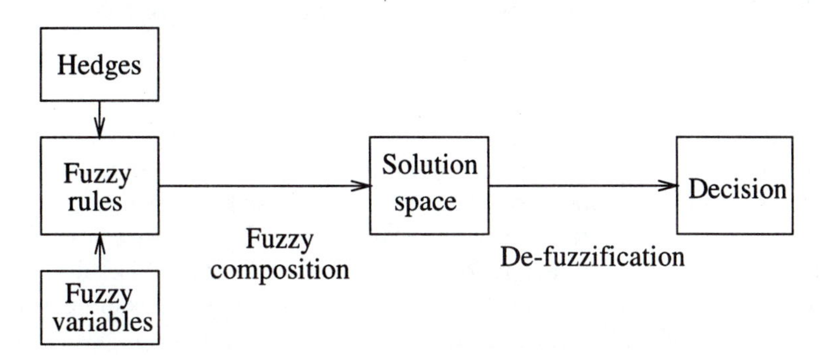

Figure 7.27: *Fuzzy reasoning—composition and de-fuzzification.*

fuzzy rules. Unconditional fuzzy rules are of the form

$$x \text{ is } A \tag{7.66}$$

and conditional fuzzy rules have the form

$$if \quad x \text{ is } A \quad then \quad w \text{ is } B \tag{7.67}$$

where A and B are linguistic variables and x and w represent scalars from their respective domains. The degree of membership associated with an unconditional fuzzy rule is simply $\mu_A(x)$. Unconditional fuzzy propositions are used either to restrict the solution space or to define a default solution space. Since these rules are unconditional, they are applied directly to the solution space by applying fuzzy set operators.

Considering conditional fuzzy rules, there are several approaches to arrive at the decision. **Monotonic fuzzy reasoning** is the simplest approach that can produce a solution directly without composition and de-fuzzification. Let again x represent a scalar gray-level describing darkness of a cloud, and w the severity of a thunderstorm. The following fuzzy rule may represent our knowledge of thunderstorm severity:

$$\text{if } x \text{ is DARK then } w \text{ is SEVERE} \tag{7.68}$$

The algorithm for monotonic fuzzy reasoning is shown in Figure 7.28. Based on determination of the cloud gray-level (x=80 in our case), the membership value $\mu_{DARK}(80) = 0.35$ is determined. This value is used to represent the membership value $\mu_{SEVERE}(w) = \mu_{DARK}(x)$ and the decision is made about the expected severity of the thunderstorm; in our case severity $w = 4.8$ on a scale between 0 and 10. This approach may also be applied to complex predicates of the form

$$if \quad (x \text{ is } A) \bullet (y \text{ is } B) \bullet \ldots \bullet (u \text{ is } F) \quad then \quad w \text{ is } Z \tag{7.69}$$

where $\bullet$ represents the conjunctive AND or disjunctive OR operations. Fuzzy intersection and union operators can be used to combine the complex predicates; AND corresponds to fuzzy intersection and OR corresponds to fuzzy union. While the monotonic approach shows the fundamental concept of fuzzy reasoning, it can only be used for a monotonic single fuzzy variable controlled by a single fuzzy rule (possibly with a complex predicate). As the complexity of the predicate proposition increases, the validity of the decision tends to decrease.

Fuzzy Composition

Knowledge related to the decision-making process is usually contained in more than one fuzzy rule. A large number of fuzzy rules may take part in the decision-making process and all fuzzy rules are fired in parallel during that process. Clearly, not all fuzzy rules contribute equally to the final solution, and rules that have no degree of truth in their premises do not contribute to the outcome at all. Several composition mechanisms facilitate rule combination; the most frequently used approach, called the **min–max rule**, will be discussed.

In the min–max composition approach, a sequence of minimizations and maximizations is applied. First, the minimum of the predicate truth (**correlation minimum**) $\mu_{A_i}(x)$ is

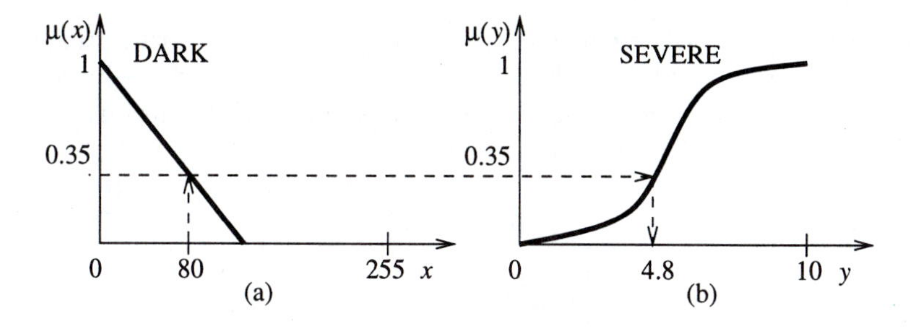

Figure 7.28: *Monotonic fuzzy reasoning based on a single fuzzy rule: If the gray-level of the cloud is* DARK, *then the thunderstorm will be* SEVERE.

used to restrict the consequent fuzzy membership function $\mu_{B_i}(w)$. Let the rules be in the form specified in equation (7.67), and let i represent the ith rule. Then, the consequent fuzzy membership functions B_i are updated in a pointwise fashion and the fuzzy membership functions B_i^+ are formed (Figure 7.29).

$$\mu_{B_i^+}(w) = \min[\mu_{B_i}(w), \mu_{A_i}(x)] \tag{7.70}$$

Second, the pointwise maxima of these minimized fuzzy sets form the solution fuzzy membership function.

$$\mu_S(w) = \max_i[\mu_{B_i^+}(w)] \tag{7.71}$$

Figure 7.29 demonstrates the min–max composition process; again, complex predicates may be considered.

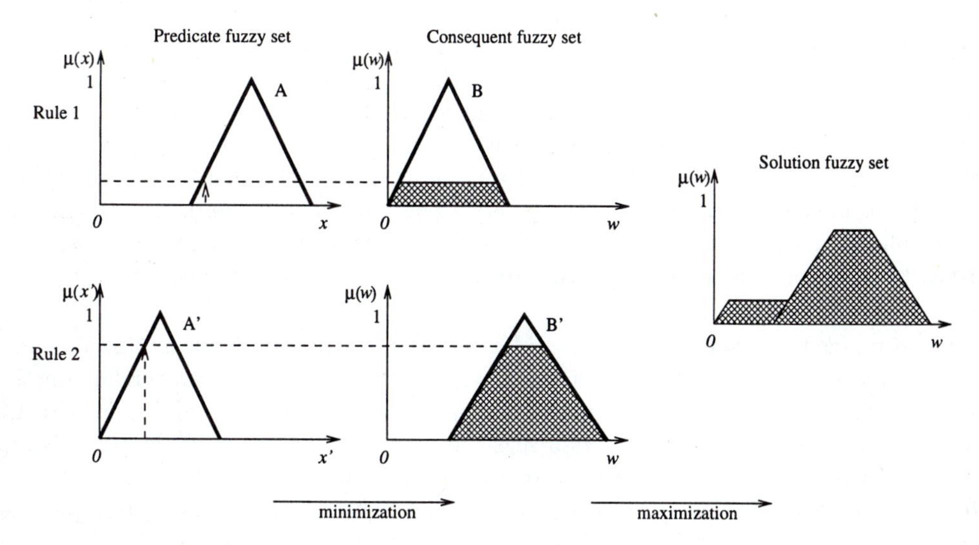

Figure 7.29: *Fuzzy min–max composition using correlation minimum.*

The correlation minimum described above is the most common approach to performing the first step of the min–max composition. An alternative approach called **correlation product** scales the original consequent fuzzy membership functions instead of truncating them. While correlation minimum is computationally less demanding and easier to de-fuzzify, correlation product represents in many ways a better method of minimization, since the original shape of the fuzzy set is retained (Figure 7.30).

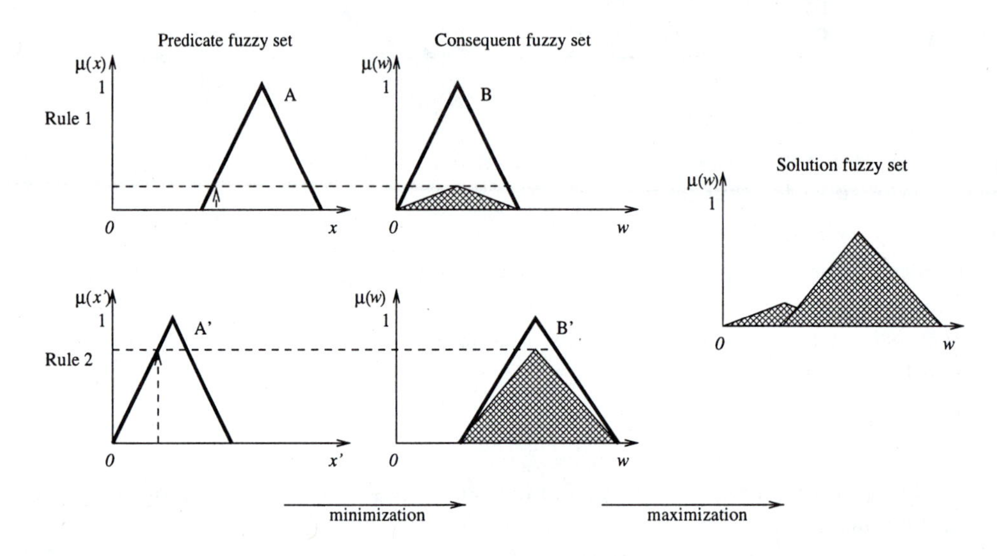

Figure 7.30: *Fuzzy min–max composition using correlation product.*

De-fuzzification

Fuzzy composition produces a single solution fuzzy membership function for each solution variable. To find the actual crisp solution that will be used for decision making, it is necessary to find a vector of scalar values (one value for each solution variable) that best represents the information contained in the solution fuzzy sets. This process is performed independently for each solution variable and is called de-fuzzification. Two de-fuzzification methods, called **composite moments** and **composite maximum**, are commonly used; many other varieties exist.

Composite moments look for the centroid c of the solution fuzzy membership function—Figure 7.31a shows how the centroid method converts the solution fuzzy membership function into a crisp solution variable c. Composite maximum identifies the domain point with the highest membership value in the solution fuzzy membership function. If this point is ambiguous (on a plateau or if there are two or more equal global maxima), the center of the plateau (or the point halfway between the leftmost and rightmost global maximum) provides the crisp solution c' (Figure 7.31b). The composite moments approach produces a result that is sensitive to all the rules, while solutions determined using the composite maximum method are sensitive to the membership function produced by the single rule that has the highest predicate truth. While composite moments are used mostly in control applications, recognition applications usually use the composite maximum method.

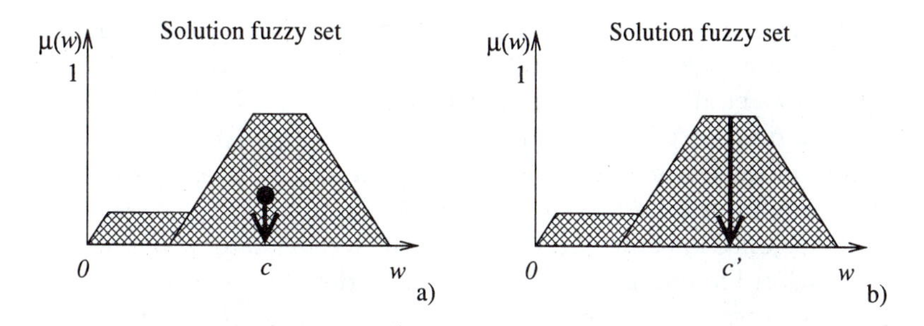

Figure 7.31: *De-fuzzification: (a) composite moments, (b) composite maximum.*

7.7.4 Fuzzy system design and training

Fuzzy system design consists of several main steps that are outlined in the following algorithm.

Algorithm 7.12: Fuzzy system design

1. Design functional and operational characteristics of the system—determine the system inputs, basic processing approaches, and system outputs. In object recognition, the inputs are patterns and the output represents the decision.

2. Define fuzzy sets by decomposing each input and output variable of the fuzzy system into a set of fuzzy membership functions. The number of fuzzy membership functions associated with each variable depends on the task at hand. Typically, an odd number of three to nine fuzzy membership functions is created for each variable. It is recommended that the neighboring fuzzy membership functions overlap by 10–50%. The sum of the membership values of the overlap are recommended to be less than one.

3. Convert problem-specific knowledge into the fuzzy *if-then* rules that represent a fuzzy associative memory. The number of designed rules is related to the number of input variables. For N variables each of which is divided into M fuzzy membership functions, M^N rules are required to cover all possible input combinations.

4. Perform fuzzy composition and de-fuzzification as described in Section 7.7.3.

5. Using a training set, determine the system's performance. If the fuzzy system's behavior does not meet the requirements, modify the fuzzy set descriptions and/or fuzzy rules and/or the fuzzy composition and/or de-fuzzification approaches. The speed and success of this fine-tuning step depend on the problem complexity, the designer's level of understanding of the problem, and the level of the designer's experience.

As can be seen from the description of steps 3 and 5 of the previous algorithm, the design of fuzzy rules may be a tedious and time-consuming process if the rules are to be designed from human experts, as has been typical in most existing applications. Recently, several approaches have been reported that generate fuzzy *if–then* rules automatically using a training set as the source of knowledge and/or for automated adjusting membership functions of fuzzy sets [Horikawa et al. 92, Simpson 92, Ishibuchi et al. 92, Ishibuchi et al. 95, Abe and Lan 95, Homaifar and McCormick 95]. Some of the approaches use neural networks or genetic algorithms to control the learning process. A genetic algorithm-based method for selecting a small number of significant rules from a training set of examples is given in [Ishibuchi et al. 95], in which the rule-selection problem is formulated as a combinatorial optimization problem and uses genetic algorithm optimization. The optimization process is designed to maximize the classification correctness and minimize the number of fuzzy *if–then* rules. An approach based on iterative splitting of the feature space that was introduced in [Park 96] generates a smaller number of fuzzy rules compared to the approach of Ishibuchi. While Ishibuchi's approach uses equal spacing for partitioning of the feature space, Park proposes using an adaptive grid defined by minimum and maximum values of individual features for each class. This adaptivity is mostly responsible for the more efficient feature space partitioning and is reflected in the smaller number of generated fuzzy rules.

Many applications of fuzzy systems exist in pattern recognition and image processing. In the field of pattern recognition, fuzzy logic has been used for supervised and nonsupervised recognition, sequential learning, fuzzy decision theoretic and syntactic classifiers, feature extraction, etc. [Bezdek 81, Kandel 82, Pal 91, Zadeh and Kacprzyk 92]. In image processing and vision, fuzzy logic has been applied to image quality assessment, edge detection, image segmentation, color image segmentation, etc. [Pal 91]. Extensive work has been done in developing fuzzy geometry approaches [Rosenfeld 79a, Rosenfeld 83, Rosenfeld 84a, Rosenfeld 84b, Rosenfeld 85]. Recently, a fuzzy approach to object definition and connectedness and its application to image segmentation was presented in [Dellepiane and Fontana 95, Udupa and Samarasekera 96]; fuzzy connectivity in mathematical morphology is discussed in [Bloch 93]; good performance of fuzzy systems in medical image segmentation and interpretation applications was reported in [Udupa and Samarasekera 96, Park et al. 96].

7.8 Summary

- **Object recognition, pattern recognition**
  - Pattern recognition is used for region and object **classification**, and represents an important building block of complex machine vision processes.
  - No recognition is possible without **knowledge**. Specific knowledge about both the objects being processed and hierarchically higher and more general knowledge about object classes is required.
- **Knowledge representation**
  - Descriptions and features
  - Grammars and languages
  - Predicate logic
  - Production rules
  - Fuzzy logic

  - Semantic nets
  - Frames, scripts

- **Statistical pattern recognition**
  - **Object recognition** is based on assigning classes to objects, and the device that does these assignments is called the **classifier**. The number of classes is usually known beforehand, and typically can be derived from the problem specification.
  - The classifier does not decide about the class from the object itself—rather, sensed object properties called **patterns** are used.
  - For statistical pattern recognition, **quantitative** description of objects is characteristic, elementary numerical descriptions—**features**—are used. The set of all possible patterns forms the **pattern space** or **feature space**. The **classes** form clusters in the feature space, which can be separated by **discrimination hypersurfaces**.
  - A **statistical classifier** is a device with n inputs and 1 output. Each input is used to enter the information about one of n features measured from an object to be classified. An R-class classifier generates one of R symbols ω_r, the **class identifiers**.
  - Classification parameters are determined from a **training set** of examples during **classifier learning**. Two common learning strategies are **probability density estimation** and **direct loss minimization**.
  - Some classification methods do not need training sets for learning. **Cluster analysis** methods divide the set of processed patterns into subsets (clusters) based on the mutual similarity of subset elements.

- **Neural nets**
  - Most neural approaches are based on combinations of elementary processors (**neurons**), each of which takes a number of inputs and generates a single output. Associated with each input is a weight, and the output is a function of the weighted sum of inputs. Pattern recognition is one of many application areas of neural networks.
  - **Feed-forward** networks are common in pattern recognition problems. Their training uses a training set of examples and is often based on the **back-propagation** algorithm.
  - **Self-organizing** networks do not require a training set to cluster the processed patterns.
  - **Hopfield** neural networks do not have designated inputs and outputs, but rather the current configuration represents the state. The Hopfield net acts as an associative memory where the exemplars are stored.

- **Syntactic pattern recognition**
  - For syntactic pattern recognition, **qualitative** description of objects is characteristic. The elementary properties of the syntactically described objects are called **primitives**. **Relational structures** are used to describe relations between the object primitives.
  - The set of all primitives is called the **alphabet**. The set of all words in the alphabet that can describe objects from one class is named the **description language**. A **grammar** represents a set of rules that must be followed when words of the specific language are constructed from the alphabet.
  - Grammar construction usually requires significant human interaction. In simple cases, an automated process of grammar construction from examples called **grammar inference** can be applied.
  - The recognition decision of whether or not the word can be generated by a particular grammar is made during **syntactic analysis**.
- **Recognition as graph matching**
  - Matching of a model and an object graph description can be used for recognition. An exact match of graphs is called graph **isomorphism**. Determination of graph isomorphism is computationally expensive.
  - In the real world, the object graph usually does not match the model graph exactly. Graph isomorphism cannot assess the level of mismatch. To identify objects represented by similar graphs, **graph similarity** can be determined.
- **Optimization techniques in recognition**
  - Optimization problems seek minimization or maximization of an **objective function**. Design of the objective function is a key factor in the performance of optimization algorithms.
  - Most conventional approaches to optimization use calculus-based **hill climbing** methods. For these, the search can easily end in a local maximum, and the global maximum can be missed.
  - **Genetic algorithms** use natural evolution mechanisms of the survival of the fittest to search for the maximum of an objective function. Potential solutions are represented as strings. Genetic algorithms search from a population of potential solutions, not a single solution. The sequence of **reproduction**, **crossover**, and **mutation** generates a new population of strings from the previous population. The fittest string represents the final solution.
  - **Simulated annealing** combines two basic optimization principles, **divide and conquer** and **iterative improvement** (hill climbing). This combination avoids getting stuck in local optima.

- **Fuzzy systems**
  - Fuzzy systems are capable of representing diverse, non-exact, uncertain, and inaccurate knowledge or information. They use qualifiers that are very close to the human way of expressing knowledge.
  - Fuzzy reasoning is performed in the context of a **fuzzy system model** that consists of control, solution, and working data variables; fuzzy sets; hedges; fuzzy rules; and a control mechanism.
  - **Fuzzy sets** represent properties of fuzzy spaces. **Membership functions** represent the fuzziness of the description and assess the degree of certainty about the membership of an element in the particular fuzzy set. Shape of fuzzy membership functions can be modified using **fuzzy set hedges**. A hedge and its fuzzy set constitute a single semantic entity called a **linguistic variable**.
  - **Fuzzy** *if–then* **rules** represent fuzzy associative memory in which knowledge is stored.
  - In fuzzy reasoning, information carried in individual fuzzy sets is combined to make a decision. The functional relationship determining the degree of membership in related fuzzy regions is called the **method of composition** and results in definition of a **fuzzy solution space**. To arrive at the decision, **de-fuzzification** is performed. Processes of composition and de-fuzzification form the basis of fuzzy reasoning.

7.9 Exercises

Short-answer questions

1. Define the *syntax* and *semantics* of knowledge representation.
2. Describe the following knowledge representations, giving for each one at least one example that is different from examples given in the text.
   (a) Descriptions (features)
   (b) Grammars
   (c) Predicate logic
   (d) Production rules
   (e) Fuzzy logic
   (f) Semantic nets
   (g) Frames (scripts)
3. Define the following terms:
   (a) Pattern
   (b) Class
   (c) Classifier
   (d) Feature space
4. Describe the main steps of pattern recognition.

5. Define the following terms:
   (a) Class identifier
   (b) Decision rule
   (c) Discrimination function
6. Explain the main concepts and derive a mathematical representation of the discrimination functions for:
   (a) A minimum distance classifier
   (b) A minimum error classifier
7. What is a training set? How is it designed? What influences its desired size?
8. Explain why learning should be inductive and sequential.
9. Describe the conceptual differences between supervised and unsupervised learning.
10. Draw schematic diagrams of a feed-forward and Hopfield neural networks. Discuss their major architectural differences.
11. For what is the back-propagation algorithm used? Explain its main steps.
12. What is the reason for including the momentum constant in back-propagation learning?
13. Explain the functionality of Kohonen neural networks. How can they be used for unsupervised pattern recognition?
14. Explain how Hopfield networks can be used for pattern recognition.
15. Compare classification approaches used by statistical pattern recognition and neural networks.
16. Define the following terms:
    (a) Primitive
    (b) Alphabet
    (c) Description language
    (d) Grammar
17. Describe the main steps of syntactic pattern recognition.
18. Give a formal definition of a grammar.
19. When are two grammars equivalent?
20. True or false? A regular grammar is a context-free grammar.
21. Name the main approaches to syntactic analysis.
22. What is grammar inference? Give its block diagram.
23. Formally define:
    (a) A graph
    (b) Graph isomorphism
    (c) Sub-graph isomorphism
    (d) Double sub-graph isomorphism
24. Define Levenshtein distance. Explain its application to assessing string similarity.

25. Explain why hill-climbing optimization approaches may converge to local instead of global optima.
26. Explain the concept and functionality of genetic algorithm optimization. What are the roles of reproduction, crossover, and mutation in genetic algorithms?
27. Explain the concept of optimization based on simulated annealing. What is the annealing schedule?
28. List the advantages and disadvantages of genetic algorithms and simulated annealing compared to optimization approaches based on derivatives.
29. Define the following terms:
    (a) Fuzzy set
    (b) Fuzzy membership function
    (c) Minimum normal form of a fuzzy membership function
    (d) Maximum normal form of a fuzzy membership function
    (e) Fuzzy system
    (f) Domain of a fuzzy set
    (g) Hedge
    (h) Linguistic variable
30. Use Zadeh's definitions to define formally:
    (a) Fuzzy intersection
    (b) Fuzzy union
    (c) Fuzzy complement
31. Explain fuzzy reasoning based on composition and de-fuzzification. Draw a block diagram of fuzzy reasoning.

Problems

1. Let a minimum distance classifier be used to recognize two-dimensional patterns from three classes K_1, K_2, K_3. The training set consists of five patterns from each class:

$$K_1 := \left\{ \begin{pmatrix} 0 \\ 6 \end{pmatrix}, \begin{pmatrix} 1 \\ 6 \end{pmatrix}, \begin{pmatrix} 2 \\ 6 \end{pmatrix}, \begin{pmatrix} 1 \\ 5 \end{pmatrix}, \begin{pmatrix} 1 \\ 7 \end{pmatrix} \right\}$$

$$K_2 := \left\{ \begin{pmatrix} 4 \\ 1 \end{pmatrix}, \begin{pmatrix} 5 \\ 1 \end{pmatrix}, \begin{pmatrix} 6 \\ 1 \end{pmatrix}, \begin{pmatrix} 5 \\ 0 \end{pmatrix}, \begin{pmatrix} 5 \\ 2 \end{pmatrix} \right\}$$

$$K_3 := \left\{ \begin{pmatrix} 8 \\ 6 \end{pmatrix}, \begin{pmatrix} 9 \\ 6 \end{pmatrix}, \begin{pmatrix} 10 \\ 6 \end{pmatrix}, \begin{pmatrix} 9 \\ 5 \end{pmatrix}, \begin{pmatrix} 9 \\ 7 \end{pmatrix} \right\}$$

   Determine (sketch) the discrimination functions in the two-dimensional feature space.

2. Let a minimum error classifier be used to recognize two-dimensional patterns from two classes, each having a normal distribution $N(\boldsymbol{\mu}_r, \boldsymbol{\Psi}_r)$:

$$\boldsymbol{\mu}_1 = \begin{pmatrix} 2 \\ 5 \end{pmatrix} \quad \boldsymbol{\Psi}_1 = \begin{pmatrix} 1 & 0 \\ 0 & 1 \end{pmatrix} \quad \boldsymbol{\mu}_2 = \begin{pmatrix} 4 \\ 3 \end{pmatrix} \quad \boldsymbol{\Psi}_2 = \begin{pmatrix} 1 & 0 \\ 0 & 1 \end{pmatrix}$$

   Assume unit loss functions and equal a priori probabilities of classes $P(\omega_1) = P(\omega_2) = 0.5$. Determine (sketch) the discrimination function in the two-dimensional feature space.

3. Repeat Problem 7.2 with $P(\omega_1) = P$, $P(\omega_2) = 1 - P$. Show how the discrimination function locations in the two-dimensional feature space change as a function of P.

4. Repeat Problem 7.2 considering modified parameters of the normal distributions:

$$\mu_1 = \begin{pmatrix} 2 \\ 5 \end{pmatrix} \quad \Psi_1 = \begin{pmatrix} 1 & 0 \\ 0 & 3 \end{pmatrix} \quad \mu_2 = \begin{pmatrix} 4 \\ 3 \end{pmatrix} \quad \Psi_2 = \begin{pmatrix} 1 & 0 \\ 0 & 3 \end{pmatrix}$$

5. Repeat Problem 7.4 with $P(\omega_1) = P$, $P(\omega_2) = (1 - P)$. Show how the discrimination function locations in the two-dimensional feature space change as a function of P.

6. Consider the training set specified in Problem 7.1. Assume that the three pattern classes have normal distributions and that a priori probabilities of classes are equal $P(\omega_1) = P(\omega_2) = P(\omega_3) = 1/3$. Determine (sketch) the discrimination functions of the minimum error classifier in the two-dimensional feature space. Discuss under what circumstances the discrimination functions of a minimum distance classifier are identical to discrimination functions of the minimum error classifier if both were trained using the same training set.

7. Create the following training and testing sets of feature vectors named TRAIN1, TEST1; TRAIN2, TEST2. The training and testing sets will be used in the experiments below.

TRAIN1

ω_i	ω_1	ω_1	ω_1	ω_1	ω_1	ω_2	ω_2	ω_2	ω_2	ω_2
x_1	2	4	3	3	4	10	9	8	9	10
x_2	3	2	3	2	3	7	6	6	7	6

TEST1

ω_i	ω_1	ω_1	ω_1	ω_1	ω_1	ω_2	ω_2	ω_2	ω_2	ω_2
x_1	3	6	5	5	6	13	12	11	11	13
x_2	5	3	4	3	5	10	8	8	9	8

TRAIN2

ω_i	ω_1	ω_1	ω_1	ω_1	ω_1	ω_2	ω_2	ω_2	ω_2	ω_2
x_1	2	6	-2	7	5	-2	-6	2	-4	-5
x_2	400	360	520	-80	180	-200	-200	-400	-600	-400

ω_i	ω_3	ω_3	ω_3	ω_3	ω_3
x_1	-10	-8	-15	-12	-14
x_2	200	140	100	50	300

TEST2

ω_i	ω_1	ω_1	ω_1	ω_1	ω_1	ω_2	ω_2	ω_2	ω_2	ω_2
x_1	4	8	-3	9	6	-1	-4	3	-2	-3
x_2	600	540	780	-120	270	-250	-250	-470	-690	-470

ω_i	ω_3	ω_3	ω_3	ω_3	ω_3
x_1	-15	-13	-6	-17	-16
x_2	230	170	130	80	450

8. Develop a program for training and classification using the minimum distance classifier. Assess classification correctness.

   (a) Train and test using data sets TRAIN1 and TEST1.

   (b) Train and test using data sets TRAIN2 and TEST2.

9. Develop a program for training and classification using the minimum error classifier, considering unit loss functions. Assume the training data have normal distribution. Assess classification correctness.

   (a) Train and test using data sets TRAIN1 and TEST1.

   (b) Train and test using data sets TRAIN2 and TEST2.

10. Develop a program for cluster analysis using the k-means approach. Vary the initialization of cluster starting points and explore its influence on clustering results. Vary the number of classes and discuss the results.

    (a) Use a combined data set TRAIN1 and TEST1.

    (b) Use a combined data set TRAIN2 and TEST2.

11. Create a training set and a testing set of feature vectors using the shape description program developed in Problem 6.14 to determine shape feature vectors. Use simple shapes (e.g., triangles, squares, rectangles, circles, etc.) of different sizes. Select up to five discriminative features to form the feature vectors of analyzed shapes. The training as well as the testing sets should consist of at least 10 patterns from each class. The training and testing sets will be used in the experiments below.

12. Apply the program developed in Problem 7.8 to the training and testing sets created in Problem 7.11. Assess classification correctness.

13. Apply the program developed in Problem 7.9 to the training and testing sets created in Problem 7.11. Assume normal distributions, and that you have a sufficient number to determine representative dispersion matrices and mean values from the training set. Assess classification correctness in the testing set. Compare with the performance of the minimum distance classifier used in Problem 7.12.

14. Apply the program developed in Problem 7.10 to the testing set created in Problem 7.11. First, assume that the number of classes is known. Assess clustering correctness and compare it with that of the supervised methods used in Problems 7.12 and 7.13. Then, vary the initialization of cluster starting points and explore the influence on clustering results.

15. Develop a program for back-propagation training and classification using a three-layer feed-forward neural network. Train and test using artificial data from a two-dimensional feature space representing patterns from at least three separable classes.

16. Apply the program developed in Problem 7.15 to the testing set created in Problem 7.11. Assess classification correctness and compare it with that of the statistical classification methods assessed in Problems 7.12 and 7.13.

17. Choice of width of layers is often a problem. Repeat Problems 7.15 and 7.16, paying attention to the size of the hidden layer. Draw some conclusions about network performance and training time as the size of this layer varies.

18. Implement Algorithm 7.5. For some datasets devised by you, or extracted from some known application, run the algorithm. Compare its performance for different sizes and topologies of output layer, and various choices of parameters.

19. Implement a Hopfield network. Train it on digitized patterns of the digits 0–9; study its performance at pattern recall (e.g., of noisy examples of digits) for various resolutions of the patterns.

20. Design a grammar G that can generate a language $L(G)$ of equilateral triangles of any size; primitives with 0°, 60°, and 120° orientation form the set of terminal symbols $V_t = \{a, b, c\}$.

21. Design three different grammars producing the language $L(G) = \{ab^n\}$ for $n = 1, 2, \ldots$.

22. Design a grammar that generates all characters P or d of the following properties:

    - Character P is represented by a square with an edge length equal to one; a vertical line is attached to the bottom left corner of the square and may have any length.
    - Character d is represented by a square with an edge length equal to one; a vertical line is attached to the top right corner of the square and may have any length.

    Obviously, there are infinitely many such characters. Use the following set of terminal symbols $V_t = \{N, W, S, E\}$; terminal symbols correspond to directions of the chain code – north, west, south, east. Design your set of non-terminal symbols, using a start symbol s. Validate your grammar design on examples. Show all steps in generating at least two P and two d characters.

23. Using Algorithm 7.8 prove or disprove isomorphism of the graphs shown in Figure 7.32.

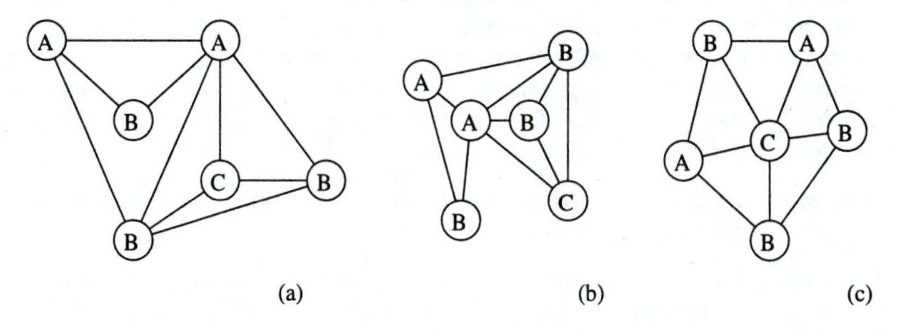

Figure 7.32: *Problem 7.23.*

24. Determine the Levenshtein distance for the following string pairs:

    (a) $S_1 = abadcdefacde$ $S_2 = abadddefacde$

    (b) $S_1 = abadcdefacde$ $S_3 = abadefaccde$

    (c) $S_1 = abadcdefacde$ $S_4 = cbadcacdae$

25. Using genetic algorithm optimization, determine the maximum of the following function (this function has several local maxima; its visualization is available in, e.g., Matlab by typing peaks, see Figure 7.33):

$$z(x,y) = 3(1-x)^2 \exp\left[-x^2 - (y+1)^2\right] - 10[(x/5) - x^3 - y^5] \exp\left[-x^2 - y^2\right]$$
$$-(1/3) \exp\left[-(x+1)^2 - y^2\right]$$

    Develop a program for genetic algorithm-based optimization following Algorithm 7.10. (Alternatively, use one of the many genetic algorithm programs freely available on the World Wide Web.) Design the code string as consisting of n bits for the x value and n bits for the y value, the value of $z(x,y)$ represents the string fitness. Limit your search space to $x \in (-4,4)$ and $y \in (-4,4)$. Explore the role of the starting population, population size S, mutation rate M, and string bit length $2n$ on the speed of convergence and solution accuracy. For several values of S, M, n, plot the function values of maximum string fitness, average population fitness, and minimum string fitness as a function of the generation number.

26. Using the definition of the fuzzy set SEVERE as given in Figure 7.28b, sketch the following fuzzy membership functions:

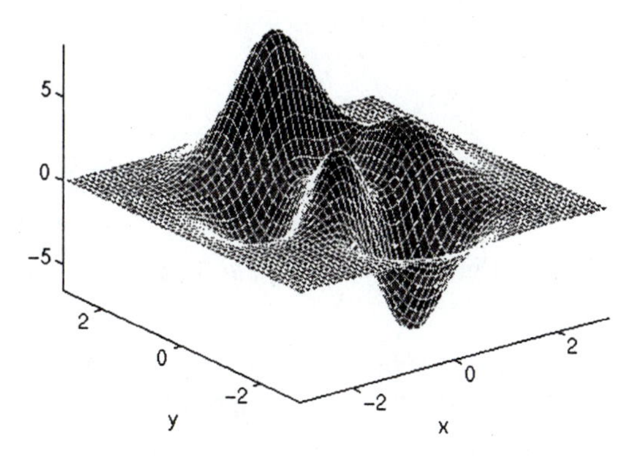

Figure 7.33: *Problem 7.25.*

(a) VERY_SEVERE

(b) SOMEWHAT_SEVERE

(c) SOMEWHAT_NOT_SEVERE

27. Using the fuzzy sets DARK and BRIGHT as given in Figure 7.25, sketch the single fuzzy membership function (NOT_VERY_DARK AND NOT_VERY_BRIGHT).

28. Considering the fuzzy sets A and B given in Figure 7.34, derive intersection, union, and complement of the two fuzzy sets.

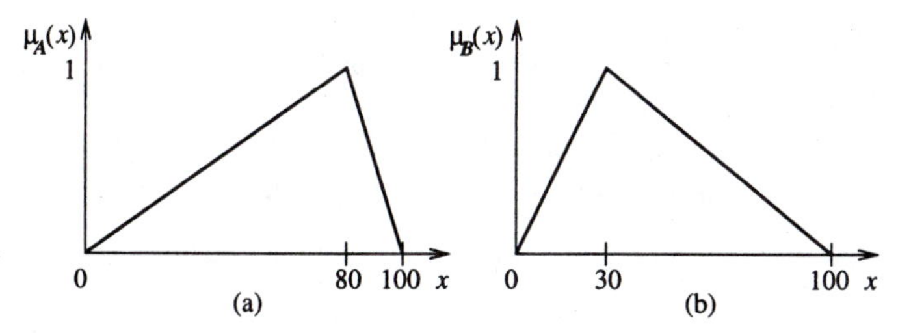

Figure 7.34: *Problem 7.28.*

29. Use the composite moments and composite maximum approaches to de-fuzzification to find the representative value of the fuzzy set provided in Figure 7.35.

30. Flash flood represents a potential danger in many areas, and its prediction is an important part of meteorological forecasting. Clearly, the following conditions increase the flood danger:

    (a) Rain amount in the past three days.

    (b) Water saturation of soil.

    (c) The rainfall expected in the next 24 hours.

    Assuming the specified information is available, design a fuzzy logic system to determine the expected flood danger within the next 24 hours.

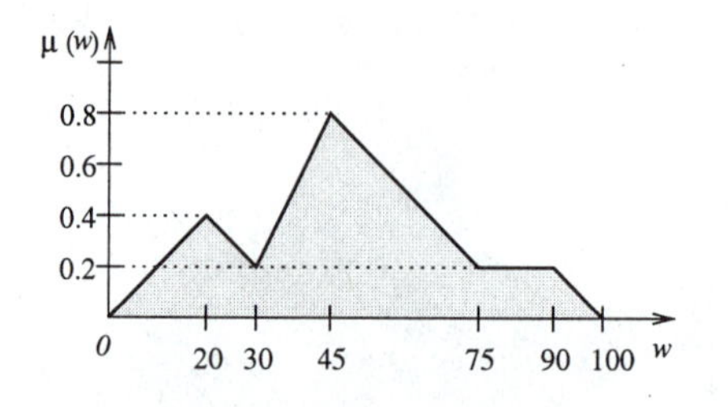

Figure 7.35: *Problem 7.29.*

31. Develop a program implementation of the fuzzy logic system designed in Problem 7.30. Explore, how different membership function shapes, and fuzzy logic composition and decomposition methods, influence the achieved results.

7.10 References

[Aarts and van Laarhoven 86] E H L Aarts and P J M van Laarhoven. Simulated annealing: A pedestrian review of the theory and some applications. In P A Devijver and J Kittler, editors, *Pattern Recognition Theory and Applications*, pages 179–192. Springer Verlag, Berlin–New York–Tokyo, 1986.

[Abe and Lan 95] S Abe and M Lan. A method for fuzzy rules extraction directly from numerical data and its application to pattern classification. *IEEE Transactions on Fuzzy Systems*, 3(2):129–139, 1995.

[Adeli and Hung 95] H Adeli and S L Hung. *Machine Learning: Neural Networks, Genetic Algorithms, and Fuzzy Systems.* Wiley, New York, 1995.

[Ambler 75] A P H Ambler. A versatile system for computer controlled assembly. *Artificial Intelligence*, 6(2):129–156, 1975.

[Amit 89] D J Amit. *Modeling Brain Function: The World of Attractor Neural Networks.* Cambridge University Press, Cambridge, England; New York, 1989.

[Arabie et al. 96] P Arabie, L J Hubert, and G De Soete, editors. *Clustering and Classification.* World Scientific, River Edge, NJ, 1996.

[Azencott 92] R Azencott, editor. *Simulated Annealing: Parallelization Techniques.* Wiley, New York, 1992.

[Baird 84] H S Baird. *Model-Based Image Matching Using Location.* MIT Press, Cambridge, MA, 1984.

[Ballard and Brown 82] D H Ballard and C M Brown. *Computer Vision.* Prentice-Hall, Englewood Cliffs, NJ, 1982.

[Barnard 87] Barnard. Stereo matching by hierarchical microcanonical annealing. *Perception*, 1:832, 1987.

[Barrero 91] A Barrero. Inference of tree grammars using negative samples. *Pattern Recognition*, 24(1):1–8, 1991.

[Barrow and Popplestone 71] H G Barrow and R J Popplestone. Relational descriptions in picture processing. *Machine Intelligence*, 6, 1971.

[Berge 76] C Berge. *Graphs and Hypergraphs.* American Elsevier, New York, 2nd edition, 1976.

[Bezdek 81] L C Bezdek. *Pattern Recognition with Fuzzy Objective Function Algorithm.* Plenum Press, New York, 1981.

[Bittner and Reingold 75] J R Bittner and E M Reingold. Backtrack programming techniques. *Communications of the ACM*, 18(11):651–656, 1975.

[Blashfield et al. 82] R K Blashfield, M S Aldenderfer, and L C Morey. Cluster analysis software. In P R Krishniah and L N Kanal, editors, *Handbook of Statistics*, pages 245–266. North Holland, Amsterdam, 1982.

[Bloch 93] I Bloch. Fuzzy connectivity and mathematical morphology. *Pattern Recognition Letters*, 14:483–488, 1993.

[Blum and Rivest 88] A Blum and R L Rivest. Training a three node neural network is NP-complete. In *Proceedings of IEEE Conference on Neural Information Processing Systems*, page 494, 1988.

[Bouman and Liu 91] C Bouman and B Liu. Multiple resolution segmentation of textured images. *IEEE Transactions on Pattern Analysis and Machine Intelligence*, 13(2):99–113, 1991.

[Braspenning et al. 95] P J Braspenning, F Thuijsman, and A J M M Weijters, editors. *Artificial Neural Networks: An Introduction to ANN Theory and Practice.* Springer Verlag, Berlin; New York, 1995.

[Bron and Kerbosch 73] C Bron and J Kerbosch. Finding all cliques of an undirected graph. *Communications of the ACM*, 16(9):575–577, 1973.

[Buckley 90] F Buckley. *Distance in Graphs.* Addison-Wesley, Redwood City, CA, 1990.

[Carling 92] A Carling. *Introducing Neural Networks.* Sigma, Wilmslow, England, 1992.

[Carpenter and Grossberg 87a] G A Carpenter and S Grossberg. ART2: Self organization of stable category recognition codes for analog input patterns. *Applied Optics*, 26:4919–4930, 1987.

[Carpenter and Grossberg 87b] G A Carpenter and S Grossberg. A massively parallel architecture for a self-organizing neural pattern recognition machine. *Computer Vision, Graphics, and Image Processing*, 37:54–115, 1987.

[Carpenter and Grossberg 91] G A Carpenter and S Grossberg. *Pattern Recognition by Self-organizing Neural Networks.* MIT Press, Cambridge, MA, 1991.

[Cerny 85] V Cerny. Thermodynamical approach to the travelling salesman problem: An efficient simulation algorithm. *Journal of Optimization Theory and Applications*, 45:41–51, 1985.

[Chambers 95] L Chambers, editor. *Practical Handbook of Genetic Algorithms.* CRC Press, Boca Raton, FL, 1995.

[Chen 76] C H Chen, editor. *Pattern Recognition and Artificial Intelligence.* Academic Press, New York, 1976.

[Chen et al. 93] C H Chen, L F Pau, and P S P Wang, editors. *Handbook of Pattern Recognition and Computer Vision.* World Scientific, Singapore; River Edge, NJ, 1993.

[Cherkassky et al. 94] V Cherkassky, J H Friedman, and H Wechsler, editors. *From Statistics to Neural Networks.* Springer Verlag, Berlin; New York, 1994.

[Chomsky 66] N Chomsky. *Syntactic Structures.* Mouton, Hague, 6th edition, 1966.

[Chomsky et al. 71] N Chomsky, J P B Allen, and P Van Buren. *Chomsky: Selected Readings.* Oxford University Press, London–New York, 1971.

[Clocksin and Mellish 81] W F Clocksin and C S Mellish. *Programming in Prolog.* Springer Verlag, Berlin–New York–Tokyo, 1981.

[Cox 94] E Cox. *The Fuzzy Systems Handbook.* AP Professional, Cambridge, England, 1994.

[Dasarathy 91] B V Dasarathy. *Nearest Neighbor (NN) Norms: NN Pattern Classification Techniques.* IEEE Computer Society Press, Los Alamitos, CA, 1991.

[Dellepiane and Fontana 95] S Dellepiane and F Fontana. Extraction of intensity connectedness for image processing. *Pattern Recognition Letters*, 16:313–324, 1995.

[Devijver and Kittler 82] P A Devijver and J Kittler. *Pattern Recognition: A Statistical Approach.* Prentice-Hall, Englewood Cliffs, NJ, 1982.

[Devijver and Kittler 86] P A Devijver and J Kittler. *Pattern Recognition Theory and Applications.* Springer Verlag, Berlin–New York–Tokyo, 1986.

[Dubes and Jain 76] R C Dubes and A K Jain. Clustering techniques: The user's dilemma. *Pattern Recognition*, 8:247–260, 1976.

[Dubes and Jain 80] R C Dubes and A K Jain. Clustering methodologies in exploratory data analysis. In M Yovits, editor, *Advances in Computers*, pages 113–228. Academic Press, New York, 1980.

[Duda and Hart 73] R O Duda and P E Hart. *Pattern Classification and Scene Analysis.* Wiley, New York, 1973.

[Even 79] S Even. *Graph Algorithms.* Computer Science Press, Rockville, MD, 1979.

[Everitt and Brian 93] B Everitt and S E Brian. *Cluster Analysis.* E Arnold, Halsted Press, New York; London, 3rd edition, 1993.

[Fausett 94] L Fausett. *Fundamentals of Neural Networks.* Prentice-Hall, Englewood Cliffs, NJ, 1994.

[Fischler and Elschlager 73] M A Fischler and R A Elschlager. The representation and matching of pictorial structures. *IEEE Transactions on Computers*, C-22(1):67–92, 1973.

[Fu 68] K S Fu. *Sequential Methods in Pattern Recognition and Machine Learning.* Academic Press, New York, 1968.

[Fu 74] K S Fu. *Syntactic Methods in Pattern Recognition.* Academic Press, New York, 1974.

[Fu 77] K S Fu. *Syntactic Pattern Recognition—Applications.* Springer Verlag, Berlin, 1977.

[Fu 80] K S Fu. Picture syntax. In S K Chang and K S Fu, editors, *Pictorial Information Systems*, pages 104–127. Springer Verlag, Berlin, 1980.

[Fu 82] K S Fu. *Syntactic Pattern Recognition and Applications.* Prentice-Hall, Englewood Cliffs, NJ, 1982.

[Fukunaga 90] K Fukunaga. *Introduction to Statistical Pattern Recognition.* Academic Press, Boston, 2nd edition, 1990.

[Furuhashi 95] T Furuhashi, editor. *Advances in Fuzzy Logic, Neural Networks, and Genetic Algorithms.* Springer Verlag, Berlin; New York, 1995.

[Geman et al. 90] D Geman, S Geman, C Graffigne, and P Dong. Boundary detection by constrained optimisation. *IEEE Transactions on Pattern Analysis and Machine Intelligence*, 12(7):609–628, 1990.

[Goldberg 89] D E Goldberg. *Genetic Algorithms in Search, Optimization, and Machine Learning.* Addison-Wesley, Reading, MA, 1989.

[Gonzalez and Thomason 74] R C Gonzalez and M G Thomason. On the inference of tree grammars for pattern recognition. In *Proceedings of the IEEE International Conference on System, Man and Cybernetics*, pages 2–4, IEEE, Piscataway, NJ, 1974.

[Haralick and Elliott 79] R M Haralick and G L Elliott. Increasing tree search efficiency for constraint satisfaction problems. In *6th International Joint Conference on Artificial Intelligence,* Tokyo, Japan, pages 356–364, 1979.

[Harary 69] F Harary. *Graph Theory.* Addison-Wesley, Reading, MA, 1969.

[Hayes 77] P J Hayes. In defense of logic. In *5th International Joint Conference on Artificial Intelligence,* Cambridge, CA, 1977.

[Haykin 94] S Haykin. *Neural Networks.* Macmillan, New York, 1994.

[Hecht-Nielsen 90] R Hecht-Nielsen. *Neurocomputing.* Addison-Wesley, Reading, MA, 1990.

[Hecht-Nielson 87] R Hecht-Nielson. Kolmogorov's mapping neural network existence theorem. In *Proceedings of the First IEEE International Conference on Neural Networks*, volume 3, pages 11–14, IEEE, San Diego, 1987.

[Hlavac and Sara 95] V Hlavac and R Sara, editors. *Computer Analysis of Images and Patterns.* Springer Verlag, Berlin; New York, 1995.

[Homaifar and McCormick 95] A Homaifar and E McCormick. Simultaneous design of memebership functions and rule sets for fuzzy controllers using genetic algorithms. *IEEE Transactions on Fuzzy Systems*, 3(2):129–139, 1995.

[Hopfield and Tank 85] J J Hopfield and D W Tank. Neural computation of decisions in optimization problems. *Biological Cybernetics*, 52:141–152, 1985.

[Hopfield and Tank 86] J J Hopfield and D W Tank. Computing with neural circuits: A model. *Science*, 233:625–633, 1986.

[Horikawa et al. 92] S Horikawa, T Furuhashi, and Y Uchikawa. On fuzzy modeling using fuzzy neural networks with the back-propagation algorithm. *IEEE Transactions on Neural Networks*, 3(5):801–806, 1992.

[Ishibuchi et al. 92] H Ishibuchi, K Nozaki, and H Tanaka. Distributed representation of fuzzy rules and its application to pattern classification. *Fuzzy Sets and Systems*, 52:21–32, 1992.

[Ishibuchi et al. 95] H Ishibuchi, K Nozaki, N Yamamoto, and H Tanaka. Selecting fuzzy if-then rules for classification problems using genetic algorithms. *IEEE Transactions on Fuzzy Systems*, 3:260–270, 1995.

[Johnson and Wichern 90] R A Johnson and D W Wichern. *Applied Multivariate Statistical Analysis.* Prentice-Hall, Englewood Cliffs, NJ, 2nd edition, 1990.

[Judd 90] J S Judd. *Neural Network Design and the Complexity of Learning.* MIT Press, Cambridge, MA, 1990.

[Kandel 82] A Kandel. *Fuzzy Techniques in Pattern Recognition.* Wiley, New York, 1982.

[Kaufman and Rousseeuw 90] L Kaufman and P J Rousseeuw. *Finding Groups in Data: An Introduction to Cluster Analysis.* Wiley, New York, 1990.

[Kaufmann 75] A Kaufmann. *Introduction to the Theory of Fuzzy Subsets-Fundamental Theoretical Elements, Vol 1.* Academic Press, New York, 1975.

[Kirkpatrick et al. 83] S Kirkpatrick, C D Gelatt, and M P Vecchi. Optimization by simulated annealing. *Science*, 220:671–680, 1983.

[Kohonen 88] T Kohonen. The "neural" phonetic typewriter. *Computer*, pages 11–22, March 1988.

[Kohonen 95] T Kohonen. *Self-organizing Maps.* Springer Verlag, Berlin; New York, 1995.

[Kolmogorov 63] A N Kolmogorov. On the representation of continuous functions of many variables by superposition of continuous functions of one variable and addition. *Doklady Akademii Nauk SSSR*, 144:679–681, 1963. (AMS Translation, 28, 55-59).

[Kosko 91] B Kosko. Adaptive bidirectional associative memories. In G A Carpenter and S Grossberg, editors, *Pattern Recognition by Self-organizing Neural Networks*, pages 425–450. MIT Press, Cambridge, MA, 1991.

[Kosko 92] B Kosko. *Neural Networks and Fuzzy Systems.* Prentice-Hall, Englewood Cliffs, NJ, 1992.

[Kowalski 79] R Kowalski. *Logic for Problem Solving.* North Holland, Amsterdam, 1979.

[Lakemeyer and Nebel 94] G Lakemeyer and B Nebel, editors. *Foundations of Knowledge Representation and Reasoning.* Springer Verlag, Berlin; New York, 1994.

[Lau 89] H T Lau. *Algorithms on Graphs.* TAB Professional and Reference Books, Blue Ridge Summit, PA, 1989.

[Linggard et al. 92] R Linggard, C Nightingale, and D Myers, editors. *Neural Networks for Vision, Speech and Natural Language.* Chapman & Hall, London, 1992.

[MacQueen 67] J MacQueen. Some methods for classification and analysis of multivariate observations. In *Proceedings of the 5th Berkeley Symposium—1*, pages 281–297, 1967.

[Masters 95] T Masters. *Advanced Algorithms for Neural Networks: A C++ Sourcebook.* Wiley, New York, 1995.

[Masuch and Polos 94] M Masuch and L Polos, editors. *Knowledge Representation and Reasoning under Uncertainty: Logic at Work.* Springer Verlag, Berlin; New York, 1994.

[McCulloch and Pitts 43] W S McCulloch and W Pitts. A logical calculus of ideas immanent in nervous activity. *Bulletin of Mathematical Biophysics*, 5:115–133, 1943.

[McEliece et al. 87] R J McEliece, E C Posner, E R Rodemich, and S S Venkatesh. The capacity of the Hopfield associative memory. *IEEE Transactions on Information Theory*, 33:461, 1987.

[McHugh 90] J A McHugh. *Algorithmic Graph Theory.* Prentice-Hall, Englewood Cliffs, NJ, 1990.

[McQuitty 87] L L McQuitty. *Pattern-Analytic Clustering: Theory, Method, Research, and Configural Findings.* University Press of America, Lanham, NY, 1987.

[Metropolis et al. 53] N Metropolis, A W Rosenbluth, M N Rosenbluth, A H Teller, and E Teller. Equation of state calculation by fast computing machines. *Journal of Chemical Physics*, 21:1087–1092, 1953.

[Michalewicz 94] Z Michalewicz. *Genetic Algorithms + Data Structures = Evolution Programs.* Springer Verlag, Berlin; New York, 2nd edition, 1994.

[Michalski et al. 83] R S Michalski, J G Carbonell, and T M Mitchell. *Machine Learning I, II.* Morgan Kaufmann Publishers, Los Altos, CA, 1983.

[Minsky 88] M L Minsky. *Perceptrons: An Introduction to Computational Geometry.* MIT Press, Cambridge, MA, 2nd edition, 1988.

[Mitchell 96] M Mitchell. *An Introduction to Genetic Algorithms.* MIT Press, Cambridge, MA, 1996.

[Mohring 91] R H Mohring, editor. *Graph-Theoretic Concepts in Computer Science—16th WG'90*, Berlin–New York–Tokyo, 1991. Springer Verlag.

[Mozer 91] M C Mozer. *The Perception of Multiple Objects: A Connectionist Approach.* MIT Press, Cambridge, MA, 1991.

[Nagl 90] M Nagl, editor. *Graph-Theoretic Concepts in Computer Science—15th WG'89*, Berlin–New York–Tokyo, 1990. Springer Verlag.

[Niemann 90] H Niemann. *Pattern Analysis and Understanding.* Springer Verlag, Berlin–New York–Tokyo, 2nd edition, 1990.

[Nigrin 93] A Nigrin. *Neural Networks for Pattern Recognition.* MIT Press, Cambridge, MA, 1993.

[Nilsson 71] N J Nilsson. *Problem Solving Methods in Artificial Intelligence.* McGraw-Hill, New York, 1971.

[Nilsson 82] N J Nilsson. *Principles of Artificial Intelligence.* Springer Verlag, Berlin, 1982.

[Oja 83] E Oja. *Subspace Methods of Pattern Recognition.* Research Studies Press, Letchworth, England, 1983.

[Otten and van Ginneken 89] R H J M Otten and L P P P van Ginneken. *The Annealing Algorithm.* Kluwer, Norwell, MA, 1989.

[Pal 91] S K Pal. Fuzzy tools for the management of uncertainty in pattern recognition, image analysis, vision, and expert systems. *International Journal of System Science*, 22:511–548, 1991.

[Pal and Majumder 86] S K Pal and D D Majumder. *Fuzzy Mathematical Approach to Pattern Recognition.* Wiley, New York, 1986.

[Park 96] W Park. Automated determination of fuzzy rules and membership functions: Application to analysis of pulmonary CT images. PhD thesis, University of Iowa, 1996.

[Park et al. 96] W Park, E A Hoffman, and M Sonka. Fuzzy logic approach to extraction of intrathoracic airway trees from three-dimensional CT images. In *Image Processing, Proceedings SPIE Vol. 2710*, pages 210–219, SPIE, Bellingham, WA, 1996.

[Patrick and Fattu 86] E A Patrick and J M Fattu. *Artificial Intelligence with Statistical Pattern Recognition.* Prentice-Hall, Englewood Cliffs, NJ, 1986.

[Pavel 93] M Pavel. *Fundamentals of Pattern Recognition.* Marcel Dekker, New York, 2nd edition, 1993.

[Pavlidis 77] T Pavlidis. *Structural Pattern Recognition.* Springer Verlag, Berlin, 1977.

[Pavlidis 80] T Pavlidis. Structural descriptions and graph grammars. In S K Chang and K S Fu, editors, *Pictorial Information Systems*, pages 86–103, Springer Verlag, Berlin, 1980.

[Pedrycz 95] W Pedrycz. *Fuzzy Sets Engineering.* CRC Press, Boca Raton, FL, 1995.

[Pospesel 76] H Pospesel. *Predicate Logic.* Prentice-Hall, Englewood Cliffs, NJ, 1976.

[Pudil et al. 94a] P Pudil, J Novovicova, and J Kittler. Floating search methods in feature selection. *Pattern Recognition Letters*, 15:1119–1125, 1994.

[Pudil et al. 94b] P Pudil, J Novovicova, and J Kittler. Simultaneous learning of decision rules and important atrtributes for classification problems in image analysis. *Image and Vision Computing*, 12:193–198, 1994.

[Rao 65] C R Rao. *Linear Statistical Inference and Its Application.* Wiley, New York, 1965.

[Rawlins 91] G J E Rawlins. *Foundations of Genetic Algorithms.* Morgan Kaufmann, San Mateo, CA, 1991.

[Reichgelt 91] H Reichgelt. *Knowledge Representation: An AI Perspective.* Ablex, Norwood, NJ, 1991.

[Rogers and Kabrisky 91] S K Rogers and M Kabrisky. *An Introduction to Biological and Artificial Neural Networks for Pattern Recognition.* SPIE, Bellingham, WA, 1991.

[Romesburg 84] H C Romesburg. *Cluster Analysis for Researchers.* Lifetime Learning Publications, Belmont, CA, 1984.

[Rosenblatt 62] R Rosenblatt. *Principles of Neurodynamics.* Spartan Books, Washington, DC, 1962.

[Rosenfeld 79a] A Rosenfeld. Fuzzy digital topology. *Information Control*, 40:76–87, 1979.

[Rosenfeld 79b] A Rosenfeld. *Picture Languages—Formal Models for Picture Recognition.* Academic Press, New York, 1979.

[Rosenfeld 83] A Rosenfeld. On connectivity properties of grayscale pictures. *Information Control*, 16:47–50, 1983.

[Rosenfeld 84a] A Rosenfeld. The diameter of a fuzzy set. *Fuzzy Sets and Systems*, 13:241–246, 1984.

[Rosenfeld 84b] A Rosenfeld. The fuzzy geometry of image subsets. *Pattern Recognition Letters*, 2:311–317, 1984.

[Rosenfeld 85] A Rosenfeld. The perimeter of a fuzzy set. *Pattern Recognition*, 18:125–130, 1985.

[Rumelhart and McClelland 86] D Rumelhart and J McClelland. *Parallel Distributed Processing.* MIT Press, Cambridge, MA, 1986.

[Schalkoff 92] R J Schalkoff. *Pattern Recognition: Statistical, Structural and Neural Approaches.* Wiley, New York, 1992.

[Schutzer 87] D Schutzer. *Artificial Intelligence, An Application-Oriented Approach.* Van Nostrand Reinhold, New York, 1987.

[Sedgewick 84] R Sedgewick. *Algorithms.* Addison-Wesley, Reading, MA, 2nd edition, 1984.

[Sejnowski and Rosenberg 87] T J Sejnowski and C R Rosenberg. Parallel systems that learn to pronounce English text. *Complex Systems*, 1:145–168, 1987.

[Sethi and Jain 91] I K Sethi and A K Jain, editors. *Artificial Neural Networks and Statistical Pattern Recognition: Old and New Connections.* North-Holland, Amsterdam; New York, 1991.

[Shapiro and Haralick 80] L G Shapiro and R M Haralick. Algorithms for inexact matching. In *5th International Conference on Pattern Recognition,* Los Alamitos, CA, pages 202–207, IEEE, Piscataway, NJ, 1980.

[Sharples et al. 89] M Sharples, D Hogg, C Hutchinson, S Torrance, and D Young. *Computers and Thought, A Practical Introduction to Artificial Intelligence.* MIT Press, Cambridge, MA, 1989.

[Simons 84] G L Simons. *Introducing Artificial Intelligence.* NCC Publications, Manchester, England, 1984.

[Simpson 90] P K Simpson. *Artificial Neural Systems: Foundations, Paradigms, Applications, and Implementations.* Pergamon Press, New York, 1990.

[Simpson 92] P K Simpson. Fuzzy min-max neural networks—Part 1: Classification. *IEEE Transactions on Fuzzy Systems*, 3(2):129–139, 1992.

[Sklansky 81] J Sklansky. *Pattern Classifiers and Trainable Machines.* Springer Verlag, New York, 1981.

[Sonka 86] M Sonka. A new texture recognition method. *Computers and Artificial Intelligence*, 5(4):357–364, 1986.

[Tan et al. 92] H K Tan, S B Gelfand, and E J Delp. A cost minimization approach to edge detection using simulated annealing. *IEEE Transactions on Pattern Analysis and Machine Intelligence*, 14(1), 1992.

[Tucker 95] A Tucker. *Applied Combinatorics.* Wiley, New York, 3rd edition, 1995.

[Udupa and Samarasekera 96] J K Udupa and S Samarasekera. Fuzzy connectedness and object definition: Theory, algorithms, and applications in image segmentation. *Graphical Models and Image Processing*, 58:246–261, 1996.

[Ullmann 76] J R Ullmann. An algorithm for subgraph isomorphism. *Journal of the Association for Computing Machinery*, 23(1):31–42, 1976.

[van Laarhoven 88] P J M van Laarhoven. *Theoretical and Computational Aspects of Simulated Annealing.* Centrum voor Wiskunde en Informatik, Amsterdam, 1988.

[van Laarhoven and Aarts 87] P J M van Laarhoven and E H L Aarts. *Simulated Annealing: Theory and Applications.* Kluwer and Dordrecht, Norwell, MA, 1987.

[Wasserman 89] P D Wasserman. *Neural Computing—Theory and Practice.* Van Nostrand Rheinhold, New York, 1989.

[Wechsler 90] H Wechsler. *Computational Vision.* Academic Press, London–San Diego, 1990.

[Winston 75] P H Winston, editor. *The Psychology of Computer Vision.* McGraw-Hill, New York, 1975.

[Winston 84] P H Winston. *Artificial Intelligence.* Addison-Wesley, Reading, MA, 2nd edition, 1984.

[Yang et al. 89] B Yang, W E Snyder, and G L Bilbro. Matching oversegmented 3D images to models using association graphs. *Image and Vision Computing*, 7(2):135–143, 1989.

[Young and Calvert 74] T Y Young and T W Calvert. *Classification, Estimation, and Pattern Recognition.* American Elsevier, New York–London–Amsterdam, 1974.

[Zadeh 65] L A Zadeh. Fuzzy sets. *Information and Control*, 8:338–353, 1965.

[Zadeh and Kacprzyk 92] L A Zadeh and J Kacprzyk, editors. *Fuzzy Logic for the Management of Uncertainty.* Wiley, New York, 1992.

[Zdrahal 81] Z Zdrahal. A structural method of scene analysis. In *7th International Joint Conference on Artificial Intelligence,* Vancouver, Canada, pages 680–682, 1981.

[Zhou 92] Y T Zhou. *Artificial Neural Networks for Computer Vision.* Springer Verlag, New York, 1992.

[Zimmermann 87] H J Zimmermann. *Fuzzy sets, decision making and expert systems.* Kluwer, Boston, 1987.

[Zimmermann 91] H J Zimmermann. *Fuzzy Set Theory and Its Applications.* Kluwer, Boston, 1991.

[Zimmermann et al. 84] H J Zimmermann, L A Zadeh, and B R Gaines. *Fuzzy Sets and Decision Analysis.* North Holland, Amsterdam–New York, 1984.

Chapter 8

Image understanding

Image understanding requires mutual interaction of processing steps. The necessary building blocks for image understanding have been presented in earlier chapters—now an internal image model must be built that represents the machine vision system's concept about the processed image of the world.

Consider a typical human approach: A human being is well prepared to do image processing, analysis, and understanding. Despite this fact, it may sometimes be difficult to recognize what is seen if what to expect is not known. If a microscopic image of some tissue is presented to an observer who has never had a chance to study tissue structure or morphology, the question of the location of diseased tissue may be unanswerable. A similar problem can result if an observer is required to understand an aerial or satellite image of some urban area, even if the data correspond to a city with which the observer is familiar. Further, we can require the observer to watch the scene on a 'per part' basis—like using a telescope; this is an approach similar to a machine vision system's abilities. If a human observer solves the problem of orientation in such a scene, probably a start is made by trying to locate some known object. The observer constructs an image model of the city starting with the object believed to be recognized. Consider an aerial city view (see the simplified map of Prague, Figure 8.1), and suppose our observer sees two Gothic towers. They may be the towers of Prague castle, of the Vysehrad castle, or of some other Gothic churches. Let our observer begin with a hypothesis that the towers belong to the Vysehrad castle; a model of Vysehrad consists of the adjacent park, closely located river, etc. The observer attempts to verify the hypothesis with the model—Does the model match the reality? If it matches, the hypothesis is supported; if it does not, the hypothesis is weakened and finally rejected. The observer constructs a new hypothesis describing the scene, builds another model, and again tries to verify it. Two main forms of knowledge are used when the internal model is constructed—the general knowledge of placement of streets, houses, parks, etc., in cities, and specific knowledge of the order of specific houses, streets, rivers, etc., in the specific city.

A machine vision system can be asked to solve similar problems. The main difference between a human observer and an artificial vision system is in a lack of widely applicable, general, and modifiable knowledge of the real world in the latter. Machine vision systems construct internal models of the processed scene, verify them, and update them, and an appropriate sequence of processing steps must be performed to fulfill the given task. If the internal model matches the reality, image understanding is achieved. On the other hand,

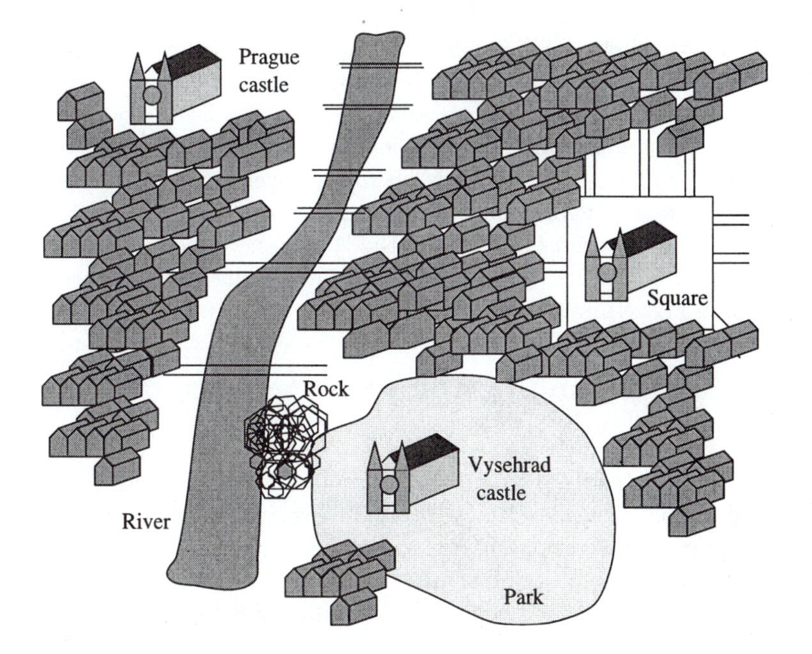

Figure 8.1: *Simulated orientation problem.*

the example described above showed that existence of an image model is a prerequisite for perception; there is no inconsistency in this. The image representation has an incremental character; new data or perceptions are compared with an existing model, and are used for model modification. Image data interpretation is not explicitly dependent on image data alone. The variations in starting models, as well as differences in previous experience, cause the data to be interpreted differently, even if always consistently with the constructed model; any final interpretation can be considered correct if just a match between a model and image data is evaluated [Levine 78, Zucker 78, Rosenfeld 79, Tsosos 84, Li and Uhr 87, Basu 87, Mulder 88, Niemann 90, Grimson 90].

We have said that machine vision consists of lower and upper processing levels, and image understanding is the highest processing level in this classification. The main task of this processing level is to define control strategies that ensure an appropriate sequence of processing steps. Moreover, a machine vision system must be able to deal with a large number of interpretations that are hypothetical and ambiguous. Generally viewed, the organization of the machine vision system consists of a weak hierarchical structure of image models.

Many important results have been achieved in image understanding in recent years. Despite that, the image understanding process remains an open area of computer vision and is under continued investigation. Image understanding is one of the most complex challenges of AI, and to cover this complicated area of computer vision in detail it would be necessary to discuss relatively independent branches of AI—knowledge representation, relational structures, semantic networks, general matching, inference, production systems, problem solving, planning, control, feedback, and learning from experience, a difficult and not fully understood area. These areas are used and described in various AI references [Bajcsy and Rosenthal

80, Nilsson 82, Michalski et al. 83, Simons 84, Devijver and Kittler 86, Wechsler 90, Reichgelt 91, Marik et al. 92, Winston 92] and their application to computer vision is an active area of research. Nevertheless, to cover these topics in detail exceeds the frame of this book; therefore, we present here an overview of basic image understanding control strategies and describe contextual and semantic methods of image understanding. Image understanding control is a crucial problem in machine vision, and the control strategies described give a better rationale for the application of various methods of image processing, object description, and recognition described earlier. At the same time, it explains why the specific AI methods are incorporated in image understanding processes.

8.1 Image understanding control strategies

Image understanding can be achieved only as a result of cooperation of complex information processing tasks and appropriate control of these tasks. Biological systems include a very complicated and complex control strategy incorporating parallel processing, dynamic sensing sub-system allocation, behavior modifications, interrupt-driven shifts of attention, etc. As in other AI problems, the main goal of computer vision is to achieve machine behavior similar to that of biological systems by applying technically available procedures.

8.1.1 Parallel and serial processing control

Both parallel and serial approaches can be applied to image processing, although sometimes it is not obvious which steps should be processed in parallel and which serially. Parallel processing makes several computations simultaneously (e.g., several image parts can be processed simultaneously), and an extremely important consideration is the synchronization of processing actions—that is, the decision of when, or if, the processing should wait for other processing steps to be completed [Ghosh and Harrison 90, Prasanna Kumar 91, Fischer and Niemann 93, Hwang and Wang 94].

Operations are always sequential in serial processing. A serial control strategy is natural for conventional von Neumann computer architectures, and the large numbers of operations that biological organisms process in parallel often cannot be done serially at the required speed. Pyramid image representations, and corresponding pyramid processor architectures, resulted from speed requirements (including implementation of cognitive processes in lower processing levels, etc.). Parallel computers have become generally available, and despite substantial difficulties with their programming, the parallel processing option is now a reality. The feasibility of parallel processing implementation of various approaches and algorithms has been mentioned throughout this book, and it has been made clear that almost all low-level image processing can be done in parallel. However, high-level processing using higher levels of abstraction is usually serial in essence. There is an obvious comparison with the human strategy of solving complex sensing problems: A human always concentrates on a single topic during later phases of vision, even if the early steps are done in parallel.

8.1.2 Hierarchical control

Image information is stored in different representations during processing. There is one crucial question related to processing control: Should the processing be controlled by the image data information or by higher-level knowledge? These different approaches can be described as follows.

1. **Control by the image data (bottom-up control)**: Processing proceeds from the raster image to segmented image, to region (object) description, and to their recognition.

2. **Model-based control (top-down control)**: A set of assumptions and expected properties is constructed from applicable knowledge. The satisfaction of those properties is tested in image representations at different processing levels in a top-down direction, down to the original image data. The image understanding is an internal model verification, and the model is either accepted or rejected.

The two basic control strategies do not differ in the types of operation applied, but do differ in the sequence of their application, in the application either to all image data or just to selected image data, etc. The control mechanism chosen is not only a route to the processing goal, it influences the whole control strategy. Neither top-down nor bottom-up control strategies can explain the vision process or solve complex vision sensing problems in their standard forms. However, their appropriate combination can yield a more flexible and powerful vision control strategy.

8.1.3 Bottom-up control strategies

A general bottom-up algorithm is

Algorithm 8.1: Bottom-up control

1. Pre-processing: Transform the raster image data (pre-process the image) to highlight information that may be useful in further processing steps. Appropriate transformations are applied throughout the image.

2. Segmentation: Detect and segment image regions that can correspond to real objects or object parts.

3. Understanding: If region descriptions were not used in step 2, determine an appropriate description for regions found in the segmented image. Compare the detected objects with real objects that are present in the solution domain (i.e., using pattern recognition techniques).

It is obvious that the bottom-up control strategy is based on the construction of data structures for the processing steps that follow. Note that each algorithm step can consist of several substeps; however, the image representation remains unchanged in the substeps. The bottom-up control strategy is advantageous if a simple and efficient processing method is available that is independent of the image data content. Bottom-up control yields good results if unambiguous data are processed and if the processing gives reliable and precise representations for later processing steps. The recognition of well-illuminated objects in robotic applications is an example—in this case, bottom-up control results in fast and reliable processing. If the input data are of low quality, bottom-up control can yield good results only if unreliability of the data causes just a limited number of insubstantial errors in each processing step. This implies that the main image understanding role must be played by a control strategy that is not only a concatenation of processing operations in the bottom-up direction, but that also uses an internal model goal specifications, planning, and complex cognitive processes.

A good example of a bottom-up control strategy is Marr's image understanding approach [Marr 82]. The processing begins with a two-dimensional intensity image and tries to achieve a three-dimensional image understanding through a sequence of intermediate image representations. Marr's understanding strategy is based on a pure bottom-up data flow using only very general assumptions about the objects to be identified—a more detailed description of this approach is given in Section 9.1.1.

8.1.4 Model-based control strategies

There is no general form of top-down control as was presented in the bottom-up control algorithm. The main top-down control principle is the construction of an internal model and its verification, meaning that the main principle is **goal-oriented processing**. Goals at higher processing levels are split into sub-goals at lower processing levels, which are split again into sub-goals etc., until the sub-goals can be either accepted or rejected directly.

An example will illustrate this principle. Imagine that you are in a large hotel, and your spouse parked your white Ford Escort somewhere in the large car park in front of the hotel. You are trying to find your car, looking from the hotel room window. The first-level goal is to find the car park. A sub-goal might be to detect all white cars in the car park and to decide which of those white cars are Ford Escorts. All the given goals can be fulfilled by looking from the window and using general models (general knowledge) of cars, colors, and Escorts.

If all the former goals are fulfilled, the last goal is to decide if the detected white Ford Escort really is your car and not some other white Escort; to satisfy this goal, specific knowledge of your car is necessary. You have to know what makes your car special—the differences between your car and others. If the test of the specific properties of the detected car is successful, the car is accepted as yours; the model you built for your white Escort is accepted, the car is located, and the search is over. If the test of specific properties is not successful, you have to resume testing at some higher level, for instance, to detect another, as yet untested white Ford Escort.

The general mechanism of top-down control is hypothesis generation and its testing. The internal model generator predicts what a specific part of the model must look like in lower image representations. The image understanding process consists of sequential hypothesis generation and testing. The internal model is updated during the processing according to the results of the hypothesis tests. The hypothesis testing relies on a (relatively small) amount of information acquired from lower representation levels, and the processing control is based on the fact that just the necessary image processing is required to test each hypothesis. The model-based control strategy (top-down, hypothesize and verify) seems to be a way of solving computer vision tasks by avoiding brute-force processing; at the same time, it does not mean that parallel processing should not be applied whenever possible.

Not surprisingly, real-world models play a substantial role in model vision. Many approaches presented throughout this book may be considered either models of a part of an image or object models. However, to represent a variety of real-world domains, to be able to model complex image objects, their physical properties must be included in the representation. This is especially true in modeling natural objects—human faces together with their mimics serve as a good example. Physical modeling is a fast-developing branch of computer vision and image understanding [Kanade and Ikeuchi 91] in which four main techniques appear: reflection models for vision, relations between shape and reflection, statistical and stochastic modeling, and modeling deformable shapes (**elastics in vision**). Clearly, all these techniques may significantly increase the knowledge available in the image understanding process. From the point of view of the context being discussed here, deformable models of non-rigid objects seem to widen substantially the rank of feasible applications.

Elastics in vision [Witkin et al. 87, Terzopoulos and Fleischer 88, Terzopoulos 91] represent deformable bodies mostly using **dynamically moving splines—snakes** [Kass et al. 87a] and **superquadrics** [Terzopoulos and Metaxas 91, Metaxas and Terzopoulos 91] to fit complex three-dimensional shapes, with a potential for sub-pixel accuracy [Zucker 88]. Snakes and balloons as active contour models are discussed in Section 8.2.

8.1.5 Combined control strategies

Combined control mechanisms that use both data- and model-driven control strategies are widely used in modern vision applications, and usually give better results than any of the previously discussed, separately applied, basic control strategies. Higher-level information is used to make the lower-level processing easier, but alone is insufficient to solve the task. Seeking cars in aerial or satellite image data is a good example; data-driven control is necessary to find the cars, but at the same time, higher-level knowledge can be used to simplify the problem since cars appear as rectangular objects of specific size, and the highest probability of their appearance is on roads.

An example of a robust approach to automated coronary border detection in angiographic images illustrates the combined control strategy [Collins et al. 91, Sonka et al. 93, Sonka et al. 95]. X-ray images are acquired after injecting a radio-opaque dye into the arteries of a human heart. An example of a successful bottom-up detection of coronary borders using a graph search approach is given in Section 5.2.4, Figure 5.26.

Unfortunately, the bottom-up graph search often fails in more complicated images, in the presence of closely parallel, branching, or overlapping vessels, and in low-quality images. Image data representing such a difficult case are shown in Figure 8.2 together with the result of the bottom-up graph search (the same method that worked so well for a single-vessel case). To achieve a reliable border detection in difficult images, a hybrid control strategy was designed combining bottom-up and top-down control steps; the following principles are incorporated in the process (see Section 16.2).

1. **Model-based approach:** The model favors symmetric left and right borders as those most typical in coronary imagery.

2. **Hypothesize and verify approach:** Based on multi-resolution processing, the approximate vessel border is detected at low resolution and the precision is increased at full resolution (also, multi-resolution speeds up the border detection process).

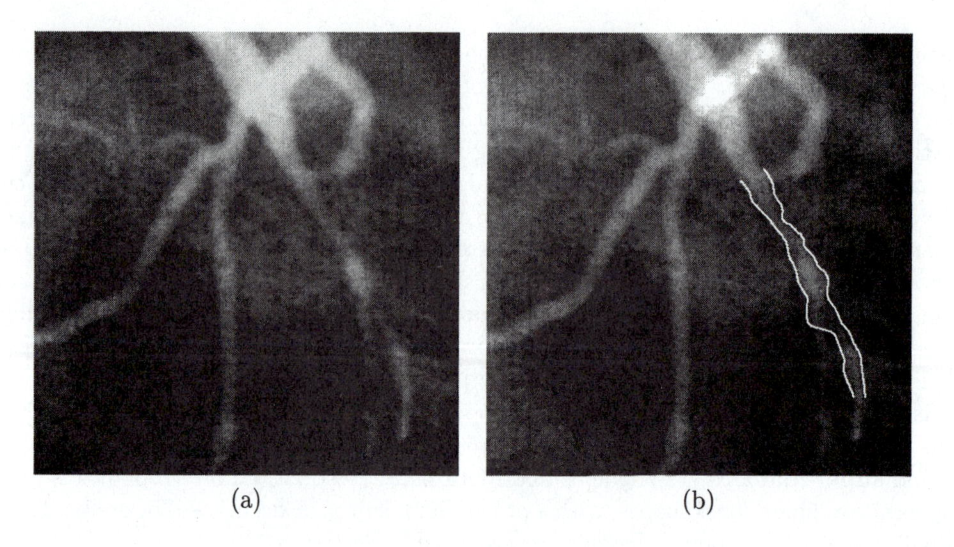

Figure 8.2: *Coronary angiogram: (a) original X-ray image; (b) borders detected by a bottom-up graph search approach.*

3. **A priori knowledge:** Knowledge about directions of edges forming the vessel border is used to modify a graph search cost function.

4. **Multi-stage approach:** Models of different strength are applied throughout the processing.

The method searches for left and right coronary borders simultaneously, performing a three-dimensional graph search and the border symmetry model is thus incorporated in the search process. The three-dimensional graph results from combining two conventional edge detection graphs of the left and the right coronary borders (see Sections 5.2.4, 5.5.1). The model guides the search in regions of poor data, and where the image data have an acceptable quality, the search is guided by the image data.

A frequent problem of model-based control strategies is that the model control necessary in some parts of the image is too strong in other parts (the symmetry requirements of the model have a larger influence on the final border than a non-symmetric reality), corrupting the border detection results. This is the rationale for a multi-stage approach where a strong model is applied at low resolution, and a weaker model leaves enough freedom for the search to be guided predominantly by image data at full-resolution, thereby achieving higher overall accuracy. Nevertheless, the low-resolution coronary borders detected by cooperation with the model guarantee that the full-resolution search will not get lost—the low-resolution border is used as a model border in the full-resolution search.

A block algorithm of the control steps is now given, accompanied by a label showing whether the particular step is done in a bottom-up or top-down manner.

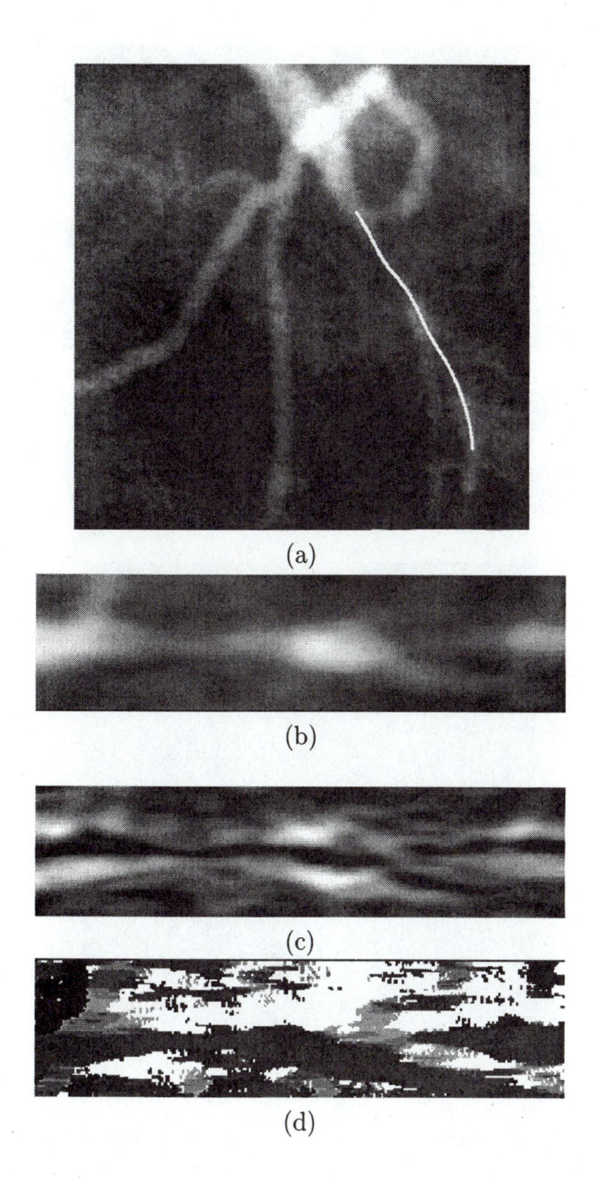

(a)

(b)

(c)

(d)

Figure 8.3: *Steps of coronary border detection (I): (a) centerline definition; (b) straightened image data; (c) edge detection; (d) edge direction detection.*

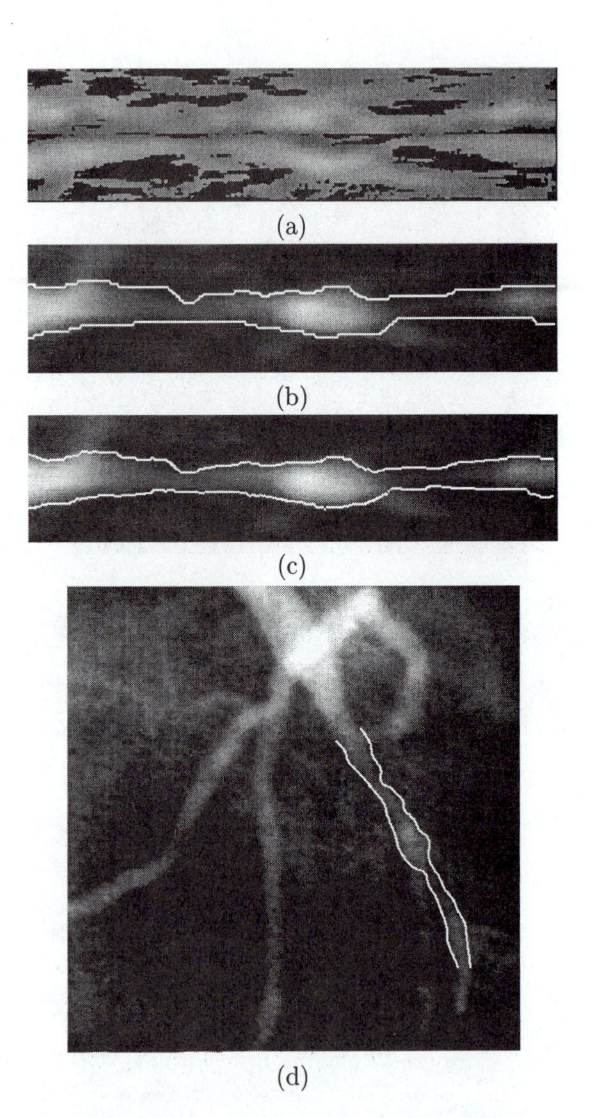

Figure 8.4: *Steps of coronary border detection (II): (a) modified cost function—note that the cost increases in non-probable border locations image areas, where edge direction does not support location of the border; (b) approximate coronary borders acquired in low resolution; (c) precise full-resolution border in straightened image; (d) full-resolution coronary borders in original image data.*

Algorithm 8.2: Coronary border detection—a combined control strategy

1. *(top-down)* Detect a vessel centerline in interaction with an operator (show which vessel is to be processed), and straighten the vessel image, Figure 8.3a,b.

2. *(bottom-up)* Detect image edges in full resolution, Figure 8.3c.

3. *(bottom-up)* Detect local edge directions in the straightened intensity image, Figure 8.3d.

4. *(top-down)* Modify the cost matrix using a priori knowledge about edge directions and the directional edge image, Figure 8.4a.

5. *(bottom-up)* Construct a low-resolution image and a low-resolution cost matrix.

6. *(top-down)* Search for the low-resolution pair of approximate borders using the vessel symmetry model Figure 8.4b.

7. *(top-down)* Find an accurate position of the full-resolution border using the low-resolution border as a model to guide the full-resolution search, Figure 8.4c. The symmetry model is much weaker than in the low-resolution search.

8. *(bottom-up)* Transform the results from the straightened image to the original image, Figure 8.4d.

9. *(top-down)* Evaluate the coronary disease severity.

Results of this strategy applied to coronary vessel data are given in Figure 8.5.

It is obvious that a combined control strategy can improve processing efficiency. Further, some of the steps are not sequential in principle (such as edge image construction) and can be computed in parallel.

8.1.6 Non-hierarchical control

There is always an upper and a lower level in hierarchical control. Conversely, non-hierarchical control can be seen as a cooperation of competing experts at the same level.

Non-hierarchical control can be applied to problems that can be separated into a number of sub-problems, each of which requires some expertise. The order in which the expertise should be deployed is not fixed. The basic idea of non-hierarchical control is to ask for assistance from the expert that can help most to obtain the final solution. The chosen expert may be known, for instance, for high reliability, high efficiency, or for the ability to provide the most information under given conditions. Criteria for selection of an expert from the set may differ; one possibility is to let the experts calculate their own abilities to contribute to the solution in particular cases—the choice is based on these local and individual evaluations. Another option is to assign a fixed evaluation to each expert beforehand and help is then requested from the expert with the highest evaluation under given conditions [Ambler 75].

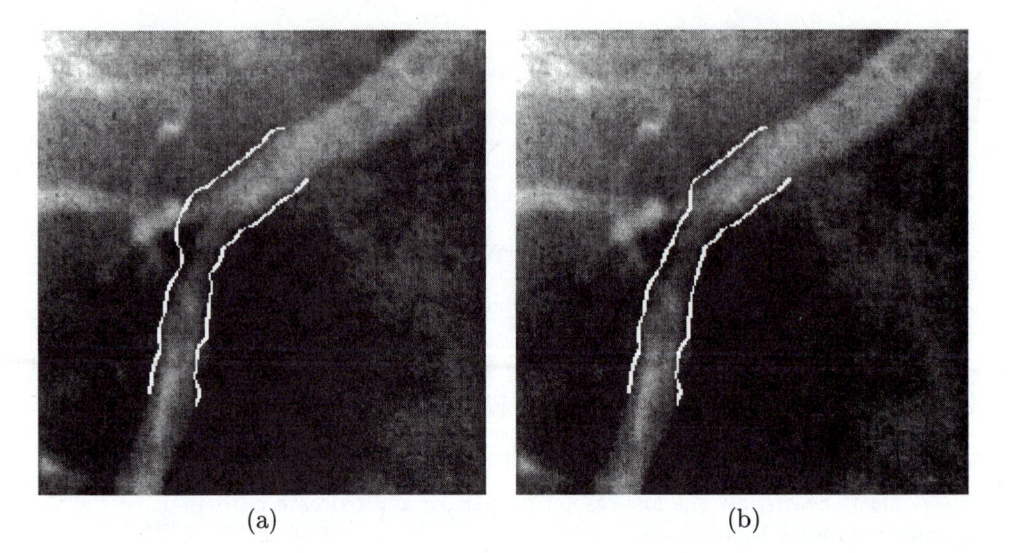

Figure 8.5: *Coronary border detection: (a) borders resulting from the pure bottom-up graph search approach follow borders of the vessel branch; (b) results of the combined control graph search strategy follow the coronary borders correctly.*

The criterion for expert choice may be based on some appropriate combination of empirically detected evaluations computed by experts, and evaluations dependent on the actual state of the problem solution. Non-hierarchical control strategies can be illustrated by the following algorithm outline.

Algorithm 8.3: Non-hierarchical control

1. Based on the actual state and acquired information about the solved problem, decide on the best action, and execute it.

2. Use the results of the last action to increase the amount of acquired information about the problem.

3. If the goals of the task are met, stop. Otherwise, return to step 1.

A system for analysis of complex aerial photographs [Nagao and Matsuyama 80] is an example of a successful application of non-hierarchical control—the blackboard principle was used for competing experts. To explain the main idea of a blackboard, imagine a classroom full of experts. If any of them wants to share knowledge or observations with others, a note is made on the blackboard. Therefore, all others can see the results and use them. A blackboard is a specific data structure that can be accessed by all the experts, and is a data structure first used in speech recognition—computer vision applications followed (e.g., VISIONS [Hanson and Riseman 78], COBIUS [Kuan et al. 89]). The blackboard usually includes a mechanism that retrieves specialized sub-systems which can immediately affect the standard control. These sub-systems are very powerful and are called **daemons**. The blackboard must include a mechanism that synchronizes the daemon activity. Programming with daemons is not easy, and the design of daemon behavior is based on general knowledge of the problem domain. Therefore, the programmer can never be absolutely sure if the daemon procedure based on some specific property will be activated or not; moreover, there is no guarantee that the daemon will be activated in the correct way. To limit the uncertainty of daemon behavior, the following additional rules are usually added.

- The blackboard represents a continuously updated part of the internal model that corresponds to image data.
- The blackboard includes a set of rules that specify which daemon sub-system should be used in specific cases.

The blackboard is sometimes called the **short-term memory**—it contains information about interpretation of the processed image. The **long-term memory**, the knowledge base, consists of more general information that is valid for (almost) all representations of the problems to be solved [Hanson and Riseman 78].

In a system for the analysis of complex aerial photographs [Nagao and Matsuyama 80], all the information about a specific image is stored in the blackboard (segmented region properties and their relations). The blackboard can activate 13 sub-systems of region detection, all of which communicate with the blackboard in a standard way, and the only way the sub-systems can communicate with each other is via the blackboard. The blackboard data structure depends on the application; in this particular case, the structure takes advantage of a priori global knowledge of the domain such as the physical size of pixels, the direction of the sun, etc. Additionally, the blackboard maintains a property table in which all the observations on image regions is stored, together with the information about the region class (resulting from recognition). An integral part of the blackboard is represented by the symbolic region image that provides information about relations between regions.

The primary aim of the blackboard system is to identify places of interest in the image that should be processed with higher accuracy, to locate places with a high probability of a target region being present. The approximate region borders are found first, based on a fast computation of just a few basic characteristics—saving computational time and making the detailed analysis easier. The control process follows the **production system** principle [Nilsson 82], using the information that comes from the region detection sub-systems via the blackboard. The blackboard serves as a place where all the conflicts between region labeling are solved (one region can be marked by two or more region detection sub-systems at the same time and it is necessary to decide which label is the best one). Furthermore, the labeling errors are detected in the blackboard, and are corrected using back-tracking principles.

The principal image understanding control strategies have been presented here—it was noted that a wide variety of knowledge representation techniques, object description methods, and processing strategies must co-exist in any image understanding system. The role of knowledge and control is reviewed in [Rao and Jain 88] within a frame of image and speech understanding systems such as ACRONYM [Brooks et al. 79], HEARSAY [Lesser et al. 75], and

VISIONS [Hanson and Riseman 78]. More recent progress in image understanding approaches generally following the bottom-up control strategy and allowing the use of semantic networks can be found in [Puliti and Tascini 93], knowledge-based composition of image interpretation processes is discussed in [Higgins et al. 94, Jurie and Gallice 95, Gong and Kulikowski 95], use of Bayesian networks for image interpretation is described in [Kumar and Desai 96], and machine learning strategies for image understanding are discussed critically in [Kodratoff and Moscatelli 94]. Further, neural networks and fuzzy logic are increasingly considered suitable vehicles for image interpretation [Zheng 95, Nakagawa and Hirota 95, Ralescu and Shanahan 95, Udupa and Samarasekera 96, Hata et al. 97].

8.2 Active contour models—snakes

The development of active contour models results from the work of Kass, Witkin, and Terzopoulos [Kass et al. 87a, Witkin et al. 87, Terzopoulos et al. 87], and they offer a solution to a variety of tasks in image analysis and machine vision. This section is based on the paper [Kass et al. 87b] in which the energy-minimization approach to achieve computer vision goals was first presented; the original notation is used.

Active contour models may be used in image segmentation and understanding, and are also suitable for analysis of dynamic image data or 3D image data. The active contour model, or **snake**, is defined as an energy-minimizing spline—the snake's energy depends on its shape and location within the image. Local minima of this energy then correspond to desired image properties. Snakes may be understood as a special case of a more general technique of matching a deformable model to an image by means of energy minimization. Snakes do not solve the entire problem of finding contours in images; rather, they depend on other mechanisms such as interaction with a user, interaction with some higher-level image understanding process, or information from image data adjacent in time or space. This interaction must specify an approximate shape and starting position for the snake somewhere near the desired contour. A priori information is then used to push the snake toward an appropriate solution—see Figures 8.6 and 8.7. Unlike most other image models, the snake is *active*, always minimizing its energy functional, therefore exhibiting dynamic behavior.

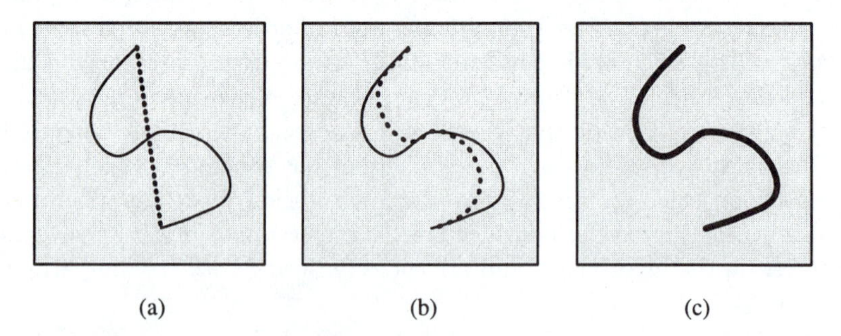

Figure 8.6: *Active contour model—snake: (a) initial snake position (dotted) defined interactively near the true contour; (b), (c) iteration steps of snake energy minimization: the snake is pulled toward the true contour.*

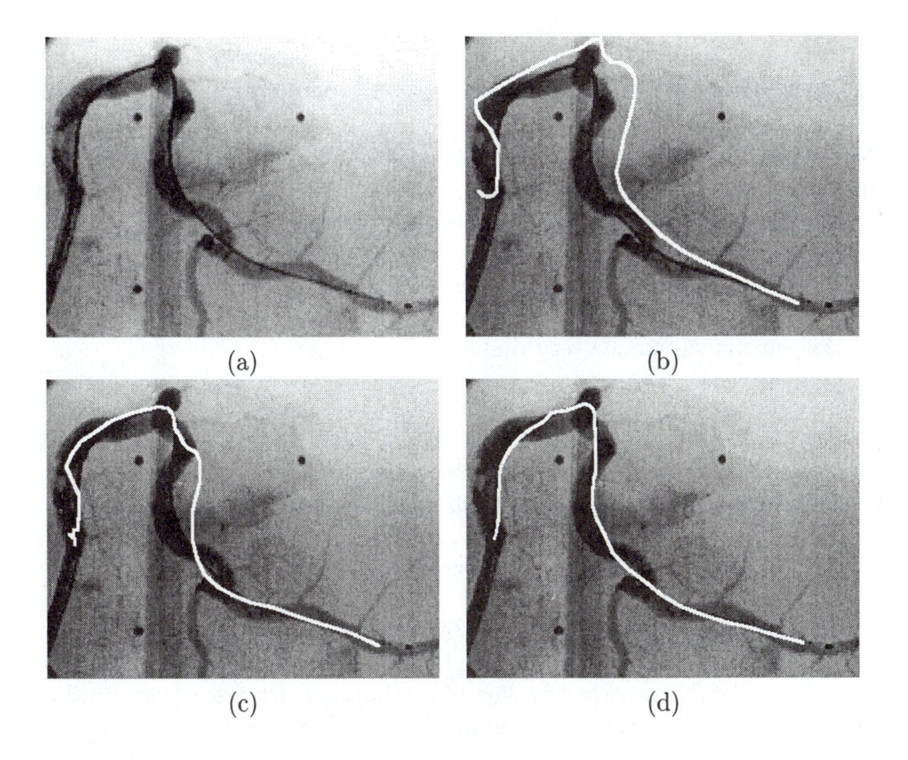

Figure 8.7: *Snake-based detection of the untravascular ultrasound catheter (dark line positioned inside the coronary artery lumen) in an angiographic X-ray image of a pig heart: (a) original angiogram; (b) initial position of the snake; (c) snake deformation after 4 iterations; (d) final position of the snake after 10 iterations.*

The energy functional which is minimized is a weighted combination of internal and external forces. The internal forces emanate from the shape of the snake, while the external forces come from the image and/or from higher-level image understanding processes. The snake is defined parametrically as $\mathbf{v}(s) = [x(s), y(s)]$, where $x(s), y(s)$ are x, y co-ordinates along the contour and $s \in [0, 1]$ (see Figure 6.11a). The energy functional to be minimized may be written as

$$\begin{aligned} E^*_{\text{snake}} &= \int_0^1 E_{\text{snake}}(\mathbf{v}(s))\, ds \\ &= \int_0^1 \{[E_{\text{int}}(\mathbf{v}(s)] + [E_{\text{image}}(\mathbf{v}(s)] + [E_{\text{con}}(\mathbf{v}(s)]\}\, ds \end{aligned} \tag{8.1}$$

where E_{int} represents the internal energy of the spline due to bending, E_{image} denotes image forces, and E_{con} external constraint forces. Usually, $\mathbf{v}(s)$ is approximated as a spline to ensure desirable properties of continuity.

The internal spline energy can be written

$$E_{\text{int}} = \alpha(s) |\frac{d\mathbf{v}}{ds}|^2 + \beta(s) |\frac{d^2\mathbf{v}}{ds^2}|^2 \tag{8.2}$$

where $\alpha(s), \beta(s)$ specify the *elasticity* and *stiffness* of the snake. Note that setting $\beta(s_k) = 0$ at a point s_k allows the snake to become second-order discontinuous at that point, and develop a corner.

The second term of the energy integral (8.1) is derived from the image data over which the snake lies. As an example, a weighted combination of three different functionals is presented which attracts the snake to lines, edges, and terminations:

$$E_{\text{image}} = w_{\text{line}} E_{\text{line}} + w_{\text{edge}} E_{\text{edge}} + w_{\text{term}} E_{\text{term}} \tag{8.3}$$

The line-based functional may be very simple

$$E_{\text{line}} = f(x, y) \tag{8.4}$$

where $f(x, y)$ denotes image gray-levels at image location (x, y). The sign of w_{line} specifies whether the snake is attracted to light or dark lines. The edge-based functional

$$E_{\text{edge}} = -|\text{grad} f(x, y)|^2 \tag{8.5}$$

attracts the snake to contours with large image gradients—that is, to locations of strong edges. Line terminations and corners may influence the snake using a weighted energy functional E_{term}: Let g be a slightly smoothed version of the image f, let $\psi(x, y)$ denote the gradient directions along the spline in the smoothed image g, and let

$$\mathbf{n}(x, y) = (\cos\psi(x, y), \sin\psi(x, y)) \quad \mathbf{n}_R(x, y) = (-\sin\psi(x, y), \cos\psi(x, y))$$

be unit vectors along and perpendicular to the gradient directions $\psi(x, y)$. Then the curvature of constant-gray-level contours in the smoothed image can be written as [Kass et al. 87a]

$$\begin{aligned} E_{\text{term}} &= \frac{\partial \psi}{\partial \mathbf{n}_R} = \frac{\partial^2 g / \partial \mathbf{n}_R^2}{\partial g / \partial \mathbf{n}} \\ &= \frac{(\partial^2 g/\partial y^2)(\partial g/\partial x)^2 - 2(\partial^2 g/\partial x \partial y)(\partial g/\partial x)(\partial g/\partial y) + (\partial^2 g/\partial x^2)(\partial g/\partial y)^2}{((\partial g/\partial x)^2 + (\partial g/\partial y)^2)^{3/2}} \end{aligned} \tag{8.6}$$

The snake behavior may be controlled by adjusting the weights $w_{\text{line}}, w_{\text{edge}}, w_{\text{term}}$. A snake attracted to edges and terminations is shown in Figure 8.8.

The third term of the integral (8.1) comes from external constraints imposed either by a user or some other higher-level process which may force the snake toward or away from particular features. If the snake is near to some desirable feature, the energy minimization will pull the snake the rest of the way. However, if the snake settles in a local energy minimum that a higher-level process determines as incorrect, an area of energy peak may be made at this location to force the snake away to a different local minimum.

A contour is defined to lie in the position in which the snake reaches a local energy minimum. From equation (8.1), the functional to be minimized is

$$E^*_{\text{snake}} = \int_0^1 E_{\text{snake}}[\mathbf{v}(s)] ds$$

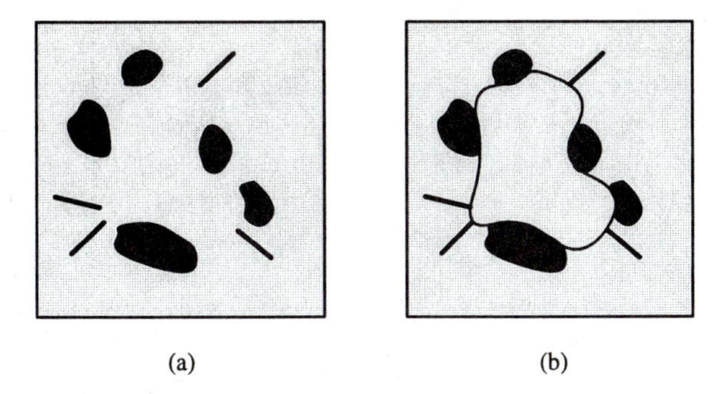

Figure 8.8: *A snake attracted to edges and terminations: (a) contour illusion; (b) a snake attracted to the subjective contour. Adapted from [Kass et al. 87b].*

Then, from the calculus of variations, the Euler-Lagrange condition states that the spline $\mathbf{v}(s)$ which minimizes E^*_{snake} must satisfy

$$\frac{d}{ds}E_{\mathbf{v}_s} - E_{\mathbf{v}} = 0 \tag{8.7}$$

where $E_{\mathbf{v}_s}$ is the partial derivative of E with respect to $d\mathbf{v}/ds$ and $E_{\mathbf{v}}$ is the partial derivative of E with respect to $\mathbf{v}$. Using equation (8.2) and denoting $E_{\text{ext}} = E_{\text{image}} + E_{\text{con}}$, the previous equation reduces to

$$-\frac{d}{ds}[\alpha(s)\frac{d\mathbf{v}}{ds}] + \frac{d^2}{ds^2}[\beta(s)\frac{d^2\mathbf{v}}{ds^2}] + \nabla E_{\text{ext}}[\mathbf{v}(s)] = 0 \tag{8.8}$$

To solve the Euler-Lagrange equation, suppose an initial estimate of the solution is available. An evolution equation is formed:

$$\frac{\partial\mathbf{v}(s,t)}{\partial t} - \frac{\partial}{\partial s}[\alpha(s)\frac{\partial\mathbf{v}(s,t)}{\partial s}] + \frac{\partial^2}{\partial s^2}[\beta(s)\frac{\partial^2\mathbf{v}(s,t)}{\partial s^2}] + \nabla E_{\text{ext}}[\mathbf{v}(s,t)] = 0 \tag{8.9}$$

The solution is found if $\partial\mathbf{v}(s,t)/\partial t = 0$. Nevertheless, minimization of the snake energy integral is still problematic; numerous parameters must be designed (weighting factors, iteration steps, etc.), a reasonable initialization must be available, and, moreover, the solution of the Euler-Lagrange equation suffers from numerical instability.

Originally, a resolution minimization method was proposed [Kass et al. 87a]; partial derivatives in s and t were estimated by the finite-differences method. Later [Amini et al. 88, Amini et al. 90], a dynamic programming approach was proposed which allows 'hard' constraints to be added to the snake. Further, a requirement that the internal snake energy must be a continuous function may thus be eliminated and some snake configurations may be prohibited (that is, have infinite energy), allowing more a priori knowledge to be incorporated.

Difficulties with the numerical instability of the original method were overcome by Berger [Berger and Mohr 90] by incorporating an idea of **snake growing**. In this method a single primary snake may begin which later divides itself into pieces. The pieces of very low energy are allowed to grow in directions of their tangents, while higher-energy pieces are eliminated. After each growing step, the energy of each snake piece is minimized (the ends are pulled to the true contour, see Figure 8.9) and the snake growing process is repeated. Further, the snake growing method may overcome the initialization problem. The primary snake may fall into an unlikely local minimum, but parts of the snake may still lie on salient features. The very-low-energy parts (the probable pieces) of the primary snake are used to initialize the snake growing in later steps. This iterative snake growing always converges and the numerical solution is therefore stable—the robustness of the method is paid for by an increase in the processing cost of the algorithm.

Algorithm 8.4: Snake growing

1. Based on a priori knowledge, estimate the desired contour position and shape as a curve S^0.
2. Use this curve S^0 to initialize the conventional snake algorithm [equations (8.1)—(8.9)]. This yields a contour C^0.
3. Segment the contour C^0 to eliminate high-energy segments, resulting in a number of initial shorter low-energy contour segments C_i^0, where i is a segment identifier.
4. Repeat steps 5 and 6 while lengthening is needed.
5. Each contour segment C_i^k is allowed to grow in the direction of tangents (see Figure 8.9). This yields a new estimate S_i^{k+1} for each contour segment C_i^k.
6. For each contour segment C_i^k, run the conventional snake algorithm using S_i^{k+1} as an initial estimate to get a new (longer) contour segment C_i^{k+1}.

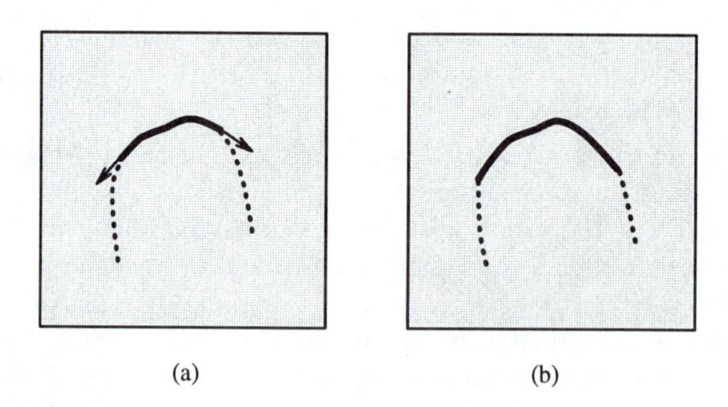

(a) (b)

Figure 8.9: *Snake growing: (a) lengthening in tangent direction; (b) energy minimization after a growing step.*

A different approach to the energy integral minimization that is based on a Galerkin solution of the finite-element method was proposed in [Cohen 91b] and has the advantage of greater numerical stability and better efficiency. This approach is especially useful in the case of closed or nearly closed contours. An additional pressure force is added to the contour interior by considering the curve as a **balloon** which is inflated. This allows the snake to overcome isolated energy valleys resulting from spurious edge points, giving better results all over (see Figures 8.10 and 8.11). Another approach using a finite-element method [Karaolani et al. 92] also significantly improves the solution's efficiency; forces are scaled by the size of an element, preventing very small contributions (which may be noise) from contributing to the global solution as much as longer elements.

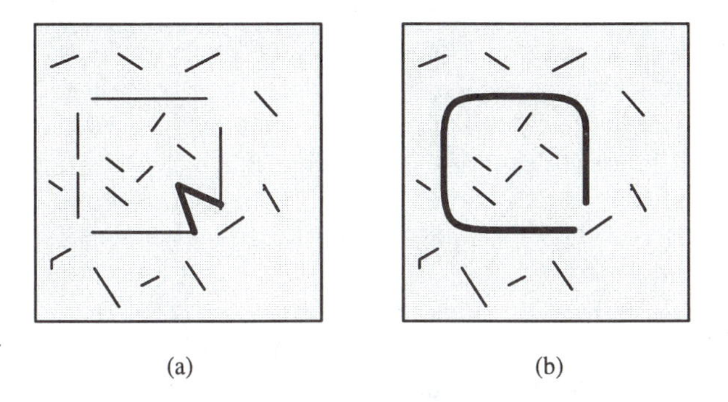

(a) (b)

Figure 8.10: *Active contour model—balloon: (a) initial contour; (b) final contour after inflation and energy minimization. Adapted from [Cohen and Cohen 92].*

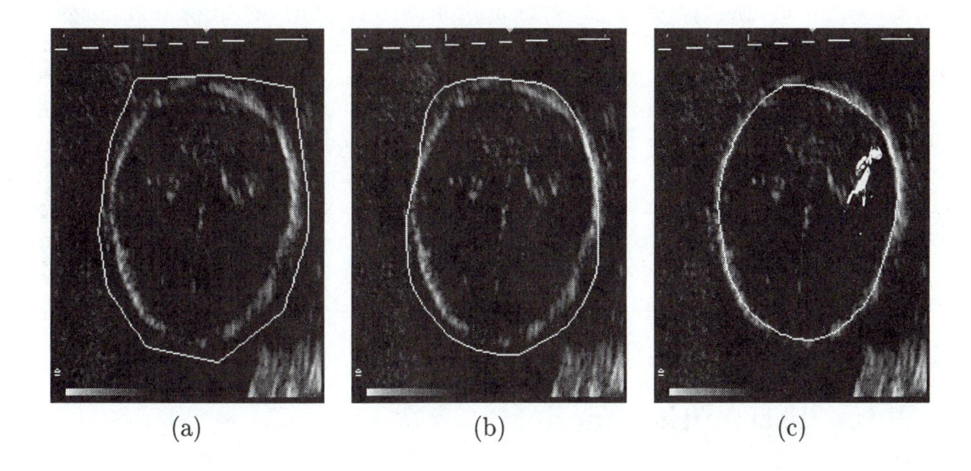

(a) (b) (c)

Figure 8.11: *B alloon-based image segmentation of an ultrasound image of a fetal head: (a) initial position of the balloon; (b) balloon deformation after 10 iterations, (c) final position of the balloon after 25 iterations. Courtesy V. Chalana, MathSoft, Seattle, WA.*

Deformable models based on active contours were generalized to three dimensions by Terzopoulos [Terzopoulos et al. 87, Terzopoulos et al. 88, McInerney and Terzopoulos 93], and 3D balloons were introduced in [Cohen and Cohen 92]; more recently [Bulpitt and Efford 96], a 3D generalization using an adaptive mesh that can refine and decimate has been successfully applied to modeling MR images of human heads and hands. In [Zhang and Braun 97], a fully three-dimensional active surface model with self-inflation and self-deflation forces has been introduced. Further, fast algorithms for active contour models are beginning to appear [Williams and Shah 92, Olstad and Tysdahl 93, Lam and Yan 94]. Several problems often accompany the snake-based approach: Snakes tend to be attracted to spurious edges, they sometimes degenerate in shape by shrinking and flattening, and the convergence and stability of the contour deformation process may be unpredictable. Additionally, while their initialization is not straightforward, it may affect their performance substantially. Many approaches have been designed to overcome these limitations, some of them using more complex knowledge and models [Radeva et al. 95, Lobregt and Viergever 95, Hoch and Litwinowicz 96] and employing neural networks [Tsai and Sun 92, Chiou and Jenq-Neng 95], adaptively modifying the number of control points in snakes represented by B-splines [Fugeiredo et al. 97], designing internal forces independent of contour shape [Xu et al. 94], determining external forces from edge images [Xu and Prince 98], limiting the region of interest by a pair of snakes [Gunn and Nixon 95], simplifying the initialization process [Neuenschwander et al. 94], considering region-based information [Etoh et al. 93, Ronfard 94], and allowing topological snake adaptation so that the snake can flow into complex shapes including branches [McInerney and Terzopoulos 95]. Applications can be found in many areas of machine vision, medical image analysis being a very promising field because living organisms and organs are naturally deformable and their shapes vary considerably; for applications to magnetic resonance images of a human head, to coronary vessel detection, etc., see [Cohen and Cohen 92, Hyche et al. 92, Lobregt and Viergever 95].

Active contour models represent a recent approach to contour detection and image interpretation. They differ substantially from classical approaches, where features are extracted from an image and higher-level processes try to interpolate sparse data to find a representation that matches the original data—active contour models start from an initial estimate based on higher-level knowledge, and an optimization method is used to refine the initial estimate. During the optimization, image data, an initial estimate, desired contour properties, and knowledge-based constraints are considered. Feature extraction and knowledge-based constrained grouping of these features are integrated into a single process, which seems to be the biggest advantage. Active contour models search for local energy minima, which is another significant difference from most other image interpretation techniques, which usually seek a global minimum of some objective function.

8.3 Point distribution models

The **point distribution model** (PDM) is a powerful shape description technique that may subsequently be used in locating new instances of such shapes in other images. It is most useful for describing features that have well understood 'general' shape, but which cannot be easily described by a rigid model (that is, particular instances are subject to variation). Examples which have been the successful subject of applications of this approach include electrical resistors [Cootes et al. 92, Cootes and Taylor 92], faces [Lanitis et al. 94, Cootes et al. 94], and bones within the hand [Efford 93]; each of these exhibits properties of 'shape' that the human can comprehend and describe easily, but which do not permit rigid model-based description. The PDM is a relatively recent development that has seen enormous application in a short time.

The PDM approach assumes the existence of a set of M examples (a training set) from which to derive a statistical description of the shape and its variation. In our context, we take

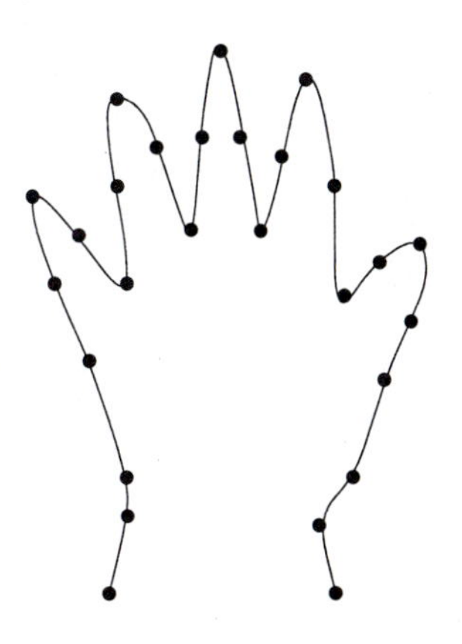

Figure 8.12: *A contour representing a hand, with possible landmark points marked.*

this to mean some number of instances of the shape represented by a boundary (a sequence of pixel co-ordinates). In addition, some number N of **landmark** points is selected on each boundary; these points are chosen to correspond to a feature of the underlying object—for example (see Figure 8.12), if the shape represents a hand, we might choose 27 points that include the fingertips, points that 'divide' the fingers, and some suitable number of intermediates.

It is intuitively clear that if the hands so represented are in 'about the same place', so will the N landmark points be. Variations in the positions of these points would then be attributable to natural variation between individuals. We may expect, though, that these differences will be 'small' measured on the scale of the overall shape. The PDM approach allows us to model these 'small' differences (and, indeed, to identify which are truly small, and which are more significant).

Aligning the training data

In order to develop this idea, it is necessary first to align all the training shapes in an approximate sense (otherwise comparisons are not 'like with like'). This is done by selecting for each example a suitable translation, scaling, and rotation to ensure that they all correspond as closely as possible—informally, the transformations are chosen to reduce (in a least-squares sense) the difference between an aligned shape and a 'mean' shape derived from the whole set. Specifically, suppose we wish to align just two shapes—each of these is described by a vector of N co-ordinate pairs:

$$\begin{aligned}\mathbf{x}^1 &= (x_1^1, y_1^1, x_2^1, y_2^1, \ldots, x_N^1, y_N^1)^T \\ \mathbf{x}^2 &= (x_1^2, y_1^2, x_2^2, y_2^2, \ldots, x_N^2, y_N^2)^T\end{aligned}$$

A transformation $\mathcal{T}$ of $\mathbf{x}^2$ composed of a translation (t_x, t_y), rotation θ, and scaling s may be represented by a matrix R applied to $\mathbf{x}^2$ using standard techniques,

$$\mathcal{T}(\mathbf{x}^2) = R\begin{pmatrix} x_i^2 \\ y_i^2 \end{pmatrix} + \begin{pmatrix} t_x \\ t_y \end{pmatrix} = \begin{pmatrix} x_i^2 s\cos\theta x_i^2 - y_i^2 s\sin\theta \\ x_i^2 s\sin\theta x_i^2 + y_i^2 s\cos\theta \end{pmatrix} + \begin{pmatrix} t_x \\ t_y \end{pmatrix}$$

and the 'best' such may be found by minimizing the expression

$$E = [\mathbf{x}^1 - R\mathbf{x}^2 - (t_x, t_y)^T]^T [\mathbf{x}^1 - R\mathbf{x}^2 - (t_x, t_y)^T] \tag{8.10}$$

This minimization is a routine application of a least-squares approach (see, for example, [Cootes et al. 92])—partial derivatives of E are calculated with respect to the unknowns (θ, s, t_x, and t_y) and set to zero, leaving simultaneous linear equations to solve.

This general idea is used to co-align all M shapes using the following algorithm.

Algorithm 8.5: Approximate alignment of similar training shapes

1. In a pairwise fashion, rotate, scale, and align each $\mathbf{x}^i$ with $\mathbf{x}^1$, for $i = 2, 3, \ldots, M$ to give the set $\{\mathbf{x}^1, \hat{\mathbf{x}}^2, \hat{\mathbf{x}}^3, \ldots, \hat{\mathbf{x}}^M\}$.

2. Calculate the mean of the transformed shapes (the details of this procedure are outlined in Section 8.3).

3. Rotate, scale, and align the mean shape to align to $\mathbf{x}^1$.

4. Rotate, scale, and align $\hat{\mathbf{x}}^2, \hat{\mathbf{x}}^3, \ldots, \hat{\mathbf{x}}^M$ to match to the adjusted mean.

5. If the mean has not converged, go to step 2.

Step 3 of this algorithm is necessary because otherwise it is ill-conditioned (underconstrained); without doing this, convergence will not occur. Final convergence may be tested by examining the differences involved in realigning the shapes to the mean.

This approach assumes that each of the landmark points is of equal significance, but that may not be the case. If for some reason one of them moves around the shape less than others, it has a desirable stability that we might wish to exploit during the alignment. This can be done by introducing a (diagonal) weight matrix W into equation (8.10):

$$E = [\mathbf{x}^1 - R\mathbf{x}^2 - (t_x, t_y)^T]^T W [\mathbf{x}^1 - R\mathbf{x}^2 - (t_x, t_y)^T] \tag{8.11}$$

The elements of W indicate the relative 'stability' of each of the landmarks, by which a high number indicates high stability (so counts for more in the error computation), and a low number the opposite. There are various ways of measuring this; one [Cootes et al. 92] is to compute for each shape the distance between landmarks k and l, and to let V_{kl} be the variance in these distances. A high variance indicates high mobility, and so setting the weight for the kth point to

$$w_k = \frac{1}{\sum_{l=1}^{N} V_{kl}}$$

has the desired weighting effect.

Deriving the model

The outcome of the alignment will be M (mutually aligned) boundaries $\hat{\mathbf{x}}^1, \hat{\mathbf{x}}^2, \ldots, \hat{\mathbf{x}}^M$, and we now proceed to determine the mean such, $\bar{\mathbf{x}}$. Each shape is given by N co-ordinate pairs,

$$\hat{\mathbf{x}}^i = (\hat{x}_1^i, \hat{y}_1^i, \hat{x}_2^i, \hat{y}_2^i, \ldots, \hat{x}_N^i, \hat{y}_N^i)^T$$

and so the mean shape is given by

$$\bar{\mathbf{x}} = (\bar{x}_1, \bar{y}_1, \bar{x}_2, \bar{y}_2, \ldots, \bar{x}_N, \bar{y}_N)$$

where

$$\bar{x}_j = \frac{1}{M} \sum_{i=1}^{M} \hat{x}_j^i \quad \text{and} \quad \bar{y}_j = \frac{1}{M} \sum_{i=1}^{M} \hat{y}_j^i$$

Knowledge of this mean allows explicit measurement of the variation and co-variation exhibited by each landmark and landmark pair; we can write

$$\mathbf{d}x^i = \hat{\mathbf{x}}^i - \bar{\mathbf{x}}$$

Doing this for each training vector, we can calculate the $2N \times 2N$ covariance matrix

$$S = \frac{1}{M} \sum_{i=1}^{M} \mathbf{d}x^i \mathbf{d}x^{iT}$$

This matrix has some particularly useful properties. If we imagine the aligned training set plotted in $2N$ dimensions, it will exhibit variation more in some directions than others (these directions will not, of course, in general align with the co-ordinate axes)—these variations are important properties of the shape we are describing. What these directions are, and their (relative) importance, may be derived from an eigen-decomposition of S—that is, solving the equation

$$S\mathbf{p}_i = \lambda_i \mathbf{p}_i \tag{8.12}$$

Solutions to equation (8.12) provide the **eigenvectors** $\mathbf{p}_i$ and **eigenvalues** λ_i of S; conventionally, we assume $\lambda_i \geq \lambda_{i+1}$. It can be shown that the eigenvectors associated with larger eigenvalues correspond to the directions of larger variation in the underlying data—they provide the **modes of variation**. Thus solving the equation and finding the highest eigenvalues tells us where the variation in the model is most likely to occur. We have performed here a *principal components analysis* [Jolliffe 86] of the data variation.

It is well known that a set of eigenvectors provides a basis, meaning that we can represent any vector $\mathbf{x}$ as a linear combination of the $2N$ different $\mathbf{p}^i$. If we write

$$P = (\mathbf{p}^1 \mathbf{p}^2 \mathbf{p}^3 \ldots \mathbf{p}^{2N})$$

then for any vector $\mathbf{x}$ a vector $\mathbf{b}$ exists such that

$$\mathbf{x} = \bar{\mathbf{x}} + P\mathbf{b}$$

where the components of $\mathbf{b}$ indicate how much variation is exhibited with respect to each of the eigenvectors.

Using the observation that the eigenvectors of lower index describe most of the changes in the training set, we may expect that the contributions from $\mathbf{p}^{2N}$, $\mathbf{p}^{2N-1}$, ..., play a small role in describing how far 'valid' shapes deviate from $\bar{\mathbf{x}}$; therefore, if we write

$$P_t = (\mathbf{p}^1 \mathbf{p}^2 \mathbf{p}^3 \dots \mathbf{p}^t)$$
$$\mathbf{b}_t = (b_1, b_2, \dots, b_t)^T \tag{8.13}$$

then the approximation

$$\mathbf{x} \approx \bar{\mathbf{x}} + P_t \mathbf{b}_t \tag{8.14}$$

will be good for sufficiently high $t \leq 2N$, if $\mathbf{x}$ is a valid shape with respect to the training set. This permits a dimensional compression of the representation—if there is a lot of structure in the data, t will be low (relative to $2N$) and good shape description will be possible very compactly by representing the shape as $\mathbf{b}_t$ rather than $\mathbf{x}$. One approach to this is to calculate λ_{total}, the sum of the λ_i, and choose t such that

$$\sum_{i=1}^{t} \lambda_i \geq \alpha \lambda_{\text{total}} \qquad 0 \leq \alpha \leq 1$$

The choice of α here will govern how much of the variation seen in the training set can be recaptured by the compacted model.

Further, it can be shown that the variance of b_i over the training set will be the associated eigenvalue λ_i; accordingly, for 'well-behaved' shapes we might expect

$$-3\sqrt{\lambda_i} \leq b_i \leq 3\sqrt{\lambda_i}$$

—that is, most of the population is within 3σ of the mean. This allows us to generate, from knowledge of P and λ_i, plausible shapes that are not part of the training set.

Example—metacarpal analysis

We can illustrate this theory with an example taken from automatic hand X-ray analysis. The finger bones (metacarpals) have characteristic long, thin shape with bulges near the ends—precise shape differs from individual to individual, and as an individual ages. Scrutiny of bone shape is of great value in diagnosing bone aging disorders and is widely used by pediatricians [Tanner et al. 83].

From a collection of X-rays, 40 landmarks (so vectors are 80-dimensional) were picked out by hand on a number (approximately 50) of segmented metacarpals. Figure 8.13 illustrates (after alignment, as described in Section 8.3) the mean shape, together with the actual positions of the landmark points from the entire data set.

Following the procedure outlined in Section 8.3, the covariance matrix and its eigenvectors associated with the variation are extracted; the relative contribution of the most influential components is illustrated in Table 8.1. From this we see that more than 95% of the shape variation is captured by the first eight modes of variation. Figure 8.14 illustrates the effect of varying the first mode of the mean shape by up to $2.5\sqrt{\lambda_1}$. This mode, which accounts for more than 60% of the variation seen in the data, captures the (asymmetric) thickening and thining of bones (relative to their length), which is an obvious characteristic of maturity. In

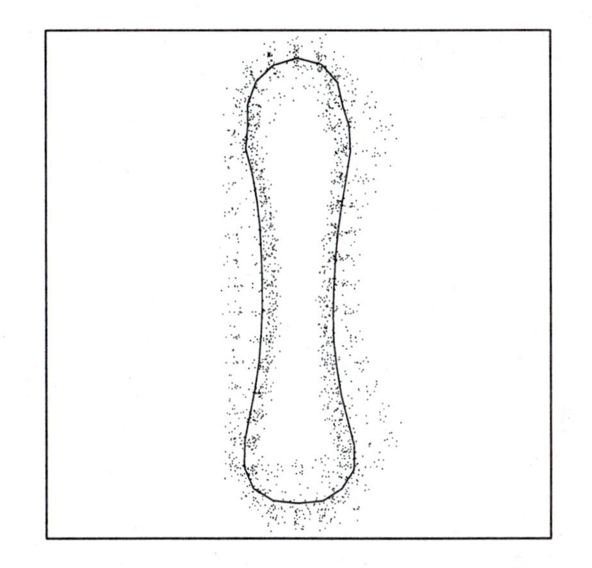

Figure 8.13: *PDM of a metacarpal. Dots mark the possible positions of landmarks, and the line denotes the mean shape. Courtesy N.D. Efford, School of Computer Studies, University of Leeds.*

this example, it is clear that 2.5 is an unlikely factor for $\sqrt{\lambda_1}$, since the resulting shapes are too extreme—thus we may expect b_1 to be smaller in magnitude for this application.

Figure 8.15 similarly illustrates extremes of the third mode. The shape change here is somewhat subtler; part of what is captured is a bending (in banana fashion) of the bone. Both extremes have a plausible 'bone-like' look about them.

Fitting models to data

A strength of this approach is that it permits plausible shapes to be fitted to new data. Given an image in which we wish to locate an instance of a modeled shape (specifically, given an edge map of the image, so having information about where boundaries are most likely to lie),

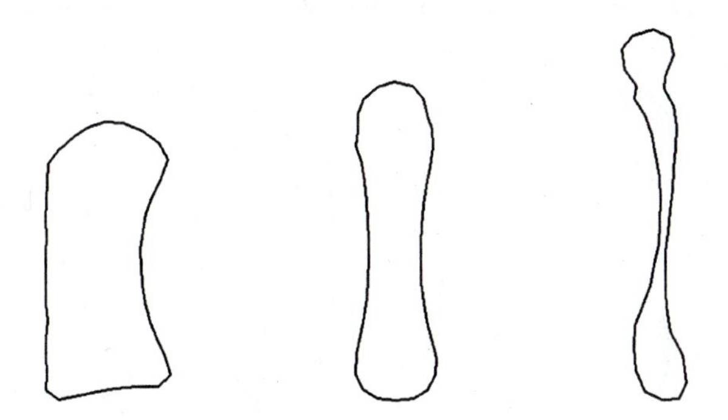

Figure 8.14: *The first mode of variation; $-2.5\lambda_1$, mean shape, $2.5\lambda_1$. Courtesy N.D. Efford, School of Computer Studies, University of Leeds.*

Index i	$(\lambda_i/\lambda_{\text{total}}) \times 100$	Cumulative total
1	63.3	63.3
2	10.8	74.1
3	9.5	83.6
4	3.4	87.1
5	2.9	90.0
6	2.5	92.5
7	1.7	94.2
8	1.2	95.4
9	0.7	96.1
10	0.6	96.7
11	0.5	97.2
12	0.4	97.6
13	0.3	97.9
14	0.3	98.2
15	0.3	98.5
16	0.2	98.7

Table 8.1: *Relative contributions to total data variance for the first 16 principal components.*

we require to know

- The mean shape $\bar{\mathbf{x}}$
- The transformation matrix P_t
- The particular shape parameter vector $\mathbf{b}_t$
- The particular pose (translation, rotation and scale)

Here, $\bar{\mathbf{x}}$ and P_t are known from the model construction. The identification of $\mathbf{b}_t$ and the pose is an optimization problem—locate the parameters that best fit the data at hand, subject to certain constraints. These constraints would include the known limits on reasonable values for the components of $\mathbf{b}_t$, and might also include domain knowledge about plausible positions

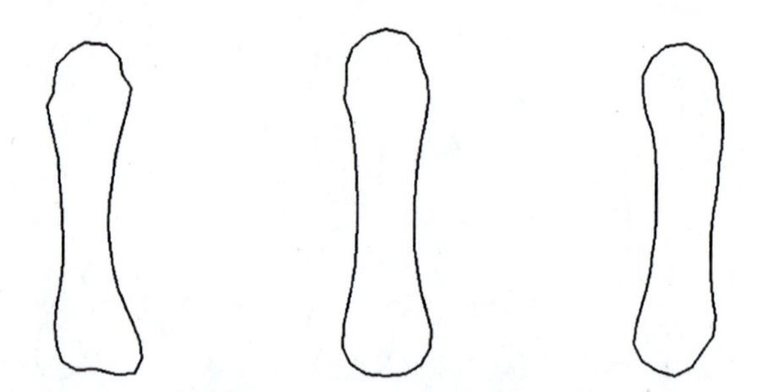

Figure 8.15: *The third mode of variation; $-2.5\lambda_3$, mean shape, $2.5\lambda_3$. Courtesy N.D. Efford, School of Computer Studies, University of Leeds.*

for the object to constrain the pose. In the metacarpal example (Section 8.3), this would include knowledge that a bone lies within the hand silhouette, is aligned with the finger and is of a known approximate size.

This approach may be used successfully with a number of well-known optimization algorithms, some of which are described in Section 7.6. It is likely, however, that convergence would be slow. An alternative, quicker approach [Cootes and Taylor 92, Smyth et al. 97] is to use the PDM as the basis of an active shape model (ASM) (sometimes referred to as a 'smart snake'—snakes, which represent a different approach to boundary fitting, are described in Section 8.2). Here, we iterate toward the best fit by examining an approximate fit, locating improved positions for the landmark points, then recalculating pose and parameters.

Algorithm 8.6: Fitting an ASM

1. Initialize an approximate fit to image data; this may be done in any suitable way but is likely to depend on geometric constraints provided by the application, together with crude image properties. This gives in local (model) co-ordinates a shape description,

$$\hat{\mathbf{x}} = (x_1, y_1, x_2, y_2, \dots, x_N, y_N)$$

2. At each landmark point, inspect the boundary normal close to the boundary, and locate the pixel of highest intensity gradient; mark this as the best target position to which to move this landmark point. This is illustrated in Figure 8.16. If there is no clear new target, the landmark is left where it is.

   We derive thereby a desired displacement vector $\mathbf{d}x$.

3. Adjust the pose parameters to provide the best fit to the target points of the current landmarks.

   There are various ways of doing this, but Algorithm 8.5 provides one approach; a quicker approximation, which is adequate since the iteration will seek out a good solution in time, is given in [Cootes and Taylor 92].

4. Determine the displacement vector $\tilde{\mathbf{d}x}$ that adjusts the model in the new pose to the target points (details follow the end of the algorithm).

5. Determine the model adjustment $\mathbf{d}b_t$ that best approximates $\tilde{\mathbf{d}x}$. From equation (8.14) we have

$$\tilde{\mathbf{x}} \approx \bar{\mathbf{x}} + P_t \mathbf{b}_t$$

   and we seek $\mathbf{d}b_t$ such that

$$\tilde{\mathbf{x}} + \tilde{\mathbf{d}x} = \bar{\mathbf{x}} + P_t(\mathbf{b}_t + \mathbf{d}b_t)$$

   Hence

$$\tilde{\mathbf{d}x} \approx P_t \mathbf{d}b_t$$

   With the properties of eigenmatrices, we can deduce

$$\mathbf{d}b_t = P_t^T \tilde{\mathbf{d}x}$$

as the best approximation. Note that since the modes of variation $t+1, t+2, \ldots,$ are discounted, this is necessarily only an approximation. Note also that we can at this stage prevent components of the vector $\mathbf{b}_t$ from growing in magnitude beyond any limits we may set by limiting them as we see appropriate—that is, should this equation generate a component deemed to be too large in magnitude, it would be set to the appropriate limit. Thus the re-fitted model will (probably) not match the targets precisely.

6. Iterate from step 2 until changes become negligible.

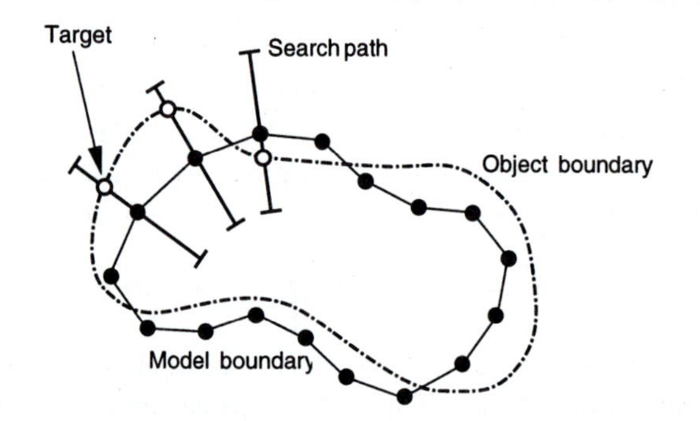

Figure 8.16: *Searching an approximate model fit for target points to which landmarks may move. Courtesy N.D. Efford, School of Computer Studies, University of Leeds.*

Step 2 assumes that a suitable target can be found, which may not always be true. If there is none, the landmark can be left where it is, and the model constraints will eventually pull it into a reasonable position. Alternatively, outlier landmarks can be automatically identified and replaced with model landmarks [Duta and Sonka 97]. There is also the option of locating targets by more sophisticated means than simple intensity gradient measurements.

Step 4 requires a calculation for $\tilde{\mathbf{d}x}$. To perform this, note that we commence with a vector $\tilde{\mathbf{x}}$ (in the 'local' frame), which is updated by a pose matrix and translation to provide $\mathbf{x}$ in the image frame,

$$\mathbf{x} = M(\theta, s)\tilde{\mathbf{x}} + (t_x, t_y)$$

where

$$M = M(\theta, s) = \begin{pmatrix} s\cos\theta & -s\sin\theta \\ s\sin\theta & s\cos\theta \end{pmatrix}$$

New pose parameters $t_x + dt_x$, $t_y + dt_y$, $\theta + d\theta$, $s(1 + ds)$ (from step 3) and a displacement $\mathbf{d}x$ (from step 2) have been calculated, giving the equation

$$\mathbf{x} + \mathbf{d}x = M[\theta + d\theta, s(1 + ds)](\tilde{\mathbf{x}} + \tilde{\mathbf{d}x}) + (t_x + dt_x, t_y + dt_y)$$

Since

$$M^{-1}(\theta, s) = M(-\theta, s^{-1})$$

we obtain

$$\tilde{\mathbf{d}x} = M\{-(\theta + d\theta), [s(1 + ds)]^{-1}\}[M(\theta, s)\tilde{\mathbf{x}} + \mathbf{d}x - (dt_x, dt_y)] - \tilde{\mathbf{x}}$$

This adjustment is 'raw' in the sense that it takes no account of the model; the next algorithm step compensates for this.

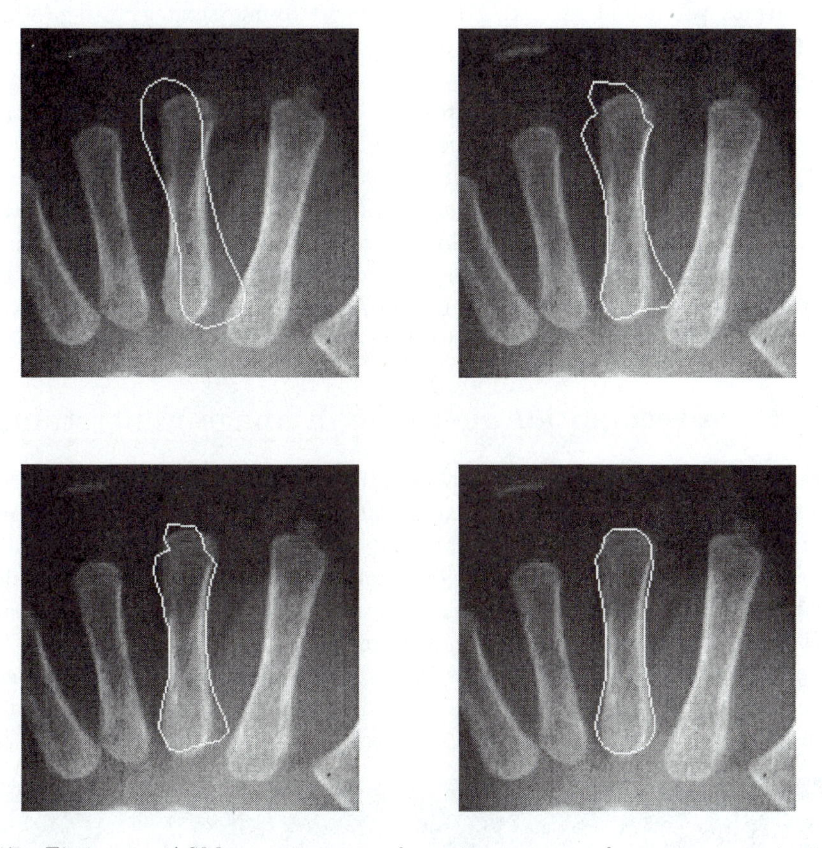

Figure 8.17: *Fitting an ASM to a metacarpal; various stages of convergence—initialization, 3, 6, and 10 iterations. Courtesy N.D. Efford, School of Computer Studies, University of Leeds.*

Algorithm 8.6 is illustrated in Figure 8.17, where we see an initialization and the position of the model after 3, 6, and 10 iterations as a metacarpal is located. Note in this figure that the model locates the correct position despite the proximity of strong boundaries that could distract it—this does not occur because the shape of the boundary is tightly bound in.

An example of the application of these algorithms is given as part of Section 16.4.

Extensions

In a short time, the literature on PDMs and ASMs has become very extensive—the technique lends itself to a very wide range of problems, but has some drawbacks.

The placing of the landmark points for construction of the training set is clearly very labor intensive, and in some applications it is error prone. Automatic placing of these points has been addressed [Hill and Taylor 94, Kotcheff and Taylor 97, Hill et al. 97, Rangarajan et al. 97, Cornic 97]. Another approach to the same task is described in Section 16.4 [Baumberg

95].

Efficiency of the approach has also been enhanced by the common idea of a multi-resolution attack [Cootes et al. 94]. Using a coarse-to-fine strategy can produce benefits in both quality of final fit and reduction of computational load.

As presented, the approach is strictly linear in the sense that control points may only move along a straight line (albeit with respect to directions of maximum variation); non-linear effects are produced by combining contributions from different modes; aside from being imperfect, this results in a representation that is not as compact as it might be if the non-linear aspects were explicitly modeled. This problem has been addressed in two ways: [Sozou et al. 94] introduces the **Polynomial Regression** PDM, which assumes dependence between the modes, with minor modes being polynomial combinations of major ones; and Heap [Heap and Hogg 96] extends the linear model by permitting polar relationships between modes, thereby efficiently capturing the ability of (parts of) objects to rotate around one another.

8.4 Pattern recognition methods in image understanding

Pattern recognition methods (Chapter 7) appear frequently in image understanding—classification-based segmentation of multi-spectral images (satellite images, magnetic resonance medical images, etc.) is a typical example.

The basic idea of classification-based segmentation is the same as that of statistical pattern recognition. Consider a magnetic resonance image (MRI) of the brain, and suppose the problem is to find areas of white matter, gray matter, and cerebro-spinal fluid (WM, GM CSF). Suppose the image data is available in two image modalities of multi-spin-echo images as $T2$-weighted and PD-weighted images (see Figure 8.18). As can be seen, neither single image can be used to detect the required areas reliably.

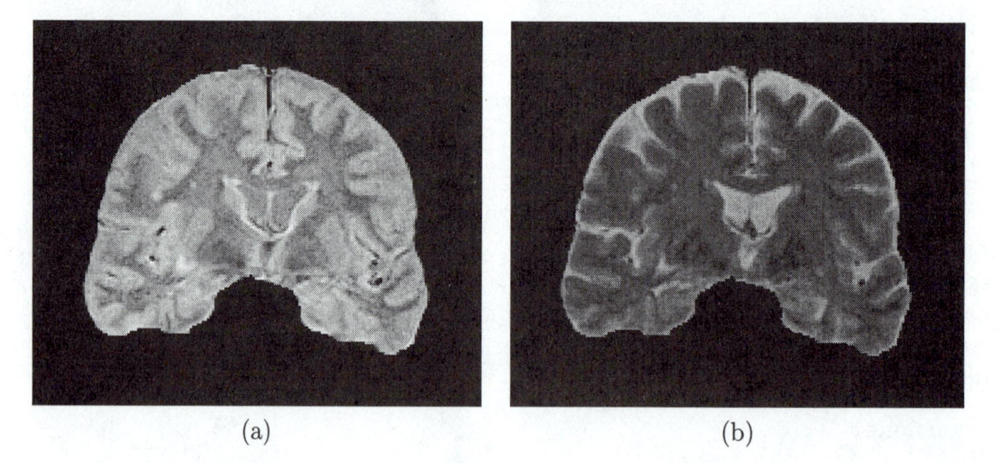

(a) (b)

Figure 8.18: *Magnetic resonance multi-spin-echo images: (a) PD-weighted, (b) T2-weighted. Courtesy N. Andreasen, G. Cohen, The University of Iowa.*

Gray-level values of pixels in particular image channels, their combinations, local texture features, etc., may be considered elements of a feature vector, one of which is assigned to each pixel. If an MR brain image is considered, four features, $PD, T2, PD - T2, PD \times T2$ may be used to construct the vector; subsequent classification-based understanding may be supervised or unsupervised.

If supervised methods are used for classification, a priori knowledge is applied to form a training set (see Figure 8.19a); classifier learning based on this training set was described in Section 7.2.3. In the image understanding stage, feature vectors derived from local multi-spectral image values of image pixels are presented to the classifier, which assigns a label to each pixel of the image. Image understanding is then achieved by pixel labeling; labels assigned to the MR brain image pixels can be seen in Figure 8.19b. Thus the understanding process segments a multi-spectral image into regions of known labels; in this case areas of white matter, gray matter, and cerebro-spinal fluid are detected and labeled.

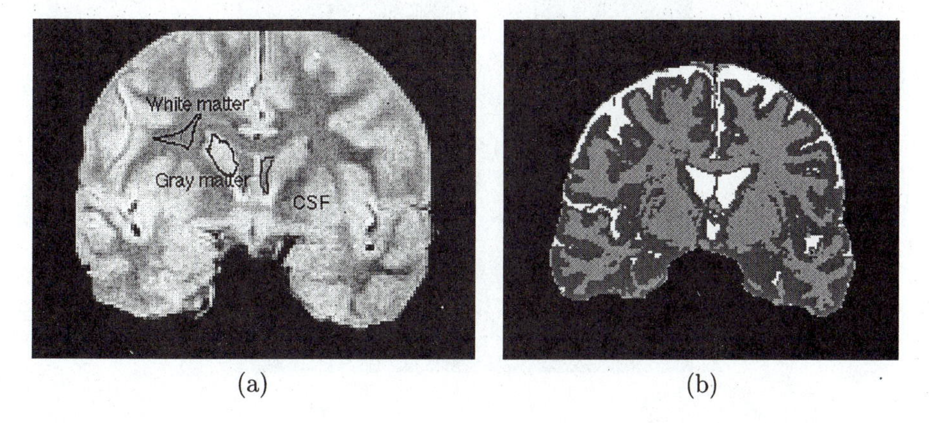

Figure 8.19: *MR brain image labeling: (a) training set construction; (b) result of supervised classification labeling. Courtesy J. Parkkinen, University of Kuopio, G. Cohen, N. Andreasen, The University of Iowa.*

Training set construction, and therefore human interaction, is necessary for supervised classification methods, but if unsupervised classification is used, training set construction is avoided (see Section 7.2.4). As a result, the clusters and the pixel labels do not have a one-to-one correspondence with the class meaning. This implies the image is segmented, but labels are not available to support image understanding. Fortunately, a priori information can often be used to assign appropriate labels to the clusters without direct human interaction. In the case of MR brain images, cerebro-spinal fluid is known always to form the brightest cluster, and gray matter to form the darkest cluster in $T2$ pixel values. Based on this information, clusters can be assigned appropriate labels. Cluster formation in feature space and results of unsupervised labeling are shown in Figure 8.20 [Parkkinen et al. 91].

In the supervised classification of MR brain images, the Bayes minimum error classification method was applied, and the ISODATA method of cluster analysis was used for unsupervised labeling. Validation of results proved a high accuracy of the method; further, the supervised and the unsupervised methods give almost identical results [Cohen 91a, Gerig et al. 92]. Many other segmentation approaches use local and global image features; promising approaches use graph partitioning and data clustering [Wu and Leahy 93, Shi and Malik 97, Comaniciu and

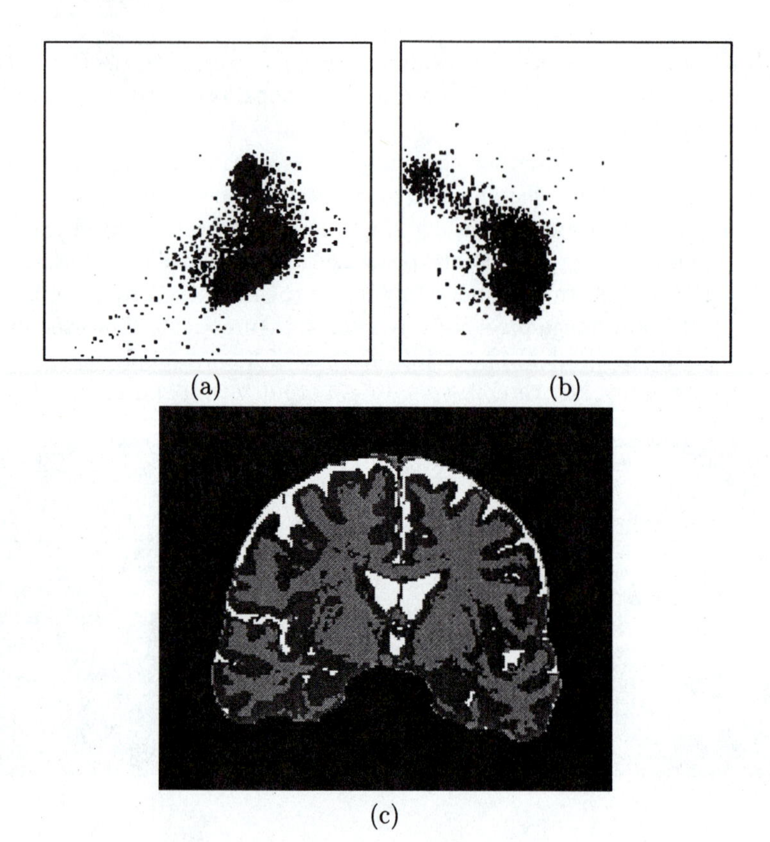

Figure 8.20: *MR brain image labeling: (a) clusters in feature space,* $(PD, T2)$ *plane; (b) clusters in feature space,* $(PD, PD \times T2)$ *plane; (c) result of unsupervised classification labeling. Courtesy J. Parkkinen, University of Kuopio, G. Cohen, N. Andreasen, The University of Iowa.*

Meer 97].

8.4.1 Contextual image classification

The method presented above works well with non-noisy data, and if the spectral properties determine classes sufficiently well. If noise or substantial variations in in-class pixel properties are present, the resulting image segmentation may have many small (often one-pixel) regions, which are misclassified. Several standard approaches can be applied to avoid this misclassification, which is very common in classification-based labeling. All of them use contextual information to some extent [Kittler and Foglein 84a].

- The first approach is to apply a post-processing filter to a labeled image. Small or single-pixel regions then disappear as the most probable label from the local neighborhood is assigned to them. This approach works well if the small regions are caused by noise. Unfortunately, the small regions can result from true regions with different properties in the original multi-spectral image, and in this case such filtering would worsen labeling results. Post-processing filters are widely used in remote sensing applications (see Figure 8.21).

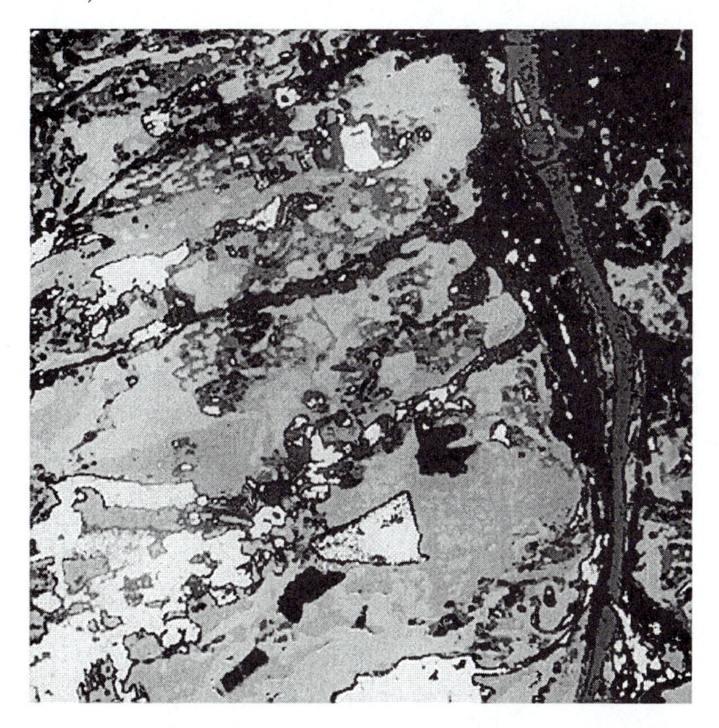

Figure 8.21: *Remotely sensed data of Prague, Landsat Thematic Mapper. Unsupervised classification, post-processing filter applied: White—no vegetation (note the sport stadium), green (varying hue)—vegetation types, red—urban areas. A color version of this picture may be seen in the color inset. Courtesy V. Cervenka, K. Charvat, Geodetic Institute Prague.*

- A slightly different post-processing classification improvement is introduced in [Wharton 82]. Pixel labels resulting from pixel classification in a given neighborhood form a new feature vector for each pixel, and a second-stage classifier based on the new feature vectors assigns final pixel labels. The contextual information is incorporated into the labeling process of the second-stage classifier learning.

- Context may also be introduced in earlier stages, merging pixels into homogeneous regions and classifying these regions (see Chapter 5).

- Another contextual pre-processing approach is based on acquiring pixel feature descriptions from a pixel neighborhood. Mean values, variances, texture description, etc., may be added to (or may replace) original spectral data. This approach is very common in textured image recognition (see Chapter 14).

- The most interesting option is to combine spectral and spatial information in the same stage of the classification process [Kittler and Foglein 84a, Kittler and Foglein 84b,

Kittler and Pairman 85]. The label assigned to each image pixel depends not only on multi-spectral gray-level properties of the particular pixel but also considers the context in the pixel neighborhood.

This section will discuss the last approach.

Contextual classification of image data is based on the Bayes minimum error classifier [Section 7.2.2, equation (7.20)]. For each pixel $\mathbf{x}_0$, a vector consisting of (possibly multi-spectral) values $f(\mathbf{x}_i)$ of pixels in a specified neighborhood $N(\mathbf{x}_0)$ is used as a feature representation of the pixel $\mathbf{x}_0$. Each pixel is represented by the vector

$$\xi = (f(\mathbf{x}_0), f(\mathbf{x}_1), \dots, f(\mathbf{x}_k)) \tag{8.15}$$

where

$$\mathbf{x}_i \in N(\mathbf{x}_0) \qquad i = 0, \dots, k$$

Some more vectors are defined which will be used later. Let labels (classification) of pixels in the neighborhood $N(\mathbf{x}_0)$ be represented by a vector (see Figure 8.22)

$$\eta = (\theta_0, \theta_1, \dots, \theta_k) \tag{8.16}$$

where

$$\theta_i \in \{\omega_1, \omega_2, \dots, \omega_R\}$$

and ω_s denotes the assigned class. Further, let the labels in the neighborhood excluding the pixel $\mathbf{x}_0$ be represented by a vector

$$\tilde{\eta} = (\theta_1, \theta_2, \dots, \theta_k) \tag{8.17}$$

Theoretically, there may be no limitation on the neighborhood size, but the majority of contextual information is believed to be present in a small neighborhood of the pixel $\mathbf{x}_0$. Therefore, a 3×3 neighborhood in 4-connectivity or in 8-connectivity is usually considered appropriate (see Figure 8.22); also, computational demands increase exponentially with growth of neighborhood size.

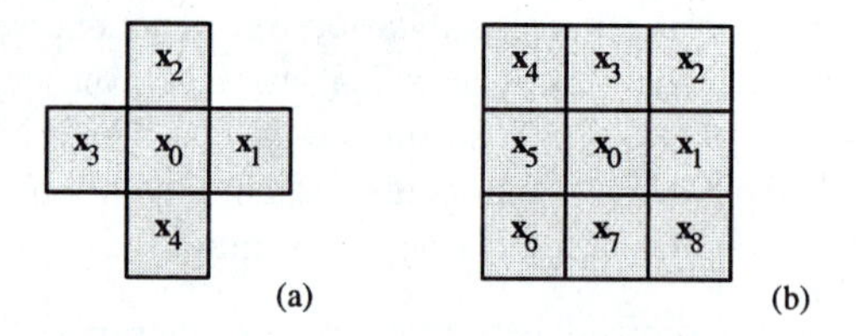

Figure 8.22: *Pixel neighborhoods used in contextual image classification, pixel indexing scheme: (a) 4-neighborhood; (b) 8-neighborhood.*

A conventional minimum error classification method assigns a pixel $\mathbf{x}_0$ to a class ω_r if the probability of $\mathbf{x}_0$ being from the class ω_r is the highest of all possible classification probabilities [as given in equation (7.20)]:

$$\theta_0 = \omega_r \quad \text{if} \quad P[\omega_r | f(\mathbf{x}_0)] = \max_{s=1,\dots,R} P[\omega_s | f(\mathbf{x}_0)] \tag{8.18}$$

A contextual classification scheme uses the feature vector $\boldsymbol{\xi}$ instead of $\mathbf{x}_0$, and the decision rule remains similar:

$$\theta_0 = \omega_r \quad \text{if} \quad P(\omega_r|\boldsymbol{\xi}) = \max_{s=1,\dots,R} P(\omega_s|\boldsymbol{\xi}) \tag{8.19}$$

The a posteriori probability $P(\omega_s|\boldsymbol{\xi})$ can be computed using the Bayes formula

$$P(\omega_s|\boldsymbol{\xi}) = \frac{p(\boldsymbol{\xi}|\omega_s)P(\omega_s)}{p(\boldsymbol{\xi})} \tag{8.20}$$

Note that each image pixel is classified using a corresponding vector $\boldsymbol{\xi}$ from its neighborhood, and so there are as many vectors $\boldsymbol{\xi}$ as there are pixels in the image. Many accompanying details, and a formal proof that contextual information increases classification reliability, are given in [Kittler and Foglein 84a]. The basic contextual classification algorithm can be summarized as follows.

Algorithm 8.7: Contextual image classification

1. For each image pixel, determine a feature vector $\boldsymbol{\xi}$ [equation (8.15)].
2. From the training set, determine parameters of probability distributions $p(\boldsymbol{\xi}|\omega_s)$ and $P(\omega_s)$.
3. Compute maximum a posteriori probabilities $P(\omega_r|\boldsymbol{\xi})$ and label (classify) all pixels in the image according to Equation (8.19). An image classification results.

A substantial limitation in considering larger contextual neighborhoods is exponential growth of computational demands with increasing neighborhood size. A **recursive contextual classification** overcomes these difficulties [Kittler and Foglein 84a, Kittler and Foglein 84b, Kittler and Pairman 85]. The main trick of this method is in propagating contextual information through the image although the computation is still kept in small neighborhoods. Spectral and neighborhood pixel labeling information are both used in classification. Therefore, context from a distant neighborhood can propagate to the labeling θ_0 of the pixel $\mathbf{x}_0$; this is illustrated in Figure 8.23.

The vector $\tilde{\boldsymbol{\eta}}$ of labels in the neighborhood may further improve the contextual representation. Clearly, if the information contained in the spectral data in the neighborhood is unreliable (e.g., based on spectral data, the pixel $\mathbf{x}_0$ may be classified into a number of classes with similar probabilities), the information about labels in the neighborhood may increase confidence in one of those classes. If a majority of surrounding pixels are labeled as members of a class ω_i, the confidence that the pixel $\mathbf{x}_0$ should also be labeled ω_i increases.

More complex dependencies may be found in the training set—for instance, imagine a thin striped noisy image. Considering labels in the neighborhood of the pixel $\mathbf{x}_0$, the decision rule becomes

$$\theta_0 = \omega_r \quad \text{if} \quad P(\omega_r|\boldsymbol{\xi}, \tilde{\boldsymbol{\eta}}) = \max_{s=1,\dots,R} P(\omega_s|\boldsymbol{\xi}, \tilde{\boldsymbol{\eta}}) \tag{8.21}$$

After several applications of the Bayes formula [Kittler and Pairman 85], the decision rule transforms into

$$\theta_0 = \omega_r \quad \text{if} \quad p(\boldsymbol{\xi}|\boldsymbol{\eta}_r)P(\omega_r|\tilde{\boldsymbol{\eta}}) = \max_{s=1,\dots,R} p(\boldsymbol{\xi}|\boldsymbol{\eta}_s)P(\omega_s|\tilde{\boldsymbol{\eta}}) \tag{8.22}$$

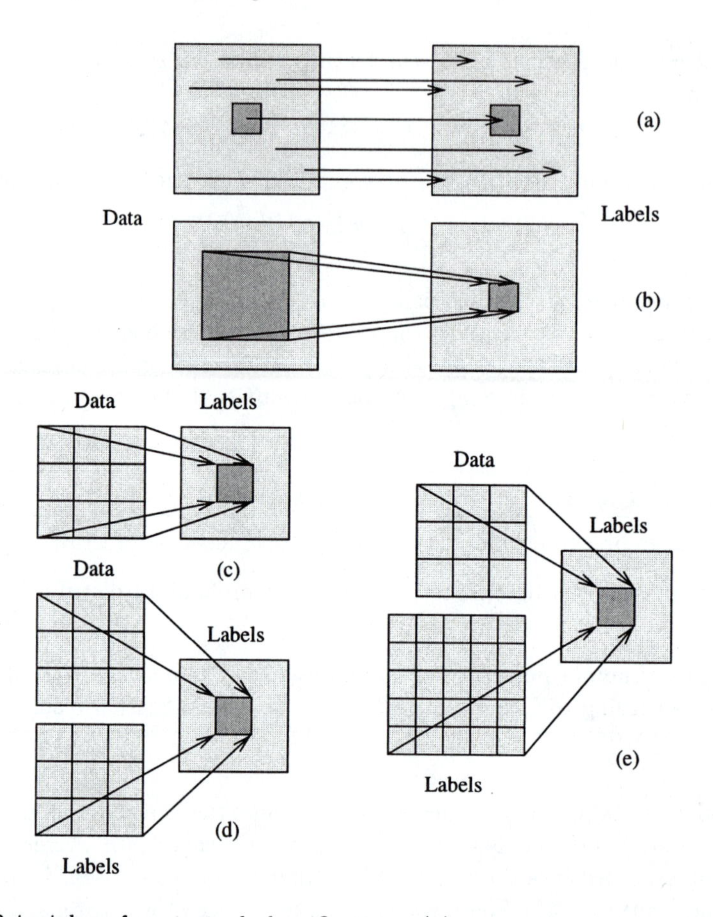

Figure 8.23: *Principles of contextual classification: (a) conventional non-contextual method; (b) contextual method; (c) recursive contextual method—step 1 of Algorithm 8.7; (d) first application of step 2; (e) second application of step 2.*

where $\boldsymbol{\eta}_r$ is a vector $\boldsymbol{\eta}$ with $\theta_0 = \omega_r$. Assuming all necessary probability distribution parameters were determined in the learning process, the recursive contextual classification algorithm follows.

Algorithm 8.8: Recursive contextual image classification

1. Determine an initial image pixel labeling using the non-contextual classification scheme, equation (8.18)

2. Update labels in each image pixel $\mathbf{x}_0$, applying the current label vectors $\boldsymbol{\eta}$, $\tilde{\boldsymbol{\eta}}$, and local spectral vector $\boldsymbol{\xi}$ to the decision rule equation (8.22).

3. Terminate the algorithm if the labels of all pixels in the image are stable; repeat step 2 otherwise.

Only a general outline of the contextual classification methods has been given; for more details, discussion of convergence, other techniques, and specific algorithms, see [Kittler and Foglein 84a, Kittler and Foglein 84b, Kittler and Pairman 85, Haralick et al. 88, Watanabe and Suzuki 89, Zhang et al. 90]. A comparison of contextual classifiers is given in [Mohn et al. 87, Watanabe and Suzuki 88], and a parallel implementation is described in [Tilton 87]. Applications are mostly related to remote sensing and medical images [Gonzalez and Lopez 89, Moller-Jensen 90, Franklin 90, Wilkinson and Megier 90, Algorri et al. 91]. Contextual classification of textures based on the context of feature vectors is described in [Fung et al. 90], and the application of neural networks to contextual image segmentation is given in [Toulson and Boyce 92].

A crucial idea is incorporated in the algorithm of recursive contextual image classification that will be seen several times throughout this chapter; this is the idea of information propagation from distant image locations without the necessity for expensive consideration of context in large neighborhoods. This is a standard approach used in image understanding.

8.5 Scene labeling and constraint propagation

Context plays a significant role in image understanding; the previous section was devoted to context present in pixel data configurations, and this section deals with semantic labeling of regions and objects. Assume that regions have been detected in an image that correspond to objects or other image entities, and let the objects and their inter-relationships be described by a region adjacency graph and/or a semantic net (see Sections 3.2.3 and 7.1). Object properties are described by unary relations, and inter-relationships between objects are described by binary (or n-ary) relations. The goal of scene labeling is to assign a label (a meaning) to each image object to achieve an appropriate image interpretation.

The resulting interpretation should correspond with available scene knowledge. The labeling should be consistent, and should favor more probable interpretations if there is more than one option. Consistency means that no two objects of the image appear in an illegal configuration—e.g., an object labeled *house* in the middle of an object labeled *lake* will be considered inconsistent in most scenes. Conversely, an object labeled *house* surrounded by an object labeled *lawn* in the middle of a *lake* may be fully acceptable.

Two main approaches may be chosen to achieve this goal.

- **Discrete** labeling allows only one label to be assigned to each object in the final labeling. Effort is directed to achieving a consistent labeling all over the image.

- **Probabilistic** labeling allows multiple labels to co-exist in objects. Labels are probabilistically weighted, with a label confidence being assigned to each object label.

The main difference is in interpretation robustness. Discrete labeling always finds either a consistent labeling or detects the impossibility of assigning consistent labels to the scene. Often, as a result of imperfect segmentation, discrete labeling fails to find a consistent interpretation even if only a small number of local inconsistencies is detected. Probabilistic labeling always gives an interpretation result together with a measure of confidence in the interpretation. Even if the result may be locally inconsistent, it often gives a better scene interpretation than a consistent and possibly very unlikely interpretation resulting from a discrete labeling. Note that discrete labeling may be considered a special case of probabilistic labeling with one label probability always being 1 and all the others being 0 for each object.

The scene labeling problem is specified by

- A set of objects R_i, $i = 1, \ldots, N$
- A finite set of labels Ω_i for each object R_i (without loss of generality, the same set of labels will be considered for each object; $\Omega_i = \Omega_j$ for any $i, j \in [1, \ldots, N]$)
- A finite set of relations between objects
- The existence of a compatibility function (reflecting constraints) between interacting objects

To solve the labeling problem considering direct interaction of all objects in an image is computationally very expensive and approaches to solving labeling problems are usually based on **constraint propagation**. This means that local constraints result in local consistencies (local optima), and by applying an iterative scheme the local consistencies adjust to global consistencies (global optima) in the whole image.

Many types of relaxation exist, some of them being used in statistical physics, for example, simulated annealing (Section 7.6.2), and stochastic relaxation [Geman and Geman 84], etc. Others, such as **relaxation labeling**, are typical in image understanding. To provide a better understanding of the idea, the discrete relaxation approach is considered first.

8.5.1 Discrete relaxation

Consider the scene shown in Figure 8.24a. Six objects are present in the scene, including the background. Let the labels be *background (B), window (W), table (T), drawer (D), phone (P)*, and let the unary properties of object interpretations be (the example is meant to be illustrative only)

- A window is rectangular.
- A table is rectangular.
- A drawer is rectangular.

Let the binary constraints be

- A window is located above a table.
- A phone is above a table.
- A drawer is inside a table.
- Background is adjacent to the image border.

Given these constraints, the labeling in Figure 8.24b is inconsistent. Discrete relaxation assigns all existing labels to each object and iteratively removes all the labels which may not be assigned to an object without violating the constraints. A possible relaxation sequence is shown in Figure 8.25.

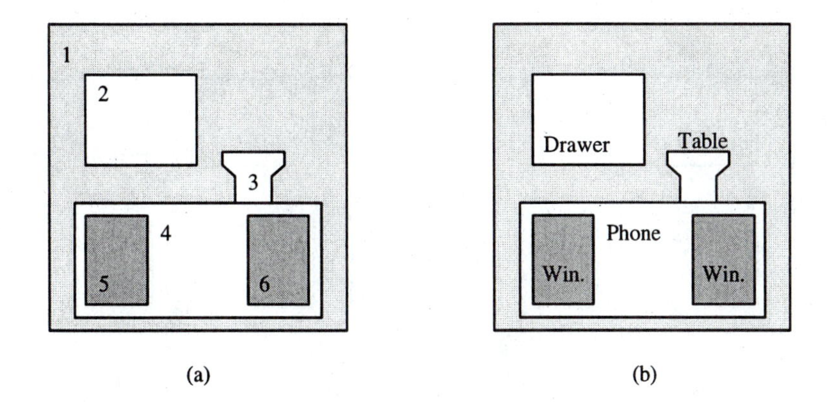

Figure 8.24: *Scene labeling: (a) scene example; (b) inconsistent labeling.*

At the beginning (Figure 8.25a), all labels are assigned to each object, and for each object all its labels are tested for consistency. Therefore, the label B can immediately be removed as inconsistent in objects 2, 3, 4, 5, and 6. Similarly, object 3 is not rectangular, therefore it violates the unary relation that must hold for T, W, D, etc.

The final consistent labeling is given in Figure 8.25c; note the mechanism of constraint propagation. The distant relations between objects may influence labeling in distant locations of the scene after several steps, making it possible to achieve a global labeling consistency of the scene interpretation although all the label-removing operations are local.

Algorithm 8.9: Discrete relaxation

1. Assign all possible labels to each object, considering the unary constraints.

2. Repeat steps 3–5 until global consistency is achieved or is found to be impossible.

3. Choose one object to update its labels.

4. Modify (delete inconsistent) labels of the chosen object by considering relations with other interacting objects.

5. If any object has no label, stop—a consistent labeling was not found.

The algorithm may be implemented in parallel with one difference: step 4 disappears as all objects are treated in parallel.

For a more detailed survey of discrete relaxation techniques, their properties, and technical difficulties that limit their applicability, see [Hancock and Kittler 90a]. Although discrete relaxation is naturally parallel, a study of the complexity of discrete relaxation given in [Kasif 90] shows that a parallel solution is unlikely to improve known sequential solutions much. An interesting discrete relaxation control strategy using asynchronous activation of object updating actions (**daemons**) was introduced in [Barrow and Tenenbaum 76].

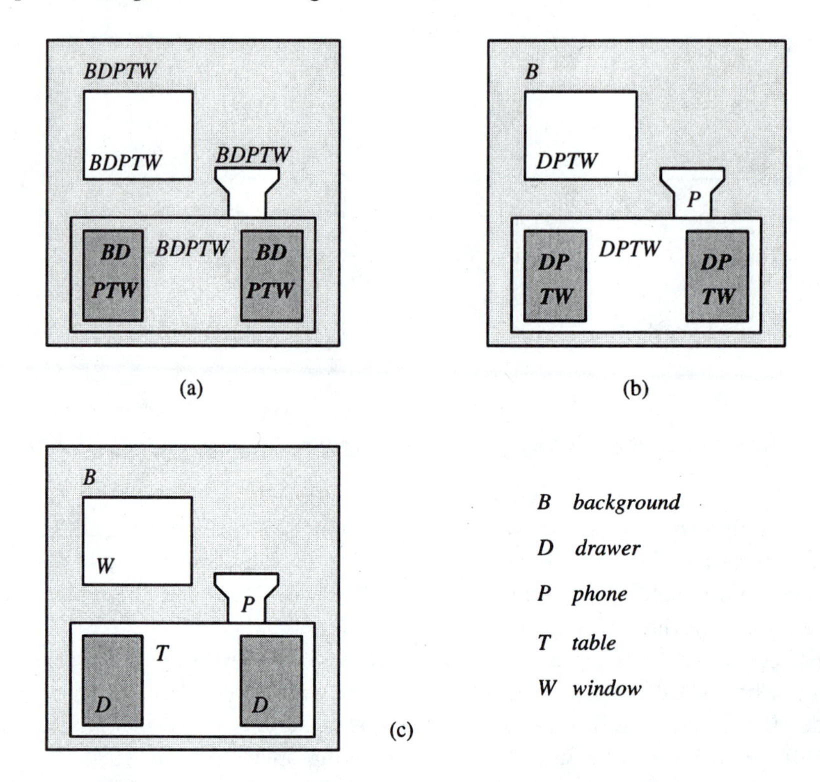

Figure 8.25: *Discrete relaxation: (a) all labels assigned to each object; (b) locally inconsistent labels are removed; (c) final consistent labeling.*

8.5.2 Probabilistic relaxation

Constraints are a typical tool in image understanding. The classical problem of discrete relaxation labeling was first introduced in [Waltz 57] in understanding perspective line drawings depicting 3D objects, and this problem is discussed briefly in Chapter 9. Discrete relaxation results in an unambiguous labeling; in a majority of real situations, however, it represents an oversimplified approach to image data understanding—it cannot cope with incomplete or imprecise segmentation. Using semantics and knowledge, image understanding is supposed to solve segmentation problems which cannot be solved by bottom-up interpretation approaches. Probabilistic relaxation may overcome the segmentation problems of missing objects or extra regions in the scene, but it results in an ambiguous image interpretation which is often inconsistent. It has been noted that a locally inconsistent but very probable (global) interpretation may be more valuable than a consistent but unlikely interpretation (e.g., a non-rectangular window located far above the table would be considered a phone in our example; this labeling would be consistent, even if very unlikely—see Figure 8.26).

Probabilistic relaxation was introduced in [Rosenfeld et al. 76] and has been used extensively in image understanding ever since. Consider the relaxation problem as specified above (regions R_i and sets of labels Ω_i) and, in addition, let each object R_i be described by a set of unary properties X_i. Similarly to discrete relaxation, object labeling depends on the object

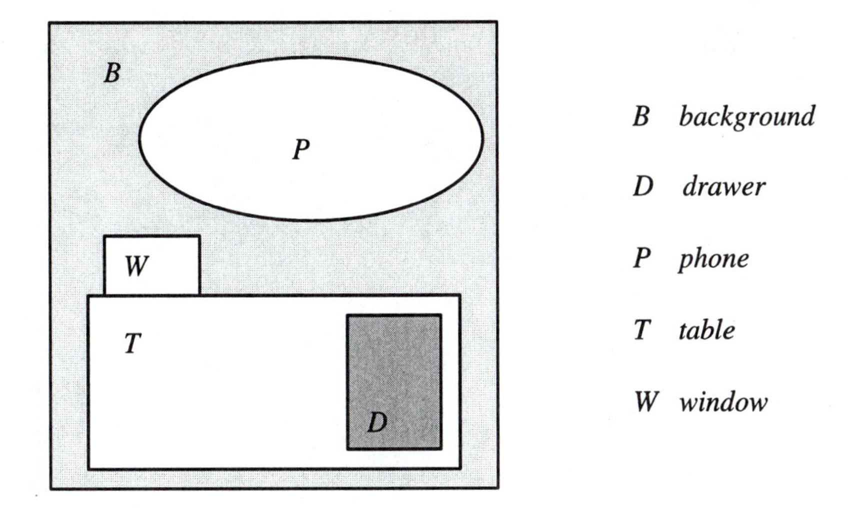

Figure 8.26: *Consistent but unlikely labeling.*

properties and on a measure of compatibility of the potential object labels with the labeling of other, directly interacting objects. All the image objects may be considered directly interacting, and a general form of the algorithm will be given assuming this. Nevertheless, only adjacent objects are usually considered to interact directly, to reduce computational demands of the relaxation. However, as before, more distant objects still interact with each other as a result of the constraint propagation. A region adjacency graph is usually used to store the adjacency information.

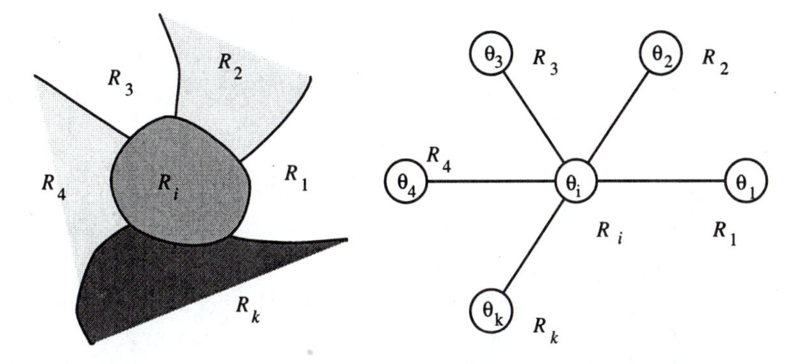

Figure 8.27: *Local configuration of objects in an image—part of a region adjacency graph.*

Consider the local configuration of objects given in Figure 8.27; let the objects R_j be labeled by θ_j; $\theta_j \in \Omega$; $\Omega = \{\omega_1, \omega_2, \ldots, \omega_R\}$. Confidence in the label θ_i of an object R_i depends on the configuration of labels of directly interacting objects. Let $r(\theta_i = \omega_k, \theta_j = \omega_l)$ represent the value of a compatibility function for two interacting objects R_i and R_j with labels θ_i and θ_j (the probability that two objects with labels θ_i and θ_j appear in a specific relation). The relaxation algorithm [Rosenfeld et al. 76] is iterative and its goal is to achieve the locally best consistency in the entire image. The **support** q_j^s for a label θ_i of the object R_i

resulting from the binary relation with the object R_j at the sth step of the iteration process is

$$q_j^s(\theta_i = \omega_k) = \sum_{l=1}^{R} r(\theta_i = \omega_k, \theta_j = \omega_l) P^s(\theta_j = \omega_l) \tag{8.23}$$

where $P^s(\theta_j = \omega_l)$ is the probability that region R_j should be labeled ω_l. The support Q^s for the same label θ_i of the same object R_i resulting from all N directly interacting objects R_j and their labels θ_j at the sth step is

$$\begin{aligned} Q^s(\theta_i = \omega_k) &= \sum_{j=1}^{N} c_{ij} q_j^s(\theta_i = \omega_k) \\ &= \sum_{j=1}^{N} c_{ij} \sum_{l=1}^{R} r(\theta_i = \omega_k, \theta_j = \omega_l) P^s(\theta_j = \omega_l) \end{aligned} \tag{8.24}$$

where c_{ij} are positive weights satisfying $\sum_{j=1}^{N} c_{ij} = 1$. The coefficients c_{ij} represent the strength of interaction between objects R_i and R_j. Originally [Rosenfeld et al. 76], an updating formula was given which specified the new probability of a label θ_i according to the previous probability $P^s(\theta_i = \omega_k)$ and probabilities of labels of interacting objects,

$$P^{s+1}(\theta_i = \omega_k) = \frac{1}{K} P^s(\theta_i = \omega_k) Q^s(\theta_i = \omega_k) \tag{8.25}$$

where K is a normalizing constant

$$K = \sum_{l=1}^{R} P^s(\theta_i = \omega_l) Q^s(\theta_i = \omega_l) \tag{8.26}$$

This form of the algorithm is usually referred to as a **non-linear relaxation scheme**. A **linear scheme** [Rosenfeld et al. 76] looks for probabilities such as

$$P(\theta_i = \omega_k) = Q(\theta_i = \omega_k) \quad \text{for all } i, k \tag{8.27}$$

with a non-contextual probability

$$P^0(\theta_i = \omega_k) = P(\theta_i = \omega_k | X_i) \tag{8.28}$$

being used only to start the relaxation process [Blake 82, Elfving and Eklundh 82].

A relaxation algorithm can also be treated as an optimization problem, the goal being maximization of the global confidence in the labeling [Berthod and Faugeras 80, Hummel and Zucker 83]. The global objective function is

$$F = \sum_{k=1}^{R} \sum_{i=1}^{N} P(\theta_i = \omega_k) \sum_{j=1}^{N} c_{ij} \sum_{l=1}^{R} r(\theta_i = \omega_k, \theta_j = \omega_l) P(\theta_j = \omega_l) \tag{8.29}$$

subject to the constraint that the solution satisfies

$$\sum_{k=1}^{R} P(\theta_i = \omega_k) = 1 \quad \text{for any } i, \quad P(\theta_i = \omega_k) > 0 \quad \text{for any } i, k \tag{8.30}$$

Optimization approaches to relaxation can be generalized to allow n-ary relations among objects. A projected gradient ascent method [Hummel and Zucker 83, Illingworth and Kittler 87] may be used to optimize equation (8.29), and an efficient version of this updating principle is introduced in [Parent and Zucker 89].

Convergence is an important property of iterative algorithms; as far as relaxation is concerned, convergence problems have not yet been satisfactorily solved. Although convergence of a discrete relaxation scheme can always be achieved by an appropriate design of the label updating scheme (e.g., to remove the inconsistent labels), convergence of more complex schemes where labels may be added, or of probabilistic relaxation, often cannot be guaranteed mathematically. Despite this fact, the relaxation approach may still be quite useful. Relaxation algorithms are one of the cornerstones of the high-level vision understanding processes, and applications can also be found outside the area of computer vision.

Relaxation algorithms are naturally parallel, since the label updating may be done on all objects at the same time. Many parallel implementations exist, and parallel relaxation does not differ in essence from the serial version. A general version is the following algorithm.

Algorithm 8.10: Probabilistic relaxation

1. Define conditional probabilities of interpretations (labels) for all objects R_i in the image [e.g., using equation (8.28)].

2. Repeat steps 3 and 4 until the best scene interpretation (a maximum of the objective function F) is reached.

3. Compute the objective function F [equation (8.29)], which represents the quality of the scene labeling.

4. Update probabilities of object interpretations (labels) to increase the value of the objective function F.

Parallel implementations of relaxation algorithms can be found in [Kamada et al. 88, Millin and Ni 89, Dew et al. 89, Bhandarker and Suk 90, Zen et al. 90, Lau and Hancock 94].

Relaxation algorithms are still being developed. One existing problem with their behavior is that the labeling improves rapidly during early iterations, followed by a degradation which may be very severe. The reason is that the search for the global optimum over the image may cause highly non-optimal local labeling. A possible treatment that allows spatial consistency to be developed while avoiding labeling degradation is based on decreasing the neighborhood influence with the iteration count [Lee et al. 89]. For a survey and an extensive list of references, see [Kittler and Illingworth 85, Kittler and Foglein 86, Kittler and Hancock 89, Hancock and Kittler 90b]. A compact theoretical basis for probabilistic relaxation and close relations to the contextual classification schemes is given in [Kittler 87]. Recent improvements of algorithms for probabilistic relaxation can be found in [Hancock 93, Lu and Chung 94, Hatef and Kittler 95, Stoddart et al. 95, Sharp and Hancock 95, Christmas et al. 96, Pelillo and Fanelli 97]. Application of the relaxation scheme to image segmentation is described in the next section.

8.5.3 Searching interpretation trees

Note that relaxation is not the only way to solve discrete labeling problems, and classical methods of **interpretation tree** searching may be applied. A tree has as many levels as there are objects present in the scene; nodes are assigned all possible labels, and a depth-first search based on back-tracking is applied. Starting with a label assigned to the first object node (tree root), a consistent label is assigned to the second object node, to the third object node, etc. If a consistent label cannot be assigned, a back-tracking mechanism changes the label of the closest node at the higher level. All the label changes are done in a systematic way.

An interpretation tree search tests all possible labelings, and therefore computational inefficiency is common, especially if an appropriate tree pruning algorithm is not available. An efficient method for searching the interpretation trees was introduced in [Grimson and Lozano-Perez 87]. The search is heuristically guided towards a *good* interpretation based on a *quality of match* that is based on constraints and may thus reflect feasibility of the interpretation. Clearly, an infeasible interpretation makes all interpretations represented down the tree infeasible also. To represent the possibility of discarding the evaluated patch, an additional interpretation tree branch is added to each node. The general search strategy is based on a depth-first approach in which the search is for the *best* interpretation. However, the search for the best solution can be very time consuming.

There have been many attempts to improve on the basic idea of the Grimson Lozano-Perez algorithm—a recent summary may be found in [Fisher 94]. Typically, correct interpretations are determined early on, and considerable time is spent by attempts to improve them further. Thus, a **cut-off** threshold is used to discontinue the search for an interpretation when the cut-off threshold is exceeded. This approach was found to be highly significant in improving the search times without adversely affecting the search results [Grimson and Lozano-Perez 87, Grimson 90]. [Fisher 93] divides a model into a tree of progressively smaller sub-models which are combined to produce an overall match, while [Fletcher et al. 96] uses a coarse-to-fine approach to describing features of surfaces to be matched (in this case, 3D data derived from MR head scans).

Yet another approach has recently demonstrated practical applicability for assessing similarity of medical images for database retrieval. Here, Voronoi diagrams representing arrangement of regions in an image (Section 3.2.3) were used together with a tree-based metric representing Voronoi diagram similarity [Tagare et al. 95].

8.6 Semantic image segmentation and understanding

This section presents a higher-level extension of region growing methods which were discussed in Section 5.3. These ideas fall under the heading of this chapter, rather than simple segmentation, for a number of reasons: Semantic approaches represent a significantly advanced field of image segmentation and as such require a well-developed set of techniques not fully covered until this point; while from another viewpoint, semantic segmentation includes image region interpretation and may result in image understanding and therefore should be included in this chapter. Whichever, it is considered appropriate to present semantic segmentation methods at this point, after the reader is comfortable with the necessary background material: region growing, object description, minimum error classification, contextual classification, image understanding strategies, etc.

Algorithms already discussed in Section 5.3 merge regions on the basis of general heuristics using local properties of regions, and may be referred to as syntactic information-based methods. Conversely, semantic information representing higher-level knowledge was included in [Feldman and Yakimovsky 74] for the first time. It is intuitively clear that including more information, especially information about assumed region interpretation, can have a beneficial effect on the merging process, and it is also clear that context and criteria for global optimization of region interpretation consistency will also play an important role. Further, the approaches described in this section are meant to serve as examples of incorporating context, semantics, applying relaxation methods to propagate constraints, and to show how the global consistency function may be optimized—for applications, see also [Roberto et al. 90, Cabello et al. 90, Strat and Fischler 91, Gelgon and Bouthemy 97].

The first issue in semantic region growing is the representation of image regions and their inter-relationships. The concept of the region adjacency graph, in which nodes represent regions and arcs connect nodes of adjacent regions, was introduced in Section 3.2.3. An artificial region may surround the image in order to treat all regions consistently. A dual graph can be constructed from the region adjacency graph in which nodes correspond to intersecting points of boundary segments of different regions and arcs correspond to boundary segments. An example of a region adjacency graph and its dual is shown in Figure 8.28. Each time two regions are merged, both graphs change— the following algorithm [Ballard and Brown 82]

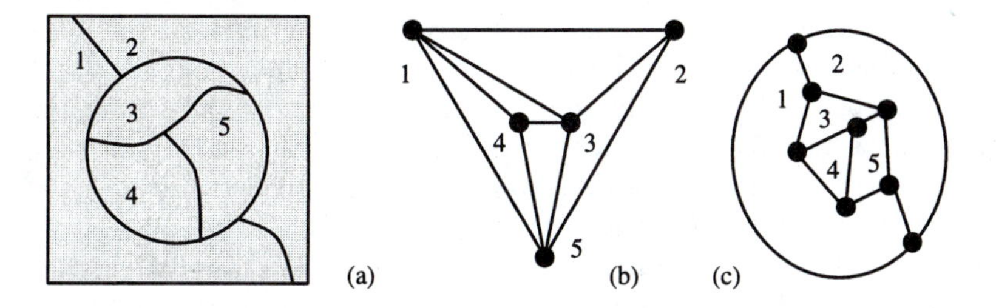

Figure 8.28: *Region adjacency graphs: (a) segmented image; (b) region adjacency graph; (c) dual graph.*

describes how to update the region adjacency graph and its dual after merging two regions R_i and R_j.

Algorithm 8.11: Updating a region adjacency graph and dual to merge two regions

1. *Region adjacency graph*

   (a) Add all nonexisting arcs connecting region R_i and all regions adjacent to R_j.

   (b) Remove the node R_j and all its arcs from the graph.

2. *Dual graph*

   (a) Remove all arcs corresponding to the boundaries between regions R_i and R_j from the graph.

(b) For each node associated with these arcs:
- If the number of arcs associated with the node is equal to 2, remove this node and combine the arcs into a single one.
- If the number of arcs associated with the node is larger than 2, update the labels of arcs that corresponded to parts of borders of region R_j to reflect the new region label R_i.

The region adjacency graph is one in which costs are associated with both nodes and arcs, implying that an update of these costs must be included in the given algorithm as node costs change due to the connecting two regions R_i and R_j.

8.6.1 Semantic region growing

A classical method of semantic region growing is now presented [Feldman and Yakimovsky 74]. Consider remotely sensed photographs, in which regions can be defined with interpretations such as *field, road, forest, town*, etc. It then makes sense to merge adjacent regions with the same interpretation into a single region. The problem is that the interpretation of regions is not known and the region description may give unreliable interpretations. In such a situation, it is natural to incorporate context into the region merging using a priori knowledge about relations (unary, binary) among adjacent regions, and then to apply constraint propagation to achieve globally optimal segmentation and interpretation throughout the image.

A region merging segmentation scheme is now considered in which semantic information is used in later steps, with the early steps being controlled by general heuristics similar to those given in Section 5.3. Only after the preliminary heuristics have terminated are semantic properties of existing regions evaluated, and further region merging is either allowed or restricted; these are steps 4 and 6 of the next algorithm. The same notation is used as in the previous section: A region R_i has properties X_i, its possible labels are denoted $\theta_i \in \{\omega_1, \dots, \omega_R\}$, and $P(\theta_i = \omega_k)$ represents the probability that the interpretation of the region R_i is ω_k.

Algorithm 8.12: Semantic region merging

1. Initialize a segmentation with many small regions.
2. Merge all adjacent regions that have at least one weak edge on their common boundary.
3. For preset constants c_1 and c_2, and threshold T_1, merge neighboring regions R_i and R_j if $S_{ij} \le T_1$, where

$$S_{ij} = \frac{c_1 + a_{ij}}{c_2 + a_{ij}} \qquad a_{ij} = \frac{(area_i)^{1/2} + (area_j)^{1/2}}{perimeter_i \; perimeter_j} \tag{8.31}$$

4. For all adjacent regions R_i and R_j, compute the conditional probability P that their mutual border B_{ij} separates them into two regions of the same interpretation ($\theta_i = \theta_j$), equation (8.34). Merge regions R_i and R_j if P is larger than a threshold T_2. If no two regions can be so merged, continue with step 5.

5. For each region R_i, compute the initial conditional probabilities

$$P(\theta_i = \omega_k | X_i) \quad k = 1, \dots, R \tag{8.32}$$

6. Repeat this step until all regions are labeled as *final*. Find a *non-final* region with the highest confidence C_i in its interpretation [equation (8.36)]; label the region with this interpretation and mark it as *final*. For each *non-final* region R_j and each of its possible interpretations $\omega_k, k = 1, \dots, R$, update the probabilities of its interpretations according to equation (8.37).

The first three steps of the algorithm do not differ in essence from Algorithm 5.19, but the final two steps, where semantic information has been incorporated, are very different and represent a variation of a serial relaxation algorithm combined with a depth-first interpretation tree search. The goal is to maximize an objective function,

$$F = \prod_{i,j=1,\dots,R} P(B_{ij} \text{ is between } \theta_i, \theta_j | X(B_{ij})) \prod_{i=1,\dots,R} P(\theta_i | X_i) \prod_{j=1,\dots,R} P(\theta_j | X_j) \tag{8.33}$$

for a given image partition.

The probability that a border B_{ij} between two regions R_i and R_j is a false one must be found in step 4. This probability P can be found as a ratio of conditional probabilities; let P_t denote the probability that the boundary should remain, and P_f denote the probability that the boundary is false (i.e., should be removed and the regions should be merged), and $X(B_{ij})$ denote properties of the boundary B_{ij}: Then

$$P = \frac{P_f}{P_t + P_f} \tag{8.34}$$

where

$$P_f = \sum_{k=1}^{R} P[\theta_i = \theta_j | X(B_{ij})] \, P(\theta_i = \omega_k | X_i) \, P(\theta_j = \omega_k | X_j)$$

$$P_t = \sum_{k=1}^{R} \sum_{l=1;k \neq l}^{R} P[\theta_i = \omega_k \text{ and } \theta_j = \omega_l | X(B_{ij})] \, P(\theta_i = \omega_k | X_i) \, P(\theta_j = \omega_l | X_j) \tag{8.35}$$

The confidence C_i of interpretation of the region R_i (step 6) can be found as follows. Let θ_i^1, θ_i^2 represent the two most probable interpretations of region R_i. Then

$$C_i = \frac{P(\theta_i^1 | X_i)}{P(\theta_i^2 | X_i)} \tag{8.36}$$

After assigning the final interpretation θ_f to a region R_f, interpretation probabilities of all its neighbors R_j (with *non-final* labels) are updated to maximize the objective function (8.33):

$$P_{\text{new}}(\theta_j) = P_{\text{old}}(\theta_j) \, P(B_{fj} \text{ is between regions labeled } \theta_f, \theta_j | X(B_{fj})) \tag{8.37}$$

The computation of these conditional probabilities is very expensive in terms of time and memory. It may be advantageous to compute them beforehand and refer to table values during processing; this table must have been constructed with suitable sampling.

It should be understood that appropriate models of the inter-relationship between region interpretations, the collection of conditional probabilities, and methods of confidence evaluation must be specified to implement this approach.

8.6.2 Genetic image interpretation

The previous section described the first historical semantic region growing method, which is still conceptually up to date. However, there is a fundamental problem in the region growing segmentation approach—the results are sensitive to the split/merge order (see Section 5.3). The conventional split-and-merge approach usually results in an undersegmented or an oversegmented image. It is practically impossible to stop the region growing process with a high confidence that there are neither too many nor too few regions in the image.

A method [Liow and Pavlidis 88, Pavlidis and Liow 90] was mentioned in Section 5.3.3 in which region growing always resulted in an oversegmented image and post-processing steps were used to remove false boundaries. Similar approach of removing false oversegmented regions can be found in a conceptually very different knowledge-based morphological region growing algorithm based on watersheds for graphs [Vincent and Soille 91]. Further, conventional region growing approaches are based on evaluation of homogeneity criteria and the goal is either to split a non-homogeneous region or to merge two regions, which may form a homogeneous region. Remember that the result is sensitive to the merging order; therefore, even if a merge results in a homogeneous region, it may not be optimal. In addition, there is no mechanism for seeking the optimal merges. Consequently, the semantic region growing approach to segmentation and interpretation starts with an oversegmented image in which some merges were not best possible. The semantic process is then trying to locate the maximum of some objective function by grouping regions which may already be incorrect and is therefore trying to obtain an optimal image interpretation from partially processed data where some significant information has already been lost. Further, conventional semantic region growing merges regions in an interpretation level only and does not evaluate properties of newly merged regions. It also very often ends in a local optimum of region labeling; the global optimum is not found because of the character of the optimization. Unreliability of image segmentation and interpretation of complex images results. The genetic image interpretation method solves these basic problems in the following manner.

- Both region merging and splitting is allowed; no merge or split is ever final, a better segmentation is looked for even if the current segmentation is already good.
- Semantics and higher-level knowledge are incorporated into the main segmentation process, not applied as post-processing after the main segmentation steps are over.
- Semantics are included in an objective evaluation function (that is similar to conventional semantic-based segmentation).
- In contrast to conventional semantic region growing, any merged region is considered a contiguous region in the semantic objective function evaluation, and all its properties are measured.

- The genetic image interpretation method does not look for local maxima; its search is likely to yield an image segmentation and interpretation specified by a (near) global maximum of an objective function.

The genetic image interpretation method is based on a **hypothesize and verify** principle. An objective function (similar to the objective functions used in previous sections) which evaluates the quality of a segmentation and interpretation is optimized by a genetic algorithm (the basics of which were presented in Section 7.6.1). The method is initialized with an oversegmented image called a **primary segmentation**, in which starting regions are called **primary regions**. Primary regions are repeatedly merged into current regions during the segmentation process. The genetic algorithm is responsible for generating new populations of feasible image segmentation and interpretation hypotheses.

An important property of genetic algorithms is that the whole population of segmentations is tested in a single processing step, in which better segmentations survive and others die (see Section 7.6.1). If the objective function suggests that some merge of image regions was a good merge, it is allowed to survive into the next generation of image segmentation (the code string describing that particular segmentation survives), while bad region merges are removed (their description code strings die).

The **primary region adjacency graph** is the adjacency graph describing the primary image segmentation. The **specific region adjacency graph** represents an image after the merging of all adjacent regions of the same interpretation into a single region (collapsing the primary region adjacency graph). The genetic algorithm requires any member of the processed population to be represented by a code string. Each primary region corresponds to one element in the code string; this correspondence is made once at the beginning of the segmentation/interpretation process. A region interpretation is given by the current code string in which each primary region of the image corresponds uniquely to some specific position. Each feasible image segmentation defined by a generated code string (segmentation hypothesis) corresponds to a unique specific region adjacency graph. The specific region adjacency graphs serve as tools for evaluating objective segmentation functions. The specific region adjacency graph for each segmentation is constructed by collapsing a primary region adjacency graph.

Design of a segmentation optimization function (the fitness function in genetic algorithms) is crucial for a successful image segmentation. The genetic algorithm is responsible for finding an optimum of the objective function. Nevertheless, the optimization function must really represent segmentation optimality. To achieve this, the function must be based on properties of image regions and on relations between the regions—a priori knowledge about the desired segmentation must be included in the optimization criterion.

An applicable objective function may be similar to that given in equation (8.29), keeping in mind that the number of regions N is not constant since it depends on the segmentation hypothesis.

The conventional approach evaluates image segmentation and interpretation confidences of all possible region interpretations. Based on the region interpretations and their confidences, the confidences of neighboring interpretations are updated, some being supported and others becoming less probable. This conventional method can easily end at a consistent but sub-optimal image segmentation and interpretation. In the genetic approach, the algorithm is fully responsible for generating new and increasingly better hypotheses about image segmentation.

Only these hypothetical segmentations are evaluated by the objective function (based on a corresponding specific region adjacency graph). Another significant difference is in the region property computation—as mentioned earlier, a region consisting of several primary regions is treated as a single region in the property computation process which gives a more appropriate region description.

Optimization criteria consist of three parts. Using the same notation as earlier, the objective function consists of

- A confidence in the interpretation θ_i of the region R_i according to the region properties X_i,

$$C(\theta_i|X_i) = P(\theta_i|X_i) \tag{8.38}$$

- A confidence in the interpretation θ_i of a region R_i according to the interpretations θ_j of its neighbors R_j,

$$C(\theta_i) = \frac{C(\theta_i|X_i) \sum_{j=1}^{N_A} [r(\theta_i, \theta_j) C(\theta_j|X_j)]}{N_A} \tag{8.39}$$

  where $r(\theta_i, \theta_j)$ represents the value of a compatibility function of two adjacent objects R_i and R_j with labels θ_i and θ_j, N_A is the number of regions adjacent to the region R_i (confidences C replace the probabilities P used in previous sections because they do not satisfy necessary conditions which must hold for probabilities; however, the intuitive meaning of interpretation confidences and interpretation probabilities remains unchanged).

- An evaluation of interpretation confidences in the whole image,

$$C_{\text{image}} = \frac{\sum_{i=1}^{N_R} C(\theta_i)}{N_R} \tag{8.40}$$

  or

$$C'_{\text{image}} = \sum_{i=1}^{N_R} \left(\frac{C(\theta_i)}{N_R} \right)^2 \tag{8.41}$$

  where $C(\theta_i)$ is computed from equation (8.39) and N_R is the number of regions in the corresponding specific region adjacency graph.

The genetic algorithm attempts to optimize the objective function C_{image}, which represents the confidence in the current segmentation and interpretation hypothesis.

As presented, the segmentation optimization function is based on both unary properties of hypothesized regions and on binary relations between these regions and their interpretations. A priori knowledge about the characteristics of processed images is used in evaluation of the local region confidences $C(\theta_i|X_i)$, and the compatibility function $r(\theta_i, \theta_j)$ represents the confidence that two regions with their interpretations can be present in an image in the existing configuration.

The method is described by the following algorithm.

Algorithm 8.13: Genetic image segmentation and interpretation

1. Initialize the segmentation into primary regions, and define a correspondence between each region and the related position of its label in the code strings generated by a genetic algorithm.

2. Construct a primary region adjacency graph.

3. Pick the starting population of code strings at random. If a priori information is available that can help to define the starting population, use it.

4. *Genetic optimization.* Collapse a region adjacency graph for each code string of the current population (Algorithm 8.11). Using the current region adjacency graphs, compute the value of the optimization segmentation function for each code string from the population.

5. If the maximum of the optimization criterion does not increase significantly in several consecutive steps, go to step 7.

6. Let the genetic algorithm generate a new population of segmentation and interpretation hypotheses. Go to step 4.

7. The code string with the maximum confidence (the best segmentation hypothesis) represents the final image segmentation and interpretation.

A simple example

Consider an image of a ball on a lawn (see Figure 8.29). Let the interpretation labeling be B for *ball* and L for *lawn*, and let the following higher-level knowledge be included: *There is a circular ball in the image* and *the ball is inside the green lawn region.* In reality, some more a priori knowledge would be added even in this simple example, but this knowledge will be sufficient for our purposes. The knowledge must be stored in appropriate data structures.

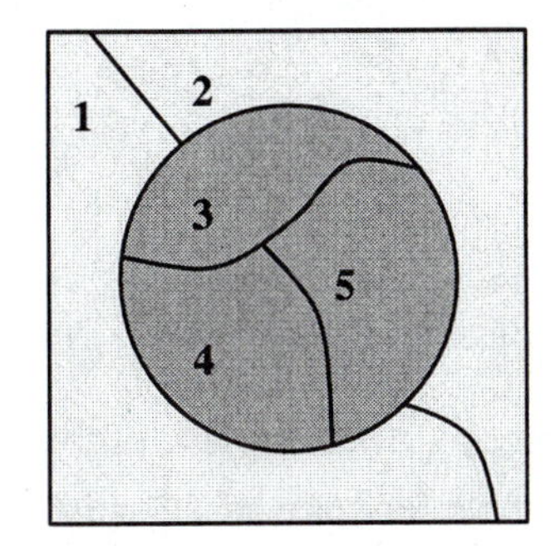

Figure 8.29: *A simulated scene 'ball on the lawn'.*

- Unary condition: Let the confidence that a region is a *ball* be based on its compactness (see Section 6.3.1),
$$C(\theta_i = B|X_i) = compactness(R_i) \tag{8.42}$$
and let the confidence that a region is *lawn* be based on its greenness,
$$C(\theta_i = L|X_i) = greenness(R_i) \tag{8.43}$$
Let the confidences for regions forming a perfect ball and perfect lawn be equal to one
$$C(B|circular) = 1 \qquad\qquad C(L|green) = 1$$
- Binary condition: Let the confidence that one region is positioned inside the other be given by a compatibility function
$$r(B \text{ is inside } L) = 1 \tag{8.44}$$
and let the confidences of all other positional combinations be equal to zero.

The unary condition says that the more compact a region is, the better its circularity, and the higher the confidence that its interpretation is a ball. The binary condition is very strict and claims that a ball can only be completely surrounded by a lawn.

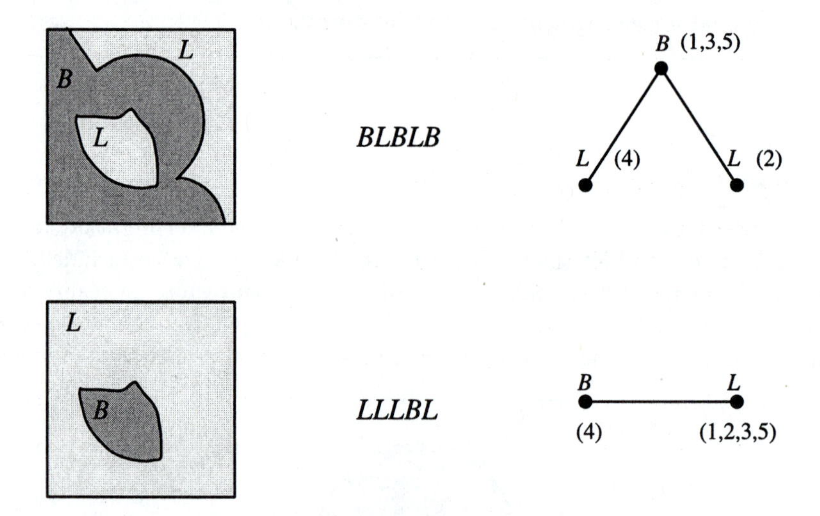

Figure 8.30: *Starting hypotheses about segmentation and interpretation: interpretation, corresponding code strings, and corresponding region adjacency graphs.*

Suppose the primary image segmentation consists of five primary regions $R_1, \ldots, R_5$ (see Figure 8.29); the primary region adjacency graph and its dual are in shown in Figure 8.28. Let the region numbers correspond to the position of region labels in code strings which are generated by the genetic algorithm as segmentation hypotheses and assume, for simplicity, that the starting population of segmentation hypotheses consists of just two strings (in any practical application the starting population would be significantly larger). Let the starting population be picked at random:

$$BLBLB$$
$$LLLBL$$

—this represents segmentation hypotheses as shown in Figure 8.30. After a random crossover between second and third positions, the population is as follows; confidences reflect the circularity of the region labeled *ball* and the positioning of the region labeled *ball* inside the *lawn* region—the confidence computation is based on equation (8.40):

$$\begin{array}{ll} BL|BLB & C_{\text{image}} = 0.00 \\ LL|LBL & C_{\text{image}} = 0.12 \\ LLBLB & C_{\text{image}} = 0.20 \\ BLLBL & C_{\text{image}} = 0.00 \end{array}$$

The second and the third segmentation hypotheses are the best ones, so they are reproduced and another crossover is applied; the first and the fourth code strings die (see Figure 8.31):

$$\begin{array}{ll} LLL|BL & C_{\text{image}} = 0.12 \\ LLB|LB & C_{\text{image}} = 0.20 \\ LLLLB & C_{\text{image}} = 0.14 \\ LLBBL & C_{\text{image}} = 0.18 \end{array}$$

After one more crossover,

$$\begin{array}{ll} LLBL|B & C_{\text{image}} = 0.20 \\ LLBB|L & C_{\text{image}} = 0.18 \\ LLBLL & C_{\text{image}} = 0.10 \\ LLBBB & C_{\text{image}} = 1.00 \end{array}$$

The code string (segmentation hypothesis) $LLBBB$ has a high (the highest achievable) confidence. If the genetic algorithm continues generating hypotheses, the confidence of the best hypothesis will not be any better and so it stops. The optimum segmentation/interpretation is shown in Figure 8.32.

Brain segmentation example

The previous example illustrated only the basic principles of the method. Practical applications require more complex a priori knowledge, the genetic algorithm has to work with larger string populations, the primary image segmentation has more regions, and the optimum solution is not found in three steps. Nevertheless, the principles remain the same as was demonstrated when the method is applied to more complex problems, and interpretation of human magnetic resonance brain images [Sonka et al. 96] is given here as such a complex example.

The genetic image interpretation method was trained on two-dimensional MR images depicting anatomically corresponding slices of the human brain. Knowledge about the unary properties of the specified neuroanatomic structures and about the binary properties between the structure pairs was acquired from manually traced contours in a training set of brain images (Figure 8.34a).

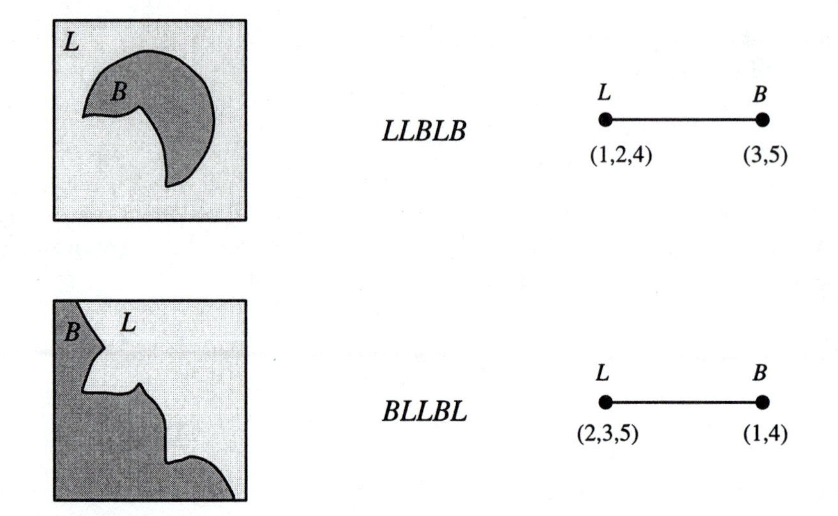

Figure 8.31: *Hypotheses about segmentation and interpretation: interpretations, corresponding code strings, and corresponding region adjacency graphs.*

Figure 8.32: *Optimal segmentation and interpretation: interpretation, corresponding code string, and region adjacency graph.*

As has been apparent from the definition of the global objective function C_{image} [equation (8.40)], the unary properties of individual regions, hypothesized interpretations of the regions, and binary relationships between regions contribute to the computation of the confidence C_{image}.

In our case, the unary region confidences $C(\theta_i|X_i)$ and the compatibility functions $r(\theta_i, \theta_j)$ were calculated based on the brain anatomy and MR image acquisition parameters. The following approach to the confidence calculations was used in the brain interpretation task [Sonka et al. 96]:

Unary confidences: The unary confidence of a region was calculated by matching the region's shape and other characteristic properties with corresponding properties representing the hypothesized interpretation (i.e., matching with the a priori knowledge).

Let the set of properties of region R_i be $X_i = \{x_{i1}, x_{i2}, \ldots, x_{iN}\}$. Matching was done for each characteristic of the region $\{x_{ij}\}$, and the unary confidence $C(\theta_i|X_i)$ was calculated as follows:

$$C(\theta_i|X_i) = P(x_{i1}) * P(x_{i2}) * \ldots * P(x_{iN}) \tag{8.45}$$

The feature confidences $P(x_{ik})$ were calculated by using the piecewise linear function shown in Figure 8.33. For example, let x_{ik} be the area of region R_i in the specific RAG and let R_i

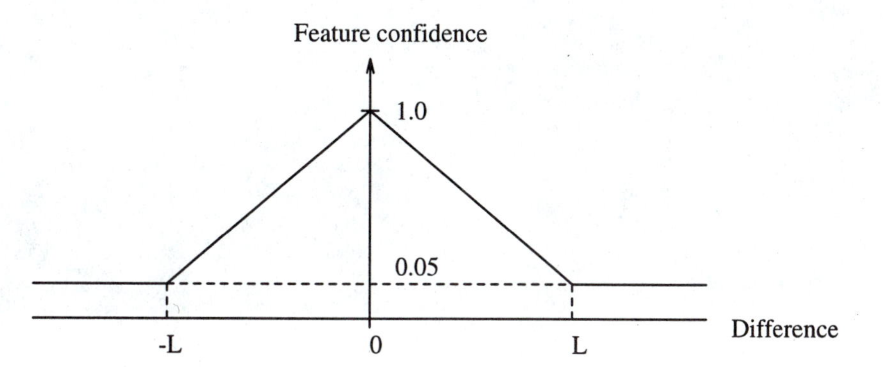

Figure 8.33: *Piecewise linear function for calculating unary confidences. L is a limit which depends on the a priori knowledge.*

be labeled θ_i. According to a priori knowledge, assume that an object labeled θ_i has an area y_{ik}. Then

$$P(x_{ik}) = \begin{cases} 1.0 - (0.95 * |x_{ik} - y_{ik}|)/L & : \; |x_{ik} - y_{ik}| < L \\ 0.05 & : \; \text{otherwise} \end{cases}$$

The limit L depends on the strength of the a priori knowledge for each particular feature.

Binary confidences: Binary confidences were defined between two regions based on their interrelationships.

The value of the compatibility function $r(\theta_i, \theta_j)$ was assigned to be in the range [0,1], depending on the strength of the a priori knowledge about the expected configuration of regions R_i and R_j.

For example, if a region R_i, labeled θ_i, is known always to be inside region R_j, labeled θ_j, then $r(\theta_i \text{ is\_inside } \theta_j) = 1$ and $r(\theta_j \text{ is\_outside } \theta_i) = 1$, whereas $r(\theta_j \text{ is\_inside } \theta_i) = 0$ and $r(\theta_i \text{ is\_outside } \theta_j) = 0$. Thus, low binary confidences serve to penalize infeasible configurations of pairs of regions.

Similarly to the calculation of the unary confidence, the compatibility function was calculated as follows:

$$r(\theta_i, \theta_j) = r(\theta_{ij1}) * r(\theta_{ij2}) * \ldots * r(\theta_{ijN}) \tag{8.46}$$

Here, $r(\theta_{ijk})$ is a binary relation (example: larger than/smaller than) between regions labeled θ_i and θ_j.

After the objective function C_{image} was designed using a number of brain images from the training set, the genetic brain image interpretation method was applied to testing brain images. For illustration, the primary region adjacency graph typically consisted of approximately 400 regions; a population of 20 strings and a mutation rate $\mu = 1/string\_length$ were used during the genetic optimization. The method was applied to a testing set of MR brain images and offered good image interpretation performance (Figure 8.34).

Conventional semantic region growing methods start with a non-semantic phase and use semantic post-processing to assign labels to regions. Based on the segmentation achieved in

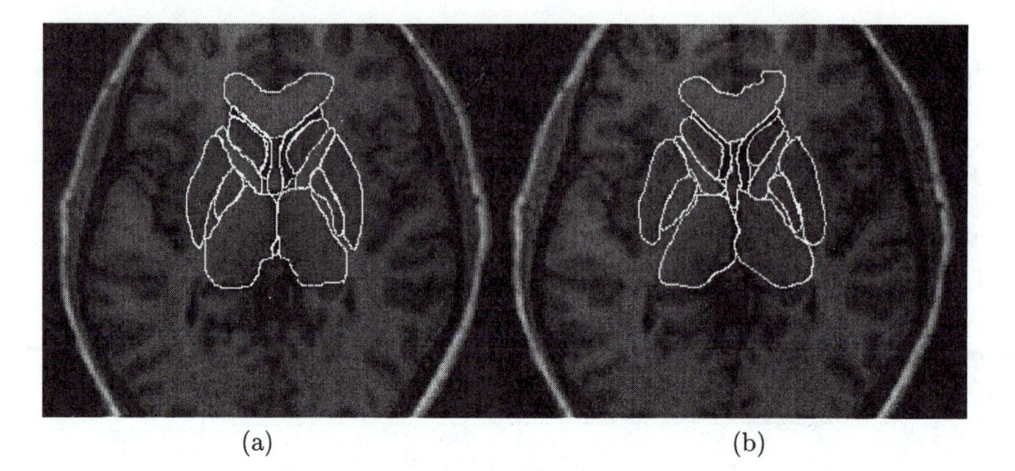

Figure 8.34: *Automated segmentation and interpretation of MR brain images: (a) observer-defined borders of the neuroanatomic structures correspond closely with (b) computer-defined borders.*

the region growing phases, the labeling process is trying to find a consistent set of interpretations for regions. The genetic image interpretation approach functions in a quite different way.

First, there are not separate phases. The semantics are incorporated into the segmentation/interpretation process. Second, segmentation hypotheses are generated first, and the optimization function is used only for evaluation of hypotheses. Third, a genetic algorithm is responsible for generating segmentation hypotheses in an efficient way.

The method can be based on any properties of region description and on any relations between regions. The basic idea of generating segmentation hypotheses solves one of the problems of split-and-merge region growing—the sensitivity to the order of region growing. The only way to re-segment an image in a conventional region growing approach if the semantic post-processing does not provide a successful segmentation is to apply feedback control to change region growing parameters in a particular image part. There is no guarantee that a global segmentation optimum will be obtained even after several feedback re-segmentation steps.

In the genetic image interpretation approach, no region merging is ever final. Natural and constant feedback is contained in the genetic interpretation method because it is part of the general genetic algorithm—this gives a good chance that a (near) global optimum segmentation/interpretation will be found in a single processing stage.

Note that throughout this chapter, the methods cannot and do not guarantee a correct segmentation—all the approaches try to achieve optimality according to the chosen optimization function. Therefore, a priori knowledge is essential to designing a good optimization function. A priori knowledge is often included into the optimization function in the form of heuristics, and moreover, may affect the choice of the starting population of segmentation hypotheses, which can affect computational efficiency.

An important property of this method is the possibility of parallel implementation. Similarly to the relaxation algorithm, this method is also naturally parallel. Moreover, there is a straightforward generalization leading to a genetic image segmentation and interpretation in three dimensions. Considering a set of image planes forming a three-dimensional image (such as MR or CT images), a primary segmentation can consist of regions in all image planes and can be represented by a 3D primary relational graph. The interesting possibility is to look for a global 3D segmentation and interpretation optimum using 3D properties of generated 3D regions in a single complex processing stage. In such an application, the parallel implementation would be a necessity.

8.7 Hidden Markov models

It is often possible when attempting image understanding to model the patterns being observed as a transitionary system. Sometimes these are transitions in time, but they may also be transitions through another pattern; for example, the patterns of individual characters when connected in particular orders represent another pattern that is a word. If the transitions are well understood, and we know the system state at a certain instant, they can be used to assist in determining the state at a subsequent point. This is a well-known idea, and one of the simplest examples is the **Markov model.**

A Markov model assumes a system may occupy one of a finite number of states $X_1, X_2, X_3, \ldots, X_n$ at times $t_1, t_2, \ldots$, and that the probability of occupying a state is determined solely by recent history. More specifically, a first-order Markov model assumes these probabilities depend only on the preceding state; thus a matrix $A = a_{ij}$ will exist in which

$$a_{ij} = P(\text{system is in state } j \mid \text{system was in state } i)$$

Thus $0 \leq a_{ij} \leq 1$ and $\sum_{j=1}^{n} a_{ij} = 1$ for all $1 \leq i \leq n$. The important point is that these parameters are time independent—the a_{ij} do not vary with t. A second-order model makes similar assumptions about probabilities depending on the last two states, and the idea generalizes obviously to order-k models for $k = 3, 4, \ldots$.

A trivial example might be to model weather forecasting: Suppose that the weather on a given day may be *sunny* (1), *cloudy* (2), or *rainy* (3) and that the day's weather depends probabilistically on the preceding day's weather *only.* We might be able to derive a matrix A,

$$A = \begin{array}{r} \\ \mathit{sun} \\ \mathit{cloud} \\ \mathit{rain} \end{array} \begin{array}{c} \begin{array}{ccc} \mathit{sun} & \mathit{cloud} & \mathit{rain} \end{array} \\ \begin{pmatrix} 0.5 & 0.25 & 0.25 \\ 0.375 & 0.125 & 0.375 \\ 0.125 & 0.625 & 0.375 \end{pmatrix} \end{array} \tag{8.47}$$

so the probability of rain after a sunny day is 0.125, the probability of cloud after a rainy day is 0.375, and so on.

In many practical applications, the states are not directly observable, and instead we observe a different set of states $Y_1, \ldots, Y_m$ (where possibly $n \neq m$), where we can only guess the exact state of the system from the probabilities

$$b_{jk} = P(Y_k \text{ observed} \mid \text{system is in state } j)$$

so $0 \leq b_{jk} \leq 1$ and $\sum_{k=1}^{m} b_{jk} = 1$. The $n \times m$ matrix B that is so defined is also time independent; that is, the observation probabilities do not depend on anything except the current state, and in particular not on how that state was achieved, or when.

Extending the weather example, it is widely believed that the moistness of a piece of seaweed is an indicator of weather; if we conjecture four states, *dry* 1, *dryish* 2, *damp* 3 or *soggy* 4,, and that the actual weather is probabilistically connected to the seaweed state, we might derive a matrix such as

$$B = \begin{array}{c} \\ dry \\ dryish \\ damp \\ soggy \end{array} \begin{array}{c} \begin{array}{ccc} sun & cloud & rain \end{array} \\ \begin{pmatrix} 0.6 & 0.25 & 0.05 \\ 0.2 & 0.25 & 0.1 \\ 0.15 & 0.25 & 0.35 \\ 0.05 & 0.25 & 0.5 \end{pmatrix} \end{array} \tag{8.48}$$

so the probability of observing dry seaweed when the weather is sunny is 0.6, the probability of observing damp seaweed when the weather is cloudy is 0.25, and so on.

A first-order **hidden Markov model (HMM)** $\lambda = (\pi, A, B)$ is specified by the matrices A and B together with an n-dimensional vector π to describe the probabilities of the state at time $t = 1$. The time-independent constraints are quite strict and in many cases unrealistic, but HMMs have seen significant practical application. In particular, they are successful in the area of speech processing [Rabiner 89], wherein the A matrix might represent the probability of a particular phoneme following another phoneme, and the B matrix refers to a feature measurement of a spoken phoneme (the Fourier spectrum, for example), where it is recognized that the fuzziness of speech means we cannot be certain which feature will be generated by which phoneme. Recently, the same ideas have seen wide application in optical character recognition (OCR) and related areas, where the A matrix might refer to letter successor probabilities, and again the B matrix is a probabilistic description of which features are generated by which letters.

A HMM poses three questions.

Evaluation: Given a model and a sequence of observations, what is the probability that the model actually generated those observations? If two different models are available, $\lambda_1 = (\pi_1, A_1, B_1)$ and $\lambda_2 = (\pi_2, A_2, B_2)$, this question indicates which one better describes some given observations. For example, if we have two models, a known weather sequence and a known sequence of seaweed observations, which model is the best description of the data?

Decoding: Given a model $\lambda = (\pi, A, B)$ and a sequence of observations, what is the most likely underlying state sequence? For pattern analysis, this is the most interesting question, since it permits an optimal estimate of what is happening on the basis of a sequence of feature measurements. For example, if we have a model and a sequence of seaweed observations, what is most likely to have been the underlying weather sequence?

Learning: Given knowledge of the set $X_1, X_2, X_3, \ldots, X_n$ and a sequence of observations, what are the best parameters π, A, B if the system is indeed a HMM? For example, given a known weather sequence and a known sequence of seaweed observations, what model parameters best describe them?

HMM Evaluation

To determine the probability that a particular model generated an observed sequence, it is straightforward to evaluate all possible sequences, calculate their probabilities, and multiply by the probability that the sequence in question generated the observations in hand. If

$$Y^k = (Y_{k_1}, Y_{k_2}, \ldots, Y_{k_T})$$

is a T long observation, and

$$X^i = (X_{i_1}, X_{i_2}, \ldots, X_{i_T})$$

is a state sequence, we require

$$P(Y^k) = \sum_{X^i} P(Y^k|X^i)P(X^i)$$

This quantity is given by summing over all possible sequences X^i, and for each such, determining the probability of the given observations; these probabilities are available from the B matrix, while the transition probabilities of X^i are available from the A matrix. Thus

$$P(Y^k) = \sum_{X^i} \pi(i_1) b_{i_1 k_1} \prod_{j=2}^{T} a_{i_{j-1} i_j} b_{i_j k_j}$$

Exhaustive evaluation over the X^i is possible since A, B, π are all available, but the load is exponential in T and clearly not in general computationally realistic. The assumptions of the model, however, permit a short cut by defining a recursive definition of partial, or intermediate, probabilities. Suppose

$$\alpha_t(j) = P(\text{state } X_j \text{ at time } t) \qquad 1 < t < T$$

Here t is between 1 and T, so this is an intermediate probability. Time independence allows us to write immediately

$$\alpha_{t+1}(j) = \sum_{i=1}^{n} [\alpha_t(i) a_{ij}] b_{jk_{t+1}}$$

since a_{ij} represents the probability of moving to state j and $b_{jk_{t+1}}$ is the probability of observing what we do at this time. Thus α is defined recursively; it may be initialized from our knowledge of the initial states,

$$\alpha_1(j) = \pi(j) b_{jk_1}$$

At time T, the individual quantities $\alpha_T(j)$ give the probability of the observed sequence occurring, with the actual system terminating state being X_j; therefore the total probability of the model generating the observed sequence Y_k is

$$P(Y^k) = \sum_{j=1}^{n} \alpha_T(j)$$

The recursive definition permits the calculation of this quantity in 'synchronous steps' without the need for exhaustively evaluating all sequences X^i individually. Any number of models

$\lambda_1 = (\pi_1, A_1, B_1)$, λ_2, λ_3, ... can be subjected to this **forward algorithm** [Baum and Eagon 63], and we would adopt the one with the maximal probability of causing the sequence observed:

$$\max_i [P(Y^k|\lambda_i)]$$

In particular, in OCR word recognition the individual patterns may be features extracted from characters or groups of characters, and an individual model may represent an individual word. We would determine which word was most likely to have generated an observed feature sequence.

HMM Decoding

Given that a particular model (π, A, B) generated an observation sequence of length T, $Y^k = (Y_{k_1}, \ldots, Y_{k_T})$, it is often not obvious what precise states the system passed through, $X^i = (X_{i_1}, \ldots, X_{i_T})$, and we therefore need an algorithm that will determine the most probable (or optimal in some sense) X^i given Y^k.

A simple approach might be to start at time $t = 1$ and ask what the most probable X_{i_1} would be, given the observation Y_{k_1}. Formally,

$$\begin{aligned} i_t &= \text{argmax}_j [P(X_j|Y_{k_t})] \\ &= \text{argmax}_j [P(Y_{k_t}|X_j)P(X_j)] \\ &= \text{argmax}_j [b_{jk_t} P(X_j)] \end{aligned} \tag{8.49}$$

which may be calculated given the probabilities of the X_j (or, more likely, some estimate thereof). This approach will generate an answer, but in the event of one or more observations being poor, a wrong decision may be taken for some t. It also has the possibility of generating illegal sequences (for example, a transition for which $a_{ij} = 0$). This frequently occurs in observation of noisy patterns, where an isolated best guess for a pattern may not be the same as the best guess taken in the context of a stream of patterns.

We do not decide on the value of i_t during the examination of the tth observation, but instead record how likely it is that a particular state *might* be reached, and if it were to be correct, which state was likely to have been its predecessor. Then at the Tth column, a decision can be taken about the final state X_T based on the entire history, which is fed back to the earlier stages —this is the **Viterbi algorithm** [Viterbi 67]. The approach is similar to that developed for dynamic programming (Section 5.2.5); we reconstruct the system evolution by imagining an $N \times T$ lattice of states; at time t we occupy one of the N possible X_i in the tth column. States in neighboring columns are connected by transition probabilities from the A matrix, but our view of this lattice (see Figure 8.35, cf. Figure 5.29) is attenuated by the observation probabilities B. The task is to find the route from the first to the Tth column of maximal probability, given the observation set.

Formally, we set

$$\delta_1(i) = \pi(i) b_{ik_1} \tag{8.50}$$

$$\delta_t(i) = \max_j [\delta_{t-1}(j) a_{ji} b_{ik_t}] \tag{8.51}$$

$$\phi_t(i) = \text{argmax}_j [\delta_{t-1}(j) a_{ji}] \tag{8.52}$$

$$i_T = \text{argmax}_i[\delta_T(i)] \tag{8.53}$$

$$i_t = \phi_{t+1}(i_{t+1}) \qquad t = T-1, \ldots, 1 \tag{8.54}$$

Here, equation (8.50) initializes the first lattice column, combining the π vector with the first observation. Equation (8.51) is a recursion relation to define the subsequent column from the predecessor, the transition probabilities, and the observation; this gives the ith element of the tth column, and informally is the probability of the 'most likely' way of being in that position, given events at time $t-1$. Equation (8.52) is a back pointer, indicating where one is most likely to have come from at time $t-1$ if currently in state i at time t (see Figure 8.36). Equation (8.53) indicates what the most likely state is at time T, given the preceding $T-1$ states and the observations. Equation (8.54) traces the back pointers through the lattice, initializing from the most likely final state.

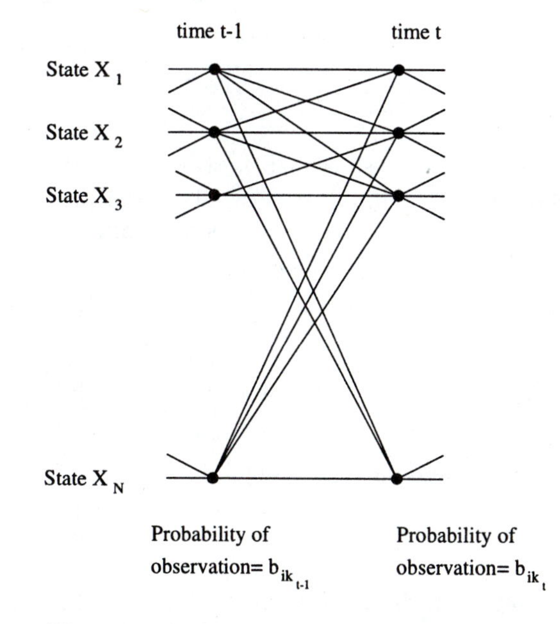

Figure 8.35: *Part of a Markov model lattice.*

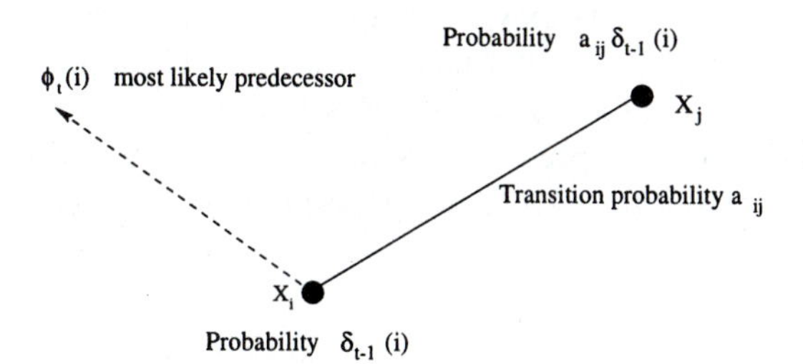

Figure 8.36: *A close-up of the HMM lattice; moving from state i at time $t-1$ to state j at time t.*

A simple example will illustrate this, considering the weather transition probabilities [equation (8.47)] and the seaweed observation probabilities [equation (8.48)], we might conjecture, without prior information, that the weather states on any given start day have equal probabilities, so $\pi = (\frac{1}{3}, \frac{1}{3}, \frac{1}{3})$. Suppose now we imagine a weather observer in a closed, locked room with a piece of seaweed—if on four consecutive days the seaweed is *dry, dryish, soggy, soggy*, the observer wishes to calculate the most likely sequence of weather states that have caused these observations. Starting with the observation *dry*, the first column of probabilities becomes [equation (8.50)];

$$\begin{aligned}
P(\text{dry observation and sunny weather}) &= \delta_1(1) \\
&= 0.333 \times 0.6 = 0.2 \\
P(\text{dry observation and cloudy weather}) &= \delta_1(2) \\
&= 0.333 \times 0.25 = 0.0833 \\
P(\text{dry observation and rainy weather}) &= \delta_1(3) \\
&= 0.333 \times 0.05 = 0.0167
\end{aligned} \tag{8.55}$$

—as expected, the *sunny* state is most probable. Now reasoning about the second day, $\delta_2(1)$ gives the probability of observing *dryish* seaweed on a *sunny* day, given the preceding day's information. For each of the three possible preceding states, we calculate the explicit probability and select the largest [equation (8.51)];

$$\begin{aligned}
&P(\text{seaweed is dryish and day 2 is sunny} \mid \text{day 1 is sunny}) \\
&\qquad = 0.2 \times 0.5 \times 0.2 = 0.02 \\
&P(\text{seaweed is dryish and day 2 is sunny} \mid \text{day 1 is cloudy}) \\
&\qquad = 0.0833 \times 0.25 \times 0.2 = 0.00417 \\
&P(\text{seaweed is dryish and day 2 is sunny} \mid \text{day 1 is rainy}) \\
&\qquad = 0.0167 \times 0.25 \times 0.2 = 0.000833
\end{aligned} \tag{8.56}$$

Thus the most probable way of reaching the *sunny* state on day 2 is from day 1 being sunny too; accordingly, we record $\delta_2(1) = 0.02$ and store the back pointer $\phi_2(1) = 1$ [equation (8.52)]. In a similar way, we find $\delta_2(2) = .0188$, $\phi_2(2) = 1$ and $\delta_2(3) = .00521$, $\phi_2(3) = 2$.

δ probabilities and back pointers may be computed similarly for the third and fourth days; we discover $\delta_4(1) = 0.00007$, $\delta_4(2) = 0.00055$, $\delta_4(3) = 0.0011$—thus the most probable final state *given all preceding information*, is rainy. We select this [equation (8.53)], and follow the ϕ back pointers of most probable predecessors to determine the optimal sequence [equation (8.54)]. In this case, it is *sunny, sunny, rainy, rainy*, which accords well with expectation given the model.

HMM Learning

The task of learning the best model to fit a given observation sequence is the hardest of the three associated with HMMs, but an estimate (often sub-optimal) can be made. An initial model is guessed, and this is refined to give a higher probability of providing the observations in hand via the **forward-backward** or **Baum-Welch** algorithm. This is essentially a gradient descent of an error measure of the current best model, and is a special case of the EM (estimate-maximize) algorithm [Baum et al. 70]. Details to assist in implementation are available in the literature [Rabiner and Juang 86].

Applications

Early uses of the HMM approach were predominantly in speech recognition, where it is not hard to see how a different model may be used to represent each word, how features may be extracted, and how the global view of the Viterbi algorithm would be necessary to recognize phoneme sequences correctly through noise and garble [Rabiner 89, Huang et al. 90]. HMMs are actively used in commercial speech recognizers [Green 95]. Wider applications in natural language processing have also been seen [Sharman 89].

The same ideas translate naturally into the related language recognition domain of OCR and handwriting recognition. One use [Hanlon and Boyle 92a, Hanlon and Boyle 92b, Hanlon 94] has been to let the underlying state sequence be grammatical tags, while the observations are features derived from segmented words in printed and handwritten text; the patterns of English grammar (which are not, of course, a first-order Markov model) closely restrict which words may follow which others, and this reduction of the size of candidate sets can be seen to assist enormously in recognition.

Similarly, HMMs lend themselves to analysis of letter sequences in text [Kundu et al. 89]; here, the transition probabilities are derived empirically from letter frequencies and patterns (so for example, 'q' is nearly always followed by 'u', and very rarely by 'j'), and the observation probabilities are the output of an OCR system—these are derived from a range of pattern features such as those described in Chapter 6. This system is seen to improve in performance when a second-order Markov model is deployed.

At a lower level, HMMs can be used to recognize individual characters. This may be done by skeletonizing characters and considering the sequence of stroke primitives to be a Markov process [Vlontzos and Kung 92]. Alternatively, vertical and horizontal projections (see Section 6.3.1) of binarized character images may be considered [Elms and Illingworth 94]. Observed through noise, a Fourier transform of the projections is derived as a feature vector, and a HMM for each possible character is trained using the Baum-Welch algorithm. Unknown characters are then identified by determining the best scoring model for features derived from an unseen image.

More recently, HMMs have found favor in analysis of visual sequences. Recognition of sign language from video has proved possible [Yamato et al. 92, Schlenzig et al. 94, Starner and Pentland 97], and Markov models have significant success in describing transitions between sub-models generated by PDMs (see Section 8.3) [Heap 97, Heap 98]. HMMs have also been applied successfully to lip and face tracking in real time [Oliver et al. 97], lip reading [Harvey et al. 97], and recognition of complex actions [Brand et al. 97].

8.8 Summary

- **Image understanding**
  - Machine vision consists of **lower** and **upper processing levels**, and image understanding is the highest processing level in this classification.

  - The main **computer vision goal** is to achieve machine behavior similar to that of biological systems by applying technically available procedures.

- Image understanding **control strategies**

  - **Parallel** and **serial** processing control

    * Parallel processing makes several computations simultaneously.
    * Serial processing operations are sequential.
    * Almost all low-level image processing can be done in parallel. High-level processing using higher levels of abstraction is usually serial in essence.

  - **Hierarchical** control

    * **Control by the image data (bottom-up control)**: Processing proceeds from the raster image to segmented image, to region (object) description, and to recognition.
    * **Model-based control (top-down control)**: A set of assumptions and expected properties is constructed from applicable knowledge. The satisfaction of those properties is tested in image representations at different processing levels in a top-down direction, down to the original image data. The image understanding is an internal model verification, and the model is either accepted or rejected.
    * **Combined** control uses both data driven and model driven control strategies.

  - **Non-hierarchical** control does not distinguish between upper and lower processing levels; non-hierarchical control can be seen as a cooperation of competing experts at the same level (often) using the **blackboard** principle. The blackboard is a shared data structure that can be accessed by multiple experts.

- **Active contour models—snakes**

  - A snake is an energy minimizing spline—the snake's energy depends on its shape and location within the image. Local minima of this energy then correspond to desired image properties.
  - The energy functional which is minimized is a weighted combination of **internal** and **external forces**.

- **Point distribution models (PDMs)**

  - A PDM is a shape description technique applicable to locating new instances of related shapes in other images. It is most useful for describing features that have well-understood 'general' shape, but which cannot be easily described by a rigid model due to shape variability.
  - The PDM approach requires the existence of a training set of examples (often shape landmarks), from which to derive a statistical description of the shape and its variation.

  - PDMs describe the **modes of variation**—directions of maximum variation are ordered so that it is known where the variation in the model is most likely to occur. Most of the shape variations are usually captured in a small number of modes of variation.

- **Pattern recognition** in image understanding
  - Supervised or unsupervised pattern recognition methods may be used for pixel classification. In the image understanding stage, feature vectors derived from local multi-spectral image values of image pixels are presented to the classifier, which assigns a label to each pixel of the image. Image understanding is then achieved by pixel labeling.
  - The resulting labeled image may have many small regions, which may be misclassified. **Context-based post-processing** approaches can be applied to avoid this misclassification.

- **Scene labeling, constraint propagation**
  - **Discrete** labeling allows only one label to be assigned to each object in a final labeling. Effort is directed to achieving a consistent labeling throughout the image. Discrete labeling always finds either a consistent labeling or detects the impossibility of assigning consistent labels to the scene.
  - **Probabilistic** labeling allows multiple labels to co-exist in objects. Labels are weighted probabilistically, with a label confidence being assigned to each object label. Probabilistic labeling always gives an interpretation result together with a measure of confidence in the interpretation.
  - The **constraint propagation** principle facilitates local consistencies adjusting to global consistencies (global optima) in the whole image.
  - Object labeling depends on the **object properties** and on a **measure of compatibility** of the potential object labels with the labeling of other directly interacting objects. Distant objects still interact with each other due to constraint propagation.
  - When **searching interpretation trees**, tree nodes are assigned all possible labels, and a depth-first search based on back-tracking is applied. An interpretation tree search tests all possible labelings.

- **Semantic image segmentation and understanding**
  - **Semantic region growing** techniques incorporate context into the region merging using a priori knowledge about relations among adjacent regions, and then to apply constraint propagation to achieve globally optimal segmentation and interpretation throughout the image.
  - **Genetic image interpretation** is based on a **hypothesize and verify** principle. An objective function which evaluates the quality of a segmentation and interpretation is optimized by a genetic algorithm that is responsible for generating new populations of image segmentation and interpretation hypotheses to be tested.

- **Hidden Markov models**
  - When attempting image understanding, patterns being observed may be modeled as a **transitionary system**. If the transitions are well understood, and the system state is known at a certain instant, transitions can be used to assist in determining the state at a subsequent point. **Markov models** represent one of the simplest examples of this idea
  - **Hidden Markov models** pose three questions: **evaluation**, **decoding**, and **learning**.
  - The **Viterbi algorithm** can be used to reconstruct the system evolution from possibly inaccurate observations.

8.9 Exercises

Short-answer questions

1. Explain how human vision differs from computer vision. Why is the problem of image understanding so difficult when every small child "knows how to do it"?
2. Explain the differences between lower and upper processing levels.
3. Explain the main ideas of the following image understanding control strategies; if possible, provide a block diagram. Specify their primary applicability within the image understanding process.
   (a) Serial control
   (b) Parallel control
   (c) Bottom-up control
   (d) Top-down control
   (e) Combined control
   (f) Non-hierarchical blackboard control
4. What is the difference between short-term and long-term memory?
5. Give a real-world example of an image understanding application (different from those in text) in which a bottom-up control strategy may be used.
6. Give a real-world example of an image understanding application (different from those in text) in which a model-based (top-down) control strategy may be used.
7. Give a real-world example of an image understanding application (different from those in text) in which a combined control strategy may be used.
8. Give a general expression for the energy functional that is minimized during snake convergence.
9. Explain the conceptual difference between snakes and balloons.
10. What kind of information can be represented by point distribution models?
11. Explain the process of determining the modes of variation represented by point distribution models. How is it possible that only a small number of modes of variation can be used to cover most of the shape variations represented by the model?
12. Discuss the importance of context in image understanding.

13. Explain the main strategy of discrete labeling. Give a real-world example of image interpretation to which discrete labeling can be applied, providing details.
14. Explain why discrete relaxation is a special case of probabilistic relaxation.
15. Define an order-k Markov model.
16. Define an order-k Hidden Markov model.
17. Define in the context of HMMs:
    - Evaluation
    - Decoding
    - Learning

Problems

1. Design an internal energy term of a snake to facilitate closely tracing sharp corners.
2. Design an image energy term of a snake to trace lines of gray-level $G = 128$.
3. Develop a program for snake-based line detection; use the energy functional specified by equation (8.1). Test its behavior in
    (a) Ideal images of contrasting lines on a homogeneous background.
    (b) Images of incomplete contrasting lines on a homogeneous background.
    (c) Images with superimposed impulse noise of varying severity—can the adverse effect of noise be overcome be adjusting the parameters of the snake functional?
4. Considering information provided in Table 8.1, how many principal components must be used in the point distribution model to leave less than 5% variation unexplained?
5. Develop a function (subroutine) for shape alignment as described in Algorithm 8.5. Test its functionality using a variety of artificial shapes.
6. Obtain a database of examples of some non-rigid shape (these may be synthetic). Choose some suitable number of landmark points and determine their 'best' placing on your dataset (if you are able to describe and develop an *automatic* placer, do so). Align all your examples using the program developed in Problem 8.5.
7. Using the function written in Problem 8.5, develop a program determining mean shape and all modes of variation of a set of shapes. Apply the program to five sets consisting of at least 10 shapes each. Give the results in a form of a table similar to Table 8.1.
8. Implement Algorithm 8.6, and test it on some instances of your shape that do not appear in the training dataset. How good is the fitting algorithm? Is it possible to characterize any errors made? If so, how might they be circumvented?
9. Suppose a sequence of images of a running cat are obtained from video. The silhouette of the animal may be extracted, and a PDM constructed. Suppose now the velocity of the silhouette's centroid (direction and speed) are also estimated, and these two features are used to augment the set of landmarks points; how might the PDM be influenced? (Hint: Consider what happens if the speed is recorded in pixels/second, miles/hour, inches/second or miles/year).
10. Give an example of how statistical pattern recognition can be used in classification of multi-spectral satellite image data. Specify possible features, classes, training set, testing set, type of classifier, and post-processing steps.

11. Implement a classifier-based image interpretation system to recognize objects of different colors in RGB color images. First, test the system using artificial color images. Next, test the system's performance in color images digitized from a scanner or a color TV camera. If the system's performance in digitized color images is unsatisfactory, implement some form of contextual post-processing. Discuss the improvements achieved.
12. Considering the recursive contextual classification approach, after how many recursive steps will image information at (x, y) location (53,145) influence the labeling at location (45,130)?
13. Implement the method for contextual image classification as given in Algorithm 8.7. Test in artificial and real-world images.
14. Implement the method for recursive contextual image classification given in Algorithm 8.8. Test in artificial and real-world images.
15. Develop a program for image interpretation using discrete relaxation as described in Algorithm 8.9. Design a complete set of unary and binary properties needed to interpret scenes similar to that given in Figure 8.24. Test in computer-generated images that do and do not belong to the set of office scenes described.
16. Explain why the objective function given in equation (8.29) is appropriate for image interpretation. Explain how each of the terms contributes to image interpretation.
17. Develop a program for region adjacency graph construction from a segmented and labeled image.
18. Using the program developed in Problem 8.17, create a program for region adjacency graph updating after two or more regions are merged.
19. Sketch a detailed block diagram of the genetic image interpretation method discussed in Section 8.6.2. How is the genetic algorithm used for generating segmentation and interpretation hypotheses?
20. Considering the 'ball on the lawn' example (Figure 8.29), sketch a specific region adjacency graph for the image segmentation and interpretation hypotheses represented by the following genetic strings: (a) LLBLB, (b) LLBBL, (c) BLLLB.
21. Implement the Viterbi algorithm.
22. Determine, either from literature or empirically, the transition probabilities of the characters of English text (it will help to considers only letters and spaces, not case and punctuation). Define some feature measurements (simple ones may be based on bay and lake counts—see Section 2.3.1). Thereby construct a first-order HMM for letter transitions.

    Use your model to decode a symbol stream. Where does it make mistakes?

    If possible, refine your feature set to improve the performance.

8.10 References

[Algorri et al. 91] M E Algorri, D R Haynor, and Y Kim. Contextual classification of multiple anatomical tissues in tomographic images. In *Proceedings of the Annual International Conference IEEE EMBS, Vol.13, 1991,* Orlando, FL, pages 106–107, IEEE, Piscataway, NJ, 1991.

[Ambler 75] A P H Ambler. A versatile system for computer controlled assembly. *Artificial Intelligence,* 6(2):129–156, 1975.

[Amini et al. 88] A Amini, S Tehrani, and T Weymouth. Using dynamic programming for minimizing the energy of active contours in the presence of hard constraints. In *2nd International Conference on Computer Vision,* Tarpon Springs, FL, pages 95–99, IEEE, Piscataway, NJ, 1988.

[Amini et al. 90] A Amini, T Weymouth, and R Jain. Using dynamic programming for solving variational problems in vision. *IEEE Transactions on Pattern Analysis and Machine Intelligence*, 12(9):855–867, 1990.

[Bajcsy and Rosenthal 80] R Bajcsy and D A Rosenthal. Visual and conceptual focus of attention. In S Tanimoto and A Klinger, editors, *Structured Computer Vision*, pages 133–154. Academic Press, New York, 1980.

[Ballard and Brown 82] D H Ballard and C M Brown. *Computer Vision.* Prentice-Hall, Englewood Cliffs, NJ, 1982.

[Barrow and Tenenbaum 76] H G Barrow and I M Tenenbaum. MSYS: A system for reasoning about scenes. Technical Report 121, Stanford Research Institute, Menlo Park, CA, 1976.

[Basu 87] S Basu. Image segmentation by semantic method. *Pattern Recognition*, 20(5):497–511, 1987.

[Baum and Eagon 63] L E Baum and J Eagon. An inequality with applications to statistical prediction for functions of Markov processes and to a model for ecology. *Bulletin of the American Mathematical Society*, 73:360–363, 1963.

[Baum et al. 70] L E Baum, T Petrie, G Soules, and M Weiss. A maximization technique occurring in the statistical analysis of probabilistic functions of Markov chains. *Annals of Mathematical Statistics*, 41:164–167, 1970.

[Baumberg 95] A M Baumberg. Learning deformable models for tracking human motion. PhD thesis, School of Computer Studies, University of Leeds, Leeds, UK, 1995.

[Berger and Mohr 90] M O Berger and R Mohr. Towards autonomy in active contour models. In *10th International Conference on Pattern Recognition,* Atlantic City, NJ, pages 847–851, IEEE, Piscataway, NJ, 1990.

[Berthod and Faugeras 80] M Berthod and O D Faugeras. Using context in the global recognition of a set of objects: An optimisation approach. In *Proceedings of the 8th World Computing Congress (IFIP),* Tokyo, Japan, pages 695–698, 1980.

[Bhandarker and Suk 90] S Bhandarker and M Suk. Computer vision as a coupled system. In *Applications of Artificial Intelligence VIII,* Orlando, FL, pages 43–54, SPIE, Bellingham, WA, 1990.

[Blake 82] A Blake. A convergent edge relaxation algorithm. Technical Report MIP-R-135, Machine Intelligence Unit, University of Edinburgh, 1982.

[Brand et al. 97] M Brand, N Oliver, and A P Pentland. Coupled hidden Markov models for complex action recognition. In *Computer Vision and Pattern Recognition*, pages 994–999, IEEE Computer Society, Los Alamitos, CA, 1997.

[Brooks et al. 79] R A Brooks, R Greiner, and T O Binford. The ACRONYM model-based vision system. In *Proceedings of the International Joint Conference on Artificial Intelligence, IJCAI-6,* Tokyo, Japan, pages 105–113, 1979.

[Bulpitt and Efford 96] A J Bulpitt and N D Efford. An efficient 3D deformable model with a self optimising mesh. *Image and Vision Computing*, 14(8):573–580, August 1996.

[Cabello et al. 90] D Cabello, A Delgado, M J Carreira, J Mira, R Moreno-Diaz, J A Munoz, and S Candela. On knowledge-based medical image understanding. *Cybernetics and Systems*, 21(2-3):277–289, 1990.

[Chiou and Jenq-Neng 95] G I Chiou and H Jenq-Neng. A neural network-based stochastic active contour model (NNS-SNAKE) for contour finding of distinct features. *IEEE Transactions on Image Processing*, 4:1407–1416, 1995.

[Christmas et al. 96] W J Christmas, J Kittler, and M Petrou. Labelling 2-D geometric primitives using probabilistic relaxation: reducing the computational requirements. *Electronics Letters*, 32:312–314, 1996.

[Cohen 91a] G A Cohen. Optimization of radiologic imaging through anatomic classification: An application to magnetic resonance imaging. PhD thesis, University of Iowa, 1991.

[Cohen 91b] L D Cohen. On active contour models and balloons. *CVGIP – Image Understanding*, 53(2):211–218, 1991.

[Cohen and Cohen 92] L D Cohen and I Cohen. Deformable models for 3D medical images using finite elements & balloons. In *Proceedings, IEEE Conference on Computer Vision and Pattern Recognition,* Champaign, IL, pages 592–598, IEEE, Los Alamitos, CA, 1992.

[Collins et al. 91] S M Collins, C J Wilbricht, S R Fleagle, S Tadikonda, and M D Winniford. An automated method for simultaneous detection of left and right coronary borders. In *Computers in Cardiology 1990,* Chicago, IL, page 7, IEEE, Los Alamitos, CA, 1991.

[Comaniciu and Meer 97] D Comaniciu and P Meer. Robust analysis of feature spaces: Color image segmentation. In *Computer Vision and Pattern Recognition*, pages 750–755, IEEE Computer Society, Los Alamitos, CA, 1997.

[Cootes and Taylor 92] T F Cootes and C J Taylor. Active shape models—'smart snakes'. In D C Hogg and R D Boyle, editors, *Proceedings of the British Machine Vision Conference,* Leeds, UK, pages 266–275, Springer Verlag, London, 1992.

[Cootes et al. 92] T F Cootes, C J Taylor, D H Cooper, and J Graham. Training models of shape from sets of examples. In D C Hogg and R D Boyle, editors, *Proceedings of the British Machine Vision Conference,* Leeds, UK, pages 9–18, Springer Verlag, London, 1992.

[Cootes et al. 94] T F Cootes, C J Taylor, and A Lanitis. Active shape models: Evaluation of a multi-resolution method for improving image search. In E Hancock, editor, *Proceedings of the British Machine Vision Conference,* York, UK, volume 1, pages 327–336. BMVA Press, 1994.

[Cornic 97] P Cornic. Another look at the dominant point detection of digital curves. *Pattern Recognition Letters*, 18:13–25, 1997.

[Devijver and Kittler 86] P A Devijver and J Kittler. *Pattern Recognition Theory and Applications.* Springer Verlag, Berlin–New York–Tokyo, 1986.

[Dew et al. 89] P M Dew, R A Earnshaw, and T R Heywood, editors. *Parallel Processing for Computer Vision and Display.* Addison-Wesley, Reading, MA, 1989.

[Duta and Sonka 97] N Duta and M Sonka. Segmentation and interpretation of MR brain images using an improved knowledge-based active shape model. In J Duncan and G Gindi, editors, *Information Processing in Medical Imaging,* pages 375–380, Springer Verlag, Berlin; New York, 1997.

[Efford 93] N D Efford. Knowledge-based segmentation and feature analysis of hand and wrist radiographs. *Proceeding of the SPIE*, 1905:596–608, 1993.

[Elfving and Eklundh 82] T Elfving and J O Eklundh. Some properties of stochastic labeling procedures. *Computer Graphics and Image Processing*, 20:158–170, 1982.

[Elms and Illingworth 94] A J Elms and J Illingworth. Combining HMMs for the recognition of noisy printed characters. In E Hancock, editor, *Proceedings of the British Machine Vision Conference,* York, UK, volume 2, pages 185–194. BMVA Press, 1994.

[Etoh et al. 93] M Etoh, Y Shirai, and M Asada. Active contour extraction based on region descriptions obtained from clustering. *Systems and Computers in Japan*, 24:55–65, 1993.

[Feldman and Yakimovsky 74] J A Feldman and Y Yakimovsky. Decision theory and artificial intelligence: A semantic–based region analyzer. *Artificial Intelligence*, 5:349–371, 1974.

[Fischer and Niemann 93] V Fischer and H Niemann. Parallelism in a semantic network for image understanding. In A Bode and M Dal Cin, editors, *Parallel Computer Architectures Theory,* pages 203–218, 1993.

[Fisher 93] R B Fisher. Hierarchical matching beats the non-wildcard and interpretation tree model matching algorithms. In J Illingworth, editor, *Proceedings of the British Machine Vision Conference,* Surrey, UK, volume 1, pages 589–598. BMVA Press, 1993.

[Fisher 94] R B Fisher. Performance comparison of ten variations of the interpretation tree matching algorithm. In J-O Eklundh, editor, *3rd European Conference on Computer Vision,* Stockholm, Sweden, volume 1, pages 507–512, Springer Verlag, Berlin, 1994.

[Fletcher et al. 96] S Fletcher, A Bulpitt, and D Hogg. Global alignment of MR images using a scale based hierarchical model. In B Buxton and R Cipolla, editors, *4th European Conference on Computer Vision,* Cambridge, England, volume 2, pages 283–292, Springer Verlag, Berlin, 1996.

[Franklin 90] S E Franklin. Topographic context of satellite spectral response. *Computers & Geosciences*, 16(7):1003–1010, 1990.

[Fugeiredo et al. 97] M A T Fugeiredo, J M N Leitao, and A K Jain. Adaptive B-splines and boundary estimation. In *Computer Vision and Pattern Recognition*, pages 724–730, IEEE Computer Society, Los Alamitos, CA, 1997.

[Fung et al. 90] P W Fung, G Grebbin, and Y Attikiouzel. Contextual classification and segmentation of textured images. In *Proceedings of the 1990 International Conference on Acoustics, Speech, and Signal Processing—ICASSP 90,* Albuquerque, NM, pages 2329–2332, IEEE, Piscataway, NJ, 1990.

[Gelgon and Bouthemy 97] M Gelgon and P Bouthemy. A region-level graph labeling approach to motion-based segmentation. In *Computer Vision and Pattern Recognition*, pages 514–519, IEEE Computer Society, Los Alamitos, CA, 1997.

[Geman and Geman 84] S Geman and D Geman. Stochastic relaxation, Gibbs distributions, and the Bayesian restoration of images. *IEEE Transactions on Pattern Analysis and Machine Intelligence*, 6(6):721–741, 1984.

[Gerig et al. 92] G Gerig, J Martin, R Kikinis, O Kubler, M Shenton, and F A Jolesz. Unsupervised tissue type segmentation of 3D dual-echo MR head data. *Image and Vision Computing*, 10(6):349–360, 1992.

[Ghosh and Harrison 90] J Ghosh and C G Harrison, editors. *Parallel Architectures for Image Processing,* Santa Clara, CA, Bellingham, WA, 1990. SPIE.

[Gong and Kulikowski 95] L Gong and C A Kulikowski. Composition of image analysis processes through object-centered hierarchical planning. *IEEE Transactions on Pattern Analysis and Machine Intelligence*, 17:997–1009, 1995.

[Gonzalez and Lopez 89] A F Gonzalez and S S Lopez. Classification of satellite images using contextual classifiers. *Digest—International Geoscience and Remote Sensing Symposium (IGARSS)*, 2:645–648, 1989.

[Green 95] T Green. A word in your ear. *Personal Computer World*, pages 354–370, April 1995.

[Grimson 90] W E L Grimson. *Object Recognition by Computer: The Role of Geometric Constraints.* MIT Press, Cambridge, MA, 1990.

[Grimson and Lozano-Perez 87] W E L Grimson and T Lozano-Perez. Localizing overlapping parts by searching the interpretation tree. *IEEE Transactions on Pattern Analysis and Machine Intelligence*, 9(4):469–482, 1987.

[Gunn and Nixon 95] S R Gunn and M S Nixon. Improving snake performance via a dual active contour. In V Hlavac and R Sara, editors, *International Conference on Computer Analysis of Images and Patterns,* Prague, Czech Republic, pages 600–605, Springer Verlag, Heidelberg, September 6th–8th, 1995.

[Hancock 93] E R Hancock. Resolving edge-line ambiguities using probabilistic relaxation. In *Proceedings Computer Vision and Pattern Recognition*, pages 300–306, IEEE, Los Alamitos, CA, 1993.

[Hancock and Kittler 90a] E R Hancock and J Kittler. Discrete relaxation. *Pattern Recognition*, 23(7):711–733, 1990.

[Hancock and Kittler 90b] E R Hancock and J Kittler. Edge-labeling using dictionary-based relaxation. *IEEE Transactions on Pattern Analysis and Machine Intelligence*, 12(2):165–181, 1990.

[Hanlon 94] S J Hanlon. A computational theory of contextual knowledge in machine reading. PhD thesis, School of Computer Studies, University of Leeds, Leeds, UK, 1994.

[Hanlon and Boyle 92a] S J Hanlon and R D Boyle. Evaluating a Hidden Markov Model of syntax in a text recognition system. In D C Hogg and R D Boyle, editors, *Proceedings of the British Machine Vision Conference,* Leeds, UK, pages 462–471, Springer Verlag, London, 1992.

[Hanlon and Boyle 92b] S J Hanlon and R D Boyle. Syntactic knowledge in word level text recognition. In R Beale and J Finlay, editors, *Neural Networks and Pattern Recognition in HCI.* Ellis Horwood, 1992.

[Hanson and Riseman 78] A R Hanson and E M Riseman. VISIONS—a computer system for interpreting scenes. In A R Hanson and E M Riseman, editors, *Computer Vision Systems*, pages 303–333. Academic Press, New York, 1978.

[Haralick et al. 88] R M Haralick, M C Zhang, and R W Ehrich. Dynamic programming approach for context classification using the Markov random field. In *9th International Conference on Pattern Recognition,* Rome, Italy, pages 1169–1181, IEEE, New York, 1988.

[Harvey et al. 97] R Harvey, I Matthews, J A Bangham, and S Cox. Lip reading from scale-space measurements. In *Computer Vision and Pattern Recognition*, pages 582–587, IEEE Computer Society, Los Alamitos, CA, 1997.

[Hata et al. 97] Y Hata, S Kobashi, N Kamiura, and M Ishikawa. Fuzzy logic approach to 3D magnetic resonance image segmentation. In J Duncan and G Gindi, editors, *Information Processing in Medical Imaging*, pages 387–392, Springer Verlag, Berlin; New York, 1997.

[Hatef and Kittler 95] M Hatef and J Kittler. Constraining probabilistic relaxation with symbolic attributes. In V Hlavac and R Sara, editors, *International Conference on Computer Analysis of Images and Patterns,* Prague, Czech Republic, pages 862–867, Springer Verlag, Heidelberg, September 6th–8th, 1995.

[Heap 97] A Heap. Learning deformable shape models for object tracking. PhD thesis, School of Computer Studies, University of Leeds, Leeds, UK, 1997.

[Heap 98] A Heap. Wormholes in shape space: Tracking through discontinuous changes in shape. In N Ahuja, editor, *International Conference on Computer Vision,* Bombay, India, pages 344–349, Narosa, Bombay, 1998.

[Heap and Hogg 96] A J Heap and D C Hogg. Extending the Point Distribution Model using polar coordinates. *Image and Vision Computing*, 14(8):589–600, 1996.

[Higgins et al. 94] W E Higgins, J M Reinhardt, and W L Sharp. Semi-automatic construction of 3D medical image-segmentation processes. In *Proceedings of the SPIE Conference on Visualization in Biomedical Computing*, volume SPIE 2359, pages 59–71, 1994.

[Hill and Taylor 94] A Hill and C J Taylor. Automatic landmark generation for PDMs. In E Hancock, editor, *Proceedings of the British Machine Vision Conference,* York, UK, volume 2, pages 429–438. BMVA Press, 1994.

[Hill et al. 97] A Hill, A D Brett, and C J Taylor. Automatic landmark identification using a new method of non-rigid correspondence. In J Duncan and G Gindi, editors, *Information Processing in Medical Imaging*, pages 483–488, Springer Verlag, Berlin; New York, 1997.

[Hoch and Litwinowicz 96] M Hoch and P C Litwinowicz. A semi-automatic system for edge tracking with snakes. *Visual Computer*, 12:75–83, 1996.

[Huang et al. 90] X D Huang, Y Akiri, and M A Jack. *Hidden Markov Models for Speech Recognition.* Edinburgh University Press, Edinburgh, Scotland, 1990.

[Hummel and Zucker 83] R A Hummel and S W Zucker. On the foundation of relaxation labeling proceses. *IEEE Transactions on Pattern Analysis and Machine Intelligence*, 5(3):259–288, 1983.

[Hwang and Wang 94] S Y Hwang and T P Wang. The design and implementation of a distributed image understanding system. *Journal of Systems Integration*, 4:107–125, 1994.

[Hyche et al. 92] M E Hyche, N F Ezquerra, and R Mullick. Spatiotemporal detection of arterial structure using active contours. In *Proceedngs of Visualization in Biomedical Computing '92 Proceedings,* Chapel Hill, NC, pages 52–62, 1992.

[Illingworth and Kittler 87] J Illingworth and J Kittler. Optimisation algorithms in probabilistic relaxation labelling. In *Pattern Recognition Theory and Applications*, pages 109–117. Springer Verlag, Berlin–New York–Tokyo, 1987.

[Jolliffe 86] I T Jolliffe. *Principal Components Analysis.* Springer Verlag, New York, 1986.

[Jurie and Gallice 95] F Jurie and J Gallice. A recognition network model-based approach to dynamic image understanding. *Annals of Mathematics and Artificial Intelligence*, 13:317–345, 1995.

[Kamada et al. 88] M Kamada, K Toraichi, R Mori, K Yamamoto, and H Yamada. Parallel architecture for relaxation operations. *Pattern Recognition*, 21(2):175–181, 1988.

[Kanade and Ikeuchi 91] T Kanade and K Ikeuchi. Special issue on physical modeling in computer vision. *IEEE Transactions on Pattern Analysis and Machine Intelligence*, 13:609–742, 1991.

[Karaolani et al. 92] P Karaolani, G D Sullivan, and K D Baker. Active contours using finite elements to control local scale. In D C Hogg and R D Boyle, editors, *Proceedings of the British Machine Vision Conference,* Leeds, UK, pages 472–480, Springer Verlag, London, 1992.

[Kasif 90] S Kasif. On the parallel complexity of discrete relaxation in constraint satisfaction networks. *Artificial Intelligence*, 45(3):275–286, 1990.

[Kass et al. 87a] M Kass, A Witkin, and D Terzopoulos. Snakes: Active contour models. *International Journal of Computer Vision*, 1(4):321–331, 1987.

[Kass et al. 87b] M Kass, A Witkin, and D Terzopoulos. Snakes: Active contour models. In *1st International Conference on Computer Vision*, London, England, pages 259–268, IEEE, Piscataway, NJ, 1987.

[Kittler 87] J Kittler. Relaxation labelling. In *Pattern Recognition Theory and Applications*, pages 99–108. Springer Verlag, Berlin–New York–Tokyo, 1987.

[Kittler and Foglein 84a] J Kittler and J Foglein. Contextual classification of multispectral pixel data. *Image and Vision Computing*, 2(1):13–29, 1984.

[Kittler and Foglein 84b] J Kittler and J Foglein. Contextual decision rules for objects in lattice configuration. In *7th International Conference on Pattern Recognition,* Montreal, Canada, pages 270–272, IEEE, Piscataway, NJ, 1984.

[Kittler and Foglein 86] J Kittler and J Foglein. On compatibility and support functions in probabilistic relaxation. *Computer Vision, Graphics, and Image Processing*, 34:257–267, 1986.

[Kittler and Hancock 89] J Kittler and E R Hancock. Combining evidence in probabilistic relaxation. *International Journal on Pattern Recognition and Artificial Intelligence*, 3:29–52, 1989.

[Kittler and Illingworth 85] J Kittler and J Illingworth. Relaxation labelling algorithms—a review. *Image and Vision Computing*, 3(4):206–216, 1985.

[Kittler and Pairman 85] J Kittler and D Pairman. Contextual pattern recognition applied to cloud detection and identification. *IEEE Transactions on Geoscience and Remote Sensing*, 23(6):855–863, 1985.

[Kodratoff and Moscatelli 94] Y Kodratoff and S Moscatelli. Machine learning for object recognition and scene analysis. *International Journal of Pattern Recognition and Artificial Intelligence*, 8:259–304, 1994.

[Kotcheff and Taylor 97] A C W Kotcheff and C J Taylor. Automatic construction of eigenshape models by genetic algorithms. In J Duncan and G Gindi, editors, *Information Processing in Medical Imaging*, pages 1–14, Springer Verlag, Berlin; New York, 1997.

[Kuan et al. 89] D Kuan, H Shariat, and K Dutta. Constraint-based image understanding system for aerial imagery interpretation. In *Proceedings of the Annual AI Systems in Government Conference,* Washington, DC, pages 141–147, 1989.

[Kumar and Desai 96] V P Kumar and U B Desai. Image interpretation using Bayesian networks. *IEEE Transactions on Pattern Analysis and Machine Intelligence*, 18:74–77, 1996.

[Kundu et al. 89] A Kundu, Y He, and P Bahi. Recognition of handwritten word: First and second order HMM based approach. *Pattern Recognition*, 22(3):283–297, 1989.

[Lam and Yan 94] K M Lam and H Yan. Fast greedy algorithm for active contours. *Electronics Letters*, 30:21–23, 1994.

[Lanitis et al. 94] A Lanitis, C J Taylor, and T F Cootes. An automatic face identification systems using flexible appearance models. In E Hancock, editor, *Proceedings of the British Machine Vision Conference,* York, UK, volume 1, pages 65–74. BMVA Press, 1994.

[Lau and Hancock 94] W H Lau and E Hancock. Pyramidal hierarchical relaxation. In *International Symposium on Speech, Image Processing and Neural Networks*, pages 768–771, IEEE, Los Alamitos, CA, 1994.

[Lee et al. 89] D Lee, A Papageorgiou, and G W Wasilkowski. Computing optical flow. In *Proceedings, Workshop on Visual Motion,* Irvine, CA, pages 99–106, IEEE, Piscataway, NJ, 1989.

[Lesser et al. 75] V R Lesser, R D Fennell, L D Erman, and D R Reddy. Organisation of the HEARSAY II speech understanding system. *IEEE Transactions on Acoustics, Speech and Signal Processing*, 23(1):11–24, 1975.

[Levine 78] M D Levine. A knowledge based computer vision system. In A R Hanson and E M Riseman, editors, *Computer Vision Systems*, pages 335–352. Academic Press, New York, 1978.

[Li and Uhr 87] Z Li and L Uhr. Pyramid vision using key features to integrate image-driven bottom-up and model-driven top-down processes. *IEEE Transactions on Systems, Man and Cybernetics*, 17(2):250–263, 1987.

[Liow and Pavlidis 88] Y Liow and T Pavlidis. Enhancements of the split-and-merge algorithm for image segmentation. In *1988 IEEE International Conference on Robotics and Automation,* Philadelphia, PA, pages 1567–1572, Computer Society Press, Washington, DC, 1988.

[Lobregt and Viergever 95] S Lobregt and M A Viergever. A discrete dynamic contour model. *IEEE Transactions on Medical Imaging*, 14:12–24, 1995.

[Lu and Chung 94] C S Lu and P C Chung. Fuzzy-based probabilistic relaxation for textured image segmentation. In *Proceedings of the International Conference on Fuzzy Systems*, pages 77–82. IEEE, 1994.

[Marik et al. 92] V Marik, O Stepankova, and R Trappl, editors. *Advanced Topics in Artificial Intelligence,* LNAI No. 617. Springer Verlag, Heidelberg, 1992.

[Marr 82] D Marr. *Vision—A Computational Investigation into the Human Representation and Processing of Visual Information.* Freeman, San Francisco, 1982.

[McInerney and Terzopoulos 93] T McInerney and D Terzopoulos. A finite element based deformable model for 3D biomedical image segmentation. In *Proceedings SPIE, Vol. 1905, Biomedical Image Processing and Biomedical Visualization,* San Jose, CA, SPIE, Bellingham, WA, 1993.

[McInerney and Terzopoulos 95] T McInerney and D Terzopoulos. Topologically adaptable snakes. In *5th International Conference on Computer Vision,* MIT, pages 840–845, IEEE, Piscataway, NJ, 1995.

[Metaxas and Terzopoulos 91] D Metaxas and D Terzopoulos. Constrained deformable superquadrics and nonrigid motion tracking. In *CVPR '91: Computer Society Conference on Computer Vision and Pattern Recognition,* Lahaina, HI, pages 337–343, IEEE, Los Alamitos, CA, 1991.

[Michalski et al. 83] R S Michalski, J G Carbonell, and T M Mitchell. *Machine Learning I, II.* Morgan Kaufmann Publishers, Los Altos, CA, 1983.

[Millin and Ni 89] B M Millin and L M Ni. A reliable parallel algorithm for relaxation labeling. In P M Dew, R A Earnshaw, and T R Heywood, editors, *Parallel Processing for Computer Vision and Display,* pages 190–207. Addison-Wesley, Reading, MA, 1989.

[Mohn et al. 87] E Mohn, N L Hjort, and G O Storvik. Simulation study of some contextual classification methods for remotely sensed data. *IEEE Transactions on Geoscience and Remote Sensing*, 25(6):796–804, 1987.

[Moller-Jensen 90] L Moller-Jensen. Knowledge-based classification of an urban area using texture and context information in Landsat-TM imagery. *Photogrammetric Engineering and Remote Sensing*, 56(6):899–904, 1990.

[Mulder 88] J A Mulder. Discrimination vision. *Computer Vision, Graphics, and Image Processing*, 43:313–336, 1988.

[Nagao and Matsuyama 80] M Nagao and T Matsuyama. *A Structural Analysis of Complex Aerial Photographs.* Plenum Press, New York, 1980.

[Nakagawa and Hirota 95] Y Nakagawa and K Hirota. Fundamentals of fuzzy knowledge base for image understanding. In *Proceedings of the International Conference on Fuzzy Systems*, pages 1137–1142, IEEE, Los Alamitos, CA, 1995.

[Neuenschwander et al. 94] W Neuenschwander, P Fua, G Szekely, and O Kubler. Initializing snakes (object delineation). In *Proceedings Computer Vision and Pattern Recognition*, pages 658–663, IEEE, Los Alamitos, CA, 1994.

[Niemann 90] H Niemann. *Pattern Analysis and Understanding.* Springer Verlag, Berlin–New York–Tokyo, 2nd edition, 1990.

[Nilsson 82] N J Nilsson. *Principles of Artificial Intelligence.* Springer Verlag, Berlin, 1982.

[Oliver et al. 97] N Oliver, A P Pentland, and F Berard. LAFTER: Lips and face real time tracker. In *Computer Vision and Pattern Recognition*, pages 123–129, IEEE Computer Society, Los Alamitos, CA, 1997.

[Olstad and Tysdahl 93] B Olstad and H E Tysdahl. Improving the computational complexity of active contour algorithms. In *8th Scandinavian Conference on Image Analysis,* Tromso, pages 257–263, International Association for Pattern Recognition, Oslo, 1993.

[Parent and Zucker 89] P Parent and S W Zucker. Radial projection: An efficient update rule for relaxation labeling. *IEEE Transactions on Pattern Analysis and Machine Intelligence*, 11(8):886–889, 1989.

[Parkkinen et al. 91] J Parkkinen, G Cohen, M Sonka, and N Andreasen. Segmentation of MR brain images. In *Proceedings of the Annual International Conference of the IEEE Engineering in Medicine and Biology Society, Volume 13,* Orlando, FL, pages 71–72, IEEE, Piscataway, NJ, 1991.

[Pavlidis and Liow 90] T Pavlidis and Y Liow. Integrating region growing and edge detection. *IEEE Transactions on Pattern Analysis and Machine Intelligence*, 12(3):225–233, 1990.

[Pelillo and Fanelli 97] M Pelillo and A M Fanelli. Autoassociative learning in relaxation labeling networks. *Pattern Recognition Letters*, 18:3–12, 1997.

[Prasanna Kumar 91] V K Prasanna Kumar. *Parallel Architectures and Algorithms for Image Understanding.* Academic Press, Boston, 1991.

[Puliti and Tascini 93] P Puliti and G Tascini. Knowledge-based approach to image interpretation. *Image and Vision Computing*, 11:122–128, 1993.

[Rabiner 89] L R Rabiner. A tutorial on Hidden Markov Models and selected applications in speech recognition. *Proceedings of the IEEE*, 77(2):257–286, 1989.

[Rabiner and Juang 86] L R Rabiner and B H Juang. An introduction to Hidden Markov Models. *IEEE ASSP Magazine*, 3:4–16, January 1986.

[Radeva et al. 95] P Radeva, J Serrat, and E Marti. A snake for model-based segmentation. In *5th International Conference on Computer Vision,* MIT, pages 816–821, IEEE, Piscataway, NJ, 1995.

[Ralescu and Shanahan 95] A L Ralescu and J G Shanahan. Fuzzy perceptual organization in image understanding. In *Proceedings of the International Conference on Fuzzy Systems*, pages 25–26, IEEE, Los Alamitos, CA, 1995.

[Rangarajan et al. 97] A Rangarajan, H Chui, and F L Bookstein. The Softassign Procrustes matching algorithm. In J Duncan and G Gindi, editors, *Information Processing in Medical Imaging*, pages 29–42, Springer Verlag, Berlin; New York, 1997.

[Rao and Jain 88] A R Rao and R Jain. Knowledge representation and control in computer vision systems. *IEEE Expert*, 3(1):64–79, 1988.

[Reichgelt 91] H Reichgelt. *Knowledge Representation: An AI Perspective.* Ablex, Norwood, NJ, 1991.

[Roberto et al. 90] V Roberto, L Gargiulo, A Peron, and C Chiaruttini. A knowledge-based system for geophysical interpretation. In *Proceedings of the 1990 International Conference on Acoustics, Speech, and Signal Processing—ICASSP 90,* Albuquerque, NM, pages 1945–1948, IEEE, Piscataway, NJ, 1990.

[Ronfard 94] R Ronfard. Region-based strategies for active contour models. *International Journal of Computer Vision*, 13:229–251, 1994.

[Rosenfeld 79] A Rosenfeld. *Picture Languages—Formal Models for Picture Recognition.* Academic Press, New York, 1979.

[Rosenfeld et al. 76] A Rosenfeld, R A Hummel, and S W Zucker. Scene labelling by relaxation operations. *IEEE Transactions on Systems, Man and Cybernetics*, 6:420–433, 1976.

[Schlenzig et al. 94] J Schlenzig, E Hunter, and R Jain. Recursive identification of gesture inputers using HMMs. In *Proceedings of the 2nd Annual Conference on Computer Vision,* Sarasota, FL, pages 187–194, IEEE Computer Society Press, New York, 1994.

[Sharman 89] R A Sharman. Hidden Markov Model methods for word tagging. Technical Report IBM UKSC 214, IBM UK Scientific Centre, Winchester, UK, December 1989.

[Sharp and Hancock 95] N G Sharp and E R Hancock. Multi-frame feature tracking by probabilistic relaxation. In D Dori and A Bruckstein, editors, *Shape, Structure and Pattern Recognition*, pages 211–220, 1995.

[Shi and Malik 97] J Shi and J Malik. Normalized cuts and image segmentation. In *Computer Vision and Pattern Recognition*, pages 731–737, IEEE Computer Society, Los Alamitos, CA, 1997.

[Simons 84] G L Simons. *Introducing Artificial Intelligence.* NCC Publications, Manchester, England, 1984.

[Smyth et al. 97] P P Smyth, C J Taylor, and J E Adams. Automatic measurement of vertebral shape using active shape models. In J Duncan and G Gindi, editors, *Information Processing in Medical Imaging*, pages 441–446, Springer Verlag, Berlin; New York, 1997.

[Sonka et al. 93] M Sonka, C J Wilbricht, S R Fleagle, S K Tadikonda, M D Winniford, and S M Collins. Simultaneous detection of both coronary borders. *IEEE Transactions on Medical Imaging*, 12(3):588–599, 1993.

[Sonka et al. 95] M Sonka, M D Winniford, and S M Collins. Robust simultaneous detection of coronary borders in complex images. *IEEE Transactions on Medical Imaging*, 14(1):151–161, 1995.

[Sonka et al. 96] M Sonka, S K Tadikonda, and S M Collins. Knowledge-based interpretation of MR brain images. *IEEE Transactions on Medical Imaging*, 15:443–452, 1996.

[Sozou et al. 94] P D Sozou, T F Cootes, C J Taylor, and A C Di-Mauro. A non-linear generalization of PDMs using polynomial regression. In E Hancock, editor, *Proceedings of the British Machine Vision Conference,* York, UK, volume 2, pages 397–406. BMVA Press, 1994.

[Starner and Pentland 97] T Starner and A Pentland. Real-time American sign language recognition from video using HMMs. In M Shah and R Jain, editors, *Motion Based Recognition*, pages 227–243, Kluwer, Boston, 1997.

[Stoddart et al. 95] A J Stoddart, M Petrou, and J Kittler. A new algorithm for probabilistic relaxation based on the Baum Eagon theorem. In V Hlavac and R Sara, editors, *International Conference on Computer Analysis of Images and Patterns,* Prague, Czech Republic, pages 674–679, Springer Verlag, Heidelberg, September 6th–8th, 1995.

[Strat and Fischler 91] T M Strat and M A Fischler. Context-based vision: Recognizing objects using information from both 2D and 3D imagery. *IEEE Transactions on Pattern Analysis and Machine Intelligence*, 13(10):1050–1065, 1991.

[Tagare et al. 95] H D Tagare, F M Vos, C C Jaffe, and J S Duncan. Arrangement: A spatial relation between parts for evaluating similarity of tomographic sections. *IEEE Transactions on Pattern Analysis and Machine Intelligence*, 17:880–893, 1995.

[Tanner et al. 83] J M Tanner, R H Whitehouse, N Cameron, W A Marshall, M J R Healy, and H Goldstein. *Assessment of Skeletal Maturity and Prediction of Adult Height.* Academic Press, London, 1983.

[Terzopoulos 91] D Terzopoulos. Visual modeling. In *Proceedings of the British Machine Vision Conference,* Glasgow, Scotland, pages 9–11, Springer Verlag, London–Berlin–New York, 1991.

[Terzopoulos and Fleischer 88] D Terzopoulos and K Fleischer. Deformable models. *The Visual Computer*, 4(6):306–331, 1988.

[Terzopoulos and Metaxas 91] D Terzopoulos and D Metaxas. Dynamic 3D models with local and global deformations: Deformable superquadrics. *IEEE Transactions on Pattern Analysis and Machine Intelligence*, 13(7):703–714, 1991.

[Terzopoulos et al. 87] D Terzopoulos, A Witkin, and M Kass. Symmetry-seeking models for 3D object reconstruction. In *1st International Conference on Computer Vision,* London, England, pages 269–276, IEEE, Piscataway, NJ, 1987.

[Terzopoulos et al. 88] D Terzopoulos, A Witkin, and M Kass. Constraints on deformable models: Recovering 3D shape and nonrigid motion. *Artificial Intelligence*, 36(1):91–123, 1988.

[Tilton 87] J C Tilton. Contextual classification on the massively parallel processor. In *Frontiers of Massively Parallel Scientific Computation,* Greenbelt, MD, pages 171–181, NASA, Washington, DC, 1987.

[Toulson and Boyce 92] D L Toulson and J F Boyce. Segmentation of MR images using neural nets. *Image and Vision Computing*, 10(5):324–328, 1992.

[Tsai and Sun 92] C T Tsai and Y N Sun. Minimizing the energy of active contour by using a Hopfield network. In *IEEE International Conference on Systems Engineering*, pages 495–498. IEEE; Wright State University; Pascal Res., 1992.

[Tsosos 84] J K Tsosos. Knowledge and the visual process. *Pattern Recognition*, 17(1):13–28, 1984.

[Udupa and Samarasekera 96] J K Udupa and S Samarasekera. Fuzzy connectedness and object definition: Theory, algorithms, and applications in image segmentation. *Graphical Models and Image Processing*, 58:246–261, 1996.

[Vincent and Soille 91] L Vincent and P Soille. Watersheds in digital spaces: An efficient algorithm based on immersion simulations. *IEEE Transactions on Pattern Analysis and Machine Intelligence*, 13(6):583–598, 1991.

[Viterbi 67] A J Viterbi. Convolutional codes and their performance in communication systems. *IEEE Transactions on Communications Technology*, 13(2):260–269, 1967.

[Vlontzos and Kung 92] J A Vlontzos and S Y Kung. HMMs for character recognition. *IEEE Transactions on Image Processing*, IP-1(4):539–543, 1992.

[Waltz 57] D L Waltz. Understanding line drawings of scenes with shadows. In *The Psychology of Computer Vision*. McGraw-Hill, New York, 1957.

[Watanabe and Suzuki 88] T Watanabe and H Suzuki. An experimental evaluation of classifiers using spatial context for multispectral images. *Systems and Computers in Japan*, 19(4):33–47, 1988.

[Watanabe and Suzuki 89] T Watanabe and H Suzuki. Compound decision theory and adaptive classification for multispectral image data. *Systems and Computers in Japan*, 20(8):37–47, 1989.

[Wechsler 90] H Wechsler. *Computational Vision*. Academic Press, London–San Diego, 1990.

[Wharton 82] S Wharton. A contextual classification method for recognising land use patterns in high resolution remotely sensed data. *Pattern Recognition*, 15:317–324, 1982.

[Wilkinson and Megier 90] G G Wilkinson and J Megier. Evidential reasoning in a pixel classification hierarchy. A potential method for integrating image classifiers and expert system rules based on geographic context. *International Journal of Remote Sensing*, 11(10):1963–1968, 1990.

[Williams and Shah 92] D J Williams and M Shah. A fast algorithm for active contours and curvature estimation. *CVGIP - Image Understanding*, 55(1):14–26, 1992.

[Winston 92] P H Winston. *Artificial Intelligence*. Addison-Wesley, Reading, MA, 3rd edition, 1992.

[Witkin et al. 87] A Witkin, D Terzopoulos, and M Kass. Signal matching through scale space. *International Journal of Computer Vision*, 1(2):133–144, 1987.

[Wu and Leahy 93] Z Wu and R Leahy. An optimal graph theoretic approach to data clustering: Theory and its application to image segmentation. *IEEE Transactions on Pattern Analysis and Machine Intelligence*, 15:1101–1113, 1993.

[Xu and Prince 98] C Xu and J L Prince. Snakes, shapes, and gradient vector flow. *IEEE Transactions on Image Processing*, 7:359–369, 1998.

[Xu et al. 94] G Xu, E Segawa, and S Tsuji. Robust active contours with insensitive parameters. *Pattern Recognition*, 27:879–884, 1994.

[Yamato et al. 92] J Yamato, J Ohya, and K Ishii. Recognising human actions in time-sequential images using HMMs. In *Proceedings, Second International Conference on Computer Vision*, Tampa, FL, pages 379–385, IEEE, Piscataway, NJ, 1992.

[Zen et al. 90] C Zen, S Y Lin, and Y Y Chen. Parallel architecture for probabilistic relaxation operation on images. *Pattern Recognition*, 23(6):637–645, 1990.

[Zhang and Braun 97] Z Zhang and M Braun. Fully 3D active surface models with self-inflation and self-deflation forces. In *Computer Vision and Pattern Recognition*, pages 85–90, IEEE Computer Society, Los Alamitos, CA, 1997.

[Zhang et al. 90] M C Zhang, R M Haralick, and J B Campbell. Multispectral image context classification using stochastic relaxation. *IEEE Transactions on Systems, Man and Cybernetics*, 20(1):128–140, 1990.

[Zheng 95] Y J Zheng. Feature extraction and image segmentation using self-organizing networks. *Machine Vision and Applications*, 8:262–274, 1995.

[Zucker 78] S W Zucker. Vertical and horizontal processes in low level vision. In A R Hanson and E M Riseman, editors, *Computer Vision Systems*, pages 187–195, Academic Press, New York, 1978.

[Zucker 88] S W Zucker. Organization of curve detection: Coarse tangent fields and fine spline coverings. *Neural Networks*, 1(1):534, 1988.

Chapter 9

3D vision, geometry, and radiometry

A number of image analysis techniques aiming at 2D images have been presented in earlier chapters. What has been overlooked hitherto, though, is the observation that the best vision system, our own, and so far unbeatable by machines, is geared to deal with the 3D world. In this chapter about 3D vision we shall fill the gap; we shall concentrate on intermediate-level vision tasks in which 3D scene properties are inferred from 2D image representations. Methods for extracting 3D information and interpreting 3D scenes will be presented.

There are several serious reasons why 3D vision using intensity images as input is regarded as difficult.

- The imaging system of a camera and the human eye performs perspective projection, which leads to considerable loss of information. All points along a line pointing from the optical center towards a scene point are projected to a single image point. We are interested in the inverse task that aims to derive 3D co-ordinates from image measurements—this task is under-constrained, and some additional information must be added to solve it unambiguously.

- The relationship between image intensity and the 3D geometry of the corresponding scene point is very complicated. The pixel intensity depends on surface reflectivity parameters, surface orientation, type and position of illuminants, and the position of the viewer. Attempting to learn about 3D geometry—surface orientation and depth—represents another ill-conditioned task.

- The mutual occlusion of objects in the scene, and even self-occlusion of one object, further complicates the vision task.

- The presence of noise in images, and the high time complexity of many algorithms, contributes further to the problem, although this is not specific to 3D vision.

The chapter is organized as follows: In Section 9.1, we shall consider various 3D vision paradigms, and Marr's theory of 3D vision from the late 1970s will be explained in more detail, since even with its known limitations it is still the most generally accepted paradigm. Section 9.2 explains the geometrical issues that constitute important mathematical machinery needed to solve 3D vision tasks. We present here recent research material in a uniform fashion; the geometry of one, two, and three cameras and related applications are sketched. Section 9.3 tackles the relation between the intensity of a pixel in a 2D image and the 3D shape of the corresponding scene point.

9.1 3D vision tasks

The field of 3D vision is young and still developing, and no unified theory is available; different research groups may have different understandings of the task. Several 3D vision tasks and related paradigms illustrate the variety of opinions:

- Marr [Marr 82] defines 3D vision as '*From an image (or a series of images) of a scene, derive an accurate three-dimensional geometric description of the scene and quantitatively determine the properties of the object in the scene*'. Here, 3D vision is formulated as a 3D object reconstruction task, i.e., description of the 3D shape in a co-ordinate system independent of the viewer. One rigid object, whose separation from the background is straightforward, is assumed, and the control of the process is strictly bottom-up from an intensity image through intermediate representations. Treating 3D vision as scene recovery seems reasonable. If vision cues give us a precise representation of a 3D scene then almost all visual tasks may be carried out; the navigation of an autonomous vehicle, parts inspection, or object recognition are examples. The recovery paradigm needs to know the relation between an image and the corresponding 3D world, and thus image formation needs to be described.

- Aloimonos and Shulman [Aloimonos and Shulman 89] see the central problem of computer vision as: '*...from one or the sequence of images of a moving or stationary object or scene taken by a monocular or polynocular moving or stationary observer, to* understand *the object or the scene and its three-dimensional properties*'. In this definition, it is the concept *understand* that makes this approach to vision different. If little a priori knowledge is available, as in human vision, then understanding is complicated. This might be seen as one limiting case; the other extreme in the complexity spectrum is, e.g., a simple object matching problem in which there are only several known possible interpretations.

- Wechsler [Wechsler 90] stresses the control principle of the process: '*The visual system casts most visual tasks as minimization problems and solves them using distributed computation and enforcing nonaccidental, natural constraints*'. Computer vision is seen as a parallel distributed representation, plus parallel distributed processing, plus active perception. The understanding is carried in the 'perception—control—action' cycle.

- Aloimonos [Aloimonos 93] asks what principles might enable us to (i) understand vision of living organisms, and (ii) equip machines with visual capabilities. There are several types of related questions:

  - *Empirical questions* (What is?) determine how existing visual systems are designed.

- *Normative questions* (What should be?) deal with classes of animals or robots that would be desirable.
- *Theoretical questions* (What could be?) are interested in mechanisms that could exist in intelligent visual systems.

System theory [Klir 91] provides a general framework that allows us to treat understanding of complex phenomena using the machinery of mathematics. The inherent complexity of the vision task is solved here by distinguishing the object (or system or phenomenon) from the background, where 'objects' mean anything of interest to solve the task at hand. The objects and their properties need to be characterized, and a formal mathematical model is typically used for this abstraction. The model is specified by a relatively small number of parameters, which are typically estimated from the (image) data.

This methodology allows us to describe the same object using qualitatively different models (e.g., algebraic or differential equations) when **varying resolution** is used during observation. Studying changes of models with respect to several resolutions may give deeper insight into the problem.

An attempt to create a computer-based vision system comprises three intertwined problems:

1. *Feature observability in images:* We need to determine whether task-relevant information will be present in the primary image data.
2. *Representation:* This problem is related to the choice of model for the observed world, at various levels of interpretation complexity.
3. *Interpretation:* This problem tackles the semantics of the data—in other words, how are data mapped to the (real) world. The task is to make certain information explicit from a mathematical model storing it in an implicit form.

Two main approaches to artificial vision, according to the flow of information and the amount of a priori knowledge, are typically considered (see Chapter 8).

1. *Reconstruction, bottom-up:* The aim is to reconstruct the 3D shape of the object from an image or set of images, which might be either intensity or range images. One extreme is given by Marr's theory [Marr 82], which is strictly bottom-up with very little a priori knowledge about the objects needed. Some, more practical, approaches aim to create a 3D model from real objects using range images [Flynn and Jain 91, Flynn and Jain 92, Soucy and Laurendeau 92, Bowyer 92].
2. *Recognition, top-down, model-based vision:* The a priori knowledge about the objects is expressed by means of the models of the objects, where 3D models are of particular interest [Brooks et al. 79, Goad 86, Besl and Jain 85, Farshid and Aggarwal 93]. Recognition based on CAD models is of practical importance [Newman et al. 93]. Additional constraints embedded in the model make under-determined vision tasks possible in many cases.

Some authors propose object recognition systems in which 3D models are avoided. The *priming-based (geons)* approach is based on the idea that 3D shapes can be inferred directly from 2D drawings [Biederman 87]—the qualitative features are called *geons*. This mimics the human recognition process in which constituents of a single object (geons) and their spatial arrangement are pointers to a human memory.

The *alignment of 2D views* is another option—lines or points in 2D views can be used for aligning different 2D views. The correspondence of points, lines, or other features must be made first. A linear combination of views has been used [Ullman and Basri 91] for recognition, and various issues related to image-based scene representations in which a collection of images with established correspondences is stored instead of a 3D model is considered in [Beymer and Poggio 96]. How this approach can be used for displaying a 3D scene from any viewpoint is considered in [Werner et al. 95, Hlaváč et al. 96].

9.1.1 Marr's theory

Marr was a pioneer in the study of computer vision whose influence has been, and continues to be, considerable despite his early death. Critical of earlier work that, while successful in limited domains or image classes, was either empirical or unduly restrictive of the images with which it could deal, Marr proposed a more abstract and theoretical approach that permitted work to be put into a larger context. Restricting himself to the 3D interpretation of single, static scenes, Marr proposed that a computer vision system was just an example of an information processing device that could be understood at three levels:

1. *Computational theory:* The theory describes what the device is supposed to do—what information it provides from other information provided as input. It should also describe the logic of the strategy that performs this task.

2. *Representation and algorithm*: These address precisely how the computation may be carried out—in particular, information representations and algorithms to manipulate them.

3. *Implementation*: The physical realization of the algorithm—specifically, programs and hardware.

It is stressed that it is important to be clear about which level is being addressed in attempting to solve or understand a particular problem. Marr illustrates this by noting that the effect of an after-image (induced by staring at a light bulb) is a physical effect, while the mental confusion provoked by the well-known Necker cube illusion (see Figure 9.1) appears to be at a different theoretical level entirely.

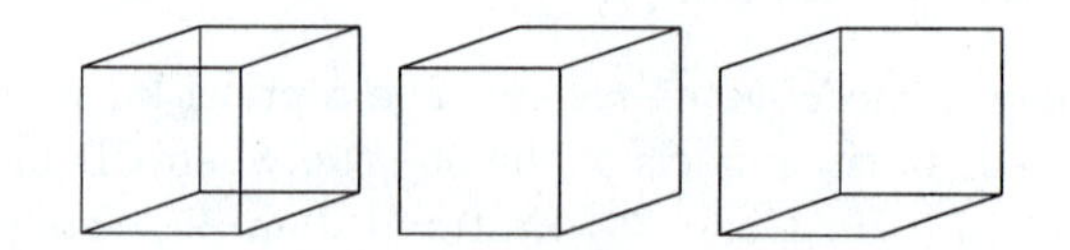

Figure 9.1: *The Necker cube, and two possible interpretations.*

The point is then made that the lynchpin of success is addressing the theory rather than algorithms or implementation—any number of edge detectors may be developed, each one specific to particular problems, but we would be no nearer any general understanding of how edge detection should or might be achieved. Marr remarks that the complexity of the vision task dictates a sequence of steps refining descriptions of the geometry of visible surfaces. Having derived some such description, it is then necessary to remove the dependence on the vantage point and to transform the description into an **object-centered** one. The requirement, then, is to move from pixels to surface delineation, then to surface characteristic description (orientation), then to a full 3D description. These transformations are effected by moving from the 2D image to a **primal sketch**, then to a **2.5D sketch**, and thence to a full 3D representation.

The primal sketch

The primal sketch aims to capture, in as general a way as possible, the significant intensity changes in an image. Hitherto, such changes have been referred to as 'edges', but Marr makes the observation that this word implies a physical meaning that cannot at this stage be inferred. The first stage is to locate these changes at a range of scales (see Section 4.3.4)—informally, a range of blurring filters are passed across the image, after which second-order zero-crossings (see Section 4.3.2) are located for each scale of blur [Marr and Hildreth 80]. The blurring recommended is a standard Gaussian filter [see equation (4.52)], while the zero-crossings are located with a Laplacian operator [see equation (4.39)]. The various blurring filters isolate features of particular scales; then zero-crossing evidence in the same locality at many scales provides strong evidence of a genuine physical feature in the scene.

To complete the primal sketch, these zero-crossings are grouped, according to their location and orientations, to provide information about tokens in the image (edges, bars, and blobs) that may help provide later information about (3D) orientation of scene surfaces. The grouping phase, paying attention to the evidence from various scales, extracts tokens that are likely to represent surfaces in the real world.

It is of interest to note that there is strong evidence for the existence of the various components used to build the primal sketch in the human visual system—we too engage in detection of features at various scales, the location of sharp intensity changes, and their subsequent grouping into tokens.

The 2.5D sketch

The 2.5D sketch reconstructs the relative distances from the viewer of surfaces detected in the scene, and may be called a **depth map**. Observe that the output of this phase uses as input features detected in the preceding one, but that in itself it does not give us a 3D reconstruction. In this sense it is midway between 2D and 3D representations, and in particular, nothing can be said about the 'other side' of any objects in view. Instead, it may be the derivation of a surface normal associated with each likely surface detected in the primal sketch, and there may be an implicit improvement in the quality of this information.

There are various routes to the 2.5D sketch, but their common thread is the continuation of the bottom-up approach in that they do not exploit any knowledge about scene contents, but rather employ additional clues such as knowledge about the nature of lighting or motion effects, and are thus generally applicable and not domain specific. The main approaches are known as **'shape from X'** techniques, and are described in Section 10.1. At the conclusion of this phase, the representation is still in viewer-centered co-ordinates.

The 3D representation

At this stage the Marr paradigm overlaps with top-down, model-based approaches. It is required to take the evidence derived so far and identify objects within it. This can only be achieved with some knowledge about what 'objects' are, and, consequently, some means of describing them. The important point is that this is a transition to an object-centered co-ordinate system, allowing object descriptions to be viewer independent.

This is the most difficult phase and successful implementation is remote, especially compared to the success seen with the derivation of the primal and 2.5D sketches—specifying what is required, however, has been very successful in guiding computer vision research since the paradigm was formulated. Unlike earlier stages, there is little physiological guidance that can be used to design algorithms since this level of human vision is not well understood. Marr observes that the target co-ordinate system(s) should be modular in the sense that each 'object' should be treated differently, rather than employing one global co-ordinate system (usually viewer centered). This prevents having to consider the orientation of model components with respect to the whole. It is further observed that a set of **volumetric** primitives is likely to be of value in representing models (in contrast to surface-based descriptions). Representations based on an object's 'natural' axes, derived from symmetries, or the orientation of stick features, are likely to be of greater use.

The Marr paradigm advocates a set of relatively independent modules; the low-level modules aim to recover a meaningful description of the input intensity image, the middle-level modules use different cues such as intensity changes, contours, texture, motion to recover shape, or location in space. It was shown later [Bertero et al. 88, Aloimonos and Rosenfeld 94] that most low-level and middle-level tasks are ill-posed, with no unique solution; one popular way developed in the 1980s to make the task well-posed is **regularization** [Tichonov and Arsenin 77, Poggio et al. 85]. A constraint requiring continuity and smoothness of the solution is often added.

9.1.2 Other vision paradigms: Active and purposive vision

When consistent geometric information has to be explicitly modeled (as for manipulation of the object), an object-centered co-ordinate system seems to be appropriate. It is not certain that Marr's attempt to create object-centered co-ordinates is confirmed in biological vision; for example, Koenderink shows that the global human visual space is viewer centered and non-Euclidean [Koenderink 90]. For small objects, the existence of an object-centered reference frame has not been confirmed in psychological studies.

There are currently two schools trying to explain the vision mechanism.

- The first and older one tries to use explicit metric information in the early stages of the visual task (lines, curvatures, normals, etc.). Geometry is typically extracted in a bottom-up fashion without any information about the purpose of this representation. The output is a geometric model.

- The second and younger school does not extract metric (geometric) information from visual data until needed for a specific task. Data are collected in a systematic way to ensure that all the object's features are present in the data, but may remain uninterpreted until a specific task is involved. A database or collection of intrinsic images (or views) is the model.

Many traditional computer vision systems and theories capture data with cameras with fixed characteristics. The same holds for traditional theories; e.g., Marr's observer is static. Some researchers advocate **active perception** [Bajcsy 88, Landy et al. 96] and purposive vision [Aloimonos 93]: In an active vision system, the characteristics of the data acquisition are dynamically controlled by the scene interpretation—many visual tasks tend to be simpler if the observer is active and controls its visual sensors. Controlled eye (or camera) movement is an example, where if there are not enough data to interpret the scene the camera can look at it from another viewpoint. In other words, active vision is intelligent data acquisition controlled by the measured, partially interpreted scene parameters and their errors from the scene. Active vision is an area of much current research.

The active approach can make most ill-posed vision tasks tractable. To provide an overview, we summarize in tabular form [Aloimonos and Rosenfeld 94] how an active observer can change ill-posed tasks to well-posed ones—see Table 9.1.

Task	*Passive observer*	*Active observer*
Shape from shading	Ill-posed. Regularization helps but a unique solution is not guaranteed due to non-linearities.	Well-posed. Stable. Unique solution. Linear equations.
Shape from contour	Ill-posed. Regularization solution not formulated yet. Solution exists only for very special cases.	Well-posed. Unique solution for monocular or binocular observer.
Shape from texture	Ill-posed. Assumptions about texture needed.	Well-posed without assumptions.
Structure from motion	Well-posed but unstable.	Well-posed and stable. Quadratic constraints. simple solution.

Table 9.1: *Active vision makes vision tasks well-posed.*

It has been generally accepted in the vision community that accurate shape recovery from intensity images is difficult. The Marr paradigm is a nice theoretic framework, but unfortunately does not lead to successful vision applications performing, e.g., recognition and navigation tasks.

There is no established theory that provides a mathematical (computational) model explaining the 'understanding' aspects of human vision; a recent account of the topic is [Ullman 96]. Two recent developments towards new vision theory are:

- *Qualitative vision*, which looks for a qualitative description of objects or scenes [Aloimonos 94]. The motivation is not to represent geometry that is not needed for qualitative (non-geometric) tasks or decisions. Further, qualitative information is more invariant to various unwanted transformations (e.g., slightly differing viewpoints) or noise than quantitative ones. Qualitativeness (or invariance) enables interpretation of observed events at several levels of complexity. Note that the human eye does not give extremely precise measurements either; a vision algorithm should look for qualities in images, e.g., convex and concave surface patches in range data [Besl and Jain 88].

- The *purposive vision* paradigm, which may help to come up with simpler solutions [Aloimonos 92]. The key question is to identify the goal of the task, the motivation being to ease the task by making explicit just that piece of information that is needed. Collision avoidance for autonomous vehicle navigation is an example where precise shape description is not needed. The approach may be heterogeneous, and a qualitative answer may be sufficient in some cases. The paradigm does not yet have a solid theoretical basis, but the study of biological vision is a rich source of inspiration. This shift of research attention resulted in many successful vision applications where no precise geometric description is necessary. Examples are collision avoidance, autonomous vehicle navigation, object tracking, etc. [Howarth 94, Buxton and Howarth 95, Fernyhough 97].

There are other vision tasks that need complete geometric 3D models, for example, to create a 3D CAD model from a real object, say, a clay model created by a human designer. Other applications are in virtual reality systems where interaction among real and virtual objects is needed. Some object recognition tasks use full 3D models as well.

9.2 Geometry for 3D vision

9.2.1 Basics of projective geometry

The basic sensor that provides computer vision with information about the surrounding 3D world is a television camera. Here, stressing the geometric aspect, we will explain how to use 2D image information for automated measurement of the 3D world, where measurements of 3D co-ordinates of points or distances from 2D images are of importance. We require to study **perspective projection** (called also central projection), which describes image formation by a pinhole camera or a thin lens. Parallel lines in the world do not remain parallel in a perspective image—consider, for example, a view along a railway or into a long corridor. Figure 9.2 illustrates this, where also some commonly used terms are introduced.

We begin with a concise introduction to basic notation and the definitions of projective space [Semple and Kneebone 63, Faugeras 93, Mohr 93]. Consider $(n+1)$-dimensional space without its origin $\mathcal{R}^{n+1} - \{(0,\dots,0)\}$, and define an equivalence relation

$$\begin{aligned} [x_1,\dots,x_{n+1}]^T &\equiv [x'_1,\dots,x'_{n+1}]^T \\ \text{iff } \exists\, \alpha \neq 0 :\ [x_1,\dots,x_{n+1}]^T &= \alpha\, [x'_1,\dots,x'_{n+1}]^T \end{aligned} \tag{9.1}$$

The projective space $\mathcal{P}^n$ is the quotient space of this equivalence relation. Points in the projective space are expressed in **homogeneous** (also projective) co-ordinates, which we will denote in bold with a tilde, e.g., $\tilde{\mathbf{x}}$. Such points are often shown with the number 1 in the rightmost position, e.g., $[x'_1,\dots,x'_n,1]^T$. This point is equivalent to any point that differs only by non-zero scaling.

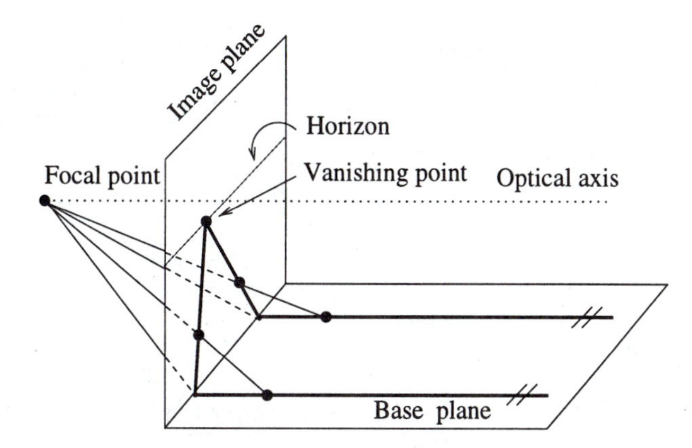

Figure 9.2: *Perspective projection of parallel lines.*

We are more accustomed to n-dimensional Euclidean space $\mathcal{R}^n$. The one-to-one mapping from the $\mathcal{R}^n$ into $\mathcal{P}^n$ is given by

$$[x_1, \ldots, x_n]^T \rightarrow [x_1, \ldots, x_n, 1]^T \tag{9.2}$$

Only the points $[x_1, \ldots, x_n, 0]^T$ do not have an Euclidean counterpart. It is easy to demonstrate that they represent points at infinity in a particular direction. Consider $[x_1, \ldots, x_n, 0]^T$ as a limiting case of $[x_1, \ldots, x_n, \alpha]^T$ that is projectively equivalent to $[x_1/\alpha, \ldots, x_n/\alpha, 1]^T$, and assume that $\alpha \rightarrow 0$. This corresponds to a point in $\mathcal{R}^n$ going to infinity in the direction of the radius vector $[x_1/\alpha, \ldots, x_n/\alpha] \in \mathcal{R}^n$.

A **co-lineation**, or projective transformation, is any mapping $\mathcal{P}^n \rightarrow \mathcal{P}^n$ that is defined by a regular $(n+1) \times (n+1)$ matrix $\mathbf{A}$, $\tilde{\mathbf{y}} = A\, \tilde{\mathbf{x}}$. Note that the matrix A is defined up to a scale factor. Co-lineations map hyper-planes to hyper-planes; a special case is the mapping of lines to lines that is often used in computer vision.

9.2.2 The single perspective camera

Consider the case of one camera with a thin lens. This pinhole model is the simplest approximation that is suitable for many computer vision applications. The pinhole camera performs perspective projection. The geometry of the device is depicted in Figure 9.3; the plane on the bottom is an **image plane** π to which the real world projects, and the vertical dotted line is the **optical axis**. The lens is positioned perpendicularly to the optical axis at the **focal point C** (also called the **optical center**). The focal length f (sometimes called the principal axis distance [Mohr 93]) is a parameter of the lens.

The projection is performed by an optical ray (also a light beam) reflected from a scene point $\mathbf{X}$ (top left in Figure 9.3) or originated from a light source. The optical ray passes through the optical center $\mathbf{C}$ and hits the image plane at the point $\mathbf{U}$.

For further explanation, we need to define four co-ordinate systems.

1. The *world Euclidean co-ordinate system* (subscript $_w$) has origin at the point $\mathbf{O}_w$. Points $\mathbf{X}$, $\mathbf{U}$ are expressed in the world co-ordinate system.

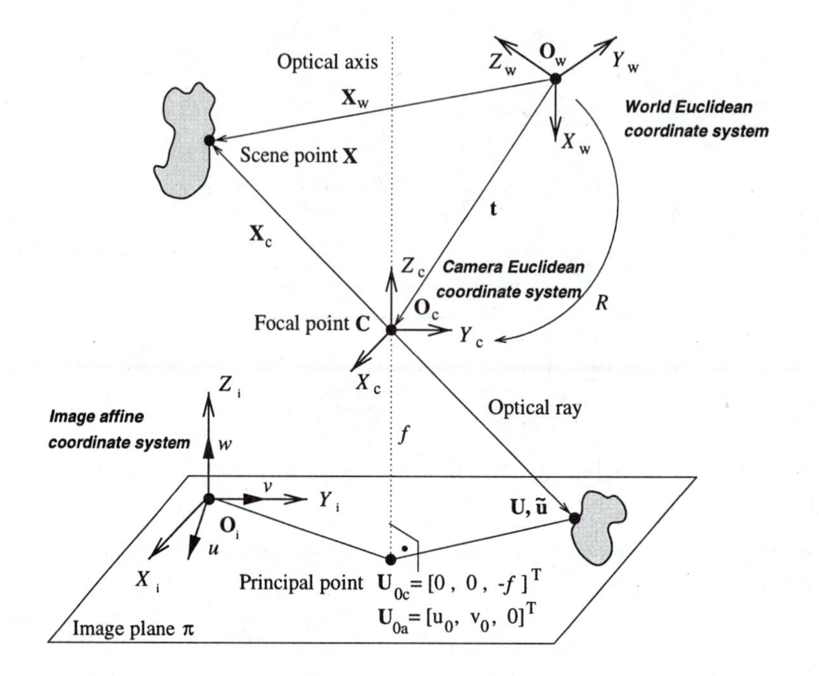

Figure 9.3: *The geometry of a linear perspective camera.*

2. The *camera Euclidean co-ordinate system* (subscript $_c$) has the focal point $\mathbf{C} \equiv \mathbf{O}_c$ as its origin. The co-ordinate axis Z_c is aligned with the optical axis and points away from the image plane.

   There is a unique relation between world and camera co-ordinate systems. We can align the world to camera co-ordinates by performing a Euclidean transformation consisting of a translation $\mathbf{t}$ and a rotation R.

3. The *image Euclidean co-ordinate system* (subscript $_i$) has axes aligned with the camera co-ordinate system, with X_i, Y_i lying in the image plane.

4. The *image affine co-ordinate system* (subscript $_a$) has co-ordinate axes u, v, w, and origin $\mathbf{O}_i$ co-incident with the origin of the image Euclidean co-ordinate system. The axes w, v are aligned with the axes Z_i, Y_i, but the axis u may have a different orientation to the axis X_i.

   The reason for introducing these co-ordinates is the fact that in general pixels need not be perpendicular, and axes can be scaled differently. The affine co-ordinate system is induced by the arrangement of the retina.

A camera performs a linear transformation from the 3D projective space $\mathcal{P}^3$ to the 2D projective space $\mathcal{P}^2$.

A scene point $\mathbf{X}$ is expressed in the world Euclidean co-ordinate system as a 3×1 vector. To express the same point in the camera Euclidean co-ordinate system, i.e., $\mathbf{X}_c$, we have to translate it by subtracting vector $\mathbf{t}$ and rotating it as specified by the matrix R.

$$\mathbf{X}_c = \begin{bmatrix} x_c \\ y_c \\ z_c \end{bmatrix} = R\,(\mathbf{X}_w - \mathbf{t}) \tag{9.3}$$

The point $\mathbf{X}_c$ is projected to the image plane π as point $\mathbf{U}_c$. The x and y co-ordinates of

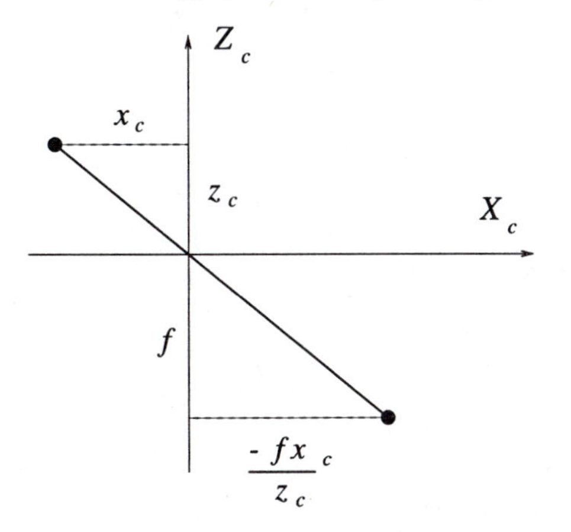

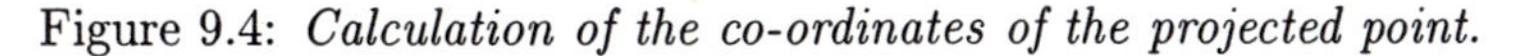

Figure 9.4: *Calculation of the co-ordinates of the projected point.*

the projected point can be derived from the similar triangles illustrated in Figure 9.4.

$$\mathbf{U}_c = \left[\ \frac{-f x_c}{z_c},\ \ \frac{-f y_c}{z_c},\ \ -f\ \right]^T \tag{9.4}$$

It remains to derive where the projected point $\mathbf{U}_c$ is positioned in the image affine co-ordinate system, i.e., to determine the co-ordinates which the real camera actually delivers.

The image affine co-ordinate system, with origin at the top left corner of the image, represents a shear and rescaling (often called the aspect ratio) of the image Euclidean co-ordinate system. The principal point $\mathbf{U}_0$—sometimes called the center of the image in camera calibration procedures—is the intersection of the optical axis with the image plane π. The principal point $\mathbf{U}_0$ is expressed in the image affine co-ordinate system as $\mathbf{U}_{0a} = [u_0, v_0, 0]^T$.

The projected point can be represented in the 2D image plane π in homogeneous co-ordinates as $\tilde{\mathbf{u}} = [U, V, W]^T$, and its 2D Euclidean counterpart is $\mathbf{u} = [u, v]^T = [U/W, V/W]^T$. Homogeneous co-ordinates allow us to express the affine transformation as a multiplication by a single 3×3 matrix where unknowns a, b, c describe the shear together with scaling along co-ordinate axes, and u_0 and v_0 give the affine co-ordinates of the principal point in the image.

$$\tilde{\mathbf{u}} = \begin{bmatrix} U \\ V \\ W \end{bmatrix} = \begin{bmatrix} a & b & -u_0 \\ 0 & c & -v_0 \\ 0 & 0 & 1 \end{bmatrix} \begin{bmatrix} \frac{-f x_c}{z_c} \\ \frac{-f y_c}{z_c} \\ 1 \end{bmatrix} = \begin{bmatrix} -fa & -fb & -u_0 \\ 0 & -fc & -v_0 \\ 0 & 0 & 1 \end{bmatrix} \begin{bmatrix} \frac{x_c}{z_c} \\ \frac{y_c}{z_c} \\ 1 \end{bmatrix} \tag{9.5}$$

We aim to collect all constants in this matrix, sometimes called the **camera calibration matrix** K. Since homogeneous co-ordinates are in use, the equation can be multiplied by any non-zero constant; thus we multiply by z_c to remove it, and can rewrite

$$\begin{aligned} z_c\,\tilde{\mathbf{u}} &= z_c \begin{bmatrix} -fa & -fb & -u_0 \\ 0 & -fc & -v_0 \\ 0 & 0 & 1 \end{bmatrix} \begin{bmatrix} \frac{x_c}{z_c} \\ \frac{y_c}{z_c} \\ 1 \end{bmatrix} = \begin{bmatrix} -fa & -fb & -u_0 \\ 0 & -fc & -v_0 \\ 0 & 0 & 1 \end{bmatrix} \begin{bmatrix} x_c \\ y_c \\ z_c \end{bmatrix} \\ &= \begin{bmatrix} -fa & -fb & -u_0 \\ 0 & -fc & -v_0 \\ 0 & 0 & 1 \end{bmatrix} R\left(\mathbf{X}_w - \mathbf{t}\right) = KR\left(\mathbf{X}_w - \mathbf{t}\right) \end{aligned} \tag{9.6}$$

The **extrinsic parameters** of the camera depend on the orientation of the camera Euclidean co-ordinates with respect to the world Euclidean co-ordinate system (see Figure 9.3). This relation is given in equation (9.6) by matrices R and $\mathbf{t}$. The rotation matrix R expresses three elementary rotations of the co-ordinate axes—rotations along the axes x, y, and z are termed pan, tilt, and roll, respectively. The translation vector $\mathbf{t}$ gives three elements of the translation of the origin of the world co-ordinate system with respect to the camera co-ordinate system. Thus there are six extrinsic camera parameters, three rotations and three translations.

The camera calibration matrix K is upper triangular as can be seen from equation (9.6). The coefficients of this matrix are called **intrinsic parameters** of the camera, and describe the specific camera independent on its position and orientation in space. If the intrinsic parameters are known, a metric measurement can be performed from images. Assume momentarily the simple case in which the world co-ordinates co-incide with the camera co-ordinates, meaning that $\mathbf{X}_w = \mathbf{X}_c$. Then equation (9.6) simplifies to

$$z_c\,\tilde{\mathbf{u}} = z_c \begin{bmatrix} U \\ V \\ W \end{bmatrix} = \begin{bmatrix} -fa & -fb & -u_0 \\ 0 & -fc & -v_0 \\ 0 & 0 & 1 \end{bmatrix} \begin{bmatrix} x_c \\ y_c \\ z_c \end{bmatrix} \tag{9.7}$$

We can write two separate equations for u and v

$$\begin{aligned} u &= \frac{U}{W} = -fa\frac{x_c}{z_c} - fb\frac{y_c}{z_c} - u_0 = \alpha_u\frac{x_c}{z_c} + \alpha_{\text{shear}}\frac{y_c}{z_c} - u_0 \\ v &= \frac{U}{W} = \quad\quad\;\; - fc\frac{y_c}{z_c} - v_0 = \quad\quad\;\; \alpha_v\frac{y_c}{z_c} - v_0 \end{aligned} \tag{9.8}$$

where we make the substitutions $\alpha_u = -fa$, $\alpha_{\text{shear}} = -fb$, and $\alpha_v = -fc$. Thus we have five intrinsic parameters, all given in pixels. The formulae also give the interpretation of the intrinsic parameters: α_u represents scaling in the u axis, measuring f in pixels along the u axis, and α_v similarly specifies f in pixels along the v axis. α_{shear} measures in pixels the degree of slant of the co-ordinate axes in the camera image plane, giving in the v axis direction how far the focal length f co-incident with the u axis is slanted from the Y_i axis.

This completes the description of the extrinsic and intrinsic camera parameters, and we can return to the general case given by equation (9.6). If we express the scene point in homogeneous co-ordinates $\tilde{\mathbf{X}}_w = [\mathbf{X}_w, 1]^T$, we can write the perspective projection using a single 3×4 matrix. The leftmost 3×3 submatrix describes a rotation and the rightmost column a translation The delimiter | denotes that the matrix is composed of two submatrices.

$$\tilde{\mathbf{u}} = \begin{bmatrix} U \\ V \\ W \end{bmatrix} = [KR\,|\,-K\;R\mathbf{t}] \begin{bmatrix} \mathbf{X}_w \\ 1 \end{bmatrix} = M \begin{bmatrix} \mathbf{X}_w \\ 1 \end{bmatrix} = M\tilde{\mathbf{X}}_w \tag{9.9}$$

where $\tilde{\mathbf{X}}$ is the 3D scene point in homogeneous co-ordinates. The matrix M is called the **projective matrix** (or camera matrix). It can be seen that the camera performs a linear projective transformation from the 3D projective space $\mathcal{P}^3$ to the 2D projective plane $\mathcal{P}^2$; notice that the introduction of projective space and homogeneous co-ordinates made the expressions simpler. Instead of the non-linear equation (9.4), we obtained the linear equation (9.9).

The 3×3 submatrix of the projective matrix M consisting of the three leftmost columns is regular, i.e., its determinant is non-zero. The scene point $\tilde{\mathbf{X}}_w$ is expressed up to scale in homogeneous co-ordinates and thus all αM are equivalent for $\alpha \neq 0$.

Sometimes the simplest form of the projection matrix M is used.

$$M = \begin{bmatrix} 1 & 0 & 0 & 0 \\ 0 & 1 & 0 & 0 \\ 0 & 0 & 1 & 0 \end{bmatrix} \tag{9.10}$$

This special matrix corresponds to the *normalized camera co-ordinate system* [Faugeras 93], in which the specific parameters of the camera can be ignored. This is useful when the properties of stereo and motion are to be explained in a simple way and independently of the specific camera.

9.2.3 An overview of single camera calibration

The calibration of one camera is a procedure that allows us to set numeric values in the camera calibration matrix K [equation (9.6)] or the projective matrix M [equation (9.9)]. The first case is applicable when we want the intrinsic camera parameters only. If the camera is calibrated and a point in the image is known, the corresponding line (ray) in camera-centered space is uniquely determined. The second case covers both intrinsic and extrinsic parameters.

We first consider basic approaches to the calibration of a single camera to give an overview of the state of the art of this developing branch of computer vision. Then we will consider some basic techniques in more detail. There are two main cases, as follows.

1. *Known scene:* Here, a set of n non-degenerate (not co-planar) points lies in the 3D world, and the corresponding 2D image points are known (see Figure 9.5[1]). Each correspondence between a 3D scene and 2D image point provides one equation,

$$\alpha_j \tilde{\mathbf{u}}_j = M \begin{bmatrix} \mathbf{X}_j \\ 1 \end{bmatrix} \tag{9.11}$$

   The solution [Faugeras 93] solves an over-determined system of linear equations. The main disadvantage is that the scene must be known, for which special calibration objects are often used.

2. *Unknown scene:* If the scene is 'unknown', more views are needed to calibrate the camera (see Figure 9.6). The intrinsic camera parameters will not change for different views, and the correspondence between image points in different views must be established.

[1]Here and in some further figures the image plane is positioned in front of the focal point—this differs from earlier figures where the image plane was behind the focal point. Such a presentation makes figures easier to comprehend and should not cause any confusion.

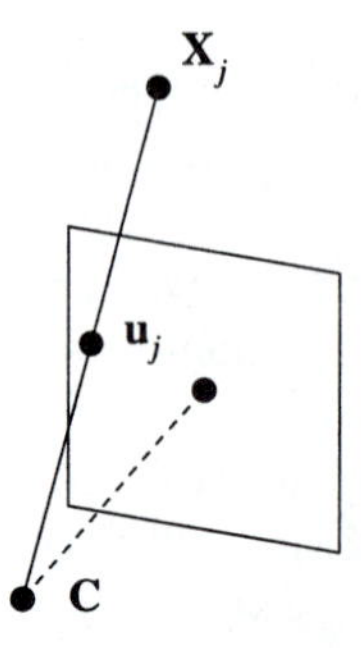

Figure 9.5: *Camera calibration from a known scene. A minimum of six corresponding pairs of scene points* $\mathbf{X}_j$ *and image points* $\mathbf{u}_j$ *are needed to calibrate the camera.*

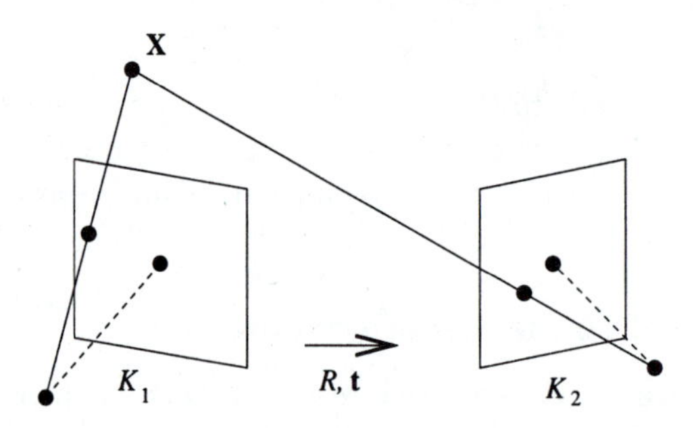

Figure 9.6: *Camera calibration from an unknown scene. At least two views are needed. It is assumed that the intrinsic parameters of the camera do not change, so* $K_1 = K_2$.

There are two cases.

(a) *Known camera motion:* Three cases can be distinguished according to the known motion constraint.

  i. *Both rotation and translation:* This general case of arbitrary known motion from one view to another has been solved [Horaud et al. 95].

  ii. *Pure rotation:* If camera motion is restricted to pure rotation, the solution is given by [Hartley 94].

  iii. *Pure translation:* The linear solution (pure translation) is due to [Pajdla and Hlaváč 95].

(b) *Unknown camera motion:* This is the most general case when there is no a priori knowledge about motion, sometimes called *camera self-calibration.* At least three views are needed, and the solution is non-linear [Maybank and Faugeras 92]. Calibration from an unknown scene is still considered numerically hard, and will not be considered here (although see, for example, [Butterfield 97] for a consideration of this problem).

9.2.4 Calibration of one camera from a known scene

Considering the case of camera calibration from a known scene in more detail, note that this is typically a two-stage process. First, the projection matrix M is estimated from the co-ordinates of points with known scene positions. Second, the extrinsic and intrinsic parameters are estimated from M. The second step is not always needed—the case of stereo vision is an example.

To obtain M, observe that each known scene point $\mathbf{X} = [x, y, z]^T$ and its corresponding 2D image point $[u, v]^T$ give one equation (9.11)—we seek the numerical values m_{ij} in the 3×4 projection matrix M. Expanding from equation (9.11),

$$\begin{bmatrix} \alpha u \\ \alpha v \\ \alpha \end{bmatrix} = \begin{bmatrix} m_{11} & m_{12} & m_{13} & m_{14} \\ m_{21} & m_{22} & m_{23} & m_{24} \\ m_{31} & m_{32} & m_{33} & m_{34} \end{bmatrix} \begin{bmatrix} x \\ y \\ z \\ 1 \end{bmatrix} \tag{9.12}$$

$$\begin{bmatrix} \alpha u \\ \alpha v \\ \alpha \end{bmatrix} = \begin{bmatrix} m_{11}x + m_{12}y + m_{13}z + m_{14} \\ m_{21}x + m_{22}y + m_{23}z + m_{24} \\ m_{31}x + m_{32}y + m_{33}z + m_{34} \end{bmatrix} \tag{9.13}$$

$$\begin{aligned} u(m_{31}x + m_{32}y + m_{33}z + m_{34}) &= m_{11}x + m_{12}y + m_{13}z + m_{14} \\ v(m_{31}x + m_{32}y + m_{33}z + m_{34}) &= m_{21}x + m_{22}y + m_{23}z + m_{24} \end{aligned} \tag{9.14}$$

Thus we obtain two linear equations, each in 12 unknowns $m_{11}, \ldots, m_{34}$, for each known corresponding scene and image point. If n such points are available, we can write equations (9.14) as a $2n \times 12$ matrix,

$$\begin{bmatrix} x & y & z & 1 & 0 & 0 & 0 & 0 & -ux & -uy & -uz & -u \\ 0 & 0 & 0 & 0 & x & y & z & 1 & -vx & -vy & -vz & -v \\ & & & & & & \vdots & & & & & \end{bmatrix} \begin{bmatrix} m_{11} \\ m_{12} \\ \vdots \\ m_{34} \end{bmatrix} = 0 \tag{9.15}$$

The matrix M actually has only 11 unknown parameters due to the unknown scaling factor, since homogeneous co-ordinates were used [Faugeras 93]. To generate a solution, at least six known corresponding scene and image points are required. Typically, more points are used and the over-determined equation (9.15) is solved using a robust least-squares method to correct for noise in measurements. The result of the calculation is the projective matrix M.

To separate the extrinsic parameters (the rotation R and translation $\mathbf{t}$) from the estimated projection matrix M, recall that the projection matrix can be written as

$$M = [KR \,|\, -KR\mathbf{t}] = [A \,|\mathbf{b}] \tag{9.16}$$

The 3×3 submatrix is denoted as A, and the rightmost column as $\mathbf{b}$.

Determining the translation vector is easy; we substituted $A = KR$ in equation (9.16), and so can write $\mathbf{t} = -A^{-1}\mathbf{b}$.

To determine R, note that the calibration matrix is upper triangular and the rotation matrix is orthogonal. The matrix factorization method called QR decomposition [Press et al. 92, Golub and Loan 89] will decompose A into a product of two such matrices, and hence recover K and R.

Alternatively, we can use **singular value decomposition (SVD)**[2]. SVD is a general tool that we shall refer to again in the solution of geometrical problems associated with 3D vision.

So far, we have assumed that the lens performs ideal central projection, as a pinhole camera does, but this is not the case with real lenses. A typical lens performs distortion of several pixels which a human observer does not notice looking at a general scene. However, when an image is used for measurements, compensation for the distortion is necessary.

When calibrating a real camera, the more realistic model of the lens includes two distortion components. First, **radial distortion** bends the ray more or less than in the ideal case; and second, **de-centering** displaces the principal point from the optical axis.

Recall the five intrinsic camera parameters introduced in equation (9.8). Here, we shall replace the focal length f of the lens by a parameter called the **camera constant**. Ideally, this is equal to the focal length, but in reality this is true only when the lens is focused at infinity; otherwise, the camera constant is slightly less than the focal length. Similarly, the co-ordinates of the principal point can change slightly from the ideal intersection of the optical axis with the image plane.

The idea behind calibration of intrinsic parameters is to observe a known calibration image with some regular pattern, for example, blobs or lines covering the whole image. Distortions observed in the pattern allow estimation of the parameters.

Both radial distortion and de-centering can in most cases be treated as **rotationally symmetric**; they are often modeled as polynomials. Let u, v denote the correct image co-ordinates, and $\tilde{u}$, $\tilde{v}$ denote the measured uncorrected image co-ordinates that come from the actual pixel co-ordinates x, y and the estimate of the position of the principal point $\hat{u}_0$, $\hat{v}_0$.

$$\begin{aligned} \tilde{u} &= x - \hat{u}_0 \\ \tilde{v} &= y - \hat{v}_0 \end{aligned} \tag{9.18}$$

The correct image co-ordinates u, v are obtained if compensations for errors δu, δv are added to the measured uncorrected image co-ordinates $\tilde{u}$, $\tilde{v}$.

$$u = \tilde{u} + \delta u$$

[2]SVD is a powerful linear algebra technique for solving linear equations in the least-square sense, and works even for singular matrices or matrices numerically close to singular. The basic information needed to use SVD can be found in [Press et al. 92], and a rigorous mathematical treatment is given in [Golub and Loan 89]. Most software packages for numerical calculations such as MATLAB (trademark of MathWorks, Inc.) contain SVD.

SVD proceeds by noting that any $m \times n$ matrix A, $m \geq n$ can be decomposed into a product of three matrices,

$$A = UDV^T \tag{9.17}$$

in which U has orthonormal columns, D is non-negative diagonal, and V^T has orthonormal rows. SVD can be used to find a solution of a set of linear equations corresponding to a singular matrix that has no exact solution—it locates the closest possible solution in a least-square sense.

Sometimes it is required to find the 'closest' singular matrix to the original matrix A—this decreases the rank from n to $n-1$. This is done by replacing the smallest diagonal element of D by zero—this new matrix is closest to the old one with respect to the Frobenius norm (which is calculated as a sum of the squared values of all matrix elements).

$$v = \tilde{v} + \delta v \tag{9.19}$$

Compensations for errors are often modeled as even power polynomials to secure rotational symmetry. Typically, polynomial degrees up to six are considered,

$$\begin{aligned} \delta u &= (\tilde{u} - u_p)(\kappa_1 r^2 + \kappa_2 r^4 + \kappa_3 r^6) \\ \delta v &= (\tilde{v} - v_p)(\kappa_1 r^2 + \kappa_2 r^4 + \kappa_3 r^6) \end{aligned} \tag{9.20}$$

where u_p, v_p is the correction to the position of the principal point. r^2 is the square of the radial distance from the center of the image,

$$r^2 = (\tilde{u} - u_p)^2 + (\tilde{u} - u_p)^2 \tag{9.21}$$

Recall that $\hat{u}_0$, $\hat{v}_0$ were used in equation (9.18). u_p, v_p are corrections to $\hat{u}_0$, $\hat{v}_0$ that can be applied after calibration to get the proper position of the principal point:

$$\begin{aligned} u_0 &= \hat{u}_0 + u_p \\ v_0 &= \hat{v}_0 + v_p \end{aligned} \tag{9.22}$$

We can visualize typical lens radial distortion for the simple second-order model as a special case of equation (9.20), i.e., no de-centering is assumed and a second-order approximation is considered:

$$\begin{aligned} u &= \tilde{u}[1 \pm \kappa_1(\tilde{u}^2 + \tilde{v}^2)] \\ v &= \tilde{v}[1 \pm \kappa_1(\tilde{u}^2 + \tilde{v}^2)] \end{aligned} \tag{9.23}$$

The original image was a square pattern, and the distorted images are shown in Figure 9.7. On the left is pincushion-like distortion [a minus sign in equation (9.23)], and the the right part depicts barrel-like distortion corresponding to a plus sign.

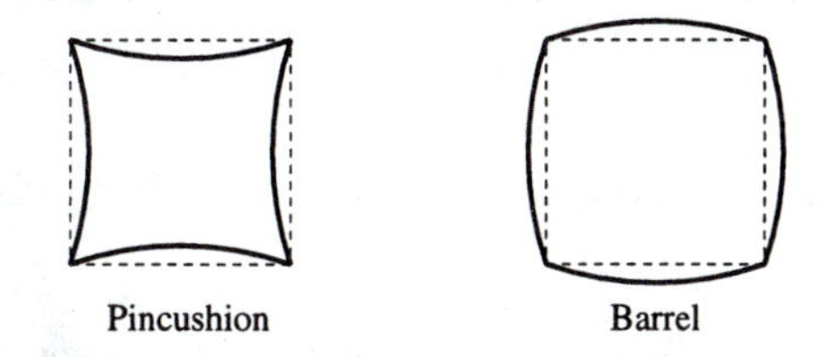

Figure 9.7: *Radial distortion of an off-the-shelf lens.*

More complicated lens models cover tangential distortions that model such effects as lens de-centering [Jain et al. 95], which we shall not describe in detail here. The reader can consult the original paper [Tsai 87] or the treatment in [Jain et al. 95]. An alternative procedure was proposed in [Prescott and McLean 97].

9.2.5 Two cameras, stereopsis

To the uneducated observer, the most obvious difference between the human visual system and most of the material presented thus far in this book is that we have two eyes and therefore (a priori, at any rate) twice as much input as a single image. From Victorian times, the use of two slightly different views to provide an illusion of 3D has been common, culminating in the '3D movies' of the 1950s. Conversely, we might hope that a 3D scene, if presenting two different views to two eyes, might permit the recapture of depth information when the information therein is combined with some knowledge of the sensor geometry (eye locations).

Stereo vision has enormous importance to us—humans. It has provoked a great deal of research into vision systems with two inputs that exploit the knowledge of their own relative geometry to derive depth information from the two views they receive.

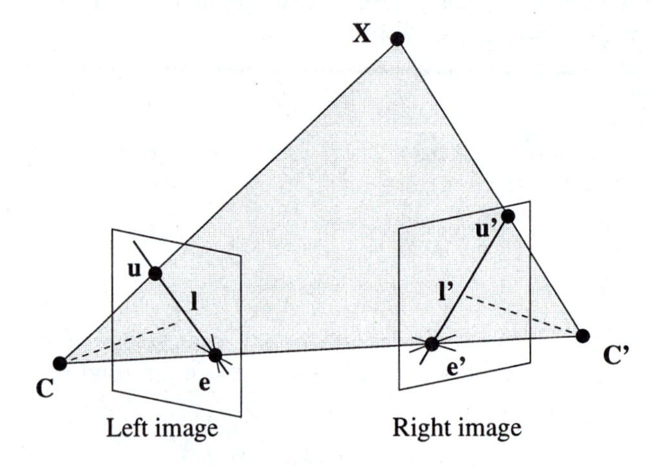

Figure 9.8: *Epipolar geometry in stereopsis.*

Calibration of one camera and knowledge of the co-ordinates of one image point allows us to determine a ray in space uniquely. If two calibrated cameras observe the same scene point $\mathbf{X}$, its 3D co-ordinates can be computed as the intersection of two such rays. This is the basic principle of **stereo vision** that typically consists of three steps:

- Camera calibration
- Establishing point correspondences between pairs of points from the left and the right images
- Reconstruction of 3D co-ordinates of the points in the scene

The geometry of the system with two cameras is given in Figure 9.8. The line connecting optical centers $\mathbf{C}$ and $\mathbf{C}'$ is called the **baseline**. Any scene point $\mathbf{X}$ observed by the two cameras, and the two corresponding rays from optical centers $\mathbf{C}$, $\mathbf{C}'$ define an **epipolar plane**. This plane intersects the image planes in the **epipolar lines** $\mathbf{l}$, $\mathbf{l}'$. When the scene point $\mathbf{X}$ moves in space, all epipolar lines pass through **epipoles** $\mathbf{e}$, $\mathbf{e}'$—the epipoles are the intersections of the baseline with the respective image planes.

Let $\mathbf{u}$, $\mathbf{u}'$ be projections of the scene point $\mathbf{X}$ in the left and right images, respectively. The ray $\mathbf{CX}$ represents all possible positions of the point $\mathbf{X}$ for the left image, and is also projected into the epipolar line $\mathbf{l}'$ in the right image. The point $\mathbf{u}'$ in the right image that corresponds to the projected point $\mathbf{u}$ in the left image must thus lie on the epipolar line $\mathbf{l}'$

in the right image. This geometry provides a strong **epipolar constraint** that reduces the dimensionality of the search space for a correspondence between $\mathbf{u}$ and $\mathbf{u}'$ in the right image from 2D to 1D.

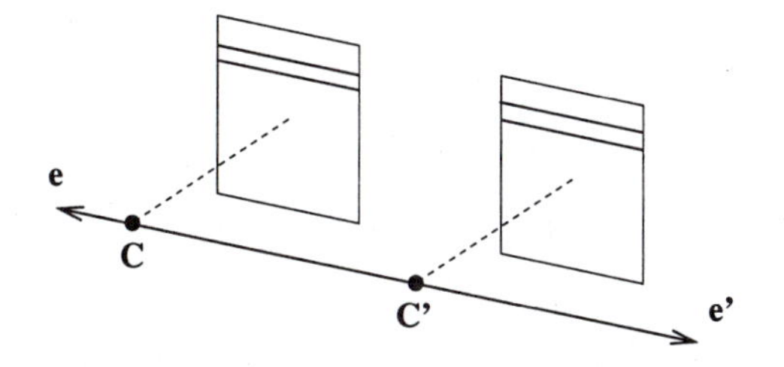

Figure 9.9: *The canonical stereo configuration where the epipolar lines are parallel in the image, and epipoles move to infinity.*

A special arrangement of the stereo camera rig, called the **canonical configuration**, is often used. The baseline is aligned to the horizontal co-ordinate axis, the optical axes of the cameras are parallel, the epipoles move to infinity, and the epipolar lines in the image planes are parallel (see Figure 9.9). For this configuration, the computation is slightly simpler; it is often used when stereo correspondence is to be determined by a human operator who will find matching points linewise to be easier (this non-automatic approach is still used in photogrammetry and remote sensing). A similar conclusion holds for computer programs too; it is easier to move along horizontal lines (rasters) than along general lines. The geometric transformation that changes a general camera configuration with non-parallel epipolar lines to the canonical one is called **image rectification**. Formulae for image rectification will be given in Section 9.2.9.

On the other hand, some authors [Mohr 93] report practical problems with the canonical stereo configuration, which adds unnecessary technical constraints to the vision hardware. If high precision of reconstruction is an issue, it is better to use general stereo geometry since rectification induces re-sampling that causes loss of resolution.

Considering first an easy canonical configuration, we shall see how to recover depth. The optical axes are parallel, which leads to the notion of disparity that is often used in stereo literature. A simple diagram demonstrates how we proceed. In Figure 9.10, which is purely schematic, we have a bird's-eye view of two cameras with parallel optical axes separated by a distance $2h$. The images they provide, together with one point P with co-ordinates (x, y, z) in the scene, show this point's projection onto left (P_l) and right (P_r) images. The co-ordinates in Figure 9.10 have the z axis representing distance from the cameras (at which $z = 0$) and the x axis representing 'horizontal' distance (the y co-ordinate, into the page, does not appear). $x = 0$ will be the position midway between the cameras; each image will have a local co-ordinate system (x_l on the left, x_r on the right) which for the sake of convenience we measure from the center of the respective images—that is, a simple translation from the global x co-ordinate. Without fear of confusion, P_l will be used simultaneously to represent the position of the projection of P onto the left image, and its x_l co-ordinate—its distance from the center of the left image (and similarly for P_r).

It is clear that there is a **disparity** between x_l and x_r as a result of the different camera positions (that is, $|P_l - P_r| > 0$); we can use elementary geometry to deduce the z co-ordinate

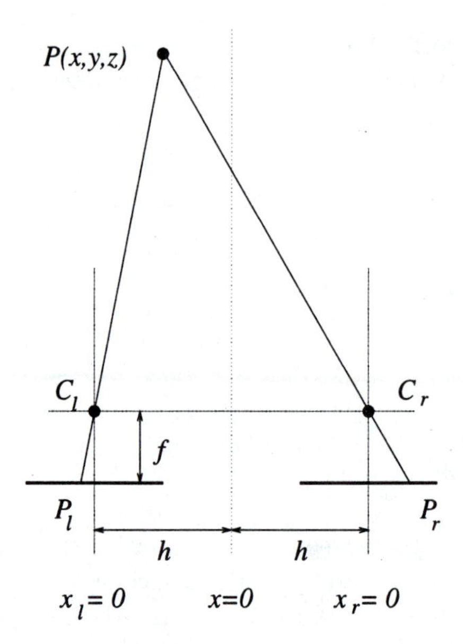

Figure 9.10: *Elementary stereo geometry in canonical configuration.*

of P.

Note that P_l, C_l and C_l, P are the hypotenuses of similar right-angled triangles. Noting further that h and f are (positive) numbers, z is a positive co-ordinate and x, P_l, P_r are co-ordinates that may be positive or negative, we can then write

$$\frac{P_l}{f} = -\frac{h+x}{z} \tag{9.24}$$

and similarly from the right-hand side of Figure 9.10,

$$\frac{P_r}{f} = \frac{h-x}{z} \tag{9.25}$$

Eliminating x from these equations gives

$$z\,(P_r - P_l) = 2hf \tag{9.26}$$

and hence

$$z = \frac{2hf}{P_r - P_l} \tag{9.27}$$

Notice in this equation that $P_r - P_l$ is the detected disparity in the observations of P. If $P_r - P_l = 0$, then $z = \infty$. Zero disparity indicates that the point is (effectively) at an infinite distance from the viewer.

9.2.6 The geometry of two cameras; the fundamental matrix

We proceed to derive a mathematical description for the general stereo rig with non-parallel optical axes, see Figure 9.11; the symbol $\simeq$ will be used to denote projection up to unknown

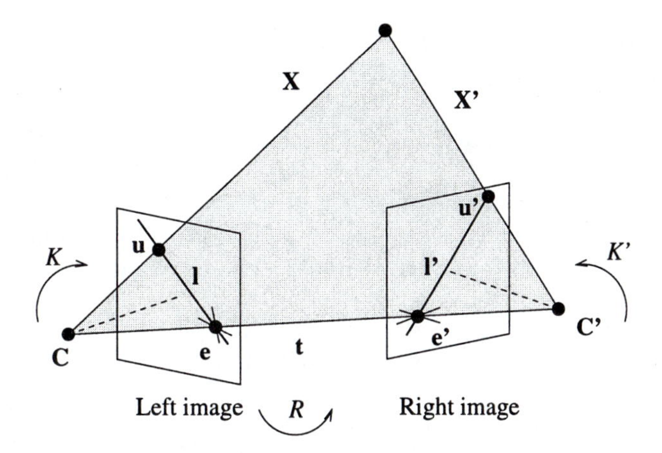

Figure 9.11: *Stereo with non-parallel axes.*

scale. The co-ordinate system of the left view can be transformed to the right view by a translation $\mathbf{t}$ from the left camera center $\mathbf{C}$ to the right camera center $\mathbf{C}'$, and the co-ordinate systems can then be transformed by the rotation R. We shall use a co-ordinate system with the origin in the left camera center $\mathbf{C}$. If K, K' are the calibration matrices of the left and right cameras, we can apply equation (9.9) to get the left projection $\mathbf{u}$ and the right projection $\mathbf{u}'$ of the scene point $\mathbf{X}$:

$$\begin{aligned} \mathbf{u} &\simeq [K|\mathbf{0}] \begin{bmatrix} \mathbf{X} \\ 1 \end{bmatrix} = K\,\mathbf{X} \\ \mathbf{u}' &\simeq [K'R\,|-K'R\mathbf{t}] \begin{bmatrix} \mathbf{X} \\ 1 \end{bmatrix} = K'(R\mathbf{X} - R\mathbf{t}) = K'\mathbf{X}' \end{aligned} \tag{9.28}$$

We know that vectors $\mathbf{X}$, $\mathbf{X}'$, and $\mathbf{t}$ are co-planar. Distinguish co-ordinates of the left and right cameras by the subscripts $_L$, $_R$, respectively—the co-ordinate vector $\mathbf{X}'$ is expressed with respect to the right camera co-ordinate system and therefore is denoted $\mathbf{X}'_R$. We shall express the epipolar constraint using the vector product $\times$, and will do this by expressing the free vector $\mathbf{X}'_R$ with respect to the left camera. The co-ordinate rotation can be written as $\mathbf{X}'_R = R\mathbf{X}'_L$, and hence $\mathbf{X}'_L = R^{-1}\mathbf{X}'_R$. The equation expressing co-planarity can be written as

$$\mathbf{X}_L^T(\mathbf{t} \times \mathbf{X}'_L) = 0 \tag{9.29}$$

Substituting from the equations $\mathbf{X}_L = K^{-1}\mathbf{u}$, $\mathbf{X}'_R = (K')^{-1}\mathbf{u}'$, and $\mathbf{X}'_L = R^{-1}(K')^{-1}\mathbf{u}'$ we get

$$(K^{-1}\mathbf{u})^T(\mathbf{t} \times R^{-1}\,(K')^{-1}\mathbf{u}') = 0 \tag{9.30}$$

This equation (9.30) is homogeneous with respect to $\mathbf{t}$, so the scale is not determined. Absolute scale cannot be recovered if a 'yardstick', i.e., the distance between two points known in advance, is not seen in the scene.

It is helpful to replace the vector product by matrix multiplication. The translation vector is $\mathbf{t} = [t_x, t_y, t_z]^T$, and a skew symmetric[3] matrix $S(\mathbf{t})$ can be created from it if $\mathbf{t} \neq \mathbf{0}$.

$$S(\mathbf{t}) = \begin{bmatrix} 0 & -t_z & t_y \\ t_z & 0 & -t_x \\ -t_y & t_x & 0 \end{bmatrix} \tag{9.31}$$

Recall that rank(S) is a number of the linearly independent lines in matrix S. Note that rank$(S) = 2$ if and only if $\mathbf{t} \neq \mathbf{0}$; the vector product can be replaced by the multiplication of two matrices; for any regular matrix A, we have

$$\mathbf{t} \times A = S(\mathbf{t})\, A \tag{9.32}$$

Thus we can rewrite equation (9.30) as

$$(K^{-1}\mathbf{u})^T \left(S(\mathbf{t})\, R^{-1}\, (K')^{-1}\mathbf{u}'\right) = 0$$

which may be re-arranged to

$$\mathbf{u}^T (K^{-1})^T S(\mathbf{t}) R^{-1} (K')^{-1} \mathbf{u}' = 0 \tag{9.33}$$

The middle part of this equation can be concentrated into a single matrix F called the **fundamental matrix** of two views.

$$F = (K^{-1})^T S(\mathbf{t}) R^{-1} (K')^{-1} \tag{9.34}$$

With the substitution for F in equation (9.33), we finally get the bilinear relation (sometimes called the Longuet-Higgins equation after the inventor [Longuet-Higgins 81] of a similar idea) between any two views,

$$\mathbf{u}^T\, F\, \mathbf{u}' = 0 \tag{9.35}$$

It can be seen that the fundamental matrix F captures all information that can be recovered from a pair of images if the correspondence problem is solved. We shall consider the properties of the fundamental matrix further in due course.

9.2.7 Relative motion of the camera; the essential matrix

A case of practical interest is a single camera moving in space, or two cameras with known calibration—this is known as **relative motion of the camera**. Knowledge of the camera calibration matrices K, K' allows us to normalize measurement in left and right images; we denote the normalized measurements by $\breve{\mathbf{u}}$, $\breve{\mathbf{u}}'$. The camera calibration matrices give the relations

$$\breve{\mathbf{u}} = K^{-1}\mathbf{u} \quad \breve{\mathbf{u}}' = (K')^{-1}\mathbf{u}' \tag{9.36}$$

If these relations are used in equation (9.33), we get a simplified version,

$$\breve{\mathbf{u}}^T\, S(\mathbf{t}) R^{-1}\, \breve{\mathbf{u}}' = 0 \tag{9.37}$$

[3] S is skew symmetric if $S^T = -S$.

Substituting $E = S(\mathbf{t})R^{-1}$, where E is called the **essential matrix**, we get

$$\breve{\mathbf{u}}^T \, E \, \breve{\mathbf{u}}' = 0 \tag{9.38}$$

Again, a bilinear relation between two views in correspondence has been obtained. The essential matrix E captures all the information about the relative motion from the first to the second position of the calibrated camera. E can be estimated from image measurements.

We summarize important *properties of the essential matrix.*

- The essential matrix E has rank 2.
- Let $\mathbf{t}$ be the translational vector, and $\mathbf{t}' = R\mathbf{t}$. Then $E\,\mathbf{t}' = 0$ and $\mathbf{t}^T E = 0$.
- SVD decomposes E as $E = UDV^T$ for a diagonal D; then

$$D = \begin{bmatrix} k & 0 & 0 \\ 0 & k & 0 \\ 0 & 0 & 0 \end{bmatrix} \tag{9.39}$$

Assuming that the essential matrix E has already been estimated, we might be interested in the rotation R and translation $\mathbf{t}$ between these two views. We present without proof a procedure to accomplish this [Hartley 92]. Equation (9.37) shows that the essential matrix is a product of the matrices $S(\mathbf{t})$ and R^{-1}. As

$$\breve{\mathbf{u}}^T \, S(\mathbf{t})R^{-1} \, \breve{\mathbf{u}}' = 0 \quad \breve{\mathbf{u}}'^T \, RS(\mathbf{t}) \, \breve{\mathbf{u}} = 0 \tag{9.40}$$

we can see also that $E = RS(\mathbf{t})$. Recall that SVD provides a similar factorization of a matrix, $E = UDV^T$. The matrix $D = \mathrm{diag}[k, k, 0]$ (where $\mathrm{diag}[x, y, \ldots]$ describes a diagonal matrix, with diagonal $x, y, \ldots$). Let

$$G = \begin{bmatrix} 0 & 1 & 0 \\ -1 & 0 & 0 \\ 0 & 0 & 1 \end{bmatrix} \quad Z = \begin{bmatrix} 0 & -1 & 0 \\ 1 & 0 & 0 \\ 0 & 0 & 0 \end{bmatrix} \tag{9.41}$$

The rotation matrix R can be calculated as

$$R = UG\,V^T \quad \text{or} \quad R = UG^T\,V^T \tag{9.42}$$

and the components of the translation vector can be derived from the matrix $S(\mathbf{t})$, remembering equation (9.31). $S(\mathbf{t})$ itself can be estimated as

$$S(\mathbf{t}) = VZ\,V^T \tag{9.43}$$

We consider now the *properties of the fundamental matrix.*

- We have seen that the rank of the essential matrix E is 2. As $F = (K^{-1})^T E K'^{-1}$ and the calibration matrices are regular, we see that the fundamental matrix F has rank 2 as well.

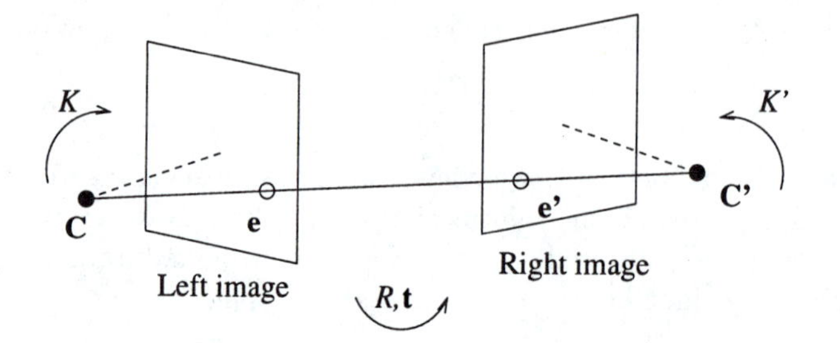

Figure 9.12: *Epipoles* $\mathbf{e}, \mathbf{e}'$, *and the fundamental matrix* F.

- Consider two epipoles $\mathbf{e}, \mathbf{e}'$, depicted in Figure 9.12. Then

$$\mathbf{e}^T F = 0 \quad \text{and} \quad F\,\mathbf{e}' = 0 \tag{9.44}$$

- SVD of the fundamental matrix gives $F = UDV^T$, where

$$D = \begin{bmatrix} k_1 & 0 & 0 \\ 0 & k_2 & 0 \\ 0 & 0 & 0 \end{bmatrix} \quad k_1 \neq k_2 \neq 0 \tag{9.45}$$

9.2.8 Fundamental matrix estimation from image point correspondences

Epipolar geometry has seven degrees of freedom [Mohr 93]: The epipoles $\mathbf{e}$, $\mathbf{e}'$ in the image have two co-ordinates each (giving 4 degrees of freedom), while another three come from the mapping of any three epipolar lines in the first image to the second. Thus the correspondence of seven points in left and right images enables the establishment of the fundamental matrix F using a non-linear algorithm [Faugeras et al. 92]. Unfortunately, this computation is numerically unstable.

If there are eight non-co-planar corresponding points available, a linear method called the **eight-point algorithm** can be used and if more points are at hand the estimation might be robust to noise and mismatches. The method was originally proposed by Longuet-Higgins [Longuet-Higgins 81] for essential matrix estimation.

The eight-point algorithm was supposed to be numerically unstable, but this is not the case if normalization (i.e., translation and scaling) of values is performed first [Hartley 95, Butterfield 97]. The algorithm is easy to implement and is fast; proper normalization is needed in most 3D geometry algorithms to obtain numerical stability.

Recall the fundamental matrix F,

$$\mathbf{u}_i{}^T\, F\, \mathbf{u}'_i = 0 \tag{9.46}$$

An image vector in homogeneous co-ordinates can be written $\mathbf{u}^T = [u_i, v_i, 1]$. The 3×3 fundamental matrix F has only eight unknowns, as it is only known up to scale; eight correspondences will generate eight matrix equations,

$$[u_i,\ v_i,\ 1]\, F \begin{bmatrix} u'_i \\ v'_i \\ 1 \end{bmatrix} = 0 \tag{9.47}$$

Rewriting the elements of the fundamental matrix as a column vector with nine elements $\mathbf{f}^T = [f_{11}, f_{12}, \ldots, f_{33}]$, equation (9.47) can be rewritten as a system of linear equations:

$$\begin{bmatrix} u_i u'_i & u_i v'_i & u_i & v_i u'_i & v_i v'_i & v_i & u'_i & v'_i & 1 \\ & & & & \vdots & & & & \end{bmatrix} \begin{bmatrix} f_{11} \\ f_{12} \\ \vdots \\ f_{33} \end{bmatrix} = 0 \qquad (9.48)$$

If the left-hand matrix in the equation (9.48) is denoted by A, we get

$$A\,\mathbf{f} = 0 \qquad (9.49)$$

The matrix A has rank 8 in a perfect case without noise. With data from real image measurements, an over-determined system of linear equations is obtained, and a least-squares solution to this set is sought [Hartley 95]. The vector $\mathbf{f}$ is determined that minimizes the Frobenius norm $||A\,\mathbf{f}||$ fulfilling the constraint $||f|| = 1$. Principal component analysis gives the solution, and $\mathbf{f}$ is the unit eigenvector of $A^T A$ corresponding to the smallest eigenvalue of A. An appropriate algorithm for achieving this is SVD. Note that another numerically plausible solution to the overdetermined systems of linear equations (9.49) is given in [Faugeras 93].

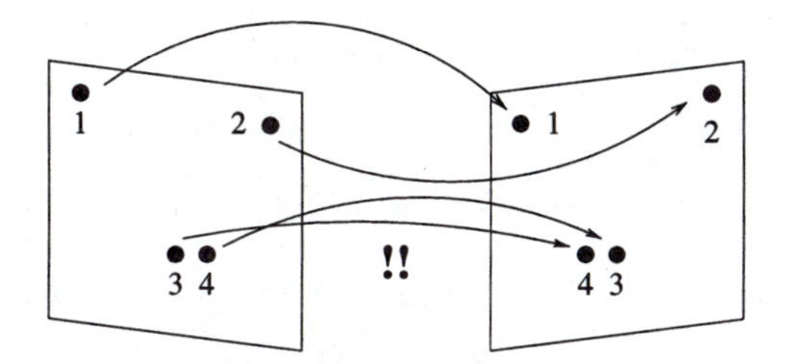

Figure 9.13: *Problem with mismatches in stereo correspondence.*

Estimation of the fundamental matrix can be corrupted by gross errors caused by **mismatches** in stereo correspondence, illustrated in Figure 9.13. The obvious solution to the problem is to attempt to drop out erroneous matches. One approach uses the least median of squares method for robust estimation instead of standard least-squares:

$$\min_{\mathbf{f}}(\mathbf{f}^T\,A^T\,A\;\mathbf{f}) \longrightarrow \min_{\mathbf{f}}\,[\mathrm{median}(||A\,\mathbf{f}||^2)]\,. \qquad (9.50)$$

The eight-point algorithm based on SVD presented above does a very similar job.

We have already seen that the fundamental matrix F should have rank 2, but a solution of equation (9.49) will not in general give such a matrix. F should be replaced by the matrix $\hat{F}$ that minimizes the Frobenius norm of $||F - \hat{F}||$ fulfilling the condition $\mathrm{rank}(A) = 2$. SVD decomposes as $F{=}UDV^T$, $D{=}\mathrm{diag}[r,s,t]$, $r \geq s \geq t$, and the solution we seek is $\hat{F}{=}U\,\mathrm{diag}[r,s,0]\,V^T$.

9.2.9 Applications of epipolar geometry in vision

Image rectification to ease the search for correspondences

We have seen that stereo geometry implies that corresponding points can be sought in 1D space along epipolar lines. In general, epipolar lines in the left image are not parallel to epipolar lines in the right image (non-parallel optical axes). Parallel epipolar lines are preferred, as they ease the search for correspondence, whether by computer or human eye. It is always possible to apply image rectification to images captured by a stereo rig with non-parallel optical axes, resulting in a new set of images with typically parallel epipolar lines.

Image rectification recalculates pixel co-ordinates using a linear transformation in projective space. This is illustrated in Figure 9.14, where C, C' are optical centers. Image planes with dashed borders show input before rectification, and image planes with solid borders and parallel horizontal epipolar lines (dotted lines in rectified images) are the desired result. Points in the left and right images are bilinearly related through the fundamental matrix F,

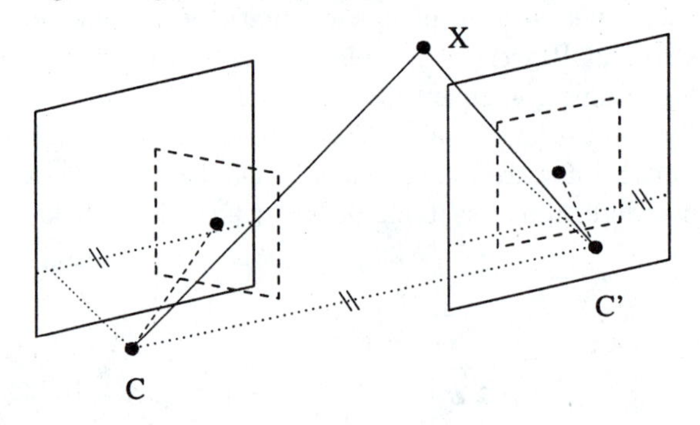

Figure 9.14: *Image rectification to get parallel epipolar lines.*

$\mathbf{u}^T F \mathbf{u}' = 0$. We seek the two 3×3 transformation matrices A, B that rectify co-ordinates (denoted with $\breve{}$) of points in left and right images, respectively, $\breve{\mathbf{u}} = A\mathbf{u}$, $\breve{\mathbf{u}}' = B\mathbf{u}'$. The fundamental matrix of the rectified images $\breve{F} = (A^{-1})^T F B^{-1}$ should correspond to epipoles that moved along horizontal axes to $-\infty$ or ∞ for the left and right images, respectively.

To complete this task we need to set values in the transformation matrices A and B—a solution is given in [Ayache and Hansen 88] which we summarize here. Since the transformation is constrained, the number of unknowns is reduced; image co-ordinate transformations using matrices A, B should not change the position of optical centers, but should align two distinct image planes into one common image plane that is parallel to the line joining C and C', and perpendicular to both newly recalculated optical axes. Moreover, it is desired that the corresponding epipolar lines have the same vertical co-ordinate, as this simplifies calculations.

Image co-ordinates after rectification are

$$\breve{\mathbf{u}} = \begin{bmatrix} U \\ V \\ W \end{bmatrix} = A \begin{bmatrix} u \\ v \\ 1 \end{bmatrix} \quad \breve{\mathbf{u}}' = \begin{bmatrix} U' \\ V' \\ W' \end{bmatrix} = B \begin{bmatrix} u' \\ v' \\ 1 \end{bmatrix} \tag{9.51}$$

Recall the projection matrix from equation (9.12). We have two such matrices M, M' for left and right images before rectification. The 3×3 submatrix on the left side of M is composed of three rows (1×3 vectors) that we denote as $\mathbf{m}_1$, $\mathbf{m}_2$, and $\mathbf{m}_3$; similarly for the right image, the matrix M' gives three vectors $\mathbf{m}'_1$, $\mathbf{m}'_2$, and $\mathbf{m}'_3$. C and C' are co-ordinates of the optical centers. The transformation matrices that perform rectification are then

$$A = \begin{bmatrix} ((C \times C') \times C)^T \\ (C \times C')^T \\ ((C - C') \times (C \times C'))^T \end{bmatrix} [\mathbf{m}_2 \times \mathbf{m}_3,\ \mathbf{m}_3 \times \mathbf{m}_1,\ \mathbf{m}_1 \times \mathbf{m}_2] \tag{9.52}$$

$$B = \begin{bmatrix} ((C \times C') \times C')^T \\ (C \times C')^T \\ ((C - C') \times (C \times C'))^T \end{bmatrix} [\mathbf{m}'_2 \times \mathbf{m}'_3,\ \mathbf{m}'_3 \times \mathbf{m}'_1,\ \mathbf{m}'_1 \times \mathbf{m}'_2] \tag{9.53}$$

This procedure is computationally inexpensive. Only two 3×3 transformation matrices need be stored, and only six multiplications, six additions and two divisions are needed per rectified pixel. Notice that the rectification is a linear transformation in projective space that preserves straight lines. If an image consists of linear segments, then it is sufficient to rectify end points of these segments. The procedure can be easily generalized to three and more images [Ayache and Hansen 88].

Ego-motion estimation from calibrated camera measurements

Camera **ego-motion** estimation of a calibrated camera considers the case of unknown movement of the camera, where rotation R and translation $\mathbf{t}$ need to be learned from point correspondences between two images (see Figure 9.15).

Suppose a point $\mathbf{u}_i$ from the first image corresponds to the point $\mathbf{u}'_i$. The following algorithm [Hartley 92] allows the computation of an unknown rotation R and translation $\mathbf{t}$ of the camera.

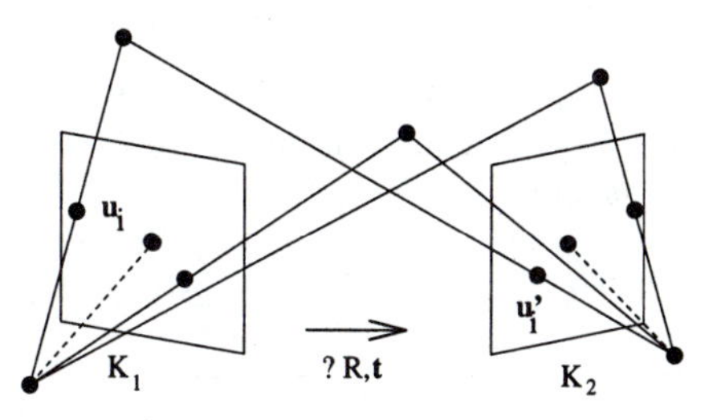

Figure 9.15: *Ego-motion estimation.*

Algorithm 9.1: Ego-motion estimation

1. Find correspondences between points $\mathbf{u}_i$ and $\mathbf{u}'_i$; these will be used to estimate a fundamental matrix.

2. The data should be normalized—this helps to minimize numerical errors.

$$\breve{\mathbf{u}} = H_1\,\mathbf{u} \quad \breve{\mathbf{u}}' = H_2\,\mathbf{u}' \tag{9.54}$$

$$H_1 = \begin{bmatrix} a_1 & 0 & c_1 \\ 0 & b_1 & d_1 \\ 0 & 0 & 1 \end{bmatrix} \qquad H_2 = \begin{bmatrix} a_2 & 0 & c_2 \\ 0 & b_2 & d_2 \\ 0 & 0 & 1 \end{bmatrix} \tag{9.55}$$

After normalization, the data should have similar order, i.e., $\text{mean}(\check{\mathbf{u}}) = 0$ and $\text{var}(\check{\mathbf{u}}) = [1,1]^T$.

3. Compute an estimate of the fundamental matrix $\hat{F}$ using the linear algorithm given in Section 9.2.8. Numerical inaccuracies may cause the estimate not to have the property that after SVD, $D = \text{diag}(k,k,0)$.

4. Compute the estimated essential matrix $\hat{E}$. This is easy, as calibration matrices K, K' are known:
$$\hat{E} = K^T \hat{F} K' \tag{9.56}$$

5. Determine a rotation R and translation $\mathbf{t}$ from the estimated essential matrix $\hat{E}$ using SVD. The translation $\mathbf{t}$ is given up to scale only.
$$\hat{E} = UDV^T \quad D = \begin{bmatrix} r & 0 & 0 \\ 0 & s & 0 \\ 0 & 0 & t \end{bmatrix} \tag{9.57}$$
Notice that we expect three different singular values due to numerical inaccuracies. We know that the essential matrix E should have two equal singular values, and the third must be zero. We can adjust singular values by zeroing t and averaging r and s
$$E = u \begin{bmatrix} \frac{r+s}{2} & 0 & 0 \\ 0 & \frac{r+s}{2} & 0 \\ 0 & 0 & 0 \end{bmatrix} V^T \tag{9.58}$$
This matrix E can be decomposed into rotation R and translation $\mathbf{t}$ in the same way as was used in Section 9.2.7. Recall that matrices G and Z were defined by equations (9.41); then we can calculate
$$R = UGV^T \text{ or } UG^TV^T \qquad S(\mathbf{t}) = VZ\,V^T \tag{9.59}$$

Notice that the translation $\mathbf{t}$ is obtained up to unknown scale only, which is to be expected. As nothing was known in advance about the scene, the same images could be seen when half-size objects are observed from half the distance.

3D similarity reconstruction from two cameras with known intrinsic calibration

3D similarity reconstruction aims to measure 3D co-ordinates of a scene point $\mathbf{X}$ from two image measurements $\mathbf{u}$ and $\mathbf{u}'$ (see Figure 9.16). We assume that the cameras are calibrated; that is, their intrinsic calibration parameters are known and are available as calibration matrices K and K'. The extrinsic parameters are unknown. This case differs from standard

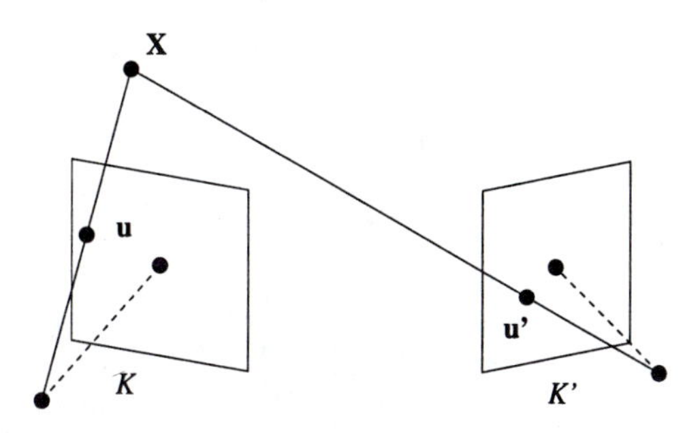

Figure 9.16: *3D similarity reconstruction from two cameras.*

stereo (full 3D Euclidean reconstruction), where the relative position of the cameras is known. Common sense suggests that less will be measured in an unknown scene compared to the standard stereo case; the reconstruction of the unknown scene is achieved up to a similarity.

The image measurements are

$$\mathbf{u} \simeq [K\,|\mathbf{0}]\,\mathbf{X} \quad \mathbf{u}' \simeq [K'R\,|-K'R\mathbf{t}]\,\mathbf{X} \tag{9.60}$$

Algorithm 9.2: 3D similarity reconstruction from two cameras

1. Find correspondences between two images.
2. Compute the essential matrix E.
3. Obtain the rotation R and translation $\mathbf{t}$ from the essential matrix E.
4. Solve equations (9.60) to get $\mathbf{X}$.

Notice that $\mathbf{X}$ is found up to scale only, meaning that we do not get a Euclidean reconstruction but a similarity reconstruction. A full Euclidean reconstruction (as in stereo vision) is unavailable because the distance between the cameras is unknown in this case.

3D projective reconstruction from two uncalibrated cameras

We now consider the most general 3D reconstruction case when the point correspondence in two uncalibrated cameras can be established, meaning that both intrinsic and extrinsic camera calibration parameters are unknown. We shall see that a 3D projective reconstruction can be obtained, which is practically appealing as we can learn something about the geometry of the scene even from a video sequence where nothing is known about the conditions under which it was captured; the camera position is unknown, and a zoom lens may be used, and we do not know the actual focal length.

The perspective projection performed by the first camera is expressed using the projective matrix M [recall equation (9.9)], which is divided into the three row vectors $\mathbf{m}_1^T$, $\mathbf{m}_2^T$, $\mathbf{m}_3^T$.

Similarly for the second camera where primed symbols are used:

$$\text{1st image} \qquad \mathbf{u} = \begin{bmatrix} U \\ V \\ W \end{bmatrix} \simeq M\,\mathbf{X} = \begin{bmatrix} \mathbf{m}_1^T \\ \mathbf{m}_2^T \\ \mathbf{m}_3^T \end{bmatrix} \mathbf{X} \tag{9.61}$$

$$\text{2nd image} \qquad \mathbf{u}' = \begin{bmatrix} u' \\ v' \\ w' \end{bmatrix} \simeq M'\,\mathbf{X} = \begin{bmatrix} \mathbf{m'}_1^T \\ \mathbf{m'}_2^T \\ \mathbf{m'}_3^T \end{bmatrix} \mathbf{X} \tag{9.62}$$

To eliminate the unknown scale factor, consider the ratio between the three rows in the projection matrix M [Faugeras and Mourrain 95].

$$\begin{aligned} u : v : w &= \mathbf{m}_1^T\mathbf{X} : \mathbf{m}_2^T\mathbf{X} : \mathbf{m}_3^T\mathbf{X} \\ u' : v' : w' &= \mathbf{m'}_1^T\mathbf{X} : \mathbf{m'}_2^T\mathbf{X} : \mathbf{m'}_3^T\mathbf{X} \end{aligned} \tag{9.63}$$

Thus three equations hold for both the first and the second camera:

$$\begin{aligned} u\mathbf{m}_2^T\mathbf{X} &= v\mathbf{m}_1^T\mathbf{X} & \qquad u'\mathbf{m'}_2^T\mathbf{X} &= v'\mathbf{m'}_1^T\mathbf{X} \\ u\mathbf{m}_3^T\mathbf{X} &= w\mathbf{m}_1^T\mathbf{X} & u'\mathbf{m'}_3^T\mathbf{X} &= w'\mathbf{m'}_1^T\mathbf{X} \\ v\mathbf{m}_3^T\mathbf{X} &= w\mathbf{m}_2^T\mathbf{X} & v'\mathbf{m'}_3^T\mathbf{X} &= w'\mathbf{m'}_2^T\mathbf{X} \end{aligned} \tag{9.64}$$

Equations (9.64) can be written in a matrix form. We present this for the first camera only; a similar expression holds for the second camera.

$$\begin{bmatrix} u\mathbf{m}_2^T & - & v\mathbf{m}_1^T \\ u\mathbf{m}_3^T & - & w\mathbf{m}_1^T \\ v\mathbf{m}_3^T & - & w\mathbf{m}_2^T \end{bmatrix} \mathbf{X} = 0 \tag{9.65}$$

If the first row in the matrix is multiplied by w and second row by $-v$ and added, we get

$$(uw\mathbf{m}_2^T - vw\mathbf{m}_1^T - uv\mathbf{m}_3^T + vw\mathbf{m}_1^T)\,\mathbf{X} = (uw\mathbf{m}_2^T - uv\mathbf{m}_3^T)\,\mathbf{X} = 0 \tag{9.66}$$

Extracting the equation corresponding to the third row of the matrix in equation (9.65), we get

$$(-w\mathbf{m}_2^T + v\mathbf{m}_3^T)\mathbf{X} = 0 \tag{9.67}$$

We see that equations (9.66) and (9.67) are linearly dependent, and the same reasoning holds for measurements from the second image. Since only two equations are linearly independent, we use the second and the third equations.

$$\begin{aligned} (u\mathbf{m}_3^T - w\mathbf{m}_1^T)\mathbf{X} &= 0 & \qquad (u'\mathbf{m'}_3^T - w'\mathbf{m'}_1^T)\mathbf{X} &= 0 \\ (v\mathbf{m}_3^T - w\mathbf{m}_2^T)\mathbf{X} &= 0 & (v'\mathbf{m'}_3^T - w'\mathbf{m'}_2^T)\mathbf{X} &= 0 \end{aligned} \tag{9.68}$$

This can be rewritten in matrix form:

$$\begin{bmatrix} u\mathbf{m}_3^T - w\mathbf{m}_1^T \\ v\mathbf{m}_3^T - w\mathbf{m}_2^T \\ u'\mathbf{m'}_3^T - w'\mathbf{m'}_1^T \\ v'\mathbf{m'}_3^T - w'\mathbf{m'}_2^T \end{bmatrix} \mathbf{X} = A\,\mathbf{X} = 0 \tag{9.69}$$

The matrix A has dimension 4×4 and $\mathbf{X}$ is a 4×1 vector.

We are interested in a non-trivial solution of equation (9.69), and therefore consider the case $\det(A) = 0$. This implies that the matrix A should have rank 3 if $\mathbf{u}$ and $\mathbf{u}'$ are really corresponding points in the first and second images.

There are two important cases to consider in reconstructing a 3D point from two corresponding 2D points in two images:

1. *Scene reconstruction with calibrated cameras.*
   This is a special case (called stereopsis, 3D Euclidean reconstruction) that has been already considered in Section 9.2.5. The current formalism concentrates all knowns into the matrix A; the projective matrices M, M' and image measurements $\mathbf{u}$, $\mathbf{u}'$ are known. The equation (9.69) can easily be replaced by an inverse mapping.

2. *Scene reconstruction with uncalibrated cameras.*
   If the calibration of a stereo rig is unknown, it can be shown that the reconstructed co-ordinates $\tilde{\mathbf{X}}$ differ from the correct Euclidean reconstruction by some (unknown) projective transformation H.
   $$\tilde{\mathbf{X}} = H\,\mathbf{X} \tag{9.70}$$
   H is a regular 4×4 matrix. The transformation H ranges from Euclidean through affine to the general projective case according to how much calibration knowledge is at hand. H is the same for all scene points for one position and calibration of the camera. Of course, the same algorithm with a different scene gives a different H.
   $$\mathbf{u} = \begin{bmatrix} u \\ v \\ w \end{bmatrix} \simeq M\,\mathbf{X} = M\,H^{-1}\,H\,\mathbf{X} = \tilde{M}\,\tilde{\mathbf{X}} \tag{9.71}$$
   $$\mathbf{u}' = \begin{bmatrix} u' \\ v' \\ w' \end{bmatrix} \simeq M'\,\mathbf{X} = M'\,H^{-1}\,H\,\mathbf{X} = \tilde{M}'\,\tilde{\mathbf{X}} \tag{9.72}$$
   Notice that $M\,\mathbf{X}$ and $\tilde{M}\,\tilde{\mathbf{X}}$ give the same measurement. The measurements $\tilde{\mathbf{X}}$ differs from the correct Euclidean measurement $\mathbf{X}$ by a projective transformation H.

   The projective transformation is determined by at least five corresponding points. The projective matrix $\tilde{M}$ should be created in such a way that it differs from the matrix M only projectively. See [Faugeras 93, Faugeras and Mourrain 95] for more details.

9.2.10 Three and more cameras

In this section we will consider the case of three or more cameras observing the same scene, assuming mutually corresponding points can be found in all views. We have already seen that views of two cameras are described using a bilinear relation expressed by the fundamental matrix, and it is natural to ask what more can be learned if three or more views are available.

Three cameras looking at the same point are sketched in Figure 9.17. The relations between projected image points $\mathbf{u}$, $\mathbf{u}'$, $\mathbf{u}''$ and their respective 3D counterparts $\mathbf{X}$, $\mathbf{X}'$, $\mathbf{X}''$ are

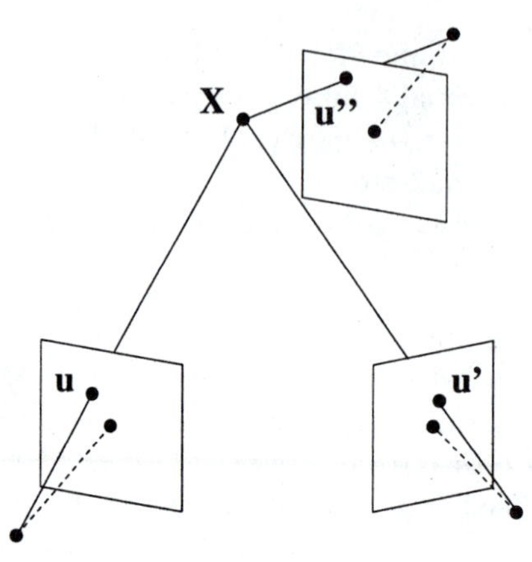

Figure 9.17: *Geometry of three cameras,* $\mathbf{u} \simeq M\,\mathbf{x}$, $\mathbf{u}' \simeq M'\,\mathbf{x}$, $\mathbf{u}'' \simeq M''\,\mathbf{x}$.

given by the projection matrices M, M', M''. using a similar approach to that given when computing 3D projective reconstruction from two uncalibrated cameras, we aim to obtain a set of linear equations that relate image measurements to their 3D counterparts.

$$\mathbf{u} = \begin{bmatrix} U \\ V \\ W \end{bmatrix} \simeq M\,\mathbf{X} = \begin{bmatrix} \mathbf{m}_1^T \\ \mathbf{m}_2^T \\ \mathbf{m}_3^T \end{bmatrix} \mathbf{X} \quad \mathbf{u}' = \begin{bmatrix} u' \\ v' \\ w' \end{bmatrix} \simeq M'\,\mathbf{X} = \begin{bmatrix} \mathbf{m'}_1^T \\ \mathbf{m'}_2^T \\ \mathbf{m'}_3^T \end{bmatrix} \mathbf{X} \tag{9.73}$$

$$\mathbf{u}'' = \begin{bmatrix} u'' \\ v'' \\ w'' \end{bmatrix} \simeq M''\,\mathbf{X} = \begin{bmatrix} \mathbf{m''}_1^T \\ \mathbf{m''}_2^T \\ \mathbf{m''}_3^T \end{bmatrix} \mathbf{X} \tag{9.74}$$

Using similar manipulations to equations (9.62)–(9.69) eliminates unknown scale factors and provides the desired relation in matrix form. We will assign labels `1:` to `6:` in the following for convenience of future reference,

$$\begin{matrix} \mathtt{1:} \\ \mathtt{2:} \\ \mathtt{3:} \\ \mathtt{4:} \\ \mathtt{5:} \\ \mathtt{6:} \end{matrix} \begin{bmatrix} u\mathbf{m}_3^T - w\mathbf{m}_1^T \\ v\mathbf{m}_3^T - w\mathbf{m}_2^T \\ u'\mathbf{m'}_3^T - w'\mathbf{m'}_1^T \\ v'\mathbf{m'}_3^T - w'\mathbf{m'}_2^T \\ u''\mathbf{m''}_3^T - w''\mathbf{m''}_1^T \\ v''\mathbf{m''}_3^T - w''\mathbf{m''}_2^T \end{bmatrix} \mathbf{X} = A\,\mathbf{X} = 0 \tag{9.75}$$

We shall follow (but simplify) an explanation given in [Faugeras and Mourrain 95], and shall use the reference numbers of equation (9.75). We are interested in the non-trivial solution to this equation, meaning that the matrix A should have rank 3. This means that the determinant of all its 4×4 submatrices must be zero; there are $C_4^6 = 6!/(4!\,2!) = 15$ such submatrices. Consider these 15 quadruples of equations and classify then according to whether they involve two or three cameras.

Three sets of equations express a bilinear relation between two cameras that are given by the fundamental matrix F as we already know. These are equations `[1234]`, `[1256]`, and

[3456]; notice that even squares of the same variable do not appear in these bilinear equations. Consider now sets of equations that express a trilinear relation among images of the same point as seen by three cameras. Of the 12 trilinearities, only four are linearly independent; three possibilities for linearly independent quadruples of equations are the following—[notice that there are always two rows of equation (9.75) corresponding to one camera, with one row for each of the remaining two cameras]:

$$\begin{array}{cccc} [1235] & [1245] & [1236] & [1246] \\ [1345] & [2345] & [1346] & [2346] \\ [1356] & [1456] & [2356] & [2456] \end{array} \qquad (9.76)$$

A geometric interpretation of the rows of equation (9.75) assists understanding; each row defines a plane passing through optical center **C** and the point **X** for which correspondence in all three views was established, see Figure 9.18.

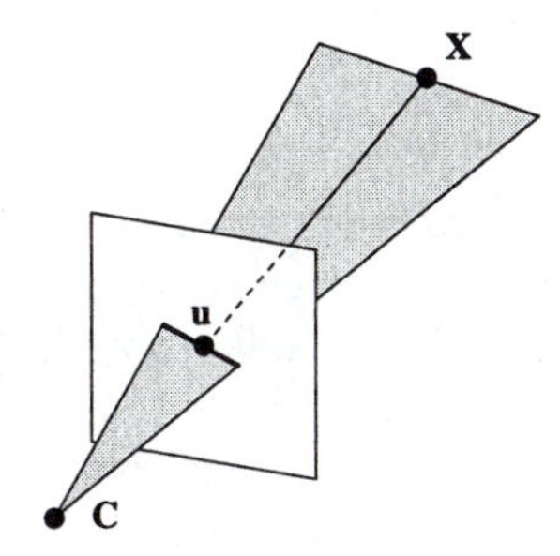

Figure 9.18: *Each of six rows defines a plane passing through the optical center* **C** *and the point* **X**.

Notice that one trilinearity relation does not ensure that observed points $\mathbf{u}$, $\mathbf{u}'$, and $\mathbf{u}''$ correspond to only one scene point **X**. Only one of the views plays a role of the ray; this is illustrated in Figure 9.19. Two rows of equation (9.75) corresponded to the measurement **u** taken by the first camera, and the corresponding ray points to **X**. The other two views constrain the point position in the space to a plane only. One trilinearity relation ensures that the ray and two planes have a common point in the projective space $\mathcal{P}^3$. In other words, **X**, **A**, and **B** are colinear but need not be co-incident.

Consider now what happens if we had four cameras. In equation (9.75) we would have two more equations. Now we can consider 4×4 sub-determinants which contain one row arising from one camera. This is called a **quadrilinear constraint** which is a polynomial of degree 4 in the co-ordinates of the points $\mathbf{m}_i$ and linear in the co-ordinates of each of them. Assuming that all the bilinear and trilinear constraints are satisfied, it is possible to show that the quadrilinear constraint can be obtained as a linear combination of bilinear and trilinear constraints. This means that *the fourth view does not contribute any additional information* if exact measurements in the image are assumed. To summarize, the relations among corresponding projections of a single point in two, three, and four images are completely understood under orthographic, similarity, and perspective projection. There is no relation involving five and more cameras that cannot be factored into relations of fewer cameras [Weinshall et al. 95].

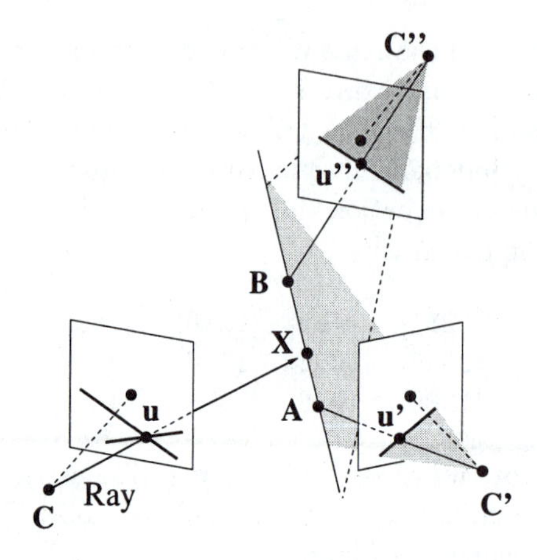

Figure 9.19: *Illustration that one trilinear relation does not assure that three measured points are co-incident as expected.*

The case of two, three, and four cameras is illustrated in Figure 9.20. The upper row shows the case of the bilinear constraint (given by the fundamental matrix F) that relates corresponding points in two images. The middle row illustrates the trilinearity constraint where correspondence between one point and two lines is established. The bottom row shows the quadrilinear constraint where correspondence of four lines is taken into account.

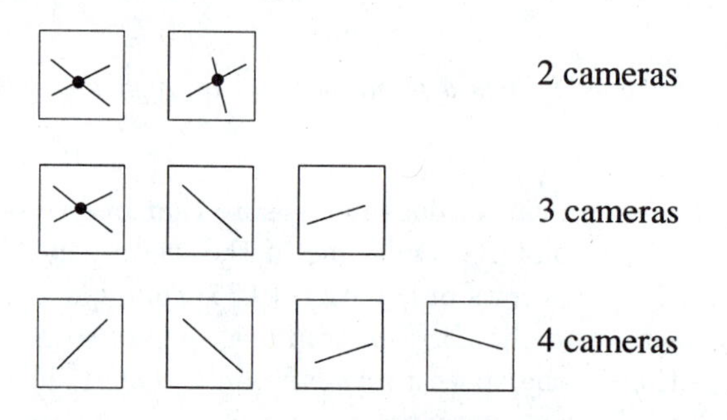

Figure 9.20: *Geometric interpretation of bilinear, trilinear, and quadrilinear constraint.*

Given this intuitive geometric understanding of the trilinear relation among views, we shall proceed to an algebraic derivation as well. Assume that the first camera is in a canonical configuration, i.e., its projective matrix is in the simplest form:

$$\begin{aligned}
\mathbf{u} &\simeq M\,\tilde{\mathbf{X}} = M\,H^{-1}\,H\,\mathbf{X} = [I|0]\,\mathbf{X} \\
\mathbf{u}' &\simeq M'\,\tilde{\mathbf{X}} = M'\,H^{-1}\,H\,\mathbf{X} = [a_{ij}]\,\mathbf{X} \quad i = 1,\dots,3 \\
\mathbf{u}'' &\simeq M''\,\tilde{\mathbf{X}} = M''\,H^{-1}\,H\,\mathbf{X} = [b_{ij}]\,\mathbf{X} \quad j = 1,\dots,4
\end{aligned} \tag{9.77}$$

The scale in the image measurement is unknown, as $\mathbf{u} \simeq [I|0]\,\mathbf{X}$.

$$\tilde{\mathbf{X}} = \begin{bmatrix} \mathbf{u} \\ \rho \end{bmatrix} \tag{9.78}$$

The scale factor ρ is to be determined. Project the scene point $\mathbf{X}$ into the second camera.

$$\mathbf{u}' \simeq [a_{ij}]\,\tilde{\mathbf{X}} = \begin{bmatrix} \mathbf{u} \\ \rho \end{bmatrix} \tag{9.79}$$

$$u_i' \simeq a_i^k u_k + a_{i4}\rho \qquad k = 1, \ldots, 3 \tag{9.80}$$

Here we have adopted Einstein's convention of omitting the summation symbol for compactness of representation. Thus $a_i^k u_k$ would originally be written as $\sum_{k=1}^{k=3} a_{ik} u_k$.

The scale factor ρ needs to be eliminated. We get three equations, of which two are independent:

$$u_i'\,(a_j^k\, u_k + a_{j4}\rho) = u_j'\,(a_i^k\, u_k + a_{i4}\rho) \tag{9.81}$$

This yields three estimates of ρ:

$$\rho = \frac{u_k(u_i'\, a_j^k - u_j'\, a_i^k)}{u_j'\, a_{i4} - u_i'\, a_{j4}} \tag{9.82}$$

The scale factor ρ is substituted back to $\tilde{\mathbf{X}}$:

$$\tilde{\mathbf{X}} = \begin{bmatrix} \mathbf{u} \\ \frac{u_k(u_i'\, a_j^k - u_j'\, a_i^k)}{u_j'\, a_{i4} - u_i'\, a_{j4}} \end{bmatrix} \simeq \begin{bmatrix} (u_j'\, a_{i4} - u_i'\, a_{j4})\,\mathbf{u} \\ u_k(u_i'\, a_j^k - u_j'\, a_i^k) \end{bmatrix} \tag{9.83}$$

Now $\tilde{\mathbf{X}}$ is projected by the third camera:

$$\mathbf{u}'' \simeq [b_{lk}]\mathbf{X} = b_l^k x_k \tag{9.84}$$

Notice that

$$\begin{aligned} u_l'' &\simeq b_l^k u_k(u_j' a_{i4} - u_i' a_{j4}) + b_{l4} u_k(u_i' a_j^k - u_j' a_i^k) \\ &\simeq u_k u_i(a_j^k b_{l4} - a_{j4} b_l^k) - u_k u_j'(a_i^k b_{l4} - a_{i4} b_l^k) \\ &\simeq u_k(u_i' T_{kjl} - u_j' T_{kil}) \end{aligned} \tag{9.85}$$

T_{ijk}, $i, j, k = 1, 2, 3$ is an algebraic entity called a tensor that depends on three indices. This can be imagined as a 'three-dimensional matrix', i.e., a $3 \times 3 \times 3$ cube consisting of 27 numbers.

The unknown scale can be eliminated if all three views are combined:

$$u_k(u_i' u_m'' T_{kjl} - u_j' u_m'' T_{kil}) = u_k(u_i' u_l'' T_{kjm} - u_j' u_l'' T_{kim}) \tag{9.86}$$

This equation is symmetric with respect to i, j and l, m; thus $i < j$ and $l < m$. There are nine equations, but only four of them are linearly independent. Assume $j = m = 3$ and for simplicity $u_3 = u_3' = u_3'' = 1$. After some manipulations we get the **trilinear constraint** among three views:

$$u_k(u_i' u_l'' T_{k33} - u_l'' T_{ki3} - u_i' T_{k3l} + T_{kil}) = 0 \tag{9.87}$$

As indices i, l can have values 1 or 2, we have four linearly independent equations.

The tensor T_{ijk} has 27 unknowns that can be estimated from at least seven corresponding points in three images.

The use of the trilinear constraint yields three practical advantages [Shashua and Werman 95].

1. The trilinear tensor can be recovered linearly from seven corresponding points in three views, while the fundamental matrix calculated from a pair of views needs at least eight points for linear solution. Practically, an overdetermined system of equations is solved using some robust estimation method.

2. The tensor can be used instead of three fundamental matrices. This is possible even in the case in which some of the fundamental matrices are singular.

3. The estimate of the constraint among three views should be numerically more stable than the estimate through three fundamental matrices.

One of the important applications of the trilinear tensor is **epipolar transfer**. Assuming that the trilinear tensor has been estimated, if two images are known, any third image can be computed using equation (9.87).

The other application of the trilinear tensor is in reconstruction and recognition. So far we have studied how one point is seen in one, two, three, or four images. The dual problem, i.e., the geometry of N 3D points in one image, allows an approach to shape under perspective projection with uncalibrated cameras [Weinshall et al. 95].

9.2.11 Stereo correspondence algorithms

We have seen in Section 9.2.6 that much can be learned about the geometry of a 3D scene if it is known which point from one image corresponds to a point in a second image. The solution of this **correspondence problem** is a key step in any photogrammetric, stereo vision, or motion analysis task. Here we describe how the same point can be found in two images if the same scene is observed from two different viewpoints. Of course, it is assumed that two images overlap and thus the corresponding points are sought in this overlapping area.

In image analysis, some methods are based on the assumption that images constitute a linear (vector) space (e.g., eigenimages or linear interpolation in images [Werner et al. 95, Ullman and Basri 91]); this linearity assumption[4] is not valid for images in general [Beymer and Poggio 96], but some authors have overlooked this fact. The structure of a vector space assumes that the ith component of one vector must refer to the ith component of another; this assumes that the correspondence problem has been solved.

Automatic solution of the correspondence problem is an evergreen computer vision topic, and the pessimistic conclusion is that it is not soluble in the general case at all. The trouble is that the correspondence problem is inherently ambiguous. Imagine an extreme case, e.g., a scene containing a white, nontextured, flat object; its image constitutes a large region with uniform brightness. When corresponding points are sought in left and right images of the flat object there are not any features that could distinguish them. Another unavoidable difficulty in searching for corresponding points is the **self-occlusion** problem, which occurs in images of non-convex objects. Some points that are visible by the left camera are not visible by the right camera and vice versa (see Figure 9.21).

Fortunately, uniform intensity and self-occlusion are rare, or at least uncommon, in scenes of practical interest. Establishing correspondence between projections of the same point in

[4]Informally, the sum of any two points from a linear space must belong to the linear space; similarly for any point multiplied by any real number.

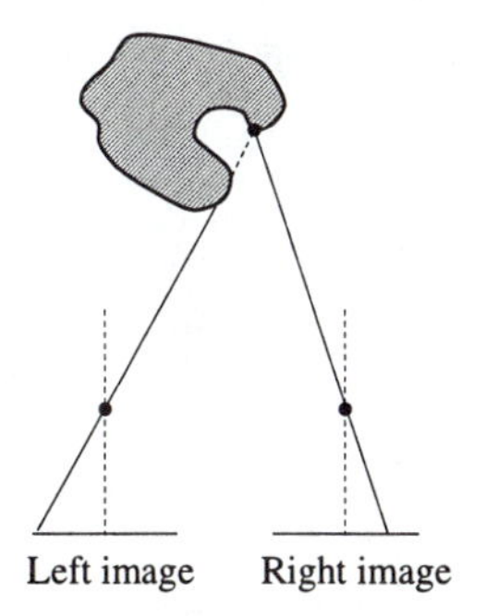

Figure 9.21: *Self-occlusion makes search for some corresponding points impossible.*

different views is based on finding image characteristics that are similar in both views, and the local similarity is calculated.

The inherent ambiguity of the *correspondence problem* can in practical cases be reduced using several **constraints**. Some of these follow from the geometry of the image capturing process, some from photometric properties of a scene, and some from prevailing object properties in our natural world. A vast number of different stereo correspondence algorithms have been proposed. We will give here only a concise taxonomy of approaches to finding correspondence—not all the constraints are used in all of them. There follows a list of constraints commonly used [Klette et al. 96] to provide insight into the correspondence problem.

The first group of constraints depends mainly on the geometry and the photometry of the image capturing process.

Epipolar constraint: This says that the corresponding point can only lie on the epipolar line in the second image. This reduces the potential 2D search space into 1D. The epipolar constraint was explained in detail in Section 9.2.5.

Uniqueness constraint: This states that, in most cases, a pixel from the first image can correspond to only one pixel in the second image. The exception arises when two or more points lie on one ray coming from the first camera and can be seen as separate points from the second. This case, which arises in the same way as self-occlusion, is illustrated in Figure 9.22.

Photometric compatibility constraint: This states that intensities of a point in the first and second images are likely to differ only a little. They are unlikely to be exactly the same due to the mutual angle between the light source, surface normal, and viewer differing, but the difference will typically be small and the views will not differ much. Practically, this constraint is very natural to image-capturing conditions. The advantage is that intensities in the left image can be transformed into intensities in the right image using very simple transformations.

Geometric similarity constraints: These build on the observation that geometric characteristics of the features found in the first and second images do not differ much (e.g., length or orientation of the line segment, region, or contour).

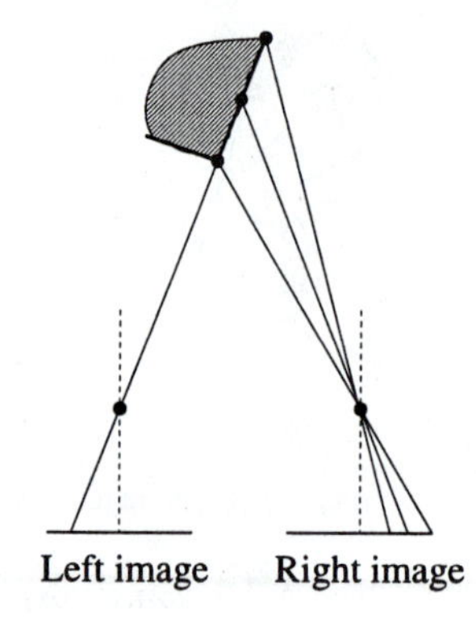

Figure 9.22: *Exception from the uniqueness constraint.*

The second group of constraints exploits some common properties of objects in typical scenes.

Disparity smoothness constraint: This claims that disparity changes slowly almost everywhere in the image. Assume two scene points $\mathbf{p}$ and $\mathbf{q}$ are close to each other, and denote the projection of $\mathbf{p}$ into the left image as $\mathbf{p}_L$ and into the right image as $\mathbf{p}_R$, and $\mathbf{q}$ similarly. If we assume that the correspondence between $\mathbf{p}_L$ and $\mathbf{p}_R$ has been established, then the quantity

$$\left|\left(|\mathbf{p}_L - \mathbf{p}_R| - |\mathbf{q}_L - \mathbf{q}_R|\right)\right|$$

(the absolute disparity difference) should be small.

Figural disparity constraint: This says that corresponding points should lie on an edge element in both right and left images, as well as fulfilling the disparity smoothness constraint.

Feature compatibility constraint: This place a restriction on possible matches on the physical origin of matched points. Points can match only if they have the same physical origin – for example, object surface discontinuity, border of a shadow cast by some objects, occluding boundary or specularity boundary. Notice that edges in an image caused by specularity or self-occlusion cannot be used to solve the correspondence problem, as they move with changing viewpoint. On the other hand, self-occlusion caused by abrupt discontinuity of the surface can be identified—see Figure 9.23.

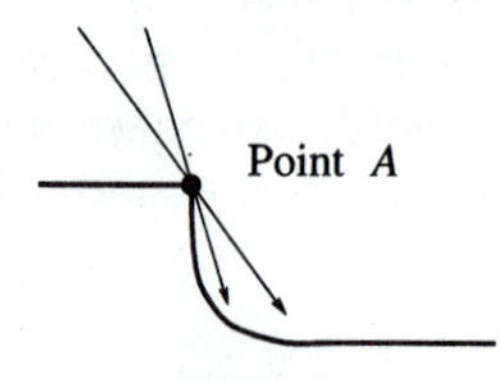

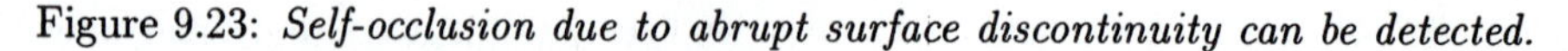

Figure 9.23: *Self-occlusion due to abrupt surface discontinuity can be detected.*

Disparity limit constraint: This originates from psycho-physical experiments in which it is demonstrated that the human vision system can only fuse stereo images if the disparity is smaller than some limit. This constrains the lengths of the search in artificial methods that seek correspondence.

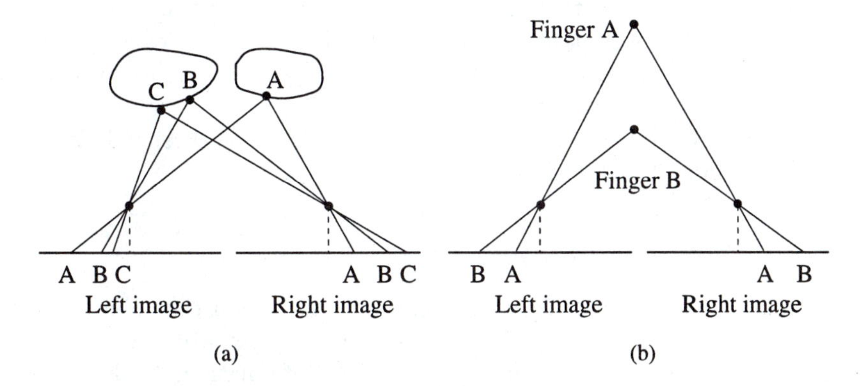

Figure 9.24: *(a) Corresponding points lie in the same order on epipolar lines. (b)This rule does not hold if there is a big discontinuity in depths.*

Ordering constraint: This says that for surfaces of similar depth, corresponding feature points typically lie in the same order on the epipolar line (see Figure 9.24a). If there is a narrow object much closer to the camera than its background, the order can be changed (see Figure 9.24b). It is easy to demonstrate violation of this ordering constraint: Hold two forefingers vertically, almost aligned but at different depths in front of your eyes. Closing the left eye and then the right eyes interchanges the left/right order of the fingers.

Mutual correspondence constraint: This helps to rule out points that do not have a corresponding counterpart due to occlusion, highlight, or noise. Assume the search started from the left image point $\mathbf{p}_L$ and a corresponding $\mathbf{p}_R$ was found. If the task is reversed, and a search starting from the point $\mathbf{p}_R$ fails to find the point $\mathbf{p}_L$, then the match is not reliable and should be ruled out.

All these constraints have been of use in one or more existing stereo correspondence algorithms; we present here a taxonomy of such algorithms. From the historical point of view, correspondence algorithms for stereopsis were and still are driven by two main paradigms:

1. Low-level, correlation-based, bottom-up methods

2. High-level, feature-based, top-down methods

Initially, it was believed that higher-level features such as corners and straight line segments should be automatically identified, and then matched. This was a natural development from photogrammetry, which has been using feature points identified by human operators since the beginning of the twentieth century.

Psychological experiments with **random dot stereograms** performed by Julesz [Julesz 90] generated a new view: These experiments show that humans do not need to create monocular features before binocular depth perception can take place. A random dot stereogram is created in the following way: A left image is entirely random, and the right image is created from it in a consistent way such that some part of it is shifted according to disparity of the desired stereo effect. The viewer must glare at the random dot stereogram from a given distance of about 20 centimeters. Such 'random dot stereograms' have been widely published under the name '3D images' in many popular magazines.

Recent developments in this area use a combination of both low-level and high-level stereo correspondence methods [Tanaka and Kak 90].

Correlation-based stereo correspondence

Correlation-based correspondence algorithms use the assumption that pixels in correspondence have very similar intensities (recall the photometric compatibility constraint). The intensity of an individual pixel does not give sufficient information, as there are typically many potential candidates with similar intensity and thus intensities of several neighboring pixels are considered. Typically, a 5×5 or 7×7 or 3×9 window may be used. These methods are sometimes called **area-based stereo**.

We shall illustrate the approach with a simple algorithm called **block matching** [Klette et al. 96]. Assuming the canonical stereo setup with parallel optical axes of both cameras, the basic idea of the algorithm is that all pixels in the window (called a block) have the same disparity, meaning that one and only one disparity is computed for each block. One of the images, say the left, is tiled into blocks, and a search for correspondence in the right image is conducted for each of these blocks in the right image. The measure of similarity between blocks can be, e.g., the mean square error of the intensity, and the disparity is accepted for the position where the mean square error is minimal. Maximal change of position is limited by the disparity limit constraint. The mean square error can have more than one minimum, and in this case an additional constraint is used to cope with ambiguity.

The result of the block matching algorithm is a sparse matrix of disparities, where disparity is calculated only for a representative point of the block; various methods allow us to refine the result to a dense disparity matrix. Block matching algorithms are typically slow, and regular pyramid implementations are often used to speed up the process.

Another relevant approach is that of Nishihara [Nishihara 84], who observes that an algorithm attempting to correlate individual pixels (by, e.g., matching zero crossings [Marr and Poggio 79]) is inclined towards poor performance when noise causes the detected location of such features to be unreliable. A secondary observation is that such pointwise correlators are very heavy on processing time in arriving at a correspondence. Nishihara notes that the sign (and magnitude) of an edge detector response is likely to be a much more stable property to match than the edge or feature locations, and devises an algorithm that simultaneously exploits a scale-space matching attack.

The approach is to match large patches at a large scale, and then refine the quality of the match by reducing the scale, using the coarser information to initialize the finer-grained match. An edge response is generated at each pixel of both images at a large scale (see Section 4.3.4), and then a large area of the left (represented by, say, its central pixel) is correlated with a large area of the right. This can be done quickly and efficiently by using the fact that the correlation function peaks very sharply at the correct position of a match, and so a small number of tests permits an ascent to a maximum of a correlation measure. This coarse area match may then be refined to any desired resolution in an iterative manner, using the knowledge from the coarser scale as a clue to the correct disparity at a given position. At any stage of the algorithm, therefore, the surfaces in view are modeled as square prisms of varying height; the area of the squares may be reduced by performing the algorithm at a finer scale—for tasks such as obstacle avoidance it is possible that only coarse scale information is necessary, and there will be a consequent gain in efficiency.

This algorithm is enhanced by casting random-dot light patterns on the scene to provide patterns to match even in areas of the scene that are texturally uniform. The resulting system has been demonstrated in use in robot guidance and bin-picking applications, and has been implemented robustly in real time.

Feature-based stereo correspondence

Feature-based correspondence methods use points or set of points that are striking and easy to find. Characteristically, these are pixels on edges, lines, corners, etc., and correspondence is sought according to properties of such features as, e.g., orientation along edges, or lengths of line segments. The advantages of feature-based methods over intensity-based correlation are:

- Feature-based methods are less ambiguous since the number of potential candidates for correspondence is smaller.
- The resulting correspondence is less dependent on photometric variations in images.
- Disparities can be computed with higher precision; features can be sought in the image to sub-pixel precision.

We shall present one example of a feature-based correspondence method—the **PMF algorithm**, named after its inventors [Pollard et al. 85]. It proceeds by assuming that a set of feature points (for example, detected edges) has been extracted from each image by some interest operator. The output is a correspondence between pairs of such points. In order to do this, three constraints are applied: the epipolar constraint, the uniqueness constraint, and the disparity gradient limit constraint.

The first two constraints are not peculiar to this algorithm (for example, they are also used by Marr [Marr and Poggio 79])—the third, however, of stipulating a disparity gradient limit, is its novelty. The **disparity gradient** measures the relative disparity of two pairs of matching points. Suppose (Figure 9.25) that a point A (B) in 3D appears as $A_l = (a_{xl}, a_y)$ [$B_l = (b_{xl}, b_y)$] in the left image and $A_r = (a_{xr}, a_y)$ [$B_r = (b_{xr}, b_y)$] in the right (the epipolar constraint requires the y co-ordinates to be equal); the **cyclopean** image is defined as that given by their average co-ordinates,

$$A_c = \left(\frac{a_{xl} + a_{xr}}{2}, a_y \right) \tag{9.88}$$

$$B_c = \left(\frac{b_{xl} + b_{xr}}{2}, b_y \right) \tag{9.89}$$

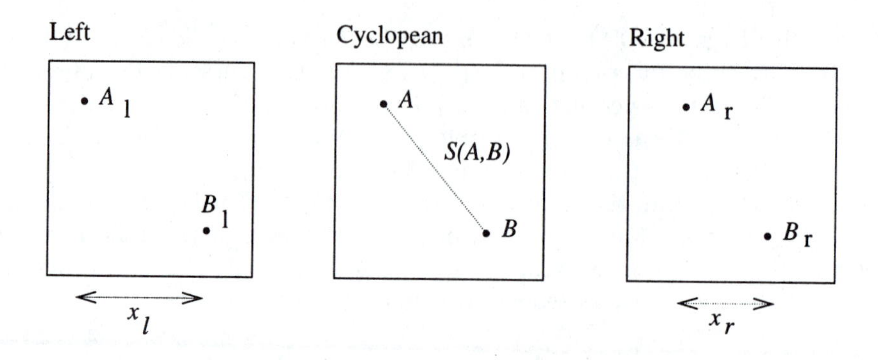

Figure 9.25: *Definition of the disparity gradient.*

and their **cyclopean separation** S is given by their distance apart in this image,

$$\begin{aligned} S(A,B) &= \sqrt{\left[\left(\frac{a_{xl}+a_{xr}}{2}\right)-\left(\frac{b_{xl}+b_{xr}}{2}\right)\right]^2+(a_y-b_y)^2} \\ &= \sqrt{\frac{1}{4}[(a_{xl}-b_{xl})+(a_{xr}-b_{xr})]^2+(a_y-b_y)^2} \\ &= \sqrt{\frac{1}{4}(x_l+x_r)^2+(a_y-b_y)^2} \end{aligned} \tag{9.90}$$

The difference in disparity between the matches of A and B is

$$\begin{aligned} D(A,B) &= (a_{xl}-a_{xr})-(b_{xl}-b_{xr}) \\ &= (a_{xl}-b_{xl})-(a_{xr}-b_{xr}) \\ &= x_l-x_r \end{aligned} \tag{9.91}$$

The disparity gradient of the pair of matches is then given by the ratio of the disparity difference to the cyclopean separation:

$$\begin{aligned} \Gamma(A,B) &= \frac{D(A,B)}{S(A,B)} \\ &= \frac{x_l-x_r}{\sqrt{\frac{1}{4}(x_l+x_r)^2+(a_y-b_y)^2}} \end{aligned} \tag{9.92}$$

Given these definitions, the constraint exploited is that, in practice, the disparity gradient Γ can be expected to be limited; in fact, it is unlikely to exceed 1. This means that very small differences in disparity are not acceptable if the corresponding points are extremely close to each other in 3D—this seems an intuitively reasonable observation, and it is supported by a good deal of physical evidence [Pollard et al. 85]. A solution to the correspondence problem is then extracted by a relaxation process in which all possible matches are scored according to whether they are supported by other (possible) matches that do not violate the stipulated disparity gradient limit. High-scoring matches are regarded as correct, permitting firmer evidence to be extracted about subsequent matches.

Algorithm 9.3: PMF stereo correspondence

1. Extract features to match in left and right images. These may be, for example, edge pixels.

2. For each feature in the left (say) image, consider its possible matches in the right; these are defined by the appropriate epipolar line.

3. For each such match, increment its likelihood score according to the number of other possible matches found that do not violate the chosen disparity gradient limit.

4. Any match which is highest scoring for *both* the pixels composing it is now regarded as correct. Using the uniqueness constraint, these pixels are removed from all other considerations.

5. Return to step 2 and re-compute the scores taking account of the definite match derived.

6. Terminate when all possible matches have been extracted

Note here that the epipolar constraint is used in step 2 to limit to one dimension the possible matches of a pixel, and the uniqueness constraint is used in step 4 to ensure that a particular pixel is never used more than once in the calculation of a gradient.

The scoring mechanism has to take account of the fact that the more remote two (possible) matches are, the more likely they are to satisfy the disparity gradient limit. This is catered for by:

- Considering only matches that are 'close' to the one being scored. In practice it is typically adequate to consider only those inside a circle of radius equal to 7 pixels, centered at the matching pixels (although this number depends on the precise geometry and scene in hand).

- Weighting the score by the reciprocal of its distance from the match being scored. Thus more remote pairs, which are more likely to satisfy the limit by chance, count for less.

The PMF algorithm has been demonstrated to work relatively successfully. It is also attractive because it lends itself to parallel implementation and could be extremely fast on suitably chosen hardware. It has a drawback (along with a number of similar algorithms) in that horizontal line segments are hard to match; they often move across adjacent rasters and, with parallel camera geometry, any point on one such line can match any point on the corresponding line in the other image.

9.2.12 Active acquisition of range images

It is extremely difficult to extract 3D shape information from intensity images of real scenes directly. Another approach—'shape from shading'—will be explained in Section 9.3.

One way to circumvent these problems is to measure distances from the viewer to points on surfaces in the 3D scene explicitly; such measurements are called **geometric signals**, i.e., a collection of 3D points in a known co-ordinate system. If the surface relief is measured from a single viewpoint, it is called a **range image** or a **depth map**. Such explicit 3D information, being closer to the geometric model that is sought, makes geometry recovery easier.[5]

Two steps are needed to obtain geometric information from a range image:

1. The range image must be captured; this procedure is discussed in this section.

2. Geometric information must be extracted from the range image. Features are sought and compared to a selected 3D model. The selection of features and geometric models leads to one of the most fundamental problems in computer vision: how to represent a solid shape [Koenderink 90].

The term **active sensor** refers to a sensor that uses and controls its own images—the term 'active' means that the sensor uses and controls electromagnetic energy, or more specifically illumination, for measuring a distance between scene surfaces and the 'observer'. An active sensor should not be confused with the active perception strategy, where the sensing subject plans how to look at objects from different views.

RADAR (RAdio Detecting And Ranging) and **LIDAR** (LIght Detecting And Ranging) in one measurement yield the distance between the sensor and a particular point in a measured scene. The sensor is mounted on an assembly that allows movement around two angles, azimuth Θ and tilt Φ, corresponding to spherical co-ordinates. The distance is proportional to the time interval between the emission of energy and the echo reflected from the measured scene object. The elapsed time intervals are very short, so very high precision is required. For this reason, the phase difference between emitted and received signals is often used.

RADAR emits electromagnetic waves in meter, centimeter, or millimeter wavelength bands. Aside from military use, it is frequently used for navigation of autonomous guided vehicles.

LIDAR often uses laser as a source of a focused light beam. The higher the power of the laser, the stronger is the reflected signal and the more precise the measured range. If LIDAR is required to work in an environment together with humans, then the energy has an upper limit, due to potential harm to the unprotected eye. Another factor that influences LIDAR safety is the diameter of the laser beam: If it is to be safe, it should not be focused too much. LIDARs have trouble when the object surface is almost tangential to the beam, as very little energy reflects back to the sensor in this case. Measurements of specular surfaces are not very accurate, as they scatter the reflected light; while transparent objects (obviously) cannot be measured with optical lasers. The advantage of LIDAR is a wide range of measured distances, from a tenth of a millimeter to several kilometers; the accuracy of the measured range is typically around 0.01 millimeter. LIDAR provides one range in an instant. If the whole-range image is to be captured, the measurement takes several tenths of a seconds as the whole scene is scanned.

Another principle of active range imaging is **structured light triangulation**, where we employ a geometric arrangement similar to that used for stereo vision, with optical axes. One camera is replaced by an illuminant that yields a light plane perpendicular to the epipolars;

[5]There are techniques that measure full 3D information directly, such as mechanical co-ordinate-measuring machines (considered in Chapter 10) or computer tomography.

the image-capturing camera is at a fixed distance from the illuminant. Since there is only one significantly bright point on each image line, the correspondence problem that makes passive stereo so problematic is avoided, although there will still be problems with self-occlusion in the scene. Distance from the observer can easily be calculated as in Figure 9.10. To capture a whole-range image, the rod with camera and illuminant should be made to move mechanically relative to the scene, and the trace of the laser should gradually illuminate all points to be measured. The conduct of the movement, together with the processing of several hundred images (i.e., one image for each distinct position of the laser-stripe) takes some time, typically from a couple of seconds to about a minute. Faster laser stripe range finders find a bright point corresponding to the intersection of a current image line using special-purpose electronics.

We shall illustrate an example of such a scanner built in the Center for Machine Perception of the Czech Technical University in Prague. Figure 9.26 shows a view of the scanner together with a target object (a wooden toy—a rabbit). The image seen by the camera with the distinct bright laser stripe is in Figure 9.27a, and the resulting range image is shown in Figure 9.27b.

Figure 9.26: *Laser plane range finder. The camera is on the left side, the laser diode on the bottom left. Courtesy T. Pajdla, Czech Technical University, Prague.*

In some applications, a range image is required in an instant, typically meaning one TV frame; this is especially useful for capturing range images of moving objects, e.g., moving humans. One possibility is to illuminate the scene by several stripes at once and code them; Figure 9.28a shows a human hand lit by a binary pattern that codes light stripes using a cyclic code such that the local configuration of squares in the image allows to us to decode which stripe it is. In this case, the pattern with coded stripes is projected from a 36×24 mm slide using a standard slide projector. The resulting range image does not provide as many samples as in the case of a moving laser stripe—in our case only 64×80, see Figure 9.28b.

It is possible to acquire a dense range sample as in the laser stripe case in one TV frame; individual stripes can be encoded using spectral colors and the image captured by a color TV

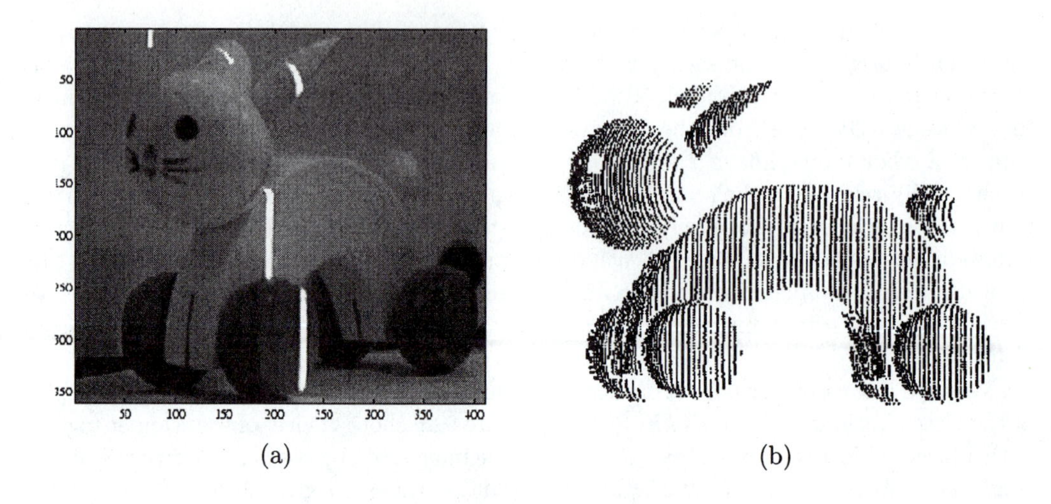

Figure 9.27: *Measurement using a laser-stripe range finger: (a) the image seen by a camera with a bright laser stripe; (b) reconstructed range image displayed as a point cloud. Courtesy T. Pajdla, Czech Technical University, Prague.*

camera [Smutný 93].

Two further measuring principles will conclude this discussion of active range sensors. One is sonar, which uses ultrasonic waves as an energy source. Sonars are used in robot navigation for close-range measurements. Their disadvantage is that measurements are typically very noisy. The second principle is Moiré interferometry [Klette et al. 96], in which two periodic patterns, typically stripes, are projected on the scene. Due to interference, the object is covered by a system of closed, non-intersecting curves, each of which lies in a plane of constant distance from the viewer. Distance measurements obtained are only relative, and absolute distances are unavailable. The properties of Moiré curves are very similar to level curves on maps.

9.3 Radiometry and 3D vision

9.3.1 Radiometric considerations in determining gray-level

A TV camera and most other artificial vision sensors measure the amount of received light energy in individual pixels as the result of interaction among various materials and light source(s); the value measured is informally called gray-level (or brightness). **Radiometry** is a branch of physics that deals with the measurement of the flow and transfer of radiant energy, and is the appropriate tool to consider the mechanism of image creation. The gray-level corresponding to a point on a 3D surface depends, informally speaking, on the shape of the object, its reflectance properties, the position of the viewer, and properties and position of the illuminants [Nicodemus et al. 77]. We will later use these concepts to consider derivation of 3D shape from shading.

The radiometric approach to understanding gray-levels is very often avoided in practical

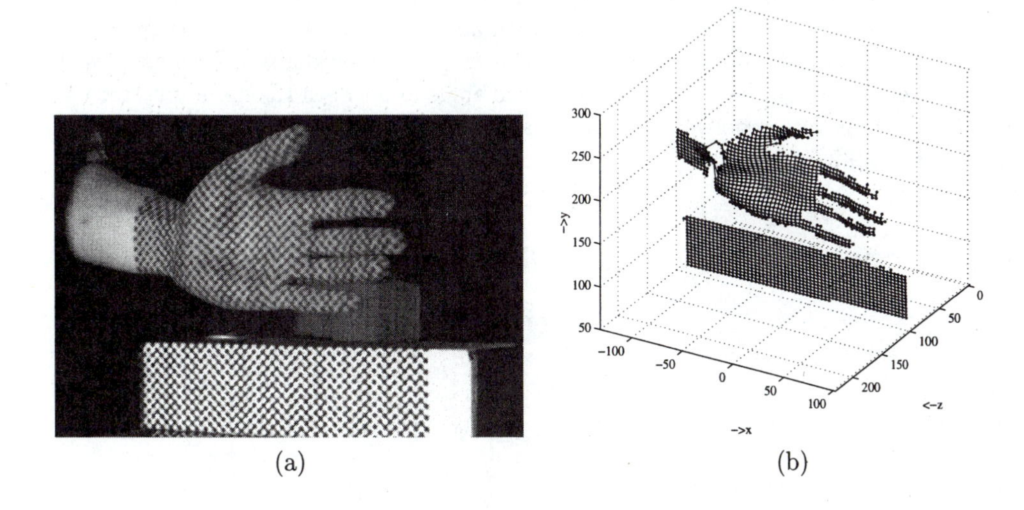

Figure 9.28: *Binary-coded range finder: (a) the captured image of a hand; (b) reconstructed surface. Courtesy T. Pajdla, Czech Technical University, Prague.*

applications because of its complexity and numerical instability. The gray-level measured typically does not provide a precise quantitative measurement (one reason is that CCD cameras are much more precise geometrically than radiometrically; another, more serious reason is that the relation between gray-level and shape is too complicated). One way to circumvent this is to use task-specific illumination that allows the location objects of interest on a qualitative level, and their separation from the background.

Photometry is a discipline closely related to radiometry that studies the sensation of radiant light energy in the human eye; both disciplines describe similar phenomena using similar quantities. Herein, we shall describe physical units using square brackets; when there is a danger of confusion we shall denote photometric quantities using the subscript $_{ph}$, and leave radiometric ones with no subscript.

The basic radiometric quantity is **radiant flux** Φ[W], and its photometric counterpart is **luminous flux** Φ_{ph} [lm (= lumen)]. For light of wavelength $\lambda = 555$ and daylight vision, we can convert between these quantities with the relation 1[W] = 680lm. Different people have different abilities to perceive light, and photometric quantities depend on the spectral characteristic of the radiation source and on the sensitivity of the photoreceptive cells of a human retina. For this reason, the international standardization body Commission International de l'Eclairage (CIE) defined a 'standard observer' corresponding to average abilities. Let $K(\lambda)$ be the **luminous efficacy** [lm W^{-1}], $S(\lambda)$[W] the spectral power of the light source, and λ[W], the wavelength. Then luminous flux Φ_{ph} is proportional to the intensity of perception and is given by

$$\Phi_{ph} = \int_{\lambda} K(\lambda) S(\lambda)\, d\lambda \tag{9.93}$$

Since photometric quantities are too observer dependent, we shall consider radiometric ones. From a viewer's point of view, the surface of an object can reflect energy into a half-sphere, differently into different directions. The **spatial angle** is given by the area on the surface of the unit sphere that is bounded by a cone with an apex in the center of the sphere. The whole half-sphere corresponds to the spatial angle 2π [sr (= steradians)]. A small area A at distance R from the origin (i.e., $R^2 \gg A$) and with angle Θ between the normal vector to the area and the radius vector between the origin and the area corresponds to the spatial angle Ω[sr] (see Figure 9.29).

$$\Omega = \frac{A \cos \Theta}{R^2} \tag{9.94}$$

Irradiance E [$\mathrm{W\,m^{-2}}$] describes the power of the light energy that falls onto a unit area of

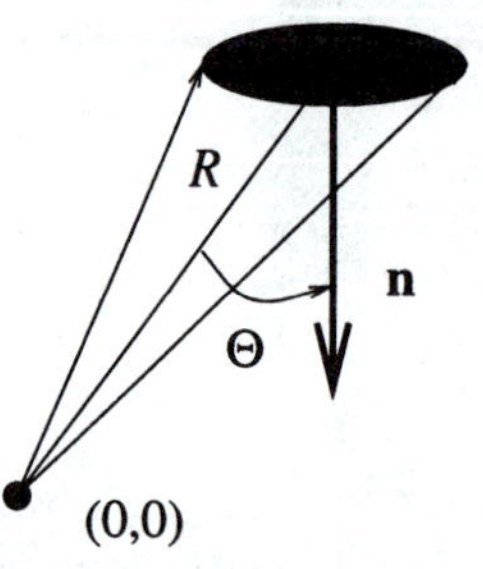

Figure 9.29: *Spatial angle for an elementary surface area.*

the object surface, $E = \delta\Phi\,\delta A$, where δA is an infinitesimal element of the surface area; the corresponding photometric quantity is **illumination** [$\mathrm{lm\,m^{-2}}$]. **Radiance** L [$\mathrm{W\,m^{-2}\,sr^{-1}}$] is the power of light that is emitted from a unit surface area into some spatial angle, and the corresponding photometric quantity is called **brightness** L_{ph} [$\mathrm{lm\,m^{-2}\,sr^{-1}}$]. Brightness is used informally in image analysis to describe the quantity that the camera measures.

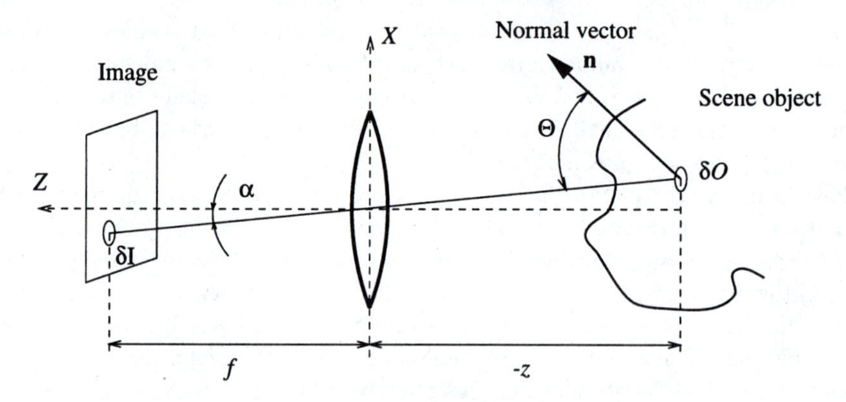

Figure 9.30: *The relation between irradiance E and radiance L.*

Irradiance is given by the amount of energy that an image-capturing device gets per unit of an efficient sensitive area of the camera [Horn 86]—then gray-levels of image pixels are quantized estimates of image irradiance. The efficient area copes with foreshortening that is caused by the mutual rotation between the elementary patch on the emitting surface and the elementary surface patch of the sensor. We shall consider the relationship between the irradiance E measured in the image and the radiance L produced by a small patch on the object surface. Only part of this radiance is captured by the lens of the camera.

The geometry of the capturing setup is given in Figure 9.30. The optical axis is aligned with the horizontal axis Z, and a lens with focal length f is placed at the co-ordinate origin (the optical center). The elementary object surface patch δO is at distance z. We are interested in how much light energy reaches an elementary patch of the sensor surface δI. The off-axis angle α spans between the axis Z and the line connecting δO with δI; as we are considering a perspective projection, this line must pass through the origin. The elementary object surface patch δO is tilted by the angle Θ measured between the object surface normal $\mathbf{n}$ at the patch and a line between δO and δI.

Light rays passing through the lens origin are not refracted; thus the spatial angle attached to the elementary surface patch in the scene is equal to the spatial angle corresponding to the elementary patch in the image. The foreshortened elementary image patch as seen from the optical center is $\delta I \cos\alpha$, and its distance from the optical center is $f/\cos\alpha$. The corresponding spatial angle is

$$\frac{\delta I \cos\alpha}{(f/\cos\alpha)^2}$$

Analogously, the spatial angle corresponding to the elementary patch δO on the object surface is

$$\frac{\delta O \cos Theta}{(z/\cos\alpha)^2}$$

As the spatial angles are equal,

$$\frac{\delta O}{\delta I} = \frac{\cos\alpha}{\cos\Theta}\frac{z^2}{f^2} \tag{9.95}$$

Consider how much light energy passes through the lens if its aperture has diameter d; the spatial angle Ω_L that sees the lens from the elementary patch on the object is

$$\Omega_L = \frac{\pi}{4}\frac{d^2\cos\alpha}{(z/\cos\alpha)^2} = \frac{\pi}{4}\left(\frac{d}{z}\right)^2 \cos^3\alpha \tag{9.96}$$

Let L be the radiance of the object surface patch that is oriented towards the lens. Then the elementary contribution to the radiant flux Φ falling at the lens is

$$\delta\Phi = L\delta O\ \Omega_L \cos\Theta = \pi L\delta O \left(\frac{d}{z}\right)^2 \frac{\cos^3\alpha\cos\Theta}{4} \tag{9.97}$$

The lens concentrates the light energy into the image. If energy losses in the lens are neglected and no other light falls on the image element, we can express the irradiation E of the elementary image patch as

$$E = \frac{\delta\Phi}{\delta I} = L\frac{\delta O}{\delta I}\frac{\pi}{4}\left(\frac{d}{z}\right)^2 \cos^3\alpha\cos\Theta \tag{9.98}$$

If we substitute for $\delta O/\delta I$ from equation (9.95), we obtain an important equation that reveals how scene radiance influences irradiation in the image:

$$E = L\frac{\pi}{4}\left(\frac{d}{f}\right)^2 \cos^4\alpha \tag{9.99}$$

The term $\cos^4 \alpha$ describes a systematic lens optical defect called **vignetting**[6], which notes that optical rays with larger span-off angle α are attenuated more; this means that pixels closer to image borders are darker. This effect is more severe with wide-angle lenses than with tele-lenses. Since vignetting is a systematic error, it can be compensated for with a radiometrically calibrated lens. The term d/f is called the f-number of the lens and describes how much the lens differs from a pinhole model.

9.3.2 Surface reflectance

In many applications, pixel gray-level is constructed as an estimate of image irradiance as a result of light reflection from scene objects. Consequently, it is necessary to understand different mechanisms involved in reflection. Here we give just a brief overview that will later permit us to explain the main ideas behind shape from shading. Consult [Ikeuchi 94, Foley et al. 90, Klette et al. 96] for more detailed explanations.

The radiance of an opaque object that does not emit its own energy depends on irradiance caused by other energy sources. The illumination that the viewer perceives depends on the strength, position, orientation, type (point or diffuse) of the light sources, and ability of the object surface to reflect energy and the local surface orientation (given by its normal vector).

An important concept now is **gradient space** which is a way of describing surface orientations (and has also been used in the analysis of line labeling problems [Mackworth 73]). Let $z(x, y)$ be the surface height. We proceed by noting that at nearly every point a surface has a unique normal $\mathbf{n}$. The components of surface gradient

$$p = \frac{\partial z}{\partial x} \qquad \text{and} \qquad q = \frac{\partial z}{\partial y} \tag{9.100}$$

can be used to specify the surface orientation. We shall express the unit surface normal using surface gradient components; if we move a small distance ∂x in the x direction, the change of height is $\partial z = p\, \partial x$. Thus the vector $[1, 0, p]^T$ is the tangent to the surface, and analogously $[0, 1, q]^T$ is also tangent to the surface. The surface normal is perpendicular to all its tangents, and may be computed using the vector product as

$$\begin{bmatrix} 1 \\ 0 \\ p \end{bmatrix} \times \begin{bmatrix} 0 \\ 1 \\ q \end{bmatrix} = \begin{bmatrix} -p \\ -q \\ 1 \end{bmatrix} \tag{9.101}$$

The unit surface normal $\mathbf{n}$ can be written as

$$\mathbf{n} = \frac{1}{\sqrt{1 + p^2 + q^2}} \begin{bmatrix} -p \\ -q \\ 1 \end{bmatrix} \tag{9.102}$$

Here we suppose that the z component of the surface normal is positive, as only the surface part oriented towards the viewer is visible.

The pair $[p, q]$ is the two-dimensional gradient space representation of the surface orientation. Gradient space has a number of attractive properties that allow elegant description

[6] One of the meanings of **vignette** is a photograph or drawing with edges that are shaded off.

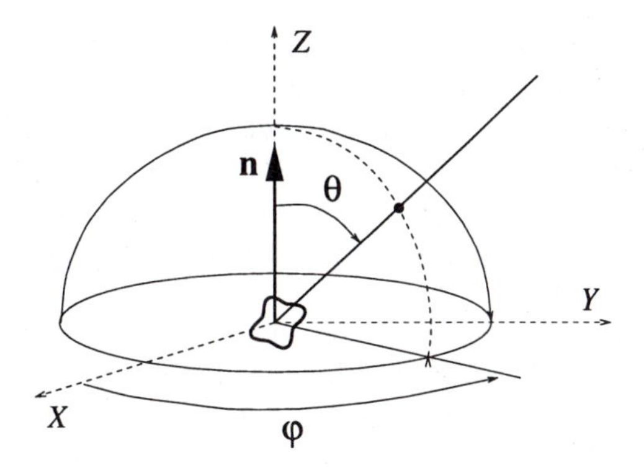

Figure 9.31: *Polar and spherical angles used to describe orientation of a surface patch.*

of the surface. Interpreting the image plane as $z = 0$, we see that the origin of gradient space corresponds to the vector $[p, q]=[0, 0]$, which is normal to the image plane. Thus $[p, q]=[0, 0]$ implies that the surface is parallel to the image plane. The more remote a vector is from the origin of gradient space, the steeper its corresponding surface patch is inclined to the image plane.

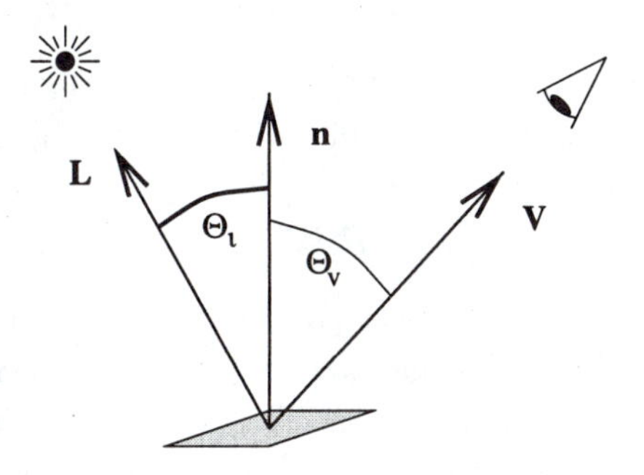

Figure 9.32: *Directions towards the viewer and the light source.*

Consider now spherical co-ordinates used to express the geometry of an infinitesimal surface patch—see Figure 9.31. The **polar angle** (also called zenith angle) is Θ and the **azimuth** is φ. We shall attempt to describe the ability of different materials to reflect light. The direction towards a point light source is denoted by subscript $_i$ (i.e., Θ_i and φ_i), while subscript $_v$ identifies the direction toward the viewer (Θ_v and φ_v)—see Figure 9.32. The irradiance of the elementary surface patch from the light source is $\delta E(\Theta_i, \varphi_i)$, and the elementary contribution of the radiance in the direction towards the viewer is $\delta L(\Theta_v, \varphi_v)$. In general, the ability of the body to reflect light is described using a **bi-directional reflectance distribution function** f_r $[sr^{-1}]$, abbreviated BDRF [Nicodemus et al. 77],

$$f_r(\Theta_i, \varphi_i; \Theta_v, \varphi_v) = \frac{\delta L(\Theta_v, \varphi_v)}{\delta E(\Theta_i, \varphi_i)} \tag{9.103}$$

The BDRF f_r describes the brightness of an elementary surface patch for a specific material, light source, and viewer directions. Modeling of the BDRF is also important for realistic rendering in computer graphics [Foley et al. 90]. The BDRF in its full complexity [equation (9.103)] is used for modeling reflection properties of materials with oriented microstructure (e.g., tiger's eye—a semi-precious golden-brown stone, peacock's feather, rough cut of aluminum).

Fortunately, for most practically applicable surfaces, the BDRF remains constant if the elementary surface patch rotates along the normal vector to the surface. In this case it is simplified and depends on $\varphi_i - \varphi_v$, i.e., $f_r[\Theta_i, \Theta_v, (\varphi_i - \varphi_v)]$. This simplification holds for both ideal diffuse (Lambertian) surfaces and for ideal mirrors.

Let $E_i(\lambda)$ denote the irradiance caused by the illumination of the surface element, and $E_r(\lambda)$ the energy flux per unit area scattered by the surface element back to the whole half-space. The ratio

$$\rho(\lambda) = \frac{E_r(\lambda)}{E_i(\lambda)} \tag{9.104}$$

is called the **reflectance coefficient** or **albedo**. Albedo describes what proportion of incident energy is reflected back to the half-space. For simplicity, assume that we may neglect color properties of the surface, and suppose that albedo does not depend on the wavelength λ. This proportion is then an integral of the surface radiance L over the solid angle Ω representing the half-space,

$$E_r = \int_\Omega L \, d\Omega \tag{9.105}$$

Now define a **reflectance function** $R(\Omega)$ that models the influence of the local surface geometry into the spatial spread of the reflected energy. Ω is an infinitesimal solid angle around the viewing direction.

$$\int_\Omega R \, d\Omega = 1 \tag{9.106}$$

In general, surface reflectance properties depend on three angles between the direction to the light source $\mathbf{L}$, the direction towards the viewer $\mathbf{V}$, and the local surface orientation given by the surface normal $\mathbf{n}$ (recall Figure 9.32). The cosines of these angles can be expressed as scalar (dot) products of vectors; thus the reflectance function is a scalar function of the following three dot products:

$$R = R(\mathbf{nL}, \mathbf{nV}, \mathbf{VL}) \tag{9.107}$$

A **Lambertian surface** (also ideally opaque, with ideal diffusion) reflects light energy in all directions, and thus the radiance is constant in all all directions. The BDRF f_{Lambert} is constant:

$$f_{\text{Lambert}}(\Theta_i, \Theta_v, \varphi_i - \varphi_v) = \frac{\rho(\lambda)}{\pi} \tag{9.108}$$

If constant albedo $\rho(\lambda)$ is assumed, then the Lambertian surface reflectance can be expressed as

$$R = \frac{1}{\pi} \mathbf{nL} = \frac{1}{\pi} \cos \Theta_i \tag{9.109}$$

Because of its simplicity, the Lambertian reflectance function has been widely accepted as a reasonable reflectance model for shape from shading. Notice that the reflectance function for the Lambertian surface is independent of the viewing direction $\mathbf{V}$.

The dependence of the surface radiance on the local surface orientation can be expressed in gradient space, and the **reflectance map** $R(p, q)$ is used for this purpose. The $R(p, q)$

can be visualized in gradient space as nested iso-contours corresponding to the same observed irradiance.

Values of the reflectance map may be:

1. Measured experimentally on a device called a goniometer stage that is able to set angles Θ and φ mechanically. A sample of the surface is attached to the goniometer and its reflectance measured for different orientations of viewer and light sources.

2. Set experimentally if a calibration object is used. Typically a half-sphere is used for this purpose.

3. Derived from a mathematical phenomenonical model describing surface reflecting properties

The best-known surface reflectance models are the Lambertian model for ideal opaque surfaces, the **Phong** model which models reflection from dielectric materials, the **Torrance-Sparrow** model which describes surfaces as a collection of planar mirror-like micro-facets with normally distributed normals, and the wave theory-based **Beckmann-Spizzichino** model. A survey of surface reflection models from the point of view of computer vision, and their recent modifications, can be found in [Ikeuchi 94].

The irradiance $E(x, y)$ of an infinitely small light sensor located at position x, y in the image plane is equal to the surface radiance at a corresponding surface patch given by its surface parameters u, v if the light is not attenuated in the optical medium between the surface and the sensor. This important relation between surface orientation and perceived image intensity is called the **image irradiance equation**:

$$E(x, y) = \rho(u, v) R(\mathbf{N}(u, v)\mathbf{L}, \mathbf{N}(u, v)\mathbf{V}, \mathbf{VL}) \tag{9.110}$$

In an attempt to reduce complexity, several simplifying assumptions [Horn 90] are usually made to ease the shape from shading task. It is assumed that:

- The object has uniform reflecting properties; i.e., $\rho(u, v)$ is constant.

- The light sources are distant; then irradiation in different places in the scene is approximately the same and the incident direction towards the light sources is the same.

- The viewer is very distant; then the radiance emitted by scene surfaces does not depend on position but only on orientation. The perspective projection is simplified to an orthographic one.

We present the simplified version of the image irradiance equation for the Lambertian surface, constant albedo, single distant illuminant, distant viewer in the same direction as illuminant, and the reflectance function R expressed in gradient space (p, q):

$$E(x, y) = \beta R[p(x, y), q(x, y)] \tag{9.111}$$

$R(p, q)$ gives the radiance of the corresponding point in the scene; the proportionality constant β comes from equation (9.99) and depends on the f-number of the lens. The vignetting degradation of the lens is negligible, as the viewer is aligned to the illuminant. The measured irradiance E can be normalized and the factor β omitted; this permits us to write the **image irradiance equation** in the simplest form as

$$E(x,y) = R[p(x,y), q(x,y)] = R(\frac{\partial z}{\partial x}, \frac{\partial z}{\partial y}) \tag{9.112}$$

The image irradiance equation in its simplest form is a first-order differential equation. It is typically non-linear, as the reflectance function R in most cases depends non-linearly on the surface gradient. This is the basic equation that is used to recover surface orientation from intensity images.

9.3.3 Shape from shading

The human brain is able to make very good use of clues from shadows and shading in general. Not only do detected shadows give a clear indication of where occluding edges are, and the possible orientation of their neighboring surfaces, but general shading properties are of great value in deducing depth. A fine example of this is a photograph of a face; from a straight-on, 2D representation, our brains make good guesses about the probable lighting model, and then deductions about the 3D nature of the face—for example, deep eye sockets and protuberant noses or lips are often recognizable without difficulty.

Recall that the intensity of a particular pixel depends on the light source(s), surface reflectance properties, and local surface orientation expressed by a surface normal $\mathbf{n}$. The aim of shape from shading is to extract information about normals of surfaces in view solely on the basis of an intensity image. If simplifying assumptions are made about illumination, surface reflectance properties, and surface smoothness, the shape from shading task has proven to be solvable. The first computer-vision related formulation comes from Horn [Horn 70, Horn 75].

Techniques similar to shape from shading were earlier proposed independently in photoclinometry [Rindfleisch 66], when astro-geologists wanted to measure steepness of slopes on planets in the solar system from intensity images observed by terrestrial telescopes. There are two significant differences here from shape from shading:

- Surface normals are calculated by the integration along a space curve (called the profile; it is a 1D entity if the curve is arc-length parameterized). In shape from shading, the integration is performed on the surface area, which is a 2D entity if the surface is parametrized.

- Shape from shading is more concerned with ambiguity of solutions. The use of singular points and occluding boundaries helps to combat this ambiguity. The surface normal can be then uniquely computed.

We shall classify shape from shading methods into three categories, and proceed to describe them.

Incremental propagation from surface points of known height

The oldest, and easiest to explain, method develops a solution along a space curve. This is also called the characteristic strip method.

We can begin to analyze the problem of global shape extraction from shading information when the reflectance function and the lighting model are both known perfectly [Horn 90]. Even given these constraints, it should be clear that the mapping 'surface orientation to brightness' is many-to-one, since many orientations can produce the same point intensity. Acknowledging this, a particular brightness can be produced by an infinite number of orientations that can be plotted as a (continuous) line in gradient space. An example for the simple case of a light source directly adjacent to the viewer, incident on a matte surface, is shown in Figure 9.33—two points lying on the same curve (circles in this case) indicate two different orientations that will reflect light of the same intensity, thereby producing the same pixel gray-level.

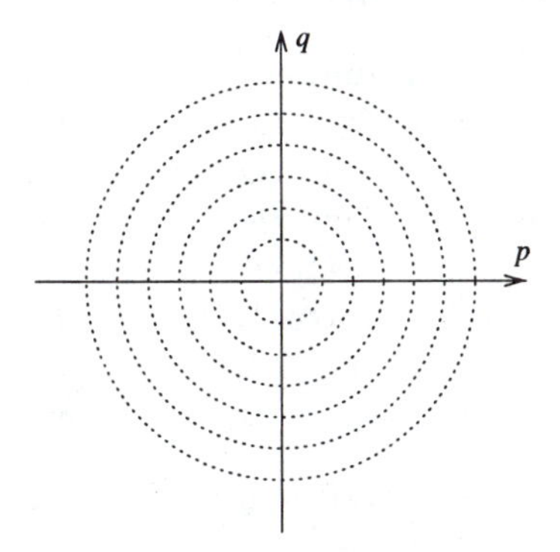

Figure 9.33: *Reflectance map for a matte surface —the light source is adjacent to the viewer.*

The original formulation [Horn 70] to the general shape from shading task assumes a Lambertian surface, one distant point light source, a distant observer, and no inter-reflections in the scene. The proposed method is based on the notion of a **characteristic strip**: Suppose that we already calculated co-ordinates of a surface point $[x, y, z]^T$ and we want to propagate the solution along an infinitesimal step on the surface, e.g., taking small steps δx and δy, then calculating the change in height δz. This can be done if the components of the surface gradient p, q are known. For compactness we use an index notation, and express $p{=}\delta z/\delta x$ as z_x, and $\delta^2 x/\delta x^2$ as z_{xx}. The infinitesimal change of height is

$$\delta z = p\,\delta x + q\,\delta y \tag{9.113}$$

The surface is followed stepwise, with values of p, q being traced along with x, y, z. Changes in p, q are calculated using second derivatives of height $r{=}z_{xx}$, $s{=}z_{xy}{=}z_{yx}$, $t{=}z_{yy}$:

$$\delta p = r\,\delta x + s\,\delta y \qquad \text{and} \qquad \delta q = s\,\delta x + t\,\delta y \tag{9.114}$$

Consider now the image irradiance equation $E(x,y) = R(p,q)$, equation (9.112), and differentiate with respect to x, y to obtain the brightness gradient

$$E_x = rR_p + sR_q \qquad \text{and} \qquad E_y = sR_p + tR_q \tag{9.115}$$

The direction of the step δx, δy can be chosen arbitrarily;

$$\delta x = R_p\xi \qquad \text{and} \qquad \delta y = R_q\xi \tag{9.116}$$

The parameter ξ changes along particular solution curves. Moreover, the orientation of the surface along this curve is known; thus it is called a characteristic strip.

We can now express changes of gradient δp, δq as dependent on gradient image intensities, which is the crucial 'trick'. A set of ordinary differential equations can be generated by considering equations (9.114) and (9.115); dot denotes differentiation with respect to ξ.

$$\dot{x} = R_p \quad \dot{y} = R_q \quad \dot{z} = p\,R_p + q\,R_q \quad \dot{p} = E_x \quad \dot{q} = E_y \tag{9.117}$$

There are points on the surface for which the surface orientation is known in advance, and these provide boundary conditions during normal vector calculations. These are

- Points of a surface **occluding boundary**; an occluding boundary is a curve on the surface due to the surface rolling away from the viewer, i.e., the set of points for which the local tangent plane co-incides with the direction towards the viewer. The surface normal at such a boundary can be uniquely determined, as it is parallel to the image plane and perpendicular to the direction towards the viewer. This normal information can be propagated into the recovered surface patch from the occluding boundary. Although the occlusion boundary uniquely constrains surface orientation, it does not constrain the solution sufficiently to recover depth uniquely [Oliensis 91].

- Singular points in the image; we have seen that at most surface points gradient is not fully constrained by image intensities. Suppose that the reflectance function $R(p, q)$ has a global maximum, so $R(p, q) < R(p_0, q_0)$ for all $[p, q] \neq p_0, q_0$. This maximum corresponds to singular points in the image:

  $$E(x_0, y_0) = R(p_0, q_0) \tag{9.118}$$

  Here, the surface normal is parallel to the direction towards the light source. Singular points are in general sources and sinks of characteristic stripes.

It is reported [Horn 90] that direct implementation of this characteristic strip method does not yield particularly good results, due to numerical instability.

Global optimization methods

These methods are formulated as a variational task in which the whole image plays a role in the chosen functional. Results obtained are in general better than those generated by incremental methods.

We already know that under the simplifying conditions for recovery of surface normals from intensities (stated in Section 9.3.2), the image irradiance equation (9.112) relates image irradiance E and surface reflection R as

$$E(x, y) = R[p(x, y), q(x, y)] \tag{9.119}$$

The task is to find the surface height $z(x, y)$ given the image $E(x, y)$ and reflectance map $R(p, q)$.

Now presented with an intensity image, a locus of possible orientations for each pixel can be located in gradient space, immediately reducing the number of possible interpretations of the scene. Of course, at this stage, a pixel may be part of any surface lying on the gradient space contour; to determine which, another constraint needs to be deployed. The key to deciding which point on the contour is correct is to note that 'almost everywhere' 3D surfaces are smooth, in the sense that neighboring pixels are very likely to represent orientations whose gradient space positions are also very close. This additional constraint allows a relaxation process to determine a best-fit (minimum-cost) solution to the problem. The details of the procedure are very similar to those used to extract optical flow, and are discussed more fully in Section 15.2.1, but may be summarized as follows.

Algorithm 9.4: Extracting shape from shading

1. For each pixel (x, y), select an initial guess to orientation $p^0(x, y)$, $q^0(x, y)$.

2. Apply two constraints:

   (a) The observed intensity $f(x, y)$ should be close to that predicted by the reflectance map $R(p, q)$ derived from foreknowledge of the lighting and surface properties.

   (b) p and q vary smoothly—therefore their Laplacians $\nabla^2 p$ and $\nabla^2 q$ should be small.

3. Apply the method of Lagrange multipliers to minimize the quantity

$$\Sigma_{(x,y)} \text{Energy}(x, y) \tag{9.120}$$

   where

$$\text{Energy}(x, y) = [f(x, y) - R(p, q)]^2 + \lambda[(\nabla^2 p)^2 + (\nabla^2 q)^2] \tag{9.121}$$

The first term of equation (9.121) is the intensity 'knowledge', while the second, in which λ is a Lagrange multiplier, represents the smoothness constraint. Deriving a suitable solution to this problem is a standard technique that iterates p and q until E falls below some reasonable bound; a closely related problem is inspected in more detail in Section 15.2.1, and the method of optimization using Lagrange multipliers is described in many books [Horn 86].

The significant work in this area is due to Horn [Horn 75] and Ikeuchi [Ikeuchi and Horn 81] and pre-dated the publication of Marr's theory, being complete in the form presented here by 1980. Shape from shading, as implemented by machine vision systems, is often not as reliable as other 'shape from' techniques, since it is so easy to confuse with reflections, or for it to fail through poorly modeled reflectance functions. This observation serves to reinforce the point that the human recognition system is very powerful, since in deploying elaborate knowledge it does not suffer these drawbacks. A review of significant developments in the area since may be found in [Horn and Brooks 89].

Local shading analysis

Local shading analysis methods use just a small neighborhood of the current point on the surface, and seek a direct relation between the differential surface structure and the local structure of the corresponding intensity image. The surface is considered as a set of small neighborhoods, each defined in some local neighborhood of one of its points. Only an estimate of local surface orientation is available, not information about the height of a particular surface point.

The main advantage of local shading analysis is that it provides surface-related information to higher-level vision algorithms from a single monocular-intensity image without any need to reconstruct the surface in explicit depth form [Sara 95]. This is possible because the intensity image is closely related to local surface orientation. The surface normal and the shape operator ('curvature matrix') form a natural shape model that can be recovered from an intensity image by local computations. This approach is, of course, much faster than the solution propagation or global variational methods.

The fundamental contribution to local shading analysis comes from Pentland [Pentland 84]; an overview can be found in [Pentland and Bichsel 94]. In addition, Šára [Sara 94] demonstrates:

1. It was known that local surface orientation and Gaussian curvature sign can be determined uniquely at occlusion boundaries. Further orientation on a self-occluding boundary can also be determined uniquely; self-occluding contours are thus a rich source of unambiguous information about the surface.

2. The differential properties of isophotes (curves of constant image intensity) are closely related to the properties of the underlying surface. Isophotes are projections of curves of constant slant from the light direction if the surface reflectance is space invariant, or the illuminant is located at the vantage point.

9.3.4 Photometric stereo

Woodham proposed **photometric stereo** as a method that recovers surface orientation unambiguously, assuming a known reflectance function [Woodham 80]. Consider a particular Lambertian surface with varying albedo ρ. The key idea of photometric stereo is to look at the surface from one fixed viewing direction while changing the direction of incident illumination. Assume we have three or more such images of the Lambertian surface; then the surface normals can be uniquely determined based on the shading variations in the observed images.

The lines of constant reflectance on the surface correspond to lines of constant irradiation E in the image (called also isophotes); these curves observed in images are second-order polynomials. The local surface orientation $\mathbf{n} = [p, q]$ is constrained along a second-order curve in the reflectance map. For different illumination directions, the surface reflectance remains the same on the surface but the observed reflectance map $R(p, q)$ changes. This provides an additional constraint on possible surface orientation that is another second-order polynomial. Two views corresponding to two distinct illumination directions are not enough to determine the surface orientation $[p, q]$ uniquely, and a third view is needed to derive a unique solution. If more than three distinct illuminations are at hand, an over-determined set of equations can be solved.

A practical setup for image capture consists of one camera and K point illumination sources, $K \geq 3$, with known intensities and illumination directions $L_1, \ldots, L_K$. Only one light source is active at any one time. The setup should be photometrically calibrated to take into account light source intensities, particular camera gain, and offset; such a calibration is described in [Haralick and Shapiro 93]. After photometric calibration, the images give K estimates of image irradiances $E_i(x, y)$; $i = 1, \ldots, K$.

If not all light is reflected from a surface, then albedo ρ, $0 \leq \rho \leq 1$, occurs in the image irradiance (as shown in equation [9.110)]. For a Lambertian surface the image irradiance equation simplifies to

$$E(x, y) = \rho\, R(p, q) \tag{9.122}$$

Recall equation (9.109) (called the cosine law), showing that the reflectance map of a Lambertian surface is given by the dot product of the surface normal $\mathbf{n}$ and the direction of the incident light L_i. If the surface reflectance map is substituted into equation (9.122), we get K image irradiance equations,

$$E_i(x, y) = \rho\, L_i\, \mathbf{n} \qquad i = 1, \ldots, K \tag{9.123}$$

For each point x, y in the image we get a vector of image irradiances $\mathbf{E} = [E_1, \ldots, E_K]^T$. The light directions can be written in the form of a $K \times 3$ matrix,

$$L = \begin{bmatrix} L_1 \\ \vdots \\ L_K \end{bmatrix} \tag{9.124}$$

At each image point, the system of image irradiance equations can be written

$$\mathbf{E} = \rho\, L\, \mathbf{n} \tag{9.125}$$

The matrix L does not depend on the pixel position in the image, and we can thus derive a vector representing simultaneously surface albedo and a local surface orientation.

If we have three light sources, $K = 3$, we can derive a solution by inverting the regular matrix L:

$$\rho\, \mathbf{n} = L^{-1}\, \mathbf{E} \tag{9.126}$$

The unit normal is then

$$\mathbf{n} = \frac{L^{-1}\, \mathbf{E}}{\|L^{-1}\, \mathbf{E}\|} \tag{9.127}$$

For more than three light sources, the pseudo-inverse of a rectangular matrix is determined to get a solution in the least-square sense:

$$\mathbf{n} = \frac{(L^T\, L)^{-1}\, L^T\, \mathbf{E}}{\|(L^T\, L)^{-1}\, L^T\, \mathbf{E}\|} \tag{9.128}$$

Note that the pseudo-inversion [or inversion in equation (9.127)] must be repeated for each image pixel x, y to derive an estimate of the corresponding normal.

9.4 Summary

- 3D vision aims at inferring 3D information from 2D scenes, a task with embedded geometric and radiometric difficulties. The geometric problem is that a single image does not provide enough information about 3D structures, and the radiometric problem is the complexity of the physical process of intensity image creation. This process is complex, and typically not all input parameters are known precisely.

- **3D vision tasks**
  - There are several approaches to 3D vision, which may be categorized as *bottom-up (or reconstruction)* or *top-down (model-based vision).*
  - *Marr's theory*, formulated in the late 1970s, is an example of the bottom-up approach. The aim is to reconstruct qualitative and quantitative 3D geometric descriptions from one or more intensity images under very weak assumptions about objects in the scene.
  - Four representations are ordered in bottom-up fashion: (1) input intensity image(s); (2) primal sketch, representing significant edges in the image in viewer-centered co-ordinates; (3) 2.5D sketch, representing depth from the observer and local orientation of the surface; and (4) 3D representation, representing object geometry in co-ordinates related to the objects themselves.
  - The 2.5D sketch is derived from the primal sketch by a variety of techniques called shape from X.
  - 3D representations are very hard to obtain; this step has not been solved in the general case.
  - More recent perception paradigms such as active, purposive, and qualitative vision try to provide a computational model explaining the 'understanding' aspects of vision.
  - None has yet led to direct practical applications, but many partial techniques (such as shape from X) are widely used in practice.
- **Radiometry and 3D vision**
  - 3D perspective geometry is the basic mathematical tool for 3D vision as it explains a pinhole camera.
  - Lines parallel in the 3D world do not project as parallel lines in 2D image.
  - The case of the single-perspective camera permits careful study of calibration of intrinsic and extrinsic camera parameters.
  - Two-perspective cameras constitute stereopsis and allow depth measurements in 3D scenes.
  - Epipolar geometry teaches us that the search for corresponding points is inherently one-dimensional. This can be expressed algebraically using the fundamental matrix.
  - This tool has several applications, such as image rectification, ego-motion estimation from calibrated cameras measurements, 3D Euclidean reconstruction from two fully calibrated cameras, 3D similarity reconstruction from two cameras with only intrinsic calibration parameters known, and 3D projective reconstruction from two uncalibrated cameras.
  - There a is trilinear relation among views of from three cameras that is expressed algebraically using a trifocal tensor.

  - The application of the trilinear relation is in epipolar transfer; if two images are known, together with the trifocal tensor, the third perspective image can be computed.
  - The correspondence problem is core to 3D vision; various passive and active techniques to solve it exist.

- **Radiometry and 3D vision**
  - Radiometry informs us about the physics of image formation.
  - If it is understood together with position of illuminants, type, surface reflectance and viewer position, something can be learned about depth and scene surface orientation from one intensity image.
  - This task is called *shape from shading.*
  - The task is ambiguous and numerically unstable. Shape from shading can be understood in the simple case of Lambertian surfaces.
  - There is a practical method that uses one camera and three known illuminants; selective illumination provides three intensity images.
  - Photometric stereo allows measure of orientation of surfaces.

9.5 Exercises

Short-answer questions

1. Explain the difference between a bottom-up approach (object reconstruction) to 3D vision as opposed to top-down (model-based).
2. Explain the basic idea of active vision, and give some examples of how this approach eases vision tasks.
3. Give examples of perspective images from everyday life. Where do parallel lines in the world not correspond to parallel ones in images?
4. What are the intrinsic and extrinsic calibration parameters of a single-perspective camera? How are they estimated from known scenes?
5. Do zoom lenses typically have worse geometric distortion compared to fixed-focal-length lenses? Is the difference significant?
6. What is the main contribution of epipolar geometry in stereopsis?
7. Where are epipoles in the case of two cameras with parallel optical axes (the *canonical* configuration for stereopsis)?
8. What is the difference between the fundamental and essential matrices in stereopsis?
9. How are mismatches in correspondences treated in stereopsis?
10. What are the applications of epipolar geometry in computer vision?
11. Explain the principle, advantages, and applications of a trilinear relation among three cameras. What is epipolar transfer?
12. Stereo correspondence algorithms are typically lost if the left and right images have large regions of uniform brightness. How can depth acquisition still be made possible?

13. Active range finders (e.g., with a laser plane) suffer from occlusions; some points are not visible by the camera and some are not lit. What are the ways of dealing with this problem?

14. What is Moiré interferometry? Does it give absolute depth?

15. Why is the relation between pixel intensity on one side and surface orientation, surface reflectance, illuminant types and position, and viewer position on the other side difficult?

16. What is the vignetting error of a lens?

17. Under which circumstances can the surface orientation be derived from intensity changes in an image?

Problems

1. This problem relates to Marr's theory, in particular the representation scheme called primal sketch (see Section 9.1.1).

   Capture an intensity image (e.g., of an office scene), and run an advanced edge detector (such as Canny's or similar) on it. Threshold the magnitude of the image gradient.

   Answer the following questions in an essay: Are the lines which you get the primal sketch? Would it be possible to derive a 2.5D sketch directly from it? How? Discuss what more would you need in the primal sketch. What about multiple scales?

2. Explain the notion of homogeneous co-ordinates. Is projective transformation linear if expressed in homogeneous co-ordinates? Why are homogeneous co-ordinates often used in robotics to express the kinematics of a manipulating arm? (Hint: Express rotation and translation of an object in 3D space using homogeneous co-ordinates.)

3. Take a camera with an off-the-shelf lens. Design and perform an experiment to find the intrinsic calibration parameters of it. Design and use an appropriate calibration object (for example, a grid like structure printed by laser printer on paper; you might capture it at different heights by placing it on a box of known height). Discuss the precision of your results. Is the pinhole model of your camera appropriate? (Hint: Look at distortions of a grid as in Figure 9.7.)

4. Consider the case of two cameras (stereopsis) with baseline $2h$. A scene point lies on the optical axis at depth d from the baseline. Assume that the precision of pixel position measurement x in the image plane is given by dispersion σ^2. Derive a formula showing the dependence of the precision of depth measurement against dispersion. (Hint: Differentiate d according to x.)

5. Conduct an experiment with stereo correspondences. For simplicity, capture a pair of stereo images using cameras in canonical configuration (epipolar lines correspond to lines in images), and cut corresponding lines from both images. Visualize the brightness profiles in those lines (for example, using MATLAB or another package). First, try to find correspondences in brightness manually. Second, decide if correlation-based or feature-based stereo techniques are more suitable for your case. Program it and test on your profiles.

6. Conduct a laboratory experiment with photometric stereo. You will need one camera and three light sources. Take some opaque object and measure its surface orientation using photometric stereo (see Section 9.3.4).

9.6 References

[Aloimonos 92] Y Aloimonos, editor. Special issue on purposive and qualitative active vision. *CVGIP B: Image Understanding*, 56, 1992.

[Aloimonos 93] Y Aloimonos, editor. *Active Perception.* Lawrence Erlbaum Associates, Hillsdale, NJ, 1993.

[Aloimonos 94] Y Aloimonos. What I have learned. *CVGIP: Image Understanding*, 60(1):74–85, 1994.

[Aloimonos and Rosenfeld 94] Y Aloimonos and A Rosenfeld. Principles of computer vision. In T Y Young, editor, *Handbook of Pattern Recognition and Image Processing: Computer Vision*, pages 1–15, Academic Press, San Diego, 1994.

[Aloimonos and Shulman 89] Y Aloimonos and D Shulman. *Integration of Visual Modules—An Extension of the Marr Paradigm.* Academic Press, New York, 1989.

[Ayache and Hansen 88] N Ayache and C Hansen. Rectification of images for binocular and trinocular stereovision. In *9th International Conference on Pattern Recognition,* Rome, Italy, pages 11–16, IEEE, Los Alamitos, CA, 1988.

[Bajcsy 88] R Bajcsy. Active perception. *Proceedings of the IEEE*, 76(8):996–1005, 1988.

[Bertero et al. 88] M Bertero, T Poggio, and V Torre. Ill-posed problems in early vision. *IEEE Proceedings*, 76:869–889, 1988.

[Besl and Jain 85] P J Besl and R C Jain. Three-dimensional object recognition. *ACM Computing Surveys*, 17(1):75–145, March 1985.

[Besl and Jain 88] P J Besl and R Jain. *Surfaces in range image understanding.* Springer Verlag, New York, 1988.

[Beymer and Poggio 96] D Beymer and T Poggio. Image representations for visual learning. *Science*, 272:1905–1909, 1996.

[Biederman 87] I Biederman. Recognition by components: A theory of human image understanding. *Psychological Review*, 94(2):115–147, 1987.

[Bowyer 92] K W Bowyer, editor. Special issue on directions in CAD-based vision. *CVGIP - Image Understanding*, 55:107–218, 1992.

[Brooks et al. 79] R A Brooks, R Greiner, and T O Binford. The ACRONYM model-based vision system. In *Proceedings of the International Joint Conference on Artificial Intelligence, IJCAI-6,* Tokyo, Japan, pages 105–113, 1979.

[Butterfield 97] S Butterfield. Reconstruction of extended environments from image sequences. PhD thesis, School of Computer Studies, University of Leeds, Leeds, UK, 1997.

[Buxton and Howarth 95] H Buxton and R J Howarth. Spatial and temporal reasoning in the generation of dynamic scene representations. In R V Rodriguez, editor, *Proceedings of Spatial and Temporal Reasoning*, pages 107–115, IJCAI-95, Montreal, Canada, 1995.

[Farshid and Aggarwal 93] A Farshid and J K Aggarwal. Model-based object recognition in dense range images—a review. *ACM Computing Surveys*, 25(1):5–43, March 1993.

[Faugeras 93] O D Faugeras. *Three-Dimensional Computer Vision: A Geometric Viewpoint.* MIT Press, Cambridge, MA, 1993.

[Faugeras and Mourrain 95] O D Faugeras and B Mourrain. On the geometry and algera of the point and line correspondences between n images. In *5th International Conference on Computer Vision,* MIT, pages 951–956, IEEE, Piscataway, NJ, 1995.

[Faugeras et al. 92] O D Faugeras, Q T Luong, and S J Maybank. Camera self-calibration: Theory and experiments. In *2nd European Conference on Computer Vision,* Santa Margherita Ligure, Italy, pages 321–333, Springer Verlag, Heidelberg, Germany, 1992.

[Fernyhough 97] J F Fernyhough. Generation of qualitative spatio-temporal representations from visual input. PhD thesis, School of Computer Studies, University of Leeds, Leeds, UK, 1997.

[Flynn and Jain 91] P J Flynn and A K Jain. CAD-based computer vision: From CAD models to relational graphs. *IEEE Transaction on Pattern Analysis and Machine Intelligence*, 13(2):114–132, February 1991.

[Flynn and Jain 92] P J Flynn and A K Jain. 3D object recognition using invariant feature indexing of interpretation tables. *CVGIP - Image Understanding*, 55(2):119–129, 1992.

[Foley et al. 90] J D Foley, A van Dam, S K Feiner, and J F Hughes. *Computer Graphics—Principles and Practice.* Addison-Wesley, Reading, MA, 2nd edition, 1990.

[Goad 86] C Goad. Fast 3D model-based vision. In A P Pentland, editor, *From Pixels to Predicates*, pages 371–374. Ablex, Norwood, NJ, 1986.

[Golub and Loan 89] G H Golub and C F Van Loan. *Matrix Computations.* Johns Hopkins University Press, Baltimore, MD, 2nd edition, 1989.

[Haralick and Shapiro 93] R M Haralick and L G Shapiro. *Computer and Robot Vision, Volume II.* Addison-Wesley, Reading, MA, 1993.

[Hartley 92] R I Hartley. Estimation of relative camera positions for uncalibrated cameras. In *2nd European Conference on Computer Vision,* Santa Margherita Ligure, Italy, pages 579–587, Springer Verlag, Heidelberg, 1992.

[Hartley 94] R I Hartley. Self-calibration from multiple views with a rotating camera. In J-O Eklundh, editor, *3rd European Conference on Computer Vision,* Stockholm, Sweden, pages A:471–478, Springer Verlag, Berlin, 1994.

[Hartley 95] R I Hartley. In defence of the 8-point algorithm. In *5th International Conference on Computer Vision,* MIT, pages 1064–1070, IEEE, Piscataway, NJ, 1995.

[Hlaváč et al. 96] V Hlaváč, A Leonardis, and T Werner. Automatic selection of reference views for image-based scene representations. In B Buxton and R Cipolla, editors, *4th European Conference on Computer Vision,* Cambridge, England, volume 1, pages 526–535, Springer Verlag, Berlin, 1996.

[Horaud et al. 95] R Horaud, R Mohr, F Dornaika, and B Boufama. The advantage of mounting a camera onto a robot arm. In *Proceedings of the Europe-China Workshop on Geometrical Modelling and Invariants for Computer Vision,* Xian, China, pages 206–213, 1995.

[Horn 70] B K P Horn. Shape from shading: A method for obtaining the shape of a smooth opaque sobject from one view. PhD thesis, Department of Electrical Engineering, MIT, Cambridge, MA, 1970.

[Horn 75] B K P Horn. Shape from shading. In P H Winston, editor, *The Psychology of Computer Vision.* McGraw-Hill, New York, 1975.

[Horn 86] B K P Horn. *Robot Vision.* MIT Press, Cambridge, MA, 1986.

[Horn 90] B K P Horn. Height and gradient from shading. *International Journal of Computer Vision*, 5(1):37–75, 1990.

[Horn and Brooks 89] B K P Horn and M J Brooks, editors. *Shape from Shading.* MIT Press, Cambridge, MA, 1989.

[Howarth 94] R J Howarth. Spatial representation and control for a surveillance system. PhD thesis, Department of Computer Science, Queen Mary and Westfield College, Univerity of London, UK, 1994.

[Ikeuchi 94] K Ikeuchi. Surface reflection mechanism. In T Y Young, editor, *Handbook of Pattern Recognition and Image Processing: Computer Vision*, pages 131–160, Academic Press, San Diego, 1994.

[Ikeuchi and Horn 81] K Ikeuchi and B K P Horn. Numerical shape from shading and occluding boundaries. *Artificial Intelligence*, 17:141–184, 1981.

[Jain et al. 95] R Jain, R Kasturi, and B G Schunk. *Machine Vision.* McGraw-Hill, New York, 1995.

[Julesz 90] B Julesz. Binocular depth perception of computer-generated patterns. *Bell Systems Technical Journal*, 39, 1990.

[Klette et al. 96] R Klette, A Koschan, and K Schlüns. *Computer Vision—Räumliche Information aus digitalen Bildern.* Friedr. Vieweg & Sohn, Braunschweig, 1996.

[Klir 91] G J Klir. *Facets of System Science.* Plenum Press, New York, 1991.

[Koenderink 90] J J Koenderink. *Solid Shape.* MIT Press, Cambridge, MA, 1990.

[Landy et al. 96] M S Landy, L T Maloney, and M Pavel, editors. *Exploratory Vision: The Active Eye.* Springer Series in Perception Engineering. Springer Verlag, New York, 1996.

[Longuet-Higgins 81] H C Longuet-Higgins. A computer algorithm for reconstruction a scene from two projections. *Nature*, 293(10):133–135, September 1981.

[Mackworth 73] A K Mackworth. Interpreting pictures of polyhedral scenes. *Artificial Intelligence*, 4(2):121–137, 1973.

[Marr 82] D Marr. *Vision—A Computational Investigation into the Human Representation and Processing of Visual Information.* Freeman, San Francisco, 1982.

[Marr and Hildreth 80] D Marr and E Hildreth. Theory of edge detection. *Proceedings of the Royal Society*, B 207:187–217, 1980.

[Marr and Poggio 79] D Marr and T A Poggio. A computational theory of human stereo vision. *Proceedings of the Royal Society*, B 207:301–328, 1979.

[Maybank and Faugeras 92] S J Maybank and O D Faugeras. A theory of self-calibration of a moving camera. *International Journal of Computer Vision*, 8(2):123–151, 1992.

[Mohr 93] R Mohr. Projective geometry and computer vision. In C H Chen, L F Pau, and P S P Wang, editors, *Handbook of Pattern Recognition and Computer Vision*, Chapter 2.4, pages 369–393. World Scientific, Singapore, 1993.

[Newman et al. 93] T S Newman, P J Flynn, and A K Jain. Model-based classification of quadric surfaces. *CVGIP: Image Understanding*, 58(2):235–249, September 1993.

[Nicodemus et al. 77] F E Nicodemus, J C Richmond, J J Hsia, I W Ginsberg, and T Limperis. Geometrical considerations and nomeclature for reflectance. US Department of Commerce, National Bureau of Standards, Washington DC, 1977.

[Nishihara 84] H K Nishihara. Practical real-time imaging stereo matcher. *Optical Engineering*, 23(5):536–545, 1984.

[Oliensis 91] J Oliensis. Shape from shading as a partially well-constrained problem. *Computer Vision, Graphics, and Image Processing: Image Understanding*, 54(2):163–183, September 1991.

[Pajdla and Hlaváč 95] Tomáš Pajdla and Václav Hlaváč. Camera calibration and Euclidean reconstruction from known translations. Presented at the workshop *Computer Vision and Applied Geometry*, Nordfjordeid, Norway, August 1–7, 1995.

[Pentland 84] A P Pentland. Local shading analysis. *IEEE Transactions on Pattern Analysis and Machine Intelligence*, 6(2):170–187, March 1984.

[Pentland and Bichsel 94] A P Pentland and M Bichsel. Extracting shape from shading. In T Y Young, editor, *Handbook of Pattern Recognition and Image Processing: Computer Vision*, pages 161–183, Academic Press, San Diego, 1994.

[Poggio et al. 85] T Poggio, V Torre, and C Koch. Computational vision and regularization theory. *Nature*, 317:314–319, 1985.

[Pollard et al. 85] S B Pollard, J E W Mayhew, and J P Frisby. PMF: A stereo correspondence algorithm using a disparity gradient limit. *Perception*, 14:449–470, 1985.

[Prescott and McLean 97] B Prescott and G F McLean. Line-based correction of radial lens distortion. *Graphical Models and Image Processing*, 59(1):39–77, January 1997.

[Press et al. 92] W H Press, , S A Teukolsky, W T Vetterling, and B P Flannery. *Numerical Recipes in C.* Cambridge University Press, Cambridge, England, 2nd edition, 1992.

[Rindfleisch 66] T Rindfleisch. Photometric method form lunar topography. *Photogrammetric Engineering*, 32(2):262–267, 1966.

[Sara 94] R Sara. Local shading analysis via isophotes properties. PhD thesis, Department of System Sciences, Johannes Kepler University Linz, March 1994.

[Sara 95] R Sara. Isophotes: The key to tractable local shading analysis. In V Hlavac and R Sara, editors, *International Conference on Computer Analysis of Images and Patterns,* Prague, Czech Republic, pages 416–423, Springer Verlag, Heidelberg, September 6th–8th, 1995.

[Semple and Kneebone 63] J G Semple and G T Kneebone. *Algebraic Projective Geometry.* Oxford University Press, London, 1963.

[Shashua and Werman 95] A Shashua and M Werman. Trilinearity of three perspective views and its associated tensor. In *5th International Conference on Computer Vision,* MIT, pages 920–925, IEEE, Piscataway, NJ, 1995.

[Smutný 93] V Smutný. Analysis of rainbow range finder errors. In V Hlaváč and T Pajdla, editors, *1st Czech Pattern Recognition Workshop*, pages 59–66. Czech Pattern Recognition Society, CTU, Prague, November 1993.

[Soucy and Laurendeau 92] M Soucy and D Laurendeau. Surface modeling from dynamic integration of multiple range views. In *11th International Conference on Pattern Recognition,* The Hague, volume I, pages 449–452, IEEE, Piscataway, NJ, September 1992.

[Tanaka and Kak 90] S Tanaka and A C Kak. A rule-based approach to binocular stereopsis. In R C Jain and A K Jain, editors, *Anaysis and Interpretation of Range Images*, Chapter 2. Springer Verlag, Berlin, 1990.

[Tichonov and Arsenin 77] A N Tichonov and V Y Arsenin. *Solution of ill-posed problems.* Winston and Wiley, Washington, DC, 1977.

[Tsai 87] R Y Tsai. A versatile camera calibration technique for high-accurancy 3D machine vision metrology using off-the-shelf cameras and lenses. *IEEE Journal of Robotics and Automation,* RA-3(4):323—344, August 1987.

[Ullman 96] S Ullman. *High-Level Vision: Object Recognition and Visual Cognition.* MIT Press, Cambridge, MA, 1996.

[Ullman and Basri 91] S Ullman and R Basri. Recognition by linear combination of models. *IEEE Transactions on Pattern Analysis and Machine Intelligence*, 13(10):992–1005, October 1991.

[Wechsler 90] H Wechsler. *Computational Vision.* Academic Press, London–San Diego, 1990.

[Weinshall et al. 95] D Weinshall, M Werman, and A Shashua. Shape tensors for efficient and learnable indexing. In *Proceedings of the IEEE Workshop Representation of Visual Scenes,* Cambridge, MA, pages 58–65, IEEE, Los Alamitos, CA, June 24, 1995.

[Werner et al. 95] T Werner, R D Hersch, and V Hlaváč. Rendering real-world objects using view interpolation. In *5th International Conference on Computer Vision,* MIT, pages 957–962, IEEE, Piscataway, NJ, 1995.

[Woodham 80] R J Woodham. Photometric method for determining surface orientation from multiple images. *Optical Engineering*, 19:139–144, 1980.

Chapter 10

Use of 3D vision

In earlier (and later) chapters, we present a constructive approach to various aspects of image processing and vision that should allow readers to reproduce ideas and build systems of their own. Most of this chapter is somewhat different; 3D vision solves complex tasks having no settled and simple theory, and here we step aside and provide an overview of recent approaches, task formulations, applications and current research. We hope thereby that the reader will learn what the current state of 3D vision is; it is possible that the material herein will be at the limit of the abilities of the vision novice, but may form a useful primer for master's or Ph.D. courses.

10.1 Shape from X

Shape from X is a generic name for techniques that aim to extracting shape from intensity images. Many of these methods estimate local surface orientation (e.g., surface normal) rather than absolute depth. If, in addition to this local knowledge, the depth of some particular point is known, the absolute depth of all other points can be computed by integrating the surface normals along a curve on a surface [Horn 86].

Several topics belonging to this category of methods have already been mentioned, i.e., *shape from stereo* (sections 9.2.5, 9.2.11), *shape from shading* (section 9.3.3), and *photometric stereo* (section 9.3.4).

10.1.1 Shape from motion

Motion is a primary property exploited by human observers of the 3D world. The real world we see is dynamic in many respects, and the relative movement of objects in view, their translation and rotation relative to the observer, the motion of the observer relative to other static and moving objects all provide very strong clues to shape and depth—consider how just moving your head from side to side provides rich information from parallax effects. It should therefore come as no surprise to learn that attempts at shape extraction are able to make use of motion. Motion, and particularly lower-level algorithms associated with its analysis, is considered in detail in Chapter 15, and in this section study is restricted to shape extraction alone.

A study of human analysis of motion is instructive and was conducted comprehensively in a computational context by Ullman [Ullman 79]. Exactly how we make deductions from moving scenes is far from clear, and several theories have come and gone in efforts to understand this matter—in particular, *Gestaltist* theories. Gestalt psychology was a revolutionary psychological paradigm proposed in Germany in the early twentieth century ('Gestalt' means 'shape' or 'form' in German). It claims that more complicated mental processes cannot be simply composed from the simpler ones, and questioned the causality of events. Its suggestion that groupings of observations are of primary importance was disproved, notably by an experiment of Ullman's. On a computer screen, he simulated two coaxial cylinders of different radii rotating about their common axis in opposite directions. The view is perpendicular to the common axis; the cylinders were not drawn, but only randomly placed dots on their surfaces. Thus what is seen (on a per-point basis) is a large number of dots moving from left to right or right to left, at varying speeds. Exactly what speed and direction depends upon which cylinder surface a dot belongs to, and at what point of rotation it is—in fact, each individual dot executes simple harmonic motion about a point that is on a line that is the projection onto the image of the axis. The striking conclusion is that the human observer is in no doubt about the nature of the scene, despite the absence of surface clues and the complete absence of structure in any single frame from the sequence.

What we exploit are particular *constraints that assist in resolving the non-uniqueness of the interpretation* of a sequence of frames as a moving 3D scene. In fact, motion may be presented to us as widely spaced (in time) discrete frames, or as (pseudo-)continuous—that is, so many frames that changes between a given pair are imperceptible. We shall examine each case separately, each time using Ullman's observation that the extraction of 3D information from moving scenes can be done as a two-phase process:

1. *Finding correspondences* or calculating the nature of the flow is a lower-level phase that operates on pixel arrays.

2. The *shape extraction* phase follows as a separate, higher-level process. This phase is examined here.

It is worth noting that researchers are not unanimous in the view that these two phases should be held separate, and approaches exist that are different from those discussed here [Negahdaripour and Horn 85].

Note that one approach to the analysis of motion is superficially similar to that of stereo vision—images that are relatively widely separated in time are taken, and correspondences between visible features made. The solution to this correspondence problem is considered in detail in Chapter 15 and Section 9.2.11. It is worth remarking here that resemblance to the stereo correspondence problem is deceptive since the scene may well contain any number of independently moving objects, which could mean that correlations may be strictly local. Two images are not of the same scene, but (more probably) of the same objects in different relative positions.

Searching for correspondence in motion analysis may be easier than when attempting it in stereo imaging. It is often possible to capture a dense sequence of images (i.e., the time separation between neighboring frames is small so that corresponding features are very close, and the search for them almost trivial). Moreover, the position of the feature in the next frame can be predicted by estimating its trajectory using techniques similar to those of control theory. The Kalman filter approach (see Section 15.4) is common.

Rigidity, and the structure from motion theorem

For now, suppose that the correspondence problem has been solved, and that it remains to extract some shape information—that is, given that a collection of points has been identified in two different views, how might they be interpreted as 3D objects? As might be expected, the large number of possible interpretations is resolved by deploying a constraint; Ullman's success in this area was based on the psycho-physical observation that the human visual system seems to assume that objects are *rigid.* This rigidity constraint prompted the proof of an elegant **structure from motion theorem** saying that *three orthographic projections of four non-co-planar points have a unique 3D interpretation as belonging to one rigid body.* We shall proceed to outline the proof of this theorem, which is constructive and therefore permits the extraction of the appropriate geometry, given point correspondences in three frames from a motion sequence. In use, the theorem allows samples of four points to be taken from an image sequence—*if* they belong to the same (rigid) body, an interpretation is generated, but if they do not, the probability of there being a chance rigid interpretation turns out to be negligibly small, meaning that the algorithm is self-verifying in the sense that it generates only answers that are 'correct'. Thus if there are N points in the correspondence, we might search for $N/4$ rigid interpretations, some of which will be invalid, and others of which will group according to the rigid object to which they belong.

The theorem proof involves a re-phrasing of the problem to permit its definition as the solution of an equivalent problem in 3D geometry. Given three orthographic views of four points that have a rigid interpretation, the correspondence allows them to be labeled as O, A, B, and C in each image. First note that the body's motion may be decomposed into translational and rotational movement; the former gives the movement of a fixed point with respect to the observer, and the latter relative rotation of the body (for example, about the chosen fixed point). *Translational movement*, as far as it is recognizable, is easy to identify. All that can be resolved is movement perpendicular to the projection, and this is given by the translation (in 2D) of an arbitrarily chosen point, say O. Observe that motion parallel to the projection cannot be identified.

It remains to identify *rotational motion*; to do this we can assume that O is a fixed point, and seek to identify an interpretation of A, B and C as belonging to the same rigid body as O. Accordingly, we transform the problem to that of knowing three pairs of (2D) co-ordinates for A, B, and C with respect to a common origin O, each a different orthographic projection; what is now required is the (3D) directions of the projections.

Formally, suppose we have in 3D an origin O and three vectors $\mathbf{a}$, $\mathbf{b}$, and $\mathbf{c}$ corresponding to A, B, and C; given projections of $\mathbf{a}$, $\mathbf{b}$, and $\mathbf{c}$ onto three planes Π_1, Π_2, and Π_3 of unknown orientation, we require to reconstruct the 3D geometry of $\mathbf{a}$, $\mathbf{b}$, and $\mathbf{c}$. Now let the co-ordinate system of the plane Π_i be defined by vectors $\mathbf{x}_i$ and $\mathbf{y}_i$; that is, $\mathbf{x}_i$ and $\mathbf{y}_i$ are orthogonal 3D unit vectors lying in the plane Π_i. With respect to these systems, suppose that on plane Π_i the points' projections have co-ordinates (a_{xi}, a_{yi}), (b_{xi}, b_{yi}), (c_{xi}, c_{yi})—these nine pairs are the input to the algorithm. Finally, let $\mathbf{u}_{ij}$ be a unit vector lying on the line defined by the intersection of planes Π_i and Π_j.

Elementary co-ordinate geometry gives

$$\begin{aligned} a_{xi} &= \mathbf{a} \cdot \mathbf{x}_i & a_{yi} &= \mathbf{a} \cdot \mathbf{y}_i \\ b_{xi} &= \mathbf{b} \cdot \mathbf{x}_i & b_{yi} &= \mathbf{b} \cdot \mathbf{y}_i \\ c_{xi} &= \mathbf{c} \cdot \mathbf{x}_i & c_{yi} &= \mathbf{c} \cdot \mathbf{y}_i \end{aligned} \tag{10.1}$$

Further, since $\mathbf{u}_{ij}$ lies on both Π_i and Π_j, there must exist scalars $\alpha_{ij}, \beta_{ij}, \gamma_{ij}, \delta_{ij}$ such that

$$\begin{aligned} \alpha_{ij}^2 + \beta_{ij}^2 &= 1 \\ \gamma_{ij}^2 + \delta_{ij}^2 &= 1 \end{aligned} \tag{10.2}$$

and

$$\begin{aligned} \mathbf{u}_{ij} &= \alpha_{ij}\mathbf{x}_i + \beta_{ij}\mathbf{y}_i \\ \mathbf{u}_{ij} &= \gamma_{ij}\mathbf{x}_j + \delta_{ij}\mathbf{y}_j \end{aligned} \tag{10.3}$$

and hence

$$\alpha_{ij}\mathbf{x}_i + \beta_{ij}\mathbf{y}_i = \gamma_{ij}\mathbf{x}_j + \delta_{ij}\mathbf{y}_j \tag{10.4}$$

We can take the scalar product of this equation with each of **a**, **b**, and **c**, and using equation (10.1) see that

$$\begin{aligned} \alpha_{ij}a_{xi} + \beta_{ij}a_{yi} &= \gamma_{ij}a_{xj} + \delta_{ij}a_{yj} \\ \alpha_{ij}b_{xi} + \beta_{ij}b_{yi} &= \gamma_{ij}b_{xj} + \delta_{ij}b_{yj} \\ \alpha_{ij}c_{xi} + \beta_{ij}c_{yi} &= \gamma_{ij}c_{xj} + \delta_{ij}c_{yj} \end{aligned} \tag{10.5}$$

—thus we have relations between unknowns $(\alpha, \beta, \gamma, \delta)$ in terms of known quantities $(a_x, a_y,$ etc.).

It is easy to show that the equations (10.5) are linearly independent (this is where the fact that O, A, B, and C are not co-planar is used). Therefore, using the constraint of equation (10.2), it is possible to solve for $\alpha_{ij}, \beta_{ij}, \gamma_{ij}, \delta_{ij}$—in fact, there are two possible solutions that differ in sign only.

This (findable) solution is important, as it means that we are able to express the vectors $\mathbf{u}_{ij}$ in terms of the co-ordinate basis vectors $\mathbf{x}_i$, $\mathbf{y}_i$, $\mathbf{x}_j$, and $\mathbf{y}_j$. To see why this is important, picture the three planes in 3D—they intersect at the common origin O and therefore define a tetrahedron; what interests us is the *relative* angles between the planes, and if the geometry of the tetrahedron can be recaptured, these angles are available. Note, though, that knowledge of $\alpha_{ij}, \beta_{ij}, \gamma_{ij}, \delta_{ij}$ allows calculation of the distances

$$\begin{aligned} d_1 &= \quad |\mathbf{u}_{12} - \mathbf{u}_{13}| \\ d_2 &= \quad |\mathbf{u}_{12} - \mathbf{u}_{23}| \\ d_3 &= \quad |\mathbf{u}_{13} - \mathbf{u}_{23}| \end{aligned} \tag{10.6}$$

For example,

$$\begin{aligned} \mathbf{u}_{12} - \mathbf{u}_{13} &= (\alpha_{12}\mathbf{x}_1 + \beta_{12}\mathbf{y}_1) - (\alpha_{13}\mathbf{x}_1 + \beta_{13}\mathbf{y}_1) \\ &= (\alpha_{12} - \alpha_{13})\mathbf{x}_1 + (\beta_{12} - \beta_{13})\mathbf{y}_1 \end{aligned} \tag{10.7}$$

and hence

$$d_1 = (\alpha_{12} - \alpha_{13})^2 + (\beta_{12} - \beta_{13})^2 \tag{10.8}$$

since $\mathbf{x}_1$ and $\mathbf{y}_1$ are orthogonal. Now the tetrahedron formed by the three intersecting planes is defined by the origin O and a triangular base—we might consider the base given by the three points at unit distance from the origin. By construction, this triangle has sides d_1, d_2, d_3, and we can thus reconstruct the required tetrahedron.

Determining the 3D structure is now possible by noting that a particular point lies at the intersection of the normals to any two of the planes drawn from the projections of the point concerned.

There is a complication in the proof not discussed here that occurs when one of the d_i is zero, and the tetrahedron is degenerate. It is possible to resolve this problem without difficulty—the full proof is given in [Ullman 79].

It is worth noting that Ullman's result is the best possible in the sense that unique reconstruction of a rigid body cannot be guaranteed with fewer than three projections of four points, or with three projections of fewer than four points. It should also be remembered that the result refers to *orthographic* projection when in general image projections are *perspective* (of which, of course, the orthographic projection is a special case). This turns out not to be a problem since a similar result is available for the perspective projection [Ullman 79]. In fact this is not necessary, since it is possible to approximate neighborhoods within a perspective projection by a number of different orthographic projections; thus in such a neighborhood, the theorem as outlined is valid. Interestingly, there seems to be evidence that the human visual system uses this sort of orthographic approximation in extracting shape information from motion.

This result is of particular value in *active* vision applications [Blake and Yuille 92, Aloimonos 93] such as a robot arm having a camera mounted upon it; when such a system finds itself unable to 'see' particular objects of interest, the arm will move for a different view, which will then need reconciling with earlier ones.

Shape from optical flow

The motion presented to human observers is not that considered in the previous section, but rather is continuous—the scene in view varies smoothly. The approach of considering widely spaced (in time) views is therefore a simplification, and it is natural to ask how to treat the 'limiting case' of separate frames being temporally very close to each other—it is well known that, in fact, the human eye perceives continuous motion from relatively few frames per second (as illustrated by cinema film). Clearly the approach of making correspondences is no longer any use since corresponding points will be separated by infinitesimally small distances—it is the apparent velocity (direction and speed) of pixels that is of interest in the study of continuous motion. In a continuous sequence, we are therefore interested in the apparent movement of each pixel (x, y) which is given by the *optical flow field* $(dx/dt, dy/dt)$. In Chapter 15, optical flow is considered at length and an algorithm is described for its extraction from observation of changes in the intensity function (gray-levels); accordingly, in this section it is assumed that the flow field is available, and we ask how it may be used to extract shape in the form of surface orientation (in fact, optical flow is useful for deducing a number of motion properties, such as the nature of the translational or rotational movement—these points are considered in Chapter 15).

Determining shape from optical flow is mathematically non-trivial, and here an early simplification of the subject is presented as an illustration [Clocksin 80]. The simplification is in two parts:

- Motion is due to the observer traveling in a straight line through a static landscape. Without loss of generality, suppose the motion is in the direction of the z axis of a viewer-centered co-ordinate system (i.e., the observer is positioned at the origin).

- Rather than being projected onto a 2D plane, the image is seen on the surface of a unit sphere, centered at the observer (a 'spherical retina'). Points in 3D are represented in spherical polar rather than Cartesian co-ordinates—spherical polar co-ordinates (r, θ, φ) (see Figure 10.1) are related to (x, y, z) by the equations

$$r^2 = x^2 + y^2 + z^2 \tag{10.9}$$

$$y = x \tan \theta \tag{10.10}$$

$$z = r \cos \varphi \tag{10.11}$$

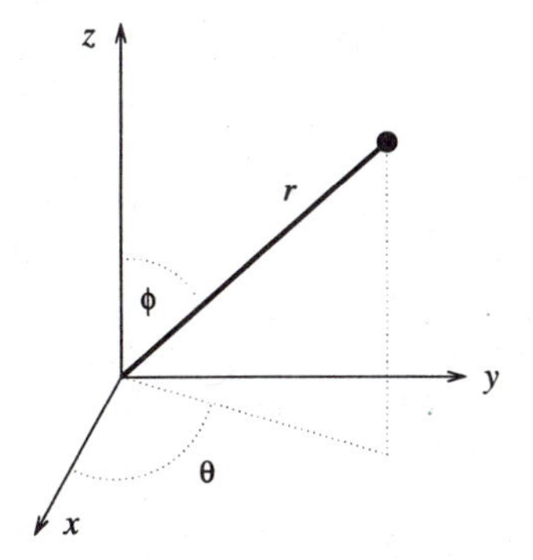

Figure 10.1: *Definition of spherical polar co-ordinates.*

Since the image is spherical, we can specify co-ordinates as (θ, φ) pairs rather than (x, y) as usual, and the optical flow is then $(d\theta/dt, d\varphi/dt)$. Supposing the observer's speed to be v (in the direction of the z axis), the motion of points in 3D is given by

$$\frac{dx}{dt} = 0 \qquad \frac{dy}{dt} = 0 \qquad \frac{dz}{dt} = -v \tag{10.12}$$

Differentiating equation (10.9) with respect to t gives

$$\begin{aligned} 2r\frac{dr}{dt} &= 2x\frac{dx}{dt} + 2y\frac{dy}{dt} + 2z\,\frac{dz}{dt} \\ &= -2vz \\ \frac{dr}{dt} &= -\frac{vz}{r} \\ &= -v\cos\varphi \end{aligned} \tag{10.13}$$

Differentiating equation (10.10) with respect to t gives

$$\begin{aligned} \frac{dy}{dt} &= \tan\theta\,\frac{dx}{dt} + x\sec^2\theta\,\frac{d\theta}{dt} \\ 0 &= 0 + x\sec^2\theta\,\frac{d\theta}{dt} \end{aligned} \tag{10.14}$$

and hence

$$\frac{d\theta}{dt} = 0 \tag{10.15}$$

Differentiating equation (10.11) with respect to t gives

$$\frac{dz}{dt} = \cos\varphi\,\frac{dr}{dt} - r\sin\varphi\,\frac{d\varphi}{dt} \tag{10.16}$$

and hence, by equations (10.12) and (10.13),

$$-v = -v\cos^2\varphi - r\sin\varphi\,\frac{d\varphi}{dt} \tag{10.17}$$

and so

$$\frac{d\varphi}{dt} = \frac{v(1-\cos^2\varphi)}{r\sin\varphi} = \frac{v\sin\varphi}{r} \tag{10.18}$$

Equations (10.15) and (10.18) are important. The former says that, for this particular motion, the rate of change of θ is zero (θ is constant). More interestingly, the latter says that given the optical flow $d\varphi/dt$, then the distance r of a 3D point from the observer can be recaptured up to a scale factor v. In particular, if v is known, then r, and a complete depth map, can be deduced from the optical flow. The depth map allows a reconstruction of the 3D scene and hence characteristics of surfaces (smoothly varying areas of r) and of edges (discontinuities in r) will be available.

In the case that v is not known, it turns out that surface information is still available directly from the flow. In particular, suppose a point P lies on a smooth surface, which at P may be specified by the direction of a normal vector $\mathbf{n}$. Such a direction may be specified by two angles α and β, where α is the angle between $\mathbf{n}$ and a plane Π_1 defined by P and the z axis, and β is the angle between $\mathbf{n}$ and a plane Π_2 which passes through P and the origin and is perpendicular to Π_1. Intuitively, it is clear that the rate of change of r with respect to θ and φ provides information about the direction of $\mathbf{n}$. Moderately straightforward co-ordinate geometry gives the relations

$$\tan\alpha = \frac{1}{r}\frac{\partial r}{\partial\varphi} \qquad \tan\beta = \frac{1}{r}\frac{\partial r}{\partial\theta} \tag{10.19}$$

These equations depend upon a knowledge of r (the depth map), but it is possible to combine them with equation (10.18) to overcome this. For convenience, write $d\varphi/dt = \dot\varphi$; then, by equation (10.18),

$$r = \frac{v\sin\varphi}{\dot\varphi} \tag{10.20}$$

and so

$$\frac{\partial r}{\partial \varphi} = v\frac{\dot{\varphi}\cos\varphi - \sin\varphi(\partial\dot{\varphi}/\partial\varphi)}{\dot{\varphi}^2}$$
$$\frac{\partial r}{\partial \theta} = -v\frac{\sin\varphi(\partial\dot{\varphi}/\partial\theta)}{\dot{\varphi}^2} \tag{10.21}$$

Substituting (10.20) and (10.21) into equations (10.19) gives

$$\tan\alpha = \cot\varphi - \frac{1}{\dot{\varphi}}\frac{\partial\dot{\varphi}}{\partial\varphi}$$
$$\tan\beta = \frac{1}{\dot{\varphi}}\frac{\partial\dot{\varphi}}{\partial\theta} \tag{10.22}$$

Thus, given the flow $\dot{\varphi}$ (which we assume), the angles α and β are immediately available, regardless of S and without any need to determine the depth map given by r.

The original reference [Clocksin 80] provides full information on this derivation, and proceeds to describe how edge information may also be extracted from knowledge of the flow. It also includes some interesting discussion of psycho-physical considerations of human motion perception in the context of a computational theory.

10.1.2 Shape from texture

A further property of which there is clear psycho-physical evidence of human use to extract depth is texture [Marr 82]. To appreciate this, it is only necessary to consider a regularly patterned object viewed in 3D. Two effects would be apparent: The angle at which the surface is seen would cause a (perspective) distortion of the **texture primitive** (**texel**), and the relative size of the primitives would vary according to distance from the observer. Simple examples, shown in Figure 10.2, are sufficient to illustrate this. Much use can be made of texture in computer vision at various levels of abstraction, and Chapter 14 examines them in some detail. Here we look briefly at the use of textural properties to assist in the extraction of shape [Bajcsy and Lieberman 76, Kanatani and Chou 89].

Considering a textured surface patterned with identical texels which have been recovered by lower-level processing, note that with respect to a viewer it has three properties at any point projected onto a retinal image: distance from the observer, **slant**; the angle at which the surface is sloping away from the viewer (the angle between the surface normal and the line of sight); and **tilt**, the direction in which the slant takes place. Attempts to re-capture some of this information is based on the **texture gradient**—that is, the direction of maximum rate of change of the perceived size of the texels, and a scalar measurement of this rate. One approach [Bajcsy and Lieberman 76] assumes a uniform texel size.

If the texture is particularly simple, the shape of the perceived texels will reveal surface orientation information. For example, if a plane is marked with identical circles, they will be seen in an image as ellipses (see Figure 10.3). The eccentricity of the ellipses provides information about the slant, while the orientation of the ellipse axes indicates the tilt [Stevens 79]. There is evidence to suggest [Stevens 79] that the human viewer uses the texture gradient as a primary clue in the extraction of tilt and (relative) distance, but that slant is inferred by processing based on estimates of the other two parameters. Tilt is indicated by the direction

Figure 10.2: *A simple texture pattern in 3D. The left side shows a vanishing brick wall and the right the shape of a woman's body perceived from texture changes.*

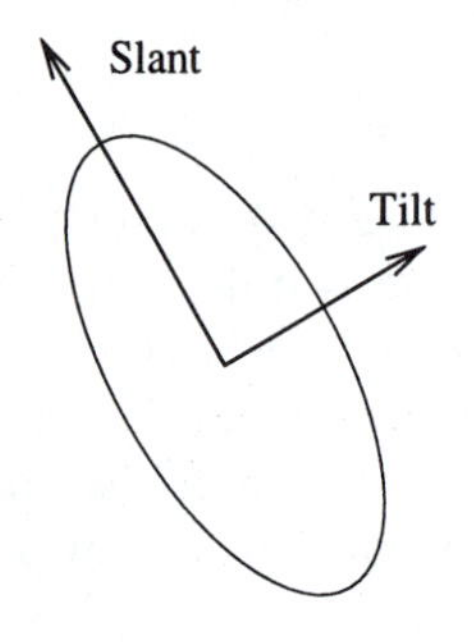

Figure 10.3: *Slant and tilt are revealed by texel properties.*

of the texture gradient (see Figure 10.4), while the apparent size of objects decreases as the reciprocal of their distance from the viewer.

Large-scale texture effects can provide information about large-scale scene geometry; in particular, strong linear effects will indicate 'vanishing points' that may be joined to give the scene horizon. Note that it is not necessary in fact for the image to contain large numbers of long straight lines for this conclusion to be available, since often such lines can be inferred by a number of segments whose co-linearity may be deduced by, for example, a Hough transform. Such co-linearity may well be a property of urban scenes in which rectangular objects, some large, can predominate.

Texture has many interpretations and there is a correspondingly large number of attempts to exploit it in shape extraction—a useful grounding may be found in [Witkin 81]. A multiple-scale approach was used in [Blostein and Ahuja 89], while [Aloimonos and Swain 85] give an interesting approach in which 'shape from texture' is shown to be equivalent to 'shape from shading', thereby allowing the use of established results in surface parameter extraction from

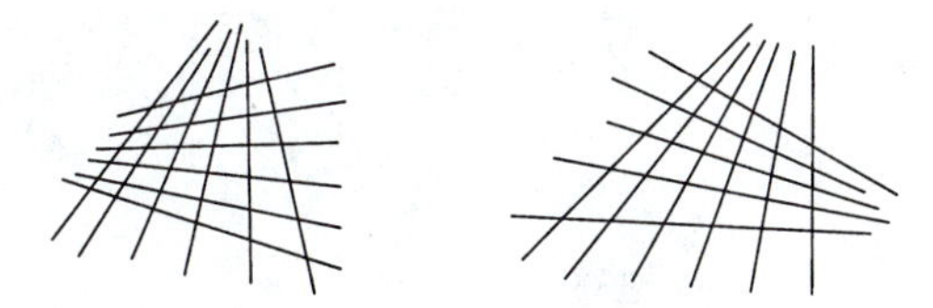

Figure 10.4: *Tilt affects the appearance of texture.*

shading. Texture is usually used as an additional or complementary feature, augmenting another, stronger clue in shape extraction.

10.1.3 Other shape from X techniques

Shape from focus/de-focus techniques are based on the fact that lenses have finite depth of field, and only objects at the correct distance are in focus; others are blurred in proportion to their distance. Two main approaches can be distinguished: shape from focus and shape from de-focus.

Shape from focus measures depth in one location in an active manner; this technique is used in 3D measuring machines in mechanical engineering. The object to be measured is fixed on a motorized table that moves along x, y, z axes. A small portion of the surface is observed by a camera through a microscopic lens, and if the surface patch in view (given by a small image window) is in focus, then the image has the maximal number of high frequencies; this qualitative information about focus serves as feedback to the z-axis servo-motor. The image is put into focus and x, y, z co-ordinates read from the motorized table. If the depth of all points in the image are to be measured, a large number of images is captured by displacing the sensor in small increments from the scene, and the image of maximum focus is detected for each image point [Krotkov 87, Nayar and Nakagawa 94].

Shape from de-focus typically estimates depth using two input images captured at different depths. The relative depth of the whole scene can be reconstructed from image blur. The image is modeled as a convolution of the image with a proper point spread function (see Section 2.1.2); the function is either known from capturing setup parameters or estimated, for example by observing a sharp depth step in the image. The depth reconstruction, which is an ill-posed problem [Pentland 87], is performed by local frequency analysis. Depth from de-focus shares an inherent problem with shape from stereo and shape from motion, in that it requires the scene to be covered by a fine texture. A real-time (30Hz) depth from de-focus sensor has been built [Nayar et al. 96]: The device uses active illumination by texture and analyzes relative blur in two images captured at two different depths. The derivation of the illumination pattern and depth estimation is posed as an optimization problem in the Fourier domain. **Shape from vergence** uses two cameras fixed on a common rod. Using two servo-mechanisms, the cameras can change the direction of their optical axes (verge) in the plane containing a line segment joining their optical centers. Such devices are called **stereo heads**; see Figure 10.5.

The aim of shape from vergence is to to ease the correspondence problem for estimating depth [Krotkov and Bajcsy 93]; vergence is used to align individual feature points in both left and right images.

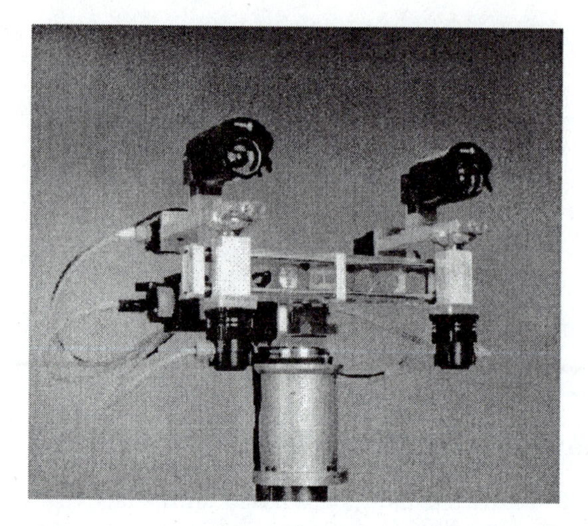

Figure 10.5: *An example of a stereo head. Courtesy J. Kittler, University of Surrey, U.K.*

Shape from contour aims to describe a 3D shape from contours seen from one or more view directions. Objects with smooth bounding surfaces are quite difficult to analyze.

Following terminology given in [Ullman 96], assume the object is observed from some view point. The set of all points on the object surface where surface normal is perpendicular to the observer's visual ray is called a **rim** [Koenderink 90]. Note that in general the rim is not a planar curve. Assuming orthographic projection, the rim points generate a **silhouette** of an object in the image. Silhouettes can be easily and reliably captured if back-light illumination is used, although there is possible complication in the special case in which two distinct rim points project to a single image point.

The most general approach considers contours as silhouettes plus images of the salient curves on the surface, e.g., those corresponding to surface curvature discontinuities. The latter are often found using an edge detector from an intensity image. The trouble is that this process is not often robust enough; the simpler—and more often used—approach explores silhouettes as contours. In Figure 10.6, silhouette and surface discontinuity are shown on an image of an apricot.

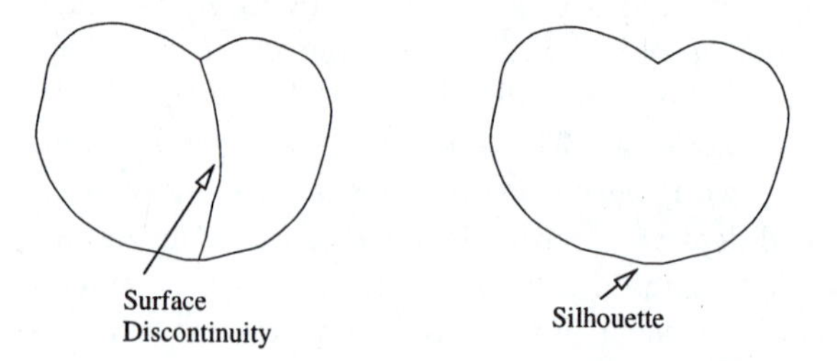

Figure 10.6: *Apricot; contour as silhouette (left) or as silhouette plus surface discontinuities (right).*

The inherent difficulty of the shape from contour comes from the loss of information in projecting 3D to 2D. We know that this projection is not invertible, since one image may result from the projection of different objects. This fact is illustrated in Figure 10.7, in which both a sphere and an ellipsoid project to the same image, the ellipse. Humans are surprisingly successful at perceiving clear 3D shapes from contours, and it seems that tremendous background knowledge is used to assist. Understanding this human ability is one of the major challenges for computer vision. Contours are used as constraints on the shape they represent, and the aim is to reduce the number of possible interpretations.

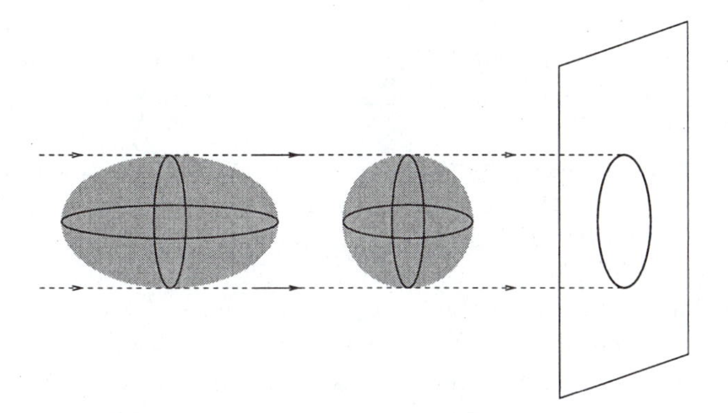

Figure 10.7: *Ambiguity of the shape from contour task.*

This section has merely formulated the shape from contour task. The reader interested in solutions should consult [Nevatia et al. 94].

10.2 Full 3D objects

10.2.1 3D objects, models, and related issues

The notion of a **3D object** allows us to consider a 3D volume as a part of the entire 3D world. This volume has a particular interpretation (semantics, purpose) for the task in hand. Such an approach accords with the way general systems theory [Klir 91] treats complex phenomena, in which objects are separated from uninteresting background. Thus far, we have treated geometric (Section 9.2) and radiometric (Section 9.3) techniques that provide intermediate 3D cues, and it was implicitly assumed that such cues help to understand the nature of a 3D object.

Shape is another informal concept that humans typically connect with a 3D object. Recall the understanding we have when thinking of the shape of a mountain, or a vase or a cup. Computer vision aims at scientific methods for 3D object description, but there are no mathematical tools yet available to express shape in its general sense. The reader interested in abstract aspects of shape might consult texts on the shapes of solid objects [Koenderink 90]. Here, however, we shall not consider 3D shape issues in their full abstract sense. Instead, the simple geometrical approach treating parts of 3D objects locally as simple volumetric or surface primitives is used. Curvilinear surfaces with no restriction on surface shape are called **free-form surfaces.**

Roughly speaking, the 3D vision task distinguishes two classes of approach:

1. *Reconstruction* of the 3D object model or representation from real-world measurements with the aim of estimating a continuous function representing the surface.

2. *Recognition* of an instance of a 3D object in the scene. It is assumed that object classes are known in advance, and that they are represented by a suitable 3D model.

The reconstruction and recognition tasks use different representations of 3D objects. Recognition may use approaches that distinguish well between distinct classes, but do not characterize an object as a whole.

Humans meet and recognize often **deformable objects** that change their shape [Terzopoulos et al. 88, Terzopoulos and Fleischer 88], an advanced topic that is too large to consider in this book.

Computer vision as well as computer graphics use **3D models** to encapsulate the shape of an 3D object. 3D models serve in computer graphics to generate detailed surface descriptions used to render realistic 2D images. In computer vision, the model is used either for reconstruction (copying, displaying an object from a different viewpoint, modifying an object slightly during animation) or for recognition purposes, where features are used that distinguish objects from different classes. There are two main classes of models: volumetric and surface. **Volumetric models** represent the 'inside' of a 3D object explicitly, while **surface models** use only object surfaces, as most vision-based measuring techniques can only see the surface of a non-transparent solid.

3D models make a transition towards an *object-centered* co-ordinate system, allowing object descriptions to be viewer independent. This is the most difficult phase within Marr's paradigm. Successful implementation is remote, especially compared to the success seen with the derivation of the primal and 2.5D sketches. Unlike earlier stages, there is little physiological guidance that can be used to design algorithms, since this level of human vision is not well understood. Marr observes that the target co-ordinate system(s) should be modular in the sense that each 'object' should be treated differently, rather than employing one global co-ordinate system—this prevents having to consider the orientation of model components with respect to the whole. Representations based on an object's 'natural' axes, derived from symmetries, or the orientation of stick features, are likely to be of greater use.

3D models of objects are common in other areas besides computer vision, notably computer-aided design (CAD) and computer graphics, where image synthesis is required—that is, an exact (2D) pictorial representation of some modeled 3D object. Use of an object representation which matches the representation generated by CAD systems has been an active research area for years, with substantial promise for industrial model-based vision. Progress in this area is presented in [Bowyer 92], with papers devoted to CAD-based models applied to pose estimation [Kriegman 92, Ponce et al. 92, Seales and Dyer 92], 3D specular object recognition [Sato et al. 92], and invariant feature extraction from range data [Flynn and Jain 92].

Various representation schemes exist, with different properties. A representation is called **complete** if two different objects cannot correspond to the same model, so a particular model is unambiguous. A representation is called **unique** if an object cannot correspond to two different models. Most 3D representation methods sacrifice either the completeness or the uniqueness property. Commercial CAD systems frequently sacrifice uniqueness; different design methodologies may produce the same object. Some solid modelers maintain multiple

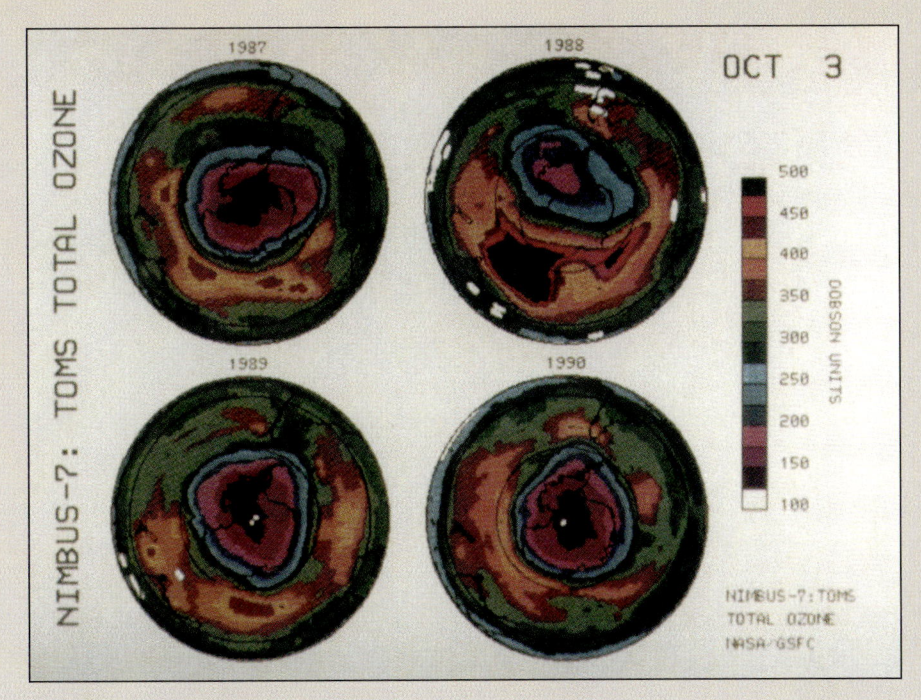

Plate 1 *The ozone layer hole.* (Figure 1.1)

Plate 2 *Pseudocolor representation of the original image.* (Figure 5.38 b)

Plate 3 *Recursive region merging.* (Figure 5.38 c)

Plate 4 *Region merging via boundary melting.* (Figure 5.38 d)

Plate 5 *Remotely sensed data of Prague, Landsat Thematic Mapper. Unsupervised classification, post-processing filter applied: White-no vegetation, green-vegetation types, red-urban areas.* (Figure 8.13)

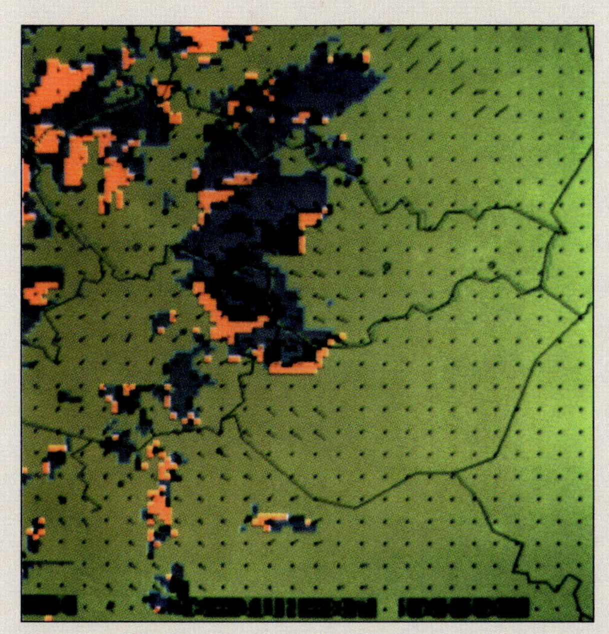

Plate 6 *Meteotrend: Cloud motion analysis in horizontal and vertical directions, vertical speed coded in color.* (Figure 14.14 a)

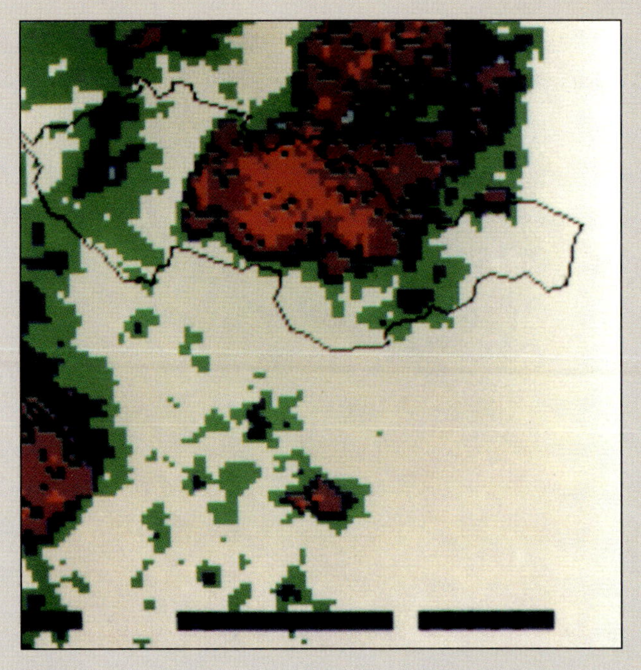

Plate 7 *Meteotrend: Cloud type classification.* (Figure 14.14 b)

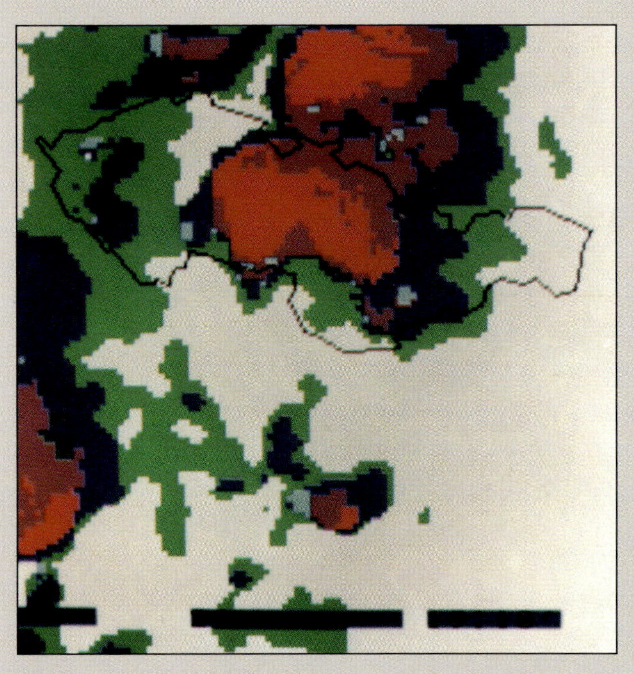

Plate 7 *Meteotrend: Cloud type prediction.* (Figure 14.14 d)

representations of objects in order to offer flexibility in design.

Due to self-occlusion of objects and to triangulation-based measuring methods, most vision-based measuring sensors inherently produce only partial 3D descriptions of objects. A fusion of several such measurements from different viewpoints is needed to obtain the shape of an object entirely. An ideal 3D sensor would provide a set of 3D uniformly sampled points on the surface together with their relation to neighboring points.

10.2.2 Line labeling

Early attempts to develop 3D vision systems tried to reconstruct a full 3D representation from a single, fully segmented view of a scene. The step between the dimensions was made by assuming that all objects in the scene had planar faces (see Figure 10.8), and that three faces met at each vertex. A perfect segmentation then provides straight-edged regions, and in general three of these will meet at a vertex. The idea was that this constraint was sufficient to permit a single 2D view to permit unambiguous reconstruction of a polyhedron. For obvious reasons, this is sometimes called a **blocks world** approach.

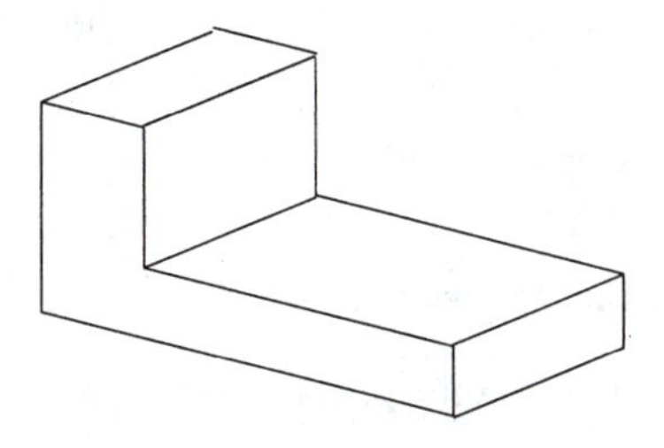

Figure 10.8: *An example blocks world object.*

The approach is clearly unrealistic for two reasons: First, the requirement for a perfect segmentation is unlikely to be met except in the most contrived situations; it is assumed that all edges are found, they are all linked into complete straight boundaries, and spurious evidence is filtered out. Second, there is a very limited number of circumstances in which objects do consist strictly of planar faces. It is perhaps possible that industrial applications exist where both conditions might be met by constraining the objects, and providing lighting of a quality that permits full segmentation.

The idea was pioneered some time ago by Roberts [Roberts 65], who made significant progress, especially considering the time at which the work was done. Independently, two other researchers built on these ideas to develop what is now a very well known **line labeling** algorithm [Clowes 71, Huffman 71]. Mindful of the limitations of the blocks world approach, research into 3D vision has left these ideas behind and they are now largely of historical interest only. What follows is only an overview of how line labeling works, but it is instructive in that first, it illustrates how the 3D reconstruction task may be naively approached, and second, it is good example of **constraint propagation** (see Chapter 8) in action. The algorithm rests on observing that, since each 3D vertex is a meeting of exactly three planar faces, only four types of junction may appear in any 2D scene (see Figure 10.9). In the 3D world, an edge may be concave or convex, and in its 2D projection the three faces meeting at a vertex may be visible or occluded. These finite possibilities permit an exhaustive listing of interpretations of a 2D vertex as a 3D vertex—there are in fact 22 of them [Clowes 71].

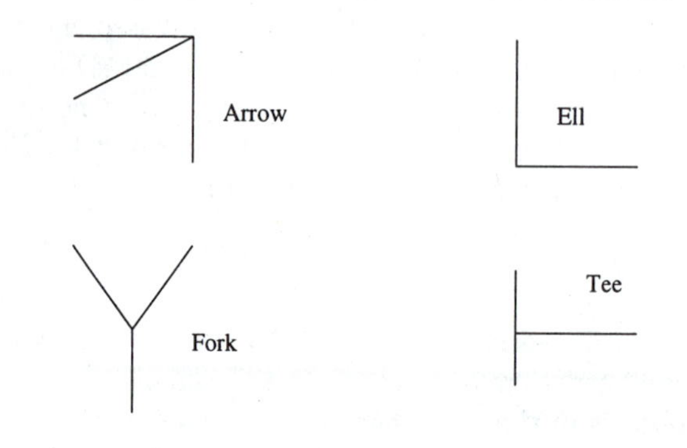

Figure 10.9: *The four possible 2D junctions.*

The problem now reduces to deriving a mutually consistent set of vertex labels; this may be done by employing constraints such as an edge interpretation (convex or concave) being the same at both ends, and that circumnavigating a region provides a coherent 3D surface interpretation. At a high level, the algorithm is as follows.

Algorithm 10.1: Line labeling

1. Extract a complete and accurate segmentation of the 2D scene projection into polygons.
2. Determine the set of possible 3D interpretations for each 2D vertex from a pre-computed exhaustive list.
3. Determine 'edge-wise' coherent labelings of vertices by enforcing either concave or convex interpretations to each end of an edge.
4. Deduce an overall interpretation by requiring a circumnavigation of a region to have a coherent 3D interpretation.

Line labeling is able to detect 'impossible' objects such as that shown in Figure 10.10a, since it would not pass the final stage of the informally described algorithm; it would not, however, register Figure 10.10b, which defies a 3D interpretation along its upper front horizontal edge, as impossible. It is also unable, in the simple form described, to cope with 'accidental' junctions which are the meeting of four or more lines (caused by chance occlusion), although these could be analyzed as special cases.

This simple approach received a lot of attention, and was extended to consider solids whose vertices may have more than three faces meeting at a vertex (such as square-based pyramids), and scenes in which regions might represent shadows of solids [Waltz 75]. Interestingly, while the number of possible junction interpretations increases enormously, the constraint satisfaction that is required for admissible candidate interpretations prevents the resulting algorithms becoming unworkable. In general, however, line labeling exists as an interesting historical idea—the way perhaps that one might approach the problem of 3D vision as a

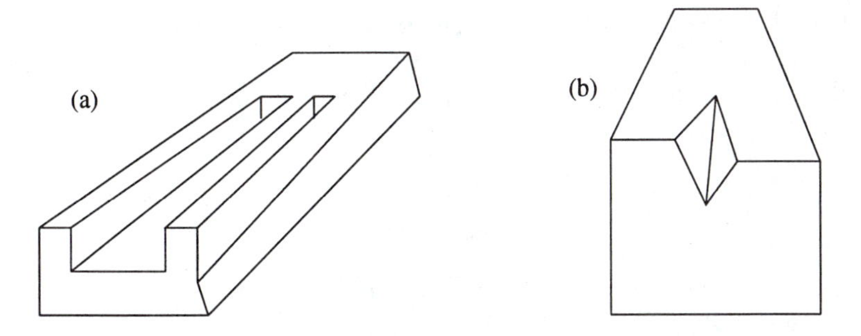

Figure 10.10: *Impossible blocks world figures.*

first attempt. It is clear, though, that its results are limited, and fraught with problems in overcoming 'special cases'.

More recent attempts at line label interpretations may be found in [Sugihara 86, Malik and Maydan 89, Shomar and Young 94].

10.2.3 Volumetric representation, direct measurements

An object is placed in some reference co-ordinate system and its volume is sub-divided into small volume elements called **voxels**—it is usual for these to be cubes. The most straightforward representation of voxel-based volumetric models is the **3D occupancy grid**, which is implemented as a 3D Boolean array. Each voxel is indexed by its x, y, z co-ordinates; if the object is present in a particular space location the voxel has value 1, and otherwise 0. Creating such a voxel-based model is an instance of discretization with similar rules to those for 2D images; for example, the Shannon sampling theorem (see Section 2.2.1) applies. An example of a voxelized toroid is shown in Figure 10.11. One way of obtaining a voxel-based volumetric model is to synthesize it using a geometric modeler, i.e., a computer graphics program. This permits composite objects to be assembled from some number of basic solids such as cubes, cylinders, and spheres.

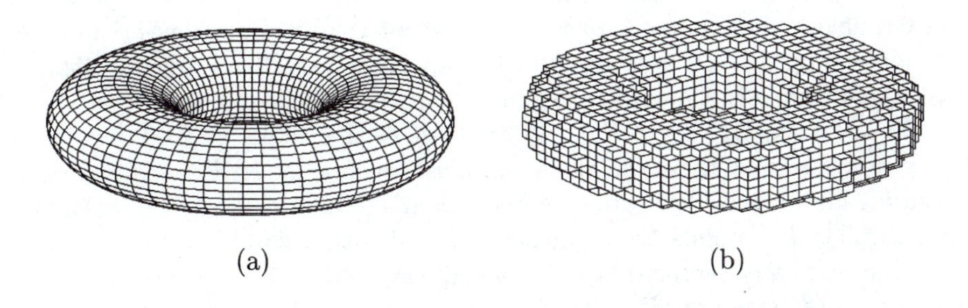

Figure 10.11: *Voxelization (discretization) in 3D: (a) continuous surface; (b) voxelized image consisting of cubes of the same size.*

Another possibility is the case in which a volumetric model is to be created from an existing real object. A simple measuring technique has been used in mechanical engineering for a long time. The object is fixed to a **measuring machine**, and an absolute co-ordinate system is attached to it. Points on the object surface are touched by a **measuring needle** which provides 3D co-ordinates; see Figure 10.12. The precision depends on the machine and the size of the objects, typically ±5 micrometers.

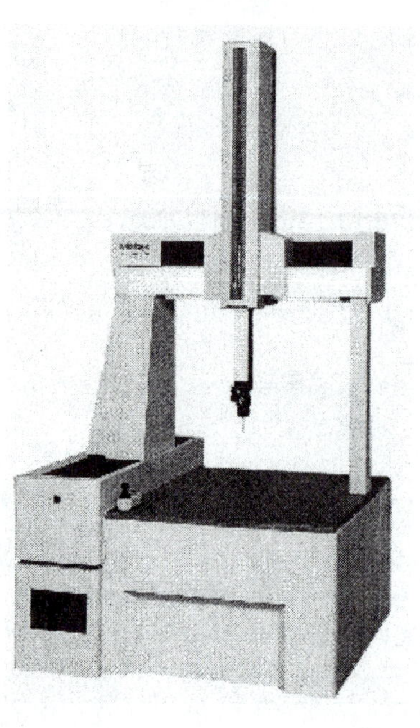

Figure 10.12: *An example of a fully motorized 3D measuring machine. Manufacturer, Mitutoyo, Inc., Japan.*

In simpler machines, navigation of the needle on the surface is performed by a human operator; x, y, z co-ordinates are recorded automatically. Such a surface representation can easily be converted it into volumetric representation.

Besides precision testing in machining or other mechanical engineering applications, this measuring technology may be used if the object is first created from clay by a designer. If computer-aided design (CAD) is to be brought into play, the 3D co-ordinates of the object are needed. An example is the automotive industry, where a clay model of a car body may be created at the scale 1:1. Actually only one half of the model is produced, as the car body is largely symmetric along the elongated axis. Such a model is measured on the 3D point measuring machine; as there are very many points to be measured, the probe navigates semi-automatically. The points are organized into strips that cover the whole surface, and the probe has a proximity sensor that automatically stops on the surface or near to it. The probe is either a needle equipped with a force sensor or a laser probe performing the same measurement but stopping at a fixed and precise distance from the surface, e.g., 3 millimeters.

Another 3D measurement technique, **computed tomography**, looks inside the object and thus yields more detailed information than the binary occupancy grid discussed so far. Tomography yields a mass density in a 2D planar slice of the object. If 3D volumetric information is required, such slices are stacked on top of one another. The resulting 3D sample space consists of voxels, the values of which are mass densities addressed by the x, y, x co-ordinates. Computed tomography is used widely in medical imaging (see Chapter 16).

10.2.4 Volumetric modeling strategies

Constructive Solid Geometry

The principal idea of Constructive Solid Geometry (CSG), which has found some success (notably with IBM's WINSOM [Quarendon 84]) is to construct 3D bodies from a selection of solid primitives. Popularly, these primitives are a cuboid, a cylinder, a sphere, a cone, and a 'half-space'—the cylinder and cone are considered to be infinite. They are scaled, positioned, and combined by union, intersection, and difference; thus a finite cone is formed by intersecting an infinite cone with an appropriately positioned half-space. A CSG model is stored as a

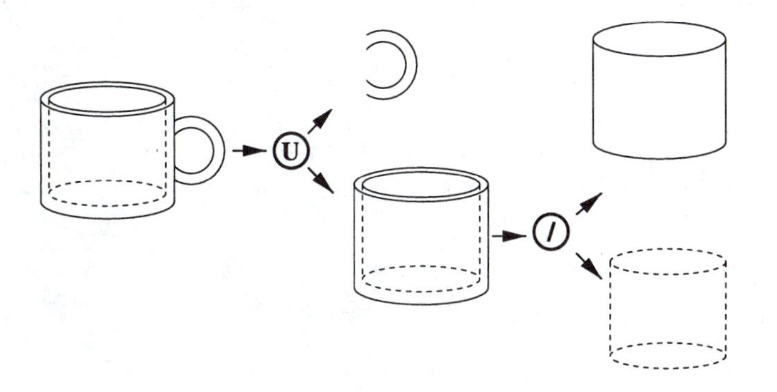

Figure 10.13: *CSG representation of a 3D object—a mug.*

tree, with leaf nodes representing the primitive solid and edges enforcing precedence among the set theoretical operations. The versatility of such a simply stated scheme is surprising. CSG models define properties such as object volume unambiguously, but suffer the drawback of being non-unique. For example, the solid illustrated in Figure 10.13—a mug—may be formed by the union of the cylinder with a hole and a handle. The cylinder with the hole is obtained from a full cylinder by subtracting (in the set sense) a smaller cylinder. Further, it is not easy to model 'natural' shapes (a human head, for instance) with CSG. A more serious drawback is that it is not straightforward to recover surfaces given a CSG description; such a procedure is computationally very expensive.

Super-quadrics

Super-quadrics are geometric bodies that can be understood as a generalization of basic quadric solids. They were introduced in computer graphics [Barr 81]. Super-ellipsoids are instances of super-quadrics used in computer vision.

The implicit equation for a super-ellipsoid is

$$\left[\left(\frac{x}{a_1}\right)^{(2/\varepsilon_{\mathrm{vert}})} + \left(\frac{y}{a_2}\right)^{(2/\varepsilon_{\mathrm{vert}})} \right]^{(\varepsilon_{\mathrm{hori}}/\varepsilon_{\mathrm{vert}})} + \left(\frac{z}{a_3}\right)^{(2/\varepsilon_{\mathrm{vert}})} = 1 \tag{10.23}$$

where a_1, a_2, and a_3 define the super-quadric size in the x, y, and z directions, respectively. $\varepsilon_{\text{vert}}$ is the squareness parameter in the latitude plane and $\varepsilon_{\text{hori}}$ is the squareness parameter in the longitude plane. The squareness values used in respective planes are 0 (i.e., square) $\leq \varepsilon \leq 2$ (i.e., deltoid), as only those are convex bodies. If squareness parameters are greater than 2, the body changes to a cross-like shape. Figure 10.14 illustrates how squareness parameters influence super-ellipsoid shape.

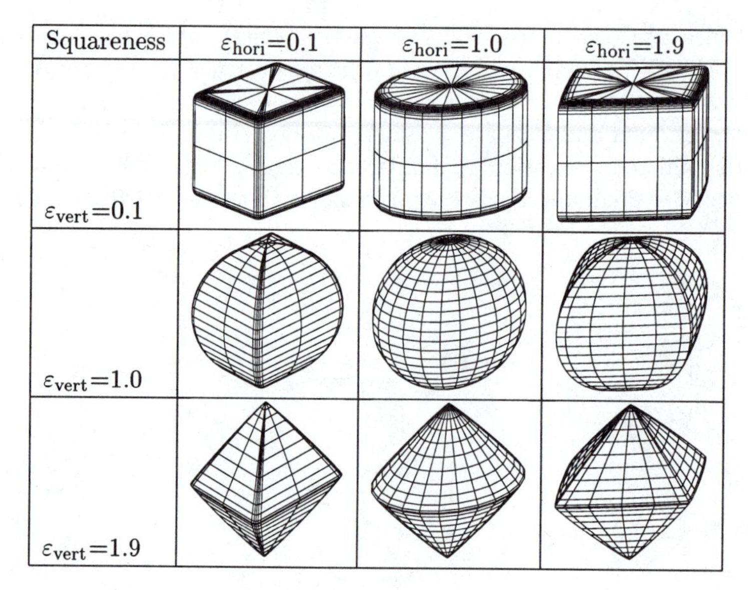

Figure 10.14: *Super-ellipses.*

Super-quadric fitting to range images is described in [Solina and Bajcsy 90, Leonardis et al. 97], and the construction of full 3D model range images taken from several views using super-quadrics is shown in [Jaklič 97]. Super-quadric volumetric primitives can be deformed by bending, twisting, and tapering, and Boolean combinations of simple entities can be used to represent more complicated shapes [Terzopoulos and Metaxas 91].

Generalized cylinders

Generalized cylinders, or **generalized cones**, are often also called **sweep representations**. Recall that a cylinder may be defined as the surface swept out by a circle whose center is traveling along a straight line (spine) normal to the circle's plane. We can generalize this idea in a number of ways—we may permit any closed curve to be 'pulled along' any line in 3-space. We may even permit the closed curve to adjust as it travels in accordance with some function, so a cone is defined by a circle whose radius changes linearly with distance traveled, moving along a straight line. Further, the closed curve section need not contain the spine. Usually it is assumed that the curve is perpendicular to the spine curve at each point. In some cases this constraint is released. Figure 10.15 illustrates two simple generalized cylinders.

These generalized cones turn out to be very good at representing some classes of solid body [Binford 71, Soroka and Bajcsy 78]. The advantage of symmetrical volumetric primitives, such

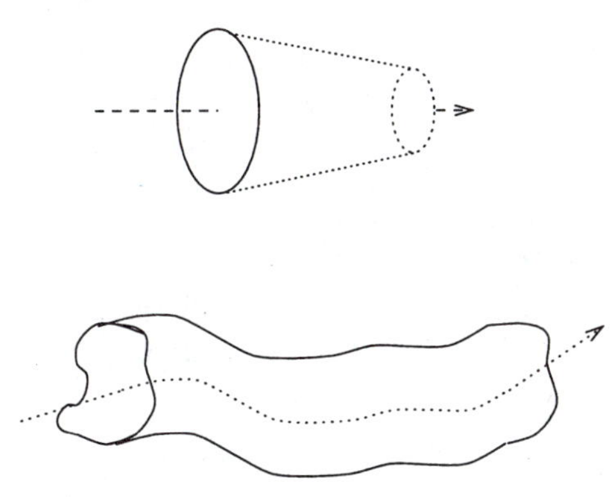

Figure 10.15: *Solids represented as generalized cylinders.*

as generalized cylinders and super-quadrics, is their ability to capture common symmetries and represent certain shapes with few parameters. They are, however, ill-suited for modeling many natural objects that do not have the set of regularities incorporated into the primitives. A well-known vision system called ACRONYM [Brooks et al. 79] used generalized cones as its modeling scheme.

There is a modification of the sweep representation called a **skeleton representation**, which stores only the spines of the objects [Besl and Jain 85].

10.2.5 Surface modeling strategies

A solid object can be represented by surfaces bounding it; such a description can vary from simple triangular patches to visually appealing structures such as non-uniform rational B-splines (NURBS) popular in geometric modeling. Computer vision solves two main problems with surfaces: First, reconstruction creates surface description from sparse depth measurements that are typically corrupted by outliers; second, segmentation aims to classify surface or surface patches into surface types.

Boundary representations (B-reps) can be viewed conceptually as a triple:

- A set of surfaces of the object
- A set of space curves representing intersections between the surfaces
- A graph describing the surface connectivity

B-reps are an appealing and intuitively natural way of representing 3D bodies in that they consist of an explicit list of the bodies' faces. In the simplest case, 'faces' are taken to be planar, so bodies are always **polyhedral**, and we are dealing the whole time with piecewise planar surfaces. A useful side effect of this scheme is that properties such as surface area and solid volume are well defined. The simplest B-rep scheme would model everything with the simplest possible 2D polygon, the triangle. By taking small enough primitives quite satisfactory representations of complex objects can be achieved, and it is an obvious generalization to consider polygons with more edges than three.

Triangulation of irregular data points (e.g., a 3D point cloud obtained from a range scanner) is an example of an interpolation method. The best-known technique is called **Delaunay triangulation**, which can be defined in two, three, or more space dimensions. Delaunay triangulation is dual to the Voronoi diagram. We assume that the Euclidean distance between data points is known; then points that are closer to each other than to other points are connected. Let $d(P,Q)$ be the Euclidean distance between points P and Q, and S be the set of points $S=\{M_1,\ldots,M_n\}$. A Voronoi diagram on the set S is a set of convex polyhedra that covers the whole space. The polyhedron V_i consists of all points that are closer to the point M_i than to other points of S.

$$V_i = \{p;\ d(p,M_i) \le d(p,M_j) \quad \text{for all } j = 1,2,\ldots,n\} \tag{10.24}$$

An algorithm to compute Delaunay triangulation can be found in [Preparata and Shamos 85]. A problem with Delaunay triangulation is that it triangulates the convex hull of the point set; constrained Delaunay triangulation [Faugeras 93] can be a solution.

We shall illustrate the idea of Delaunay triangulation on the simplest case of a 2D planar point set (see Figure 10.16). The task is to find triangles that cover all data points in such a way that the circumcircle of any one triangle contains only the three points that are vertices of that particular triangle. The triangulation has the following properties:

- The boundary of the set of points covered by triangles corresponds to the convex hull of the point set.
- The incremental algorithm constructing a triangulation of N points has expected time complexity $\mathcal{O}(N \log N)$ [Gubais et al. 92].
- The 2D Delaunay triangulation algorithm provides a unique solution if no more than three points lie on one circle.

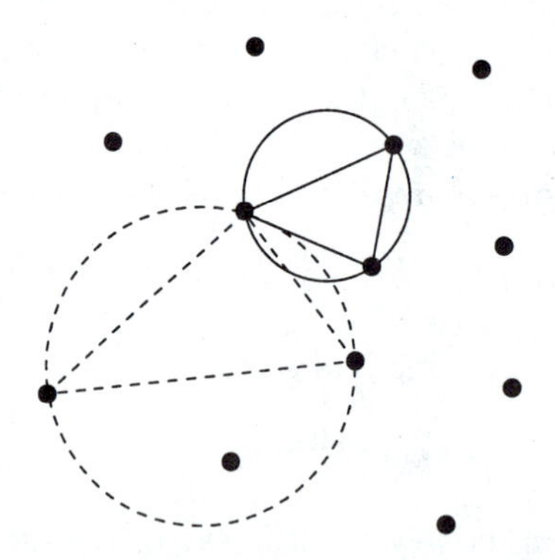

Figure 10.16: *2D Delaunay triangulation. The solid triangle belongs to the Delaunay triangulation but the dotted one does not, as its circumcircle contains an additional point.*

A drawback with polyhedral or triangulated B-reps is that the concept of 'face' may not be well defined. A face should have no 'dangling' edges, and the union of a body's faces should be its boundary. Unfortunately, the real world is not cooperative, and many (simple) bodies exist in which face boundaries are not well defined.

The next step in generalizing descriptions of surface patches is the **quadric surface model**. Quadric surfaces are defined using second-degree polynomials in three co-ordinates x, y, z. In implicit form, the equation has up to 10 coefficients and represents hyperboloids, ellipsoids, paraboloids, and cylinders.

$$\sum_{i,j,k=0\ldots2} a_{ijk}\, x^i\, y^j\, z^k = 0 \tag{10.25}$$

More complicated objects may be created from quadric surface patches. *Parametric bi-cubic surfaces* defined by bivariate cubic polynomials are used in CAD systems; the commonly used Bézier surfaces fall into this category. These surfaces have the advantage that surface patches can be smoothly joined along the intersection curves, and undesirable curvature discontinuity artifacts are thus avoided. Such an approach permits much greater flexibility in the description, but it becomes important to restrict the number of possible face edges in order to limit the complexity of the computations involved.

An independent discipline called **geometric modeling** considers object representation issues from the designer's point of view.

10.2.6 Registering surface patches and their fusion to get a full 3D model

A **range image** represents distance measurements from an observer to an object; it yields a partial 3D description of the surface from one view only. It may be visualized as a relief made by a sculptor—shape information from different views, e.g., from the other side of the object, is not available. Techniques for range image acquisition have been mentioned in Section 9.2.12.

Several range images are needed to capture the whole surface of an object. Each image yields a point cloud in the co-ordinates related to the range sensor, and successive images are taken in such a way that neighboring views overlap slightly, providing information for later fusion of partial range measurements into one global, object-centered, co-ordinate system.

A fusion of partial surface descriptions into global, object-centered co-ordinates implies known geometric transformations between the object and the sensor. The process depends on the data representing one view, e.g., from simple point clouds, triangulated surfaces, to parametric models as quadric patches.

Range image registration finds a rigid geometric transformation between two range images of the same object captured from two different viewpoints. The recovery can be based either on explicit knowledge of sensor positions, e.g., if it is held in a precise robot arm, or on geometric features measured from the overlapped parts of range data. Typically, both sources of information are used; an initial estimate of the appropriate geometric transformation can be provided by image feature correspondence, range image sensor data, an object manipulation device, or in many cases by a human operator.

This 3D model reconstruction task has been approached by several research groups in recent years, and many partial solutions have been proposed, e.g., [Hoppe et al. 92, Higuchi et al. 95, Uray 97]. We present here one of the possible approaches to the task. The method automates the construction of a 3D model of a 3D free-form object from a set of range images as follows.

1. The object is placed on a turntable and a **set of range images from different viewpoints is measured** by a structured-light (laser-plane) range finder.

2. A triangulated surface is constructed over the range images.

3. Large data sets are reduced by **decimation** of triangular meshes in each view.

4. Surfaces are registered into a common object-centered co-ordinate system and outliers in measurements are removed.

5. A full 3D model of the object is reconstructed by a surface fusion process.

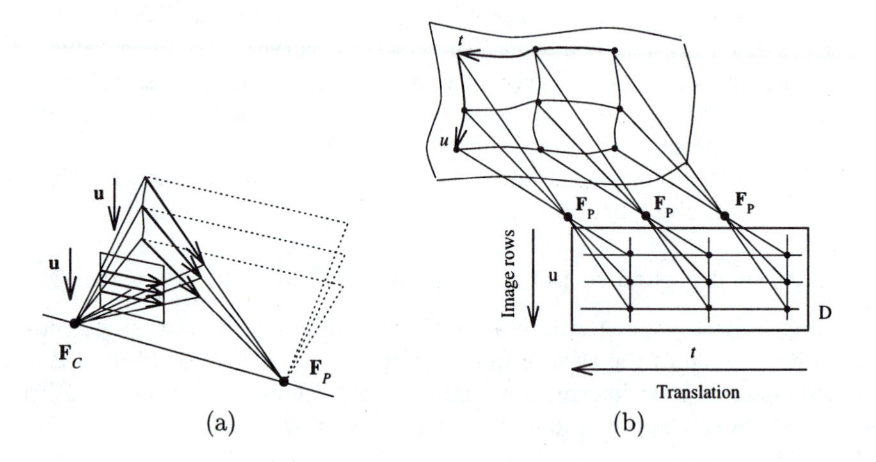

Figure 10.17: *Surface parametrization is composed from the projected laser ray and translation.*

Measurement from a laser-plane range finder provides a natural connectivity relation of points along stripes. Their parameterization allows easy construction of a 4-connected mesh following parametric curves just by connecting points found in the neighboring rows of the image and the points in neighboring scans with the same image row co-ordinate. The parameterization obtained of the measured surface is shown in Figure 10.17. The assumption of surface continuity is implemented as a restriction on the distance of neighboring points; only neighbors closer than a pre-defined ε are considered to lie on one surface next to each other. Points with no close neighbors are assumed to be outliers and are removed from data.

A 4-connected mesh cannot represent all objects; e.g., a sphere cannot be covered by a four-sided polygon. By splitting each polygon by an edge, a *triangulation* of the surface, which is able to represent any surface, is easily obtained. A polygon may be split two ways; it is preferable to choose the shortest edge because this results in triangles with larger inner angles.

Often, we wish to reduce the number of triangles representing the surface in areas where curvature is low [Soucy and Laurendeau 96]. Data reduction is particularly useful for the registration of neighboring views, since it has worst-case complexity $O(N^2)$ in the number of points. We formulate the task as a search for the best approximation of a triangulated surface by another triangulated surface that is close to the vertices of the original mesh [Hoppe et al. 92]. For instance, we might look for the closest triangulated surface with maximally n

triangles, or we might want simultaneously to minimize n and a residual error to get a consensus between the precision and space costs using the minimum description length principle (MDL) [Rissanen 89]. The surface triangulation procedure and node decimation is demonstrated on the synthetic pattern in Figures 10.18 and 10.19. Decimation of a triangulated surface from a real range image is shown in Figure 10.20.

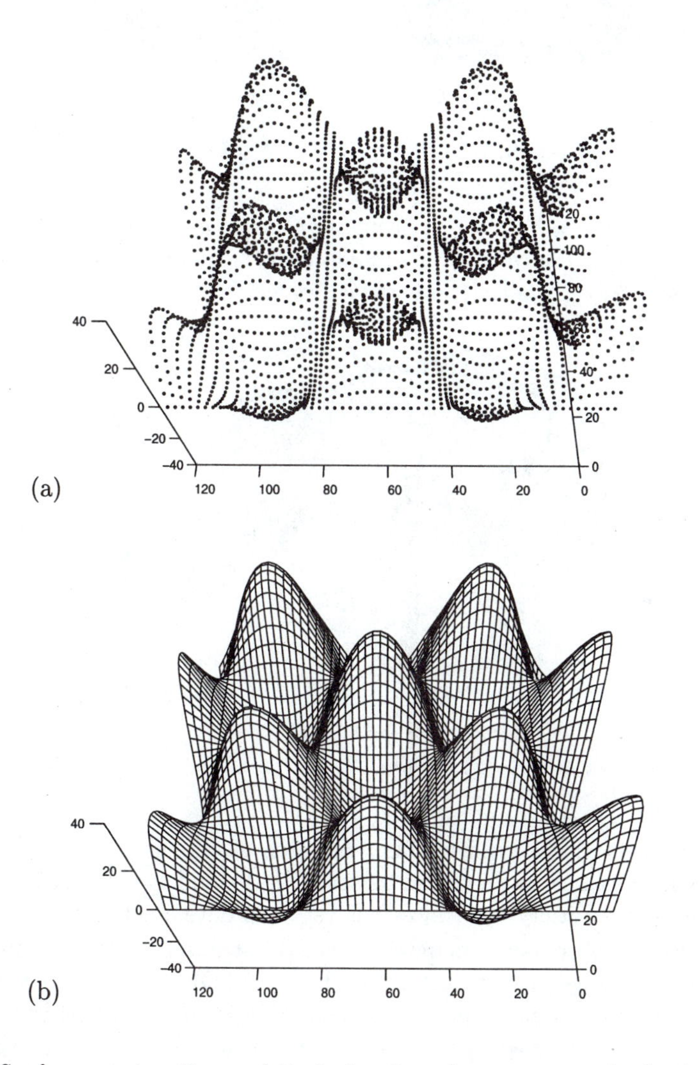

Figure 10.18: *Surface construction and its decimation shown on a synthetic sinusoidal pattern: (a) point cloud; (b) 4-connected mesh. Courtesy T. Pajdla, D. Večerka, Czech Technical University, Prague.*

Integration of partial shape descriptions attempts to get known geometric transformations among the views in order to register them and express them in a common co-ordinate system. Precise alignment of the data can be done automatically by gradient minimization, provided

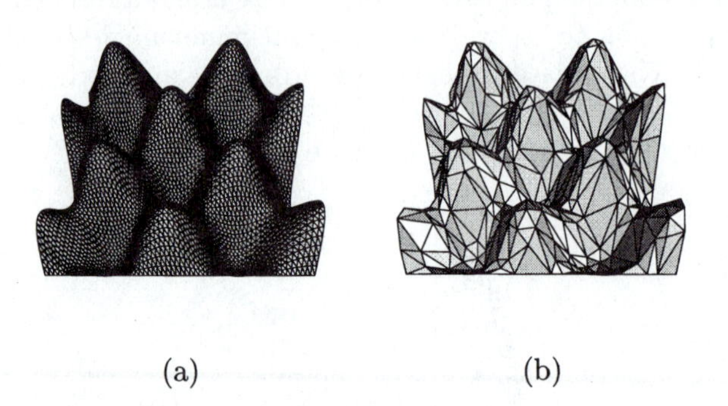

Figure 10.19: *Surface construction and its decimation shown on a synthetic sinusoidal pattern; (a) triangulated surface; (b) surface after decimation of a number of triangles. Courtesy T. Pajdla, D. Večerka, Czech Technical University, Prague.*

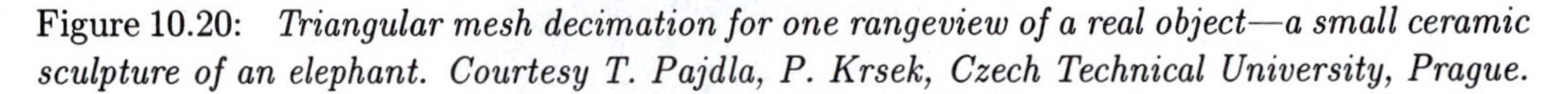

Figure 10.20: *Triangular mesh decimation for one rangeview of a real object—a small ceramic sculpture of an elephant. Courtesy T. Pajdla, P. Krsek, Czech Technical University, Prague.*

good starting transformations are available. In some cases, matching based on invariant features detected on the visual surfaces can be used [Pajdla and Van Gool 95], but no method able to cope with a large variety of surfaces has yet been developed.

Figure 10.21 shows approximate manual surface registration, with the help of user interaction. The mutual position of two surfaces is defined by aligning three pairs of matching points; the user selects a few point pairs (the minimum is three) on the surfaces. The approximate registration is obtained by moving one of the surfaces so that the sum of squared distances between the matching point pairs is minimal. An interactive program, Geomview (authored by the Geometry Center, University of Minnesota), for 3D surface viewing and manipulation, was used to let the user do the registration. Figures 10.22–10.24 illustrate this procedure.

When two partially overlapping surface patches are roughly registered, automatic refinement of the registration follows. It is assumed that two partially overlapping surfaces P and X related by a global transformation are available. In our case, all transformations are

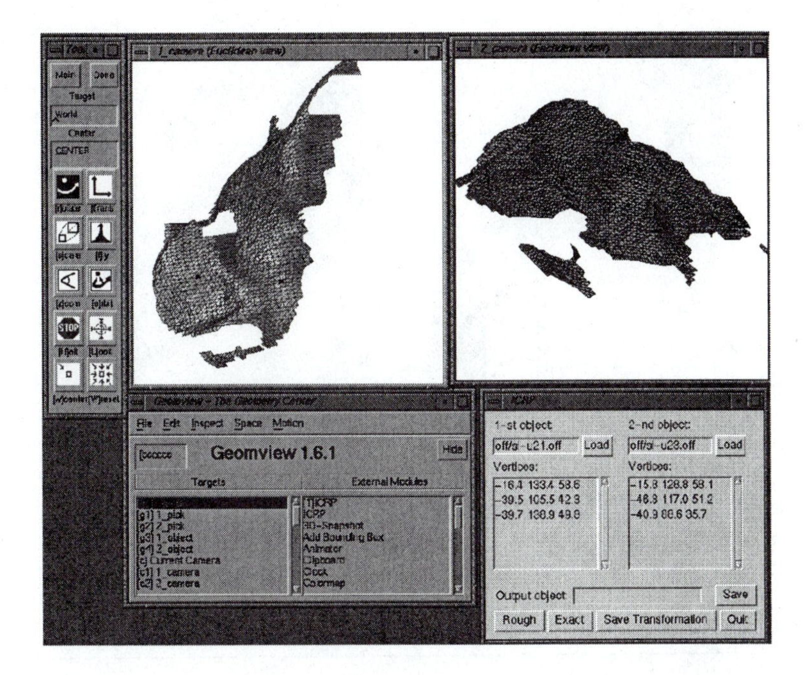

Figure 10.21: *Manual registration of surfaces in Geomview.*

subgroups of a projective group in $\mathcal{P}^3$. Surface registration looks for the best Euclidean transformation T that overlays P and X. T is found by minimization of

$$e = \min_T \rho[P, T(X)] \tag{10.26}$$

where ρ is a cost function evaluating the quality of match of two surfaces. In Euclidean geometry, it might be the distance between the points on a surface.

The **iterative closest point algorithm** (ICP) developed by Besl and McKay [Besl and McKay 92] solves the registration automatically provided a good initial estimate of T is available. The algorithm assumes that one of the surfaces is a subset of the second, meaning that only one surface can contain points without correspondence to the second surface. The ICP is an iterative optimization procedure that looks for the geometric transformation of one surface best to match the second. It is likely that the cost function will be non-convex, and so there is a consequent danger of falling into local minima—thus a good initial estimate is needed.

We present here a modification of the ICP algorithm which is able to register partial corresponding surfaces. This approach uses the idea of reciprocal points [Pajdla and Van Gool 95] to eliminate points without correspondence. Assume point $\mathbf{p}$ is on the surface P and that y is the closest point on the surface X. The closest point on the surface P to $\mathbf{y}$ is the point $\mathbf{r}$ (see Figure 10.25). Points $\mathbf{p}$, satisfying the condition that the distance is less then ϵ, are called ϵ-reciprocal—only these points are registered. Let P_ϵ denote the set of ϵ-reciprocal points on the surface P; then the iterative reciproal control point algorithm (ICRP) algorithm is as follows.

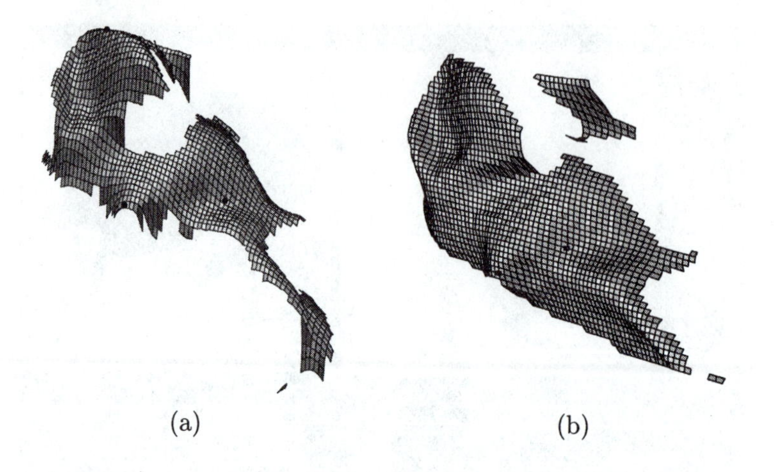

Figure 10.22: *The process of registration: (a) three points on the first surface; (b) three points on the second surface; Courtesy T. Pajdla, D. Večerka, Czech Technical University, Prague.*

Algorithm 10.2: Iterative closest reciprocal points

1. Initialize $k = 0$ and $P_0 = P$.
2. Find closest points Y_k for P_k and X .
3. Find reciprocal points $P_{\epsilon 0}$ and $Y_{\epsilon k}$.
4. Compute the mean square distance d_k between $P_{\epsilon k}$ and $Y_{\epsilon k}$.
5. Compute the transformation T between $Y_{\epsilon k}$ and $P_{\epsilon 0}$ in the least-squares sense.
6. Apply the transformation T: $P_{k+1} = T(P_0)$.
7. Compute the mean square distance $d_{k'}$ between $P_{\epsilon k+1}$ and $Y_{\epsilon k}$.
8. Terminate if the difference $d_k - d_{k'}$ is below a preset threshold or if the maximal number of iterations is exceeded; otherwise go to step 2.

When visual surfaces are properly registered, surface *integration* follows. All partial measurements will have been registered and can be expressed in one object-centered co-ordinate system, and constitute a global point. A problem is that the 3D object representation was created from overlapping surface patches corresponding to several views; these patches were integrated following one traversal around the object. All measurements are corrupted by some noise, and this noise is propagated from one surface patch to the other. The important issue is to have approximately the same error when joining the first and last patches. To ensure that the global error is minimal, registered surface points should be rearranged during the surface integration to maintain global consistency.

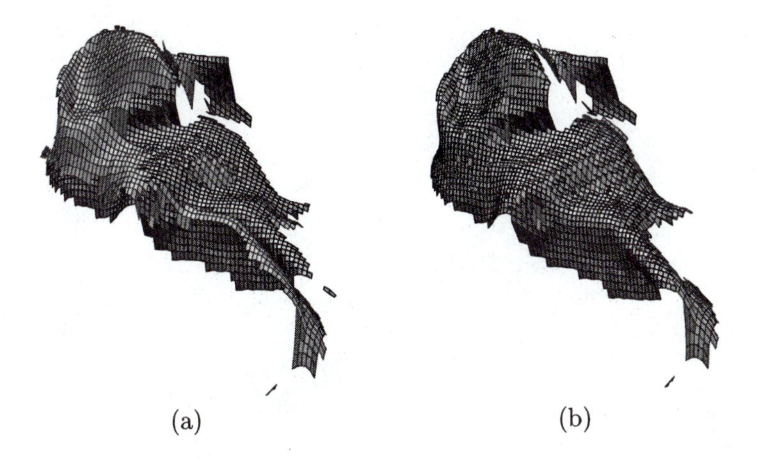

Figure 10.23: *The process of registration: (a) surfaces after rough registration; (b) after exact registration; Courtesy T. Pajdla, D. Večerka, Czech Technical University, Prague.*

The next task is to create an analytic shape description of the object's surface by approximation, using some implicit or explicit formulae; commonly, this is not done for the whole object but just for parts of it. A recent promising [Sara and Bajcsy 98] uses a multi-scale local noise model. An uncertainty ellipsoid (called a fish scale) is used to integrate local information of the point cloud on the surface into a global shape description. The fish scales can be created at multiple resolutions, and their overlap and consistency is explored when creating a 3D shape model.

10.3 3D model-based vision

10.3.1 General considerations

The recognition approach to 3D vision aims at successful recognition of real-world objects using standard pattern recognition approaches. The tremendous complexity of understanding the 3D scene which is needed for 3D object reconstruction is avoided. The key issue is foreknowledge about specific objects under inspection—the classes of objects in which we are interested are represented by **partial models**.

Object models contain more information than sensor data extracted from one view, and it is therefore not possible to transform sensor data into a complete model data representation. Fortunately, sensor data can be matched with partial model data. It is useful to work with an intermediate representation scheme computable both from sensor and model data which permits the reduction of the large amount of sensor data. A matching between an object and the model is then carried out at this intermediate representation level. The best matching occurs when the hypothetical object model configuration represents the scene world as accurately as possible. Accuracy might then be measured quantitatively by **matching errors**. This difference might be used to control the whole recognition process by closing a feedback loop.

A selection of lower-level processing techniques will generate some selection of clues and

Figure 10.24: *The process of registration: rendered result. Courtesy T. Pajdla, D. Večerka, Czech Technical University, Prague.*

features, and from some model base there will be a selection of objects which may or may not match the observations. Most model-based recognition systems are domain oriented and do not provide a general solution. This approach is particularly useful for industrial and other applications.

A typical object recognition system operates in four phases:

- Data acquisition
- Feature detection
- Hypothesis generation
- Hypothesis verification

Earlier chapters have discussed several approaches to **matching**, such as hypothesize and verify, efficient pruning of interpretation trees, graph (sub-graph) matching, properly indexed database, etc. In Section 10.2.2 the line labeling approach to the recapture of 3D was outlined, and it was explained why in general this solution is inadequate. **Verification** can be approached in two ways:

- At a data level, where synthetic data is created from a hypothesized object, and its position and orientation are compared with the actual input measurements
- At a feature level, where some metric is used to compare features from a hypothesized and a real object.

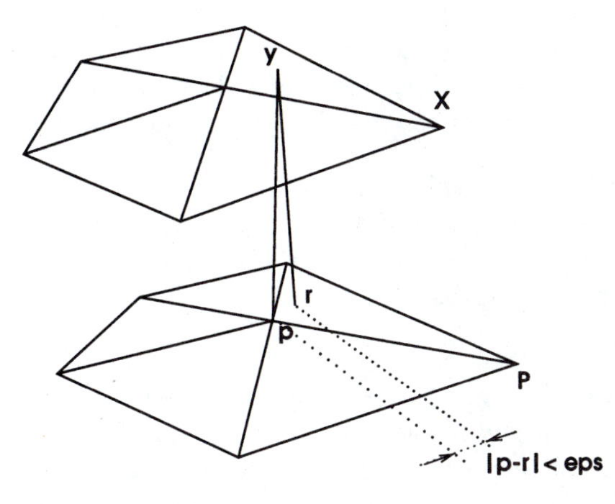

Figure 10.25: *The notion of the closest point of the surface for introducing reciprocal points into the ICP algorithm.*

Here we shall make some general remarks about 3D object recognition, and by way of example present in detail for didactic a well-known algorithm that solves the problem for the simple case of a well-understood object with planar, straight-edged faces by replacing the 3D problem by many 2D ones [Goad 86].

10.3.2 Goad's algorithm

A particular model matching algorithm due to Goad [Goad 86] has received a lot of attention. Goad's algorithm is interesting because it is simple enough in principle to show how the task may be done, but generates enough complexities in implementation to illustrate how difficult the task of model matching, even in a relatively simple case, can become.

Goad's algorithm aims to recover the 3D co-ordinates and orientation of a known polyhedral object from a single intensity image. The object is 'known' in the sense that its edges, which are assumed to be straight, and their relative position to each other, are exhaustively listed in the object model with which the algorithm is provided. This contrasts with the easier problems of locating a known object of known orientation (a common aim in industrial vision systems), and the harder one of locating imprecisely defined objects of unknown orientation (a more widely applicable, and elusive, solution).

Following the terminology of the original reference, an object edge will refer to a straight line segment in 3D that forms part of the boundary of an object face. Projections of these into 2D (the image) are referred to as lines. The algorithm proceeds on a number of assumptions.

1. It is assumed that an edge and line detector have done their work and the lines (straight boundaries) in the image have been extracted. The algorithm permits these extracted lines to be imprecise within certain bounds, and, in its full form, is able to make allowances for spurious and missing evidence (that is, lines where there should be none and no lines where there ought to be).

2. The object to be located is either fully in the field of view, or not visible at all.

3. The distance to the object is known; this permits the further assumption that the camera lies at some point on a sphere centered at the origin of an object-based co-ordinate system. Without loss of generality, we assume this is a unit sphere.

4. The field of view is sufficiently narrow to permit the assumption that changing the **orientation** of the camera at a given position only causes the features in view to undergo a simple rotation and translation. While such a change in orientation may affect which features are visible, the lengths of lines in view will not alter, within a small tolerance.

The general strategy of the algorithm is intuitively simple—an edge of the object is taken and a likely match for it is found in the image. This possible match will not constrain the position of the object completely—there will be a range (a locus) of camera positions that is consistent with what is observed. Now we select another edge from the model; the restrictions provided by the (putatively) matched edge will limit the possible position of the projection of this new edge into the image; we may thus predict the position of this edge in the image, within bounds governed by the accuracy of measurements and line finders. If this projected edge cannot be located, the supposed match is false. If it can, it may be expected to restrict further the locus of possible camera positions that is consistent with all hitherto deduced possible camera positions—the **observation** is used to **back-project** and thereby reduce the possible locus. Another edge is now selected and the procedure repeated iteratively until either the match fails or there is strong enough evidence (sufficient matched edges, and a restricted enough locus) to deduce that the match is correct and thereby specify the object's location and orientation. Early in the matching process, with few object edges matched, the bounds on the prediction may be very wide; as the match proceeds, the predictions become more precise. When the match fails, we may back-track to a point where more than one image edge was a possible match for an object edge and try again.

This 'predict-observe-back project' cycle is a simple instance of an elementary matching algorithm—sequential matching with back-tracking [Hayes-Roth et al. 83], and is a typical example of a top-down, hypothesize-and-verify approach.

Some notation and definitions are necessary to proceed. Remember we are working in object-centered co-ordinates, so the object is regarded as fixed while the camera position and orientation are the unknowns to be determined. Throughout, an object edge will be regarded as an **oriented** line segment, given by an ordered pair of co-ordinates, while an image line may be unoriented. Thus an object edge is an **ordered** pair of (3D) co-ordinates, while an image line is given by a (perhaps unordered) pair of (2D) co-ordinates.

Let $\mathbf{p}$ be a 3D (camera) position, and $\mathbf{q}$ a 3D (camera) orientation. $\mathbf{p}$ is just a 3D co-ordinate, which we are constraining to lie on the unit sphere. If e is an object edge, let $P([\mathbf{p},\mathbf{q}],e)$ denote the **oriented** image line which results from viewing e from $\mathbf{p}$ at orientation $\mathbf{q}$, using a perspective transformation. $P([\mathbf{p},\mathbf{q}],e)$ may be undefined if e is occluded, or the projection is outside the field of view. If it is defined, it is an **ordered** pair of 2D co-ordinates.

The possible positions on the surface of the unit sphere are quantized—a set of such positions will be referred to as a **locus**. A given edge e will only be visible from some of these positions, which is referred to as the **visibility locus** of e.

An assignment between object edges and image lines will be called a **match** M. For an object edge e, $M(e)$ will denote the oriented image line assigned to it by M; for some e, $M(e)$

may be undefined, so a match need not be a complete assignment of object edges. Then a match M is **consistent** with a camera position and orientation $[\mathbf{p}, \mathbf{q}]$ if for each object edge e we have $P([\mathbf{p}, \mathbf{q}], e) = M(e)$ to within errors of measurement. A match M is consistent with a camera position $\mathbf{p}$ if there is some orientation $\mathbf{q}$ such that M is consistent with $[\mathbf{p}, \mathbf{q}]$. A match is consistent with a locus L if it is consistent with every position of that locus. L is initialized from the assumptions deduced from the match of the first edge.

In overview, the algorithm is then as follows.

Algorithm 10.3: Goad's matching algorithm

1. Initialize by finding a plausible match in the image for one edge.
2. For the current match M and the current locus L, select an unmatched edge e.
3. By considering a matched edge e_0, compute bounds on the possible position of $P([\mathbf{p}, \mathbf{q}], e)$ relative to $P([\mathbf{p}, \mathbf{q}], e_0)$ as $\mathbf{p}$ ranges over L (this position depends only on $\mathbf{p}$ and not on $\mathbf{q}$ from assumption 4 above). Thus determine a range of possible positions for $M(e)$ in the image.
4. If a candidate for $M(e)$ is located, back-project—that is, restrict the locus L to L' by rejecting all points in L that are not consistent with the measured position of $M(e)$. L' is those points $\mathbf{p}$ in L from which the predicted position of $P([\mathbf{p}, \mathbf{q}], e)$ relative to $P([\mathbf{p}, \mathbf{q}], e_o)$ is the same as the position of $M(e)$ relative to $M(e_0)$ to within measurement error.
5. If more than one candidate for $M(e)$ is located, mark this as a choice point for future back-tracking.
6. If no candidate for $M(e)$ is located, back-track to the last choice point.
7. Iterate until the match is regarded as certain.

It is acknowledged that the image line detector is not going to be perfect. Some e, although expected to be in view, will not have a match $M(e)$ in the image. Two measures are maintained as the algorithm proceeds that gauge whether its oversights are 'reasonable', and whether it is likely to be 'complete'.

1. **Reliability**: The probability that the edges making up the match to date arose by chance from background information is calculated, and the inverse of this probability is called the reliability of the match. When the reliability exceeds a certain threshold, the match is regarded as correct and the algorithm terminated. These probabilities may best be computed on the basis of statistics gathered from images of the same class as that under examination.
2. **Plausibility**: Assuming the match is correct but has missed some edges, the probability that those edges would indeed have been missed by the line detector in use is calculated —these probabilities assume knowledge of the performance of the line detector which once again are best accumulated from running it on sample images.

Now, high reliability indicates that the match is correct, while low plausibility indicates that it is probably incorrect (although we must beware—high plausibility does not imply a correct match and low reliability does not imply that it is incorrect). Plausibility is introduced into the algorithm by requiring that if it falls below a certain threshold, we must back-track. In fact, this generates another possible choice point—if we assume e is visible, search for it and fail to find it, we may assume it should be visible but is absent from the image and proceed with reduced plausibility accordingly. Only if this assumption leads to no match do we back-track and consider whether the edge should be visible at all.

Edge **visibility** considers from which points of L an edge e may actually be seen (that is, whether the visibility locus V of e intersects L). This provides another possible choice point for back-tracking: First assume e is visible (that is, restrict L to its intersection with V) and proceed as described above. If a match is not found, and we need to back-track, we can assume that e is not visible and restrict L to its intersection with the complement of V. 'Visibility' needs to be defined with caution here—an edge is regarded as visible only if it is likely to be found by the line detector. Very short lines (object edges viewed nearly 'end on', for instance) would not meet this criterion.

A feature of the problem that this algorithm is designed to solve is that the object sought is modeled precisely. This fact may be exploited to speed up what would otherwise be at best a ponderous execution by going through a 'setup' phase during which the features of the object are coded into the algorithm to be exploited at run time. Goad refers to this as the 'compile time' of the algorithm.

There are several ways we may exploit this compile time;

1. From a given position $\mathbf{p}$, we require during the algorithm to determine bounds on the position of an edge e relative to a matched edge e_0. This relative position, $relpos(e, e_o, \mathbf{p})$ depends only on the object, which is fully characterized independent of the run. These relative position bounds may therefore be computed at compile time and stored for look-up. A complete, reliable, and plausible set of constraints on $relpos$ is proposed in [Bray and Hlavac 91].

   In fact we require $relpos(e, e_0, \mathbf{p})$ for all $\mathbf{p} \epsilon L$; this is easily done by taking the union of the bounds for each such $\mathbf{p}$. This table can also be used for the back-projection; given a likely $M(e)$, we need only determine for which $\mathbf{p} \epsilon L$ this $M(e)$ is within the bounds predicted by $relpos(e, e_0, \mathbf{p})$.

2. When selecting the next edge to match, care should be taken to maximize the return on the effort put into trying to match it. This may be ensured by maximizing the likelihood of the selected edge being visible (and findable by the line detector), and by requiring that the measurements of the image position of the observed edge should provide as much information as possible about the camera position (so the locus is reduced as much as possible by making a match).

   These judgments may be made at compile time. Supposing a uniform distribution of camera positions around the sphere (in fact, allowances can be made if this is an unreasonable assumption), then the probability of the visibility of a given edge over any locus can be pre-computed. Likewise, the 'value' of measuring the position of a given edge can be computed at compile time. If we determine a way of combining these factors by appropriately weighting the values determined (and this is not necessarily straightforward), the 'best next edge' to match, given a particular partial match, can be determined at compile time. Goad observes that this particular pre-computation may be very expensive.

3. The elemental contributions to the plausibility measurements can also be pre-computed.

There is no doubt that performing the compile time operations outlined will be very expensive, but this should not worry us since the expected runtime efficiency gain will make the effort well worth the cost. It is a familiar idea to pay the price of lengthy compilation in order to generate efficient running code.

Goad's algorithm is in principle quite simple—there is a variety of things we may expect to see, and they are sought in a 2D projection. Nevertheless, when examined with all its ramifications, it should be clear that the algorithm's implementation is not quite so simple.

When running, Goad managed to get respectable efficiency from his system. To deduce a complete match, it is really only necessary to make reliable matches for four object edges. He reports that on an 'average 1 MIPS machine' (remember this work is dated 1983), one matching step will take of the order of a few milliseconds, permitting several hundred attempts at matching every second. The runtime for a complete match is quoted at approximately one second, but this excludes edge and line detection. As has been remarked, much of this efficiency has been achieved at the expense of a very long 'compile time'.

The algorithm has actually been applied to several problems—single occurrences of objects such as a connecting rod or universal joint have been located in cluttered scenes, and, more interesting, key caps (the plastic keys from a keyboard or typewriter) have been located in an image of a pile of caps. In industrial terms this problem, often referred to as 'bin-picking', is unreasonably difficult—multiple occurrences of the target object at multiple orientations, many partially occluded (remember the first assumption above, that the object is visible either fully or not at all). The algorithm succeeds despite the fact that the background consists of features very similar to those composing the target, and that the target has few distinguishing features.

Goad's algorithm turns out to be quite powerful. The idea of 'pre-compiling' the object description (special-purpose automatic programming) produces systems of acceptable efficiency. Various elaborations exist which are not explored here, such as exploiting recurring patterns or symmetries in the object, or variable camera-to-object distance. Remember this object location is done by two-dimensional matching; that it works despite unknown orientation and is dependent on complete and thorough knowledge of the image and line detector properties, and the target object. Various elaborations on the ideas presented here have been developed [Lowe 85, Grimson 89, Bray and Hlavac 91]

10.3.3 Model-based recognition of curved objects from intensity images

Consider the case in which 3D curved objects are to be recognized using 2D intensity images—this is hard, as it is too ambiguous. Salient curves on the object surface are believed to be seen as identifiable curves in images. These are often silhouettes or curves corresponding to surface discontinuities as steps or crest lines. The dependence on illumination causes problems. Sometimes image curves due to shadows are more visible in the image than those due to features of interest.

The **curvature primal sketch** [Brady 84] is a systematic, domain-independent method permitting the description of curvilinear objects. Quantities derived from differential geometry, such as lines of curvature, asymptotes, bounding contours, surface intersections and planar surface patches, are used. Principal curvatures and principal directions are computed. An ad-hoc, breadth-first method is used to link the principal direction at each point into the line of curvature.

Typically, curvature is a primary feature, and there are several ways to compute it for a discrete curve [Worring and Smeulders 93]. Good success in curvature estimation based on smoothing in different scales by Gaussian filters [Lowe 89] has been observed. Such estimates are significantly biased as the smoothed curve shrinks; this shrink can be compensated for [Hlaváč et al. 94], and the resulting algorithm is easy to use. Mokhtarian [Mokhtarian 95] introduced the notion of curvature scale-space which permits the analysis and recognition of curves at multiple resolutions.

A systematic method exists which permits the use of theorems from continuous mathematics for digitized objects The basic idea is first to approximate the discrete curve by an analytic function in an explicit form; polynomials are often used for this. The differential characteristics are then computed using analytic formulae.

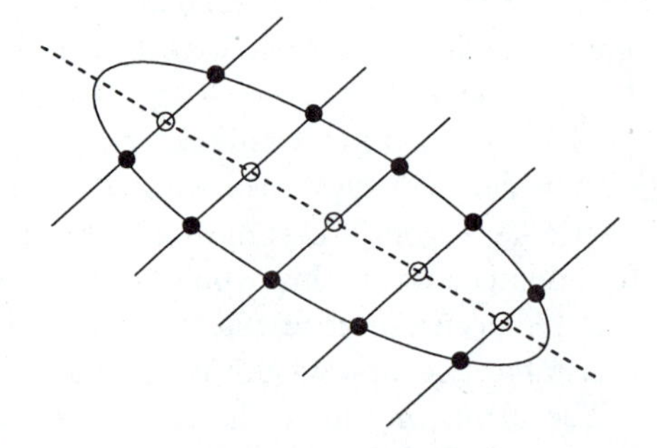

Figure 10.26: *Mid-points of parallel chords across a conic lie on a straight line*

Weiss' approximation to general curves by conic sections [Weiss 88, Weiss and Rosenfeld 96] is known to be robust; this is extendible to curved surfaces, where quadric patches are used. The approximation observes that mid-points of chords across a conic lie on a straight line (see Figure 10.26). Having a reliable approximation of the curve by conic sections, Weiss proceeds to use differential and algebraic **invariants** to perspective projection. These invariants provide features on which to base recognition of 3D models from 2D images—Weiss' invariants are based on derivatives up to the fourth order [Rivlin and Weiss 95]. The practical disadvantage of this method is that it does not work well for curves that are close to straight lines.

A very rich description may be based on properties of object surfaces; typically these will range images (Section 10.3.4). Here we present similar techniques suitable for intensity images. The term **surface features** (or characteristics) refers to descriptive features of a general smooth surface; then **surface characterization** refers to the process of partitioning a surface into regions with equivalent characteristics. A very similar process might be applied to an intensity image, and a function describing an intensity image has the same mathematical form as a surface, i.e., a function of two variables. What makes the problem difficult is the separation of illumination effects.

The symbolic scene description features should be invariant to translations and rotations of object surfaces. Differential geometry-based features are useful if they can be reliably computed from the sensor data, but this is not usually the case and surface (or curve) approximations should be adopted first.

One interesting possibility useful both for an intensity image and range images is **topographic characterization** of the surface. At each pixel in an image, a local facet-model bi-cubic polynomial surface is fitted to estimate the first three partial derivatives. The magnitude of the gradient, eigenvalues of a 2×2 Hessian matrix (the matrix of second derivatives) and directions of its eigenvectors are computed; these five numbers are then used to characterize the surface at the pixel. The following 10 labels are used [Besl and Jain 85]: peak, pit, ridge, ravine (valley), saddle, flat (planar), slope, convex hill, concave hill, and saddle hill. This is called a **topographic primal sketch**. The main advantage of these labels is the invariance to monotonic gray-level transformations such as change in brightness and contrast.

10.3.4 Model-based recognition based on range images

The recognition of simple polyhedral or piecewise quadric models has been attempted out by several researchers, e.g., [Chen and Kak 89, Newman et al. 93]; a useful review can be found in [Arman and Aggarwal 93].

We take a further step down the road of complexity and consider *free-form objects*. The common objects we meet everyday, such as the bust of a person, a car, a banana, etc., fall into this class. There are objects with which computer vision still cannot deal, such as flowers, bushes, or a hairy piece of cloth, due to complexity and problems of range data acquisition.

The traditional approach attempts to approximate a free-form surface by set of planes; for example, Faugeras and Hebert [Faugeras and Hebert 86] used planar surfaces to approximate models. Creases between planar facets and vertices were used as features for finding pairings between object and model.

The idea of the **extended Gaussian image (EGI)** has been proposed [Horn 84]; it is a distribution of surface normals on a Gaussian sphere. All surface normals are moved to the origin and positional information is ignored. A model of an 3D object is then a set of EGIs, each from a different viewing direction on the viewing sphere. The 3D recognition problem is reduced to a set of 2D problems similar to Goad's algorithm (Section 10.3.2). The principal limitation is that only convex objects can be uniquely represented using EGIs. If the connectivity information about the original surface normals is preserved [Liang and Todhunter 90], this convexity can be overcome.

Another similar surface representation is called the **simplex angle image** (SAI) [Higuchi et al. 95]; this stores various surface curvature measures in an generic representation. Starting with a pre-defined 3D mesh (e.g., an ellipsoid), the mesh is iteratively deformed until it fits the surface. The one-to-one relation between nodes of the initial and deformed meshes is preserved. Curvature features computed from the mesh are stored in the initial mesh and used to represent an object. Connectivity among mesh nodes is stored, and thus the representation copes with non-convex images.

10.4 2D view-based representations of a 3D scene

10.4.1 Viewing space

Most 3D objects or scenes representations discussed hitherto have been **object-centered**—another option is to use **viewer-centered** representations, where the set of possible appearances of a 3D object is stored as a collection of 2D images. The trouble is that there is potentially an infinite number of possible viewpoints that induce an infinite number of object appearances. To cope with the huge number of viewpoints and appearances it is necessary to sample a viewpoint space and group together similar neighboring views. The original motivation was the recognition of polyhedral objects, which was later generalized to view-based recognition of curved objects [Ullman and Basri 91]. More recent is the view-based representation of 3D scenes for display from any viewpoint [Beymer and Poggio 96].

Consider two models of the viewing space as a representation of possible views on the object or scene. The general model of the viewing space considers all points in a 3D space in which the 3D object is located at the origin. This viewpoint representation is needed if perspective projection is used. A simplified model is a **viewing sphere** model that is often used in the orthographic projection case. Then the object is enclosed by a unit sphere; a point on the sphere's surface gives a viewing direction. The surface can be densely discretized into view patches.

To simplify working with a viewing sphere it is often approximated by a regular polyhedron, of which the most common choice [Horn 86] is an icosahedron (with 20 equilateral triangular faces). Twenty viewing directions defined by the centers of the triangles are often not enough, in which case the faces are further regularly divided into four triangles in a recursive manner. This yields 80, 320, 1280, ... viewing directions.

10.4.2 Multi-view representations and aspect graphs

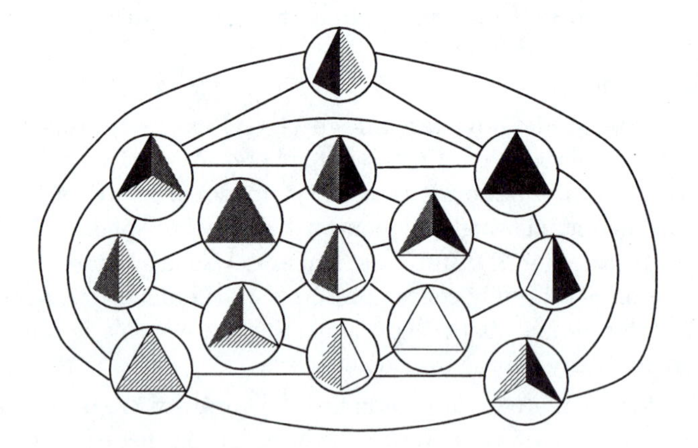

Figure 10.27: *Aspect graph for the tetrahedron.*

Other representation methods attempt to combine all the viewpoint-specific models into a single data structure. One of them is the **characteristic view technique** [Chakravarty and Freeman 82], in which all possible 2D projections of the convex polyhedral object are grouped into a finite number of topologically equivalent classes. Different views within an equivalence class are related via linear transformations. This general-purpose representation specifies the 3D structure of the object. A similar approach is based on **aspect** [Koenderink and van Doorn 79], which is defined as the topological structure of singularities in a single view of an object—aspect has useful invariance properties. Most small changes in vantage point will not affect aspect, and such vantage points (that is, most) are referred to as **stable**. Then the set of stable vantage points yielding a given aspect is contiguous in space, and the space surrounding an object may be partitioned into subspaces from within which the same aspect is generated. An example of the aspect graph for the simplest regular polyhedron—a tetrahedron—is shown in Figure 10.27. Moving from one such subspace into another generates an **event**—a change in aspect; the event is said to connect the two aspects. It is now possible to construct an **aspect graph** with respect to this connectivity relationship in which nodes are aspects and edges are events. This aspect graph is sometimes referred to as the **visual potential** of the object, and paths through it represent orbits of an observer around the object.

10.4.3 Geons as a 2D view-based structural representation

The psychologist Biederman influenced 3D computer vision by the recognition by components theory of human object recognition [Biederman 87]. He advocates structural representations in which shape is composed from a set of primitive parts. Representations used by Biederman are qualitative as opposed to the more commonly used quantitative representations of solids that are specified in terms of numerical parameters. For visual recognition tasks the quantitative description might contain redundant detail, while examining some qualitative features of the segmented primitives can ease recognition. Biederman calls 'primal access' the real-time entry-level shape classification by humans, and replication of this ability in machines is the main challenge for geons. However, qualitative representations cannot in general be used to synthesize an image of an object.

Psycho-physical experiments provide two main advantages of recognition by components.

- The basic-level representation derived from an image is invariant to scale and translation or orientation in depth.
- The object/scene is composed of parts and this decomposes a complicated task into simpler ones. The information about the scene is given by mutual relations between parts.

Biederman developed a catalogue of three dozen qualitative volumetric primitives called geons (GEOmetrical iONs). Each member of the catalogue has a unique set of four qualitative distinctions motivated by generalized cylinders:

- Edge: straight or curved
- Symmetry: rotational, reflective, asymmetric
- Size variation: constant, expanding, expanding/contracting
- Spine: straight or curved

The Cartesian product of values of these distinctions gives the final number of geons. Geons extend the idea of generalized cylinders by adding a qualitative taxonomy; 3D objects would be composed of a number of connected geons. Recent views on recognition by component theory can be found in [Dickinson et al. 97].

Biederman proposes an edge-based procedure for segmenting line drawing-like images into their geon components—places where geons join are sought using non-accidental alignments and concavities. On the other hand, the lack of quantitative information limits the use of geons, since distinguishing qualitatively similar objects (for example, parts which differ in scale) becomes difficult. One solution is to add some quantitative information. In general, it is possible to recognize an object's type using qualitative information, but it would be much more difficult to determine position if no quantitative information were available.

There is still some interest in geons in the vision community. The main drawbacks are:

- It is very difficult to extract a good line drawing from real images. Thus geons have been applied to very constrained domains in which line segmentation is often performed manually, or the objects are unrealistically simple.
- Missing depth information in drawings causes problems in understanding the 3D world.

10.4.4 Visualizing 3D real-world scenes using stored collections of 2D views

Methods which are able to capture a real object and render it from an arbitrary viewpoint usually use a 3D geometric model of the object. The bottleneck of these methods is the 3D reconstruction, which is a non-trivial problem, often failing for objects of more complex shapes. Virtual reality systems can augment real objects to the generated scenes.

Alternatively, the **image-based scene representation paradigm** (or view-based) attempts to display a real 3D scene from any viewing direction without using a 3D model. The scene is represented by a collection of 2D **reference views** instead of a full model. The actual image to be displayed is called a **virtual view** and is created by interpolation from the reference views using correspondences among them. The new bottleneck becomes the correspondence problem, being simpler than 3D reconstruction. The aim of such a procedure is to avoid the difficult problem of consistent 3D model reconstruction and thus more complex objects can be handled. In addition, faster access to a view can be achieved than by rendering the 3D model.

To succeed, the following problems must be solved.

Correspondence problem. How are correspondences between reference views found?

Image interpolation/extrapolation. Given the positions and intensities in reference views, how are they predicted in a new view?

Geometry. How is the visibility of points in the new view determined? It can be shown that without knowledge of geometry it is not possible to render scene correctly.

View selection. How is the optimal set of necessary reference views determined? A solution has been proposed [Werner et al. 96] in which the natural criterion of optimality is the minimal number of views allowing synthesis of an image that looks similar to that which would be seen when looking from the same viewpoint. It is non-trivial to determine a good measure of image similarity that corresponds well to a human understanding of image similarity.

The image based approach copes with complicated free-form surfaces—see Figure 10.28. Notice the mistake in the interpolated image in Figure 10.28b—due to a mismatch in correspondences, one of the lines in the top center is doubled. This error is likely to be overlooked, demonstrating how human understanding of an image content is not sensitive to such errors.

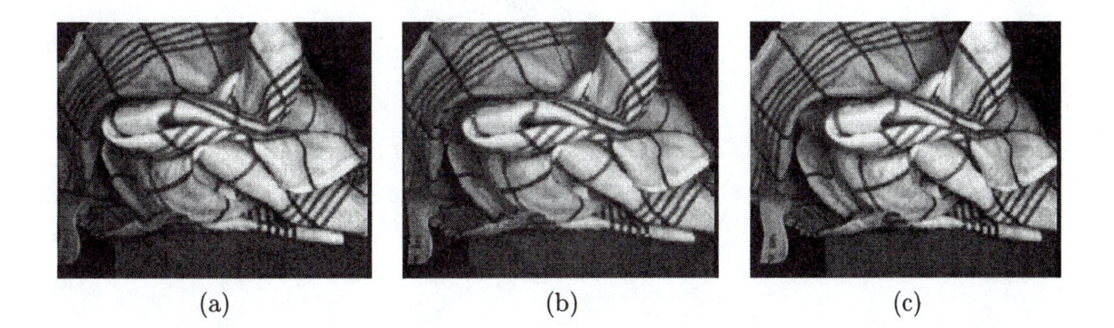

Figure 10.28: *Left and right are two reference views of a linen towel with view directions 10° apart. The virtual image in the middle was obtained using interpolation. Courtesy T. Werner, Czech Technical University, Prague.*

It has been thought [Laveau and Faugeras 94, Seitz and Dyer 95, Werner et al. 95, Kumar et al. 95, Irani et al. 95] that displaying a 3D scene from stored 2D images is quite different from rendering a 3D model. The difference seems to follow from Ullman's observation [Ullman and Basri 91] that objects could be recognized just on the basis of linear combination of corresponding points in their orthographic images. This is in contrast to recognition based on verification of a 3D model projected to an image.

Ullman's approach has attracted new attention since it has been shown that for a perspective camera a trilinear function replaces the linear one [Shashua 93]; more recently, it has been shown that any projective reconstruction of the scene suffices for visualization [Laveau and Faugeras 94, Hartley 95], and hence tedious calibration of the camera can be avoided.

It has been shown [Seitz and Dyer 95] that visualizing an object by interpolating close views is a well-posed process, and therefore for certain limited tasks a perfect correspondence algorithm is not needed. Elsewhere [Werner et al. 95, Cox et al. 92] it has been shown that even quite complicated scenes can be shown by interpolating between reference views. However, to make the visual effect realistic, many reference images may be needed [Werner et al. 95]. This is caused by a principal deficiency of image interpolation—its inability to display a general object from an arbitrary viewing angle using the images and the correspondences obtained from a sparse set of views. Surprisingly, no object, not even a convex polyhedron, can be completely viewed by interpolating between a finite number of reference views.

Consider the situation in which reference views are located around a simple polyhedron (see Figure 10.29). The images from the virtual camera C lying in the segment B_1C_3, cannot be constructed by interpolating reference views C_2 and C_3, since the camera C_2 does not see both sides of the polyhedron which are seen by the camera C. It will not help to move one of the reference cameras, e.g., C_3, closer to B_1, since then the problem moves to the segment C_3C_4. The only solution would be to increase the density of views near B_1 to infinity: Indeed, the algorithm for finding the best sparse set of reference views [Werner et al. 96] has tended to select many reference views in places where aspect changes—point B_1 is such a place in Figure 10.29.

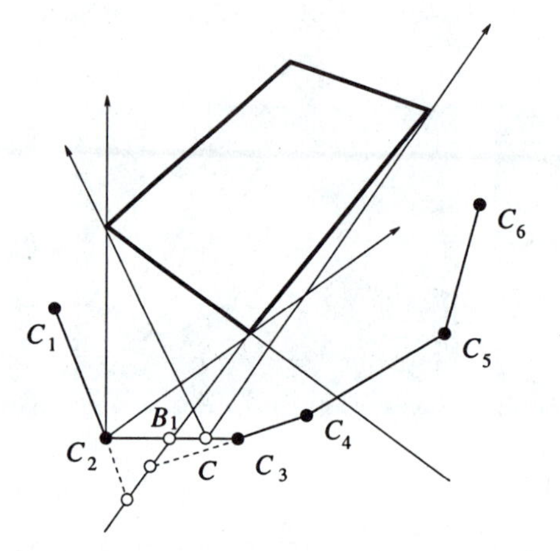

Figure 10.29: *The virtual view C cannot be constructed by interpolation from the views C_2 and C_3, but can be extrapolated from C_3 and C_4.*

This argument shows that views cannot be constructed by interpolating reference views from different aspects. On the other hand, it is certainly possible to construct view C from C_3 and C_4 by extrapolation of views. The significant difference between interpolation and extrapolation manifests itself on the border between aspects, where the virtual camera has to switch from the views C_1, C_2 to the views C_3, C_4. Unlike in the interpolation case, where switching has been done at the centers of the reference cameras, here it is not clear where to switch between the reference views until the aspect of the object is not known. Finding the changes in aspect is equivalent to finding depth discontinuities; moreover, for reference views in general positions, it is not possible to move smoothly along the object just by linear extrapolation and switching on the borders between aspects as the line C_1C_2 need not intersect the line C_3C_4 on the change of aspect. When crossing the boundary of an aspect, it is also necessary to determine the visibility of the points, because not all of them have to disappear at the same time.

Finding discontinuities and resolving visibility are problems known from 3D surface reconstruction and visualization of 3D models. In order to synthesize the images from a virtual camera moving smoothly around an object, one has to step back from pure image-based scene representation and interpolation mechanisms to partial projective reconstructions of the shape and their correct visualization.

A method for constructing virtual views regardless of the aspect of the reference views has been developed [Werner et al. 97], where a practically technology that permits the building of an image-based representation of a 3D scene routinely without human aid is proposed. The novel contribution is:

- The key step towards view extrapolation is to solve the correspondence problem. The proposed solution tracks edge features in a dense sequence of images.
- The visibility of points in virtual views is an important issue. Oriented projective geometry is used to formulate and solve the problem.
- It is shown that it is not necessary to transfer each point from reference views when creating the virtual image. Instead, it is proposed to triangulate the correspondences in reference images, to transfer only their vertices, and to fill triangle interiors.

We demonstrate the displaying of a 3D object of a ceramic doll—see Figure 10.30. The inputs are just two reference images that are captured from two view directions 10° apart. Figure 10.30e shows a virtual view from the same direction as one of the reference views; it can be seen that the rendering of the triangulated surface gives very similar results. Figures 10.30f, g demonstrate that virtual views can be extrapolated quite far from the two originals. The limit of this range of views is the viewing directions where there is not enough information in the reference images to solve the visibility. This is demonstrated in Figure 10.30h.

The original aim was to visualize a scene using the image-based approach, but it turns out that the whole chain from acquisition to representation to rendering can be made to work. Things start to resemble working with a 3D model; for example, considerations about visibility had to lean on the concept of 3D model, and techniques such as z-buffering, known from 3D model rendering could be used. The boundary separating approaches based on images and 3D models is not sharp, and one could think that this boundary is determined by whether the points in the rendered image are obtained by transfer or re-projection. However, this criterion cannot be taken seriously, for there is little difference between transfer and projective reconstruction $\rightarrow$ reprojection. There are three domains which we should distinguish in the spectrum of approaches: (i) view interpolation, (ii) view extrapolation, and (iii) displaying a consistent 3D model.

For further considerations it is necessary to divide the possible viewpoints around the scene (i.e., centers of a camera viewing the scene) into **visibility classes** as follows: A set of points belongs to the same visibility class if and only if the same portion of the scene is visible from any of these points. Note that only convex polyhedra have a finite number of visibility classes.

If two reference cameras are both in the same visibility class and the correspondences satisfy the ordering constraint, view interpolation is a well-posed problem [Seitz and Dyer 95], and solving visibility can be neglected. When the reference views are not in the same visibility class, the interpolation yields a good approximation of the correct view if the reference views are close to each other. Moreover, for close reference views, linear interpolation [Werner et al. 95] can replace the general transfer based on the trilinear relationship.

It can be shown that no scene can be displayed correctly from an **arbitrary** viewpoint by switching among view interpolation algorithms used for different sets of reference views. However, if both reference views and the virtual cameras are all in the same visibility class, view extrapolation remains also well-posed. The extrapolation becomes in this case as simple as the interpolation because no visibility has to be solved. Convex polyhedra can be displayed correctly from any viewpoint by switching among these algorithms used for the different sets of reference views.

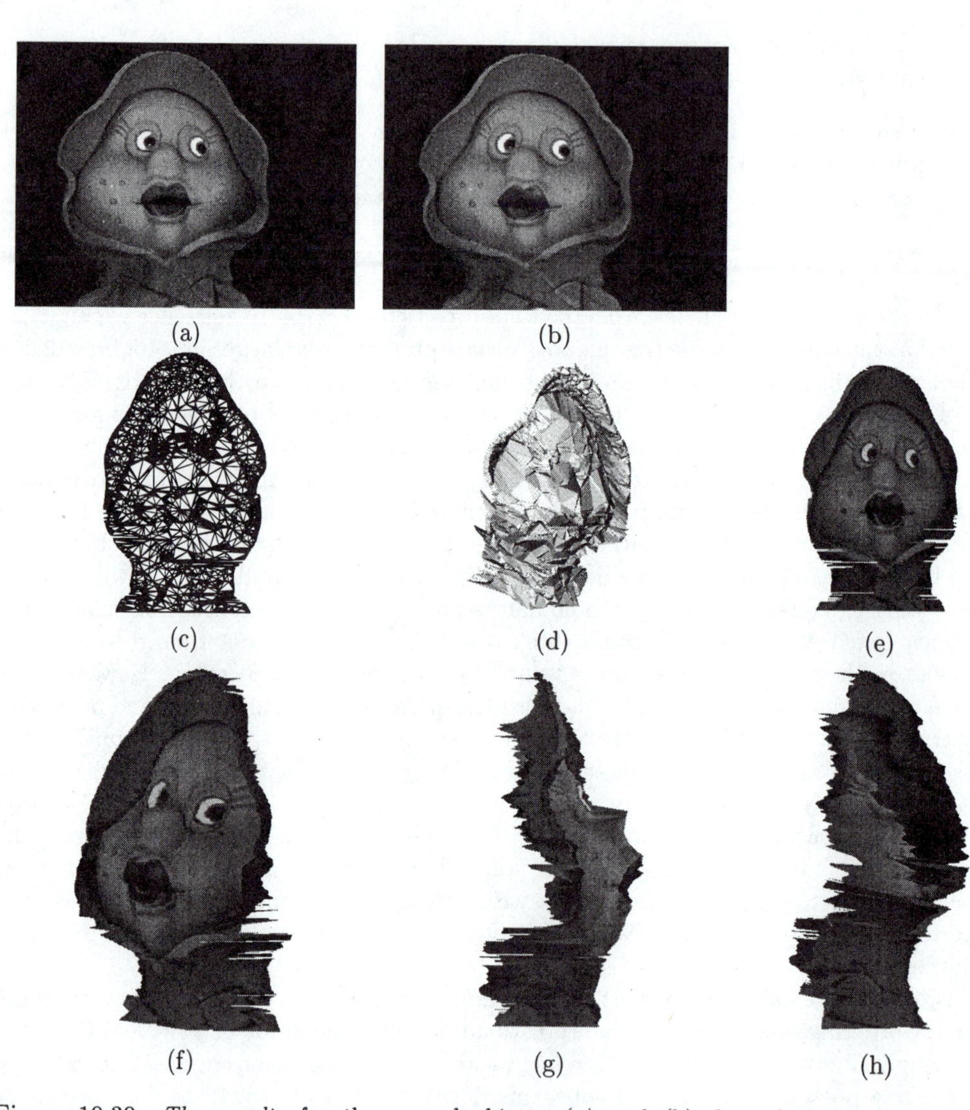

Figure 10.30: *The results for the second object: (a) and (b) the reference views; (c) the triangulation of the first reference view; (d) the projective model; (e) the virtual view from the same viewpoint as (a); (f) and (g) other virtual views; (h) shows the incorrectly solved visibility in (g)—the extrapolation extended beyond the reasonable limit (there was an attempt to look at something which was not visible in the reference images). Courtesy T. Werner, Czech Technical University, Prague.*

If the virtual camera is in a different visibility class than the reference ones, this simple algorithm cannot be used and visibility in the virtual image plane must be solved explicitly. Since determining whether the virtual camera is in the same class as the reference ones is very difficult, it is in practice always necessary to solve visibility. Actually, this is exactly the situation considered here which is called view extrapolation.

Some views of complex objects cannot be obtained from the reference cameras placed in the same visibility class, because they have in general infinitely many classes. In such a case, only part of the virtual view can be obtained correctly from any set of reference views. The whole virtual view needs to be constructed by **fusing** parts obtained from different sets of reference views. Surprisingly, it seems that there is no qualitative difference between fusing parts of images to get a physically valid virtual view and the reconstruction of a consistent projective model.

10.5 Summary

- This material is overview in nature. We have given a taxonomy of various 3D vision tasks, an overview of current approaches, formulated tasks, shown some applications, and hinted at recent research directions.

- **Shape from X**

  - Shape may be extracted from motion, optical flow, texture, focus/de-focus, vergence, and contour.
  - Each of these techniques may used to derive a 2.5D sketch for Marr's vision theory; they are also of practical use on their own.

- **Full 3D objects**

  - Line labeling is an outmoded but accessible technique for reconstructing object with planar faces.
  - Transitions to 3D objects need a co-ordinate system that is object centered.
  - 3D objects may be measured mechanically or by computed tomography.
  - Volumetric modeling strategies include constructive solid geometry, super-quadrics and generalized cylinders.
  - Surface modeling strategies include boundary representations, triangulated surfaces, and quadric patches.

- **3D model-based vision**

  - To create a full 3D model from a set of range images, the measured surfaces must first be registered—rotations and translations should be found that match one surface to another.
  - Model-based vision uses a priori knowledge about an object to ease its recognition.
  - Goad's algorithm is an illustration that searches for polyhedra in a single intensity image.

  - Techniques exist to locate curved objects from range images.

- **2D view-based representations of a 3D scene**
  - 2D view-based representations of 3D scenes may be achieved with multi-view representations or geons.
  - It is possible to select a few stored reference images, and render any view from them.
  - Interpolation of views is not enough and view extrapolation is needed. This requires knowledge of geometry, and the view-based approach does not differ significantly from 3D geometry reconstruction.

10.6 Exercises

Short-answer questions

1. For each of the following *shape from X* techniques, briefly describe the principle underlying it, explain how they can be of practical use, and give some examples.
   - Motion
   - Texture
   - Focus
   - Vergence
   - Contour
2. Give a counter-example in which apparent motion measured by an optical flow field does not correspond to actual motion of the 3D points. [Hint: Consider an opaque sphere, where (1) the sphere does not move and illumination changes; (2) the sphere rotates and the illumination is constant.]
3. Fashion designers cheat the human ability to derive shape from texture. How do they do it?
4. Give an example of a rim which is not a planar curve.
5. Are there line drawings that do not correspond to physically plausible polyhedral objects? If yes, show an example.
6. What was the contribution of image interpretation via line labeling to computer vision? Where else can it be used and why? (Hint: model-based 3D object recognition)
7. Draw a picture that functionally decomposes the human body into several generalized cylinders from the point of view of movement.
8. Describe how to create a full 3D model of an object from range images taken from different views.
9. Explain the principle of 2D view-based representation of a 3D scene.

Problems

1. Conduct a laboratory experiment that establishes depth from focus. Take a lens with small depth of focus—if you have an ordinary lens at hand, open the aperture as much as possible. Mount the camera with the lens on a stand that can adjust the distance between the camera and the object—a photo magnifier stand can be used for this purpose).

   Select a window in the image that corresponds to a point in the scene from which the distance is measured. Devise an algorithm that assists in achieving the sharpest image. (Hint: The sharpest image has the most high frequencies; is there some method to find such an image that is simpler than the Fourier transform?)

   Finally, the distance can be read from the mechanical scale on the stand.

2. Write an essay about shape. Most computer vision techniques measure co-ordinates of distinct points on a 3D object surface, e.g., those where correspondences were found; the result is a 3D point cloud. The human notion of shape is not of a point cloud. Elaborate on this problem, e.g., in the case of human face where 3D data are measured using stereopsis. How would you move from a point cloud to a surface? Is the surface the shape we seek? (Hint: Shape is treated from the geometer's point of view in [Koenderink 90]).

10.7 References

[Aloimonos 93] Y Aloimonos, editor. *Active Perception.* Lawrence Erlbaum Associates, Hillsdale, NJ, 1993.

[Aloimonos and Swain 85] J Aloimonos and M J Swain. Shape from texture. In *9th International Joint Conference on Artificial Intelligence,* Los Angeles, volume 2, pages 926–931, Morgan Kaufmann Publishers, Los Altos, CA, 1985.

[Arman and Aggarwal 93] F Arman and J K Aggarwal. Model-based object recognition in dense-range images—A review. *ACM Computing Surveys*, 25(1):5–43, 1993.

[Bajcsy and Lieberman 76] R Bajcsy and L Lieberman. Texture gradient as a depth cue. *Computer Graphics and Image Processing*, 5:52–67, 1976.

[Barr 81] A H Barr. Superquadrics and angle-preserving transformations. *IEEE Computer Graphics and Applications*, 1(1):11–23, January 1981.

[Besl and Jain 85] P J Besl and R C Jain. Three-dimensional object recognition. *ACM Computing Surveys*, 17(1):75–145, March 1985.

[Besl and McKay 92] P J Besl and N D McKay. A method for registration of 3-D shapes. *IEEE Transactions on Pattern Analysis and Machine Intelligence*, 14(2):239–256, February 1992.

[Beymer and Poggio 96] D Beymer and T Poggio. Image representations for visual learning. *Science*, 272:1905–1909, 1996.

[Biederman 87] I Biederman. Recognition by components: A theory of human image understanding. *Psychological Review*, 94(2):115–147, 1987.

[Binford 71] T O Binford. Visual perception by computer. In *Proceedings of the IEEE Conference on Systems, Science and Cybernetics*, IEEE, Miami, FL, December 1971.

[Blake and Yuille 92] A Blake and A Yuille. *Active Vision.* MIT Press, Cambridge, MA, 1992.

[Blostein and Ahuja 89] D Blostein and N Ahuja. Shape from texture: Integrating texture-element extraction and surface estimation. *IEEE Transactions on Pattern Analysis and Machine Intelligence*, 11:1233–1251, 1989.

[Bowyer 92] K W Bowyer, editor. Special issue on directions in CAD-based vision. *CVGIP - Image Understanding*, 55:107–218, 1992.

[Brady 84] M Brady. Representing shape. In M Brady, L A Gerhardt, and H F Davidson, editors, *Robotics and Artificial Intelligence*, pages 279–300. Springer + NATO, Berlin, 1984.

[Bray and Hlavac 91] A J Bray and V Hlavac. Properties of local geometric constraints. In *Proceedings of the British Machine Vision Conference,* Glasgow, Scotland, pages 95–103, Springer Verlag, London–Berlin–New York, September 1991.

[Brooks et al. 79] R A Brooks, R Greiner, and T O Binford. The ACRONYM model-based vision system. In *Proceedings of the International Joint Conference on Artificial Intelligence, IJCAI-6,* Tokyo, Japan, pages 105–113, 1979.

[Chakravarty and Freeman 82] I Chakravarty and H Freeman. Characteristic views as a basis for three-dimensional object recognition. *Proceedings of The Society for Photo-Optical Instrumentation Engineers Conference on Robot Vision*, 336:37–45, 1982.

[Chen and Kak 89] C H Chen and A C Kak. A robot vision system for recognizing 3D objects in low order polynomial time. *IEEE Transactions on Systems, Man and Cybernetics*, 19(6):1535–1563, 1989.

[Clocksin 80] W F Clocksin. Perception of surface slant and edge labels from optical flow—a computational approach. *Perception*, 9:253–269, 1980.

[Clowes 71] M B Clowes. On seeing things. *Artificial Intelligence*, 2(1):79–116, 1971.

[Cox et al. 92] I J Cox, S Hingorani, B M Maggs, and S B Rao. Stereo without disparity gradient smoothing: A Bayesian sensor fusion solution. In D C Hogg and R D Boyle, editors, *Proceedings of the British Machine Vision Conference,* Leeds, UK, pages 337–346, Springer Verlag, London, 1992.

[Dickinson et al. 97] S J Dickinson et al. Panel report: The potential of geons for generic 3D object recognition. *Image and Vision Computing*, 15:277–292, 1997.

[Faugeras 93] O D Faugeras. *Three-Dimensional Computer Vision: A Geometric Viewpoint.* MIT Press, Cambridge, MA, 1993.

[Faugeras and Hebert 86] O D Faugeras and M Hebert. The representation, recognition, locating of 3-D objects. *International Journal of Robotics Research*, 5(3):27–52, 1986.

[Flynn and Jain 92] P J Flynn and A K Jain. 3D object recognition using invariant feature indexing of interpretation tables. *CVGIP - Image Understanding*, 55(2):119–129, 1992.

[Goad 86] C Goad. Fast 3D model-based vision. In A P Pentland, editor, *From Pixels to Predicates*, pages 371–374. Ablex, Norwood, NJ, 1986.

[Grimson 89] W E L Grimson. On the recognition of curved objects. *IEEE Transactions on Pattern Analysis and Machine Intelligence*, 11(6):632–643, 1989.

[Gubais et al. 92] L J Gubais, D E Knuth, and Sharir M. Randomized incremental construction of Delunay and Voronoi diagrams. *Algorithmica*, 7:381–413, 1992.

[Hartley 95] R I Hartley. A linear method for reconstruction from lines and points. In *5th International Conference on Computer Vision,* MIT, pages 882–887, IEEE, Piscataway, NJ, 1995.

[Hayes-Roth et al. 83] F Hayes-Roth, D A Waterman, and D B Lenat. *Building Expert Systems.* Addison-Wesley, Reading, MA, 1983.

[Higuchi et al. 95] K Higuchi, M Hebert, and K Ikeuchi. Building 3-D models from unregistered range images. *Graphics Models and Image Processing*, 57(4):313–333, July 1995.

[Hlaváč et al. 94] V Hlaváč, T Pajdla, and M Sommer. Improvement of the curvature computation. In *12th International Conference on Pattern Recognition,* Jerusalem, Israel, volume 1, Computer Vision and Image Processing, pages 536–538, IEEE, Piscataway, NJ, October 1994.

[Hoppe et al. 92] H Hoppe, T DeRose, T Duchamp, J McDonald, and W Stuetzle. Surface reconstruction from unorganized points. *Computer Graphics*, 26(2):71–78, 1992.

[Horn 84] B K P Horn. Extended Gaussian image. *Proceedings of the IEEE*, 72(12):1671–1686, December 1984.

[Horn 86] B K P Horn. *Robot Vision.* MIT Press, Cambridge, MA, 1986.

[Huffman 71] D A Huffman. Impossible objects as nonsense sentences. In B Metzler and D M Michie, editors, *Machine Intelligence*, volume 6, pages 295–323. Edinburgh University Press, Edinburgh, 1971.

[Irani et al. 95] M Irani, P Anandan, and S Hsu. Mosaic-based representations of video sequences and their applications. In *5th International Conference on Computer Vision,* MIT, pages 605–611, IEEE, Piscataway, NJ, 1995.

[Jaklič 97] A Jaklič. Construction of CAD models from range images. PhD thesis, Department of Computer and Information Science, University of Ljubljana, Ljubljana, Slovenia, March 1997.

[Kanatani and Chou 89] K Kanatani and T C Chou. Shape from texture: General principle. *Artificial Intelligence*, 38(1):1–48, 1989.

[Klir 91] G J Klir. *Facets of System Science.* Plenum Press, New York, 1991.

[Koenderink 90] J J Koenderink. *Solid Shape.* MIT Press, Cambridge, MA, 1990.

[Koenderink and van Doorn 79] J J Koenderink and A J van Doorn. Internal representation of solid shape with respect to vision. *Biological Cybernetics*, 32(4):211–216, 1979.

[Kriegman 92] D J Kriegman. Computing stable poses of piecewise smooth objects. *CVGIP – Image Understanding*, 55(2):109–118, 1992.

[Krotkov 87] E Krotkov. Focusing. *International Journal of Computer Vision*, 1:223–237, 1987.

[Krotkov and Bajcsy 93] E P Krotkov and R Bajcsy. Active vision for reliable ranging: Cooperating focus, stereo, and vergence. *International Journal of Computer Vision*, 11(2):187–203, October 1993.

[Kumar et al. 95] R Kumar, P Anadan, M Irani, J Bergen, and K Hanna. Representation of scenes from collection of images. In *Proceedings of the Visual Scene Representation Workshop,* Boston, pages 10–17. IEEE, June 24th, 1995.

[Laveau and Faugeras 94] S Laveau and O D Faugeras. 3D scene representation as a collection of images. In *12th International Conference on Pattern Recognition,* Jerusalem, Israel, pages 689–691, 1994.

[Leonardis et al. 97] A Leonardis, A Jaklič, and F Solina. Superquadrics for segmenting and modeling range data. *IEEE Transactions on Pattern Analysis and Machine Intelligence*, 19(11):1289–1295, November 1997.

[Liang and Todhunter 90] P Liang and J S Todhunter. Representation and recogntion of surface shapes in range images: A differential geometry approach. *Computer Vision, Graphics, and Image Processing*, 52:78–109, 1990.

[Lowe 85] D G Lowe. *Perceptual Organisation and Visual Recognition.* Kluwer Nijhoff, Norwell, MA, 1985.

[Lowe 89] D G Lowe. Organization of smooth image curves at multiple scales. *International Journal of Computer Vision*, 1:119–130, 1989.

[Malik and Maydan 89] J Malik and D Maydan. Recovering three-dimensional shape from a single image of curved object. *IEEE Transactions on Pattern Analysis and Machine Intelligence*, 11(6):555–566, June 1989.

[Marr 82] D Marr. *Vision—A Computational Investigation into the Human Representation and Processing of Visual Information.* Freeman, San Francisco, 1982.

[Mokhtarian 95] F Mokhtarian. Silhouette-based object recognition through curvature scale space. *IEEE Transactions on Pattern Analysis and Machine Intelligence*, 17:539–544, 1995.

[Nayar and Nakagawa 94] S K Nayar and Y Nakagawa. Shape from focus. *IEEE Transactions on Pattern Analysis and Machine Intelligence*, 16(8):824–831, 1994.

[Nayar et al. 96] S K Nayar, M Watanabe, and M Hoguchi. Real-time focus range sensor. *IEEE Transactions on Pattern Analysis and Machine Intelligence*, 18(12):1186–1197, December 1996.

[Negahdaripour and Horn 85] S Negahdaripour and B K P Horn. Determining 3D motion of planar objects from image brightness measurements. In *9th International Joint Conference on Artificial Intelligence,* Los Angeles, volume 2, pages 898–901, Morgan Kaufmann Publishers, Los Altos, CA, 1985.

[Nevatia et al. 94] R Nevatia, M Zerroug, and F Ulupinar. Recovery of three-dimensional shape of curved objects from a single image. In T Y Young, editor, *Handbook of Pattern Recognition and Image Processing: Computer Vision*, pages 101–129, Academic Press, San Diego, CA, 1994.

[Newman et al. 93] T S Newman, P J Flynn, and A K Jain. Model-based classification of quadric surfaces. *CVGIP: Image Understanding*, 58(2):235–249, September 1993.

[Pajdla and Van Gool 95] T Pajdla and L Van Gool. Matching of 3-D curves using semi-differential invariants. In *5th International Conference on Computer Vision,* MIT, pages 390–395, IEEE, Piscataway, NJ, 1995.

[Pentland 87] A Pentland. A new sense for depth of field. *IEEE Transactions on Pattern Analysis and Machine Intelligence*, 9(4):523–531, 1987.

[Ponce et al. 92] J Ponce, A Hoogs, and D J Kriegman. On using CAD models to compute the pose of curved 3d objects. *CVGIP - Image Understanding*, 55(2):184–197, 1992.

[Preparata and Shamos 85] F P Preparata and M I Shamos. *Computational Geometry—An Introduction.* Springer Verlag, Berlin, 1985.

[Quarendon 84] P Quarendon. *WINSOM User's Guide.* IBM, IBM UK Scientific Centre, Winchester, England, August 1984.

[Rissanen 89] J Rissanen. *Stochastic Complexity in Statistical Inquiry.* World Scientific, Series in Computer Science, IBM Almaden Research Center, San Jose, CA, 1989.

[Rivlin and Weiss 95] E Rivlin and I Weiss. Local invariants for recognition. *IEEE Transactions on Pattern Analysis and Machine Intelligence*, 17(3):226–238, March 1995.

[Roberts 65] L G Roberts. Machine perception of three-dimensional solids. In J T Tippett, editor, *Optical and Electro-Optical Information Processing*, pages 159–197. MIT Press, Cambridge, MA, 1965.

[Sara and Bajcsy 98] R Sara and R Bajcsy. Fish-scales: Representing fuzzy manifolds. In N Ahuja, editor, *International Conference on Computer Vision,* Bombay, India, pages 811–817, Narosa, Bombay, 1998.

[Sato et al. 92] K Sato, K Ikeuchi, and T Kanade. Model based recognition of specular objects using sensor models. *CVGIP - Image Understanding*, 55(2):155–169, 1992.

[Seales and Dyer 92] W B Seales and C R Dyer. Viewpoints from occluding contour. *CVGIP - Image Understanding*, 55(2):198–211, 1992.

[Seitz and Dyer 95] S M Seitz and C R Dyer. Physically-valid view synthesis by image interpolation. In *Proceedings of the Visual Scene Representation Workshop,* Boston, pages 18–27. IEEE, June 24th, 1995.

[Shashua 93] A Shashua. On geometric and algebraic aspects of 3D affine and projective structures from perspective 2D views. Technical Report AI-1405, Massachusetts Institute of Technology, Artifical Intelligence Laboratory, Cambridge, MA, July 1993.

[Shomar and Young 94] W J Shomar and T Y Young. Three-dimensional shape recovery from line drawings. In *Handbook of Pattern Recognition and Image Processing: Computer Vision*, volume 2, pages 53–100, Academic Press, San Diego, CA, 1994.

[Solina and Bajcsy 90] F Solina and R Bajcsy. Recovery of parametric models from range images: The case for superquadrics with global deformations. *IEEE Transactions on Pattern Analysis and Machine Intelligence*, 12(2):131–147, December 1990.

[Soroka and Bajcsy 78] B I Soroka and R K Bajcsy. A program for describing complex three dimensional objects using generalised cylinders. In *Proceedings of the Pattern Recognition and Image Processing Conference* Chicago, pages 331–339, IEEE, New York, 1978.

[Soucy and Laurendeau 96] M Soucy and D Laurendeau. Multiresolution surface modeling based on hierarchical triangulation. *Computer Vision and Image Understanding*, 63(1):1–14, January 1996.

[Stevens 79] K A Stevens. Representing and analyzing surface orientation. In P A Winston and R H Brown, editors, *Artificial Intelligence: An MIT Persepctive*, volume 2. MIT Press, Cambridge, MA, 1979.

[Sugihara 86] K Sugihara. *Machine Interpretation of Line Drawings.* MIT Press, Cambridge, MA, 1986.

[Terzopoulos and Fleischer 88] D Terzopoulos and K Fleischer. Deformable models. *The Visual Computer*, 4(6):306–331, 1988.

[Terzopoulos and Metaxas 91] D Terzopoulos and D Metaxas. Dynamic 3D models with local and global deformations: Deformable superquadrics. *IEEE Transactions on Pattern Analysis and Machine Intelligence*, 13(7):703–714, July 1991.

[Terzopoulos et al. 88] Demetri Terzopoulos, Andrew Witkin, and Michael Kass. Constraints on deformable models: Recovering 3-D shape and nonrigid motion. *Artificial Intelligence*, 36:91–123, 1988.

[Ullman 79] S Ullman. *The Interpretation of Visual Motion.* MIT Press, Cambridge, MA, 1979.

[Ullman 96] S Ullman. *High-Level Vision: Object Recognition and Visual Cognition.* MIT Press, Cambridge, MA, 1996.

[Ullman and Basri 91] S Ullman and R Basri. Recognition by linear combination of models. *IEEE Transactions on Pattern Analysis and Machine Intelligence*, 13(10):992–1005, October 1991.

[Uray 97] P Uray. From 3D point clouds to surface and volumes. PhD thesis, Technische Universitaet Graz, Austria, October 1997.

[Waltz 75] D L Waltz. Understanding line drawings of scenes with shadows. In P H Winston, editor, *The Psychology of Computer Vision*, pages 19–91. McGraw-Hill, New York, 1975.

[Weiss 88] I Weiss. Projective invariants of shapes. In *Proceedings of the DARPA Image Understanding Workshop,* Cambridge, MA, volume 2, pages 1125–1134. DARPA, 1988.

[Weiss and Rosenfeld 96] I Weiss and A Rosenfeld. A simple method for ellipse detection. Technical Report CS-TR-3717, University of Maryland, Center for Automation Research, College Park, MD, 1996.

[Werner et al. 95] T Werner, R D Hersch, and V Hlaváč. Rendering real-world objects using view interpolation. In *5th International Conference on Computer Vision,* MIT, pages 957–962, IEEE, Piscataway, NJ, 1995.

[Werner et al. 96] T Werner, V Hlaváč, A Leonardis, and T Pajdla. Selection of reference views for image-based representation. In *Proceedings of the 13th International Conference on Pattern Recognition,* volume I—Track A: Computer Vision, pages 73–77, IEEE Computer Society Press, Los Alamitos, CA, Vienna, Austria, August 1996.

[Werner et al. 97] T Werner, T Pajdla, and V Hlaváč. Visualizing 3-D real-world scenes using view extrapolation. Technical Report K335-CMP-1997-137, Department of Control, Faculty of Electrical Engineering, Czech Technical University, Prague, June 1997.

[Witkin 81] A P Witkin. Recovering surface shape and orientation from texture. *Artificial Intelligence,* 17:17–45, 1981.

[Worring and Smeulders 93] M Worring and A W M Smeulders. Digital curvature estimation. *CVGIP: Image Understanding,* 58(3):366–382, November 1993.

Chapter 11

Mathematical morphology

11.1 Basic morphological concepts

Mathematical morphology, which started to develop in the late 1960s, stands as a relatively separate part of image analysis. It is based on the algebra of non-linear operators operating on object shape and in many respects supersedes the linear algebraic system of convolution. It performs in many tasks—pre-processing, segmentation using object shape, and object quantification—better and more quickly than the standard approaches. The main obstacle for the novice user of mathematical morphology tools is the slightly different algebra than is usual in standard algebra and calculus courses.

The main protagonists of mathematical morphology were Matheron [Matheron 67] and Serra [Serra 82], whose monographs are highly mathematical books, and more recent books are typically written in a similar spirit, e.g., [Giardina and Dougherty 88, Dougherty 92, Heijmans 94]. Other significant references are [Maragos and Schafer 87a, Maragos and Schafer 87b, Serra 87, Roerdink and Heijmans 88].

Our aim is to present morphology in a manner that is relatively easy to follow [Haralick and Shapiro 92, Vincent 95]. Morphological tools are implemented in most advanced image analysis packages, and we hope the reader will learn enough to apply them in a qualified way. Mathematical morphology is very often used in applications where shape of objects and speed is an issue—for example, analysis of microscopic images (in biology, material science, geology, and criminology), industrial inspection, optical character recognition, and document analysis.

The non-morphological approach to image processing is close to calculus, being based on the point-spread function concept and linear transformations such as convolution, and we have discussed image modeling and processing from this point of view in earlier chapters. Mathematical morphology uses tools of non-linear algebra and operates with point sets, their connectivity and shape. Morphological operations simplify images, and quantify and preserve the main shape characteristics of objects.

Morphological operations are used predominantly for the following purposes:

- Image pre-processing (noise filtering, shape simplification)
- Enhancing object structure (skeletonizing, thinning, thickening, convex hull, object marking)

- Segmenting objects from the background
- Quantitative description of objects (area, perimeter, projections, Euler-Poincaré characteristic)

Mathematical morphology exploits point set properties, results of integral geometry, and topology. The initial assumption states that real images can be modeled using **point sets** of any dimension (e.g., N-dimensional Euclidean space); the Euclidean 2D space $\mathcal{E}^2$ and its system of subsets is a natural domain for planar shape description. Understanding of inclusion ($\subset$ or $\supset$), intersection ($\cap$), union ($\cup$), the empty set $\emptyset$, and set complement (^{c}) is assumed. Set **difference** is defined by

$$X \setminus Y = X \cap Y^c \tag{11.1}$$

Computer vision uses the digital counterpart of Euclidean space—sets of integer pairs ($\in \mathcal{Z}^2$) for binary image morphology or sets of integer triples ($\in \mathcal{Z}^3$) for gray-scale morphology or binary 3D morphology.

We begin by considering binary images that can be viewed as subsets of the 2D space of all integers, $\mathcal{Z}^2$. A point is represented by a pair of integers that give co-ordinates with respect to the two co-ordinate axes of the digital raster; the unit length of the raster equals the sampling period in each direction. We talk about a **discrete grid** if the neighborhood relation between points is well defined. This representation is suitable for both rectangular and hexagonal grids, but a rectangular grid is assumed hereafter.

A binary image can be treated as a 2D point set. Points belonging to objects in the image represent a set X—these points are pixels with value equal to one. Points of the complement set X^c correspond to the background with pixel values equal to zero. The origin (marked as a diagonal cross in our examples) has co-ordinates $(0,0)$, and co-ordinates of any point are interpreted as (x,y) in the common way used in mathematics. Figure 11.1 shows an example of such a set—points belonging to the object are denoted by small black squares. Any point x from a discrete image $X = \{(1,0),(1,1),(1,2),(2,2),(0,3),(0,4)\}$ can be treated as a vector with respect to the origin (0,0).

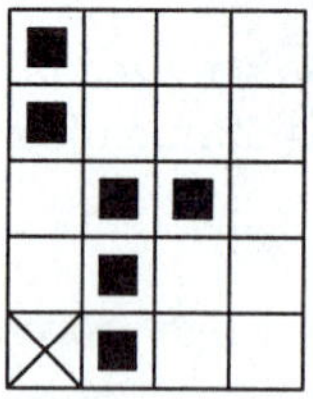

Figure 11.1: *A point set example.*

A **morphological transformation** Ψ is given by the relation of the image (point set X) with another small point set B called a **structuring element**. B is expressed with respect to a local origin $\mathcal{O}$ (called the representative point). Some typical structuring elements are shown in Figure 11.2. Figure 11.2c illustrates the possibility of the point $\mathcal{O}$ not being a member of the structuring element B.

To apply the morphological transformation $\Psi(X)$ to the image X means that the structuring element B is moved systematically across the entire image. Assume that B is positioned

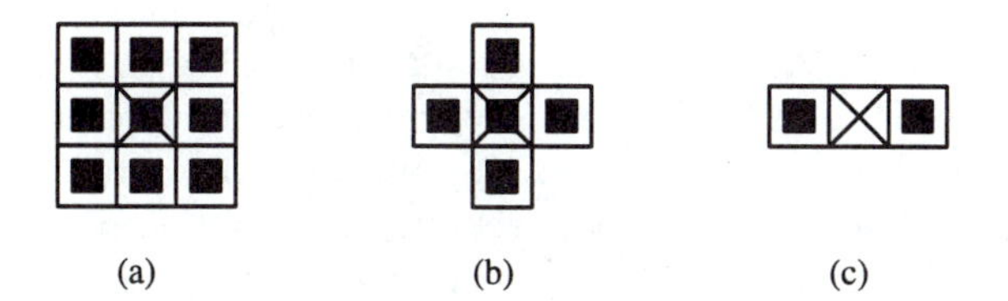

Figure 11.2: *Typical structuring elements.*

at some point in the image; the pixel in the image corresponding to the representative point $\mathcal{O}$ of the structuring element is called the *current* pixel. The result of the relation (which can be either zero or one) between the image X and the structuring element B in the current position is stored in the output image in the current image pixel position.

The **duality** of morphological operations is deduced from the existence of the set complement; for each morphological transformation $\Psi(X)$ there exists a dual transformation $\Psi^*(X)$

$$\Psi(X) = [\Psi^*(X^c)]^c \tag{11.2}$$

The **translation** of the point set X by the vector h is denoted by X_h; it is defined by

$$X_h = \left\{ p \in \mathcal{E}^2,\ p = x + h \text{ for some } x \in X \right\} \tag{11.3}$$

This is illustrated in Figure 11.3.

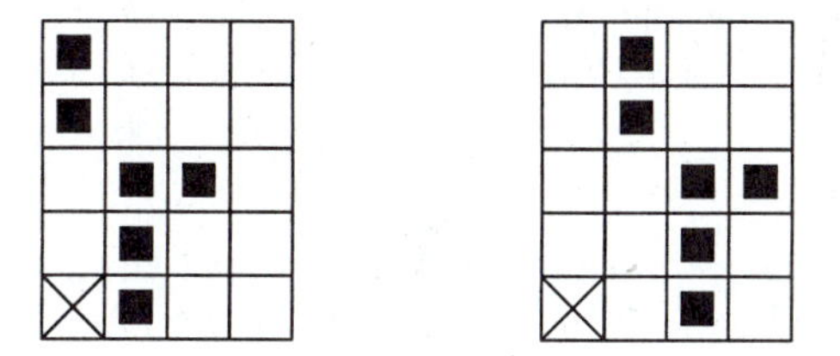

Figure 11.3: *Translation by a vector.*

11.2 Four morphological principles

It is appropriate to restrict the set of possible morphological transformations in image analysis by imposing several constraints on it; we shall briefly present here four morphological principles that express such constraints. These concepts may be difficult to understand, but an understanding of them is not essential to a comprehension of what follows, and they may be taken for granted. A detailed explanation of these matters may be found in [Serra 82].

Humans have an intuitive understanding of **spatial structure**. The structure of the Alps versus an oak tree crown is perceived as different. Besides the need for objective descriptions of such objects, the scientist requires a quantitative description. Generalization is expected as well; the interest is not in a specific oak tree, but in the class of oaks.

The morphological approach with quantified results consists of two main steps: (a) geometrical transformation and (b) the actual measurement. [Serra 82] gives two examples. The first is from chemistry, where the task is to measure the surface area of some object. First, the initial body is reduced to its surface, e.g., by marking by some chemical matter. Second, the quantity of the marker needed to cover the surface is measured. Another example is from sieving analysis, often used in geology when the distribution of sizes of milled rocks is of interest. The milled product is passed through sieves with different sizes of holes from coarse to fine. The result is a sequence of subsets of milled product. For each sieve size some oversize particles remain on it after sieving, and these are measured.

A morphological operator is (by definition) composition of a mapping Ψ (or geometrical transformation) followed by a measure μ which is a mapping $Z \times \ldots \times Z \longrightarrow R$. The geometrically transformed set $\Psi(X)$ can be the boundary, oversized particles in sieving analysis, etc., and the measure $\mu[\Psi(X)]$ yields a number (weight, surface area, volume, etc.). The discussion here is simplified just to the transformation Ψ, but the axiomatics can be transposed to measures as well.

A morphological transformation is called **quantitative** if and only if it satisfies four basic principles [Serra 82].

- Compatibility with translation: Let the transformation Ψ depend on the position of the origin $\mathcal{O}$ of the co-ordinate system, and denote such a transformation by $\Psi_{\mathcal{O}}$. If all points are translated by the vector $-h$, it is expressed as Ψ_{-h}. The *compatibility with translation* principle is given by

$$\Psi_{\mathcal{O}}(X_h) = [\Psi_{-h}(X)]_h \tag{11.4}$$

  If Ψ does not depend on the position of the origin $\mathcal{O}$, then the compatibility with translation: principle reduces to invariance under translation:

$$\Psi(X_h) = [\Psi(X)]_h \tag{11.5}$$

- Compatibility with change of scale: Let λX represent the homothetic scaling of a point set X (i.e., the co-ordinates of each point of the set are multiplied by some positive constant λ). This is equivalent to change of scale with respect to some origin. Let Ψ_λ denote a transformation that depends on the positive parameter λ (change of scale). *Compatibility with change of scale* is given by

$$\Psi_\lambda(X) = \lambda\, \Psi(\frac{1}{\lambda}\, X) \tag{11.6}$$

  If Ψ does not depend on the scale λ, then compatibility with change of scale reduces to invariance to change of scale:

$$\Psi(\lambda\, X) = \lambda\, \Psi(X) \tag{11.7}$$

- Local knowledge: The local knowledge principle considers the situation in which only a part of a larger structure can be examined—this is always the case in reality, due to the restricted size of the digital grid. The morphological transformation Ψ satisfies the *local knowledge principle* if for any bounded point set Z' in the transformation $\Psi(X)$ there exists a bounded set Z, knowledge of which is sufficient to provide Ψ. The local knowledge principle may be written symbolically as

$$[\Psi(X \cap Z)] \cap Z' = \Psi(X) \cap Z' \tag{11.8}$$

- Upper semi-continuity: The upper semi-continuity principle says that the morphological transformation does not exhibit any abrupt changes. A precise explanation needs many concepts from topology and is given in [Serra 82].

11.3 Binary dilation and erosion

The sets of black and white pixels constitute a description of a binary image. Assume that only black pixels are considered, and the others are treated as a background. The primary morphological operations are dilation and erosion, and from these two, more complex morphological operations such as opening, closing, and shape decomposition can be constituted. We present them here using Minkowski's formalism [Haralick and Shapiro 92]. The Minkowski algebra is closer to the notions taught in standard mathematics courses (an alternative is Serra's formalism based on stereological concepts [Serra 82]).

11.3.1 Dilation

The morphological transformation **dilation** $\oplus$ combines two sets using vector addition (or Minkowski set addition, e.g., $(a,b) + (c,d) = (a+c, b+d)$). The dilation $X \oplus B$ is the point set of all possible vector additions of pairs of elements, one from each of the sets X and B.

$$X \oplus B = \left\{ p \in \mathcal{E}^2 \, : \, p = x + b, \;\; x \in X \text{ and } b \in B \right\} \tag{11.9}$$

Figure 11.4 illustrates an example of dilation.

$$\begin{aligned} X &= \{(1,0),(1,1),(1,2),(2,2),(0,3),(0,4)\} \\ B &= \{(0,0),(1,0)\} \\ X \oplus B &= \{(1,0),(1,1),(1,2),(2,2),(0,3),(0,4), \\ &\quad\;\; (2,0),(2,1),(2,2),(3,2),(1,3),(1,4)\} \end{aligned}$$

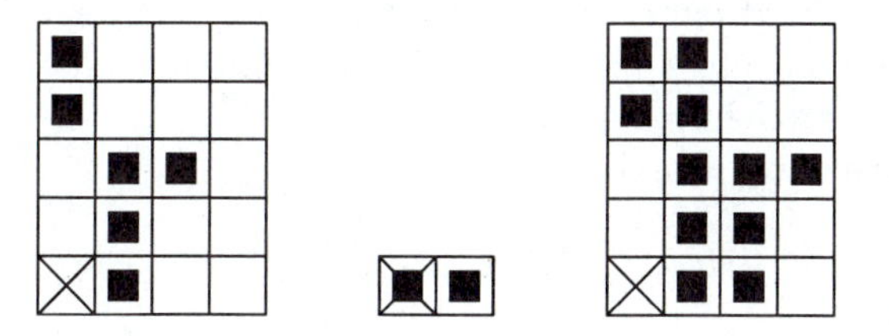

Figure 11.4: *Dilation.*

Figure 11.5 shows a 256×256 original image (the emblem of the Czech Technical University) on the left. A structuring element of size 3×3, see Figure 11.2a, is used. The result of dilation is shown on the right side of Figure 11.5. In this case the dilation is an **isotropic** expansion (it behaves the same way in all directions). This operation is also sometimes called *fill* or *grow*.

Dilation with an isotropic 3×3 structuring element might be described as a transformation which changes all background pixels neighboring the object to object pixels.

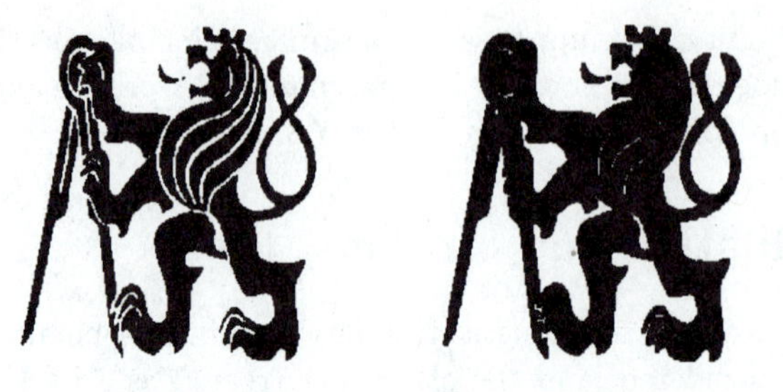

Figure 11.5: *Dilation as isotropic expansion.*

Dilation has several interesting properties that may ease its hardware or software implementation; we present some here without proof. The interested reader may consult [Serra 82] or the tutorial paper [Haralick et al. 87].

The dilation operation is commutative,

$$X \oplus B = B \oplus X \tag{11.10}$$

and is also associative,

$$X \oplus (B \oplus D) = (X \oplus B) \oplus D \tag{11.11}$$

Dilation may also be expressed as a union of shifted point sets,

$$X \oplus B = \bigcup_{b \in B} X_b \tag{11.12}$$

and is invariant to translation,

$$X_h \oplus B = (X \oplus B)_h \tag{11.13}$$

Equations (11.12) and (11.13) show the importance of shifts in speeding up implementation of dilation, and this holds for implementations of binary morphology on serial computers in general. One processor word represents several pixels (e.g., 32 for a 32-bit processor), and shift or addition corresponds to a single instruction. Shifts may also be easily implemented as delays in a pipeline parallel processor.

Dilation is an **increasing** transformation;

$$\text{If } X \subseteq Y \text{ then } X \oplus B \subseteq Y \oplus B \tag{11.14}$$

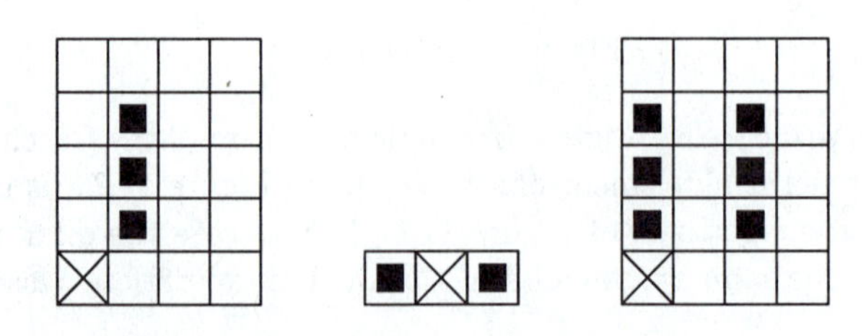

Figure 11.6: *Dilation where the representative point is not a member of the structuring element.*

Dilation is used to fill small holes and narrow gulfs in objects. It increases the object size—if the original size needs to be preserved, then dilation is combined with erosion, described in the next section.

Figure 11.6 illustrates the result of dilation if the representative point is not a member of the structuring element B; if this structuring element is used, the dilation result is substantially different from the input set. Notice that the connectivity of the original set has been lost.

11.3.2 Erosion

Erosion $\ominus$ combines two sets using vector subtraction of set elements and is the dual operator of dilation. Neither erosion nor dilation is an invertible transformation.

$$X \ominus B = \{p \in \mathcal{E}^2 : \; p + b \in X \text{ for every } b \in B\} \tag{11.15}$$

This formula says that every point p from the image is tested; the result of the erosion is given by those points p for which all possible $p + b$ are in X. Figure 11.7 shows an example of the point set X eroded by the structuring element B.

$$\begin{aligned} X &= \{(1,0),(1,1),(1,2),(0,3),(1,3),(2,3),(3,3),(1,4)\} \\ B &= \{(0,0),(1,0)\} \\ X \ominus B &= \{(0,3),(1,3),(2,3)\} \end{aligned}$$

Figure 11.7: *Erosion.*

Figure 11.8 shows the erosion by a 3×3 element (see Figure 11.2a) of the same original as in Figure 11.5. Notice that single-pixel-wide lines disappear. Erosion (such as Figure 11.8) with an isotropic structuring element is called *shrink* or *reduce* by some authors.

Basic morphological transformations can be used to find the contours of objects in an image very quickly. This can be achieved, for instance, by subtraction from the original picture of its eroded version—see Figure 11.9.

Erosion is used to simplify the structure of an object—objects or their parts with width equal to one will disappear. It might thus decompose complicated objects into several simpler ones.

There is an equivalent definition of erosion [Matheron 75]. Recall that B_p denotes B translated by p:

$$X \ominus B = \{p \in \mathcal{E}^2 : \; B_p \subseteq X\} \tag{11.16}$$

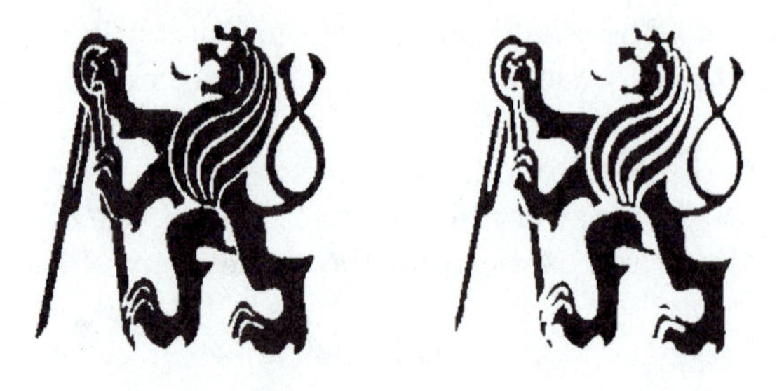

Figure 11.8: *Erosion as isotropic shrink.*

Figure 11.9: *Contours obtained by subtraction of an eroded image from an original (left).*

The erosion might be interpreted by structuring element B sliding across the image X; then, if B translated by the vector p is contained in the image X, the point corresponding to the representative point of B belongs to the erosion $X \ominus B$.

An implementation of erosion might be simplified by noting that an image X eroded by the structuring element B can be expressed as an intersection of all translations of the image X by the vector[1] $-b \in B$:

$$X \ominus B = \bigcap_{b \in B} X_{-b} \tag{11.17}$$

If the representative point is a member of the structuring element, then erosion is an anti-extensive transformation; that is, if $(0,0) \in B$, then $X \ominus B \subseteq X$. Erosion is also translation invariant,

$$X_h \ominus B = (X \ominus B)_h \tag{11.18}$$

$$X \ominus B_h = (X \ominus B)_{-h} \tag{11.19}$$

and, like dilation, is an increasing transformation.

$$\text{If } X \subseteq Y \text{ then } X \ominus B \subseteq Y \ominus B \tag{11.20}$$

If B, D are structuring elements, and D is contained in B, then erosion by B is more aggressive than by D; that is, if $D \subseteq B$, then $X \ominus B \subseteq X \ominus D$. This property enables the ordering of erosions according to structuring elements of similar shape but different sizes.

[1]This definition of erosion, $\ominus$, differs from that used in [Serra 82]. There $\ominus$ denotes Minkowski subtraction, which is an intersection of all translations of the image by the vector $b \in B$. In our case the minus sign has been added. In our notation, if convex sets are used, the dilation of erosion (or the other way around) is identity.

Denote by $\breve{B}$ the **symmetrical set** to B (called the **transpose** [Serra 82] or **rational set** [Haralick et al. 87] by some authors) with respect to the representative point $\mathcal{O}$:

$$\breve{B} = \{-b : b \in B\} \tag{11.21}$$

For example,

$$\begin{aligned} B &= \{(1,2),(2,3)\} \\ \breve{B} &= \{(-1,-2)(-2,-3)\} \end{aligned} \tag{11.22}$$

We have already mentioned that erosion and dilation are dual transformations. Formally,

$$(X \ominus Y)^C = X^C \oplus \breve{Y} \tag{11.23}$$

The differences between erosion and dilation are illustrated by the following properties. Erosion (in contrast to dilation) is not commutative:

$$X \ominus B \neq B \ominus X \tag{11.24}$$

The properties of erosion and intersection combined together are

$$\begin{aligned} (X \cap Y) \ominus B &= (X \ominus B) \cap (Y \ominus B) \\ B \ominus (X \cap Y) &\supseteq (B \ominus X) \cup (B \ominus Y) \end{aligned} \tag{11.25}$$

On the other hand, image intersection and dilation cannot be interchanged; the dilation of the intersection of two images is contained in the intersection of their dilations:

$$(X \cap Y) \oplus B = B \oplus (X \cap Y) \subseteq (X \oplus B) \cap (Y \oplus B) \tag{11.26}$$

The order of erosion may be interchanged with set union. This fact enables the structuring element to be decomposed into a union of simpler structuring elements:

$$B \oplus (X \cup Y) = (X \cup Y) \oplus B = (X \oplus B) \cup (Y \oplus B)$$

$$\begin{aligned} (X \cup Y) \ominus B &\supseteq (X \ominus B) \cup (Y \ominus B) \\ B \ominus (X \cup Y) &= (X \ominus B) \cap (Y \ominus B) \end{aligned} \tag{11.27}$$

Successive dilation (respectively, erosion) of the image X first by the structuring element B and then by the structuring element D is equivalent to the dilation (erosion) of the image X by $B \oplus D$:

$$\begin{aligned} (X \oplus B) \oplus D &= X \oplus (B \oplus D) \\ (X \ominus B) \ominus D &= X \ominus (B \oplus D) \end{aligned} \tag{11.28}$$

11.3.3 Hit-or-miss transformation

The hit-or-miss transformation is the morphological operator for finding local patterns of pixels, where *local* means the size of the structuring element. It is a variant of template matching that finds collections of pixels with certain shape properties (such as corners, or border points). We shall see later that it may be used for thinning and thickening of objects (Section 11.5.3).

Operations described hitherto used a structuring element B, and we have tested points for their membership of X; we can also test whether some points do not belong to X. An operation may be denoted by a pair of disjoint sets $B = (B_1, B_2)$, called a **composite structuring element**. The **hit-or-miss** transformation $\otimes$ is defined as

$$X \otimes B = \{x : \; B_1 \subset X \text{ and } B_2 \subset X^c\} \tag{11.29}$$

This means that for a point x to be in the resulting set, two conditions must be fulfilled simultaneously. First the part B_1 of the composite structuring element that has its representative point at x must be contained in X, and second, the part B_2 of the composite structuring element must be contained in X^c.

The hit-or-miss transformation operates as a binary matching between an image X and the structuring element (B_1, B_2). It may be expressed using erosions and dilations as well

$$X \otimes B = (X \ominus B_1) \cap (X^c \ominus B_2) = (X \ominus B_1) \setminus (X \oplus \breve{B}_2) \tag{11.30}$$

11.3.4 Opening and closing

Erosion and dilation are not inverse transformations—if an image is eroded and then dilated, the original image is not re-obtained. Instead, the result is a simplified and less detailed version of the original image.

Erosion followed by dilation creates an important morphological transformation called **opening**. The opening of an image X by the structuring element B is denoted by $X \circ B$ and is defined as

$$X \circ B = (X \ominus B) \oplus B \tag{11.31}$$

Dilation followed by erosion is called **closing**. The closing of an image X by the structuring element B is denoted by $X \bullet B$ and is defined as

$$X \bullet B = (X \oplus B) \ominus B \tag{11.32}$$

If an image X is unchanged by opening with the structuring element B, it is called *open with respect to* B. Similarly, if an image X is unchanged by closing with B, it is called *closed with respect to* B.

Opening and closing with an isotropic structuring element is used to eliminate specific image details smaller than the structuring element—the global shape of the objects is not distorted. Closing connects objects that are close to each other, fills up small holes, and smooths the object outline by filling up narrow gulfs. Meanings of 'near', 'small', and 'narrow' are related to the size and the shape of the structuring element. Opening is illustrated in Figure 11.10, and closing in Figure 11.11.

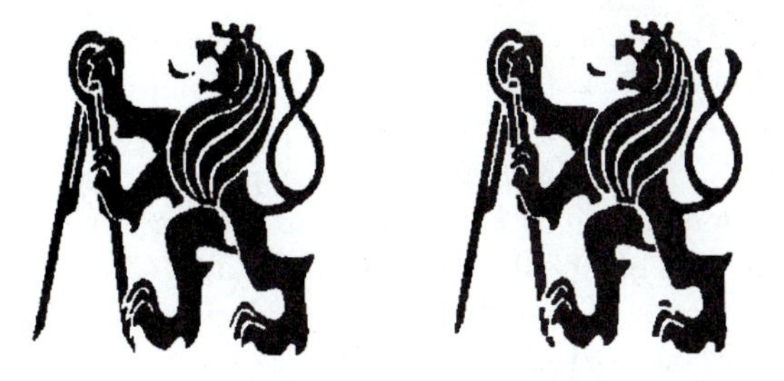

Figure 11.10: *Opening (original on the left).*

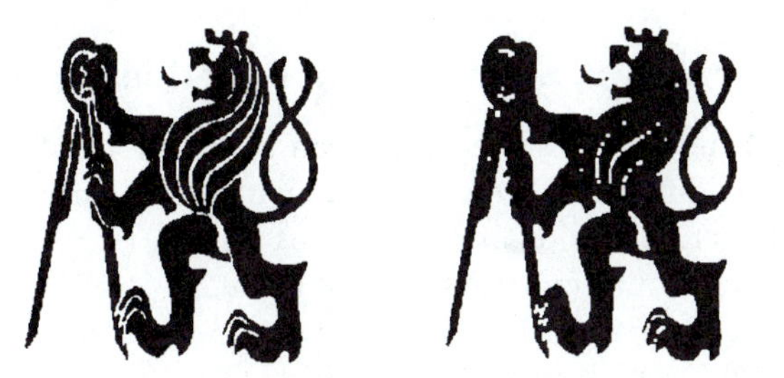

Figure 11.11: *Closing (original on the left).*

Unlike dilation and erosion, opening and closing are invariant to translation of the structuring element. Equations (11.14) and (11.20) imply that both opening and closing are increasing transformations. Opening is anti-extensive ($X \circ B \subseteq X$) and closing is extensive ($X \subseteq X \bullet B$).

Opening and closing, like dilation and erosion, are dual transformations:

$$(X \bullet B)^C = X^C \circ \breve{B} \tag{11.33}$$

Another significant fact is that iteratively used openings and closings are **idempotent**, meaning that reapplication of these transformations does not change the previous result. Formally,

$$X \circ B = (X \circ B) \circ B \tag{11.34}$$

$$X \bullet B = (X \bullet B) \bullet B \tag{11.35}$$

11.4 Gray-scale dilation and erosion

Binary morphological operations acting on binary images are easily extendible to gray-scale images using the 'min' and 'max' operations. Erosion (respectively, dilation) of an image is the operation of assigning to each pixel the minimum (maximum) value found over a neighborhood of the corresponding pixel in the input image. The structuring element is more rich than in the binary case, where it gave only the neighborhood. In the gray-scale case, the structuring element is a function of two variables that specifies the desired local gray-level property. The value of the structuring element is added (subtracted) when the maximum (or minimum) is calculated in the neighborhood.

This extension permits a **topographic view** of gray-scale images—the gray-level is interpreted as the height of a particular location of a hypothetical landscape. Light and dark spots in the image correspond to hills and hollows in the landscape. Such a morphological approach permits the location of global properties of the image, i.e., to identify characteristic topographic features on images as valleys, mountain ridges (crests), and watersheds.

We follow the explanation used in [Haralick and Shapiro 92] for gray-scale dilation and erosion, where the concept of **umbra** and **top of the point set** is introduced. Gray-scale dilation is expressed as the dilation of umbras.

11.4.1 Top surface, umbra, and gray-scale dilation and erosion

Consider a point set A in n-dimensional Euclidean space, $A \subset \mathcal{E}^n$, and assume that the first $(n-1)$ co-ordinates of the set constitute a spatial domain and the nth co-ordinate corresponds to the value of a function or functions at a point ($n = 3$ for gray-scale images). This interpretation matches the topographic view for a 2D Euclidean space, where points are given by triples of co-ordinates; the first two co-ordinates locate the position in the 2D support set and the third co-ordinate gives the height.

The **top surface** of a set A is a function defined on the $(n-1)$-dimensional support. For each $(n-1)$-tuple, the top surface is the highest value of the last co-ordinate of A, as illustrated in Figure 11.12. If the space is Euclidean the highest value means supremum.

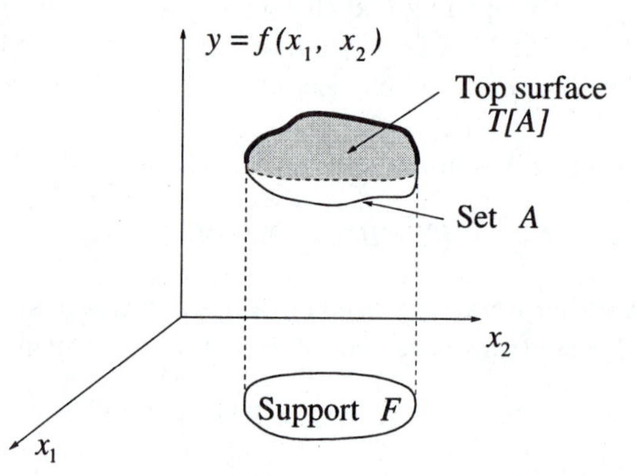

Figure 11.12: *Top surface of the set A corresponds to maximal values of the function $f(x_1, x_2)$.*

Let $A \subseteq \mathcal{E}^n$ and the support $F = \{x \in \mathcal{E}^{n-1}$ for some $y \in \mathcal{E}$, $(x, y) \in A\}$. The **top surface** of A, denoted by $T[A]$, is a mapping $F \to \mathcal{E}$ defined as

$$T[A](x) = \max\{y,\ (x, y) \in A\} \tag{11.36}$$

The next concept is the **umbra** of a function f defined on some subset F (support) of $(n-1)$-dimensional space. The usual definition of umbra is a region of complete shadow resulting from obstructing the light by a non-transparent object. In mathematical morphology, the umbra of f is a set that consists of the top surface of f and everything below it; see Figure 11.13.

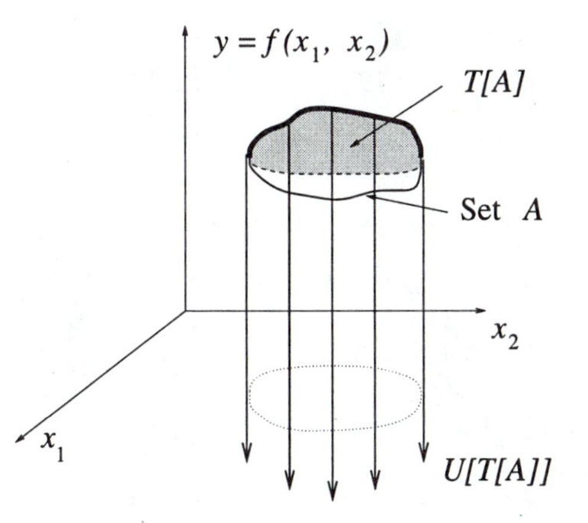

Figure 11.13: *Umbra of the top surface of a set is the whole subspace below it.*

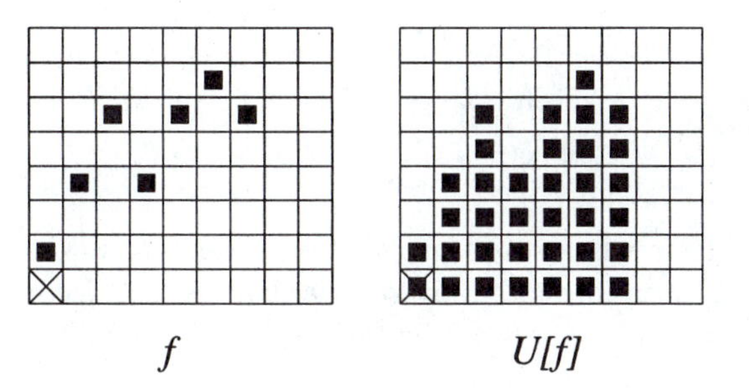

Figure 11.14: *Example of a 1D function (left) and its umbra (right).*

Formally, let $F \subseteq \mathcal{E}^{n-1}$ and $f : F \to \mathcal{E}$. The **umbra** of f, denoted by $U[f]$, $U[f] \subseteq F \times \mathcal{E}$, is defined by

$$U[f] = \{(x, y) \in F \times \mathcal{E},\ y \leq f(x)\} \tag{11.37}$$

We see that the umbra of an umbra of f is an umbra.

We can illustrate the top surface and umbra in the case of a simple 1D gray-scale image. Figure 11.14 illustrates a function f (which might be a top surface) and its umbra.

We can now define the gray-scale dilation of two functions as the top surface of the dilation of their umbras. Let $F, K \subseteq \mathcal{E}^{n-1}$ and $f : F \to \mathcal{E}$ and $k : K \to \mathcal{E}$. The **dilation** $\oplus$ of f by k, $f \oplus k : F \oplus K \to \mathcal{E}$ is defined by

$$f \oplus k = T\,\{U[f] \oplus U[k]\} \tag{11.38}$$

Notice here that $\oplus$ on the left-hand side is dilation in the gray-scale image domain, and $\oplus$ on the right-hand side is dilation in the binary image. A new symbol was not introduced since no confusion is expected here; the same applies to erosion $\ominus$ in due course.

Similarly to binary dilation, one function, say f, represents an image, and the second, k, a small structuring element. Figure 11.15 shows a discretized function k that will play the role of the structuring element. Figure 11.16 shows the dilation of the umbra of f (from the example given in Figure 11.14) by the umbra of k.

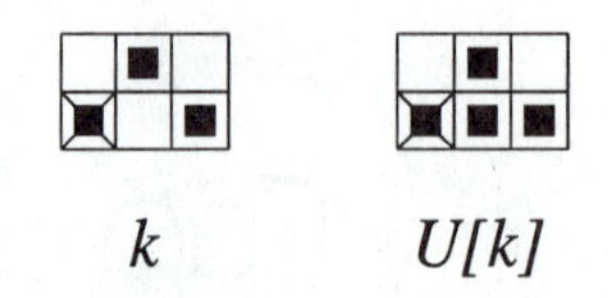

Figure 11.15: *A structuring element: 1D function (left) and its umbra (right).*

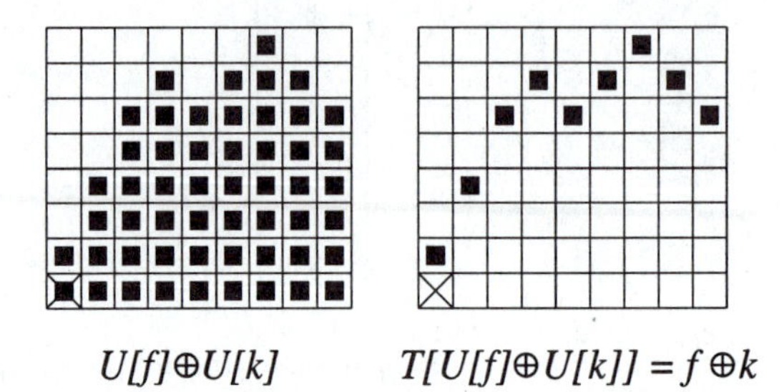

Figure 11.16: *1D example of gray-scale dilation. The umbras of the 1D function f and structuring element k are dilated first, $U[f] \oplus U[k]$. The top surface of this dilated set gives the result, $f \oplus k = T[U[f] \oplus U[k]]$.*

This definition explains what gray-scale dilation means, but does not give a reasonable algorithm for actual computations in hardware. We shall see that a computationally plausible way to calculate dilation can be obtained by taking the maximum of a set of sums:

$$(f \oplus k)(x) = \max\{f(x-z) + k(z),\ z \in K,\ x - z \in F\} \tag{11.39}$$

The computational complexity is the same as for convolution in linear filtering, where a summation of products is performed.

The definition of **gray-scale erosion** is analogous to gray-scale dilation. The gray-scale erosion of two functions (point sets)

1. Takes their umbras
2. Erodes them using binary erosion
3. Gives the result as the top surface

Let $F, K \subseteq \mathcal{E}^{n-1}$ and $f : F \to \mathcal{E}$ and $k : K \to \mathcal{E}$. The **erosion** $\ominus$ of f by k, $f \ominus k : F \ominus K \to \mathcal{E}$ is defined by

$$f \ominus k = T\,\{U[f] \ominus U[k]\} \tag{11.40}$$

Erosion is illustrated in Figure 11.17. To decrease computational complexity, the actual computations are performed in another way as the minimum of a set of differences (notice the similarity to correlation):

$$(f \ominus k)(x) = \min_{z \in K}\{f(x+z) - k(z)\} \tag{11.41}$$

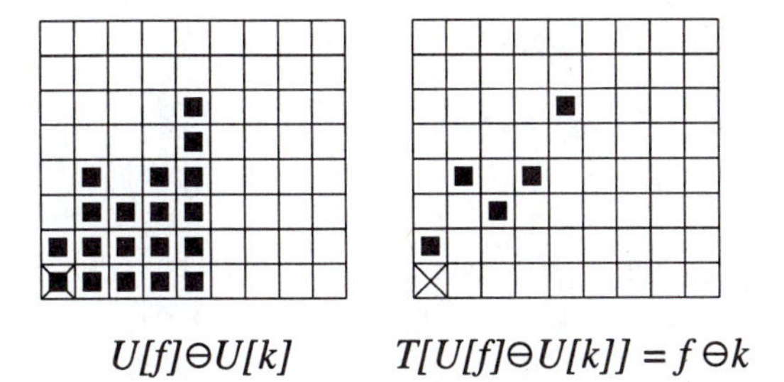

Figure 11.17: *1D example of gray-scale erosion. The umbras of 1D function f and structuring element k are eroded first, $U[f] \ominus U[k]$. The top surface of this eroded set gives the result, $f \ominus k = T[U[f] \ominus U[k]]$.*

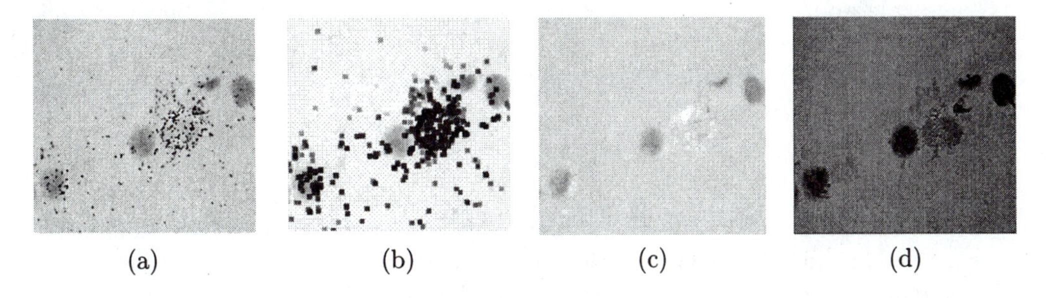

Figure 11.18: *Morphological pre-processing: (a) cells in a microscopic image corrupted by noise; (b) eroded image; (c) dilation of (b), the noise has disappeared; (d) reconstructed cells. Courtesy P. Kodl, Rockwell Automation Research Center, Prague, Czech Republic.*

We illustrate morphological pre-processing on a microscopic image of cells corrupted by noise in Figure 11.18a; the aim is to reduce noise and locate individual cells. Figure 11.18b shows erosion of the original image, and Figure 11.18c illustrates dilation of the original image. A 3×3 structuring element was used in both cases—notice that the noise has been considerably reduced. The individual cells can be located by the reconstruction operation (to be explained in Section 11.5.4). The original image is used as a mask and the dilated image in Figure 11.18c is an input for reconstruction. The result is shown in image 11.18d, in which the black spots depict the cells.

11.4.2 Umbra homeomorphism theorem, properties of erosion and dilation, opening and closing

The top surface always inverts the umbra operation; i.e., the top surface is a left inverse of the umbra, $T[U[f]] = f$. However, the umbra is not an inverse of the top surface. The strongest conclusion that can be deduced is that the umbra of the top surface of a point set A contains A (recall Figure 11.13).

The notion of top surface and umbra provides an intuitive relation between gray-scale and binary morphology. The **umbra homeomorphism theorem** states that the umbra operation is a homeomorphism from gray-scale morphology to binary morphology. Let $F, K \subseteq$

$\mathcal{E}^{n-1}$, $f : F \to \mathcal{E}$, and $k : K \to \mathcal{E}$. Then

$$\begin{aligned} &\text{(a)} \qquad U[f \oplus k] = U[f] \oplus U[k] \\ &\text{(b)} \qquad U[f \ominus k] = U[f] \ominus U[k] \end{aligned} \tag{11.42}$$

(the proof may be found elsewhere [Haralick and Shapiro 92]). The umbra homeomorphism is used for deriving properties of gray-scale operations. The operation is expressed in terms of umbra and top surface, then transformed to binary sets using the umbra homeomorphism property, and finally transformed back using the definitions of gray-scale dilation and erosion. Using this idea, properties already known from binary morphology can be derived, e.g., commutativity of dilation, the chain rule that permits decomposition of large structural elements into successive operations with smaller ones, duality between erosion and dilation.

Gray-scale opening and closing is defined in the same way as in the binary morphology. **Gray-scale opening** is defined as $f \circ k = (f \ominus k) \oplus k$. Similarly, **gray-scale closing** $f \bullet k = (f \oplus k) \ominus k$. The **duality** between opening and closing is expressed as (recall that $\breve{k}$ means the transpose, i.e., symmetric set with regards to origin of co-ordinates)

$$-(f \circ k)(x) = [(-f) \bullet \breve{k}](x) \tag{11.43}$$

There is a simple geometric interpretation of gray-scale opening; see [Haralick and Shapiro 92] for derivation and details. The opening of f by structuring element k can be interpreted as sliding k on the landscape f. The position of all highest points reached by some part of k during the slide gives the opening, and a similar interpretation exists for erosion.

Gray-scale opening and closing is often used in applications to extract parts of a gray-scale image with given shape and gray-scale structure.

11.4.3 Top hat transformation

The top hat transformation is used as a simple tool for segmenting objects in gray-scale images that differ in brightness from background, even when the background is of uneven gray-scale. The top hat transform is superseded by the watershed segmentation (to be described in Section 11.7.3) for more complicated backgrounds.

Assume a gray-level image X and a structuring element K. The residue of opening as compared to original image $X \setminus (X \circ K)$ constitutes a new useful operation called a **top hat transformation** [Meyer 78].

The top hat transformation is a good tool for extracting light objects (or, conversely, dark ones, of course) on a dark (or light) but slowly changing background. Those parts of the image that cannot fit into structuring element K are removed by opening. Subtracting the opened image from the original provides an image where removed objects stand out clearly. The actual segmentation can be performed by simple thresholding. The concept is illustrated for the 1D case in Figure 11.19, where we can see the origin of the transformation name. If an image were a hat, the transformation would extract only the top of it, provided that the structuring element is larger than the hole in the hat.

An example from visual industrial inspection provides a practical application of gray-level morphology and the top hat transformation. A factory producing glass capillaries for mercury maximal thermometers had the following problem: The thin glass tube should be narrowed

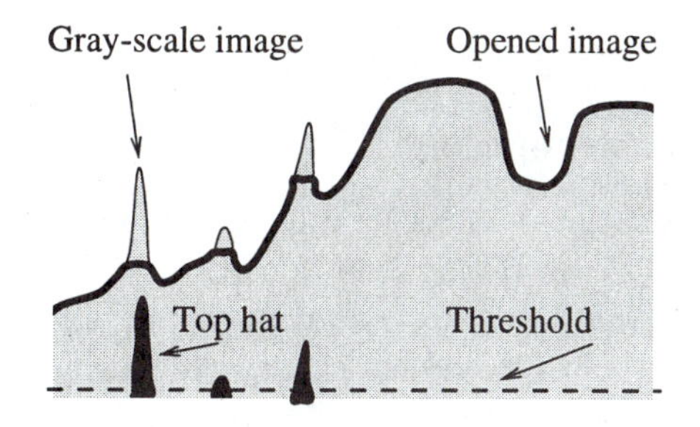

Figure 11.19: *The top hat transform permits the extraction of light objects from an uneven background.*

in one particular place to prevent mercury falling back when the temperature decreases from the maximal value. This is done by using a narrow gas flame and low pressure in the capillary. The capillary is illuminated by a collimated light beam—when the capillary wall collapses due to heat and low pressure, an instant specular reflection is observed and serves as a trigger to cover the gas flame. Originally the machine was controlled by a human operator who looked at the tube image projected optically on the screen; the gas flame was covered when the specular reflection was observed. This task had to be automated and the trigger signal learned from a digitized image. The specular reflection is detected by a morphological procedure—see Figure 11.20.

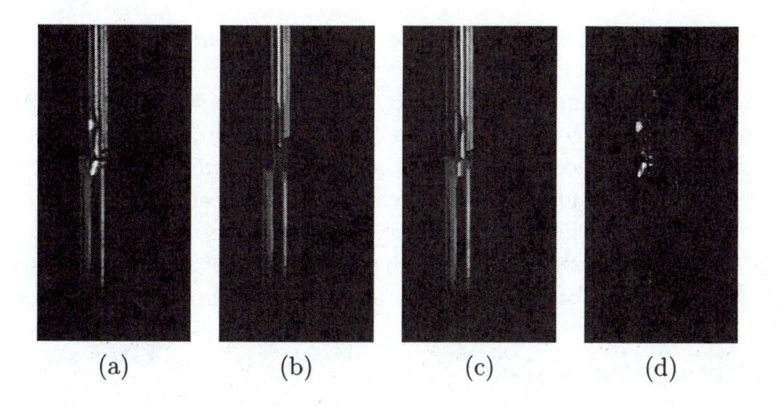

Figure 11.20: *An industrial example of gray-scale opening and top hat segmentation, i.e., image-based control of glass tube narrowing by gas flame: (a) original image of the glass tube, 512×256 pixels; (b) erosion by a one-pixel-wide vertical structuring element 20 pixels long; (c) opening with the same element; (d) final specular reflection segmentation by the top hat transformation. Courtesy V. Smutný, R. Šára, CTU Prague, P. Kodl, Rockwell Automation Research Center, Prague, Czech Republic.*

11.5 Skeletons and object marking

11.5.1 Homotopic transformations

Topological properties are associated with continuity (Section 2.3.1), and mathematical morphology can be used to study such properties of objects in images. There is an interesting group among morphological transformations called **homotopic transformations** [Serra 82].

A transformation is homotopic if it does not change the continuity relation between regions and holes in the image. This relation is expressed by the homotopic tree; its root corresponds to the background of the image, first-level branches correspond to the objects (regions), second-level branches match holes within objects, etc.

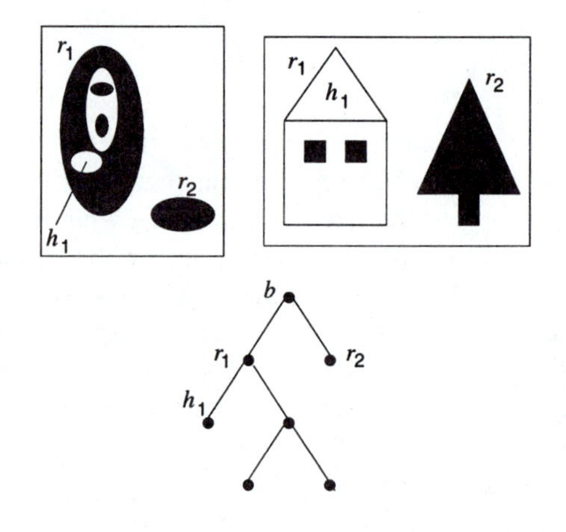

Figure 11.21: *The same homotopic tree for two different images.*

Figure 11.21 shows an example of a homotopic tree in which there are two different images with the homotopic tree below them. On the left side are some biological cells and on the right side a house and a spruce tree; both images have the same homotopic tree. Its root b corresponds to the background, node r_1 matches the larger cell (the outline of the house), and node r_2 matches the smaller cell (the spruce tree). Node h_1 corresponds to the empty hole in the cell r_1 (the hole inside the roof of the house)—the other correspondences to nodes should now be clear. A transformation is homotopic if it does not change the homotopic tree.

11.5.2 Skeleton, maximal ball

It is sometimes advantageous to convert an object to an archetypical stick figure called a **skeleton** (also considered in Section 6.3.4). We shall explain this in the context of 2D Euclidean space first, which is more illustrative than on the digital grid that we shall consider later.

The idea of skeleton was introduced by Blum under the name **medial axis transform** [Blum 67] and illustrated on the following 'grassfire' scenario. Assume a region (point set) $X \subset \mathcal{R}^2$: A grassfire is lit on the entire region boundary at the same instant, and propagates towards the region interior with constant speed. The **skeleton** $S(X)$ is the set of points where two or more firefronts meet; see Figure 11.22.

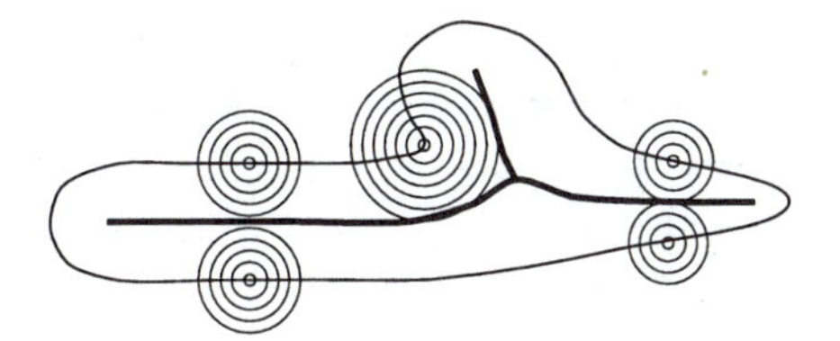

Figure 11.22: *Skeleton as points where two or more wavefronts of grassfire meet.*

A more formal definition of skeleton is based on the concept of maximal ball. A **ball** $B(p,r)$ with center p and radius r, $r \geq 0$, is the set of points with distances d from the center less than or equal to r.

The ball B included in a set X is said to be **maximal** if and only if there is no larger ball included in X that contains B, i.e., each ball B', $B \subseteq B' \subseteq X \Longrightarrow B' = B$. Balls and maximal balls are illustrated in Figure 11.23.

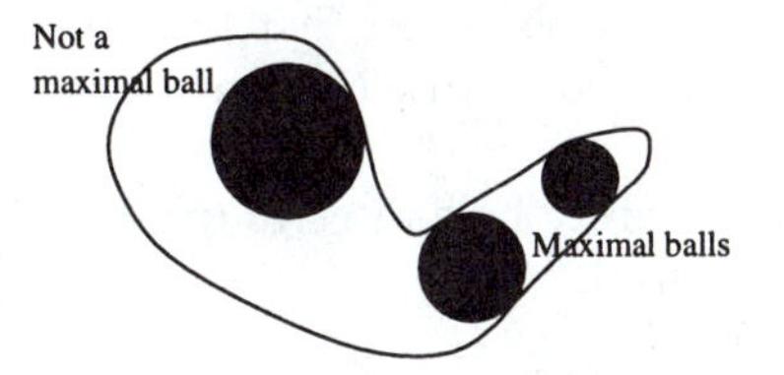

Figure 11.23: *Ball and maximal balls in Euclidean plane.*

The distance metric d that is used depends on the grid and definition of connectivity. Unit balls in a plane (i.e., unit disks) are shown in Figure 11.24.

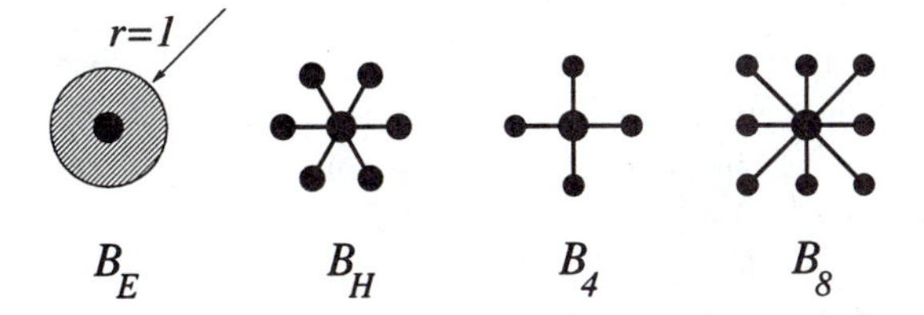

Figure 11.24: *Unit-size disk for different distances, from left side: Euclidean distance, 6-, 4-, and 8-connectivity, respectively.*

The plane $\mathcal{R}^2$ with the usual Euclidean distance gives the ball B_E. Three distances and balls are often defined in the discrete plane $\mathcal{Z}^2$. If a hexagonal grid and 6-connectivity is used, the hexagonal ball B_H is obtained. If the support is a square grid, two unit balls are possible: B_4 for 4-connectivity and B_8 for 8-connectivity.

The **skeleton by maximal balls** $S(X)$ of a set $X \subset \mathcal{Z}^2$ is the set of centers p of maximal balls:

$$S(X) = \{p \in X : \exists r \geq 0,\ B(p,r) \text{ is a maximal ball of } X\}$$

This definition of skeleton has an intuitive meaning in the Euclidean plane. The skeleton of a disk reduces to its center, the skeleton of a stripe with rounded endings is a unit thickness line at its center, etc.

Figure 11.25 shows several objects together with their skeletons—a rectangle, two touching balls, and a ring. The properties of the (Euclidean) skeleton can be seen here—in particular, the skeleton of two adjacent circles consists of two distinct points instead of a straight line joining these two points, as might be intuitively expected.

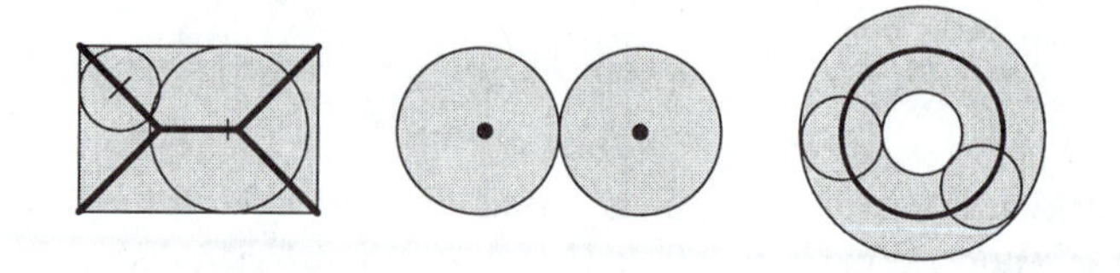

Figure 11.25: *Skeletons of rectangle, two touching balls, and a ring.*

The skeleton by maximal balls has two unfortunate properties in practical applications. First, it does not necessarily preserve the homotopy (connectivity) of the original set; and second, some of the skeleton lines may be wider than one pixel in the discrete plane. We shall see later that the skeleton is often substituted by sequential homotopic thinning that does not have these two properties.

Dilation can be used in any of the three discrete connectivities to create balls of varying radii. Let nB be the ball of radius n:

$$nB = B \oplus B \oplus \ldots \oplus B \tag{11.44}$$

The skeleton by maximal balls can be obtained as the union of the residues of opening of the set X at all scales [Serra 82]:

$$S(X) = \bigcup_{n=0}^{\infty} [(X \ominus nB) \setminus (X \ominus nB) \circ B] \tag{11.45}$$

The trouble with this is that the resulting skeleton is completely disconnected and this property is not useful in many applications. Thus **homotopic skeletons** that preserve connectivity are often preferred. We present an intuitive homotopic skeletonization algorithm based on consecutive erosions (thinning) in Section 11.5.3.

11.5.3 Thinning, thickening, and homotopic skeleton

One application of the hit-or-miss transformation (Section 11.3.3) is **thinning** and **thickening** of point sets. For an image X and a composite structuring element $B = (B_1, B_2)$ (notice that B here is not a ball), *thinning* is defined as

$$X \oslash B = X \setminus (X \otimes B) \tag{11.46}$$

and *thickening* is defined as

$$X \odot B = X \cup (X \otimes B) \tag{11.47}$$

When thinning, a part of the boundary of the object is subtracted from it by the set difference operation. When thickening, a part of the boundary of the background is added to the object. Thinning and thickening are dual transformations:

$$(X \odot B)^c = X^c \oslash B \quad B = (B_2, B_1) \tag{11.48}$$

Thinning and thickening transformations are very often used sequentially. Let $\{B_{(1)}, B_{(2)}, B_{(3)}, \ldots, B_{(n)}\}$ denote a sequence of composite structuring elements $B_{(i)} = (B_{i_1}, B_{i_2})$. **Sequential thinning** can then be expressed as a sequence of eight structuring elements for square rasters,

$$X \oslash \{B_{(i)}\} = (((X \oslash B_{(1)}) \oslash B_{(2)}) \ldots \oslash B_{(n)}) \tag{11.49}$$

and **sequential thickening** as

$$X \odot \{B_{(i)}\} = (((X \odot B_{(1)}) \odot B_{(2)}) \ldots \odot B_{(n)}) \tag{11.50}$$

There are several sequences of structuring elements $\{B_{(i)}\}$ that are useful in practice. Many of them are given by a permissible rotation of a structuring element in the appropriate digital raster (e.g., hexagonal, square, or octagonal). These sequences, sometimes called the **Golay alphabet** [Golay 69], are summarized for the hexagonal raster in [Serra 82]. We shall present structuring elements of the Golay alphabet for octagonal rasters. 3×3 matrices will be shown for the first two rotations, from which the other rotations can easily be derived.

A composite structuring element can be expressed by a single matrix only. A value of one in it means that this element belongs to B_1 (it is a subset of objects in the hit-or-miss transformation), and a value zero belongs to B_2 and is a subset of the background. An asterisk $*$ in the matrix denotes an element that is not used in the matching process, so its value is not significant.

Thinning and thickening sequential transformations converge to some image—the number of iterations needed depends on the objects in the image and the structuring element used. If two successive images in the sequence are identical, the thinning (or thickening) is stopped.

Sequential thinning by structuring element L

This sequential thinning is quite important, as it serves as the homotopic substitute of the skeleton; the final thinned image consists only of lines of width one and isolated points.

The structuring element L from the Golay alphabet is given by

$$L_1 = \begin{bmatrix} 0 & 0 & 0 \\ * & 1 & * \\ 1 & 1 & 1 \end{bmatrix} \quad L_2 = \begin{bmatrix} * & 0 & 0 \\ 1 & 1 & 0 \\ * & 1 & * \end{bmatrix} \ldots \tag{11.51}$$

(The other six elements are given by rotation). Figure 11.26 shows the result of thinning with the structuring element L, after five iterations to illustrate an intermediate result, and Figure 11.27 shows the homotopic substitute of the skeleton when the idempotency was reached (in both cases, the original is shown on the left).

Sequential thinning by structuring element E

Assume that the homotopic substitute of the skeleton by element L has been found. The skeleton is usually jagged, because of sharp points on the outline of the object, but it is possible to 'smooth' the skeleton by sequential thinning by structuring element E. Using n iterations, several points (whose number depends on n) from the lines of width one (and isolated points as well) are removed from free ends. If thinning by element E is performed until the image does not change, then only closed contours remain.

Figure 11.26: *Sequential thinning using element L after five iterations.*

Figure 11.27: *Homotopic substitute of the skeleton (element L).*

The structuring element E from the Golay alphabet is given again by eight rotated masks,

$$E_1 = \begin{bmatrix} * & 1 & * \\ 0 & 1 & 0 \\ 0 & 0 & 0 \end{bmatrix} \quad E_2 = \begin{bmatrix} 0 & * & * \\ 0 & 1 & 0 \\ 0 & 0 & 0 \end{bmatrix} \quad \dots \tag{11.52}$$

Figure 11.28 shows sequential thinning (five iterations) by the element E of the skeleton from Figure 11.27. Notice that lines have been shortened from their free ends.

Figure 11.28: *Five iterations of sequential thinning by element E.*

There are three other elements M, D, C in the Golay alphabet [Golay 69]. These are not much used in practice at present, and other morphological algorithms are used instead to find skeletons, convex hulls, and homotopic markers.

The computationally most efficient algorithm of which we are aware creates the connected skeleton as the minimal superset of the skeleton by maximal balls [Vincent 91]. Its performance is shown in Figure 11.29. The homotopy is preserved.

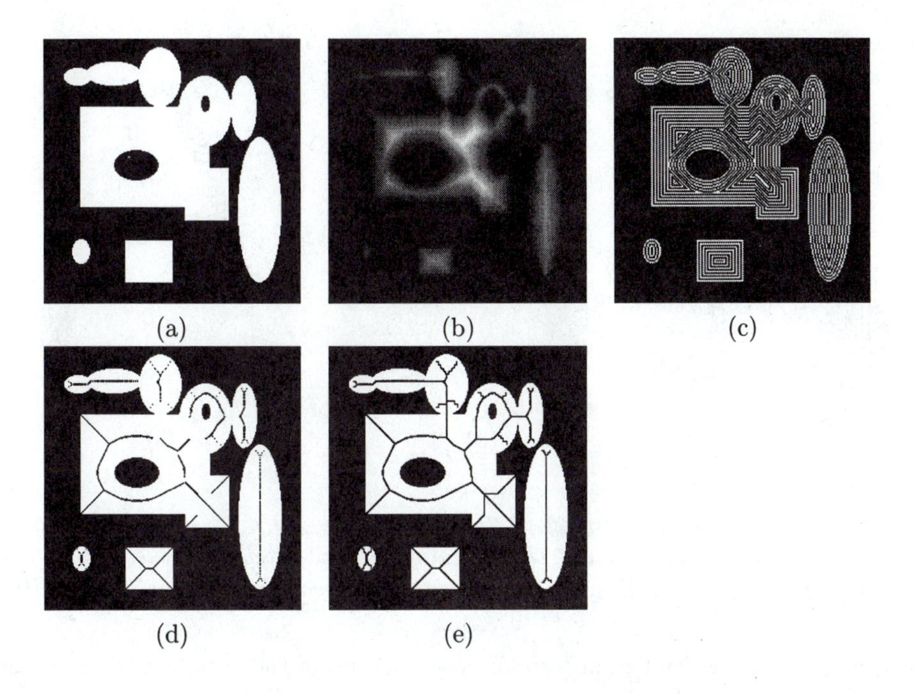

Figure 11.29: *Performance of Vincent's quick skeleton by maximal balls algorithm: (a) original binary image; (b) distance function (to be explained later); (c) distance function visualized by contouring; (d) non-continuous skeleton by maximal balls; (e) final skeleton.*

The skeleton can be applied to native 3D images as well, e.g., in the analysis of computer tomography images. Figure 11.30 illustrates examples of thinning of 3D point sets; parallel algorithms are available [Ma and Sonka 96].

11.5.4 Quench function, ultimate erosion

The binary point set X can be described using maximal balls B. Every point p of the skeleton $S(X)$ by maximal balls has an associated ball of radius $q_X(p)$; the term **quench function** is used for this association. An important property of $q_X(p)$ is that the quench function permits the reconstruction of the original set X completely as a union of its maximal balls B:

$$X = \bigcup_{p \in S(X)} [p + q_X(p) B] \tag{11.53}$$

This formula allows lossless compression of a binary image. Similar ideas are used for encoding documents in CCITT group 4 compression algorithms.

It is useful to distinguish several types of extrema, and use the analogy of a topographic view of images to make the explanation more intuitive. The **global maximum** is the pixel with highest value (lightest pixel, highest summit in the countryside); similarly, the **global minimum** corresponds to the deepest chasm in the countryside.

A pixel p of a gray-scale image is a **local maximum** if and only if for every neighboring pixel q of a pixel p, $I(p) \geq I(q)$. For example, the local maximum may mean that

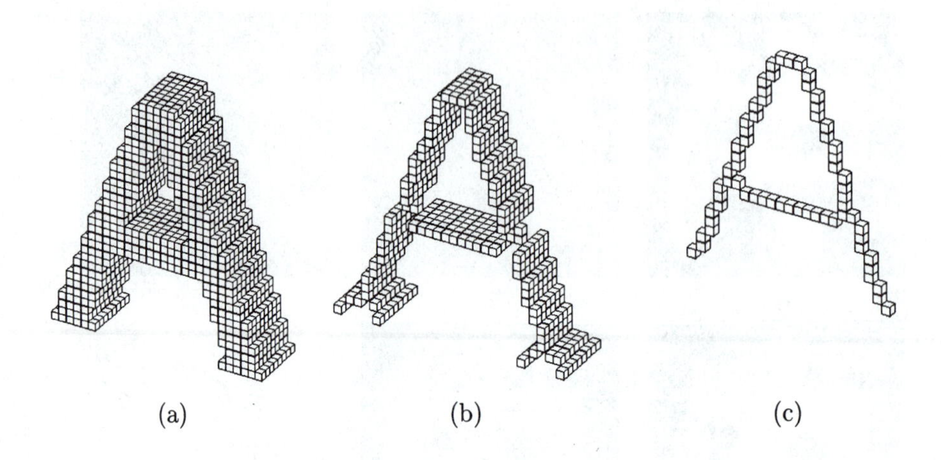

Figure 11.30: *Morphological thinning in 3D: (a) original 3D data set, a character A; (b) thinning performed in one direction; (c) resulting of one pixel thick skeleton obtained by thinning image (b) in second direction. Courtesy K. Palágyi, University of Szeged, Hungary.*

the landscape around is studied in a small neighborhood of the current position (neighborhood in morphology is defined by the structuring element). If no ascent is seen within the neighborhood, the pixel is at a local maximum.

The **regional maximum** M of a digital gray-scale image (function) I is a connected set of pixels with an associated value h (plateau at altitude h), such that every neighboring pixel of M has strictly lower value than h. There is no connected path leading upwards from a regional maximum. Topographically, regional extrema correspond to geographic summits and hollows. If M is a regional maximum of I and $p \in M$, then p is a local maximum. The converse does not hold. Global, local, and regional maxima are illustrated for the 1D case in Figure 11.31.

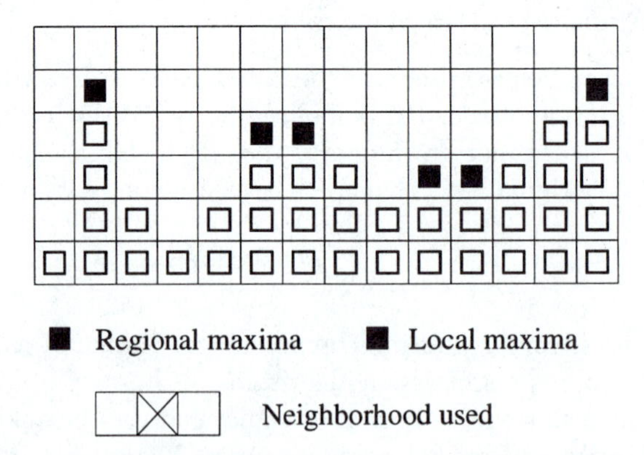

Figure 11.31: *1D illustration of global, regional, and local maxima.*

The definition of various maxima allows us to analyze the quench function. The quench function is also useful to define **ultimate erosion**, which is often used as a marker of convex

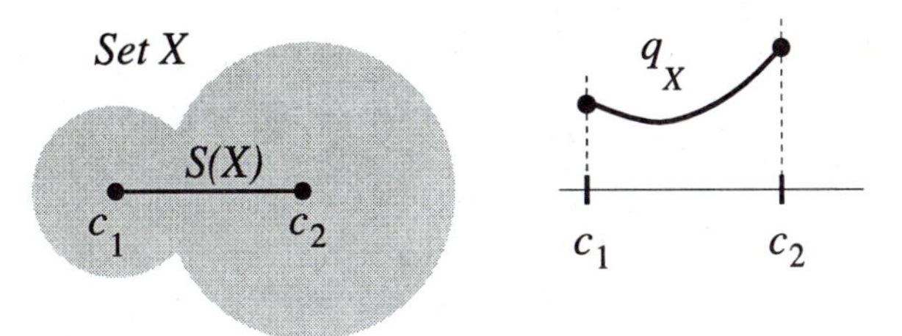

Figure 11.32: *Skeleton of a set X, and associated quench function $q_X(p)$. Regional maxima give the ultimate erosion.*

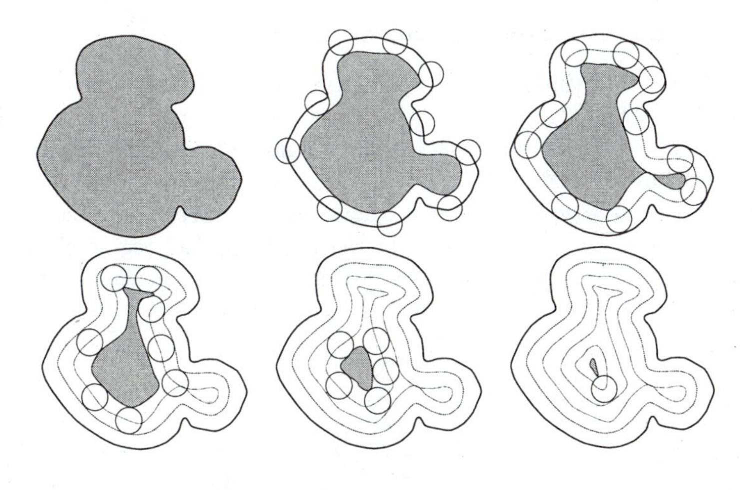

Figure 11.33: *When successively eroded, the components are first separated from the rest and finally disappear from the image. The union of residua just before disappearance gives ultimate erosion.*

objects in binary images. The **ultimate erosion** of a set X, denoted $\text{Ult}(X)$, is the set of regional maxima of the quench function. The natural markers are centers of the largest maximal balls. The trouble arises if the objects are overlapping—here ultimate erosion comes into play. Consider first the simplest case, in which the set X consists of two overlapping disks (see Figure 11.32). The skeleton is a line segment between the centers. The associated quench function has regional maxima that are located at the disk centers in this particular example. These maxima are called ultimate erosion and can be used as markers of overlapping objects. The ultimate erosion provides a tool that extracts one marker per object of a given shape, even if objects overlap. The remaining trouble is that some objects are still multiply marked.

Consider a binary image, a set X, consisting of three rounded overlapping components of different size. When iteratively eroding the set by a unit-size ball, the set is shrunk, then separates, and finally disappears, as illustrated in Figure 11.33. During the successive erosions, the residuals of connected components (just before they disappear) are stored. Their union is the ultimate erosion of the original set X (Figure 11.34).

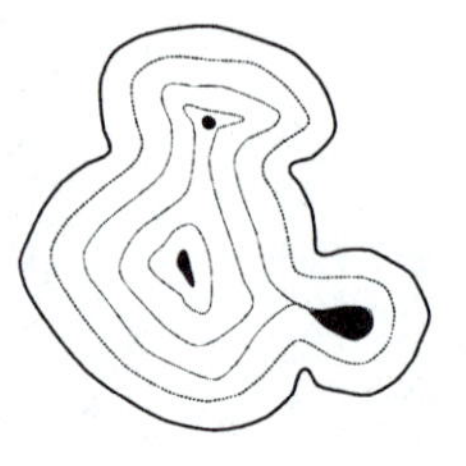

Figure 11.34: *Ultimate erosion is the union of residual connected components before they disappear during erosions.*

11.5.5 Ultimate erosion and distance functions

We seek to present this ultimate erosion procedure formally, and introduce the **morphological reconstruction** operator for this purpose. Assume two sets A, B, $B \subseteq A$. The reconstruction $\rho_A(B)$ of the set A from set B is the union of connected components of A with non-empty intersection with B (see Figure 11.35—notice that set A consists of two components). Notice that B may typically consist of markers that permit the reconstruction of the required part of the set A. Markers point to the pixel or small region that belongs to the object. Morphological reconstruction will be discussed in detail in Section 11.5.7.

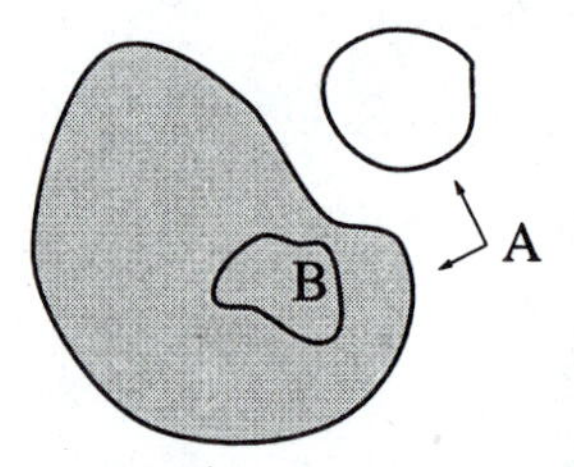

Figure 11.35: *Reconstruction $\rho_A(B)$ (in gray) of the set A from the set B. Notice that set A may consist of more than one connected component.*

Let $\mathcal{N}$ be the set of integers. Ultimate erosion can be expressed by the formula

$$\mathrm{Ult}(X) = \bigcup_{n \in \mathcal{N}} \left((X \ominus nB) \setminus \rho_{X \ominus nB}[X \ominus (n+1)B] \right) \tag{11.54}$$

There is a computationally effective ultimate erosion algorithm that uses the distance function (which is the core of several other quick morphological algorithms, as we shall see). The distance function $\mathrm{dist}_X(p)$ associated with each pixel p of the set X is the size of the first erosion of X that does not contain p, i.e.,

$$\forall p \in X \quad \mathrm{dist}_X(p) = \min\{n \in \mathcal{N}, p \text{ not in } (X \ominus nB)\} \tag{11.55}$$

This behaves as one would expect: $\mathrm{dist}_X(p)$ is the shortest distance between the pixel p and background X^C.

There are two direct applications of the distance function.

- The ultimate erosion of a set X corresponds to the union of the regional maxima of the distance function of X.

- The skeleton by maximal balls of a set X corresponds to the set of local maxima of the distance function of X.

The last concept that will be introduced here is **skeleton by influence zones**, often abbreviated **SKIZ**. Let X be composed of n connected components X_i, $i = 1, \ldots, n$. The influence zone $Z(X_i)$ consists of points which are closer to set X_i than to any other connected component of X.

$$Z(X_i) = \{p \in \mathcal{Z}^2,\ \forall i \neq j,\ d(p, X_i) \leq d(p, X_j)\} \tag{11.56}$$

The **skeleton by influence zones** denoted SKIZ(X) is the set of boundaries of influence zones $\{Z(X_i)\}$.

11.5.6 Geodesic transformations

Geodesic methods [Vincent 95] modify morphological transformations to operate only on some part of an image. For instance, if an object is to be reconstructed from a marker, say a nucleus of a cell, it is desirable to avoid growing from a marker outside the cell. Another important advantage of geodesic transformations is that the structuring element can vary at each pixel, according to the image.

The basic concept of geodesic methods in morphology is geodesic distance. The path between two points is constrained within some set. The term has its roots in an old discipline—geodesy—that measures distances on the Earth's surface. Suppose that a traveler seeks the distance between London and Tokyo—the shortest distance passes *through* the Earth, but obviously the geodesic distance that is of interest to the traveler is constrained to the Earth's surface.

The **geodesic distance** $d_X(x, y)$ is the shortest path between two points x, y while this path remains entirely contained in the set X. If there is no path connecting points x, y, we set the geodesic distance $d_X(x, y) = +\infty$. Geodesic distance is illustrated in Figure 11.36.

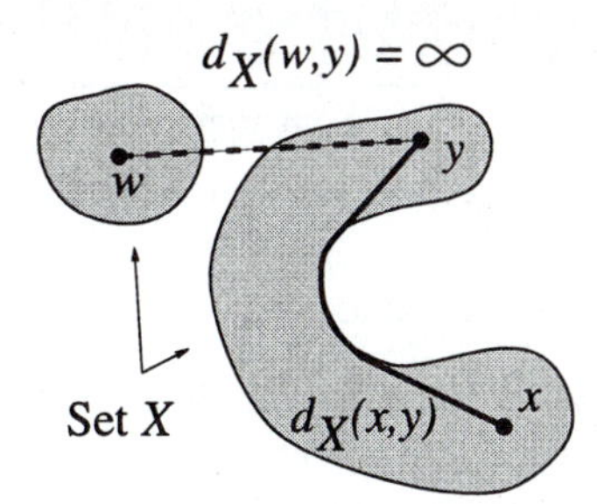

Figure 11.36: *Geodesic distance $d_X(x, y)$.*

The geodesic ball is the ball constrained by some set X. The **geodesic ball** $B_X(p, n)$ of center $p \in X$ and radius n is defined as

$$B_X(p, n) = \{p' \in X,\ d_X(p, p') \leq n\} \tag{11.57}$$

The existence of a geodesic ball permits dilation or erosion only within some subset of the image; this leads to definitions of geodetic dilations and erosions of a subset Y of X.

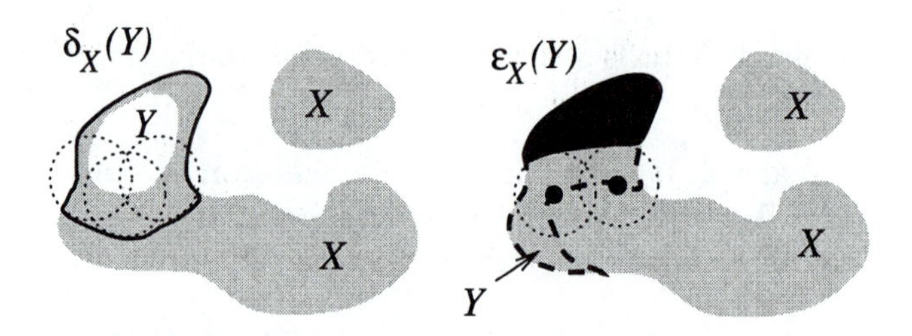

Figure 11.37: *Illustration of geodesic dilation (left) and erosion (right) of the set Y inside the set X.*

The **geodesic dilation** $\delta_X^{(n)}$ of size n of a set Y inside the set X is defined as

$$\delta_X^{(n)}(Y) = \bigcup_{p \in Y} B_X(p,n) = \{p' \in X,\ \exists p \in Y,\ d_X(p,p') \le n\} \tag{11.58}$$

Similarly the dual operation of **geodesic erosion** $\epsilon_X^{(n)}(Y)$ of size n of a set Y inside the set X can be written as

$$\epsilon_X^{(n)}(Y) = \{p \in Y,\ B_X(p,n) \subseteq Y\} = \{p \in Y,\ \forall p' \in X \setminus Y,\ d_X(p,p') > n\} \tag{11.59}$$

Geodesic dilation and erosion are illustrated in Figure 11.37.

The outcome of a geodesic operation on a set $Y \subseteq X$ is always included within the set X. Regarding implementation, the simplest geodesic dilation of size 1 ($\delta_X^{(1)}$) of a set Y inside X is obtained as the intersection of the unit-size dilation of Y (with respect to the unit ball B) with the set X.

$$\delta_X^{(1)} = (Y \oplus B) \cap X \tag{11.60}$$

Larger geodesic dilations are obtained by iteratively composing unit dilations n times

$$\delta_X^{(n)} = \underbrace{\delta_X^{(1)}(\delta_X^{(1)}[\delta_X^{(1)} \dots (\delta_X^{(1)})])}_{n\ times} \tag{11.61}$$

The fast iterative way to calculate geodesic erosion is similar.

11.5.7 Morphological reconstruction

Assume that we want to reconstruct objects of a given shape from a binary image that was originally obtained by thresholding. All connected components in the input image constitute the set X. However, only some of the connected components were marked by markers that represent the set Y. This task and its desired result are shown in Figure 11.38.

Successive geodesic dilations of the set Y inside the set X enable the reconstruction of the connected components of X that were initially marked by Y. When dilating from the marker, it is impossible to intersect a connected component of X which did not initially contain a marker Y; such components disappear.

Geodesic dilations terminate when all connected components set X previously marked by Y are reconstructed, i.e., idempotency is reached:

$$\forall n > n_0,\ \delta_X^{(n)}(Y) = \delta_X^{(n_0)}(Y) \tag{11.62}$$

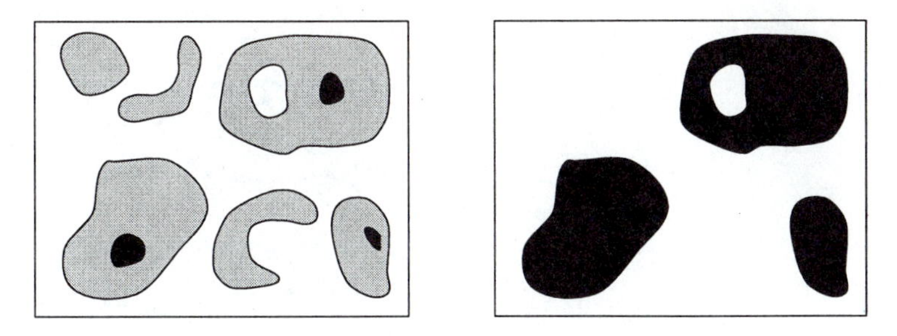

Figure 11.38: *Reconstruction of X (shown in light gray) from markers Y (black). The reconstructed result is shown in black on the right side.*

This operation is called **reconstruction** and denoted by $\rho_X(Y)$. Formally,

$$\rho_X(Y) = \lim_{n \to \infty} \delta_X^{(n)}(Y) \tag{11.63}$$

In some applications it is desirable that one connected component of X is marked by several markers Y. If it is not acceptable for the sets grown from various markers to become connected, the notion of influence zones can be generalized to **geodesic influence zones** of the connected components of set Y inside X. The idea is illustrated in Figure 11.39.

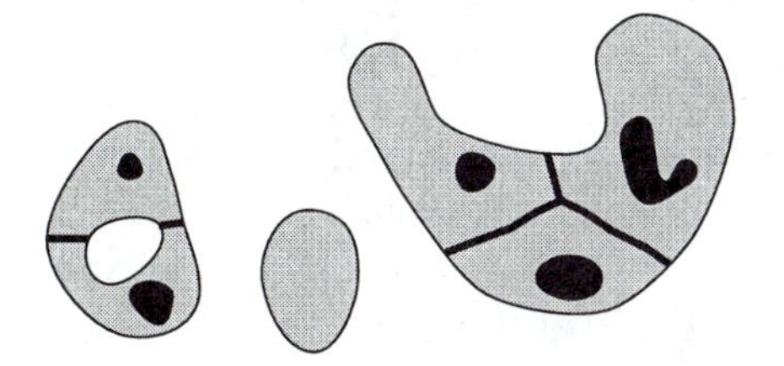

Figure 11.39: *Geodesic* SKIZ.

We are now ready to generalize the reconstruction to gray-scale images; this requires the extension of geodesy to gray-scale images. The core of the extension is the statement (which is valid for discrete images) that any increasing transformation defined for binary images can be extended to gray-level images [Serra 82]. By this transformation we mean a transformation Ψ such that

$$\forall X, Y \subset \mathcal{Z}^2, \; Y \subseteq X \Longrightarrow \Psi(Y) \subseteq \Psi(X) \tag{11.64}$$

The generalization of transformation Ψ is achieved by viewing a gray-level image I as a stack of binary images obtained by successive thresholding—this is called the threshold decomposition of image I [Maragos and Ziff 90]. Let D_I be the domain of the image I, and the gray values of image I be in $\{0, 1, \ldots, N\}$. The thresholded images $T_k(I)$ are

$$T_k(I) = \{p \in D_I, \; I(P) \geq k\} \quad k = 0, \ldots, N \tag{11.65}$$

The idea of threshold decomposition is illustrated in Figure 11.40.

Threshold-decomposed images $T_k(I)$ obey the inclusion relation

$$\forall k \in [1, N], \; T_k(I) \subseteq T_{k-1}(I) \tag{11.66}$$

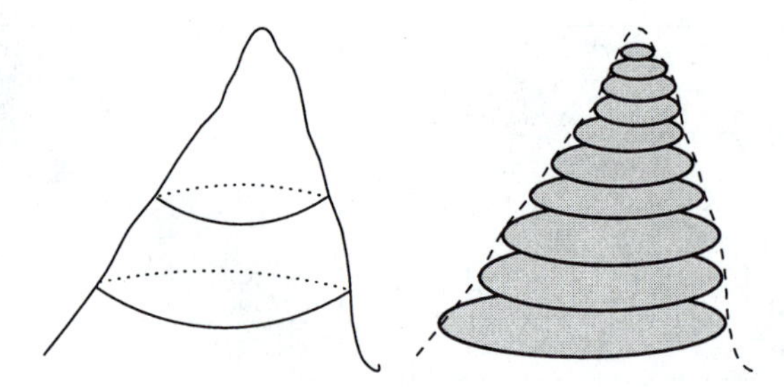

Figure 11.40: *Threshold decomposition of a gray-scale image.*

Consider the increasing transformation Ψ applied to each threshold-decomposed image; their inclusion relationship is kept. The transformation Ψ can be extended to gray-scale images using the following **threshold decomposition principle**:

$$\forall p \in D_I, \ , \Psi(I)(p) = \max\{k \in [0, \dots, N], \ p \in \Psi(T_k(I))\} \tag{11.67}$$

Returning to the reconstruction transformation, the binary geodesic reconstruction ρ is an increasing transformation, as it satisfies

$$Y_1 \subseteq Y_2, \ X_1 \subseteq X_2, \ Y_1 \subseteq X_1, \ Y_2 \subseteq X_2, \ \Longrightarrow \rho_{X_1}(Y_1) \subseteq \rho_{X_2}(Y_2) \tag{11.68}$$

We are ready to generalize binary reconstruction to **gray-level reconstruction** applying the threshold decomposition principle [equation (11.67)]. Let J, I be two gray-scale images defined on the same domain D, with gray-level values from the discrete interval $[0, 1, \dots, N]$. If, for each pixel $p \in D$, $J(p) \leq I(p)$, the gray-scale reconstruction $\rho_I(J)$ of image I from image J is given by

$$\forall p \in D, \ \rho_I(J)(p) = \max\{k \in [0, N], \ p \in \rho_{T_k}[T_K(J)]\} \tag{11.69}$$

Recall that binary reconstruction grows those connected components of the mask which are marked. The gray-scale reconstruction extracts peaks of the mask I that are marked by J (see Figure 11.41).

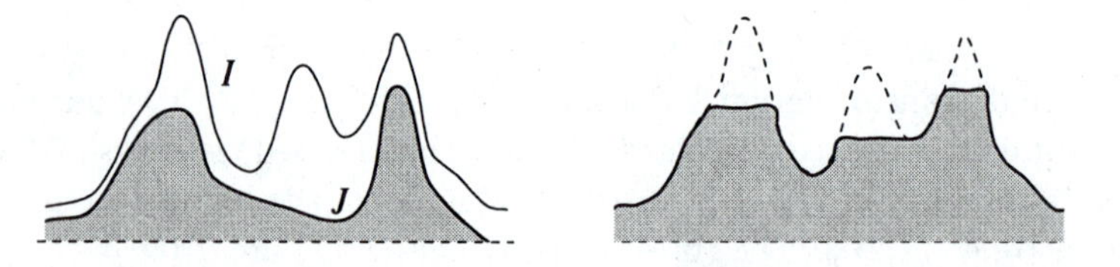

Figure 11.41: *Gray-scale reconstruction of mask I from marker J.*

The duality between dilation and erosion permits the expression of gray-scale reconstruction using erosion.

11.6 Granulometry

Granulometry was introduced by stereologists (mathematicians attempting to understand 3D shape from cross sections)—the name comes from the Latin *granulum*, meaning grain. Matheron [Matheron 67] used it as a tool for studying porous materials, where the distribution of pore sizes was quantified by a sequence of openings of increasing size. Currently, granulometry is an important tool of mathematical morphology, particularly in material science and biology applications. The main advantage is that granulometry permits the extraction of shape information without a priori segmentation.

Consider first a **sieving analysis** analogy; assume that the input is a heap of stones (or granules) of different sizes. The task is to analyze how many stones in the heap fit into several size classes. Such a task is solved by sieving using several sieves with increasing sizes of holes in the mesh. The result of analysis is a discrete function; on its horizontal axis are increasing sizes of stones and on its vertical axis the numbers of stones of that size. In morphological granulometry, this function is called a **granulometric spectrum** or **granulometric curve**.

In binary morphology, the task is to calculate a granulometric curve where the independent variable is the size of objects in the image. The value of the granulometric curve is the number of objects of given size in the image. The most common approach is that sieves with increasing hole sizes (as in the example) are replaced by a sequence of openings with structural elements of increasing size.

Granulometry plays a very significant role in mathematical morphology that is analogous to the role of frequency analysis in image processing or signal analysis. Recall that frequency analysis expands the signal into a linear combination of harmonic signals of growing frequency. The **frequency spectrum** provides the contribution of individual harmonic signals—it is clear that the granulometric curve (spectrum) is analogous to a frequency spectrum.

Let $\Psi = (\psi_\lambda)$, $\lambda \geq 0$, be a family of transformations depending on a parameter λ. This family constitutes a **granulometry** if and only if the following properties of the transformation ψ hold:

$$\begin{array}{ll} \forall \lambda \geq 0 & \psi_\lambda \text{ is increasing} \\ & \psi_\lambda \text{ is anti-extensive} \\ \forall \lambda \geq 0,\ \mu \geq 0 & \psi_\lambda \psi_\mu = \psi_\mu \psi_\lambda = \psi_{\max(\lambda\mu)} \end{array} \tag{11.70}$$

The consequence of property (11.70) is that for every $\lambda \geq 0$ the transformation ϕ_λ is idempotent. (ψ_λ), $\lambda \geq 0$ is a decreasing family of openings (more precisely, algebraic openings [Serra 82] that generalize the notion of opening presented earlier). It can be shown that for any convex structuring element B, the **family of openings** with respect to $\lambda B = \{\lambda b,\ b \in B\}$, $\lambda \geq 0$, constitutes a granulometry [Matheron 75].

Consider more intuitive granulometry acting on discrete binary images (i.e., sets). Here the granulometry is a sequence of openings ψ_n indexed by an integer $n \geq 0$—each opening result is smaller than the previous one. Recall the analogy with sieving analysis; each opening, which corresponds to one sieve mesh size, removes from the image more than the previous one. Finally, the empty set is reached. Each sieving step is characterized by some measure $m(X)$ of the set (image) X (e.g., number of pixels in a 2D image, or volume in 3D). The rate at which the set is sieved characterizes the set. The pattern spectrum provides such a characteristic.

The **pattern spectrum**, also called **granulometric curve**, of a set X with respect to the granulometry $\Psi = \psi_n$, $n \geq 0$ is the mapping

$$PS_\Psi(X)(n) = m[\psi_n(X)] - m[\psi_{n-1}(X)] \quad \forall n > 0 \tag{11.71}$$

The sequence of openings $\Psi(X)$, $n \geq 0$ is a decreasing sequence of sets, i.e., $[\psi_0(X) \supseteq \psi_1(X) \supseteq \psi_2(X) \supseteq \ldots]$. The granulometry and granulometric curve can be used.

Suppose that the granulometric analysis with family of openings needs to be computed for a binary input image. The binary input image is converted into a gray-level image using a granulometry function $G_\Psi(X)$, and the pattern spectrum PS_Ψ is calculated as a histogram of the granulometry function.

The **granulometry function** $G_\Psi(X)$ of a binary image X from granulometry $\Psi = (\psi_n)$, $n \geq 0$, maps each pixel $x \in X$ to the size of the first n such that $x \notin \psi_n(X)$:

$$x \in X, \ G_\Psi(X)(x) = \min\{n > 0, \ x \notin \psi_n(X)\} \tag{11.72}$$

The pattern spectrum PS_Ψ of a binary image X for granulometry $\Psi = (\psi_n)$, $n \geq 0$, can be computed from the granulometry function $G_\Psi(X)$ as its histogram:

$$\forall n > 0, \ PS_\Psi(X)(n) = \mathrm{card}\{p, \ G_\Psi(X)(p) = n\} \tag{11.73}$$

(where 'card' denotes cardinality). An example of granulometry is given in Figure 11.42. The input binary image with circles of different radii is shown in Figure 11.42a; Figure 11.42b shows one of the openings with a square structuring element. Figure 11.42c illustrates the granulometric power spectrum. At a coarse scale, three most significant signals in the power spectrum indicate three prevalent sizes of object. The less significant signals on the left side are caused by the artifacts that occur due to discretization. The Euclidean circles have to be replaced by digital entities (squares).

We see in this example that granulometries extract size information without the need to identify (segment) each object a priori. In applications, this is used for shape description, feature extraction, texture classification, and removal of noise introduced by image borders.

Until recently, granulometric analysis using a family of openings was too slow to be practically useful, but recent developments have made granulometries quick and useful; the reader interested in implementation may consult [Haralick et al. 95, Vincent 95]. For binary images, the basic idea towards speed-up is to use linear structuring elements for openings and more complex 2D ones derived from it, such as cross, square, or diamond (see Figure 11.43). The next source of computational saving is the fact that some 2D structuring elements can be decomposed as Minkowski addition of two 1D structuring elements. For example, the square structuring element can be expressed as Minkowski addition of horizontal and vertical lines.

Gray-scale granulometric analysis is another recent development that permits the extraction of size information directly from gray-level images. The interested reader should consult [Vincent 94].

11.7 Morphological segmentation and watersheds

11.7.1 Particles segmentation, marking, and watersheds

The concept "segmentation" commonly means finding objects of interest in the image. Mathematical morphology helps mainly to segment images of texture or images of particles—here

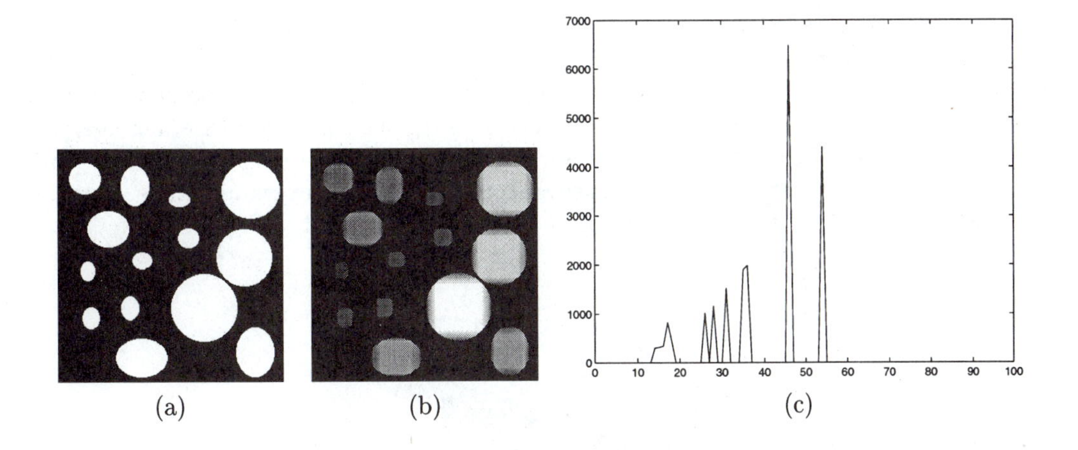

Figure 11.42: *Example of binary granulometry performance: (a) original binary image; (b) maximal square probes inscribed—the initial probe size was 2×2 pixels; (c) granulometric power spectrum as histogram of (b)—the horizontal axis gives the size of the object and the vertical axis the number of pixels in an object of given size. Courtesy P. Kodl, Rockwell Automation Research Center, Prague, Czech Republic.*

Figure 11.43: *Structural elements used for fast binary granulometry are derived from line structuring elements, e.g., cross, square, and diamond.*

we consider particle segmentation in which the input image can be either binary or gray-scale. In the binary case, the task is to segment overlapping particles; in the gray-scale case, the segmentation is the same as object contour extraction. The explanation here is inspired by Vincent's view [Vincent 95], which is intuitive and easy to understand.

Morphological particle segmentation is performed in two basic steps: (1) location of particle markers, and (2) watersheds used for particle reconstruction. The latter is explained later in this section.

Marker extraction resembles human behavior when one is asked to indicate objects; the person just points to objects and does not outline boundaries. The **marker** of an object or set X is a set M that is included in X. Markers have the same homotopy as the set X, and are typically located in the central part of the object (particle).

A robust marker-finding technique will need to know the nature of the objects sought, and thus application-specific knowledge should be used. Often combinations of non-morphological and morphological approaches are used. Moreover, object marking is in many cases left to the user, who marks objects manually on the screen. Typically, software for analysis of microscopic images has user-friendly tools for manual or semi-automatic marking.

When the objects are marked, they can be grown from the markers, e.g., using the watershed transformation (Section 11.7.3), which is motivated by the topographic view of images. Consider the analogy of a landscape and rain; water will find the swiftest descent path until it reaches some lake or sea. We already know that lakes and seas correspond to regional minima. The landscape can be entirely partioned into regions which attract water to a particular sea or lake—these will be called **catchment basins**. These regions are influence zones of the regional minima in the image. **Watersheds**, also called **watershed lines**, separate catchment basins. Watersheds and catchment basins are illustrated in Figure 11.44.

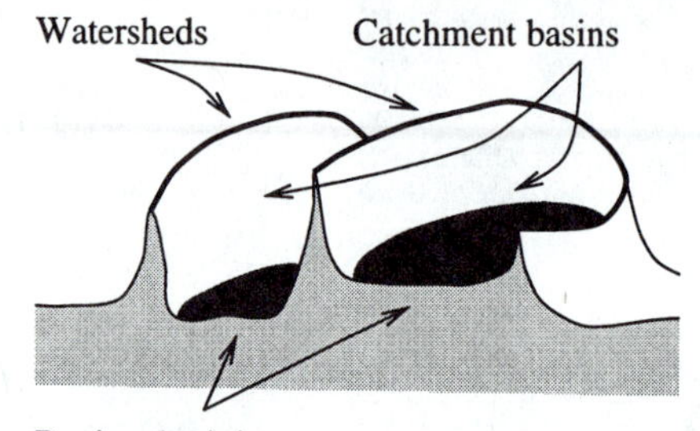

Figure 11.44: *Illustration of catchment basins and watersheds in a 3D landscape view.*

11.7.2 Binary morphological segmentation

If the task is to find objects that differ in brightness from an uneven background, the top hat transformation (Section 11.5.4) is a simple solution. The top hat approach just finds peaks in the image function that differ from the local background. The gray-level shape of the peaks does not play any role, but the shape of the structuring element does. Watershed segmentation takes into account both sources of information and supersedes the top hat method.

Morphological segmentation in binary images aims to find regions corresponding to individual overlapping objects (typically particles), and most of the tools for performing this task have already been explained. Each particle is marked first—ultimate erosion may be used for this purpose (Section 11.5.4), or markers may be placed manually. The next task is to grow objects from the markers provided they are kept within the limits of the original set and parts of objects are not joined when they come close to each other.

The oldest technique for this purpose is called **conditional dilation**. Ordinary dilation is used for growing, and the result is constrained by the two conditions (remain in the original set, and do not join particles).

Geodesic reconstruction (Section 11.5.7) is more sophisticated and performs much faster than conditional dilation. The structuring element adapts according to the neighborhood of the processed pixel.

Geodesic influence zones (Section 11.5.7) are sometimes used for segmenting particles. Figure 11.45 shows that the result can differ from our intuitive expectation.

The best solution is the **watershed transformation**. Only the basic idea will be described here—the reader interested in theory and fast implementation is referred to Section 11.7.3 and [Bleau et al. 92, Vincent 93, Vincent 95]. The original binary image is

Figure 11.45: *Segmentation by geodesic influence zones* (SKIZ) *need not lead to correct results.*

converted into gray-scale using the negative distance transform −dist [equation (11.55)]. If a drop of water falls onto a topographic surface of the −dist image, it will follow the steepest slope towards a regional minimum. This idea is illustrated in Figure 11.46.

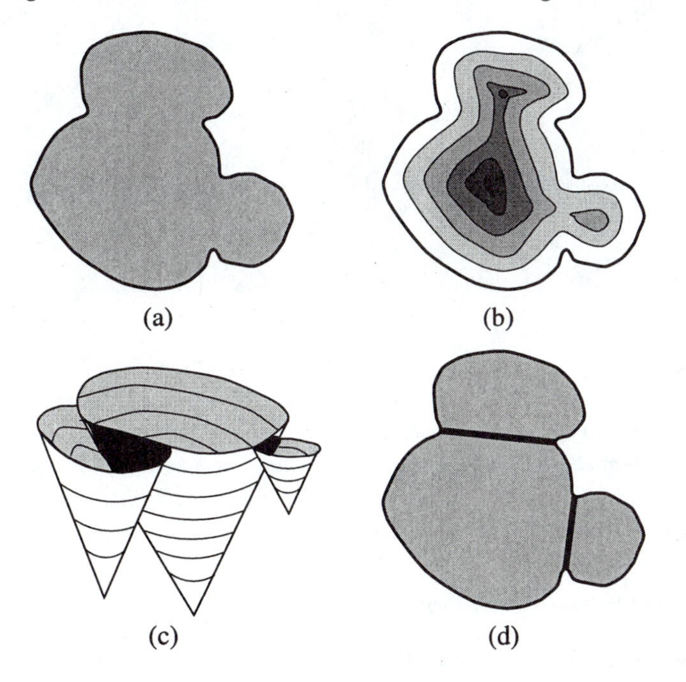

Figure 11.46: *Segmentation of binary particles: (a) input binary image, (b) gray-scale image created from (a) using the* −dist *function, (c) topographic notion of the catchment basin, (d) correctly segmented particles using watersheds of image (b).*

Application of watershed particle segmentation is shown in Figure 11.47. We selected an image of a few touching particles as an input Figure 11.47a. The distance function calculated from the background is visualized using contours in Figure 11.47b for better understanding. The regional maxima of the distance function serve as markers of the individual particles, see Figure 11.47c. The markers are dilated in Figure 11.47d. In preparation for watershed segmentation, the distance function is negated, and is shown together with the dilated markers in Figure 11.47e. The final result of particle separation is illustrated in Figure 11.47f, where particle contours are shown.

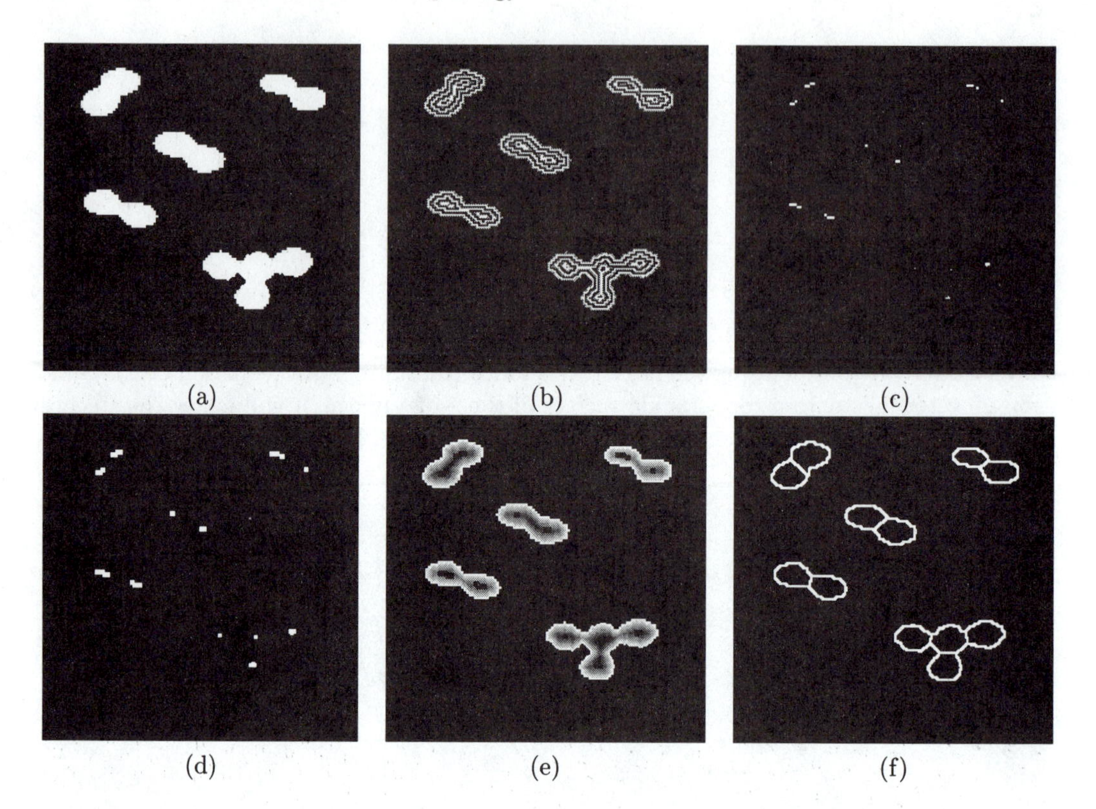

Figure 11.47: *Particle segmentation by watersheds: (a) original binary image; (b) distance function visualized using contours; (c) regional maxima of the distance function used as particle markers; (d) dilated markers; (e) inverse of the distance function with the markers superimposed; (f) resulting contours of particles obtained by watershed segmentation. Courtesy P. Kodl, Rockwell Automation Research Center, Prague, Czech Republic.*

11.7.3 Gray-scale segmentation, watersheds

The markers and watersheds method can also be applied to gray-scale segmentation. Watersheds are also used as crest-line extractors in gray-scale images. The contour of a region in a gray-level image corresponds to points in the image where gray-levels change most quickly—this is analogous to edge-based segmentation considered in Chapter 5. The watershed transformation is applied to the gradient magnitude image in this case (see Figure 11.48). There is a simple approximation to the gradient image used in mathematical morphology called Beucher's gradient [Serra 82], calculated as the algebraic difference of unit-size dilation and unit-size erosion of the input image X.

$$\mathrm{grad}(X) = (X \oplus B) - (X \ominus B) \tag{11.74}$$

The main problem with segmentation via gradient images without markers is **oversegmentation**, meaning that the image is partitioned into too many regions (Figure 11.47c). Some techniques to limit oversegmentation in watershed segmentation are given in [Vincent

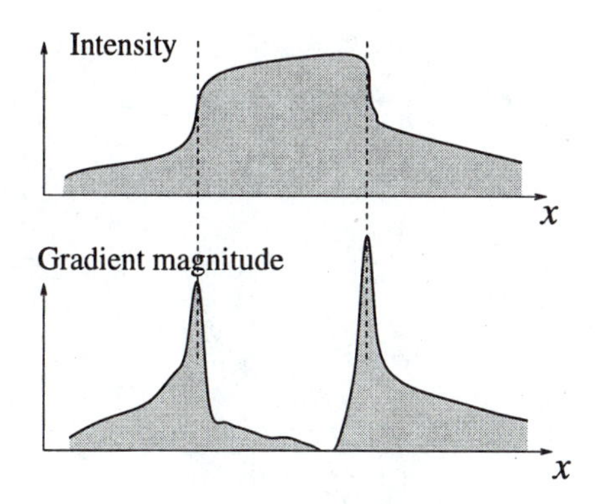

Figure 11.48: *Segmentation in gray-scale images using gradient magnitude.*

93]. The watershed segmentation methods with markers do not suffer from oversegmentation, of course.

An example from opthalmology will illustrate the application of watershed segmentation. The input image shows a microscopic picture of part of a human retina, Figure 11.49a—the task is to segment individual cells on the retina. The markers/watershed paradigm was followed, with markers being found using a carefully tuned Gaussian filter (see Figure 11.49b). The final result with the outlined contours of the cells is in Figure 11.49c.

11.8 Summary

- **Mathematical morphology**
  - Mathematical morphology stresses the role of **shape** in image pre-processing, segmentation, and object description. It constitutes a set of tools that have a solid mathematical background and lead to fast algorithms. The basic entity is a **point set**. Morphology operates using transformations that are described using operators in a relatively simple **non-linear algebra**. Mathematical morphology constitutes a counterpart to traditional signal processing based on linear operators (such as convolution).
  - Mathematical morphology is usually divided into **binary mathematical morphology** which operates on binary images (2D point sets), and **gray-level mathematical morphology** which acts on gray-level images (3D point sets).
- **Morphological operations**
  - In images, morphological operations are **relations of two sets**. One is an image and the second a small probe, called a **structuring element**, that systematically traverses the image; its relation to the image in each position is stored in the output image.
  - Fundamental operations of mathematical morphology are **dilation** and **erosion**. Dilation expands an object to the closest pixels of the neighborhood. Erosion

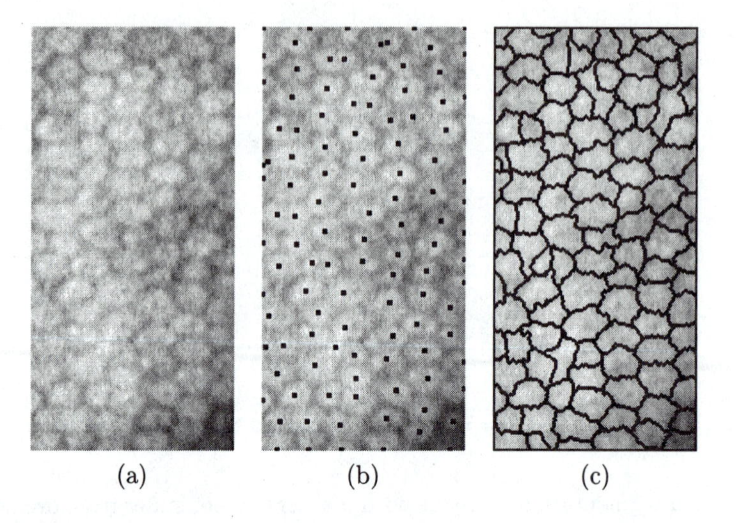

Figure 11.49: *Watershed segmentation on the image of a human retina: (a) original gray-scale image; (b) dots are superimposed markers found by nonmorphological methods; (c) boundaries of retina cells found by watersheds from markers (b). Data and markers courtesy R. Šára, Czech Technical University, Prague, segmentation courtesy P. Kodl, Rockwell Automation Research Center Prague, Czech Republic.*

shrinks the object. Erosion and dilation are not invertible operations; their combination constitutes new operations—**opening** and **closing**.

- Thin and elongated objects are often simplified using a **skeleton** that is an archetypical stick replacement of original objects. The skeleton constitutes a line that is in 'the middle of the object'.

- The **distance function** to the background constitutes a basis for many fast morphological operations. **Ultimate erosion** is often used to mark blob centers. There is an efficient **reconstruction** algorithm that grows the object from the marker to its original boundary.

- **Geodesic transformations** allow changes to the structuring element during processing and thus provide more flexibility. They provide quick and efficient algorithms for image **segmentation**. The **watershed transform** represents one of the better segmentation approaches. Boundaries of the desired regions are influence zones of regional minima (i.e., seas and lakes in the landscape). Region boundaries are watershed lines between these seas and lakes. The segmentation is often performed from **markers** chosen by a human or from some automatic procedure that takes into account semantic properties of the image.

- **Granulometry** is a quantitative tool for analyzing images with particles of different size (similar to sieving analysis). The result is a discrete **granulometric curve** (spectrum).

11.9 Exercises

Short-answer questions

1. What is mathematical morphology?
2. What is a structuring element? What is its role in mathematical morphology?
3. Give the definition of erosion and dilation for binary images.
4. Is erosion a commutative operation?
5. What is the difference between the result of opening performed once and twice? What is idempotency?
6. Dilation and erosion for gray-level images is a generalization of the same notions in binary images. Describe how they are done.
7. What is the top hat transformation and when is it used?
8. What is the skeleton by maximal balls? What does the skeleton of two touching filled circles look like?
9. What does the homotopic skeleton of two touching balls look like?
10. Explain the role of ultimate erosion for marking particles.
11. What is a regional maximum?
12. What is geodesic distance? How is it used in mathematical morphology?
13. What is granulometry?
14. What is the relation between a granulometric curve (spectrum) and the Fourier spectrum?
15. What is a watershed? How is it used for morphological segmentation?

Problems

1. Prove that dilation is commutative and associative.
2. Consider a one-dimensional gray-scale image (signal). Draw a picture that demonstrates that gray-scale dilation fills narrow bays in the image.
3. Explain how the top hat transformation permits the segmentation of dark characters on a light background with varying intensity. Draw a picture of a one-dimensional cross section through the image.
4. Explain the calculation steps of the skeleton by maximal balls in Figure 11.29.
5. Draw a one-dimensional gray-level continuous function with local, regional, and global maxima (minima). Explain the role of regional extremes in mathematical morphology.
6. Name three application areas in which granulometry analysis may be used.
7. Explain the role of markers in morphological segmentation. Why would an attempt to perform watershed segmentation without markers lead to oversegmentation?

11.10 References

[Bleau et al. 92] A Bleau, J de Guise, and R LeBlanc. A new set of fast algorithms for mathematical morphology II. Identification of topographic features on grayscale images. *CVGIP: Image Understanding*, 56(2):210–229, September 1992.

[Blum 67] H Blum. A transformation for extracting new descriptors of shape. In W Wathen-Dunn, editor, *Proceedings of the Symposium on Models for the Perception of Speech and Visual Form*, pages 362–380, MIT Press, Cambridge, MA, 1967.

[Dougherty 92] E R Dougherty. *An Introduction to Mathematical Morphology Processing.* SPIE Press, Bellingham, WA, 1992.

[Giardina and Dougherty 88] C R Giardina and E R Dougherty. *Morphological Methods in Image and Signal Processing.* Prentice-Hall, Englewood Cliffs, NJ, 1988.

[Golay 69] M J E Golay. Hexagonal parallel pattern transformation. *IEEE Transactions on Computers*, C–18:733–740, 1969.

[Haralick and Shapiro 92] R M Haralick and L G Shapiro. *Computer and Robot Vision, Volume I.* Addison-Wesley, Reading, MA, 1992.

[Haralick et al. 87] R M Haralick, S R Stenberg, and X Zhuang. Image analysis using mathematical morphology. *IEEE Transactions on Pattern Analysis and Machine Intelligence*, 9(4):532–550, 1987.

[Haralick et al. 95] R M Haralick, P L Katz, and E R Dougherty. Model-based morphology: The opening spectrum. *Graphical Models and Image Processing*, 57(1):1–12, 1995.

[Heijmans 94] H J A M Heijmans. *Morphological Image Operators.* Academic Press, Boston, 1994.

[Ma and Sonka 96] C M Ma and M Sonka. A fully parallel 3D thinning algorithm and its applications. *Computer Vision and Image Understanding*, 64:420–433, 1996.

[Maragos and Schafer 87a] P Maragos and R W Schafer. Morphological filters—part I: Their set-theoretic analysis and relations to linear shift-invariant filters. *IEEE Transactions on Acoustics, Speech and Signal Processing*, 35(8):1153–1169, August 1987.

[Maragos and Schafer 87b] P Maragos and R W Schafer. Morphological filters—part II: Their relation to median, order-statistic, and stack filters. *IEEE Transactions on Acoustics, Speech and Signal Processing*, 35(8):1170–1184, August 1987.

[Maragos and Ziff 90] P Maragos and R Ziff. Threshold superposition in morphological image analysis. *IEEE Transactions on Pattern Analysis and Machine Intelligence*, 12(5), May 1990.

[Matheron 67] G Matheron. *Eléments pour une theorie des milieux poreux* (in French). Masson, Paris, 1967.

[Matheron 75] G Matheron. *Random Sets and Integral Geometry.* Wiley, New York, 1975.

[Meyer 78] F Meyer. Contrast feature extraction. In J-L Chermant, editor, *Quantitative Analysis of Microstructures in Material Science, Biology and Medicine*, Riederer Verlag, Stuttgart, Germany, 1978. Special issue of *Practical Metalography.*

[Roerdink and Heijmans 88] J B T M Roerdink and H J A M Heijmans. Mathematical morphology for structures without translation symmetry. *Signal Processing*, 15(3):271–277, October 1988.

[Serra 82] J Serra. *Image Analysis and Mathematical Morphology.* Academic Press, London, 1982.

[Serra 87] J Serra. Morphological optics. *Journal of Microscopy*, 145(1):1–22, 1987.

[Vincent 91] L Vincent. Efficient computation of various types of skeletons. In *Proceedings of the SPIE Symposium Medical Imaging,* San Jose, CA, volume 1445, SPIE, Bellingham, WA, February 1991.

[Vincent 93] L Vincent. Morphological grayscale reconstruction in image analysis: Applications and efficient algorithms. *IEEE Transactions on Image Processing,* 2(2):176–201, April 1993.

[Vincent 94] L Vincent. Fast opening functions and morphological granulometries. In *Proceedings Image Algebra and Morphological Image Processing V,* pages 253–267, SPIE, San Diego, CA, July 1994.

[Vincent 95] L Vincent. Lecture notes on granulometries, segmentation and morphological algorithms. In K Wojciechowski, editor, *Proceedings of the Summer School on Morphological Image and Signal Processing,* Zakopane, Poland, pages 119–216. Silesian Technical University, Gliwice, Poland, September 1995.

Chapter 12

Linear discrete image transforms

Image processing and analysis based on continuous or discrete image transforms is a classic processing technique. Transforms are widely used in image filtering, image data compression, image description, etc.; they were actively studied at the end of the 1960s and in the beginning of the 1970s. This research was highly motivated by space flight achievements: First, an efficient method of image data transmission was needed; and second, transmitted images were often received either locally or globally corrupted. Both analogue and discrete methods were investigated—analogue optical methods suffered from a lack of suitable high-resolution storage material, and the rapid development of digital computer technology slowed analogue research in the middle of the 1970s. However, much research is now being devoted to optical methods again, especially with new technological achievements in developing high-resolution optical storage materials. The main focus is in possible real-time image processing.

Image transform theory is a well-known area characterized by a precise mathematical background, and image transforms represent a powerful, unified area of image processing. This chapter introduces only the basic concepts and essential references, because a number of excellent alternative books are available. There are many monographs devoted just to two-dimensional image transforms [Andrews 70, Shulman 70, Ahmed and Rao 75, Andrews and Hunt 77, Huang 81, Nussbaumer 82, Dudgeon and Mersereau 84, Yaroslavskii 85, Wolberg 90, Rao and Yip 90], and many image processing books with extensive coverage of image transforms [Hall 79, Rosenfeld and Kak 82, Gonzalez and Wintz 87, Lim 90, Pratt 91, Gonzalez and Woods 92, Sid-Ahmed 95, Bracewell 95, Castleman 96], not to mention an almost endless list of signal processing or purely mathematical references. Only general theory is discussed in this chapter, with an image processing application outline. Familiarity with the one-dimensional Fourier transform, basic operations in the frequency domain, and spatial convolution performed in the frequency domain is assumed. All of the mathematical proofs, more complicated mathematical connections, and all implementation aspects are omitted.

12.1 Basic theory

Let an image f be represented as a matrix of integer numbers. An image transform can generally process either the whole image or some (usually rectangular) subimage. We will assume that the image size is $M \times N$.

$$\mathbf{f} = \begin{bmatrix} f(0,0) & f(0,1) & \dots & f(0,N-1) \\ & \dots & & \dots \\ f(M-1,0) & f(M-1,1) & \dots & f(M-1,N-1) \end{bmatrix} \tag{12.1}$$

Transform matrices $\mathbf{P}$ and $\mathbf{Q}$ of dimension $M \times M$ and $N \times N$, respectively, are used to transform $\mathbf{f}$ into a matrix $\mathbf{F}$ ($M \times N$ matrix) of the same size,

$$\mathbf{F} = \mathbf{P} \ \mathbf{f} \ \mathbf{Q} \tag{12.2}$$

which we may write as

$$F(u,v) = \sum_{m=0}^{M-1} \sum_{n=0}^{N-1} P(u,m) f(m,n) Q(n,v) \tag{12.3}$$

$$u = 0,1,\dots,M-1 \qquad v = 0,1,\dots,N-1$$

If $\mathbf{P}$ and $\mathbf{Q}$ are non-singular (i.e., have non-zero determinants), inverse matrices $\mathbf{P}^{-1}$ and $\mathbf{Q}^{-1}$ exist and the inverse transform can be computed as

$$\mathbf{f} = \mathbf{P}^{-1} \mathbf{F} \ \mathbf{Q}^{-1} \tag{12.4}$$

Some preparatory terms and formulae are required. Let $\mathbf{M}^T$ represent the transpose of matrix $\mathbf{M}$.

- $\mathbf{M}$ is **symmetric** if $\mathbf{M} = \mathbf{M}^T$.
- $\mathbf{M}$ is **orthogonal** if $\mathbf{M}^T\mathbf{M} = \mathbf{I}$, where $\mathbf{I}$ is the identity matrix.
- For any symmetric, real, and orthogonal matrix, $\mathbf{M}^{-1} = \mathbf{M}$.
- A complex matrix $\mathbf{C}$ is **Hermitian** if $\mathbf{C}^{*T} = \mathbf{C}$, where $\mathbf{C}^*$ is derived from $\mathbf{C}$ by taking a complex conjugate of every element.
- A complex matrix $\mathbf{C}$ is **unitary** if $\mathbf{C}^{*T}\mathbf{C} = \mathbf{I}$.
- For any square, complex, Hermitian, and unitary matrix, $\mathbf{C}^{-1} = \mathbf{C}$.

If $\mathbf{P}$ and $\mathbf{Q}$ are both symmetric, real, and orthogonal, then

$$\mathbf{F} = \mathbf{P} \ \mathbf{f} \ \mathbf{Q} \qquad \qquad \mathbf{f} = \mathbf{P} \ \mathbf{F} \ \mathbf{Q} \tag{12.5}$$

and the transform is an **orthogonal transform**. If $\mathbf{P}, \mathbf{Q}$ are complex matrices, equation (12.5) still holds provided they are Hermitian and unitary.

12.2 Fourier transform

The basics of the one-dimensional continuous Fourier transform were discussed briefly in Section 2.1.3. The discrete Fourier transform is analogous to the continuous one and may be efficiently computed using the fast Fourier transform algorithm. The properties of linearity, shift of position, modulation, convolution, multiplication, and correlation are analogous to the continuous case, with the difference of the discrete periodic nature of the image and its transform. Let $\mathbf{\Phi}_{JJ}$ be a transform matrix of size $J \times J$:

$$\mathbf{\Phi}_{JJ}(k,l) = \frac{1}{J}\exp(-i\frac{2\pi}{J}kl) \qquad k,l = 0,1,\ldots,J-1 \tag{12.6}$$

The discrete Fourier transform can be defined according to equation (12.2): $\mathbf{P} = \mathbf{\Phi}_{MM}$, $\mathbf{Q} = \mathbf{\Phi}_{NN}$

$$\mathbf{F} = \mathbf{\Phi}_{MM}\ \mathbf{f}\ \mathbf{\Phi}_{NN} \tag{12.7}$$

and

$$F(u,v) = \frac{1}{MN}\sum_{m=0}^{M-1}\sum_{n=0}^{N-1} f(m,n)\exp[-2\pi i(\frac{mu}{M}+\frac{nv}{N})] \tag{12.8}$$

$$u = 0,1,\ldots,M-1 \quad v = 0,1,\ldots,N-1$$

The inverse transform matrix $\mathbf{\Phi}_{JJ}^{-1}$ is

$$\Phi_{JJ}^{-1}(k,l) = \exp(\frac{2\pi i}{J}kl) \tag{12.9}$$

and the inverse Fourier transform is given by

$$f(m,n) = \sum_{u=0}^{M-1}\sum_{v=0}^{N-1} F(u,v)\exp[2\pi i(\frac{mu}{M}+\frac{nv}{N})] \tag{12.10}$$

$$m = 0,1,\ldots,M-1 \quad n = 0,1,\ldots,N-1$$

The kernel function of the discrete transform (12.8) is

$$\exp[-2\pi i(\frac{mu}{M}+\frac{nv}{N})] \tag{12.11}$$

Considering implementation of the discrete Fourier transform, note that equation (12.8) can be modified to

$$F(u,v) = \frac{1}{M}\sum_{m=0}^{M-1}[\frac{1}{N}\sum_{n=0}^{N-1}\exp(\frac{-2\pi inv}{N})f(m,n)]\exp(\frac{-2\pi imu}{M}) \tag{12.12}$$

$$u = 0,1,\ldots,M-1 \quad v = 0,1,\ldots,N-1$$

The term in square brackets corresponds to the one-dimensional Fourier transform of the mth line and can be computed using standard fast Fourier transform (FFT) procedures (usually assuming $N = 2^k$). Each line is substituted with its Fourier transform, and the one-dimensional discrete Fourier transform of each column is computed. Algorithms for FFT computation can be found in [Nussbaumer 82, Pavlidis 82, Gonzalez and Wintz 87, Gonzalez and Woods 92, Press 88].

Periodicity is an important property of the discrete Fourier transform. We have defined the transform as a matrix with elements $F(u,v)$ for $u = 0, 1, \ldots, M-1$; $v = 0, 1, \ldots, N-1$. In arrays where other values of u, v are allowed, a periodic transform F [equation (12.8)] is derived and a periodic image f defined:

$$\begin{array}{ll} F(u,-v) = F(u, N-v) & f(-m,n) = f(M-m,n) \\ F(-u,v) = F(M-u,v) & f(m,-n) = f(m,N-n) \end{array} \tag{12.13}$$

and

$$F(aM+u, bN+v) = F(u,v) \quad f(aM+m, bN+n) = f(m,n) \tag{12.14}$$

where a and b are integers.

In image processing, image Fourier spectra play an important role. Considering equation (12.10), image values $f(m,n)$ can be interpreted as a linear combination of periodic patterns $\exp\{2\pi i[(mu/M)+(nv/N)]\}$ and $F(u,v)$ may be considered a weighting function. The Fourier transform of a real function is a complex function, that is,

$$F(u,v) = R(u,v) + iI(u,v) \tag{12.15}$$

where $R(u,v)$ and $I(u,v)$ are, respectively, the real and imaginary components of $F(u,v)$. The magnitude function $|F(u,v)|$ is called the **frequency spectrum** of image $f(m,n)$; in addition, the **phase spectrum** $\phi(u,v)$ and **power spectrum** $P(u,v)$ are used. The frequency spectrum is defined by

$$|F(u,v)| = \sqrt{R^2(u,v) + I^2(u,v)} \tag{12.16}$$

the phase spectrum by

$$\phi(u,v) = \tan^{-1}\left[\frac{I(u,v)}{R(u,v)}\right] \tag{12.17}$$

(note that care must be taken over interpretation of the signs of these quantities, since tan is π periodic, and ϕ ranges from 0 to 2π) and the power spectrum (also called **spectral density**) as

$$P(u,v) = |F(u,v)|^2 = R^2(u,v) + I^2(u,v) \tag{12.18}$$

An example of an image and its frequency spectrum is given in Figure 12.1. Note the cross that is apparent in the spectrum. Actually, two crosses are visible: One is formed by the image borders and depicts the x and y axes in the spectrum. The second is rotated by approximately 10° anti-clockwise with respect to the first. This cross comes from the image data; its directions correspond to the main edge directions present in the image. However, it is important to realize that the frequency spectrum lines seem to be rotated by 90° with respect to the image edge directions because of the perpendicular relationship between image edges and the image intensity changes (in this case, the sinusoidal basis functions) assessing the frequency character of the edges. Thus, the approximately vertical line of the cross corresponds to the approximately horizontal image edges and the approximately horizontal line corresponds to the approximately vertical image edges.

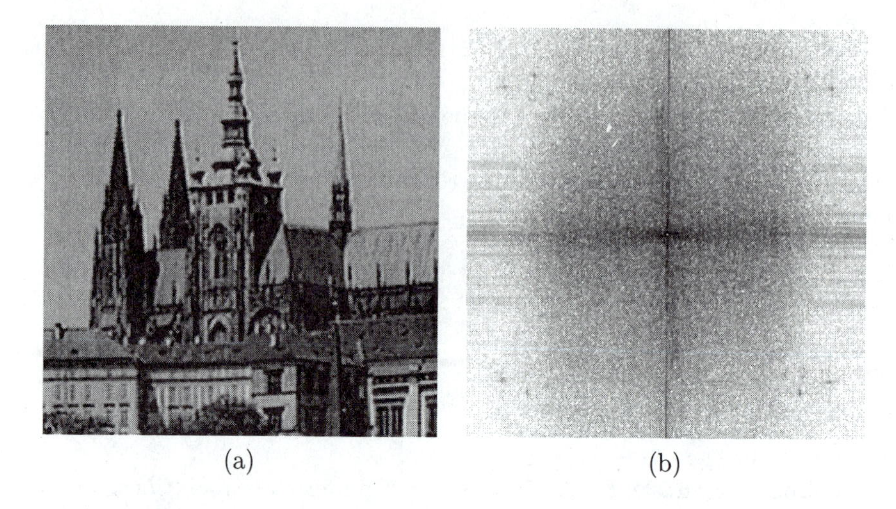

Figure 12.1: *Fourier spectrum: (a) image; (b) frequency spectrum as an intensity function—dark pixels correspond to high spectral values.*

The Fourier transform is of great use in the calculation of image convolutions (see Sections 2.1.2 and 4.3). If a convolution of two (periodic) images f and h of the same size $M \times N$ is computed, a (periodic) image g results:

$$g(a,b) = \frac{1}{MN} \sum_{m=0}^{M-1} \sum_{n=0}^{N-1} f(m,n) h(a-m, b-n) \tag{12.19}$$

Note that h is periodic, or the index calculation in $h(a-m, b-r)$ must be done modulo M, N respectively. The **convolution theorem** states that f, g, and h, and their transforms F, G, and H, are related by the equation

$$G(u,v) = F(u,v) H(u,v) \tag{12.20}$$

that is, element-by-element multiplication. Thus we may write

$$g(a,b) = \sum_{u=0}^{M-1} \sum_{v=0}^{N-1} F(u,v) H(u,v) \exp[2\pi i (\frac{au}{M} + \frac{bv}{N})] \tag{12.21}$$

Use of this relationship can reduce the computational load of calculating convolutions very significantly (usually for convolution kernels larger than 7×7).

12.3 Hadamard transform

If matrices **P** and **Q** in equation (12.2) are **Hadamard** matrices (see below), then **F** is a Hadamard transform of an image **f**. As shown in the previous section, the Fourier transform may be represented using a set of orthogonal sinusoidal waveforms; the coefficients of the Fourier representation are called frequency components and the waveforms are ordered by frequency. If the set of orthogonal functions used to represent an orthogonal transform is chosen to consist of square waves (namely, Walsh functions), analogous spectral analysis may be performed with a simple computation. The Walsh functions are ordered by the number of their zero-crossings, and the coefficients are called **sequency components** [Ahmed and Rao 75]. The Walsh functions are real (not complex) and take only the values ± 1. A Hadamard matrix $\mathbf{H}_{JJ}$ is a symmetric matrix $J \times J$, with elements all ± 1; the Hadamard matrix of the second order $\mathbf{H}_{22}$ is

$$\mathbf{H}_{22} = \begin{vmatrix} 1 & 1 \\ 1 & -1 \end{vmatrix} \tag{12.22}$$

Any Hadamard matrix of order 2^k can be written as

$$\mathbf{H}_{2J2J} = \begin{vmatrix} \mathbf{H}_{JJ} & \mathbf{H}_{JJ} \\ \mathbf{H}_{JJ} & -\mathbf{H}_{JJ} \end{vmatrix} \tag{12.23}$$

Hadamard matrices of orders other than 2^k exist, but they are not widely used in image processing. Inverse Hadamard matrices can be easily computed,

$$\mathbf{H}_{JJ}^{-1} = \frac{1}{J} \mathbf{H}_{JJ} \tag{12.24}$$

and the Hadamard transform can therefore be represented by

$$\mathbf{F} = \mathbf{H}_{MM} \mathbf{f} \ \mathbf{H}_{NN} \qquad \mathbf{f} = \frac{1}{MN} \mathbf{H}_{MM} \mathbf{F} \ \mathbf{H}_{NN} \tag{12.25}$$

It can be seen that only matrix multiplication is necessary to compute a Hadamard transform, and further, only additions are computed during it. The Hadamard transform is sometimes called a Walsh-Hadamard transform, since the base of the transform consists of Walsh functions. This transform has been found useful in early applications of image coding and pattern recognition [Hall 79].

12.4 Discrete cosine transform

There are four definitions of the discrete cosine transform, sometimes denoted DCT-I, DCT-II, DCT-III, and DCT-IV [Rao and Yip 90]. The most commonly used discrete cosine transform in image processing and compression is DCT-II—it can be defined using equation (12.2) and considering a set of basis vectors that are sampled cosine functions. Assuming a square $N \times N$ image, the discrete transform matrix (equation 12.2) can be expressed as [Hall 79]

$$\begin{aligned} C_{NN}(k,l) &= \frac{1}{\sqrt{N}} && \text{for } l = 0 \\ &= \sqrt{\frac{2}{N}} \cos[\frac{(2k+1)l\pi}{2N}] && \text{all other } k,l \end{aligned} \tag{12.26}$$

and

$$\mathbf{F} = \mathbf{C}_{NN} \ \mathbf{f} \ \mathbf{C}_{NN}^T \qquad\qquad \mathbf{f} = \mathbf{C}_{NN}^T \mathbf{F} \ \mathbf{C}_{NN} \tag{12.27}$$

In the two-dimensional case, the formula for a normalized version of the discrete cosine transform (forward cosine transform DCT-II) may be written [Rao and Yip 90]

$$F(u,v) = \frac{2c(u)c(v)}{N} \sum_{m=0}^{N-1} \sum_{n=0}^{N-1} f(m,n) \cos\left(\frac{2m+1}{2N} u\pi\right) \cos\left(\frac{2n+1}{2N} v\pi\right) \tag{12.28}$$

$$u = 0, 1, \ldots, N-1 \qquad v = 0, 1, \ldots, N-1$$

where

$$c(k) \begin{array}{ll} = \frac{1}{\sqrt{2}} & \text{for} \quad k = 0 \\ = 1 & \text{otherwise} \end{array}$$

and the inverse cosine transform is

$$f(m, n) = \frac{2}{N} \sum_{u=0}^{N-1} \sum_{v=0}^{N-1} c(u) c(v) F(u, v) \cos\left(\frac{2m+1}{2N} u\pi\right) \cos\left(\frac{2n+1}{2N} v\pi\right) \tag{12.29}$$

$$m = 0, 1, \ldots, N-1 \qquad n = 0, 1, \ldots, N-1$$

Note that the discrete cosine transform computation can be based on the Fourier transform—all N coefficients of the discrete cosine transform may be computed using a $2N$-point fast Fourier transform [Ahmed and Rao 75, Rosenfeld and Kak 82, Rao and Yip 90, Hung and Meng 94]. Discrete cosine transform forms the basis of JPEG image compression (Chapter 13).

12.5 Wavelets

Wavelets represent another approach to decomposing complex signals into sums of basis functions—in this respect they are similar to Fourier decomposition approaches, but they have an important difference. We present here a brief overview only as the theory of wavelets is well understood and documented in a number of mathematical texts; [Chui 92] provides a good example, while [Graps 95] gives a good introduction from the applications point of view for the non-mathematician; [Castleman 96] presents the topic from the point of view of image processing.

Fourier functions are localized in frequency but not in space, in the sense that they isolate frequencies, but not isolated occurrences of those frequencies (that is, not throughout the domain of interest of the signal). This means that small frequency changes in a Fourier transform will produce changes everywhere in the time domain. Wavelets are local in both frequency (via dilations) and time (via translations)—because of this they are able to analyze data at different scales or resolutions much better than simple sine and cosines can. To understand this, note that modeling a spike in a function (a noise dot, for example) with a sum of infinite functions will be hard because of its strict locality, while functions that are already local will be naturally suited to the task. This means that such functions lend themselves to more compact representation via wavelets—sharp spikes and discontinuities normally take fewer wavelet bases to represent than if sine-cosine basis functions are used.

In the same way as Fourier analysis, wavelets are derived from a basis function $\Phi(x)$, called the **Mother function** or **analyzing wavelet**. The wavelet basis is then provided by the functions

$$\Phi_{(s,l)}(x) = 2^{-(s/2)} \Phi(2^{-s} x - l)$$

Here, the scale factor s indicates the wavelet's width (a power of 2) and the location index l its position (an integer). Note that the $\Phi_{(s,l)}$ are self-similar and are selected to be orthonormal, so

$$\int \Phi_{(s_1,l_1)} \Phi_{(s_2,l_2)} = 0 \qquad \text{if } s_1 \neq s_2 \text{ or } l_1 \neq l_2$$

and it is thus possible to represent other functions as a linear combination of the $\Phi_{(s,l)}$.

The Mother function is obviously key in the application of wavelets; the simplest such is the **Haar**, which is a simple step:

$$\begin{aligned} \Phi(x) &= 1 \qquad 0 \leq x < 0.5 \\ \Phi(x) &= -1 \qquad 0.5 \leq x < 1 \end{aligned}$$

—see Figure 12.2, in which the scaling and shifting effects are clear. Other well-known basis functions provide the Meyer [Meyer 93] and the Daubechies [Daubechies 88] mother functions. An example of wavelet coefficients obtained from the original image shown in Figure 12.1a is given in Figure 12.3; the Daubechies mother function was used.

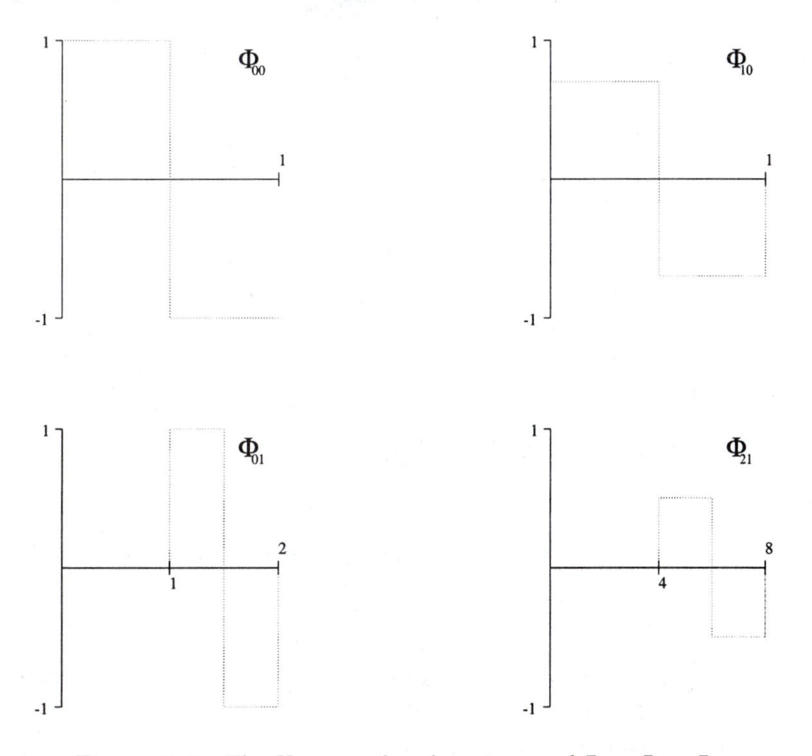

Figure 12.2: *The Haar mother function, and* Φ_{10}, Φ_{01}, Φ_{21}

Applications have proved the value of the approach. Wavelets have been used with enormous success in data compression (for example, in fingerprint image data reduction [Bradley et al. 93], see Figure 13.3) and in image noise suppression [Donoho 93]—it is possible to erase to zero the contribution of wavelet components that are 'small' and correspond to noise *without* erasing the important small detail in the underlying image. Thus, there is noise suppression without the blurring characteristic of Fourier filters (see Section 12.7).

The interested reader is referred to specialized texts for a fuller exposition on this topic [Chui 92, Meyer 93, Chui et al. 94, Castleman 96].

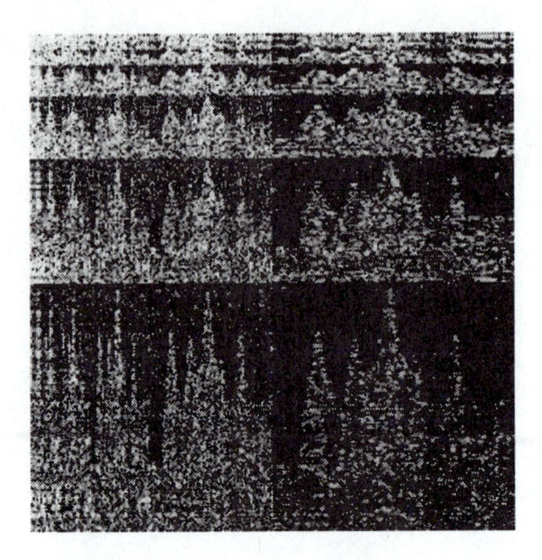

Figure 12.3: *Daubechies wavelet coefficients—dark pixels correspond to low spectral values. The coefficients in the upper left corner are related to lower image resolution; the resolution increases in both horizontal and vertical directions (right and down). For visualization purposes, absolute values of the coefficients are shown after histogram equalization. See Figure 12.1 for the original image.*

12.6 Other orthogonal image transforms

Many other orthogonal image transforms exist. **Paley** and **Walsh** transforms are both very similar to the Hadamard transform, using transformation matrices consisting of ± 1 elements only. Details can be found in [Gonzalez and Wintz 87, Gonzalez and Woods 92].

The **Haar** transform is based on non-symmetric Haar matrices whose elements are either 1, −1, or 0, multiplied by powers of $\sqrt{2}$, and may also be computed in an efficient way. The **Hadamard-Haar** transform is a combination of the Haar and Hadamard transforms, and a modified Hadamard-Haar transform is similar. The **Slant** transform and its modification the **Slant-Haar** transform represents another transform containing sawtooth waveforms or **slant** basic vectors; a fast computational algorithm is also available. The **discrete sine transform** is very similar to the discrete cosine transform. All transforms mentioned here are discussed in detail in [Hall 79, Dougherty and Giardina 87, Gonzalez and Wintz 87, Pratt 91, Gonzalez and Woods 92], where references for computational algorithms can also be found.

The significance of image reconstruction from projections can be seen in computed tomography (CT), magnetic resonance imaging (MRI), positron emission tomography (PET), astronomy, holography, etc., where image formation is based on the **Radon** transform [Dougherty and Giardina 87, Sanz et al. 88, Kak and Slaney 88, Jain 89, Bracewell 95]. In image reconstruction, projections in different directions are acquired by sensors and the two-dimensional image must be reconstructed. The inverse Radon transform is of particular interest. The main Radon inverse transform techniques are based on Fourier transforms, convolution, or algebraic formulations. Note that the Hough transform (see Section 5.2.6) has been shown to be an adaptation of the more general Radon transform.

The **Karhunen-Loeve** transform plays a slightly different role. Its transform matrices consist of eigenvectors of the covariance matrix of the original image, or of a class of original images. The power of the Karhunen-Loeve transform is in the search for non-correlated transform coefficients, which makes it very useful in data compression (see Chapter 13 and Section 8.3). The Karhunen-Loeve transform is used to order features according to their information content in statistical recognition methods (see Chapter 14). A detailed description can be found in [Andrews 70, Rosenfeld and Kak 82].

12.7 Applications of discrete image transforms

Section 12.2 noted that the Fourier transform makes convolution of two images in the frequency domain very easy. In Chapter 4 many filters used in image pre-processing were presented—the convolution masks in most cases were used for image filtering or image gradient computation. It is natural to think about processing these (and many other) convolutions in the frequency domain. Such operations are usually called **spatial frequency filtering**.

Assume that f is an input image and F is its Fourier transform. A convolution filter h can be represented by its Fourier transform H; h may be called the unit pulse response of the filter and H the frequency transfer function, and either of the representations h or H can be used to describe the filter. The Fourier transform of the filter output after an image f has been convolved with the filter h can be computed in the frequency domain [equations (12.19)–(12.21)]

$$G(u,v) = F(u,v)H(u,v) \tag{12.30}$$

Note that equation (12.30) represents a term-by-term multiplication, not a matrix multiplication. The filtered image g can be obtained by applying the inverse Fourier transform to $\mathbf{G}$—equation (12.4).

Some basic examples of spatial filtering are linear **low-pass**, **high-pass**, and **band-pass** frequency filters.

- A low-pass filter is defined by a frequency transfer function $H(u,v)$ with small values at points located far from the co-ordinate origin in the frequency domain (that is, small transfer values for high spatial frequencies) and large values at points close to the origin (large transfer values for low spatial frequencies)—see Figure 12.4a. It preserves low spatial frequencies and suppresses high spatial frequencies, and has behavior similar to smoothing by standard averaging—it blurs sharp edges.

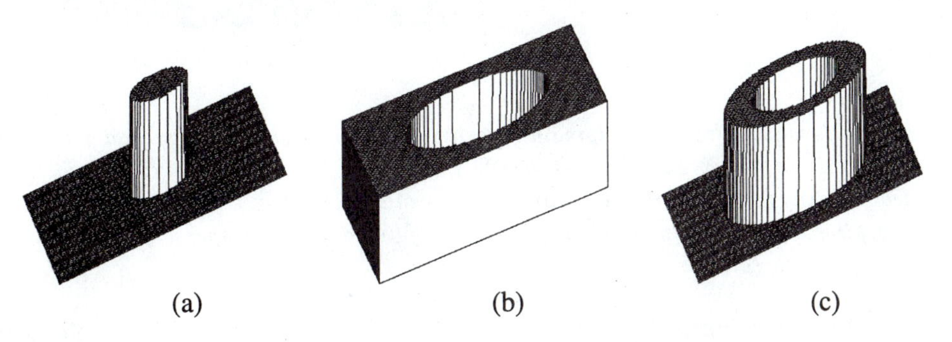

Figure 12.4: *Frequency filters displayed in 3D: (a) low-pass filter; (b) high-pass filter; (c) band-pass filter.*

- A high-pass filter is defined by small transfer function values located around the frequency co-ordinate system origin, and larger values outside this area—larger transfer coefficients for higher frequencies (Figure 12.4b).

- Band-pass filters, which select frequencies in a certain range for enhancement, are constructed in a similar way, and also filters with directional response, etc. (Figure 12.4c).

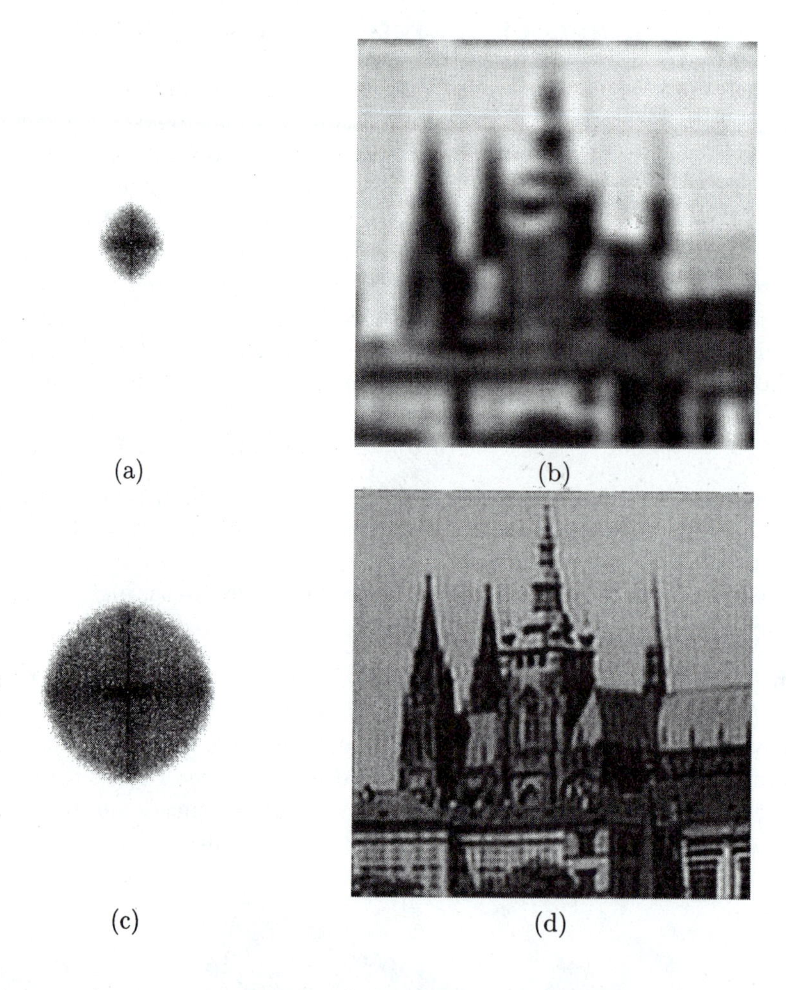

Figure 12.5: *Low-pass frequency-domain filtering—for the original image and its spectrum see Figure 12.1: (a) spectrum of a low-pass filtered image, all higher frequencies filtered out; (b) image resulting from the inverse Fourier transform applied to spectrum (a); (c) spectrum of a low-pass filtered image, only very high frequencies filtered out; (d) inverse Fourier transform applied to spectrum (c).*

The most common image enhancement problems include noise suppression, edge enhancement, and structured noise removal. Noise represents a high-frequency image component, and to suppress it, the magnitudes of image frequencies of the noise must be decreased. This can be achieved by applying a low-pass filter as shown in Figure 12.5, which demonstrates the principles of frequency filtering on Fourier image spectra; the original image spectrum is multiplied by the filter spectrum and a low-frequency image spectrum results. Unfortunately, as a result of noise suppression, all high-frequency phenomena are suppressed, including high frequencies that are not related to noise (sharp edges, lines, etc.). Low-pass filtering results in a blurred image.

Again, edges represent a high-frequency image phenomenon. Therefore, to enhance the edges, low-frequency components of the image spectrum must be suppressed, and to achieve this, a high-frequency filter must be applied.

To remove structured noise, the filter design must include a priori knowledge about the noise properties. This knowledge may be acquired either from the image data or from the corrupted image Fourier spectrum, where the structured noise usually causes notable peaks.

Some examples of frequency domain image filtering are shown in Figures 12.5–12.8. The original image and its frequency spectrum are given in Figure 12.1. Figure 12.6 shows results after application of a high-pass filter followed by an inverse Fourier transform. It can be seen that edges represent high-frequency phenomena in the image. Results of band-pass filtering can be seen in Figure 12.7. Figure 12.8 gives an even more powerful example of frequency filtering—removal of periodic noise that was present in the image. The vertical periodic noise lines in the original image are transformed into frequency spectrum peaks after the transform. To remove these frequencies from an image, a filter was designed which suppresses the periodic noise in the image, which is visible as white circular areas. Discrete image transforms are computationally expensive (although with the continuous increase in routinely available computing power, this is not as important an issue any more), but note that the Fourier transform can be obtained in real time using optical processing methods (for example, a convex lens may produce the Fourier transform) [Shulman 70, Francon 79, Stark 82].

In Section 12.5 the importance of wavelets in image noise suppression was mentioned. Because of the localized behavior of the wavelet transform, single coefficients may be erased without affecting the entire reconstructed image. This can be used for random noise suppression without the blurring characteristic of Fourier filters.

Figures 12.9 and 12.10 demonstrate such wavelet noise suppression. Figure 12.9a has been subjected to random Gaussian noise characterized by zero mean and standard deviation equal to one-half of the gray-level standard deviation of the original image; Figure 12.9b illustrates wavelet filtering based on an inverse wavelet transform that uses 15% of the largest wavelet coefficients; Figure 12.9c shows low-pass frequency filtering, with the low-frequency portion of the spectrum marked by the circle used for the inverse Fourier transform—the cut-off frequency was determined to minimize the mean square error between the reconstructed image and the original image (Figure 12.1). In the wavelet-reconstructed image (Figure 12.10a), small-size details were not blurred—compare with the result of low-pass filtering in the Fourier frequency domain shown in Figure 12.10b. The fact that wavelet filtering does not remove important detail from the image while the Fourier filtering does is demonstrated in Figures 12.10c,d. Note that there are no obvious border artifacts present in part (c) while these artifacts are clearly present in part (d).

The most substantial progress in discrete image transforms was probably motivated by image transmission and image data storage requirements. The main goal is to decrease the amount of data necessary for image representation and to decrease the amount of transferred or stored data. Compressed images are frequently used for transmission using the World Wide Web; many multimedia presentations rely on image compression to store large numbers of images and movies on CD-ROMs, and standards exist for such image compression. For example, the commonly used JPEG and MPEG image compression standards are based on the discrete cosine transform presented here (Section 12.4). The next chapter is devoted to the area of image compression.

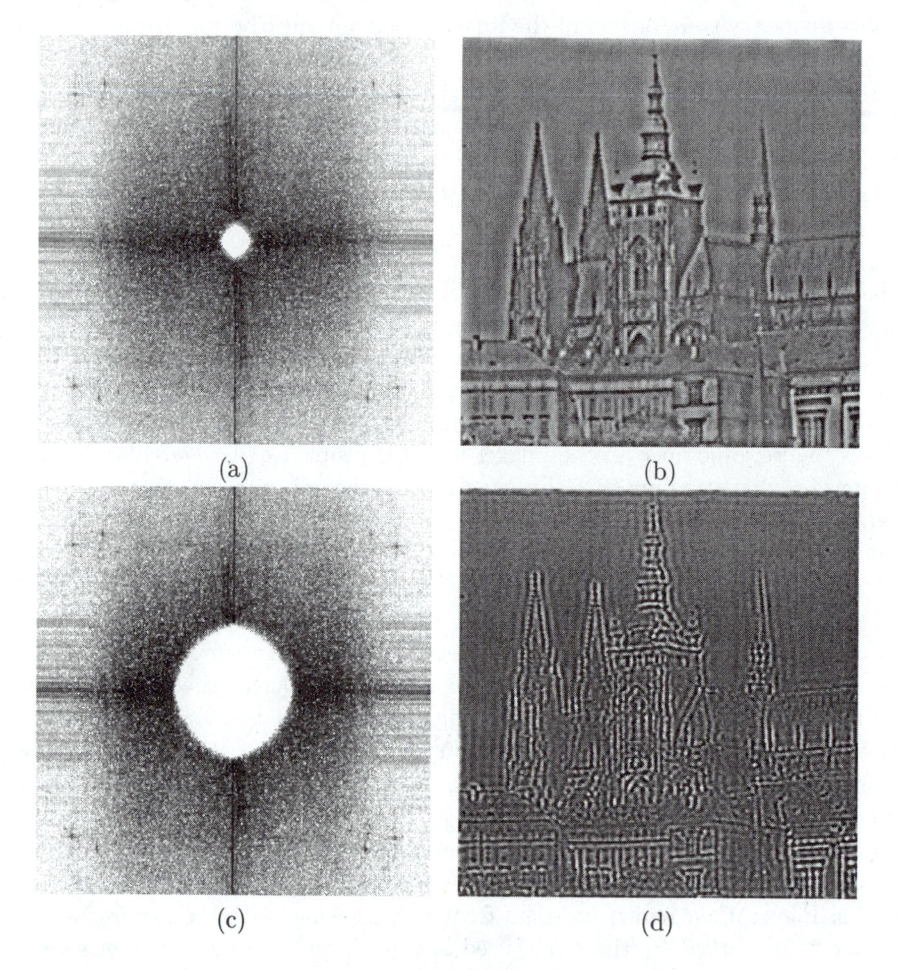

Figure 12.6: *High-pass frequency domain filtering: (a) spectrum of a high-pass filtered image, only very low frequencies filtered out; (b) image resulting from the inverse Fourier transform applied to spectrum (a); (c) spectrum of a high-pass filtered image, all lower frequencies filtered out; (d) inverse Fourier transform applied to spectrum (c).*

Another large application area can be found in image description, especially in texture description and recognition (see Section 14.1.1). Reconstruction of images corrupted by camera motion, defocusing, etc., in the image acquisition stage can also be achieved using

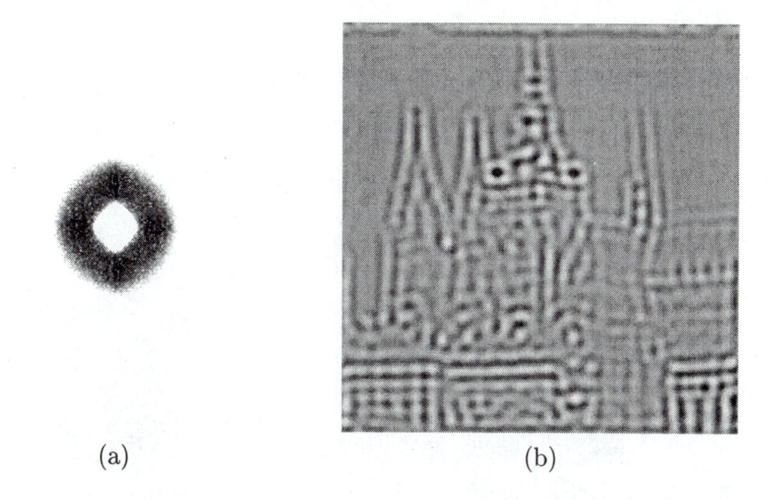

Figure 12.7: *Band-pass frequency domain filtering: (a) spectrum of a band-pass-filtered image, low and high frequencies filtered out; (b) image resulting from the inverse Fourier transform applied to spectrum (a).*

discrete image transforms [Andrews and Hunt 77, Bates and McDonnell 86].

12.8 Summary

- **Linear discrete image transforms**
  - Continuous or discrete image transforms represent a classic approach to image processing and analysis.
  - Linear discrete image transforms are widely used in **image filtering, image data compression, image description**, etc.
- **Fourier transform**
  - The Fourier transform may be represented using a set of orthogonal **sinusoidal waveforms**; the coefficients of the Fourier representation are called **frequency components** and the waveforms are ordered by frequency.
  - The discrete Fourier transform is analogous to the continuous one and may be efficiently computed using the **fast Fourier transform** algorithm.
  - The Fourier transform is of great use in the calculation of **image convolutions**.
- **Hadamard transform**
  - The base of the Hadamard transform consists of **square waves** (**Walsh functions**).
  - The Walsh functions are ordered by the number of their zero-crossings and the coefficients are called **sequency components**.

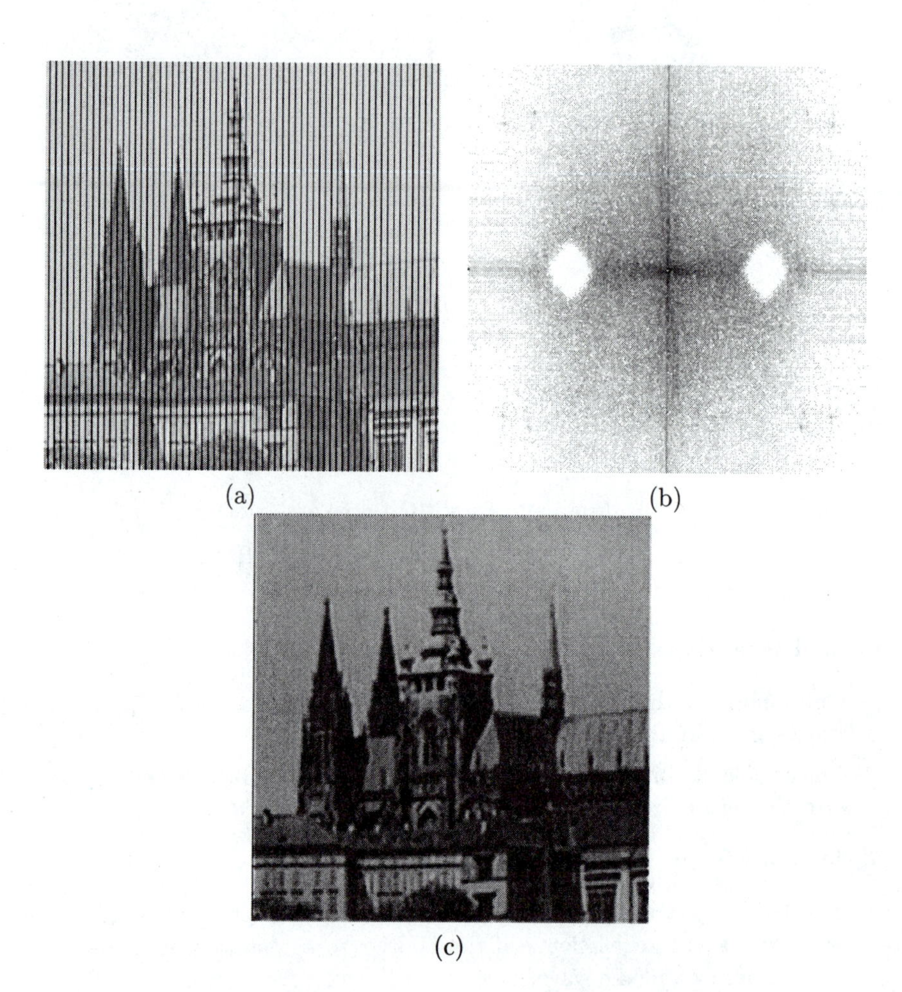

Figure 12.8: *Periodic noise removal: (a) noisy image; (b) image spectrum used for image reconstruction—note that the areas of frequencies corresponding with periodic vertical lines are filtered out; (c) filtered image.*

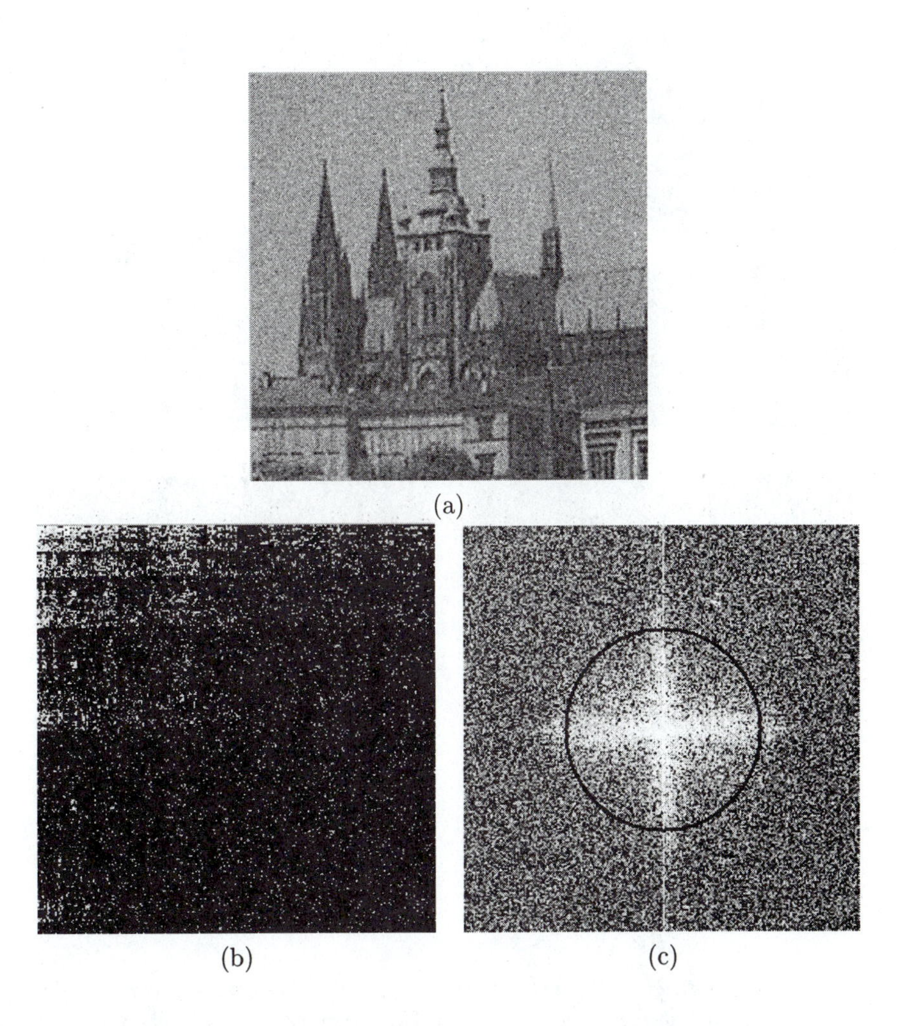

Figure 12.9: *Noise filtering: (a) noisy image; (b) wavelet filtering (unused 85% of the coefficients are shown in black, compare with all wavelet coefficients shown for the original image in Figure 12.3); (c) low-pass frequency filtering.* Courtesy G. Prause, The University of Iowa.

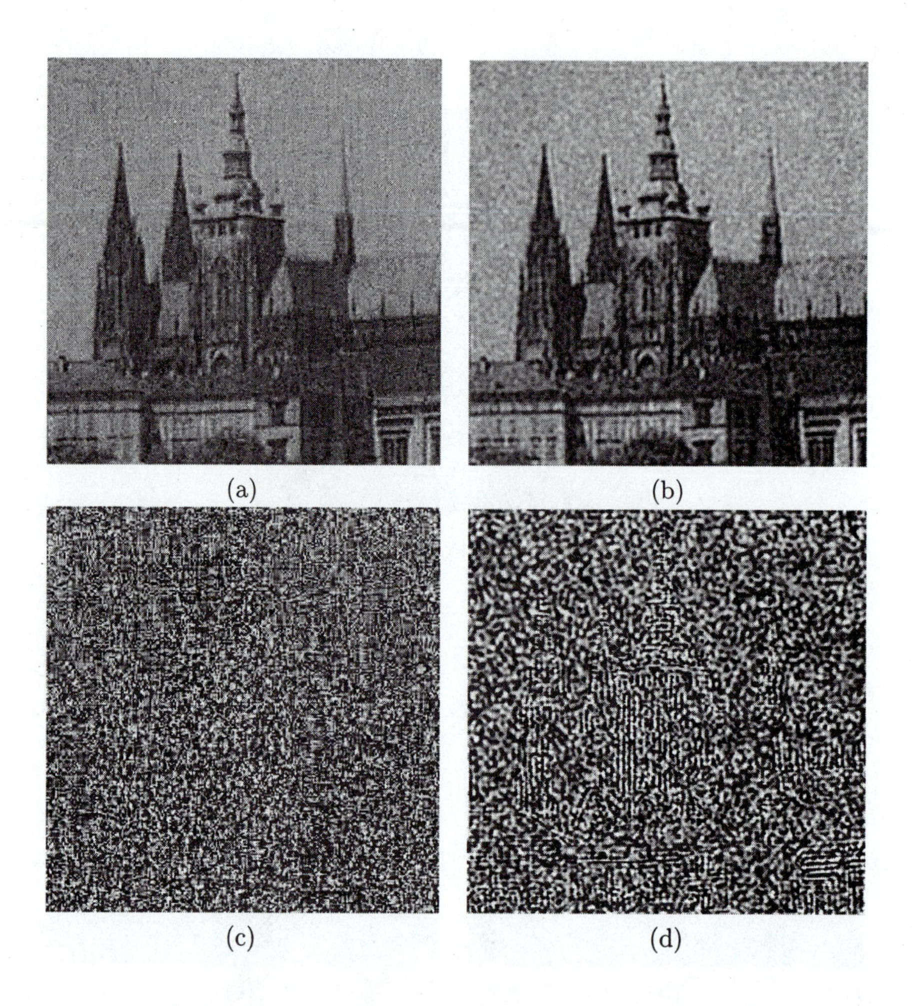

Figure 12.10: *Noise filtering: (a) wavelet filtering applied to the image shown in Figure 12.9a, using wavelet coefficients shown in Figure 12.9b; (b) Fourier frequency filtering using the low-frequency portion of the spectrum shown in Figure 12.9c; (c) absolute differences between the original and the wavelet-reconstructed images (histogram equalized); (d) absolute differences between the original and the Fourier-reconstructed images (histogram equalized). Courtesy G. Prause, The University of Iowa. (Compare with noise-reduction results of Figure 4.10.)*

- **Discrete cosine transform**
  - There are four definitions of the discrete cosine transform, denoted DCT-I, DCT-II, DCT-III, and DCT-IV. The most commonly used discrete cosine transform in image processing and compression is **DCT-II**, which forms the basis for **JPEG** image compression.
  - DCT computation can be based on the Fourier transform—all N coefficients of the discrete cosine transform may be computed using a $2N$-point fast Fourier transform.
- **Wavelets**
  - Wavelets decompose complex signals into sums of basis functions—in this respect they are similar to other discrete image transforms.
  - However, wavelets are local in both frequency and time (and/or space in image processing) and are able to analyze data at different scales or resolutions much better than simple sine and cosines can.
- **Other discrete image transforms**
  - Many other discrete image transforms exist, e.g., **Haar, Hadamard-Haar, Slant, Slant-Haar, Radon, discrete sine transform**, etc.
  - The power of the **Karhunen-Loeve** transform is in the search for non-correlated transform coefficients, which makes it very useful in data compression.
- **Applications of discrete image transforms**
  - Image convolutions can be performed in the frequency domain. Such operations are usually called **spatial frequency filtering**.
  - Basic examples of spatial filtering are linear **low-pass** and **high-pass** frequency filters, directional filters, noise reduction, etc.

12.9 Exercises

Short-answer questions

1. Define an orthogonal transform.
2. Give a definition of the two-dimensional discrete Fourier transform and its inverse.
3. Show how the fast Fourier transform can be used to compute the two-dimensional discrete Fourier transform efficiently. What are the usual image size assumptions?
4. Give a definition of the following image spectra: frequency spectrum, phase spectrum, power spectrum.
5. Show how the Fourier transform can be employed for computing image convolution. For which convolution kernel sizes is the use of Fourier transform typically more efficient than direct convolution?
6. Give a definition of the two-dimensional Hadamard transform.

7. Give a definition of the two-dimensional discrete cosine transform DCT-II.
8. Explain why wavelets are better suited to analyze image data in multiple scales than the Fourier transform.
9. Give a general definition of the mother wavelet and the specific definition of the Haar mother wavelet.
10. Which transform facilitates image reconstruction from projections?
11. Explain how the Fourier transform can be used to implement high-pass and low-pass filtering.
12. Explain how the Fourier transform can be used to remove periodic image noise.

Problems

1. Prove that a unitary matrix always has a determinant 1.
2. The power spectrum of a signal is identical to the Fourier transform of the signal's autocorrelation function. Prove this property for one-dimensional signals.
3. Determine the Hadamard matrix of order 2^3, and determine its inverse.
4. Demonstrate that Hadamard matrices of higher orders are symmetric and orthogonal.
5. Show how the fast Fourier transform can be employed to compute discrete cosine transform.
6. For the following 4×4 image, determine its forward and inverse transforms, and compare the inverse transforms with the original image data:

$$\begin{matrix} 2 & 0 & 1 & 0 \\ 1 & 1 & 0 & 1 \\ 1 & 0 & 0 & 1 \\ 2 & 1 & 2 & 3 \end{matrix}$$

   Use the following image transforms:

   (a) The discrete Fourier transform
   (b) The Hadamard transform
   (c) The discrete cosine transform

7. Using one of the many freely available functions (subroutines) [Press 88] for the one-dimensional Fourier transform and its inverse, develop a program for the computation of the two-dimensional Fourier transform and its inverse, considering real data inputs.
8. Using a program for a two-dimensional Fourier transform (e.g., developed in Problem 12.7), develop a program for high-pass, low-pass, and band-pass image filtering.
9. Develop a program for image smoothing in the frequency domain that uses the convolution property of the Fourier transform.

   (a) Write a program in which the frequency domain smoothing is equivalent to the convolution with an 11×11 averaging filter.
   (b) Implement the same smoothing by direct convolution.
   (c) Compare the efficiency of the two approaches.
   (d) Use 3×3 averaging for the efficiency comparison.

12.10 References

[Ahmed and Rao 75] N Ahmed and K R Rao. *Orthogonal Transforms for Digital Signal Processing.* Springer Verlag, Berlin, 1975.

[Andrews 70] H C Andrews. *Computer Techniques in Image Processing.* Academic Press, New York, 1970.

[Andrews and Hunt 77] H C Andrews and B R Hunt. *Digital Image Restoration.* Prentice-Hall, Englewood Cliffs, NJ, 1977.

[Bates and McDonnell 86] R H T Bates and M J McDonnell. *Image Restoration and Reconstruction.* Clarendon Press, Oxford, 1986.

[Bracewell 95] R N Bracewell. *Two-dimensional Imaging.* Prentice-Hall, Englewood Cliffs, NJ, 1995.

[Bradley et al. 93] J Bradley, C Brislawn, and T Hopper. The FBI wavelet/scalar quantization standard for gray-scale fingerprint image compression. Technical Report LA-UR-93-1659, Los Alamos National Laboratory, Los Alamos, NM, 1993.

[Castleman 96] K R Castleman. *Digital Image Processing.* Prentice-Hall, Englewood Cliffs, NJ, 1996.

[Chui 92] C K Chui. *An Introduction to Wavelets.* Academic Press, New York, 1992.

[Chui et al. 94] C K Chui, Laura Montefusco, and L Puccio. *Wavelets: Theory, Algorithms, and Applications.* Academic Press, San Diego, CA, 1994.

[Daubechies 88] I Daubechies. Orthonormal bases of compactly supported wavelets. *Communications on Pure and Applied Mathematics*, 41:906–966, 1988.

[Donoho 93] D Donoho. Nonlinear wavelet methods for recovery of signals, densities and spectra from indirect and noisy data. In I Daubechies, editor, *Proceedings of Symposia in Applied Mathematics*, pages 173–205. AMS, 1993.

[Dougherty and Giardina 87] E R Dougherty and C R Giardina. *Image Processing—Continuous to Discrete*, volume 1. Prentice-Hall, Englewood Cliffs, NJ, 1987.

[Dudgeon and Mersereau 84] D F Dudgeon and R M Mersereau. *Multidimensional Digital Signal Processing.* Prentice-Hall, Englewood Cliffs, NJ, 1984.

[Francon 79] M Francon. *Optical Image Formation and Processing.* Academic Press, New York, 1979.

[Gonzalez and Wintz 87] R C Gonzalez and P Wintz. *Digital Image Processing.* Addison-Wesley, Reading, MA, 2nd edition, 1987.

[Gonzalez and Woods 92] R C Gonzalez and R E Woods. *Digital Image Processing.* Addison-Wesley, Reading, MA, 1992.

[Graps 95] A Graps. An introduction to wavelets. *IEEE Transactions on Computational Science and Engineering*, 2(2):50–61, 1995.

[Hall 79] E L Hall. *Computer Image Processing and Recognition.* Academic Press, San Diego–New York, 1979.

[Huang 81] T S Huang. *Two-Dimensional Digital Signal Processing II: Transform and Median Filters.* Springer Verlag, Berlin–New York, 1981.

[Hung and Meng 94] A C Hung and T H Y Meng. A comparison of fast inverse discrete cosine transform algorithms. *Journal of Multimedia Systems*, 2:204–217, 1994.

[Jain 89] A K Jain. *Fundamentals of Digital Image Processing.* Prentice-Hall, Englewood Cliffs, NJ, 1989.

[Kak and Slaney 88] A C Kak and M Slaney. *Principles of Computerized Tomographic Imaging.* IEEE, Piscataway, NJ, 1988.

[Lim 90] J S Lim. *Two-Dimensional Signal and Image Processing.* Prentice-Hall, Englewood Cliffs, NJ, 1990.

[Meyer 93] Y Meyer. *Wavelets: Algorithms and Applications.* Society for Industrial and Applied Mathematics, Philadelphia, 1993.

[Nussbaumer 82] H J Nussbaumer. *Fast Fourier Transform and Convolution Algorithms.* Springer Verlag, Berlin, 2nd edition, 1982.

[Pavlidis 82] T Pavlidis. *Algorithms for Graphics and Image Processing.* Computer Science Press, New York, 1982.

[Pratt 91] W K Pratt. *Digital Image Processing.* Wiley, New York, 2nd edition, 1991.

[Press 88] W H Press. *Numerical recipes in C: The art of scientific computing.* Cambridge University Press, Cambridge, 1988.

[Rao and Yip 90] K R Rao and P Yip. *Discrete Cosine Transform, Algorithms, Advantages, Applications.* Academic Press, Boston, 1990.

[Rosenfeld and Kak 82] A Rosenfeld and A C Kak. *Digital Picture Processing.* Academic Press, New York, 2nd edition, 1982.

[Sanz et al. 88] J L C Sanz, E B Hinkle, and A K Jain. *Radon and Projection Transform-Based Computer Vision.* Springer Verlag, Berlin–New York, 1988.

[Shulman 70] A R Shulman. *Optical Data Processing.* Wiley, New York, 1970.

[Sid-Ahmed 95] M A Sid-Ahmed. *Image Processing.* McGraw-Hill, New York, 1995.

[Stark 82] H Stark. *Applications of Optical Fourier Transforms.* Academic Press, New York, 1982.

[Wolberg 90] G Wolberg. *Digital Image Warping.* IEEE, Los Alamitos, CA, 1990.

[Yaroslavskii 85] L P Yaroslavskii. *Digital Picture Processing: An Introduction.* Springer Verlag, Berlin–New York, 1985.

Chapter 13

Image data compression

Image processing is often very difficult because of the large amounts of data used to represent an image. Technology permits ever-increasing image resolution (spatially and in gray-levels), and increasing numbers of spectral bands, and there is a consequent need to limit the resulting data volume. Consider an example from the remote sensing domain, where image data compression is a very serious problem. A Landsat D satellite broadcasts 85×10^6 bits of data every second and a typical image from one pass consists of 6100×6100 pixels in seven spectral bands—260 megabytes of image data. As another example, the Japanese Advanced Earth Observing Satellite (ADEOS), which has the ability to observe the Earth's surface with a spatial resolution of 8 meters for the polychromatic band and 16 meters for the multi-spectral bands has a transmitted data rate of 120 Mbps [Arai 90]. Thus the amount of storage media needed for archiving of such remotely sensed data is enormous. One possible approach to decreasing the necessary amount of storage is to work with compressed image data.

We have seen that segmentation techniques have the side effect of image compression; by removing all areas and features that are not of interest, and leaving only boundaries or region descriptors, the reduction in data quantity is considerable. However, from this sort of representation no image reconstruction to the original uncompressed image (or only a very limited reconstruction) is possible. Conversely, image compression algorithms aim to remove redundancy in data in a way which makes image reconstruction possible; this is sometimes called *information preserving compression.* Compression is the main goal of the algorithm—we aim to represent an image using fewer bits per pixel, without losing the ability to reconstruct the image. It is necessary to find statistical properties of the image to design an appropriate compression transformation of the image; the more correlated the image data are, the more data items can be removed. In this chapter, we will discuss this group of methods which do not change image entropy or image information content. More detailed surveys of image compression techniques may be found in [Rosenfeld and Kak 82, Clarke 85, Lovewell and Basart 88, Netravali 88, Jain 89, Rabbani 91, Witten et al. 94, Furht et al. 95, Clarke 95].

A general algorithm for data compression and image reconstruction is shown in a block diagram in Figure 13.1. The first step removes information redundancy caused by high correlation of image data—transform compressions, predictive compressions, and hybrid approaches are used. The second step is coding of transformed data using a code of fixed or variable-length. An advantage of variable-length codes is the possibility of coding more fre-

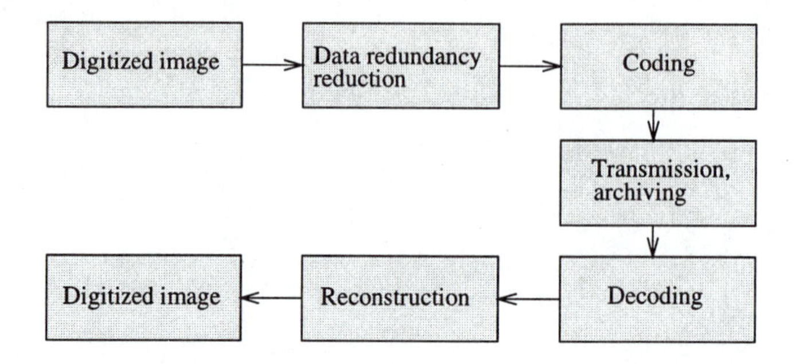

Figure 13.1: *Data compression and image reconstruction.*

quent data using shorter code words and therefore increasing compression efficiency, while an advantage of fixed length coding is a standard codeword length that offers easy handling and fast processing. Compressed data are decoded after transmission or archiving and reconstructed. Note that no non-redundant image data may be lost in the data compression process—otherwise error-free reconstruction is impossible.

Data compression methods can be divided into two principal groups: **information preserving** compressions permit error-free data reconstruction (lossless compression), while compression methods **with loss of information** do not preserve the information completely (lossy compression). In image processing, a faithful reconstruction is often not necessary in practice and then the requirements are weaker, but the image data compression must not cause significant changes in an image. Data compression success in the reconstructed image is usually measured by the mean square error (MSE), signal-to-noise ratio etc., although these global error measures do not always reflect subjective image quality.

Image data compression design consists of two parts. Image data properties must be determined first; gray-level histograms, image entropy, various correlation functions, etc., often serve this purpose. The second part yields an appropriate compression technique design with respect to measured image properties.

Data compression methods with loss of information are typical in image processing and therefore this group of methods is described in some detail. Although lossy compression techniques can give substantial image compression with very good quality reconstruction, there are considerations that may prohibit their use. For example, diagnosis in medical imaging is often based on visual image inspection, so no loss of information can be tolerated and information preserving techniques must be applied. Information preserving compression methods are mentioned briefly at the end of the chapter.

13.1 Image data properties

Information content of an image is an important property, of which **entropy** is a measure. If an image has G gray-levels and the probability of gray-level k is $P(k)$ (see Section 2.1.4), then entropy H_e, not considering correlation of gray-levels, is defined as

$$H_e = - \sum_{k=0}^{G-1} P(k) \log_2[P(k)] \tag{13.1}$$

Information **redundancy** r is defined as

$$r = b - H_e \tag{13.2}$$

where b is the smallest number of bits with which the image quantization levels can be represented. This definition of image information redundancy can be evaluated only if a good estimate of entropy is available, which is usually not so because the necessary statistical properties of the image are not known. Image data entropy however can be estimated from a gray-level histogram [Moik 80, Pratt 91]. Let $h(k)$ be the frequency of gray-level k in an image f, $0 \leq k \leq 2^b - 1$, and let the image size be $M \times N$. The probability of occurrence of gray-level k can be estimated as

$$\tilde{P}(k) = \frac{h(k)}{MN} \tag{13.3}$$

and the entropy can be estimated as

$$\tilde{H}_e = - \sum_{k=0}^{2^b-1} \tilde{P}(k) \log_2[\tilde{P}(k)] \tag{13.4}$$

The information redundancy estimate is $\tilde{r} = b - \tilde{H}_e$. The definition of the **compression ratio** K is then

$$K = \frac{b}{\tilde{H}_e} \tag{13.5}$$

Note that a gray-level histogram gives an inaccurate estimate of entropy because of gray-level correlation. A more accurate estimate can be obtained from a histogram of the first gray-level differences.

Theoretical limits of possible image compression can be found using these formulae. For example, the entropy of satellite remote sensing data may be $\tilde{H}_e \in [4,5]$, where image data are quantized into 256 gray-levels, or 8 bits per pixel. We can easily compute the information redundancy as $\tilde{r} \in [3,4]$ bits. This implies that these data can be represented by an average data volume of 4–5 bits per pixel with no loss of information, and the compression ratio would be $K \in [1.6, 2]$.

13.2 Discrete image transforms in image data compression

Image data representation by coefficients of discrete image transforms (see Chapter 12) is the basic idea of this approach. The transform coefficients are ordered according to their importance, i.e., according to their contribution to the image information contents, and the least important (low-contribution) coefficients are omitted. Coefficient importance can be judged, for instance, in correspondence to spatial or gray-level visualization abilities of the display; image correlation can then be avoided and data compression may result.

To remove correlated image data, the **Karhunen-Loeve** transform is the most important. This transform builds a set of non-correlated variables with decreasing variance. The variance of a variable is a measure of its information content; therefore, a compression strategy is based on considering only transform variables with high variance, thus representing an image by only the first k coefficients of the transform. More details about the Karhunen-Loeve transform can be found in [Rosenfeld and Kak 82, Savoji and Burge 85, Netravali 88, Jain 89, Pratt 91].

The Karhunen-Loeve transform is computationally expensive, with a two-dimensional transform of an $M \times N$ image having computational complexity $\mathcal{O}(M^2N^2)$. It is the only transform that guarantees non-correlated compressed data, and the resulting data compression is optimal in the statistical sense. This makes the transform basis vectors image dependent, which also makes this transform difficult to apply for routine image compression. Therefore, the Karhunen-Loeve transform is used mainly as a benchmark to evaluate other transforms. For example, one reason for the popularity of the discrete cosine transform DCT-II is that its performance approaches the Karhunen-Loeve transform better than others.

Other discrete image transforms (see Chapter 12) are computationally less demanding—fast algorithms of these transforms have computational complexity $\mathcal{O}[MN \log_2(MN)]$. Cosine, Fourier, Hadamard, Walsh, or binary transforms are all suitable for image data compression. If an image is compressed using discrete transforms, it is usually divided into subimages of 8×8 or 16×16 pixels to speed up calculations, and then each subimage is transformed and processed separately. The same is true for image reconstruction, with each subimage being reconstructed and placed into the appropriate image position [Anderson and Huang 71, Wintz 72]. This image segmentation into a grid of subimages does not consider any possible data redundancy caused by subimage correlation even if this correlation is the most serious source of redundancy. **Recursive block** coding [Farelle 90] is an important novel approach to reducing inter-block redundancy and tiling effects (blockiness). The most popular image transform used for image compression seems to be the discrete cosine transform, with many modifications [Chen and Bovik 89, Azadegan 90, Chitprasert and Rao 90] and variations of wavelet transforms [Ebrahimi et al. 90, Zettler et al. 90, Chui et al. 94, Hilton et al. 94] (Section 12.5).

Discrete cosine transform image compression possibilities are shown in Figure 13.2. The DCT-II applied here provides good compression with low computational demands, the compression ratios being $K = 6.2$ and $K = 10.5$. The lower compression ratio was achieved after setting 90.0% of the transform coefficients to zero; the higher compression ratio resulted after setting 94.9% of the transform coefficients to zero. Note that square blocks resulting from DCT compression and reconstruction decrease the image quality for larger compression ratios. Consequently, **wavelet image compression** is gaining acceptance, since it can be efficiently applied to the entire image [Mallat 89] and thus the square image compression artifacts are not present. Wavelet compression consists of the same steps as DCT compression, but the DCT is replaced by a wavelet transform followed by generally identical quantization and coding. Figure 13.3 shows the reconstructed image after wavelet compression with two different compression ratios, $K = 6.2$ and $K = 10.5$. The lower compression ratio (Figure 13.3a,b) was achieved after setting 89.4% of the transform coefficients to zero, the higher compression ratio (Figure 13.3a,b) resulted after setting 94.4% of the transform coefficients to zero. Note that no blocking artifacts exist.

13.3 Predictive compression methods

Predictive compressions use image information redundancy (correlation of data) to construct an estimate $\tilde{f}(i,j)$ of the gray-level value of an image element (i,j) from values of gray-levels in the neighborhood of (i,j). In image parts where data are not correlated, the estimate $\tilde{f}$ will not match the original value. The differences between estimates and reality, which may

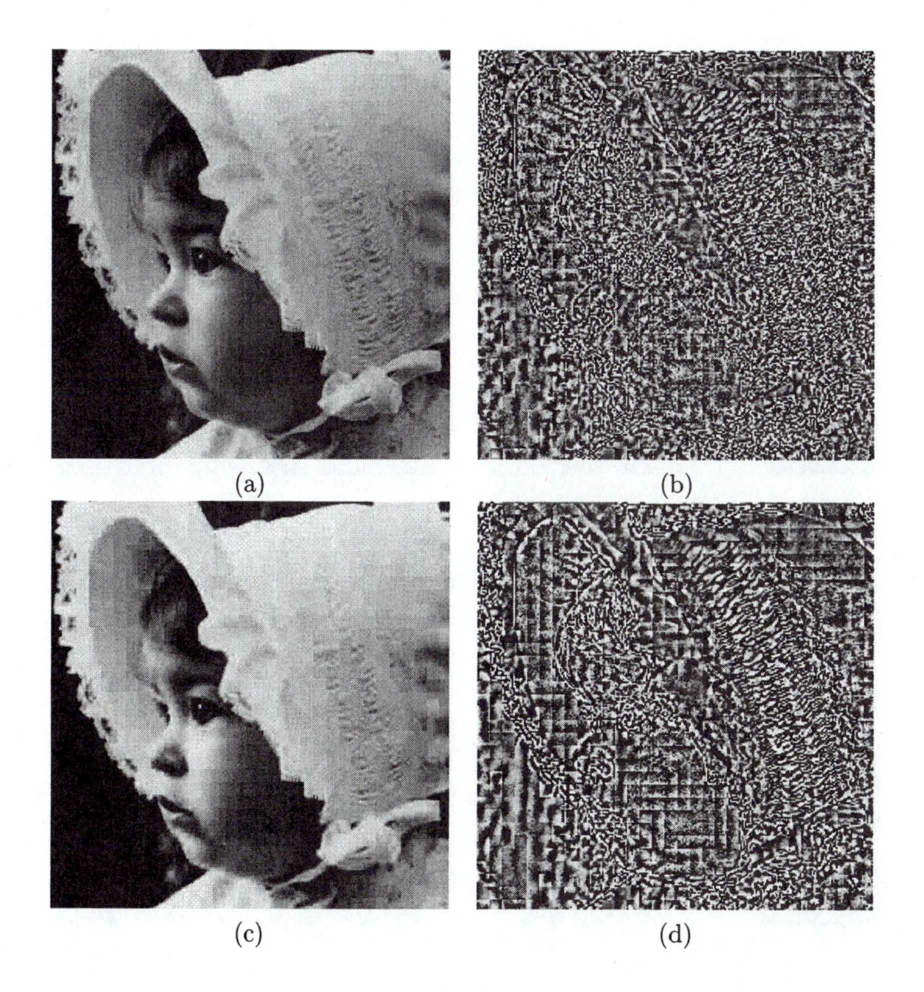

Figure 13.2: *Discrete cosine image compression applied to subblocks of* 8×8 *pixels as in JPEG. (a) Reconstructed image, compression ratio $K = 6.2$. (b) difference image—differences between pixel values in the original and the reconstructed image ($K = 6.2$); the maximum difference is 56 gray-levels, mean squared reconstruction error MSE = 32.3 gray-levels (the image is histogram equalized for visualization purposes). (c) Reconstructed image, compression ratio $K = 10.5$. (d) Difference image—differences between pixel values in the original and the reconstructed image ($K = 10.5$); the maximum difference is 124 gray-levels, MSE = 70.5 gray-levels (the image is histogram equalized for visualization purposes). Courtesy A. Kruger, G. Prause, The University of Iowa.*

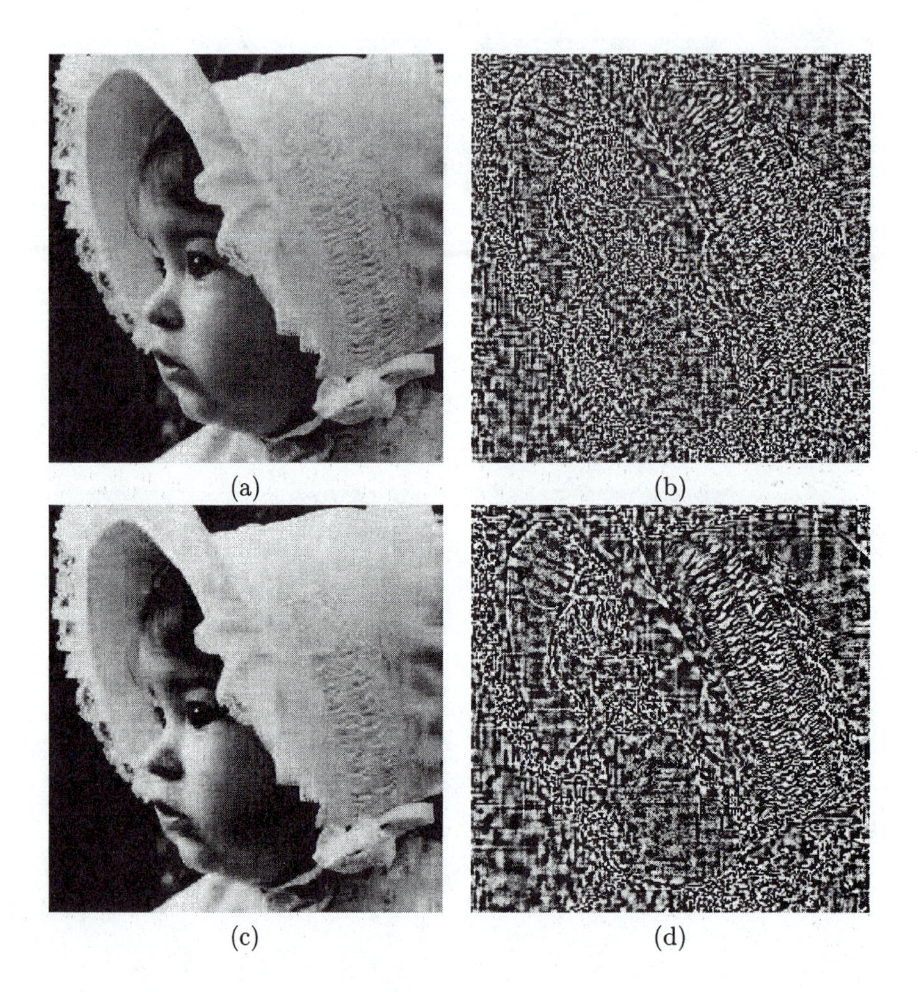

Figure 13.3: *Wavelet image compression: (a) Reconstructed image, compression ratio* $K = 6.2$. *(b) Difference image—differences between pixel values in the original and the reconstructed image (*$K = 6.2$*); the maximum difference is 37 gray-levels, mean squared reconstruction error MSE = 32.0 gray-levels (the image is histogram equalized for visualization purposes). (c) Reconstructed image, compression ratio* $K = 10.5$. *(d) Difference image—differences between pixel values in the original and the reconstructed image (*$K = 10.5$*); the maximum difference is 79 gray-levels, MSE = 65.0 gray-levels (the image is histogram equalized for visualization purposes). Courtesy G. Prause, The University of Iowa.*

be expected to be relatively small in absolute terms, are coded and transmitted (stored) together with prediction model parameters—the whole set now represents compressed image data. The gray value at location (i, j) is reconstructed from a computed estimate $\tilde{f}(i, j)$ and the stored difference $d(i, j)$

$$d(i,j) = \tilde{f}(i,j) - f(i,j) \tag{13.6}$$

This method is called differential pulse code modulation (DPCM)—its block diagram is presented in Figure 13.4. Experiments show that a linear predictor of the third order is sufficient for estimation in a wide variety of images [Habibi 71]. If the image is processed line by line, the estimate $\tilde{f}$ can be computed as

$$\tilde{f}(i,j) = a_1 f(i, j-1) + a_2 f(i-1, j-1) + a_3 f(i-1, j) \tag{13.7}$$

where a_1, a_2, a_3 are image prediction model parameters. These parameters are set to minimize the mean quadratic estimation error e,

$$e = \mathcal{E}\{[\tilde{f}(i,j) - f(i,j)]^2\} \tag{13.8}$$

and the solution, assuming f is a stationary random process with a zero mean, using a predictor of the third order, is

$$\begin{aligned} a_1 R(0,0) + a_2 R(0,1) + a_3 R(1,1) &= R(1,0) \\ a_1 R(0,1) + a_2 R(0,0) + a_3 R(1,0) &= R(1,1) \\ a_1 R(1,1) + a_2 R(1,0) + a_3 R(0,0) &= R(0,1) \end{aligned} \tag{13.9}$$

where $R(m, n)$ is the autocorrelation function of the random process f (see Chapter 2). The image data autocorrelation function is usually of exponential form and the variance of differences $d(i, j)$ is usually smaller than the variance of the original values $f(i, j)$, since the differences $d(i, j)$ are not correlated. The (probable) relatively small magnitude of the differences $d(i, j)$ makes data compression possible.

Predictive compression algorithms are described in detail in [Rosenfeld and Kak 82, Netravali 88]. A predictive method of second order with variable code length coding of the differences $d(i, j)$ was used to obtain the compressed images shown in Figure 13.5; data compression ratios $K = 3.8$ and $K = 6.2$ were achieved. Note that horizontal lines and false contours resulting from the predictive compression and reconstruction decrease the image quality for larger compression ratios.

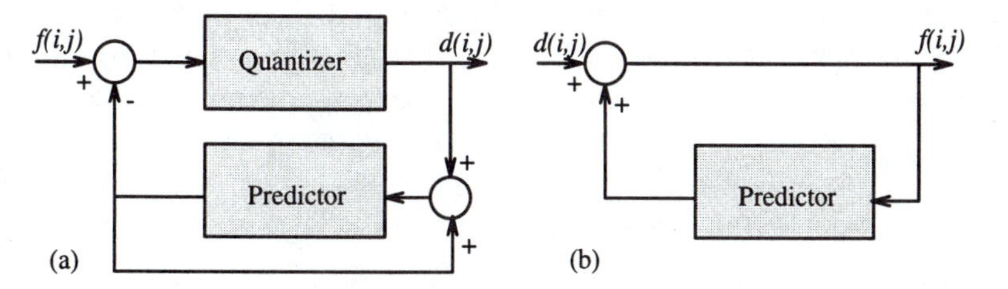

Figure 13.4: *Differential pulse code modulation: (a) compression; (b) reconstruction.*

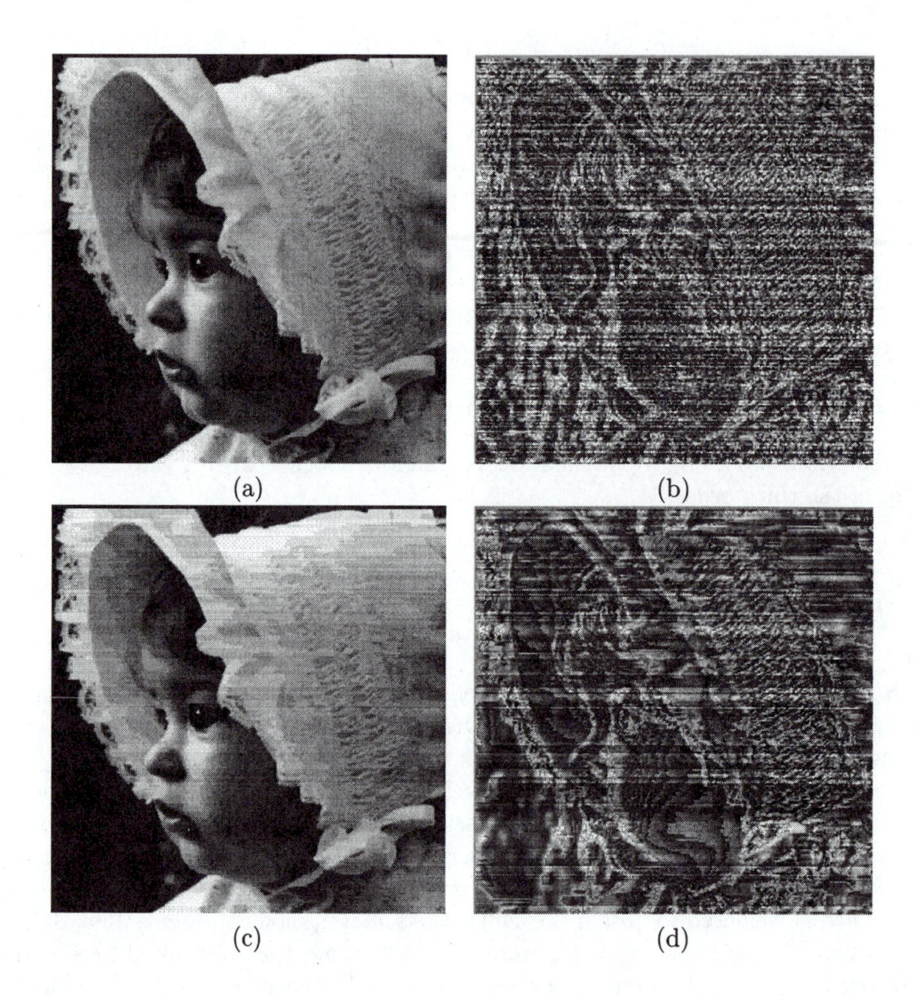

Figure 13.5: *Predictive compression: (a) Reconstructed image, compression ratio $K = 3.8$. (b) Difference image—differences between pixel values in the original and the reconstructed image ($K = 3.8$); the maximum difference is 6 gray-levels (the image is histogram equalized for visualization purposes). (c) Reconstructed image, compression ratio $K = 6.2$. (d) Difference image—differences between pixel values in the original and the reconstructed image ($K = 6.2$); the maximum difference is 140 gray-levels (the image is histogram equalized for visualization purposes). Courtesy A. Kruger, The University of Iowa.*

Many modifications of predictive compression methods can be found in the literature, some of them combining predictive compression with other coding schemes [Guha et al. 88, Daut and Zhao 90, Gonzalez et al. 90, Zailu and Taxiao 90].

13.4 Vector quantization

Dividing an image into small blocks and representing these blocks as vectors is another option [Gray 84, Chang et al. 88, Netravali 88, Gersho and Gray 92]. The basic idea for this approach comes from information theory (Shannon's rate distortion theory), which states that better compression performance can always be achieved by coding vectors instead of scalars. Input data vectors are coded using unique codewords from a codeword dictionary, and instead of vectors, the vector codes are stored or transmitted. The codeword choice is based on the best similarity between the image block represented by a coded vector and the image blocks represented by codewords from the dictionary. The code dictionary (code book) is transmitted together with the coded data. The advantage of vector quantization is a simple receiver structure consisting of a look-up table, but a disadvantage is a complex coder. The coder complexity is not caused directly by the vector quantization principle; the method can be implemented in a reasonably simple way, but the coding will be very slow. To increase the processing speed, special data structures (K-D trees) and other special treatments are needed which increase the coder complexity. Further, the necessary statistical properties of images are usually not available. Therefore, the compression parameters must be based on an image training set and the appropriate code book may vary from image to image. As a result, images with statistical properties dissimilar from images in the training set may not be well represented by the code vectors in the look-up table. Furthermore, edge degradation may be more severe than with other techniques. To decrease the coder complexity, the coding process may be divided into several levels, two being typical. The coding process is hierarchical, using two or more code books according to the number of coding levels. However, the combination of a complex coder facilitating high compression ratios and a simple decoder may be advantageous in **asymmetric** applications when the image is compressed once and decompressed many times. Within such a scenario, the higher compression ratio gained by the more complex coder and/or more time-consuming compression algorithm does not matter as long as the decompression process is simple and fast. Multimedia encyclopedias and paperless publishing serve as good examples. On the other hand, in **symmetric** applications such as video conferencing, similar complexity of coding and decoding operations is required.

A modification that allows blocks of variable size is described in [Boxerman and Lee 90], where a segmentation algorithm is responsible for detecting appropriate image blocks. The block vector quantization approach may also be applied to compression of image sequences. Identifying and processing only blocks of the image that change noticeably between consecutive frames using vector quantization and DPCM is described in [Thyagarajan and Sanchez 89]. Hybrid DPCM combined with vector quantization of colored prediction errors is presented in [De Lameillieure and Bruyland 90].

13.5 Hierarchical and progressive compression methods

Multi-resolution pyramids have been mentioned many times throughout this book, and they may also be used for efficient hierarchical image compression. **Run length** codes were introduced in Section 3.2.2, Figure 3.2; run length coding identifies long runs of the same value pixels, and stores them as this value together with a word count. If the image is characterized by such long runs, this will significantly reduce storage requirements. A similar approach may be applied to image pyramids. Recently, it was shown that a substantial reduction in bit volume can be obtained by merely representing a source as a pyramid [Rao and Pearlman 91], and even more significant reduction can be achieved for images with large areas of the same gray-level if a quadtree coding scheme is applied (see Section 3.3.2). An example is given in Figure 13.6, where the principle of quadtree image compression is presented. Large image areas of the same gray-level can be represented in higher-level quadtree nodes without the necessity of including lower-level nodes in the image representation [White 87]. Clearly, the compression ratio achieved is image dependent and, for instance, a fine checkerboard image will not be represented efficiently using quadtrees. Modifications of the basic method exist, some of them successfully applied to motion image compression [Strobach 90] or incorporating hybrid schemes [Park and Lee 91].

Nevertheless, there may be an even more important aspect connected with this compression approach—the feasibility of progressive image transmission and the idea of smart compression.

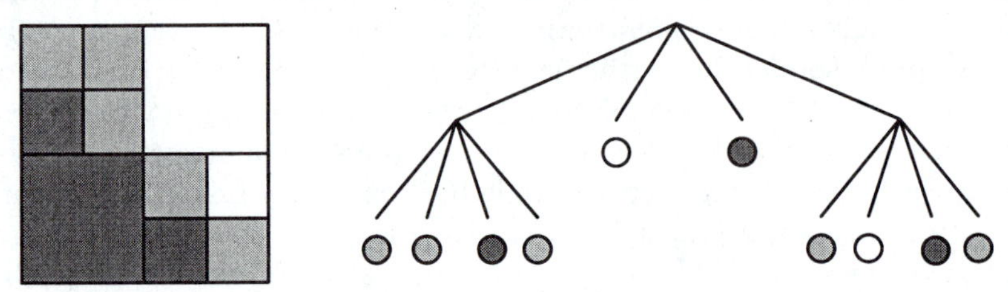

Figure 13.6: *Principle of quadtree image compression: original image and corresponding quadtree.*

Progressive image transmission is based on the fact that transmitting all image data may not be necessary under some circumstances. Imagine a situation in which an operator is searching an image database looking for a particular image. If the transmission is based on a raster scanning order, all the data must be transmitted to view the whole image, but often it is not necessary to have the highest possible image quality to find the image for which the operator is looking. Images do not have to be displayed with the highest available resolution, and lower resolution may be sufficient to reject an image and to begin displaying another one. This approach is also commonly used to decrease the waiting time needed for the image to start appearing after transmission and is used by World Wide Web image transmissions. In progressive transmission, the images are represented in a pyramid structure, the higher pyramid levels (lower resolution) being transmitted first. The number of pixels representing a lower-resolution image is substantially smaller and thus the user can decide from lower-resolution images whether further image refinement is needed. A standard M-pyramid (mean or matrix pyramid) consists of about one third more nodes than the number of image pixels.

Several pyramid encoding schemes have been designed to decrease the necessary number of nodes in pyramid representation: reduced sum pyramids, difference pyramids, and reduced difference pyramids [Wang and Goldberg 89, Wang and Goldberg 90]. The reduced difference pyramid has the number of nodes exactly equal to the number of image pixels and can be used for a lossless progressive image transmission with some degree of compression. It is shown in [Wang and Goldberg 90] that using an appropriate interpolation method in the image reconstruction stage, reasonable image quality can be achieved at a bit rate of less than 0.1 bit/pixel and excellent quality at a bit rate of about 1.2 bits/pixel. Progressive image transmission stages can be seen in Figure 2.3, where a sequence of four image resolutions is presented. Considering a hypothetical progressive image transmission, a 1/8-resolution image is transmitted first (Figure 2.3d). Next, the image is transmitted and displayed in 1/4 resolution (Figure 2.3c), followed by 1/2 resolution (Figure 2.3b) and then full resolution (Figure 2.3a).

The concept of **smart compression** is based on the sensing properties of human visual sensors [Burt 89]. The spatial resolution of the human eye decreases significantly with increasing distance from the optical axis. Therefore, the human eye can only see in high resolution in a very small area close to the point where the eye is focused. Similarly, as with image displays, where it does not make sense to display or even transmit an image in higher resolution than that of the display device, it is not necessary to display an image in full resolution in image areas where the user's eyes are not focused. This is the principle of smart image compression. The main difficulty remains in determining the areas of interest in the image on which the user will focus. When considering a smart progressive image transmission, the image should be transmitted first in higher resolution in areas of interest—this improves the subjective rating of transmission speed as sensed by a human user. The areas of interest may be obtained in a feedback control manner by tracking the user's eyes (assuming the communication channel is fast enough). The image point on which the user is focused may be used to increase the resolution in that particular image area so that the most important data are transmitted first. This smart image transmission and compression may be extremely useful if applied to dynamic image generators in driving or flight simulators, or to high-definition television [Abdel-Malek and Bloomer 90].

13.6 Comparison of compression methods

The main goal of image compression is to minimize image data volume with no significant loss of information, and all basic image compression groups have advantages and disadvantages. Transform-based methods better preserve subjective image quality, and are less sensitive to statistical image property changes both inside a single image and between images. Prediction methods, on the other hand, can achieve higher compression ratios in a much less expensive way, tend to be much faster than transform-based or vector quantization compression schemes, and can easily be realized in hardware. If compressed images are transmitted, an important property is insensitivity to transmission channel noise. Transform-based techniques are significantly less sensitive to channel noise—if a transform coefficient is corrupted during transmission, the resulting image distortion is spread homogeneously through the image or image part and is not too disturbing. Erroneous transmission of a difference value in prediction compressions causes not only an error in a particular pixel, it influences values in the neighborhood because the predictor involved has a considerable visual effect in a reconstructed image. Vector quantization methods require a complex coder, their parameters are very sensitive to image data, and they blur image edges. The advantage is in a simple decoding scheme consisting of a look-up table only. Pyramid-based techniques have a natural compression ability and show a potential for further improvement of compression ratios. They are suitable for dynamic image compression and for progressive and smart transmission approaches.

Hybrid compression methods combine good properties of the various groups [Habibi 74, Habibi and Robinson 74]. A hybrid compression of three-dimensional image data (two spatial dimensions plus one spectral dimension) is a good example. A two-dimensional discrete transform (cosine, Hadamard, ...) is applied to each mono-spectral image followed by a predictive compression in the third dimension of spectral components. Hybrid methods combine the different dimensionalities of transform compressions and predictive compressions. As a general rule, at least a one-dimensional transform compression precedes predictive compression steps. In addition to combinations of transform and predictive approaches, predictive approaches are often combined with vector quantization. A discrete cosine transform combined with vector quantization in a pyramid structure is described in [Park and Lee 91].

For more detailed comparisons of some image compression techniques refer to [Chang et al. 88, Lovewell and Basart 88, Jaisimha et al. 89, DiMento and Berkovich 90, Heer and Reinfelder 90, Hung and Meng 94].

13.7 Other techniques

Various other image data compression methods exist. If an image is quantized into a small number of gray-levels and if it has a small number of regions of the same gray-level, an effective compression method may be based on **coding region borders** [Wilkins and Wintz 71]. Image representation by its **low and high frequencies** is another method—image reconstruction is a superposition of inverse transforms of low- and high-frequency components. The low-frequency image can be represented by a significantly smaller volume of data than the original image. The high-frequency image has significant image edges only and can be represented efficiently [Graham 67, Giunta et al. 90]. The **region growing process** compression method stores an algorithm for region growing from region seed points, each region being represented by its seed point. If an image can be represented only by region seed points, significant data compression is achieved.

A technique that is gaining popularity is **block truncation** coding [Delp and Mitchell 79, Rosenfeld and Kak 82, Kruger 92], in which an image is divided into small square blocks of pixels and each pixel value in a block is truncated to one bit by thresholding and moment preserving selection of binary levels. One bit value per pixel has to be transmitted, together with information describing how to recreate the moment preserving binary levels during reconstruction. This method is fast and easy to implement. **Visual pattern image** coding is capable of high-quality compression with very good compression ratios (30:1) and is exceptionally fast [Silsbee et al. 91].

Fractal image compression is another approach offering extremely high compression ratios and high-quality image reconstruction. Additionally, because fractals are infinitely magnifiable, fractal compression is resolution independent and so a single compressed image can be used efficiently for display in any image resolution including resolution higher than the original [Furht et al. 95]. Breaking an image into pieces (fractals) and identifying self-similar ones is the main principle of the approach [Barnsley and Hurd 93, Fisher 94]. First, the image is partitioned into non-overlapping *domain* regions of any size and shape that completely cover it. Then, larger *range* regions are defined that can overlap and need not cover the entire image. These range regions are geometrically transformed using affine transforms (Section 4.2.1) to match the domain regions. Then the set of affine coefficients together with information about the selection of domain regions represents the fractal image encoding. The fractally compressed images are stored and transmitted as recursive algorithms—sets of equations with instructions on how to reproduce the image. Clearly, fractal compression is compute demanding. However, decompression is simple and fast; domain regions are iteratively replaced with appropriately geometrically transformed range regions using the affine coefficients. Thus, fractal compression represents another example of an extremely promising asymmetric compression-decompression scheme. Among others, fractal image compression has been employed in the popular multimedia encyclopedia, Microsoft's Encarta.

13.8 Coding

In addition to techniques designed explicitly to cope with 2D (or higher-dimensional) data, there is a wide range of well-known algorithms designed with serial data (e.g., simple text files) in mind. These algorithms see wide use in the compression of ordinary computer files to reduce disk consumption. Very well known is **Huffman encoding**, which can provide optimal compression and error-free decompression [Rosenfeld and Kak 82, Benelli 86]. The main idea of Huffman coding is to represent data by codes of variable length, with more frequent data being represented by shorter codes. Many modifications of the original algorithm [Huffman 52] exist, with recent adaptive Huffman coding algorithms requiring only one pass over the data [Knuth 85, Vitter 87]. More recently, the **Lempel-Ziv** (or Lempel-Ziv-Welch, LZW) algorithm of **dictionary-based** coding [Ziv and Lempel 78, Welch 84, Nelson 89] has found wide favor as a standard compression algorithm. In this approach, data are represented by pointers referring to a dictionary of symbols.

These, and a number of similar techniques, are in widespread use for de-facto standard image representations which are popular for Internet and World Wide Web image exchange. Of these, the **GIF** format (Graphics Interchange Format) is probably the most popular currently in use. GIF is a creation of Compuserve, Inc., and is designed for the encoding of RGB images (and the appropriate palette) with pixel depths between 1 and 8 bits. Blocks of data are encoded using the LZW algorithm. GIF has two versions—87a [Compuserve 87] and 89a [Compuserve 89], the latter supporting the storing of text and graphics in the same file. Additionally, **TIFF** (Tagged Image File Format) is widely encountered (and is the cause of much popular confusion). TIFF was first defined by the Aldus Corporation in 1986, and has gone through a number of versions to incorporate RGB color, compressed color (LZW), other color formats, and ultimately (in Version 6 [Aldus 92]), JPEG compression (see Section 13.9)—these versions all have backward compatibility. There are some recorded problems with the JPEG implementation, and TIFF has a reputation for being complex, although this is undeserved and it is a powerful programmer's tool. It is a particularly popular format among desktop publishers, and for scanners.

13.9 JPEG and MPEG image compression

There is an increasing effort to achieve standardization in image compression. The Joint Photographic Experts Group (JPEG) has developed an international standard for general purpose, color, still image compression. As a logical extension of JPEG still image compression, the Motion Picture Experts Group (MPEG) standard was developed for full-motion video image sequences with applications to digital video distribution and high-definition television (HDTV) in mind.

13.9.1 JPEG—still image compression

The JPEG compression system is widely used in many application areas. Four compression modes are furnished [Aravind et al. 89, Wallace 90, Pennebaker and Mitchell 93, Furht et al. 95]:

- Sequential DCT-based compression
- Progressive DCT-based compression
- Sequential lossless predictive compression
- Hierarchical lossy or lossless compression

While the lossy compression modes were designed to achieve compression ratios around 15 with very good or excellent image quality, the quality deteriorates for higher compression ratios. A compression ratio between 2 and 3 is typically achieved in the lossless mode.

Sequential JPEG compression

Following Figure 13.1, sequential JPEG compression consists of a forward DCT transform, a quantizer, and and an entropy encoder, while decompression starts with entropy decoding followed by dequantizing and inverse DCT.

In the compression stage, the unsigned image values from the interval $[0, 2^b - 1]$ are first shifted to cover the interval $[-2^{b-1}, 2^{b-1} - 1]$. The image is then divided into 8×8 blocks and each block is independently transformed into the frequency domain using the DCT-II transform [Section 12.4, equation (12.28)]. Many of the 64 DCT coefficients have zero or near-zero values in typical 8×8 blocks, which forms the basis for compression. The 64 coefficients are quantized using a quantization table $Q(u, v)$ of integers from 1 to 255 that is specified by the application to reduce the storage/transmission requirements of coefficients that contribute little or nothing to the image content. The following formula is used for quantization:

$$F_Q(u, v) = \text{round} \left[\frac{F(u, v)}{Q(u, v)} \right] \tag{13.10}$$

After quantization, the dc coefficient $F(0, 0)$ is followed by the 63 ac coefficients that are ordered in a 2D matrix in a zigzag fashion according to their increasing frequency. The dc coefficients are then encoded using predictive coding (Section 13.3), the rationale being that average gray-levels of adjacent 8×8 blocks (dc coefficients) tend to be similar.

The last step of the sequential JPEG compression algorithm is entropy encoding. Two approaches are specified by the JPEG standard. The baseline system uses simple Huffman coding, while the extended system uses arithmetic coding and is suitable for a wider range of applications.

Sequential JPEG decompression uses all the steps described above in the reverse order. After entropy decoding (Huffman or arithmetic), the symbols are converted into DCT coefficients and dequantized:

$$F'_Q(u,v) = F_Q(u,v)Q(u,v) \tag{13.11}$$

where again, the $Q(u,v)$ are quantization coefficients from the quantization table that is transmitted together with the image data. Finally, the inverse DCT transform is performed according to equation (12.29) and the image gray values are shifted back to the interval $[0, 2^b - 1]$.

The JPEG compression algorithm can be extended to color or multi-spectral images with up to 256 spectral bands [Wallace 91, Steinmetz 94].

Progressive JPEG compression

The JPEG standard also facilitates progressive image transmission (Section 13.5). In the progressive compression mode, a sequence of scans is produced, each scan containing a coded subset of DCT coefficients. Thus, a buffer is needed at the output of the quantizer to store all DCT coefficients of the entire image. These coefficients are selectively encoded.

Three algorithms are defined as part of the JPEG progressive compression standard: **progressive spectral selection**, **progressive successive approximation**, and the **combined progressive algorithm**. In the progressive spectral selection approach, the dc coefficients are transmitted first, followed by groups of low-frequency and higher-frequency coefficients. In the progressive successive approximation, all DCT coefficients are sent first with lower precision, and their precision is increased as additional scans are transmitted. The combined progressive algorithm uses both of the above principles together.

Sequential lossless JPEG compression

The lossless mode of the JPEG compression uses a simple predictive compression algorithm and Huffman coding to encode the prediction differences (Section 13.3).

Hierarchical JPEG compression

Using the hierarchical JPEG mode, decoded images can be displayed either progressively or at different resolutions. A pyramid of images is created and each lower-resolution image is used as a prediction for the next-higher-resolution pyramid level (Section 13.5). The three main JPEG modes can be used to encode the lower-resolution images—sequential DCT, progressive DCT, or lossless.

In addition to still image JPEG compression, motion JPEG (MJPEG) compression exists that can be applied to real-time full motion applications. However, MPEG compression represents a more common standard and is described below.

13.9.2 MPEG—full-motion video compression

Video and associated audio data can be compressed using MPEG compression algorithms. Using inter-frame compression, compression ratios of 200 can be achieved in full-motion, motion-intensive video applications maintaining reasonable quality. MPEG compression facilitates the following features of the compressed video; random access, fast forward/reverse searches, reverse playback, audio-visual synchronization, robustness to error, editability, format flexibility, and cost trade-off [LeGall 91, Steinmetz 94]. Currently, three standards are frequently cited:

- MPEG-1 for compression of low-resolution (320×240) full-motion video at rates of 1–1.5 Mb/s
- MPEG-2 for higher-resolution standards such as TV and HDTV at rates of 2–80 Mb/s
- MPEG-4 for small-frame full motion compression with slow refresh needs, rates of 9–40 kb/s for video telephony and interactive multimedia such as video conferencing

MPEG can be equally well used for both symmetric and asymmetric applications. Here, MPEG video compression will be described; a description of the audio compression that is also part of the MPEG standard can be found elsewhere [Pennebaker and Mitchell 93, Steinmetz 94].

The video data consist of a sequence of image frames. In the MPEG compression scheme, three frame types are defined: **intraframes I**; **predicted frames P**; and **forward, backward, or bi-directionally predicted or interpolated frames B**. Each frame type is coded using a different algorithm; Figure 13.7 shows how the frame types may be positioned in the sequence.

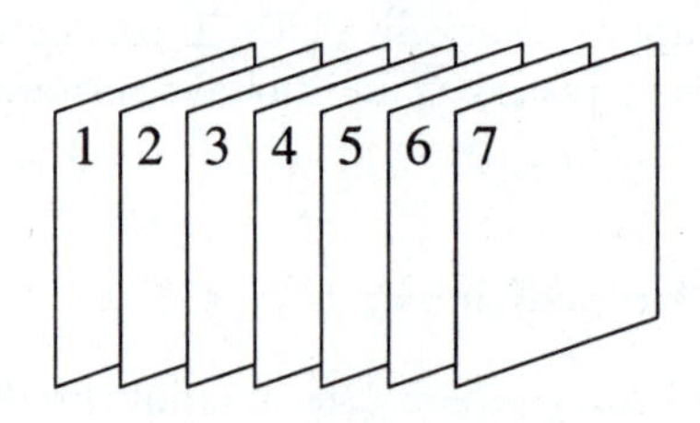

Figure 13.7: *MPEG image frames.*

I-frames are self-contained and coded using a DCT-based compression method similar to JPEG. Thus, I-frames serve as random access frames in MPEG frame streams. Consequently, I-frames are compressed with the lowest compression ratios. P-frames are coded using forward predictive coding with reference to the previous I- or P-frame, and the compression ratio for P-frames is substantially higher than that for I-frames. B-frames are coded using forward, backward, or bi-directional motion-compensated prediction or interpolation using two reference frames, closest past and future I- or P-frames, and offer the highest compression ratios.

Note that in the hypothetical MPEG stream shown in Figure 13.7, the frames must be transmitted in the following sequence (subscripts denote frame numbers): I_1–P_4–B_2–B_3–I_7–B_5–B_6– etc.; the frames B_2 and B_3 must be transmitted after frame P_4 to enable frame interpolation used for B-frame decompression. Clearly, the highest compression ratios can be achieved by incorporation of a large number of B-frames; if only I-frames are used, MJPEG compression results. The following sequence seems to be effective for a large number of applications [Steinmetz 94].

$$(\mathrm{I\ B\ B\ P\ B\ B\ P\ B\ B\ })(\mathrm{I\ B\ B\ P\ B\ B\ P\ B\ B\ })\ldots \qquad (13.12)$$

While coding the I-frames is straightforward, coding of P- and B-frames incorporates motion estimation (see also Chapter 15). For every 16×16 block of P- or B-frames, one motion vector is determined for P- and forward or backward predicted B-frames, two motion vectors are calculated for interpolated B-frames. The motion estimation technique is not specified in the MPEG standard, but block matching techniques are widely used, generally following the matching approaches presented in Section 5.4, equations (5.38)–(5.40) [Furht et al. 95]. After the motion vectors are estimated, differences between the predicted and actual blocks are determined and represent the error terms which are encoded using DCT. As usually, entropy encoding is employed as the final step.

MPEG-1 decoders are widely used in video players for multimedia applications and on the World Wide Web.

13.10 Summary

- **Image data compression**
  - The main goal of image compression is to minimize image data volume with no significant loss of information.
  - Image compression algorithms aim to remove redundancy present in data (correlation of data) in a way which makes image reconstruction possible; this is called **information preserving compression**.
  - A typical image **compression/decompression** sequence consists of data redundancy reduction, coding, transmission, decoding, and reconstruction.
  - Data compression methods can be divided into two principal groups:
    * **Information preserving** compressions permit error-free data reconstruction (**lossless compression**).
    * Compression methods **with loss of information** do not preserve the information completely (**lossy compression**).
- **Image data properties**
  - Information content of an image is an important property of which **entropy** is a measure.
  - Knowing image entropy, information **redundancy** can be determined.

- **Discrete image transforms in image data compression**
  - Image data are represented by **coefficients** of discrete image transforms. The transform coefficients are **ordered** according to their importance, i.e., according to their contribution to the image information contents, and the least important (low-contribution) coefficients are omitted.
  - To remove correlated (redundant) image data, the **Karhunen-Loeve** transform is the most effective.
  - **Cosine, Fourier, Hadamard, Walsh**, or **binary** transforms are all suitable for image data compression.
  - Performance of **discrete cosine transform DCT-II** approaches that of the Karhunen-Loeve transform better than others. The DCT is usually applied to small image blocks (typically 8×8 pixels), yielding quality-decreasing blocking artifacts for larger compression ratios.
  - Consequently, **wavelet image compression** is gaining acceptance because it does not generate square image compression artifacts.
- **Predictive compression methods**
  - Predictive compressions use image information redundancy to construct an **estimate** of the gray-level value of an image element from values of gray-levels in its neighborhood.
  - The **differences** between estimates and reality, which are expected to be relatively small in absolute terms, are coded and transmitted together with prediction model parameters.
- **Vector quantization**
  - Vector quantization compression is based on dividing an image into small blocks and representing these blocks as **vectors**.
  - Input data vectors are coded using unique **codewords** from a **codeword dictionary**; instead of vectors, the vector codes are stored or transmitted.
  - The code dictionary (code book) is transmitted together with the coded data.
- **Hierarchical and progressive compression methods**
  - Substantial reduction in bit volume can be obtained by merely representing a source as a pyramid. Even more significant reduction can be achieved for images with large areas of the same gray-level in a quadtree coding scheme.
  - Hierarchical compression facilitates **progressive** and **smart** image transmission.
  - Progressive image transmission is based on the fact that transmitting all image data may not be necessary under some circumstances.
  - Smart compression is based on the sensing properties of human visual sensors—it is not necessary to display an image in full resolution in image areas where the user's eyes are not focused.

- **Comparison of compression methods**
  - Transform-based methods better preserve subjective image quality, and are less sensitive to statistical image property changes both inside a single image and between images.
  - Prediction methods can achieve larger compression ratios in a much less expensive way, and tend to be much faster than transform-based or vector quantization compression schemes.
  - Vector quantization methods require a complex coder, their parameters are very sensitive to image data, and they blur image edges. The advantage is in a simple decoding scheme consisting of a look-up table only.

- **Other techniques**
  - Various other image data compression methods exist.
  - **Fractal image compression** offers extremely high compression ratios and high-quality image reconstruction. Breaking an image into pieces (fractals) and identifying self-similar ones is the main principle of the approach. Fractals are infinitely magnifiable, thus fractal compression is **resolution independent** and a single compressed image can be efficiently used for display in any image resolution.

- **Coding**
  - **Huffman encoding** can provide optimal compression and error-free decompression. The main idea of Huffman coding is to represent data by codes of variable length, with more frequent data being represented by shorter codes.

- **JPEG and MPEG image compression**
  - There is an increasing effort to achieve **standardization** in image compression.
  - The **Joint Photographic Experts Group (JPEG)** has developed an international standard for general-purpose, color, still image compression. This standard is widely used in many application areas. Four JPEG compression modes exist:
    * Sequential DCT-based compression
    * Progressive DCT-based compression
    * Sequential lossless predictive compression
    * Hierarchical lossy or lossless compression
  - The **MPEG standard** was developed for full-motion video image sequences.
  - Currently, three standards are frequently cited:
    * MPEG-1 for compression of low-resolution full-motion video
    * MPEG-2 for higher-resolution standards
    * MPEG-4 for small-frame full-motion compression with slow refresh needs

13.11 Exercises

Short-answer questions

1. Explain the difference between lossy and lossless image compression,
2. Give definitions of the following terms:
   (a) Image entropy
   (b) Image redundancy
   (c) Compression ratio
3. Explain the basic idea of image compression using discrete image transforms.
4. Explain the basic idea of predictive image compression.
5. Explain the basic idea of image compression based on vector quantization.
6. What are symmetric and asymmetric image compression applications?
7. Explain the concept of progressive image transmission and smart compression.
8. Explain the basic idea of fractal image compression.
9. What is the basic idea of Huffman coding?
10. What kind of image data can be compressed using the JPEG and MPEG compression algorithms?
11. What are the four compression modes available in the JPEG compression standard? Describe the algorithm employed in each of the four modes.
12. What image data are the three most frequently used MPEG standards applied to?
13. Explain the roles of the **I-**, **P-**, **B-** frames in MPEG compression.

Problems

1. Assume a gray-level image represented using 2 bits per pixel. The probabilities of the four gray-levels $\{0, 1, 2, 3\}$ are as follows: $P(0) = 0.1$, $P(1) = 0.3$, $P(2) = 0.5$, $P(3) = 0.1$.
   (a) Determine the image entropy.
   (b) Determine the information redundancy.
   (c) Determine the theoretically achievable lossless image compression ratio.
   (d) What is the minimum size of a 4-level 512×512 image after such lossless compression?
2. Repeat Problem 13.1 considering a 4-level image in which all gray-levels are equally probable.
3. Compare entropies of the following two 4-level images A and B; image A has the gray-level probabilities $P(0) = 0.2$, $P(1) = 0.0$, $P(2) = 0.0$, $P(3) = 0.8$ and image B has the probabilities $P(0) = 0.0$, $P(1) = 0.8$, $P(2) = 0.0$, $P(3) = 0.2$.
4. Develop a program to compute an estimate of the achievable compression ratio of an image from its gray-level histogram.
5. Compare image compression performance for several widely available algorithms. Using an image viewer capable of converting between different formats (for example, under Unix, XView is widely available), compare the storage requirements of a variety of image formats. Choose the image formats from the following:

(a) PBM

(b) GIF

(c) TIFF—no compression

(d) TIFF—LZW compression

(e) JPEG 50%

(f) JPEG 25%

(g) JPEG 15%

(h) JPEG 5%

By re-opening the stored images, compare the image quality visually; do you see any blocking artifacts if JPEG compression was employed? Order the image formats/compression parameters according to their storage requirements. What can be concluded if you assess the image quality considering the storage requirements?

6. Use your computer screen-grabbing tools to grab a portion of a text from a window. Store it in a gray-scale mode using GIF and JPEG (75%, 25%, and 5%). Compare the storage requirements and the resulting image quality.

7. Develop a program for transform-based image compression by applying the Fourier transform to 8×8 blocks (use the 2D discrete Fourier transform function developed in Problem 12.7 or one of the freely available Fourier transform functions). Compute a difference image between the reconstructed image and the original image. Determine the compression ratio and the maximum gray-level difference. Perform the compression in the following ways:

   (a) Retain 15% of the transform coefficients.

   (b) Retain 25% of the transform coefficients.

   (c) Retain 50% of the transform coefficients.

8. Develop a program for DPCM image compression using a third-order predictor.

9. GIF (lossless) and JPEG (lossy) compressed images are commonly used on the World Wide Web. Determine compression ratios of several GIF and JPEG compressed images; assume that the original color images were represented using 24 bits per pixel, assume 8 bits per pixel for original gray-scale images. Which compression scheme seems to provide consistently higher compression ratio? Can you determine by inspection which of the two compression approaches was used?

10. MPEG compressed image sequences are commonly used on the World Wide Web. Determine compression ratios of several MPEG compressed sequences; assume that the original color images were represented using 24 bits per pixel, assume 8 bits per pixel for original gray-scale images. (An MPEG viewer that provides information about the number of frames in the sequence will be needed.)

13.12 References

[Abdel-Malek and Bloomer 90] A Abdel-Malek and J Bloomer. Visually optimized image reconstruction. In *Proceedings of the SPIE Conference on Human Vision and Electronic Imaging,* Santa Clara, CA, pages 330–335, SPIE, Bellingham, WA, 1990.

[Aldus 92] Aldus. *TIFF Developer's Toolkit, Revision 6.0.* Aldus Corporation, Seattle, WA, 1992.

[Anderson and Huang 71] G B Anderson and T S Huang. Piecewise Fourier transformation for picture bandwidth compression. *IEEE Transactions on Communications Technology*, 19:133–140, 1971.

[Arai 90] K Arai. Preliminary study on information lossy and loss-less coding data compression for the archiving of ADEOS data. *IEEE Transactions on Geoscience and Remote Sensing*, 28(4):732–734, 1990.

[Aravind et al. 89] R Aravind, G L Cash, and J P Worth. On implementing the JPEG still-picture compression algorithm. In *Proceedings of SPIE Conference Visual Communications and Image Processing IV,* Philadelphia, PA, pages 799–808, SPIE, Bellingham, WA, 1989.

[Azadegan 90] F Azadegan. Discrete cosine transform encoding of two-dimensional processes. In *Proceedings of the 1990 International Conference on Acoustics, Speech, and Signal Processing—ICASSP 90,* Abuquerque, NM, pages 2237–2240, IEEE, Piscataway, NJ, 1990.

[Barnsley and Hurd 93] M Barnsley and L Hurd. *Fractal Image Compression.* A K Peters Ltd., Wellesley, MA, 1993.

[Benelli 86] G Benelli. Image data compression by using the Laplacian pyramid technique with adaptive Huffman codes. In V Cappellini and R Marconi, editors, *Advances in Image Processing and Pattern Recognition*, pages 229–233. North Holland, Amsterdam, 1986.

[Boxerman and Lee 90] J L Boxerman and H J Lee. Variable block-sized vector quantization of grayscale images with unconstrained tiling. In *Visual Communications and Image Processing '90,* Lausanne, Switzerland, pages 847–858, SPIE, Bellingham, WA, 1990.

[Burt 89] P J Burt. Multiresolution techniques for image representation, analysis, and 'smart' transmission. In *Visual Communications and Image Processing IV,* Philadelphia, PA, pages 2–15, SPIE, Bellingham, WA, 1989.

[Chang et al. 88] C Y Chang, R Kwok, and J C Curlander. Spatial compression of Seasat SAR images. *IEEE Transactions on Geoscience and Remote Sensing*, 26(5):673–685, 1988.

[Chen and Bovik 89] D Chen and A C Bovik. Fast image coding using simple image patterns. In *Visual Communications and Image Processing IV,* Philadelphia, PA, pages 1461–1471, SPIE, Bellingham, WA, 1989.

[Chitprasert and Rao 90] B Chitprasert and K R Rao. Discrete cosine transform filtering. *Signal Processing*, 19(3):233–245, 1990.

[Chui et al. 94] C K Chui, Laura Montefusco, and L Puccio. *Wavelets: Theory, Algorithms, and Applications.* Academic Press, San Diego, CA, 1994.

[Clarke 85] R J Clarke. *Transform Coding of Images.* Academic Press, London, 1985.

[Clarke 95] R J Clarke. *Digital Compression of Still Images and Video.* Academic Press, London, New York, 1995.

[Compuserve 87] Compuserve. *GIF Graphics Interchange Format: A Standard Defining a Mechanism for the Storage and Trasmission of Bitmap-Based Graphics Information.* CompuServe Incorporated, Columbus, OH, 1987.

[Compuserve 89] Compuserve. *Graphics Interchange Format: Version 89a.* CompuServe Incorporated, Columbus, OH, 1989.

[Daut and Zhao 90] D G Daut and D Zhao. Improved DPCM algorithm for image data compression. In *Image Processing Algorithms and Techniques,* Santa Clara, CA, pages 199–210, SPIE, Bellingham, WA, 1990.

[De Lameillieure and Bruyland 90] J De Lameillieure and I Bruyland. Single stage 280 Mbps coding of HDTV using HDPCM with a vector quantizer based on masking functions. *Signal Processing: Image Communication*, 2(3):279–289, 1990.

[Delp and Mitchell 79] E J Delp and O R Mitchell. Image truncation using block truncation coding. *IEEE Transactions on Communications*, 27:1335–1342, 1979.

[DiMento and Berkovich 90] L J DiMento and S Y Berkovich. The compression effects of the binary tree overlapping method on digital imagery. *IEEE Transactions on Communications*, 38(8):1260–1265, 1990.

[Ebrahimi et al. 90] T Ebrahimi, T R Reed, and M Kunt. Video coding using a pyramidal Gabor expansion. In *Visual Communications and Image Processing '90,* Lausanne, Switzerland, pages 489–502, SPIE, Bellingham, WA, 1990.

[Farelle 90] P M Farelle. *Recursive Block Coding for Image Data Compression.* Springer Verlag, New York, 1990.

[Fisher 94] Y Fisher. *Fractal Compression: Theory and Applications to Digital Images.* Springer Verlag, Berlin, New York, 1994.

[Furht et al. 95] B Furht, S W Smoliar, and H Zhang. *Video and Image Processing in Multimedia Systems.* Kluwer, Boston–Dordrecht–London, 1995.

[Gersho and Gray 92] A Gersho and R M Gray. *Vector Quantization and Signal Compression.* Kluwer, Norwell, MA, 1992.

[Giunta et al. 90] G Giunta, T R Reed, and M Kunt. Image sequence coding using oriented edges. *Signal Processing: Image Communication*, 2(4):429–439, 1990.

[Gonzalez et al. 90] C A Gonzalez, K L Anderson, and W B Pennebaker. DCT based video compression using arithmetic coding. In *Image Processing Algorithms and Techniques,* Santa Clara, CA, pages 305–311, SPIE, Bellingham, WA, 1990.

[Graham 67] D N Graham. Image transmission by two–dimensional contour coding. *Proceedings IEEE*, 55:336–346, 1967.

[Gray 84] R M Gray. Vector quantization. *IEEE ASSP Magazine*, 1(2):4–29, 1984.

[Guha et al. 88] R K Guha, A F Dickinson, and G Ray. Non-transform methods of picture compression applied to medical images. In *Proceedings—Twelfth Annual Symposium on Computer Applications in Medical Care,* Washington, DC, pages 483–487, IEEE, Piscataway, NJ, 1988.

[Habibi 71] A Habibi. Comparison of nth order DPCM encoder with linear transformations and block quantization techniques. *IEEE Transactions on Communications Technology*, 19(6):948–956, 1971.

[Habibi 74] A Habibi. Hybrid coding of pictorial data. *IEEE Transactions on Communications Technology*, 22(4):614–623, 1974.

[Habibi and Robinson 74] A Habibi and G S Robinson. A survey of digital picture coding. *Computer*, 7(5):22–34, 1974.

[Heer and Reinfelder 90] V K Heer and H E Reinfelder. Comparison of reversible methods for data compression. In *Medical Imaging IV: Image Processing,* Newport Beach, CA, pages 354–365, SPIE, Bellingham, WA, 1990.

[Hilton et al. 94] M L Hilton, B D Jawerth, and A Sengupta. Compressing still and moving images with wavelets. *Journal of Multimedia Systems*, 2:218–227, 1994.

[Huffman 52] D A Huffman. A method for the construction of minimum-redundancy codes. *Proceedings of IRE*, 40(9):1098–1101, 1952.

[Hung and Meng 94] A C Hung and T H Y Meng. A comparison of fast inverse discrete cosine transform algorithms. *Journal of Multimedia Systems*, 2:204–217, 1994.

[Jain 89] A K Jain. *Fundamentals of Digital Image Processing.* Prentice-Hall, Englewood Cliffs, NJ, 1989.

[Jaisimha et al. 89] M Y Jaisimha, H Potlapalli, H Barad, and A B Martinez. Data compression techniques for maps. In *Energy and Information Technologies in the Southeast,* Columbia, SC, pages 878–883, IEEE, Piscataway, NJ, 1989.

[Knuth 85] D E Knuth. Dynamic Huffman coding. *Journal of Algorithms*, 6:163–180, 1985.

[Kruger 92] A Kruger. Block truncation compression. *Dr Dobb's J Software Tools*, 17(4):48–55, 1992.

[LeGall 91] D LeGall. MPEG: A video compression standard for multimedia applications. *Communications of the ACM*, 34:45–68, 1991.

[Lovewell and Basart 88] B K Lovewell and J P Basart. Survey of image compression techniques. *Review of Progress in Quantitative Nondestructive Evaluation*, 7A:731–738, 1988.

[Mallat 89] S G Mallat. A theory of multiresolution signal decomposition: The wavelet representation. *IEEE Transactions on Pattern Analysis and Machine Intelligence*, 11:674–693, 1989.

[Moik 80] J G Moik. *Digital Processing of Remotely Sensed Images.* NASA SP–431, Washington, DC, 1980.

[Nelson 89] M R Nelson. Lzw data compression. *Dr Dobb's J Software Tools*, 14, October 1989.

[Netravali 88] A N Netravali. *Digital Pictures: Representation and Compression.* Plenum Press, New York, 1988.

[Park and Lee 91] S H Park and S U Lee. Pyramid image coder using classified transform vector quantization. *Signal Processing*, 22(1):25–42, 1991.

[Pennebaker and Mitchell 93] W B Pennebaker and J L Mitchell. *JPEG Still Image Data Compression Standard.* Van Nostrand Reinhold, New York, 1993.

[Pratt 91] W K Pratt. *Digital Image Processing.* Wiley, New York, 2nd edition, 1991.

[Rabbani 91] M Rabbani. *Digital Image Compression.* SPIE Optical Engineering Press, Bellingham, WA, 1991.

[Rao and Pearlman 91] R P Rao and W A Pearlman. On entropy of pyramid structures. *IEEE Transactions on Information Theory*, 37(2):407–413, 1991.

[Rosenfeld and Kak 82] A Rosenfeld and A C Kak. *Digital Picture Processing.* Academic Press, New York, 2nd edition, 1982.

[Savoji and Burge 85] M H Savoji and R E Burge. On different methods based on the Karhunen–Loeve expansion and used in image analysis. *Computer Vision, Graphics, and Image Processing*, 29:259–269, 1985.

[Silsbee et al. 91] P Silsbee, A C Bovik, and D Chen. Visual pattern image sequencing coding. In *Visual Communications and Image Processing '90,* Lausanne, Switzerland, pages 532–543, SPIE, Bellingham, WA, 1991.

[Steinmetz 94] R Steinmetz. Data compression in multimedia computing—standards and systems, parts i and ii. *Journal of Multimedia Systems*, 1:166–172 and 187–204, 1994.

[Strobach 90] P Strobach. Tree-structured scene adaptive coder. *IEEE Transactions on Communications*, 38(4):477–486, 1990.

[Thyagarajan and Sanchez 89] K S Thyagarajan and H Sanchez. Image sequence coding using interframe VDPCM and motion compensation. In *International Conference on Acoustics, Speech, and Signal Processing,* Glasgow, Scotland, pages 1858–1861, IEEE, Piscataway, NJ, 1989.

[Vitter 87] J S Vitter. Design and analysis of dynamic Huffman codes. *Journal of the ACM*, 34(4):825–845, 1987.

[Wallace 90] G K Wallace. Overview of the JPEG (ISO/CCITT) still image compression standard. In *Proceedings of SPIE Conference Image Processing Algorithms and Techniques,* Santa Clara, CA, pages 220–233, SPIE, Bellingham, WA, 1990.

[Wallace 91] G Wallace. The JPEG still image compression standard. *Communications of the ACM*, 34:30–44, 1991.

[Wang and Goldberg 89] L Wang and M Goldberg. Reduced-difference pyramid: A data structure for progressive image transmission. *Optical Engineering*, 28(7):708–716, 1989.

[Wang and Goldberg 90] L Wang and M Goldberg. Reduced-difference pyramid. A data structure for progressive image transmission. In *Image Processing Algorithms and Techniques,* Santa Clara, CA, pages 171–181, SPIE, Bellingham, WA, 1990.

[Welch 84] T A Welch. A technique for high performance data compression. *Computer*, 17(6):8–19, 1984.

[White 87] R G White. Compressing image data with quadtrees. *Dr Dobb's J Software Tools*, 12(3):16–45, 1987.

[Wilkins and Wintz 71] L C Wilkins and P A Wintz. Bibliography on data compression, picture properties and picture coding. *IEEE Transactions on Information Theory*, 17:180–199, 1971.

[Wintz 72] P A Wintz. Transform picture coding. *Proceedings IEEE*, 60:809–820, 1972.

[Witten et al. 94] I H Witten, A Moffat, and T C Bell. *Managing Gigabytes: Compressing and Indexing Documents and Images.* Van Nostrand Reinhold, New York, 1994.

[Zailu and Taxiao 90] H Zailu and W Taxiao. MDPCM picture coding. In *1990 IEEE International Symposium on Circuits and Systems,* New Orleans, LA, pages 3253–3255, IEEE, Piscataway, NJ, 1990.

[Zettler et al. 90] W R Zettler, J Huffman, and D C P Linden. Application of compactly supported wavelets to image compression. In *Image Processing Algorithms and Techniques,* Santa Clara, CA, pages 150–160, SPIE, Bellingham, WA, 1990.

[Ziv and Lempel 78] J Ziv and A Lempel. Compression of individual sequences via variable-rate coding. *IEEE Transactions on Information Theory*, 24(5):530–536, 1978.

Chapter 14

Texture

Texture refers to properties that represent the surface or structure of an object (in reflective or transmissive images, respectively); it is widely used, and perhaps intuitively obvious, but has no precise definition due to its wide variability. We might define texture as *something consisting of mutually related elements*; therefore we are considering a group of pixels (a **texture primitive** or **texture element**) and the texture described is highly dependent on the number considered (the texture scale) [Haralick 79]. Examples are shown in Figure 14.1; dog fur, grass, river pebbles, cork, checkered textile, and knitted fabric. Many other examples can be found in [Brodatz 66].

Texture consists of texture **primitives** or texture **elements**, sometimes called **texels**. Primitives in grass and dog fur are represented by several pixels and correspond to a stalk or a pile; cork is built from primitives that are comparable in size with pixels. It is difficult, however, to define primitives for the checkered textile or fabric, which can be defined by at least two hierarchical levels. The first level of primitives corresponds to textile checks or knitted stripes, and the second to the finer texture of the fabric or individual stitches. As we have seen in many other areas, this is a problem of **scale**; texture description is **scale dependent**.

The main aim of texture analysis is texture recognition and texture-based shape analysis. Textured properties of regions were referred to many times while considering image segmentation (Chapter 5), and derivation of shape from texture was discussed in Chapter 9. People usually describe texture as **fine**, **coarse**, **grained**, **smooth**, etc., implying that some more precise features must be defined to make machine recognition possible. Such features can be found in the **tone** and **structure** of a texture [Haralick 79]. Tone is based mostly on pixel intensity properties in the primitive, while structure is the spatial relationship between primitives.

Each pixel can be characterized by its location and tonal properties. A texture primitive is a contiguous set of pixels with some tonal and/or regional property, and can be described by its average intensity, maximum or minimum intensity, size, shape, etc. The spatial relationship of primitives can be random, or they may be pairwise dependent, or some number of primitives can be mutually dependent. Image texture is then described by the number and types of primitives and by their spatial relationship.

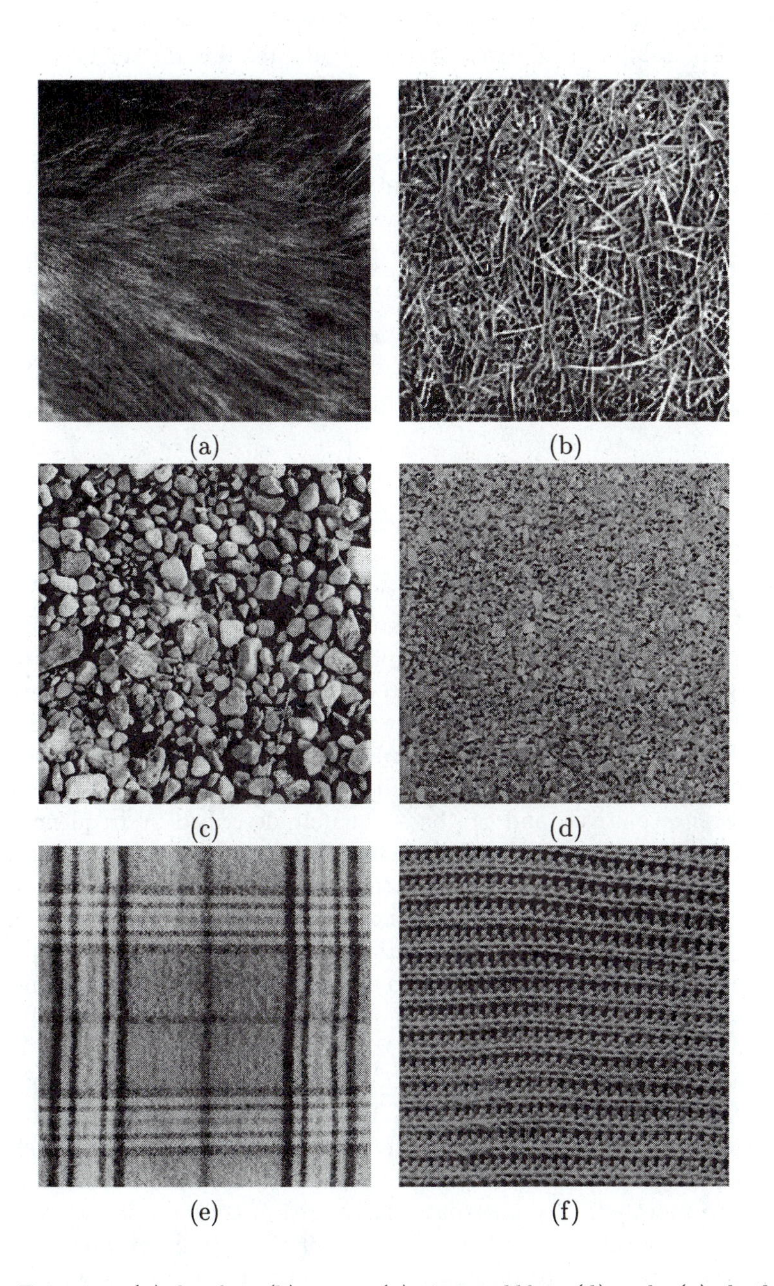

Figure 14.1: *Textures: (a) dog fur; (b) grass; (c) river pebbles; (d) cork; (e) checkered textile; (f) knitted fabric.*

Figures 14.1a,b and 14.2a,b show that the same number and the same type of primitives does not necessarily give the same texture. Similarly, Figures 14.2a and 14.2c show that the same spatial relationship of primitives does not guarantee texture uniqueness, and therefore is not sufficient for texture description. Texture tone and structure are not independent; textures always display both tone and structure even though one or the other usually dominates, and we usually speak about one or the other only. Tone can be understood as tonal properties of primitives, taking primitive spatial relationships into consideration. Structure refers to spatial relationships of primitives considering their tonal properties as well.

```
* * * * * * * * * *       ** ** ** ** **       # # # # # # # # # #
 * * * * * * * * * *      ** ** ** ** **        # # # # # # # # # #
* * * * * * * * * *       ** ** ** ** **       # # # # # # # # # #
 * * * * * * * * * *      ** ** ** ** **        # # # # # # # # # #
* * * * * * * * * *       ** ** ** ** **       # # # # # # # # # #
 * * * * * * * * * *      ** ** ** ** **        # # # # # # # # # #
* * * * * * * * * *       ** ** ** ** **       # # # # # # # # # #

        (a)                    (b)                     (c)
```

Figure 14.2: *Artificial textures.*

If the texture primitives in the image are small and if the tonal differences between neighboring primitives are large, a **fine** texture results (Figures 14.1a,b and 14.1d). If the texture primitives are larger and consist of several pixels, a **coarse** texture results (Figures 14.1c and 14.1e). Again, this is a reason for using both tonal and structural properties in texture description. Note that the fine/coarse texture characteristic depends on scale.

Further, textures can be classified according to their strength—texture strength then influences the choice of texture description method. **Weak** textures have small spatial interactions between primitives, and can be adequately described by frequencies of primitive types appearing in some neighborhood. Because of this, many statistical texture properties are evaluated in the description of weak textures. In **strong** textures, the spatial interactions between primitives are somewhat regular. To describe strong textures, the frequency of occurrence of primitive pairs in some spatial relationship may be sufficient. Strong texture recognition is usually accompanied by an exact definition of texture primitives and their spatial relationships [Chetverikov 82, Rao 90].

It remains to define a constant texture. One existing definition [Sklansky 78] claims that '*an image region has a constant texture if a set of its local properties in that region is constant, slowly changing, or approximately periodic*'. The set of local properties can be understood as primitive types and their spatial relationships. An important part of the definition is that the properties must be repeated inside the constant texture area. How many times must the properties be repeated? Assume that a large area of constant texture is available, and consider smaller and smaller parts of that texture, digitizing it at constant resolution as long as the texture character remains unchanged. Alternatively, consider larger and larger parts of the texture, digitizing it at constant raster, until details become blurred and the primitives finally disappear. We see that image resolution (scale) must be a consistent part of the texture description; if the image resolution is appropriate, the texture character does not change for any position in our window.

Two main texture description approaches exist—**statistical** and **syntactic** [Haralick 79].

Statistical methods compute different properties and are suitable if texture primitive sizes are comparable with the pixel sizes. Syntactic and **hybrid** methods (combination of statistical and syntactic) are more suitable for textures where primitives can be assigned a label—the primitive type—meaning that primitives can be described using a larger variety of properties than just tonal properties; for example, shape description. Instead of tone, brightness will be used more often in the following sections because it corresponds better to gray-level images.

Research on pre-attentive (early) vision [Julesz 81, Julesz and Bergen 87] shows that human ability to recognize texture quickly is based mostly on **textons**, which are elongated blobs (rectangles, ellipses, line segments, line ends, crossings, corners) that can be detected by pre-attentive vision, while the positional relationship between neighboring textons must be done slowly by an attentive vision sub-system. As a result of these investigations, another group of methods has begun to appear, based on texton detection and texton density computation [Voorhees and Poggio 87, Ando 88].

14.1 Statistical texture description

Statistical texture description methods describe textures in a form suitable for statistical pattern recognition. As a result of the description, each texture is described by a feature vector of properties which represents a point in a multi-dimensional feature space. The aim is to find a deterministic or probabilistic decision rule assigning a texture to some specific class (see Chapter 7).

14.1.1 Methods based on spatial frequencies

Measuring spatial frequencies is the basis of a large group of texture recognition methods. Textural character is in direct relation to the spatial size of the texture primitives; coarse textures are built from larger primitives, fine textures from smaller primitives. Fine textures are characterized by higher spatial frequencies, coarse textures by lower spatial frequencies.

One of many related spatial frequency methods evaluates the **autocorrelation function of a texture**. In an autocorrelation model, a single pixel is considered a texture primitive, and primitive tone property is the gray-level. Texture spatial organization is described by the correlation coefficient that evaluates linear spatial relationships between primitives. If the texture primitives are relatively large, the autocorrelation function value decreases slowly with increasing distance, while it decreases rapidly if texture consists of small primitives. If primitives are placed periodically in a texture, the autocorrelation increases and decreases periodically with distance.

Texture can be described using the following algorithm.

Algorithm 14.1: Autocorrelation texture description

1. Evaluate autocorrelation coefficients for several different values of parameters p, q:

$$C_{ff}(p,q) = \frac{MN}{(M-p)(N-q)} \frac{\sum_{i=1}^{M-p} \sum_{j=1}^{N-q} f(i,j) f(i+p, j+q)}{\sum_{i=1}^{M} \sum_{j=1}^{N} f^2(i,j)} \tag{14.1}$$

   where p, q is the position difference in the i, j direction, and M, N are the image dimensions.

2. Alternatively, the autocorrelation function can be determined in the frequency domain from the image power spectrum [Castleman 96]:

$$C_{ff} = \mathcal{F}^{-\infty}\{|\mathcal{F}|^{\in}\} \tag{14.2}$$

If the textures described are circularly symmetric, the autocorrelation texture description can be computed as a function of the absolute position difference not considering direction—that is, a function of one variable.

Spatial frequencies can also be determined from an **optical image transform** (recall that the Fourier transform can be realized by a convex lens—see Chapter 12) [Shulman 70], a big advantage of which is that it may be computed in real time. The Fourier transform describes an image by its spatial frequencies; average values of energy in specific wedges and rings of the Fourier spectrum can be used as textural description features (see Figure 14.3). Features evaluated from rings reflect coarseness of the texture—high energy in large-radius rings is characteristic of fine textures (high frequencies), while high energy in small radii is characteristic of coarse textures (with lower spatial frequencies). Features evaluated from wedge slices of the Fourier transform image depend on directional properties of textures—if a texture has many edges or lines in a direction ϕ, high energy will be present in a wedge in direction $\phi + \pi/2$.

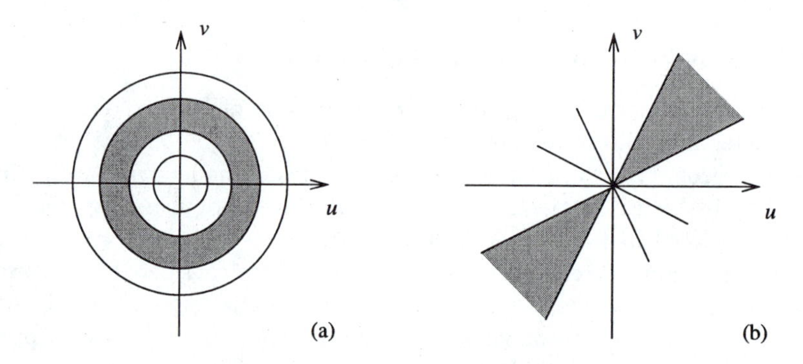

Figure 14.3: *Partitioning of Fourier spectrum: (a) ring filter; (b) wedge filter reflecting the Fourier spectrum symmetry.*

Similarly, a **discrete image transform** may be used for texture description. A textured image is usually divided into small square non-overlapping subimages. If the subimage size is $n \times n$, the gray-levels of its pixels may be interpreted as an n^2-dimensional vector, and an image can be represented as a set of vectors. These vectors are transformed applying a Fourier, Hadamard, or other discrete image transform (Chapter 12). The new co-ordinate system's basis vectors are related to the spatial frequencies of the original texture image and can be used for texture description [Rosenfeld 76]. When description of noisy texture becomes necessary, the problem becomes more difficult. From a set of 28 spatial frequency-domain features, a subset of features insensitive to additive noise was extracted (dominant peak energy, power spectrum shape, entropy) in [Liu and Jernigan 90].

Spatial frequency texture description methods are based on a well-known approach. Despite that, many problems remain—the resulting description is not invariant even to monotonic image gray-level transforms; further, it can be shown [Weszka et al. 76] that the frequency-based approach is less efficient than others. A joint spatial/spatial frequency approach is recommended; the Wigner distribution was shown to be useful in a variety of synthetic and Brodatz textures [Reed et al. 90].

14.1.2 Co-occurrence matrices

The co-occurrence matrix method of texture description is based on the repeated occurrence of some gray-level configuration in the texture; this configuration varies rapidly with distance in fine textures and slowly in coarse textures [Haralick et al. 73]. Suppose the part of a textured image to be analyzed is an $M \times N$ rectangular window. An occurrence of some gray-level configuration may be described by a matrix of relative frequencies $P_{\phi,d}(a,b)$, describing how frequently two pixels with gray-levels a, b appear in the window separated by a distance d in direction ϕ. These matrices are symmetric if defined as given below. However, an asymmetric definition may be used, where matrix values are also dependent on the direction of co-occurrence. A co-occurrence matrix computation scheme was given in Algorithm 3.1.

Non-normalized frequencies of co-occurrence as functions of angle and distance can be represented formally as

$$\begin{aligned}
P_{0^\circ,d}(a,b) &= |\{[(k,l),(m,n)] \in D : \\
&\quad k-m=0, |l-n|=d, \ f(k,l)=a, f(m,n)=b\}| \\
P_{45^\circ,d}(a,b) &= |\{[(k,l),(m,n)] \in D : \\
&\quad (k-m=d, l-n=-d) \text{ OR } (k-m=-d, l-n=d), \\
&\quad f(k,l)=a, f(m,n)=b\}| \\
P_{90^\circ,d}(a,b) &= |\{[(k,l),(m,n)] \in D : \\
&\quad |k-m|=d, l-n=0, \ f(k,l)=a, f(m,n)=b\}| \\
P_{135^\circ,d}(a,b) &= |\{[(k,l),(m,n)] \in D : \\
&\quad (k-m=d, l-n=d) \text{ OR } (k-m=-d, l-n=-d), \\
&\quad f(k,l)=a, f(m,n)=b\}|
\end{aligned} \tag{14.3}$$

where $|\{\ldots\}|$ refers to set cardinality and $D = (M \times N) \times (M \times N)$.

An example illustrates co-occurrence matrix computations for the distance $d = 1$. A 4×4 image with four gray-levels is presented in Figure 14.4. The matrix $P_{0^\circ,1}$ is constructed as

0	0	1	1
0	0	1	1
0	2	2	2
2	2	3	3

Figure 14.4: *Gray-level image.*

follows: The element $P_{0^\circ,1}(0,0)$ represents the number of times the two pixels with gray-levels 0 and 0 appear separated by distance 1 in direction 0°; $P_{0^\circ,1}(0,0) = 4$ in this case. The element $P_{0^\circ,1}(3,2)$ represents the number of times two pixels with gray-levels 3 and 2 appear separated by distance 1 in direction 0°; $P_{0°,1}(3,2) = 1$. Note that $P_{0°,1}(2,3) = 1$ due to matrix symmetry:

$$P_{0°,1} = \begin{vmatrix} 4 & 2 & 1 & 0 \\ 2 & 4 & 0 & 0 \\ 1 & 0 & 6 & 1 \\ 0 & 0 & 1 & 2 \end{vmatrix} \qquad P_{135°,1} = \begin{vmatrix} 2 & 1 & 3 & 0 \\ 1 & 2 & 1 & 0 \\ 3 & 1 & 0 & 2 \\ 0 & 0 & 2 & 0 \end{vmatrix}$$

The construction of matrices $P_{\phi,d}$ for other directions ϕ and distance values d is similar.

Texture classification can be based on criteria derived from the following co-occurrence matrices.

- **Energy**, or angular second moment (an image homogeneity measure—the more homogeneous the image, the larger the value)

$$\sum_{a,b} P^2_{\phi,d}(a,b) \tag{14.4}$$

- **Entropy:**

$$\sum_{a,b} P_{\phi,d}(a,b) \log_2 P_{\phi,d}(a,b) \tag{14.5}$$

- **Maximum probability:**

$$\max_{a,b} P_{\phi,d}(a,b) \tag{14.6}$$

- **Contrast** (a measure of local image variations; typically $\kappa = 2, \lambda = 1$):

$$\sum_{a,b} |a-b|^\kappa P^\lambda_{\phi,d}(a,b) \tag{14.7}$$

- **Inverse difference moment:**

$$\sum_{a,b;a\neq b} \frac{P^\lambda_{\phi,d}(a,b)}{|a-b|^\kappa} \tag{14.8}$$

- **Correlation** (a measure of image linearity, linear directional structures in direction ϕ result in large correlation values in this direction):

$$\frac{\sum_{a,b}[(ab)P_{\phi,d}(a,b)] - \mu_x\mu_y}{\sigma_x\sigma_y} \tag{14.9}$$

where μ_x, μ_y are means and σ_x, σ_y are standard deviations,

$$\mu_x = \sum_a a \sum_b P_{\phi,d}(a,b)$$

$$\mu_y = \sum_b b \sum_a P_{\phi,d}(a,b)$$

$$\sigma_x = \sum_a (a-\mu_x)^2 \sum_b P_{\phi,d}(a,b)$$

$$\sigma_y = \sum_b (b-\mu_x)^2 \sum_a P_{\phi,d}(a,b)$$

Following is a general algorithm for texture description based on co-occurrence matrices.

Algorithm 14.2: Co-occurrence method of texture description

1. Construct co-occurrence matrices for given directions and given distances.

2. Compute texture feature vectors for four directions ϕ, different values of d, and the six characteristics. This results in many correlated features.

The co-occurrence method describes second-order image statistics and works well for a large variety of textures (see [Gotlieb and Kreyszig 90] for a survey of texture descriptors based on co-occurrence matrices). Good properties of the co-occurrence method are the description of spatial relations between tonal pixels, and invariance to monotonic gray-level transformations. On the other hand, it does not consider primitive shapes, and therefore cannot be recommended if the texture consists of large primitives. Memory requirements represent another big disadvantage, although this is definitely not as limiting as it was a few years ago. The number of gray-levels may be set to 32 or 64, which decreases the co-occurrence matrix sizes, but loss of gray-level accuracy is a resulting negative effect (although this loss is usually insignificant in practice).

Although co-occurrence matrices give very good results in discrimination between textures, the method is computationally expensive. A fast algorithm for co-occurrence matrix computation is given in [Argenti et al. 90], and a modification of the method that is efficiently applicable to texture description of detected regions is proposed in [Carlson and Ebel 88], in which a co-occurrence array size varies with the region size.

14.1.3 Edge frequency

Methods discussed so far describe texture by its spatial frequencies, but comparison of edge frequencies in texture can be used as well. Edges can be detected either as micro-edges using small edge operator masks, or as macro-edges using large masks [Davis and Mitiche 80]. The simplest operator that can serve this purpose is Robert's gradient, but virtually any other edge detector can be used (see Section 4.3.2). Using a gradient as a function of distance between pixels is another option [Sutton and Hall 72]. The distance-dependent texture description function $g(d)$ can be computed for any subimage f defined in a neighborhood N for variable distance d:

$$\begin{aligned} g(d) = \quad & |f(i,j) - f(i+d,j)| + |f(i,j) - f(i-d,j)| \\ + \quad & |f(i,j) - f(i,j+d)| + |f(i,j) - f(i,j-d)| \end{aligned} \tag{14.10}$$

The function $g(d)$ is similar to the negative autocorrelation function; its minimum corresponds to the maximum of the autocorrelation function, and its maximum corresponds to the autocorrelation minimum.

Algorithm 14.3: Edge-frequency texture description

1. Compute a gradient $g(d)$ for all pixels of the texture.

2. Evaluate texture features as average values of gradient in specified distances d.

Dimensionality of the texture description feature space is given by the number of distance values d used to compute the edge gradient.

Several other texture properties may be derived from first-order and second-order statistics of edge distributions [Tomita and Tsuji 90].

- **Coarseness:** Edge density is a measure of coarseness. The finer the texture, the higher is the number of edges present in the texture edge image.

- **Contrast:** High-contrast textures are characterized by large edge magnitudes.

- **Randomness:** Randomness may be measured as entropy of the edge magnitude histogram.

- **Directivity:** An approximate measure of directivity may be determined as entropy of the edge direction histogram. Directional textures have an even number of significant histogram peaks, directionless textures have a uniform edge direction histogram.

- **Linearity:** Texture linearity is indicated by co-occurrences of edge pairs with the same edge direction at constant distances, and edges are positioned in the edge direction (see Figure 14.5, edges a and b).

- **Periodicity:** Texture periodicity can be measured by co-occurrences of edge pairs of the same direction at constant distances in directions perpendicular to the edge direction (see Figure 14.5, edges a and c).

- **Size:** Texture size measure may be based on co-occurrences of edge pairs with opposite edge directions at constant distance in a direction perpendicular to the edge directions (see Figure 14.5, edges a and d).

Note that the first three measures are derived from first-order statistics, the last three measures from second-order statistics.

Many existing texture recognition methods are based on texture detection. The concepts of pre-attentive vision and textons have been mentioned, which are also based mostly on edge-related information. A zero-crossing operator was applied to edge-based texture description in [Perry and Lowe 89]; the method determines image regions of a constant texture, assuming no a priori knowledge about the image, texture types, or scale. Feature analysis is performed across multiple window sizes.

A slightly different approach to texture recognition may require detection of borders between homogeneous textured regions. A hierarchical algorithm for textured image segmentation is described in [Fan 89], and a two-stage contextual classification and segmentation of textures based on a coarse-to-fine principle of edge detection is given in [Fung et al. 90].

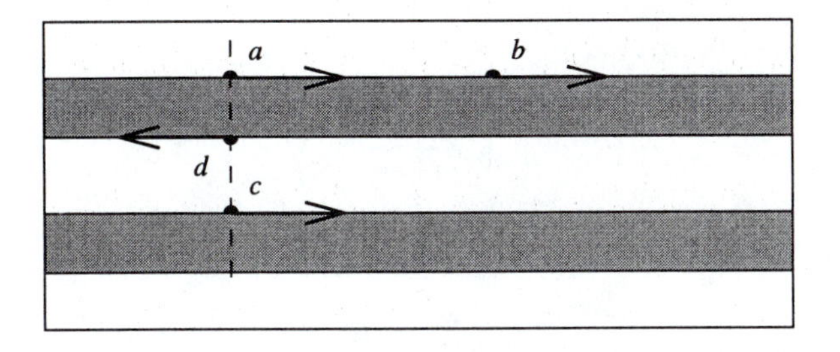

Figure 14.5: *Texture linearity, periodicity, and size measures may be based on image edges. Adapted from [Tomita and Tsuji 90].*

Texture description and recognition in the presence of noise represents a difficult problem. A noise-tolerant texture classification approach based on a Canny-type edge detector is discussed in [Kjell and Wang 91] where texture is described using periodicity measures derived from noise-insensitive edge detection.

14.1.4 Primitive length (run length)

A large number of neighboring pixels of the same gray-level represents a coarse texture, a small number of these pixels represents a fine texture, and the lengths of texture primitives in different directions can serve as a texture description [Galloway 75]. A primitive is a maximum contiguous set of constant-gray-level pixels located in a line; these can then be described by gray-level, length, and direction. The texture description features can be based on computation of continuous probabilities of the length and the gray-level of primitives in the texture [Gisolfi et al. 86].

Let $B(a,r)$ be the number of primitives of all directions having length r and gray-level a, M,N the image dimensions, and L the number of image gray-levels. Let N_r be the maximum primitive length in the image. The texture description features can be determined as follows. Let K be the total number of runs,

$$K = \sum_{a=1}^{L} \sum_{r=1}^{N_r} B(a,r) \tag{14.11}$$

Then

Short primitives emphasis:

$$\frac{1}{K} \sum_{a=1}^{L} \sum_{r=1}^{N_r} \frac{B(a,r)}{r^2} \tag{14.12}$$

Long primitives emphasis:

$$\frac{1}{K} \sum_{a=1}^{L} \sum_{r=1}^{N_r} B(a,r) r^2 \tag{14.13}$$

Gray-level uniformity:

$$\frac{1}{K} \sum_{a=1}^{L} [\sum_{r=1}^{N_r} B(a,r)]^2 \tag{14.14}$$

Primitive length uniformity:

$$\frac{1}{K}\sum_{r=1}^{N_r}[\sum_{a=1}^{L}B(a,r)]^2 \tag{14.15}$$

Primitive percentage:

$$\frac{K}{\sum_{a=1}^{L}\sum_{r=1}^{N_r} r\ B(a,r)} = \frac{K}{MN} \tag{14.16}$$

A general algorithm might then be the following.

Algorithm 14.4: Primitive-length texture description

1. Find primitives of all gray-levels, all lengths, and all directions in the texture image.
2. Compute texture description features as given in (14.12)–(14.16). These features then provide a description vector.

14.1.5 Laws' texture energy measures

Laws' texture energy measures determine texture properties by assessing average gray-level, edges, spots, ripples, and waves in texture [Laws 79, Wu et al. 92]. The measures are derived from three simple vectors: $L_3 = (1, 2, 1)$ which represents averaging; $E_3 = (-1, 0, 1)$ calculating first difference (edges); and $S_3 = (-1, 2, -1)$, corresponding to the second difference (spots). After convolution of these vectors with themselves and each other, five vectors result:

$$\begin{aligned} L_5 &= (1, 4, 6, 4, 1) \\ E_5 &= (-1, -2, 0, 2, 1) \\ S_5 &= (-1, 0, 2, 0, -1) \\ R_5 &= (1, -4, 6, -4, 1) \\ W_5 &= (-1, 2, 0, -2, -1) \end{aligned} \tag{14.17}$$

Mutual multiplying of these vectors, considering the first term as a column vector and the second term as a row vector, results in 5×5 Laws' masks. For example, the following mask can be derived

$$L_5^T \times S_5 = \begin{bmatrix} -1 & 0 & 2 & 0 & -1 \\ -4 & 0 & 8 & 0 & -4 \\ -6 & 0 & 12 & 0 & -6 \\ -4 & 0 & 8 & 0 & -4 \\ -1 & 0 & 2 & 0 & -1 \end{bmatrix} \tag{14.18}$$

By convoluting the Laws' masks with a texture image and calculating energy statistics, a feature vector is derived that can be used for texture description.

14.1.6 Fractal texture description

Fractal-based texture analysis was introduced in [Pentland 84], where a correlation between texture coarseness and fractal dimension of a texture was demonstrated. A fractal is defined [Mandelbrot 82] as a set for which the Hausdorff-Besicovich dimension [Hausdorff 19, Besicovitch and Ursell 37] is strictly greater than the topological dimension; therefore, fractional dimension is the defining property. Fractal models typically relate a metric property such as line length or surface area to the elementary length or area used as a basis for determining the metric property; measuring coast length is a frequently used example [Mandelbrot 82, Pentland 84, Lundahl et al. 86]. Suppose the coast length is determined by applying a 1-km-long ruler end to end to the coastline; the same procedure can be repeated with a 0.5-km ruler and other shorter or longer rulers. It is easy to understand that shortening of the ruler will be associated with an increase in the measured length. Importantly, the relation between the ruler length and the measured coast length can be considered a measure of the coastline's geometric properties, e.g., its roughness. The functional relationship between the ruler size r and the measured length L can be expressed as

$$L = c r^{1-D} \tag{14.19}$$

where c is a scaling constant and D is the **fractal dimension** [Mandelbrot 82]. Fractal dimension has been shown to correlate well with the function's intuitive roughness.

While equation (14.19) can be applied directly to lines and surfaces, it is often more appropriate to consider the function as a stochastic process. One of the most important stochastic fractal models is the fractional Brownian motion model described in [Mandelbrot 82], which considers naturally rough surfaces as the end results of random walks. Importantly, intensity surfaces of textures can also be considered as resulting from random walks, and the fractional Brownian motion model can be applied to texture description.

Fractal description of textures is typically based on determination of fractal dimension and **lacunarity** to measure texture roughness and granularity from the image intensity function. The topological dimension of an image is equal to three—two spatial dimensions and the third dimension representing the image intensity. Considering the topological dimension T_d, the fractal dimension D can be estimated from the Hurst coefficient H [Hurst 51, Mandelbrot 82] as

$$H = T_d - D \tag{14.20}$$

For images ($T_d = 3$), the Hurst parameter H or the fractal dimension D can be estimated from the relationship

$$E\left[(\Delta f)^2\right] = c\left[(\Delta r)^H\right]^2 = c(\Delta r)^{6-2D} \tag{14.21}$$

where $E()$ is an expectation operator, $\Delta f = f(i,j) - f(k,l)$ is the intensity variation, c is a scaling constant, and $\Delta r = ||(i,j) - (k,l)||$ is the spatial distance. A simpler way to estimate fractal dimension is to use the following equation:

$$E(|\Delta f|) = \kappa(\Delta r)^{3-D} \tag{14.22}$$

where $\kappa = E(|\Delta f|)_{\Delta r=1}$. By applying the log function and considering that $H = 3 - D$,

$$\log E(|\Delta f|) = \log \kappa + H \log(\Delta r) \tag{14.23}$$

The parameter H can be obtained by using least-squares linear regression to estimate the slope of the curve of gray-level differences $gd(k)$ versus k in log–log scales [Wu et al. 92]. Considering an $M \times M$ image f,

$$\begin{aligned} gd(k) = & \frac{\sum_{i=0}^{M-1} \sum_{j=0}^{M-k-1} |f(i,j) - f(i,j+k)|}{2M(M-k-1)} \\ + & \frac{\sum_{i=0}^{M-k-1} \sum_{j=0}^{M-1} |f(i,j) - f(i+k,j)|}{2M(M-k-1)} \end{aligned} \tag{14.24}$$

The scale k varies from 1 to the maximum selected value s. Fractal dimension D is then derived from the value of the Hurst coefficient. The approximation error of the regression line fit should be determined to prove that the analyzed texture is a fractal, and can thus be efficiently described using fractal measures. A small value of the fractal dimension D (large value of the parameter H) represents a fine texture, while large D (small H) corresponds to a coarse texture.

Single fractal dimension is not sufficient for description of natural textures. Lacunarity measures describe characteristics of textures of different visual appearance that have the same fractal dimension [Voss 86, Keller et al. 89, Wu et al. 92]. Given a fractal set A, let $P(m)$ represent the probability that there are m points within a box of size L centered about an arbitrary point of A. Let N be the number of possible points within the box, then $\sum_{m=1}^{N} P(m) = 1$ and the lacunarity λ is defined as

$$\lambda = \frac{M_2 - M^2}{M^2} \tag{14.25}$$

where

$$\begin{aligned} M &= \sum_{m=1}^{N} mP(m) \\ M_2 &= \sum_{m=1}^{N} m^2 P(m) \end{aligned} \tag{14.26}$$

Lacunarity represents a second-order statistic and is small for fine textures and large for coarse ones.

A multi-resolution approach to fractal feature extraction was introduced in [Wu et al. 92]. The multi-resolution feature vector MF that describes both texture roughness and lacunarity is defined as

$$MF = (H^{(m)}, H^{(m-1)}, \ldots, H^{(m-n+1)}) \tag{14.27}$$

where the parameters $H^{(k)}$ are estimated from pyramidal images $f^{(k)}$, where $f^{(m)}$ represents the full-resolution image of size $M = 2^m$, $f^{(m-1)}$ is the half-resolution image of size $M = 2^{m-1}$, etc., and n is the number of resolution levels considered. The multi-resolution feature vector MF can serve as a texture descriptor. Textures with identical fractal dimensions and different lacunarities can be distinguished, as was shown by classification of ultrasonic liver images in three classes—normal, hepatoma, and cirrhosis [Wu et al. 92]. Practical considerations regarding calculation of fractal-based texture description features can be found in [Sarkar and Chaudhuri 94, Huang et al. 94, Jin et al. 95].

14.1.7 Other statistical methods of texture description

A brief overview of some other texture description techniques will illustrate the variety of published methods; we present here only the basic principles of some additional approaches [Haralick 79, Ahuja and Rosenfeld 81, Davis et al. 83, Derin and Elliot 87, Tomita and Tsuji 90].

The **mathematical morphology** approach looks for spatial repetitiveness of shapes in a binary image using structure primitives (see Chapter 11). If the structuring elements consist of a single pixel only, the resulting description is an autocorrelation function of the binary image. Using larger and more complex structuring elements, general correlation can be evaluated. The structuring element usually represents some simple shape, such as a square, a line, etc. When a binary textured image is eroded by this structuring element, texture properties are present in the eroded image [Serra and Verchery 73]. One possibility for feature vector construction is to apply different structuring elements to the textured image and to count the number of pixels with unit value in the eroded images, each number forming one element of the feature vector. The mathematical morphology approach stresses the shape properties of texture primitives, but its applicability is limited due to the assumption of a binary textured image. Methods of gray-level mathematical morphology may help to solve this problem. The mathematical morphology approach to texture description is often successful in granulated materials, which can be segmented by thresholding. Using a sequence of openings and counting the number of pixels after each step, a texture measure was derived in [Dougherty et al. 89].

The **texture transform** represents another approach. Each texture type present in an image is transformed into a unique gray-level; the general idea is to construct an image g where the pixels $g(i,j)$ describe a texture in some neighborhood of the pixel $f(i,j)$ in the original textured image f. If micro-textures are analyzed, a small neighborhood of $f(i,j)$ must be used, and an appropriately larger neighborhood should be used for description of macro-textures. In addition, a priori knowledge can be used to guide the transformation and subsequent texture recognition and segmentation [Simaan 90]. In [Linnett and Richardson 90], local texture orientation is used to transform a texture image into a feature image, after which supervised classification is applied to recognize textures.

Linear estimates of gray-levels in texture pixels can also be used for texture description. Pixel gray-levels are estimated from gray-levels in their neighborhood—this method is based on the **autoregression texture model**, where linear estimation parameters are used [Deguchi and Morishita 78]. The model parameters vary substantially in fine textures, but remain mostly unchanged if coarse texture is described. The autoregression model has been compared with an approach based on second-order spatial statistics [Gagalowicz et al. 88]; it was found that even if the results are almost the same, spatial statistics performed much more quickly and reliably.

The **peak and valley** method [Mitchell et al. 77, Ehrick and Foith 78] is based on detection of local extrema of the brightness function in vertical and horizontal scans of a texture image. Fine textures have a large number of small-sized local extrema, coarse textures are represented by a smaller number of larger-sized local extrema—higher peaks and deeper valleys.

The sequence of pixel gray-levels can be considered a **Markov chain** in which the transition probabilities of an mth-order chain represent $(m+1)^{th}$-order statistics of textures [Pratt and Faugeras 78]. This approach may also be used for texture generation [Gagalowicz 79].

Texture description is highly scale dependent. To decrease scale sensitivity, a texture may be described in multiple resolutions and an appropriate scale may be chosen to achieve the maximum texture discrimination [Unser and Eden 89]. **Gabor transforms** and **wavelets** are well suited to multi-scale texture characterization [Coggins and Jain 85, Mallat 89, Bovik et al. 90, Unser 95]. Both approaches represent multi-scale spatial-spatial frequency filtering approaches, which were in the past dominated by Gabor filters. Recently, wavelets have been successfully applied to texture classification using pyramid- or tree-structured discrete wavelet transforms [Chang and Kuo 93] (Section 12.5), typically outperforming conventional texture characterization approaches. In [Unser 95], overcomplete discrete wavelet frames were shown to outperform standard critically sampled wavelet texture feature extraction. Comparison of texture classification performance of Gabor transforms and wavelets is given in [Vautrot et al. 96]. If texture segmentation is a goal, a coarse-to-fine multi-resolution strategy is often used [Gagalowicz and Graffigne 88, Bouman and Liu 91], approximate position of borders between texture regions being detected first in a low-resolution image, and accuracy being improved in higher resolutions using the low-level segmentation as a priori information.

Many of the texture description features presented so far are interrelated; the Fourier power spectrum, the autoregression model, and autocorrelation functions represent the same subset of second-order statistics. The mathematical relationships between texture description methods are summarized in [Tomita and Tsuji 90], an experimental comparison of performance between several methods can be found in [Du Buf et al. 90, Iversen and Lonnestad 94, Zhu and Goutte 95, Wang et al. 96], and criteria for comparison are discussed in [Soh and Huntsberger 91].

It has been shown that higher than second-order statistics contain little information that can be used for texture discrimination [Julesz and Caelli 79]. Nevertheless, identical second-order statistics do not guarantee identical textures; examples can be found in [Julesz and Bergen 87] together with a study of human texture perception. Texture-related research of the human visual system seems to bring useful results, and a texture analysis method based on studies of it was designed to emulate the process of texture feature extraction in each individual channel in the multi-channel spatial filtering model of perception [Rao 93]. Results of the texture recognition process were compared with co-occurrence matrix recognition, and the model-based approach gave superior results in many respects [Tan and Constantinides 90].

14.2 Syntactic texture description methods

Syntactic and hybrid texture description methods are not as widely used as statistical approaches [Tomita et al. 82]. **Syntactic** texture description is based on an analogy between the texture primitive spatial relations and the structure of a formal language. Descriptions of textures from one class form a language that can be represented by its grammar, which is inferred from a training set of words of the language (from descriptions of textures in a training set)—during a learning phase, one grammar is constructed for each texture class present in the training set. The recognition process is then a syntactic analysis of the texture description word. The grammar that can be used to complete the syntactic analysis of the description word determines the texture class (see Section 7.4).

Purely syntactic texture description models are based on the idea that textures consist of primitives located in almost regular relationships. Primitive descriptions and rules of primitive placement must be determined to describe a texture [Tsuji and Tomita 73, Lu and Fu 78]. Primitive spatial relation description methods were discussed at the beginning of this chapter. One of the most efficient ways to describe the structure of primitive relationships is using a grammar which represents a rule for building a texture from primitives, by applying transformation rules to a limited set of symbols. Symbols represent the texture primitive types and transformation rules represent the spatial relations between primitives. In Chapter 7 it was noted that any grammar is a very strict formula. On the other hand, textures of the real world are usually irregular, and structural errors, distortions, or even structural variations are frequent. This means that no strict rule can be used to describe a texture in reality. To make syntactic description of real textures possible, variable rules must be incorporated into the description grammars, and non-deterministic or stochastic grammars must be used (see Section 7.4 and [Fu 74]). Further, there is usually no single description grammar for a texture class, which might be described by an infinite number of different grammars using different symbols and different transformation rules, and different grammar types as well. We will discuss chain grammars and graph grammars in the next sections [Pavlidis 80], and other grammars suitable for texture description (tree, matrix) can be found in [Fu 80, Ballard and Brown 82, Fu 82, Vafaie and Bourbakis 88]. Another approach to texture description using generative principles is to use **fractals** [Mandelbrot 82, Barnsley 88].

14.2.1 Shape chain grammars

Shape chain grammars, whose definition matches that given in Section 7.4, are the simplest grammars that can be used for texture description. They generate textures beginning with a start symbol followed by application of transform rules, called **shape rules**. The generating process is over if no further transform rule can be applied. Texture generation consists of several steps. First, the transform rule is found. Second, the rule must be geometrically adjusted to match the generated texture exactly (rules are more general; they may not include size, orientation, etc.)

Algorithm 14.5: Shape chain grammar texture generation

1. Start a texture generation process by applying some transform rule to the start symbol.

2. Find a part of a previously generated texture that matches the left side of some transform rule. This match must be an unambiguous correspondence between terminal and non-terminal symbols of the left-hand side of the chosen transform rule with terminal and non-terminal symbols of the part of the texture to which the rule is applied. If no such part of the texture can be found, stop.

3. Find an appropriate geometric transform that can be applied to the left-hand side of the chosen rule to match it to the considered texture part exactly.

4. Apply this geometric transform to the right-hand side of the transform rule.

5. Substitute the specified part of the texture (the part that matches a geometrically transformed left-hand side of the chosen rule) with the geometrically transformed right-hand side of the chosen transform rule.

6. Continue with step 2.

We can demonstrate this algorithm on an example of hexagonal texture generation. Let V_n be a set of non-terminal symbols, V_t a set of terminal symbols, R a set of rules, S the start symbol (as in Section 7.4). The grammar [Ballard and Brown 82] is illustrated in Figure 14.6, which can then be used to generate hexagonal texture following Algorithm 14.5—note that

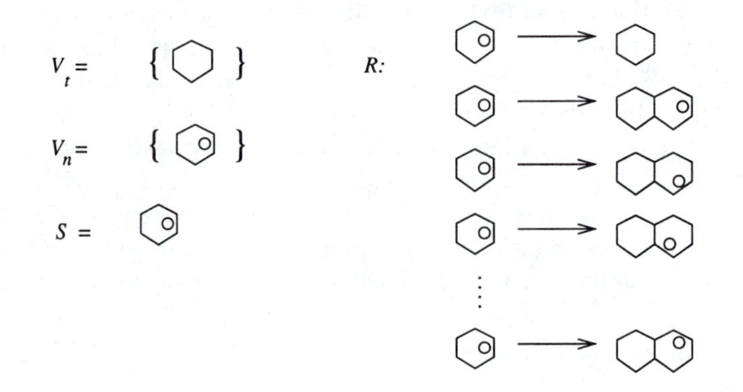

Figure 14.6: *Grammar generating hexagonal textures.*

the non-terminal symbol may appear in different rotations. Rotation of primitives here is represented by a small circle attached to one side of the primitive hexagon in Figure 14.6. Recognition of hexagonal textures is the proof that the texture can be generated by this grammar; the texture recognition uses syntactic analysis as described in Section 7.4. Note that the texture shown in Figure 14.7a will be accepted by the grammar (Figure 14.6), and recognized as a hexagonal texture. Figure 14.7b will be rejected—it is not a hexagonal texture according to the definition of Figure 14.6.

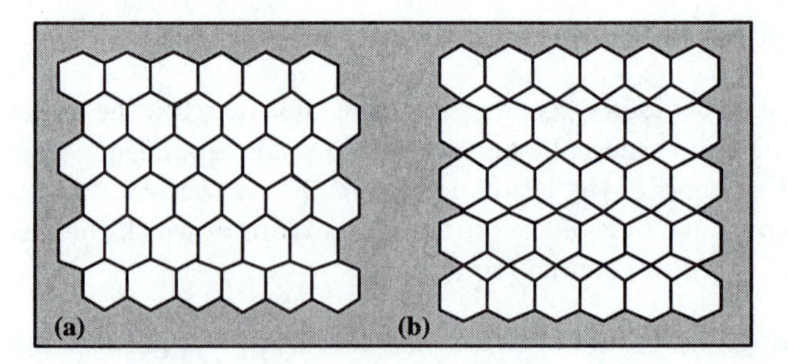

Figure 14.7: *Hexagonal textures: (a) accepted; (b) rejected.*

14.2.2 Graph grammars

Texture analysis is more common than texture synthesis in machine vision tasks (even if texture synthesis is probably more common in general, i.e., in computer graphics and computer games). The natural approach to texture recognition is to construct a planar graph (a graph which can be drawn in a plane without intersecting arcs) of primitive layout and to use it in the recognition process. (planar graph can be drawn in a plane without the presence of intersecting arcs). Primitive classes and primitive spatial relations must be known to construct such a graph; spatial relationships between texture primitives will then be reflected in the graph structure. Texture primitive classes will be coded in graph nodes, each primitive having a corresponding node in the graph, and two nodes will be connected by an arc if there is no other primitive in some specified neighborhood of these two primitives. The size of this neighborhood is the main influence on the complexity of the resulting planar graph—the larger the size of the neighborhood, the smaller the number of graph arcs. Note that choosing the neighborhood too large may result in no arcs for some nodes (the same may be true for the neighborhood being too small). Characteristic properties of some graphs used practically (relative neighborhood graphs, Gabriel graphs, Voronoi diagrams) are described in [Urquhart 82, Ahuja 82, Tuceryan and Jain 90]. These graphs are undirected since the spatial neighborhood relation is symmetric, with evaluated arcs and nodes. Each node is labeled with a primitive class to which it corresponds, and arcs are evaluated by their length and direction.

The texture classification problem is then transformed into a graph recognition problem for which the following approaches may be used.

1. Simplify the texture description by decomposition of the planar graph into a set of chains (sequences of adjacent graph nodes), and apply the algorithms discussed in the previous section. The chain descriptions of textures can represent border primitives of closed regions, different graph paths, primitive neighborhood, etc. A training set is constructed from the decomposition of several texture description planar graphs for each texture class. Appropriate grammars are inferred which represent textures in the training sets. The presence of information noise is highly probable, so stochastic grammars should be used. Texture classification consists of the following steps.

    - A classified texture is represented by a planar graph.
    - The graph is decomposed into chains.
    - The description chains are presented for syntactic analysis.
    - A texture is classified into the class whose grammar accepts all the chains of the decomposed planar graph. If more than one grammar accepts the chains, the texture can be classified into the class whose grammar accepted the chains with the highest probability.

    The main advantage of this approach is its simplicity. The impossibility of reconstructing the original planar graph from the chain decomposition is a disadvantage; it means that some portion of the syntactic information is lost during decomposition.

2. Another class of planar graph description may be represented by a stochastic graph grammar or by an extended graph grammar for description of distorted textures. This

approach is very difficult from both the implementational and algorithmic points of view; the main problem is in grammar inference.

3. The planar graphs can be compared directly using graph matching approaches. It is necessary to define a 'distance' between two graphs as a measure of their similarity; if such a distance is defined, standard methods used in statistical classifier learning can be used—exemplar computation, cluster analysis, etc.

The syntactic approach is valued for its ability to describe a texture character at several hierarchical levels. It permits a qualitative analysis of textures, for decomposition into descriptive substructures (primitive grouping), to incorporate texture descriptions into the whole description of image, scene, etc. From this point of view, it significantly exceeds the complexity of simple object classification. Not considering the implementation difficulties, the second approach from the list above is recommended; if a descriptive graph grammar is chosen appropriately, it can generate a class of graphs independently of their size. It can be used if a pattern is sought in an image at any hierarchical level. An example of a planar graph describing a texture is shown in Figure 14.8.

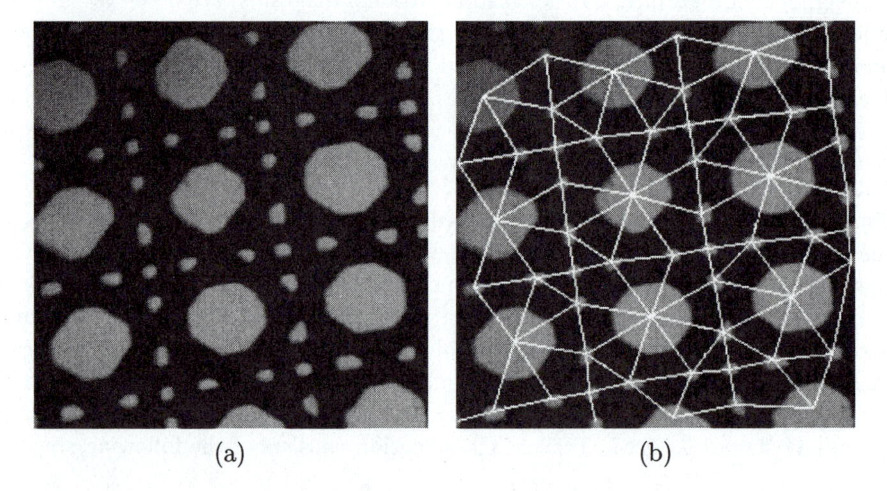

(a) (b)

Figure 14.8: *Planar graph describing a texture: (a) texture primitives; (b) planar graph overlaid.*

14.2.3 Primitive grouping in hierarchical textures

Several levels of primitives can be detected in hierarchical textures—lower-level primitives form some specific pattern which can be considered a primitive at a higher description level (Figure 14.9). The process of detecting these primitive patterns (units) in a texture is called **primitive grouping**. Note that these units may form new patterns at an even higher description level. Therefore, the grouping process must be repeated until no new units can be formed.

Grouping makes a syntactic approach to texture segmentation possible. It plays the same role as local computation of texture features in statistical texture recognition. It has been

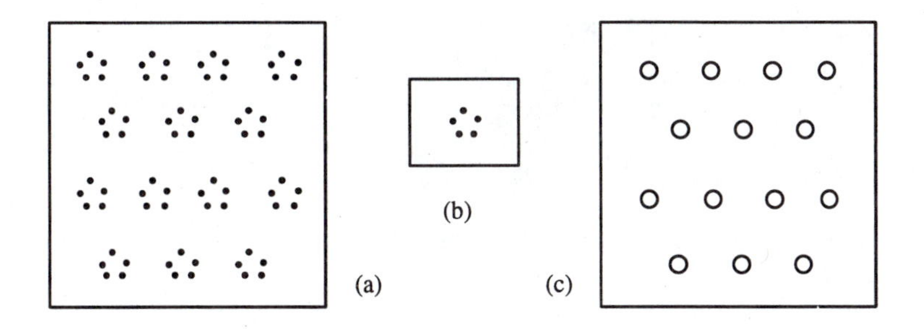

Figure 14.9: *Hierarchical texture: (a) texture; (b) a pattern formed from low-level primitives, this pattern can be considered a primitive in the higher level; (c) higher-level texture.*

claimed several times that different primitives and/or different spatial relationships represent different textures. Consider an example (Figure 14.10a) in which the primitives are the same (small circles) and textures differ in the spatial relations between primitives. If a higher hierarchical level is considered, different primitives can be detected in both textures—the textures do not consist of the same primitive types any more, see Figure 14.10b.

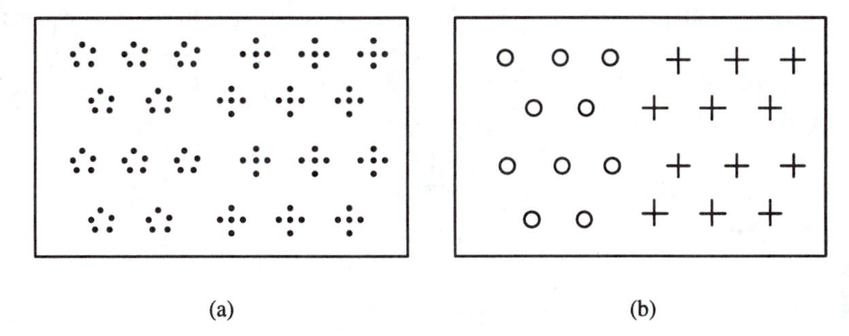

Figure 14.10: *Primitive grouping: (a) two textures, same primitives in the lowest description level; (b) the same two textures, different primitives in the higher description level.*

A primitive grouping algorithm is described in [Tomita and Tsuji 90].

Algorithm 14.6: Texture primitive grouping

1. Determine texture primitive properties and classify primitives into classes.

2. Find the nearest and the second nearest neighbor for each texture primitive. Using the primitive class and distances to the nearest two neighboring primitives d_1, d_2, classify low-level primitives into **new** classes, see Figure 14.11.

3. Primitives with the same **new** classification which are connected (close to each other), are linked together and form higher-level primitives, see Figure 14.11.

4. If any two resulting homogeneous regions of linked primitives overlap, let the overlapped part form a separate region, see Figure 14.12.

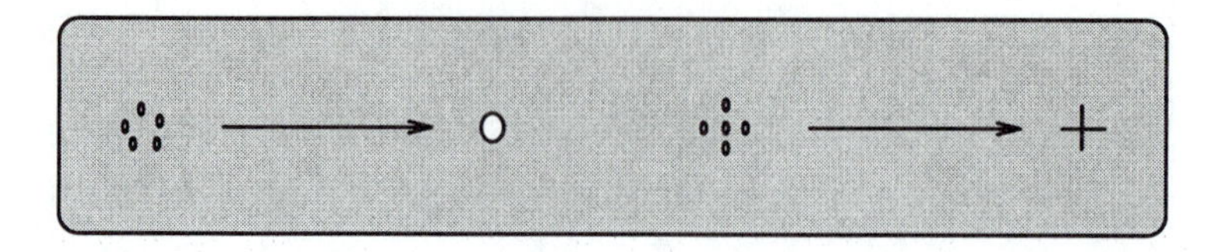

Figure 14.11: *Primitive grouping—low-level primitive patterns are grouped into single primitives at a higher level.*

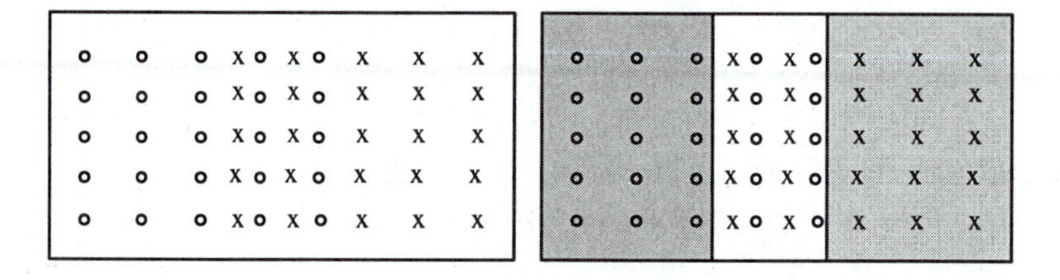

Figure 14.12: *Overlap of homogeneous regions results in their splitting.*

Regions formed from primitives of the lower level may be considered primitives in the higher level and the grouping process may be repeated for these new primitives. Nevertheless, sophisticated control of the grouping process is necessary to achieve meaningful results—it must be controlled from a high-level vision texture understanding sub-system. A recursive primitive grouping, which uses histograms of primitive properties and primitive spatial relations is presented in [Tomita and Tsuji 90] together with examples of syntactic-based texture segmentation results.

14.3 Hybrid texture description methods

Purely syntactic methods of texture description experience many difficulties with syntactic analyzer learning and with graph (or other complex) grammar inference. This is the main reason why purely syntactic methods are not widely used. On the other hand, a precise definition of primitives brings many advantages and it is not wise to avoid it completely. Hybrid methods of texture description combine the statistical and syntactic approaches; the technique is partly syntactic because the primitives are exactly defined, and partly statistical because spatial relations between primitives are based on probabilities [Conners and Harlow 80b].

The hybrid approach to texture description distinguishes between weak and strong textures. The syntactic part of weak texture description divides an image into regions based on a tonal image property (e.g., constant gray-level regions) which are considered texture primitives. Primitives can be described by their shape, size, etc. The next step constructs histograms of sizes and shapes of all the texture primitives contained in the image. If the image can be segmented into two or more sets of homogeneous texture regions, the histogram is bi-modal and each primitive is typical of one texture pattern. This can be used for texture segmentation.

If the starting histogram does not have significant peaks, a complete segmentation cannot be achieved. The histogram-based segmentation can be repeated in each hitherto segmented homogeneous texture region. If any texture region consists of more than one primitive type, the method cannot be used and spatial relations between primitives must be computed. Some methods are discussed in [Haralick 79].

Description of strong textures is based on the spatial relations of texture primitives and two-directional interactions between primitives seem to carry most of the information. The simplest texture primitive is a pixel and its gray-level property, while the maximum contiguous set of pixels of constant gray-level is a more complicated texture primitive [Wang and Rosenfeld 81]. Such a primitive can be described by its size, elongatedness, orientation, average gray-level, etc. The texture description includes spatial relations between primitives based on distance and adjacency relations. Using more complex texture primitives brings more textural information. On the other hand, all the properties of single pixel primitives are immediately available without the necessity of being involved in extensive primitive property computations.

The hybrid multi-level texture description and classification method [Sonka 86] is based on primitive definition and spatial description of inter-primitive relations. The method considers both tone and structural properties and consists of several consequent steps. Texture primitives must be extracted first, and then described and classified. As a result of this processing stage, a classifier knows how to classify texture primitives. Known textures are presented to the texture recognition system in the second stage of learning. Texture primitives are extracted from the image and the first-level classifier recognizes their classes. Based on recognized texture primitives, spatial relations between primitive classes are evaluated for each texture from the training set. Spatial relations between texture primitives are described by a feature vector used to adjust a second-level classifier. If the second-level classifier is set, the two-level learning process is over, and unknown textures can be presented to the texture recognition system. The primitives are classified by the first-level classifier, spatial primitive properties are computed and the second-level classifier assigns the texture to one of the texture classes. Another hybrid method [Bruno 87] uses Fourier descriptors for shape coding and a texture is modeled by a reduced set of joint probability distributions obtained by vector quantization.

14.4 Texture recognition method applications

The estimated yield of crops or localization of diseased forests from remotely sensed data, automatic diagnosis of lung diseases from X-ray images, recognition of cloud types from meteorological satellite data, etc., are examples of texture recognition applications. Textures are very common in our world, and application possibilities are almost unlimited. The effectiveness of various texture recognition methods is discussed in [Conners and Harlow 80a].

Texture recognition of roads, road crossings, buildings, agricultural regions, and natural objects, or classification of trees into five classes, belong to classical applications of spatial frequency-based texture description methods. An interesting proof of the role of textural information in outdoor object recognition was done by comparison of classification correctness if textural information was and was not used; spectral information-based classification achieved 74% correctly classified objects. Adding the textural information, accuracy increased to 99% [Haralick 79]. Real industrial applications of texture description and recognition are becoming more and more common. Examples can be found in almost all branches of industrial and biomedical activities—quality inspection in the motor or textile industries [Roning and Hall 87, Chen and Jain 88, Wood 90], workpiece surface monitoring [Adam and Nickolay 89], road surface skidding estimation [Heaton et al. 90, Kennedy et al. 90], micro-electronics [Rao and Jain 90], remote sensing [IGARSS-89 90, Monjoux and Rudant 91], mammography [Miller and Astley 92], MR brain imaging [Toulson and Boyce 92, Schad et al. 93], three-dimensional texture images [Ip and Lam 95], content-based data retrieval from image databases [Puzicha et al. 97], etc.

14.5 Summary

- **Texture**
  - Texture is widely used and intuitively obvious but has no precise definition due to its wide variability. One existing definition claims that '*an image region has a constant texture if a set of its local properties in that region is constant, slowly changing, or approximately periodic*'.
  - Texture consists of texture **primitives** (texture **elements**) called **texels.**
  - A texture primitive is a contiguous set of pixels with some tonal and/or regional property.
  - **Texture description** is based on **tone** and **structure**. Tone describes pixel intensity properties in the primitive, while structure reflects spatial relationships between primitives.
  - Texture description is **scale dependent**.
  - **Statistical** methods of texture description compute different texture properties and are suitable if texture primitive sizes are comparable with the pixel sizes.
  - **Syntactic** and **hybrid** methods (combination of statistical and syntactic) are more suitable for textures where primitives can be easily determined and their properties described.
- **Statistical texture description**
  - Statistical texture description methods describe textures in a form suitable for statistical pattern recognition. As a result of the description, each texture is described by a feature vector of properties which represents a point in a multi-dimensional feature space.
  - Coarse textures are built from larger primitives, fine textures from smaller primitives. Textural character is in direct relation to the spatial size of the texture primitives.
  - Fine textures are characterized by higher spatial frequencies, coarse textures by lower spatial frequencies.
  - Measuring spatial frequencies is the basis of a large group of texture recognition methods:
    * Autocorrelation function of a texture

    * Optical image transform
    * Discrete image transform

  - Texture description may be based on the repeated occurrence of some gray-level configuration in the texture; this configuration varies rapidly with distance in fine textures, slowly in coarse textures. **Co-occurrence matrices** represent such an approach.
  - **Edge frequency** approaches describe frequencies of edge appearances in texture.
  - In the **primitive length (run length)** approach, texture description features can be computed as continuous probabilities of the length and the gray-level of primitives in the texture.
  - **Laws' texture measures** determine texture properties by assessing average gray-level, edges, spots, ripples, and waves in texture.
  - **Fractal** approach to texture description is based on correlation between texture coarseness and fractal dimension and texture granularity and lacunarity.
  - Other statistical approaches exist:
    * Mathematical morphology
    * Texture transform
    * Wavelet description
  - Variety of texture properties may be derived from **first-order** and **second-order statistics** of elementary measures such as co-occurrences, edge distributions, primitive lengths, etc.
  - Higher than second-order statistics contain little information that can be used for texture discrimination.

- **Syntactic and hybrid texture description**
  - Syntactic texture description is based on an analogy between texture primitive spatial relations and structure of a formal language.
  - Hybrid methods of texture description combine the statistical and syntactic approaches; the technique is partly syntactic because the primitives are exactly defined, and partly statistical because spatial relations between primitives are based on probabilities.
  - Purely syntactic texture description models utilize the idea that textures consist of primitives located in almost regular relationships. Primitive descriptions and rules of primitive placement must be determined to describe a texture.
  - Textures of the real world are usually irregular, with frequent structural errors, distortions, and/or structural variations causing strict grammars to be inapplicable. To make syntactic description of real textures possible, variable rules must be incorporated into the description grammars, and non-deterministic or stochastic grammars must be used.
  - Syntactic texture description methods include:

    * **Shape chain grammars**, which are the simplest grammars that can be used for texture description. They generate textures beginning with a start symbol followed by application of **shape transform rules.**
    * **Graph grammars**, an approach that constructs a **planar graph of primitive layout**. Primitive classes and primitive spatial relations must be known to construct such a graph; spatial relationships between texture primitives are reflected in the graph structure. The texture classification problem is then transformed into a **graph recognition** problem

- The syntactic approach is valued for its ability to describe a texture character at several **hierarchical levels.**
- **Primitive grouping** can be performed if lower-level primitives that form some specific pattern can be considered a primitive at a higher description level.
- Syntactic and hybrid texture description methods are not as widely used as statistical approaches.

14.6 Exercises

Short-answer questions

1. What is a texel?
2. Explain the difference between weak and strong textures.
3. Explain the difference between fine and coarse textures.
4. What is the role of scale in texture description?
5. Specify the main texture description and recognition strategies employed in statistical, syntactic, and hybrid methods. For each of the three general approaches, describe the texture types for which the approach can be expected to perform well and for which it is not appropriate.
6. Define fractal dimension and lacunarity, explain how these measures can be used for texture description.
7. Determine whether the following texture features exhibit relatively higher or lower values in fine textures compared to coarse textures:
    (a) Energy in a small-radius circle of the Fourier power spectrum (Figure 14.3a)
    (b) Average edge frequency feature (equation 14.10) for a small value of d
    (c) Short primitive emphasis
    (d) Long primitive emphasis
    (e) Fractal dimension
    (f) Lacunarity
8. Name several texture description features that are well suited for characterization of directional textures.
9. Explain why deterministic grammars are too restrictive for description of real-world textures.
10. Explain the rationale for primitive grouping. What are its advantages? To which textures is this approach applicable? How can it be used for texture description?

11. Texture characteristics can be used in image segmentation. Describe how texture descriptors may be used in region growing segmentation.
12. Name several application areas in which texture description and recognition can be used.

Problems

1. Determine the co-occurrence matrices $P_{0°,2}$, $P_{45°,2}$, and $P_{90°,3}$ for the image in Figure 14.4.
2. Determine the average edge frequency function [equation (14.10)] for $d \in \{1, 2, 3, 4, 5\}$ for a 30×30 checkerboard image with 3×3 binary checkers (values of 0 and 1, 0-level checker in the upper left corner).
3. For a 30×30 checkerboard image with 3×3 binary checkers (values of 0 and 1, 0-level checker in the upper left corner) and a 30×30 image with vertical binary 3-pixel-wide stripes (0-level stripe along the left image edge), determine the following non-directional texture descriptors:
   (a) Short primitives emphasis
   (b) Long primitives emphasis
   (c) Gray-level uniformity
   (d) Primitive length uniformity
   (e) Primitive percentage
4. Repeat Problem 14.3 with modified texture descriptors that consider directionality of primitives.
5. Develop functions (subroutines) for computing the following texture features in an image of a given size:
   (a) Co-occurrence descriptors
   (b) Average edge frequency
   (c) Primitive length descriptors
   (d) Laws' energy descriptors
   (e) Fractal texture descriptors
6. Using the World Wide Web, find several images of dissimilar homogeneous textures (Brodatz textures [Brodatz 66] from a Web-based database may be a good choice). Create your personal database, which will be used in the experiments below.
7. For five highly dissimilar textures ranging from fine to coarse selected from the database created in Problem 14.6 and, using the functions developed in Problem 14.5, determine:
   (a) Co-occurrence features specified by equations (14.4)–(14.9)
   (b) Average edge frequency

   for four distinct values of d (e.g., $d = 1, 3, 5, 7$; note that you may select other values of d according to the scale of your textures). Select three texture descriptors to answer the following questions: What can be concluded from inspection of the relationship between the texture descriptors and the value of d in each texture? What can be concluded for the relationship between texture coarseness and the values of individual texture descriptors?
8. For five highly dissimilar textures ranging from fine to coarse selected from the database created in Problem 14.6 and, using the functions developed in Problem 14.5, determine:
   (a) Primitive length features specified by equations (14.12)–(14.16)

(b) Laws' energy features

(c) Fractal-based texture features—fractal dimension and lacunarity

What can be concluded for the relationship between texture coarseness and the values of individual texture descriptors?

9. Design a shape chain grammar that generates the texture shown in Figure 14.7b.

10. By using a digitizer or searching the World Wide Web (Brodatz textures [Brodatz 66] from a Web-based database may be a good choice) create a database of three dissimilar texture classes with at least ten images belonging to each class. Develop a program for statistical texture description and recognition. Texture descriptors should reflect the discriminative properties of the three texture classes. Train a simple statistical classifier (Chapter 7) using one-half of the texture images from each class. Determine texture classification correctness using the remaining texture images.

11. Develop a texture-based image segmentation program. Using the texture images from the database created in Problem 14.6, create several artificial images that contain three to five regions of different texture. Develop a texture classification program that performs texture recognition in rectangular windows. Using a moving window, the program shall classify each window into one of the texture classes. Train the classifier in single-texture images using windows of the same size as is used in the classification phase. Assess image segmentation accuracy as the percentage of correctly classified image area.

14.7 References

[Adam and Nickolay 89] W Adam and B Nickolay. Texture analysis for the evaluation of surface characteristics in quality monitoring. *SME Technical Paper (Series)*, MS, 1989.

[Ahuja 82] N Ahuja. Dot pattern processing using Voronoi neighborhood. *IEEE Transactions on Pattern Analysis and Machine Intelligence*, 4:336–343, 1982.

[Ahuja and Rosenfeld 81] N Ahuja and A Rosenfeld. Mosaic models for textures. *IEEE Transactions on Pattern Analysis and Machine Intelligence*, 3(1):1–11, 1981.

[Ando 88] S Ando. Texton finders based on Gaussian curvature of correlation with an application to rapid texture classification. In *Proceedings of the 1988 IEEE International Conference on Systems, Man, and Cybernetics,*, pages 25–28, IEEE, Beijing/Shenyang, China, 1988.

[Argenti et al. 90] F Argenti, L Alparone, and G Benelli. Fast algorithms for texture analysis using co-occurrence matrices. *IEE Proceedings, Part F: Radar and Signal Processing*, 137(6):443–448, 1990.

[Ballard and Brown 82] D H Ballard and C M Brown. *Computer Vision.* Prentice-Hall, Englewood Cliffs, NJ, 1982.

[Barnsley 88] M F Barnsley. *Fractals Everywhere.* Academic Press, Boston, 1988.

[Besicovitch and Ursell 37] A S Besicovitch and H D Ursell. Sets of fractional dimensions (V): On dimensional numbers of some continuous curves. *Journal of the London Mathematical Society*, 12:18–25, 1937.

[Bouman and Liu 91] C Bouman and B Liu. Multiple resolution segmentation of textured images. *IEEE Transactions on Pattern Analysis and Machine Intelligence*, 13(2):99–113, 1991.

[Bovik et al. 90] A C Bovik, M Clark, and W S Geisler. Multichannel texture analysis using localized spatial filters. *IEEE Transactions on Pattern Analysis and Machine Intelligence*, 12:55–73, 1990.

[Brodatz 66] P Brodatz. *Textures: A Photographic Album for Artists and Designers.* Dover, Toronto, 1966.

[Bruno 87] A Bruno. Hybrid model for the description of structured textures. In *Optical and Digital Pattern Recognition,* Los Angeles, CA, pages 20–23, SPIE, Bellingham, WA, 1987.

[Carlson and Ebel 88] G E Carlson and W J Ebel. Co-occurrence matrix modification for small region texture measurement and comparison. In *IGARSS'88—Remote Sensing: Moving towards the 21st Century,* Edinburgh, Scotland, pages 519–520, IEEE, Piscataway, NJ, 1988.

[Castleman 96] K R Castleman. *Digital Image Processing.* Prentice-Hall, Englewood Cliffs, NJ, 1996.

[Chang and Kuo 93] T Chang and C C J Kuo. Texture analysis and classification with tree-structured wavelet transforms. *IEEE Transactions on Image Processing,* 2:429–441, 1993.

[Chen and Jain 88] J Chen and A K Jain. Structural approach to identify defects in textured images. In *Proceedings of the 1988 IEEE International Conference on Systems, Man, and Cybernetics,* pages 29–32, IEEE, Beijing/Shenyang, China, 1988.

[Chetverikov 82] D Chetverikov. Experiments in the rotation-invariant texture discrimination using anisotropy features. In *Proceedings of the 6th IEEE Congress on Pattern Recognition,* pages 1071–1073, IEEE, Munich, Germany, 1982.

[Coggins and Jain 85] J M Coggins and A K Jain. A spatial filtering approach to texture analysis. *Pattern Recognition Letters,* 3:195–203, 1985.

[Conners and Harlow 80a] R W Conners and C A Harlow. A theoretical comparison of texture algorithms. *IEEE Transactions on Pattern Analysis and Machine Intelligence,* 2(3):204–222, 1980.

[Conners and Harlow 80b] R W Conners and C A Harlow. Toward a structural textural analyser based on statistical methods. *Computer Graphics and Image Processing,* 12:224–256, 1980.

[Davis and Mitiche 80] L S Davis and A Mitiche. Edge detection in textures. *Computer Graphics and Image Processing,* 12:25–39, 1980.

[Davis et al. 83] L S Davis, L Janos, and S M Dunn. Efficient recovery of shape from texture. *IEEE Transactions on Pattern Analysis and Machine Intelligence,* 5(5):485–492, 1983.

[Deguchi and Morishita 78] K Deguchi and I Morishita. Texture characterization and texture-based partitioning using two-dimensional linear estimation. *IEEE Transactions on Computers,* 27:739–745, 1978.

[Derin and Elliot 87] H Derin and H Elliot. Modelling and segmentation of noisy and textured images using Gibbs random fields. *IEEE Transactions on Pattern Analysis and Machine Intelligence,* 9(1):39–55, 1987.

[Dougherty et al. 89] E R Dougherty, E J Kraus, and J B Pelz. Image segmentation by local morphological granulometries. In *Proceedings of IGARSS '89 and Canadian Symposium on Remote Sensing,* Vancouver, Canada, pages 1220–1223, IEEE, New York, 1989.

[Du Buf et al. 90] J M H Du Buf, M Kardan, and M Spann. Texture feature performance for image segmentation. *Pattern Recognition,* 23(3–4):291–309, 1990.

[Ehrick and Foith 78] R W Ehrick and J P Foith. A view of texture topology and texture description. *Computer Graphics and Image Processing,* 8:174–202, 1978.

[Fan 89] Z Fan. Edge-based hierarchical algorithm for textured image segmentation. In *International Conference on Acoustics, Speech, and Signal Processing,* Glasgow, Scotland, pages 1679–1682, IEEE, Piscataway, NJ, 1989.

[Fu 74] K S Fu. *Syntactic Methods in Pattern Recognition.* Academic Press, New York, 1974.

[Fu 80] K S Fu. Picture syntax. In S K Chang and K S Fu, editors, *Pictorial Information Systems,* pages 104–127. Springer Verlag, Berlin, 1980.

[Fu 82] K S Fu. *Syntactic Pattern Recognition and Applications.* Prentice-Hall, Englewood Cliffs, NJ, 1982.

[Fung et al. 90] P W Fung, G Grebbin, and Y Attikiouzel. Contextual classification and segmentation of textured images. In *Proceedings of the 1990 International Conference on Acoustics, Speech, and Signal Processing—ICASSP 90,* Albuquerque, NM, pages 2329–2332, IEEE, Piscataway, NJ, 1990.

[Gagalowicz 79] A Gagalowicz. Stochatic texture fields synthesis from a priori given second order statistics. In *Proceedings, Pattern Recognition and Image Processing,* Chicago, IL, pages 376–381, IEEE, Piscataway, NJ, 1979.

[Gagalowicz and Graffigne 88] A Gagalowicz and C Graffigne. Blind texture segmentation. In *9th International Conference on Pattern Recognition,* Rome, Italy, pages 46–50, IEEE, Piscataway, NJ, 1988.

[Gagalowicz et al. 88] A Gagalowicz, C Graffigne, and D Picard. Texture boundary positioning. In *Proceedings of the 1988 IEEE International Conference on Systems, Man, and Cybernetics,* pages 16–19, IEEE, Beijing/Shenyang, China, 1988.

[Galloway 75] M M Galloway. Texture classification using gray level run length. *Computer Graphics and Image Processing,* 4:172–179, 1975.

[Gisolfi et al. 86] A Gisolfi, S Vitulano, and A Cacace. Texture and structure. In V Cappelini and R Marconi, editors, *Advances in Image Processing and Pattern Recognition,* pages 179–183. North Holland, Amsterdam, 1986.

[Gotlieb and Kreyszig 90] C C Gotlieb and H E Kreyszig. Texture descriptors based on co-occurrence matrices. *Computer Vision, Graphics, and Image Processing,* 51(1):70–86, 1990.

[Haralick 79] R M Haralick. Statistical and structural approaches to texture. *Proceedings IEEE,* 67(5):786–804, 1979.

[Haralick et al. 73] R M Haralick, K Shanmugam, and I Dinstein. Textural features for image classification. *IEEE Transactions on Systems, Man and Cybernetics,* 3:610–621, 1973.

[Hausdorff 19] F Hausdorff. Dimension und ausseres Mass. *Mathematische Annalen,* 79:157–179, 1919.

[Heaton et al. 90] B S Heaton, J J Henry, and J C Wambold. Texture measuring equipment vs skid testing equipment. In *Proceedings of the 15th ARRB Conference,* Darwin, Australia, pages 53–64, Australian Road Research Board, Nunawading, 1990.

[Huang et al. 94] Q Huang, J R Lorch, and R C Dubes. Can the fractal dimension of images be measured? *Pattern Recognition,* 27:339–349, 1994.

[Hurst 51] H E Hurst. Long-term storage capacity of reservoirs. *Transactions of the American Society of Civil Engineers,* 116:770–808, 1951.

[IGARSS-89 90] *Quantitative Remote Sensing: An Economic Tool for the Nineties, IGARSS'89,* Vancouver, Canada, 1990. IEEE.

[Ip and Lam 95] H H S Ip and S W C Lam. Three-dimensional structural texture modeling and segmentation. *Pattern Recognition,* 28:1299–1319, 1995.

[Iversen and Lonnestad 94] H Iversen and T Lonnestad. An evaluation of stochastic models for analysis and synthesis of gray-scale texture. *Pattern Recognition Letters*, 15:575–585, 1994.

[Jin et al. 95] X C Jin, S H Ong, and Jayasooriah. A practical method for estimating fractal dimension. *Pattern Recognition Letters*, 16:457–464, 1995.

[Julesz 81] B Julesz. Textons, the elements of texture perception, and their interactions. *Nature*, 290:91–97, 1981.

[Julesz and Bergen 87] B Julesz and J R Bergen. Textons, the fundamental elements in preattentive vision and perception of textures. In *Readings in Computer Vision*, pages 243–256. Morgan Kaufmann Publishers, Los Altos, CA, 1987.

[Julesz and Caelli 79] B Julesz and T Caelli. On the limits of Fourier decompositions in visual texture perception. *Perception*, 8:69–73, 1979.

[Keller et al. 89] J M Keller, S Chen, and R M Crownover. Texture description and segmentation through fractal geometry. *Computer Vision, Graphics, and Image Processing*, 45(2):150–166, 1989.

[Kennedy et al. 90] C K Kennedy, A E Young, and I C Butler. Measurement of skidding resistance and surface texture and the use of results in the UK. In *First International Symposium on Surface Characteristics,* State College, PA, pages 87–102, ASTM, Philadelphia, 1990.

[Kjell and Wang 91] B P Kjell and P Y Wang. Noise-tolerant texture classification and image segmentation. In *Intelligent Robots and Computer Vision IX: Algorithms and Techniques,* Boston, pages 553–560, SPIE, Bellingham, WA, 1991.

[Laws 79] K I Laws. Texture energy measures. In *DARPA Image Understanding Workshop,* Los Angeles, CA, pages 47–51, DARPA, Los Altos, CA, 1979.

[Linnett and Richardson 90] L M Linnett and A J Richardson. Texture segmentation using directional operators. In *Proceedings of the 1990 International Conference on Acoustics, Speech, and Signal Processing—ICASSP 90,* Albuquerque, NM, pages 2309–2312, IEEE, Piscataway, NJ, 1990.

[Liu and Jernigan 90] S S Liu and M E Jernigan. Texture analysis and discrimination in additive noise. *Computer Vision, Graphics, and Image Processing*, 49:52–67, 1990.

[Lu and Fu 78] S Y Lu and K S Fu. A syntactic approach to texture analysis. *Computer Graphics and Image Processing*, 7:303–330, 1978.

[Lundahl et al. 86] T Lundahl, W J Ohley, S M Kay, and R Siffert. Fractional Brownian motion: A maximum likelihood estimator and its application to image texture. *IEEE Transactions on Medical Imaging*, 5:152–161, 1986.

[Mallat 89] S G Mallat. A theory of multiresolution signal decomposition: The wavelet representation. *IEEE Transactions on Pattern Analysis and Machine Intelligence*, 11:674–693, 1989.

[Mandelbrot 82] B B Mandelbrot. *The Fractal Geometry of Nature.* Freeman, New York, 1982.

[Miller and Astley 92] P Miller and S Astley. Classification of breast tissue by texture analysis. *Image and Vision Computing*, 10(5):277–282, 1992.

[Mitchell et al. 77] O R Mitchell, C R Myer, and W Boyne. A max-min measure for image texture analysis. *IEEE Transactions on Computers*, 26:408–414, 1977.

[Monjoux and Rudant 91] E Monjoux and J P Rudant. Texture segmentation in aerial images. In *Image Processing Algorithms and Techniques II,* San Jose, pages 310–318, SPIE, Bellingham, WA, 1991.

[Pavlidis 80] T Pavlidis. Structural descriptions and graph grammars. In S K Chang and K S Fu, editors, *Pictorial Information Systems*, pages 86–103, Springer Verlag, Berlin, 1980.

[Pentland 84] A P Pentland. Fractal-based description of natural scenes. *IEEE Transactions on Pattern Analysis and Machine Intelligence*, 6:661–674, 1984.

[Perry and Lowe 89] A Perry and D G Lowe. Segmentation of non-random textures using zero-crossings. In *1989 IEEE International Conference on Systems, Man, and Cybernetics*, Cambridge, MA, pages 1051–1054, IEEE, Piscataway, NJ, 1989.

[Pratt and Faugeras 78] W K Pratt and O C Faugeras. Development and evaluation of stochastic-based visual texture features. *IEEE Transactions on Systems, Man and Cybernetics*, 8:796–804, 1978.

[Puzicha et al. 97] J Puzicha, T Hofmann, and J M Buhman. Non-parametric similarity measure for unsupervised texture segmentation and image retrieval. In *Computer Vision and Pattern Recognition*, pages 267–272, IEEE Computer Society, Los Alamitos, CA, 1997.

[Rao 90] A R Rao. *A Taxonomy for Texture Description and Identification.* Springer Verlag, New York, 1990.

[Rao 93] A R Rao. Identifying high level features of texture perception. *CVGIP – Graphical Models and Image Processing*, 55:218–233, 1993.

[Rao and Jain 90] A R Rao and R Jain. Quantitative measures for surface texture description in semiconductor wafer inspection. In *Integrated Circuit Metrology, Inspection, and Process Control IV,* San Jose, CA, pages 164–172, SPIE, Bellingham, WA, 1990.

[Reed et al. 90] T R Reed, H Wechsler, and M Werman. Texture segmentation using a diffusion region growing technique. *Pattern Recognition*, 23(9):953–960, 1990.

[Roning and Hall 87] J Roning and E L Hall. Shape, form, and texture recognition for automotive brake pad inspection. In *Automated Inspection and Measurement,* Cambridge, MA, pages 82–90, SPIE, Bellingham, WA, 1987.

[Rosenfeld 76] A Rosenfeld, editor. *Digital Picture Analysis.* Springer Verlag, Berlin, 1976.

[Sarkar and Chaudhuri 94] N Sarkar and B B Chaudhuri. An efficient differential box-counting approach to compute fractal dimension of image. *IEEE Transactions on Systems, Man and Cybernetics*, 24:115–120, 1994.

[Schad et al. 93] L R Schad, S Bluml, and I Zuna. Tissue characterization by magnetic resonance spectroscopy and imaging: Results of a concerted research project of the European Economic Community. IX. MR tissue characterization of intracranial tumors by means of texture analysis. *Magnetic Resonance Imaging*, 11:889–896, 1993.

[Serra and Verchery 73] J Serra and G Verchery. Mathematical morphology applied to fibre composite materials. *Film Science Technology*, 6:141–158, 1973.

[Shulman 70] A R Shulman. *Optical Data Processing.* Wiley, New York, 1970.

[Simaan 90] M Simaan. Knowledge-guided segmentation of texture images. In *Proceedings of the 1990 IEEE International Conference on Systems Engineering,* Pittsburgh, PA, pages 539–542, IEEE, Piscataway, NJ, 1990.

[Sklansky 78] J Sklansky. Image segmentation and feature extraction. *IEEE Transactions on Systems, Man and Cybernetics*, 8(4):237–247, 1978.

[Soh and Huntsberger 91] S N J Soh, Y Murthy and T L Huntsberger. Development of criteria to compare model-based texture analysis methods. In *Intelligent Robots and Computer Vision IX: Algorithms and Techniques,* Boston, pages 561–573, SPIE, Bellingham, WA, 1991.

[Sonka 86] M Sonka. A new texture recognition method. *Computers and Artificial Intelligence*, 5(4):357–364, 1986.

[Sutton and Hall 72] R Sutton and E Hall. Texture measures for automatic classification of pulmonary diseases. *IEEE Transactions on Computers*, C–21(1):667–678, 1972.

[Tan and Constantinides 90] T N Tan and A G Constantinides. Texture analysis based on a human visual model. In *Proceedings of the 1990 International Conference on Acoustics, Speech, and Signal Processing—ICASSP 90,* Albuquerque, NM, pages 2137–2140, IEEE, Piscataway, NJ, 1990.

[Tomita and Tsuji 90] F Tomita and S Tsuji. *Computer Analysis of Visual Textures.* Kluwer, Norwell, MA, 1990.

[Tomita et al. 82] F Tomita, Y Shirai, and S Tsuji. Description of textures by a structural analysis. *IEEE Transactions on Pattern Analysis and Machine Intelligence*, 4(2):183–191, 1982.

[Toulson and Boyce 92] D L Toulson and J F Boyce. Segmentation of MR images using neural nets. *Image and Vision Computing*, 10(5):324–328, 1992.

[Tsuji and Tomita 73] S Tsuji and F Tomita. A structural analyser for a class of textures. *Computer Graphics and Image Processing*, 2:216–231, 1973.

[Tuceryan and Jain 90] M Tuceryan and A K Jain. Texture segmentation using Voronoi polygons. *IEEE Transactions on Pattern Analysis and Machine Intelligence*, 12(2):211–216, 1990.

[Unser 95] M Unser. Texture classification and segmentation using wavelet frames. *IEEE Transactions on Image Processing*, 4:1549–1560, 1995.

[Unser and Eden 89] M Unser and M Eden. Multiresolution feature extraction and selection for texture segmentation. *IEEE Transactions on Pattern Analysis and Machine Intelligence*, 11(7):717–728, 1989.

[Urquhart 82] R Urquhart. Graph theoretical clustering based on limited neighbourhood sets. *Pattern Recognition*, 15(3):173–187, 1982.

[Vafaie and Bourbakis 88] H Vafaie and N G Bourbakis. Tree grammar scheme for generation and recognition of simple texture paths in pictures. In *Third International Symposium on Intelligent Control 1988,* Arlington, VA, pages 201–206, IEEE, Piscataway, NJ, 1988.

[Vautrot et al. 96] P Vautrot, N Bonnet, and M Herbin. Comparative study of different spatial/spatial frequency methods for texture segmentation/classification. In *Proceedings of the IEEE International Conference on Image Processing,* Lausanne, Switzerland, pages III:145–148, IEEE, Piscataway, NJ, 1996.

[Voorhees and Poggio 87] H Voorhees and T A Poggio. Detecting textons and texture boundaries in natural images. In *1st International Conference on Computer Vision,* London, England, pages 250–258, IEEE, Piscataway, NJ, 1987.

[Voss 86] R Voss. Random fractals: Characterization and measurement. In *Scaling Phenomena in Disordered Systems.* Plenum Press, New York, 1986.

[Wang and Rosenfeld 81] S Wang and A Rosenfeld. A relative effectiveness of selected texture primitive. *IEEE Transactions on Systems, Man and Cybernetics*, 11:360–370, 1981.

[Wang et al. 96] Z Wang, A Guerriero, and M De Sario. Comparison of several approaches for the segmentation of texture images. *Pattern Recognition Letters*, 17:509–521, 1996.

[Weszka et al. 76] J S Weszka, C Dyer, and A Rosenfeld. A comparative study of texture measures for terrain classification. *IEEE Transactions on Systems, Man and Cybernetics*, 6(4):269–285, 1976.

[Wood 90] E J Wood. Applying Fourier and associated transforms to pattern characterization in textiles. *Textile Research Journal*, 60(4):212–220, 1990.

[Wu et al. 92] C M Wu, Y C Chen, and K S Hsieh. Texture features for classification of ultrasonic liver images. *IEEE Transactions on Medical Imaging*, 11:141–152, 1992.

[Zhu and Goutte 95] Y M Zhu and R Goutte. A comparison of bilinear space/spatial-frequency representations for texture discrimination. *Pattern Recognition Letters*, 16:1057–1068, 1995.

Chapter 15

Motion analysis

In recent years, interest in motion processing has increased with advances in motion analysis methodology and processing capabilities. The usual input to a motion analysis system is a temporal image sequence, with a corresponding increase in the amount of processed data. Motion analysis is often connected with real-time analysis, for example, for robot navigation. Another common motion analysis problem is to obtain comprehensive information about moving and static objects present in a scene. Detecting 3D shape and relative depth from motion are also fast-developing fields—these issues are considered in Chapter 9.

A set of assumptions can help to solve motion analysis problems—as always, prior knowledge helps to decrease the complexity of analysis. Prior knowledge includes information about the camera motion—mobile or static—and information about the time interval between consecutive images, especially whether this interval was short enough for the sequence to represent continuous motion. This prior information about data helps in the choice of an appropriate motion analysis technique. As in other areas of machine vision, there is no foolproof technique in motion analysis, no general algorithm; furthermore, the techniques presented in this chapter work only if certain conditions are met. A very interesting aspect of motion analysis is research into visual sensing of living organisms that are extremely well adapted to motion analysis. The psycho-physiological and cognitive aspects of motion sensing can be studied in [Koenderink and Doorn 75, Ullman 79, Clocksin 80, Thompson and Barnard 81, Marr 82, Watson and Ahumada 85, Koenderink 86].

There are three main groups of motion-related problems from the practical point of view [Yachida et al. 80, Radig 84, Kanatani 85, Horn 86, Schalkoff 87, Vesecky 88, Anandan 88, Aggarwal and Nandhakumar 88, Schalkoff 89, Vega-Riveros and Jabbour 89, Murray and Buxton 89, Thompson and Pong 90, Fleet 92, Cedras and Shah 95].

1. **Motion detection** is the simplest problem. This registers any detected motion and is often used for security purposes. This group usually uses a single static camera.

2. **Moving object detection and location** represents another set of problems. A camera is usually in a static location and objects are moving in the scene, or the camera moves and objects are static. These problems are considerably more difficult in comparison with the first group. If only moving object detection is required (note the difference between motion detection and moving object detection), the solution can be based on motion-based segmentation methods. Other more complex problems include the detec-

tion of a moving object, the detection of the trajectory of its motion, and the prediction of its future trajectory. Image object-matching techniques are often used to solve this task—typically, direct matching in image data, matching of object features, matching of specific representative object points (corners, etc.) in an image sequence, or representing moving objects as graphs and consequent matching of these graphs. Practical examples of methods from this group include cloud tracking from a sequence of satellite meteorological data, including cloud character and motion prediction, motion analysis for autonomous road vehicles, automatic satellite location by detecting specific points of interest on the Earth's surface, city traffic analysis, and many military applications. The most complex methods of this group work even if both camera and objects are moving. Section 16.4 gives an example of pedestrian tracking that illustrates this area.

3. The third group is related to the **derivation of 3D object properties** from a set of 2D projections acquired at different time instants of object motion. Three-dimensional object description is covered in Chapter 9, and more information on this topic can be found in [Ullman 79, Tsai and Huang 81, Tsai and Huang 82, Webb and Aggarwal 82, Dreschler and Nagel 82, Huang 83, Costabile et al. 85, Mutch and Thomson 85, Jain et al. 87, WVM 89, Weng et al. 89, Adiv 89, Murray and Buxton 90, Zheng and Tsuji 90, Heel 90]. An excellent survey of motion-based recognition approaches is given in [Cedras and Shah 95], and additional citations can be found among references to Chapter 9.

Even though motion analysis is often called **dynamic image analysis**, it is frequently based on a small number of consecutive images, sometimes just two or three in a sequence. This case is similar to an analysis of static images, and the motion is actually analyzed at a higher level, looking for **correspondence** between pairs of points of interest in sequential images. This is the main rationale for the extensive application of matching in motion analysis. A two-dimensional representation of a (generally) three-dimensional motion is called a **motion field**, in which each point is assigned a **velocity vector** corresponding to the motion direction, velocity, and distance from an observer at an appropriate image location.

A different approach analyzes motion from an **optical flow** computation (Section 15.2), where a very small time distance between consecutive images is required, and no significant change occurs between two consecutive images. Optical flow computation results in motion direction and motion velocity determination at (possibly all) image points. The immediate aim of optical flow-based image analysis is to determine a motion field. As will be discussed later, optical flow does not always correspond to the true motion field because illumination changes are reflected in the optical flow. Object motion parameters can be derived from computed optical flow vectors. In reality, estimates of optical flow or point correspondence are noisy, but, unfortunately, three-dimensional interpretation of motion is ill-conditioned and requires high precision of optical flow or point correspondence. To overcome these difficulties, approaches that are not based on optical flow or point correspondence have begun to appear, since if the intermediate step (optical flow, point correspondence) does not have to be computed, possible errors can be avoided. Estimates of general motion of multiple moving objects in an image sequence based on gray-level and image gradient without using any higher-level information such as corners or borders is introduced in [Wu and Kittler 90]. Motion field construction using steerable filters also falls into this category [Freeman and

Adelson 91, Huang and Chen 95].

Motion field or **velocity field** computations represent a compromise technique; information similar to the optical flow is determined, but it is based on images acquired at intervals that are not short enough to ensure small changes due to motion. The velocity field can also be acquired if the number of images in a sequence is small.

Motion evaluation may or may not depend on object detection. An example of object-independent analysis is optical flow computation, whereas velocity field computation or differential methods search for points of interest or points of motion and represent object-dependent analysis. Object-dependent methods are usually based on searching for a correspondence between points of interest or between regions. A recent approach to motion analysis uses active contour models called snakes, which were discussed in Section 8.2. In motion analysis, the initial estimate necessary to start the snake energy minimization process is obtained from the detected position of the contour in the previous frame. For details, see Section 8.2.

If motion analysis is based on detection of moving objects or object feature points, the following object motion assumptions can help to localize moving objects (Figure 15.1).

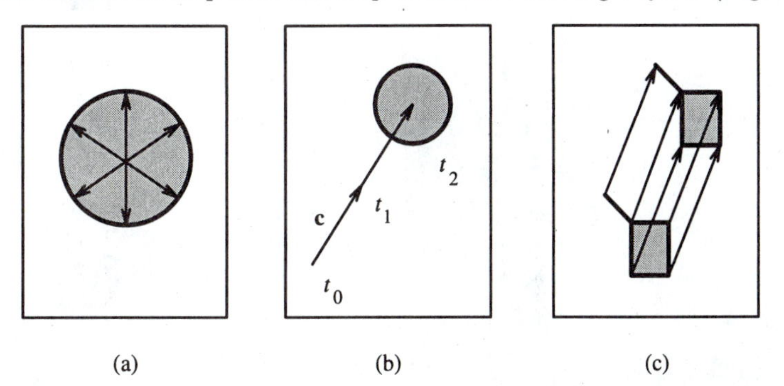

Figure 15.1: *Object motion assumptions: (a) maximum velocity (the shaded circle represents the area of possible object location); (b) small acceleration (the shaded circle represents the area of possible object location at time t_2); (c) common motion and mutual correspondence (rigid objects).*

- **Maximum velocity**: Assume that a moving object is scanned at time intervals of dt. A possible position of a specific object point in an image is inside a circle with its center at the object point position in the previous frame and its radius $c_{\max}\, dt$, where $c_{\max}$ is the assumed maximum velocity of the moving object.
- **Small acceleration**: The change of velocity in time dt is bounded by some constant.
- **Common motion** (similarity in motion): All the object points move in a similar way.
- **Mutual correspondence**: Rigid objects exhibit stable pattern points. Each point of an object corresponds to exactly one point in the next image in sequence and vice versa, although there are exceptions due to occlusion and object rotation.

As noted earlier, there are two major approaches to extracting two-dimensional motion from image sequences: optical flow and motion correspondence. We will introduce both of them, and additionally a simpler analysis based on Kalman filtering.

15.1 Differential motion analysis methods

Simple subtraction of images acquired at different instants in time makes motion detection possible, assuming a stationary camera position and constant illumination. A **difference**

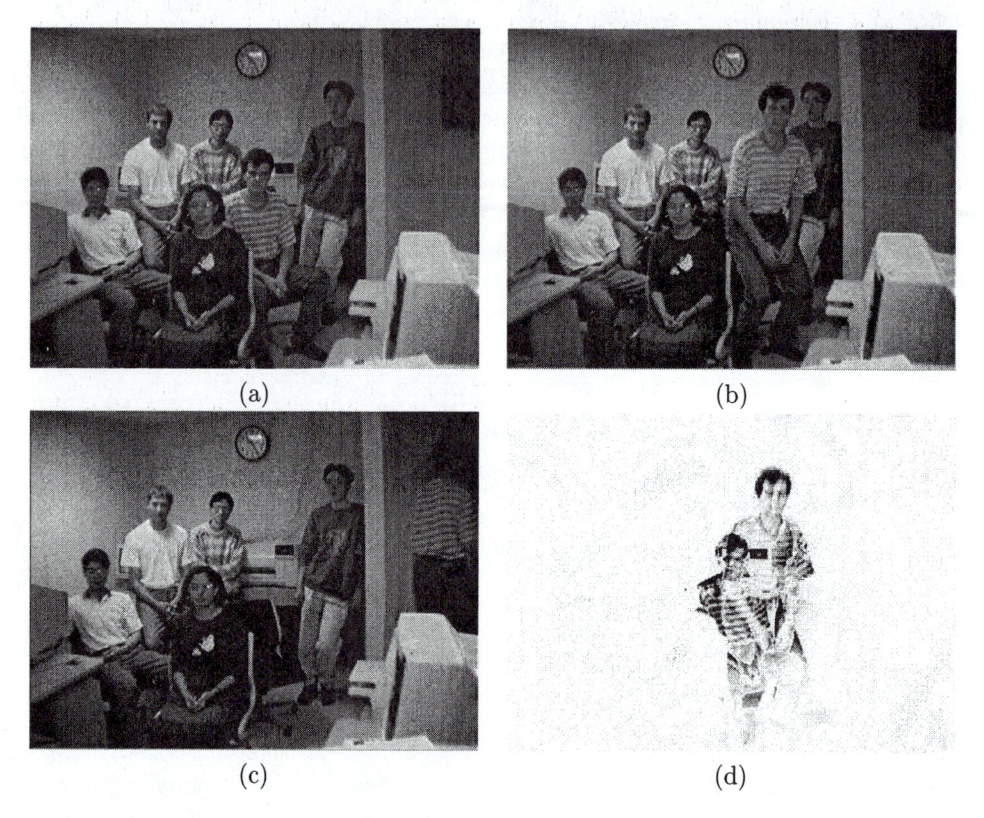

Figure 15.2: *Motion detection: (a) first frame of the image sequence; (b) frame 2 of the sequence; (c) last frame (frame 5); (d) differential motion image constructed from image frames 1 and 2 (inverted to improve visualization).*

image $d(i,j)$ is a binary image where non-zero values represent image areas with motion, that is, areas where there was a substantial difference between gray-levels in consecutive images f_1 and f_2:

$$\begin{aligned} d(i,j) &= 0 \quad \text{if } |f_1(i,j) - f_2(i,j)| \le \varepsilon \\ &= 1 \quad \text{otherwise} \end{aligned} \tag{15.1}$$

where ε is a small positive number. Figure 15.2 shows an example of motion detection using a difference image. The difference image can be based on more complex image features such as average gray-level in some neighborhood, local texture features, etc. It is clear that the motion of any object distinct from its background can be detected (considering motion detection represents motion registration only).

Let f_1 and f_2 be two consecutive images separated by a time interval. An element $d(i,j)$ of the difference image between images f_1 and f_2 may have a value one due to any one of the following reasons (Figure 15.2):

1. $f_1(i,j)$ is a pixel on a moving object
   $f_2(i,j)$ is a pixel on the static background
   (or vice versa)

2. $f_1(i,j)$ is a pixel on a moving object
   $f_2(i,j)$ is a pixel on another moving object

3. $f_1(i,j)$ is a pixel on a moving object
   $f_2(i,j)$ is a pixel on a different part of the same moving object

4. Noise, inaccuracies of stationary camera positioning, etc.

The system errors mentioned in the last item must be suppressed. The simplest solution is not to consider any regions of the difference image that are smaller than a specified threshold, although this may prevent slow motion and small object motions being detected. Further, results of this approach are highly dependent on an object–background contrast. On the other hand, we can be sure that all the resulting regions in the difference images result from motion.

Trajectories detected using differential image motion analysis may not reveal the direction of the motion. If direction is needed, construction of a **cumulative difference image** can solve this problem. Cumulative difference images contain information about motion direction and other time-related motion properties, and about slow motion and small object motion as well. The cumulative difference image d_{cum} is constructed from a sequence of n images, with the first image (f_1) being considered a reference image. Values of the cumulative difference image reflect how often (and by how much) the image gray-level was different from the gray-level of the reference image (if we do not include weight coefficients a_k):

$$d_{\text{cum}}(i,j) = \sum_{k=1}^{n} a_k |f_1(i,j) - f_k(i,j)| \tag{15.2}$$

a_k gives the **significance** of images in the sequence of n images; more recent images may be given greater weights to reflect the importance of current motion and to specify current object location. Figure 15.3 shows the cumulative difference image determined from the sequence of five image frames depicting motion analyzed in Figure 15.2.

Suppose that an image of a static scene is available, and only stationary objects are present in the scene. If this image is used for reference, the difference image suppresses all motionless areas, and any motion in the scene can be detected as areas corresponding to the actual positions of the moving objects in the scene. Motion analysis can then be based on a sequence of difference images.

A problem with this approach may be the impossibility of getting an image of a static reference scene if the motion never ends; then a learning stage must construct the reference image. The most straightforward method is to superimpose moving image objects on non-moving image backgrounds from other images taken in a different phase of the motion.

Figure 15.3: *Cumulative difference image determined from a sequence of five image frames depicting motion analyzed in Figure 15.2 (inverted for improved visualization).*

Which image parts should be superimposed can be judged from difference images, or the reference image can be constructed interactively (which can be allowed in the learning stage). Section 16.4 describes approaches to this problem.

Subsequent analysis usually determines motion trajectories; often only the center of gravity trajectory is needed. The task may be considerably simplified if objects can be segmented out of the first image of the sequence. A practical problem is the prediction of the motion trajectory if the object position in several previous images is known. There are many methods [Jain 81, Jain 84, Jain et al. 95] that find other motion parameters from difference images—whether the object is approaching or receding, which object overlaps which, etc. Note that difference motion analysis methods give good examples of motion analysis principles and present a good introduction to the problem; unfortunately, the difference images do not carry enough information to work reliably in reality. Some problems are common for most motion field detection approaches—consider just a simple example of a rectangular object moving parallel to the object boundary; differential motion analysis can only detect the motion of two sides of the rectangle (see Figures 15.4a,b). Similarly, an aperture problem may cause ambiguity of contained motion information—in the situation shown in Figure 15.4c, only part of an object boundary is visible and it is impossible to determine the motion completely. The arrows indicate three possibilities of motion, all yielding the same final position of the object boundary in the image. Differential motion analysis is often used in digital subtraction angiography, where vessel motion is estimated [Rong et al. 89, Abdel-Malek et al. 90].

While the difference image carries information about presence of motion, characteristics of motion derived from it are not very reliable. The motion parameter estimation robustness can be improved if intensity characteristics of regions or groups of pixels in two image frames are compared. A conceptually straightforward approach to **robust motion detection** is to compare corresponding areas of the image. Such corresponding **superpixels** are usually formed by non-overlapping rectangular regions, the size of which can be derived from the camera aspect ratio. Then, the superpixels may be matched in the compared frames using correlation or likelihood approaches [Jain et al. 95].

Detecting **moving edges** helps overcome several limitations of differential motion analysis methods. By combining the spatial and temporal image gradients, differential analysis can

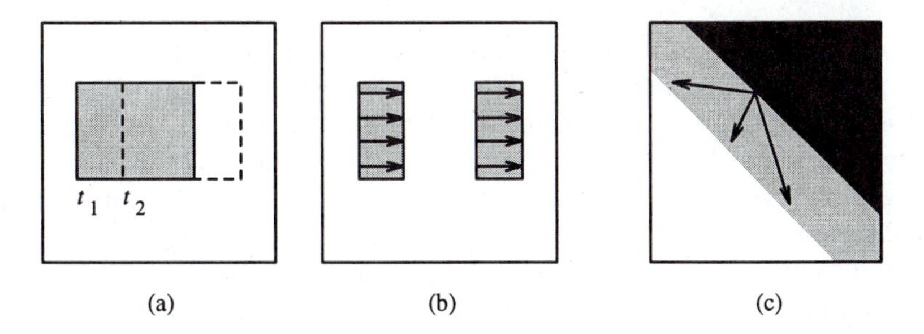

Figure 15.4: *Problems of motion field construction: (a) object position at times t_1 and t_2; (b) motion field; (c) aperture problem—ambiguous motion.*

be used reliably for detection of slow-moving edges as well as of weak edges that move with higher speed. Moving edges can be determined by logical AND operations of the spatial and temporal image edges [Jain et al. 79]. The spatial edges can be identified by virtually any edge detector from the variety given in Section 4.3.2, the temporal gradient can be approximated using the difference image, and the logical AND can be implemented through multiplication. Then, the moving edge image $d_{\text{med}}(i,j)$ can be determined as

$$d_{\text{med}}(i,j) = S(i,j)D(i,j) \tag{15.3}$$

where $S(i,j)$ represents edge magnitudes determined in one of the two analyzed image frames and $D(i,j)$ is the absolute difference image. An example of a moving-edge image determined from the first and second frames of an image sequence (Figure 15.2) is given in Figure 15.5.

Figure 15.5: *Moving-edge image determined from the first and second frames of the image sequence analyzed in Figure 15.2 (inverted for improved visualization).*

15.2 Optical flow

Optical flow reflects the image changes due to motion during a time interval dt, and the optical flow field is the velocity field that represents the three-dimensional motion of object points across a two-dimensional image [Kearney and Thompson 88]. Optical flow is an abstraction typical of the kind that computational methods are trying to achieve. Therefore, it should represent only those motion-related intensity changes in the image that are required in further processing, and all other image changes reflected in the optical flow should be considered errors of flow detection. For example, optical flow should not be sensitive to illumination changes and motion of unimportant objects (e.g., shadows). However, a non-zero optical flow is detected if a fixed sphere is illuminated by a moving source, and a smooth sphere rotating under constant illumination provides no optical flow despite the rotational motion and the true non-zero motion field [Horn 86]. Of course, the aim is to determine an optical flow that corresponds closely with the true motion field. Optical flow computation is a necessary precondition of subsequent higher-level processing that can solve motion-related problems if a camera is stationary or moving; it provides tools to determine parameters of motion, relative distances of objects in the image, etc. A simulated example of two consecutive images and a corresponding optical flow image are shown in Figure 15.6.

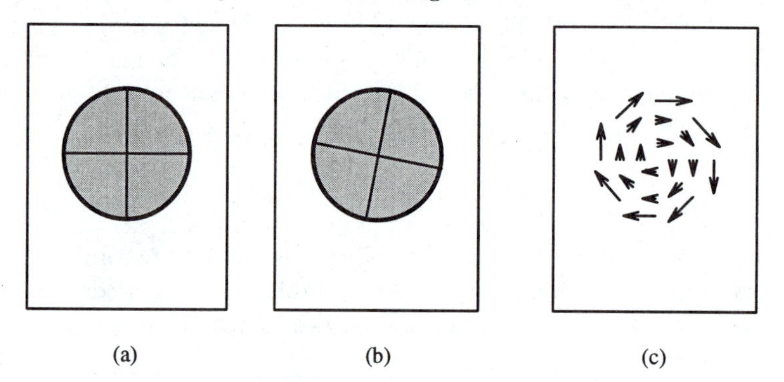

Figure 15.6: *Optical flow: (a) time t_1; (b) time t_2; (c) optical flow.*

15.2.1 Optical flow computation

Optical flow computation is based on two assumptions:

1. The observed brightness of any object point is constant over time.
2. Nearby points in the image plane move in a similar manner (the **velocity smoothness** constraint).

Suppose we have a continuous image; $f(x, y, t)$ refers to the gray-level of (x, y) at time t. Representing a dynamic image as a function of position and time permits it to be expressed as a Taylor series:

$$f(x + dx, y + dy, t + dt) = f(x, y, t) + f_x dx + f_y dy + f_t dt + O(\partial^2) \quad (15.4)$$

where f_x, f_y, f_z denote the partial derivatives of f. We can assume that the immediate neighborhood of (x, y) is translated some small distance (dx, dy) during the interval dt; that is, we can find dx, dy, dt such that

$$f(x + dx, y + dy, t + dt) = f(x, y, t) \quad (15.5)$$

If dx, dy, dt are very small, the higher-order terms in equation (15.4) vanish and

$$-f_t = f_x \frac{dx}{dt} + f_y \frac{dy}{dt} \tag{15.6}$$

The goal is to compute the velocity

$$\mathbf{c} = (\frac{dx}{dt}, \frac{dy}{dt}) = (u, v)$$

f_x, f_y, f_t can be computed, or at least approximated, from $f(x, y, t)$. The motion velocity can then be estimated as

$$-f_t = f_x u + f_y v = \text{grad}(f)\mathbf{c} \tag{15.7}$$

where grad(f) is a two-dimensional image gradient. It can be seen from equation (15.7) that the gray-level difference f_t at the same location of the image at times t and $t+dt$ is a product of spatial gray-level difference and velocity in this location according to the observer.

Equation (15.7) does not specify the velocity vector completely; rather, it only provides the component in the direction of the brightest gradient (see Figure 15.4c). To solve the problem completely, a smoothness constraint is introduced; that is, the velocity vector field changes slowly in a given neighborhood. Full details of this approach may be found in [Horn and Schunk 81], but the approach reduces to minimizing the squared error quantity

$$E^2(x, y) = (f_x u + f_y v + f_t)^2 + \lambda(u_x^2 + u_y^2 + v_x^2 + v_y^2) \tag{15.8}$$

where $u_x^2, u_y^2, v_x^2, v_y^2$ denote partial derivatives squared as error terms. The first term represents a solution to equation (15.7), the second term is the smoothness criterion, and λ is a Lagrange multiplier. Using standard techniques [Horn and Schunk 81], this reduces to solving the differential equations

$$(\lambda^2 + f_x^2)u + f_x f_y v = \lambda^2 \overline{u} - f_x f_t \tag{15.9}$$
$$f_x f_y u + (\lambda^2 + f_y^2)v = \lambda^2 \overline{v} - f_y f_t$$

where $\overline{u}, \overline{v}$ are mean values of the velocity in directions x and y in some neighborhood of (x, y). It can be shown that a solution to these equations is

$$u = \overline{u} - f_x \frac{P}{D} \tag{15.10}$$

$$v = \overline{v} - f_y \frac{P}{D} \tag{15.11}$$

where

$$P = f_x \overline{u} + f_y \overline{v}, \qquad D = \lambda^2 + f_x^2 + f_y^2 \tag{15.12}$$

Determination of the optical flow is then based on a Gauss-Seidel iteration method using pairs of (consecutive) dynamic images [Horn 86].

Algorithm 15.1: Relaxation computation of optical flow from dynamic image pairs

1. Initialize velocity vectors $\mathbf{c}(i, j) = 0$ for all (i, j).

2. Let k denote the number of iterations. Compute values u^k, v^k for all pixels (i,j)

$$\begin{aligned} u^k(i,j) &= \overline{u}^{k-1}(i,j) - f_x(i,j)\frac{P(i,j)}{D(i,j)} \\ v^k(i,j) &= \overline{v}^{k-1}(i,j) - f_y(i,j)\frac{P(i,j)}{D(i,j)} \end{aligned} \tag{15.13}$$

The partial derivatives f_x, f_y, f_t can be estimated from the pair of consecutive images.

3. Stop if

$$\sum_i \sum_j E^2(i,j) < \varepsilon$$

where ε is the maximum permitted error; return to step 2 otherwise.

If more than two images are to be processed, computational efficiency may be increased by using the results of one iteration to initialize the current image pair in sequence.

Algorithm 15.2: Optical flow computation from an image sequence

1. Evaluate starting values of the optical flow $\mathbf{c}(i,j)$ for all points (i,j).

2. Let m be the sequence number of the currently processed image. For all pixels of the next image, evaluate

$$\begin{aligned} u^{m+1}(i,j) &= \overline{u}^{m}(i,j) - f_x(i,j)\frac{P(i,j)}{D(i,j)} \\ v^{m+1}(i,j) &= \overline{v}^{m}(i,j) - f_y(i,j)\frac{P(i,j)}{D(i,j)} \end{aligned} \tag{15.14}$$

3. Repeat step 2 to process all images in the sequence.

Both these algorithms are naturally parallel. The iterations may be very slow, with a computational complexity $\mathcal{O}(n^p)$, where p is the order of the partial differential equation set (15.9). Experimentally, it is found that thousands of iterations are needed until convergence if a second-order smoothness criterion is applied [Glazer 84]. On the other hand, the first 10–20 iterations usually leave an error smaller than the required accuracy, and the rest of the iterative process is then very gradual.

If the differences dx, dy, dt are very small, all the higher-order terms vanish in the continuous derivative of equation (15.4). Unfortunately, in reality this is often not the case if subsequent images are not taken frequently enough. As a result, the higher-order terms do not vanish and an estimation error results if they are neglected. To decrease this error, the second-order terms may be considered in the Taylor series, and the problem becomes a minimization of an integral over a local neighborhood N [Nagel 83, Nagel 86, Nagel 87]:

$$\int\int_N \Big(f(x,y,t) - f(x_0,y_0,t_0) - f_x[x-u] - f_y[y-v] - \tfrac{1}{2} f_{xx}[x-u]^2 - f_{xy}[x-u][y-v] - \tfrac{1}{2} f_{yy}[y-v]^2 \Big)^2 \; dx \; dy \tag{15.15}$$

This minimization is rather complex and may be simplified for image points that correspond to corners (Section 4.3.8). Let the co-ordinate system be aligned with the main curvature direction at (x_0, y_0); then $f_{xy} = 0$ and the only non-zero second-order derivatives are f_{xx} and f_{yy}. However, at least one of them must cross zero at (x_0, y_0) to get a maximum gradient: If, say, $f_{xx} = 0$, then $f_x \to \max$ and $f_y = 0$. With these assumptions, equation (15.15) simplifies, and the following formula is minimized [Nagel 83, Vega-Riveros and Jabbour 89]:

$$\sum_{x,y \in N} \left[f(x,y,t) - f(x_0,y_0,t_0) - f_x(x-u) - \frac{1}{2} f_{yy}(y-v)^2 \right]^2 \tag{15.16}$$

A conventional minimization approach of differentiating equation (15.16) with respect to u and v and equating to zero results in two equations in the two velocity components u, v.

15.2.2 Global and local optical flow estimation

Optical flow computation will be in error to the extent that the constant brightness and velocity smoothness assumptions are violated. Unfortunately, in real imagery, their violation is quite common. Typically, the optical flow changes dramatically in highly textured regions, around moving boundaries, depth discontinuities, etc. [Kearney and Thompson 88]. A significant advantage of global relaxation methods of optical flow computation is to find the smoothest velocity field consistent with the image data; as discussed in Section 8.5, an important property of relaxation methods is their ability to propagate local constraints globally. As a result, not only constraint information but also all optical flow estimation errors propagate across the solution. Therefore, even a small number of problem areas in the optical flow field may cause widespread errors and poor optical flow estimates.

Since global error propagation is the biggest problem of the global optical flow computation scheme, local optical flow estimation appears a natural solution to the difficulties. The local estimate is based on the same brightness and smoothness assumptions, and the idea is to divide the image into small regions where the assumptions hold. This solves the error propagation problem but another problem arises—in regions where the spatial gradients change slowly, the optical flow estimation becomes ill-conditioned because of lack of motion information, and it cannot be detected correctly. If a global method is applied to the same region, the information from neighboring image parts propagates and represents a basis for optical flow computation even if the local information was not sufficient by itself. The conclusion of this comparison is that global sharing of information is beneficial in constraint sharing and detrimental with respect to error propagation [Kearney and Thompson 88].

One way to cope with the smoothness violation problem is to detect regions in which the smoothness constraints hold. Two heuristics for identifying neighboring constraint equations that differ substantially in their flow value were introduced in [Horn and Schunk 81]. The main problem is in selecting a threshold to decide which flow value difference should be considered substantial—if the threshold is set too low, many points are considered positioned along flow discontinuities, while if the threshold is too high, some points violating smoothness remain part of the computational net. The boundary between smooth subnets is not closed; paths between them remain, and the error propagation problem is not solved.

An approach of continuous adaptation to errors was introduced in [Kearney et al. 87]. As with the basic global relaxation method, optical flow is determined iteratively by combining the local average flow vector with the gradient constraint equation. However, a confidence is assigned to each flow vector based on heuristic judgments of correctness, and the local average flow is computed as a weighted average by confidence. Thus, the propagation of error-free estimates is inhibited. Details of confidence estimation, smoothness violation detection, combining partial estimates, implementation details, and discussion of results are given in [Kearney et al. 87, Kearney and Thompson 88].

Performance of the method is illustrated in Figures 15.7 and 15.8. The first image pair, shown in Figures 15.7a,b, contains a collection of toys, and the second pair of images (Figures 15.8a,b) simulates a view from an aircraft flying over a city. Optical flow resulting from a simple local optimization is shown in Figures 15.7c and 15.8c, and results of the global method of continuous adaptation to errors are given in Figures 15.7d and 15.8d. The optical flow improvement achieved by the latter method is clearly visible.

15.2.3 Optical flow computation approaches

Optical flow—its computation and use, and the many variations thereon—has generated an enormous literature, and continues so to do.

Optical flow constraints are discussed in detail in papers [Snyder 89, Girosi et al. 89, Verri et al. 89, Arnspang 93]. A recent survey can be found in [Barron et al. 92a], computational aspects are considered in [Lee et al. 89]. Several approaches to optical flow computation have been investigated which are not based on velocity smoothness; a spatio-temporal gradient method is described in [Paquin and Dubois 83], and motion of planar surfaces is considered in [Waxman and Wohn 85, Waxman et al. 87]. Smoothing the flow only along one-dimensional curves corresponding to zero-crossings was introduced in [Hildreth 83], and selective application of smoothness requirements also can be found in [Davis et al. 83, Nagel and Enkelmann 84, Fogel 88, Fogel 89]. A robust algorithm for computing image flow from object boundary information is presented in [Schunck 89]. Minimization of a penalty function as a weighted sum of the two constraining terms and the smoothness term (using a divergence-free and an incompressibility constraint) is proposed in [Song and Leahy 91], where results were tested on cine CT image sequences of a beating heart. Another approach is to compute the flow at certain locations in the image at seed points [Brady 87]; this approach has a direct link to the parallel computation of optical flow. A detailed description of local optical flow computation necessary for fully parallel implementation is discussed in [Burt et al. 83, Gong 87, Anandan 87, Gong 88, Enkelmann 88, Gong 89].

Many other approaches to computation of constrained optical flow equations can be found. A linear combination of the optical flow estimate in adjacent windows is used in [Cafforio and Rocca 83]. Optical flow computation using spline representation of the displacement field that removes the need for overlapping of adjacent correlation windows is given [Szeliski

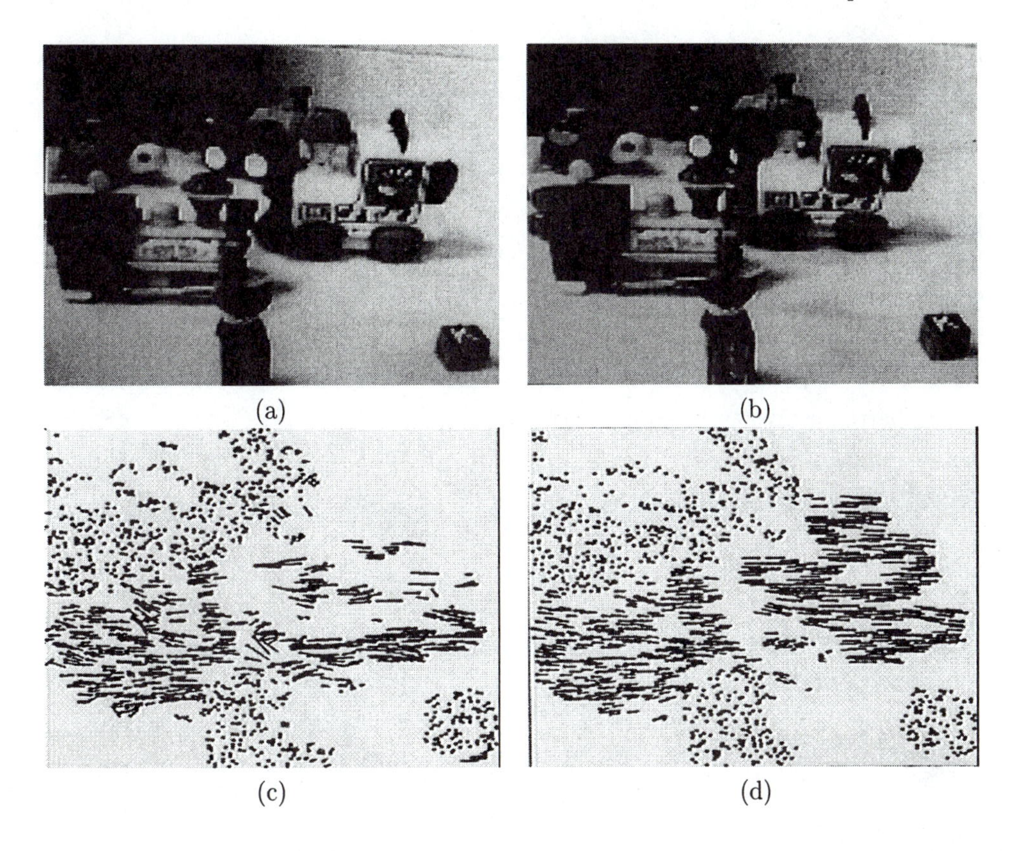

Figure 15.7: *Moving trains image sequence: (a) first frame; (b) last frame; (c) optical flow detection—local optimization method; (d) optical flow detection—method of continuous adaptation to errors. (Only 20% of vectors with moderate and high confidence shown.) Courtesy J. Kearney, The University of Iowa.*

and Coughlan 94]. A vector field smoothness constrained only in the direction perpendicular to the gray value gradient is considered in [Nagel and Enkelmann 84]. A cluster matching approach to optical flow detection is presented in [Kottke and Sun 94]. A multi-modal approach to optical flow estimation that uses several complementary constraints and relies on Bayesian estimation associated with global statistical models—specifically Markov random fields—is introduced in [Heitz and Bouthemy 93]; the constraints overcome the problem of discontinuous optical flow that is otherwise present in proximity to borders of moving objects. Another approach to avoiding the loss of information associated with the discontinuities of the velocity field can be found in [Nesi 93]. A spatio-temporal method that does not smooth the optical flow field across the edges and is based on computation of evolving curves is proposed in [Kumar et al. 96]. Frequency-based approaches to optical flow detection also exist. Some of them use different orientation Gabor or Gable filters [Watson and Ahumada 85, Heeger 87], others employ steerable filters that may have arbitrary orientation synthesized as a linear combination of basis filters [Freeman and Adelson 91, Huang and Chen 95]. Optical flow computation in pyramid image representation is described in [Glazer 84]; iterations

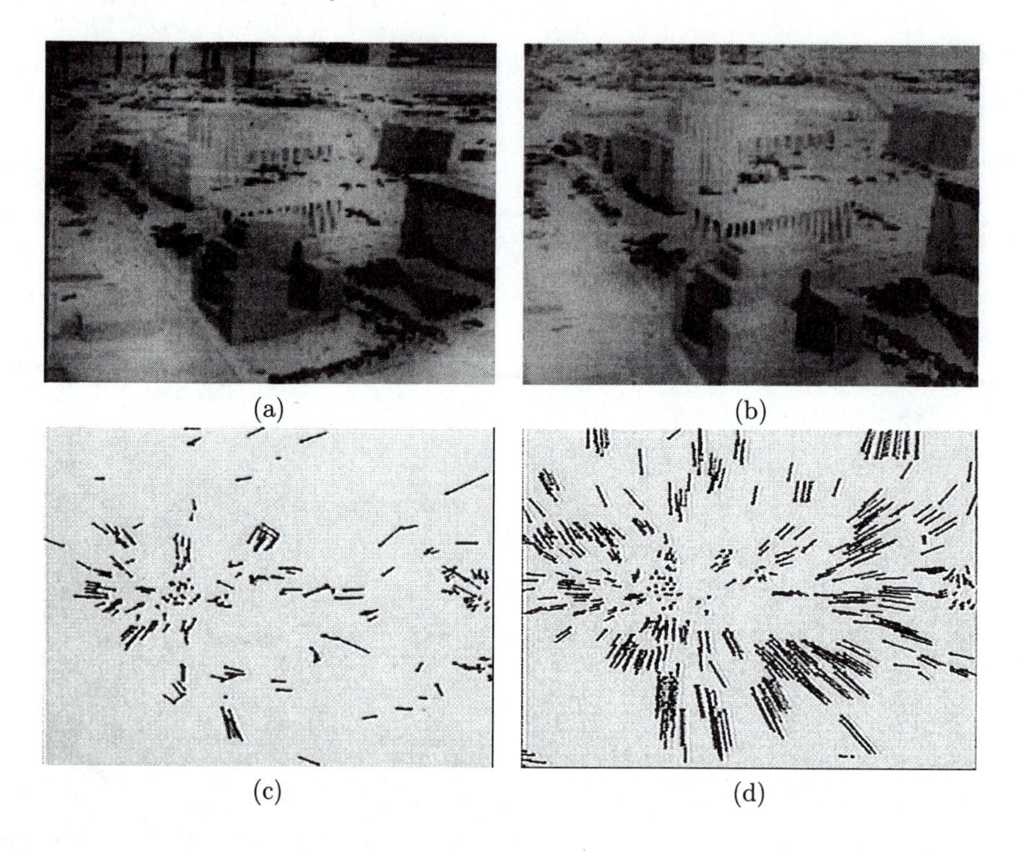

Figure 15.8: *Simulated flyover image sequence: (a) first frame; (b) last frame; (c) optical flow detection—local optimization method; (d) optical flow detection—method of continuous adaptation to errors. (Again, only 20% of vectors with moderate and high confidence shown.) Courtesy J. Kearney, The University of Iowa.*

converge significantly faster in lower-resolution pyramid levels, and the computed values can be used as starting values in higher-resolution pyramid levels. This hierarchical approach is claimed to give better results than conventional single-resolution methods because the optical flow computation assumptions can be satisfied more easily, especially if the optical flow is relatively large and the accompanying gray-level changes are small. Several other approaches are mentioned in [Vega-Riveros and Jabbour 89].

Comparison of the performance of many optical flow techniques is given in [Barron et al. 92a]—local differential approaches [Lucas and Kanade 81, Fleet and Jepson 90] were found to be most accurate and robust. Techniques using global smoothness constraints were found to produce visually attractive motion fields, but give an accuracy suitable only for qualitative use, insufficient for ego-motion computation and 3D structure from motion detection; see [Barron et al. 92a, Barron et al. 92b] for details.

15.2.4 Optical flow in motion analysis

Optical flow gives a description of motion and can be a valuable contribution to image interpretation even if no quantitative parameters are obtained from motion analysis. Optical flow can be used to study a large variety of motions—moving observer and static objects, static observer and moving objects, or both moving. Optical flow analysis does not result in motion trajectories as described in Section 15.1; instead, more general motion properties are detected that can significantly increase the reliability of complex dynamic image analysis [Thompson et al. 85, Sandini and Tistarelli 86, Kearney et al. 87, Aggarwal and Martin 88].

Motion, as it appears in dynamic images, is usually some combination of four basic elements:

- Translation at constant distance from the observer
- Translation in depth relative to the observer
- Rotation at constant distance about the view axis
- Rotation of a planar object perpendicular to the view axis

Optical-flow based motion analysis can recognize these basic elements by applying a few relatively simple operators to the flow [Thompson et al. 84, Mutch and Thompson 84]. Motion form recognition is based on the following facts (Figure 15.9):

- Translation at constant distance is represented as a set of parallel motion vectors.
- Translation in depth forms a set of vectors having a common focus of expansion.
- Rotation at constant distance results in a set of concentric motion vectors.
- Rotation perpendicular to the view axis forms one or more sets of vectors starting from straight line segments.

Exact determination of rotation axes and translation trajectories can be computed, but with a significant increase in difficulty of analysis.

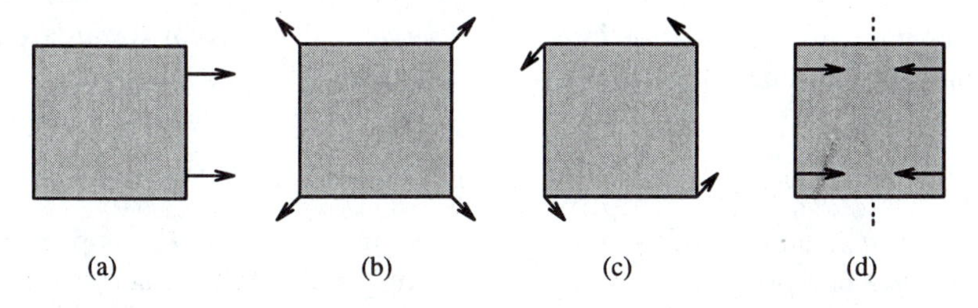

Figure 15.9: *Motion form recognition: (a) translation at constant distance; (b) translation in depth; (c) rotation at constant distance; (d) planar object rotation perpendicular to the view axis.*

Consider translational motion: If the translation is not at constant depth, then optical flow vectors are not parallel, and their directions have a single focus of expansion (FOE). If the translation is at constant depth, the FOE is at infinity. If several independently moving objects are present in the image, each motion has its own FOE—this is illustrated in Figure 15.10, where an observer moves in a car towards other approaching cars on the road.

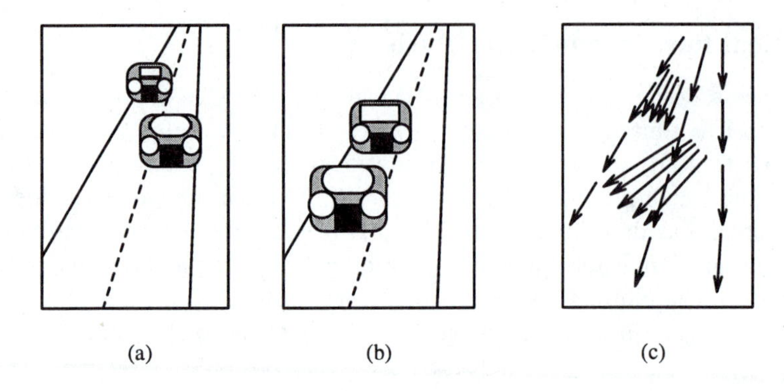

Figure 15.10: *Focus of expansion: (a) time t_1; (b) time t_2; (c) optical flow.*

Mutual velocity

The mutual velocity $\mathbf{c}$ of an observer and an object represented by an image point can be found in an optical flow representation. Let the mutual velocities in directions x, y, z be $c_x = u, c_y = v, c_z = w$, where z gives information about the depth (note that $z > 0$ for points in front of the image plane). To distinguish image co-ordinates from real-world co-ordinates in the following, let the image co-ordinates be x', y'. From perspective considerations, if (x_0, y_0, z_0) is the position of some point at time $t_0 = 0$, then the position of the same point at time t can, assuming unit focal distance of the optical system and constant velocity, be determined as follows:

$$(x', y') = (\frac{x_0 + ut}{z_0 + wt}, \frac{y_0 + vt}{z_0 + wt}) \tag{15.17}$$

FOE determination

The FOE in a two-dimensional image can be determined from this equation. Let us assume motion directed towards an observer; as $t \to -\infty$, the motion can be traced back to the originating point at infinite distance from the observer. The motion towards an observer continues along straight lines and the originating point in the image plane is

$$\mathbf{x}'_{\mathrm{FOE}} = (\frac{u}{w}, \frac{v}{w}) \tag{15.18}$$

Note that the same equation can be used for $t \to \infty$ and motion in the opposite direction. Clearly, any change of motion direction results in changes of velocities u, v, w, and the FOE changes its location in the image [Jain 83].

Distance (depth) determination

Because of the presence of a z co-ordinate in equation (15.17), optical flow can be used to determine the current distance of a moving object from the observer's position. The distance information is contained indirectly in equation (15.17). Assuming points of the same rigid object and translational motion, at least one actual distance value must be known to evaluate the distance exactly. Let $D(t)$ be the distance of a point from the FOE, measured in a two-dimensional image, and let $V(t)$ be its velocity dD/dt. The relationship between these quantities and the optical flow parameters is then

$$\frac{D(t)}{V(t)} = \frac{z(t)}{w(t)} \tag{15.19}$$

This formula is a basis for determination of distances between moving objects. Assuming an object moving towards the observer, the ratio z/w specifies the time at which an object moving at a constant velocity w crosses the image plane. Based on the knowledge of the distance of any single point in an image which is moving with a velocity w along the z axis, it is possible to compute the distances of any other point in the image that is moving with the same velocity w

$$z_2(t) = \frac{z_1(t)V_1(t)D_2(t)}{D_1(t)V_2(t)} \tag{15.20}$$

where $z_1(t)$ is the known distance and $z_2(t)$ is the unknown distance. Using the given formulae, relations between real-world co-ordinates x, y and image co-ordinates x', y' can be found related to the observer position and velocity:

$$\begin{aligned} x(t) &= \frac{x'(t)w(t)D(t)}{V(t)} \\ y(t) &= \frac{y'(t)w(t)D(t)}{V(t)} \\ z(t) &= \frac{w(t)D(t)}{V(t)} \end{aligned} \tag{15.21}$$

Note that the above equations cover both the moving objects and the moving camera as long as the motion is along the camera optical axis. Situations in which motion is not realized along the optical axis are treated in [Jain et al. 95].

Collision Prediction

A practical application is analysis of the motion of a robot in the real world, where the optical flow approach is able to detect potential collisions with scene objects. Observer motion—as seen from optical flow representation—aims into the FOE of this motion; co-ordinates of this FOE are $(u/w, v/w)$. The origin of image co-ordinates (the imaging system focal point) proceeds in the direction $\mathbf{s} = (u/w, v/w, 1)$ and follows a path in real-world co-ordinates at each time instant defined as a straight line,

$$(x, y, z) = t\mathbf{s} = t(\frac{u}{w}, \frac{v}{w}, 1) \tag{15.22}$$

where the parameter t represents time. The position of an observer $\mathbf{x}_{\mathrm{obs}}$ when at its closest point of approach to some $\mathbf{x}$ in the real world is then

$$\mathbf{x}_{\mathrm{obs}} = \frac{\mathbf{s}(\mathbf{s} \cdot \mathbf{x})}{\mathbf{s} \cdot \mathbf{s}} \tag{15.23}$$

The smallest distance $d_{\min}$ between a point $\mathbf{x}$ and an observer during observer motion is

$$d_{\min} = \sqrt{(\mathbf{x} \cdot \mathbf{x}) - \frac{(\mathbf{x} \cdot \mathbf{s})^2}{\mathbf{s} \cdot \mathbf{s}}} \tag{15.24}$$

Thus, a circular-shaped observer with radius r will collide with objects if their smallest distance of approach $d_{\min} < r$.

The analysis of motion, computation of FOE, depth, possible collisions, time to collision, etc., are all very practical problems. Interpretation of motion is discussed in [Subbarao 88], and motion analysis and computing range from an optical flow map is described in [Albus and Hong 90]. A comprehensive approach to motion parameter estimation from optical flow together with a comprehensive overview of existing techniques is given in [Hummel and Sundareswaran 93]. A robust method for extracting dense depth maps from a sequence of noisy intensity images is described in [Shahraray and Brown 88], and a method of unique determination of rigid body motion from optical flow and depth is given in [Zhuang et al. 88]. Ego-motion estimation from optical flow fields determined from multiple cameras is presented in [Tsao et al. 97]. Obstacle detection by evaluation of optical flow is presented in [Enkelmann 91]. Edge-based obstacle detection derived from determined size changes is presented in [Ringach and Baram 94]. Time to collision computation from first-order derivatives of image flow is described in [Subbarao 90], where it is shown that higher-order derivatives, which are unreliable and computationally expensive, are not necessary. Computation of FOE does not have to be based on optical flow; the spatial gradient approach and a natural constraint that an object must be in front of the camera to be imaged are used in a direct method of locating FOE in [Negahdaripour and Horn 89, Negahdaripour and Ganesan 92].

15.3 Analysis based on correspondence of interest points

The optical flow analysis method can be applied only if the intervals between image acquisitions are very short. Motion detection based on correspondence of **interest points (feature points)** works for inter-frame time intervals that cannot be considered small enough. Detection of corresponding object points in subsequent images is a fundamental part of this method—if this correspondence is known, velocity fields can easily be constructed (this does not consider the hard problem of constructing a dense velocity field from a sparse-correspondence-point velocity field).

The first step of the method is to find significant points in all images of the sequence—points least similar to their surrounding representing object corners, borders, or any other characteristic features in an image that can be tracked over time. Point detection is followed by a matching procedure, which looks for correspondences between these points. The process results in a sparse velocity field construction.

15.3.1 Detection of interest points

The Moravec operator described in Section 4.3.8 can be used as an interest-point detector which evaluates a point significance from a small neighborhood. Corners play a significant role in detection of interest points; the Kitchen-Rosenfeld and Zuniga–Haralick operators look for object vertices in images [Section 4.3.8, equation (4.75)]. The operators are almost equivalent, even though it is possible to get slightly better results applying the Zuniga–Haralick operator where a located vertex must be positioned at an edge pixel. This is represented by a term

$$\frac{1}{\sqrt{c_2^2 + c_3^2}}$$

in the facet model [Haralick and Watson 81]. This assumption has computationally important consequences: Significant edges in an edge image can be located first and a vertex function then evaluated at significant edge pixels only, a vertex being defined as a significant edge pixel with a vertex measuring function registering above some threshold.

An optimal detector of corners, which are defined as the junction points of two or more straight line edges, is described in [Rangarajan et al. 89]. The approach detects corners of arbitrary angles and performs well even in noisy images. Another definition of a corner as an intersection of two half-edges oriented in two different directions, which are not 180° apart, is introduced in [Mehrotra and Nichani 90]. In addition to the location of corner points, information about the corner angle and orientation is determined.

These methods detect significant image points whose location changes due to motion, and motion analysis works with these points only. To detect points of interest that are connected with the motion, a difference motion analysis method can be applied to two or more images of a sequence.

15.3.2 Correspondence of interest points

Assuming that interest points have been located in all images of a sequence, a correspondence between points in consecutive images is sought [Ullman 79, Shah and Jain 84]. Many approaches may be applied to seek an optimal correspondence, and several possible solutions have been presented earlier (Chapters 7 and 9). The graph matching problem, stereo matching, and 'shape from X' problems treat essentially the same problem.

One method [Barnard 79, Thompson and Barnard 81] is a very good example of the main ideas of this approach: The correspondence search process is iterative and begins with the detection of all potential correspondence pairs in consecutive images. A maximum velocity assumption can be used for potential correspondence detection, which decreases the number of possible correspondences, especially in large images. Each pair of corresponding points is assigned a number representing the probability of correspondence. These probabilities are then iteratively recomputed to get a globally optimum set of pairwise correspondences [the maximum probability of pairs in the whole image, equation (15.30)] using another motion assumption—the common motion principle. The process ends if each point of interest in a previous image corresponds with precisely one point of interest in the following image **and**

- The global probability of correspondences between image point pairs is significantly higher than other potential correspondences
- Or the global probability of correspondences of points is higher than a pre-selected threshold
- Or the global probability of correspondences gives a maximum probability (optimum) of all possible correspondences (note that $n!$ possible correspondences exist for n pairs of interest points).

Let $A_1 = \{\mathbf{x}_m\}$ be the set of all interest points in the first image, and $A_2 = \{\mathbf{y}_n\}$ the interest points of the second image. Let $\mathbf{c}_{mn}$ be a vector connecting points $\mathbf{x}_m$ and $\mathbf{y}_n$ ($\mathbf{c}_{mn}$ is thus a velocity vector; $\mathbf{y}_n = \mathbf{x}_m + \mathbf{c}_{mn}$). Let the probability of correspondence of two points $\mathbf{x}_m$

and $\mathbf{y}_n$ be P_{mn}. Two points $\mathbf{x}_m$ and $\mathbf{y}_n$ can be considered potentially corresponding if their distance satisfies the assumption of maximum velocity,

$$|\mathbf{x}_m - \mathbf{y}_n| \leq c_{\max} \tag{15.25}$$

where $c_{\max}$ is the maximum distance a point may move in the time interval between two consecutive images. Two correspondences of points $\mathbf{x}_m\mathbf{y}_n$ and $\mathbf{x}_k\mathbf{y}_l$ are termed consistent if

$$|\mathbf{c}_{mn} - \mathbf{c}_{kl}| \leq c_{\text{dif}} \tag{15.26}$$

where c_{dif} is a preset constant derived from prior knowledge. Clearly, consistency of corresponding point pairs increases the probability that a correspondence pair is correct. This principle is applied in Algorithm 15.3 [Barnard and Thompson 80].

Algorithm 15.3: Velocity field computation from two consecutive images

1. Determine the sets of interest points A_1 and A_2 in images f_1, f_2, and detect all potential correspondences between point pairs $\mathbf{x}_m \in A_1$ and $\mathbf{y}_n \in A_2$.

2. Construct a data structure in which potential correspondence information of all points $\mathbf{x}_m \in A_1$ with points $\mathbf{y}_n \in A_2$ is stored, as follows:

$$[\mathbf{x}_m, (\mathbf{c}_{m1}, P_{m1}), (\mathbf{c}_{m2}, P_{m2}), \ldots, (V^*, P^*)] \tag{15.27}$$

   P_{mn} is the probability of correspondence of points $\mathbf{x}_m$ and $\mathbf{y}_n$, and V^* and P^* are special symbols indicating that no potential correspondence was found.

3. Initialize the probabilities P^0_{mn} of correspondence based on local similarity—if two points correspond, their neighborhood should correspond as well:

$$P^0_{mn} = \frac{1}{(1 + k w_{mn})} \tag{15.28}$$

   where k is a constant and

$$w_{mn} = \sum_{\Delta\mathbf{x}} [f_1(\mathbf{x}_m + \Delta\mathbf{x}) - f_2(\mathbf{y}_n + \Delta\mathbf{x})]^2 \tag{15.29}$$

   $\Delta\mathbf{x}$ defines a neighborhood for image match testing—a neighborhood consists of all points $(\mathbf{x}+\Delta\mathbf{x})$, where $\Delta\mathbf{x}$ may be positive or negative and usually defines a symmetric neighborhood around $\mathbf{x}$.

4. Iteratively determine the probability of correspondence of a point $\mathbf{x}_m$ with all potential points $\mathbf{y}_n$ as a weighted sum of probabilities of correspondence of all consistent pairs $\mathbf{x}_k\mathbf{y}_l$, where $\mathbf{x}_k$ are neighbors of $\mathbf{x}_m$ and the consistency of $\mathbf{x}_k\mathbf{y}_l$ is evaluated according to $\mathbf{x}_m, \mathbf{y}_n$. A quality q_{mn} of the correspondence pair is

$$q^{(s-1)}_{mn} = \sum_k \sum_l P^{(s-1)}_{kl} \tag{15.30}$$

   where s denotes an iteration step, k refers to all points $\mathbf{x}_k$ that are neighbors of $\mathbf{x}_m$, and l refers to all points $\mathbf{y}_l \in A_2$ that form pairs $\mathbf{x}_k\mathbf{y}_l$ consistent with the pair $\mathbf{x}_m\mathbf{y}_n$.

5. Update the probabilities of correspondence for each point pair $\mathbf{x}_m$, $\mathbf{y}_n$

$$\hat{P}_{mn}^{(s)} = P_{mn}^{(s-1)} \; (a + b q_{mn}^{(s-1)}) \tag{15.31}$$

where a and b are preset constants. Normalize

$$P_{mn}^{(s)} = \frac{\hat{P}_{mn}^{(s)}}{\sum_j \hat{P}_{mj}^{(s)}} \tag{15.32}$$

6. Repeat steps 4 and 5 until the best correspondence $\mathbf{x}_m\mathbf{y}_n$ is found for all points $\mathbf{x}_m \in A_1$
7. Vectors $\mathbf{c}_{ij}$ of the correspondence form a velocity field of the analyzed motion.

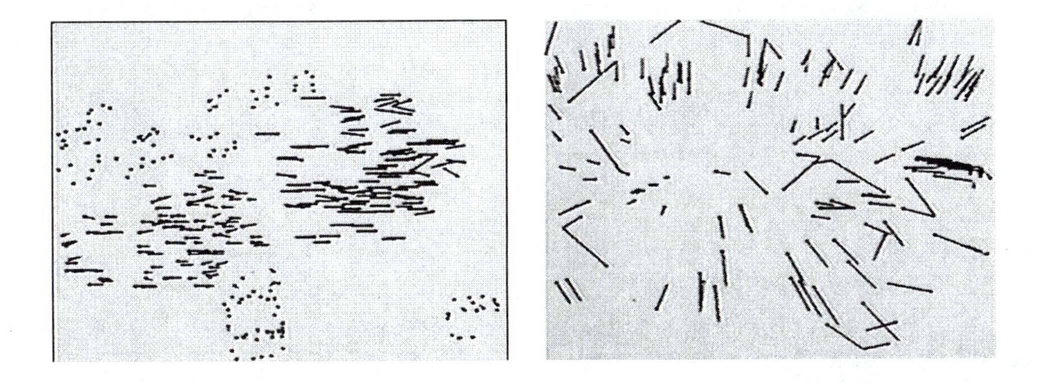

Figure 15.11: *Velocity fields of the train sequence (left) and flyover (right) (original images shown in Figures 15.7a,b and 15.8a,b). Courtesy J. Kearney, The University of Iowa.*

The velocity field resulting from this algorithm applied to the image pairs given in Figures 15.7a,b and 15.8a,b are shown in Figure 15.11. Note that the results are much better for the train sequence; compare the flyover velocity field with the optical flow results given in Figure 15.8d.

Velocity fields can be applied in position prediction tasks as well as optical flow. A good example of interpretation of motion derived from detecting interest points is given in [Scott 88]. Detection of moving objects from a moving camera using point correspondence in two orthographic views is discussed in [Thompson et al. 93]. Fluid motion analysis using particle correspondence and dynamic programming is described in [Shapiro et al. 95]. Two algorithms applicable to motion analysis in long monocular image sequences were introduced in [Hu and Ahuja 93]; one of the two algorithms uses inter-frame correspondence, the other is based on analysis of point trajectories.

Approaches that allow object registration without determination of explicit point correspondences have begun to appear. In [Fua and Leclerc 94], a method using full three-dimensional surface models is presented that may be used together with shape from motion analysis. An accurate and fast method for motion analysis that seeks correspondence of moving objects via a multi-resolution Hough transform and employing robust statistics [Hampel et al. 86] is introduced in [Bober and Kittler 94].

15.3.3 Object tracking

Tracking of object motion in a sequence of frames is a very common application. If there is only one object in the image sequence, the task can often be solved using approaches already described, but if there are many objects moving simultaneously and independently, more complex approaches are needed to incorporate individual object motion-based constraints. In such situations, motion assumptions/constraints described earlier should be examined (maximum velocity, small acceleration, common motion, mutual correspondence, smoothness of motion). Consequently, it is possible to formulate the notion of path coherence which implies that the motion of an object at any point in an image sequence will not change abruptly [Jain et al. 95].

The **path coherence function** Φ represents a measure of agreement between the derived object trajectory and the motion constraints. Path coherence functions should follow the following four principles [Sethi and Jain 87, Jain et al. 95]:

- The function value is always positive.
- The function reflects local absolute angular deviations of the trajectory.
- The function should respond equally to positive and negative velocity changes.
- The function should be normalized $[\Phi(\cdot) \in (0,1]]$.

Let the trajectory T_i of an object i be represented by a sequence of points in the projection plane,

$$T_i = (X_i^1, X_i^2, \dots, X_i^n) \tag{15.33}$$

where X_i^k represents a (three-dimensional) trajectory point in image k of the sequence (see Figure 15.12). Let $\mathbf{x}_i^k$ be the projection image co-ordinates associated with the point X_i^k. Then the trajectory can be expressed in vector form:

$$T_i = (\mathbf{x}_i^1, \mathbf{x}_i^2, \dots, \mathbf{x}_i^n) \tag{15.34}$$

Deviation function

Deviation in the path can be used to measure the path coherence. Let d_i^k be the deviation in the path of the point i in the image k:

$$d_i^k = \Phi(\overline{\mathbf{x}_i^{k-1}\mathbf{x}_i^k}, \overline{\mathbf{x}_i^k\mathbf{x}_i^{k+1}}) \quad \text{or} \quad d_i^k = \Phi(X_i^{k-1}, X_i^k, X_i^{k+1}) \tag{15.35}$$

where $\overline{\mathbf{x}_i^{k-1}\mathbf{x}_i^k}$ represents the motion vector from point X_i^{k-1} to point X_i^k and Φ is the path coherence function. The deviation D_i of the entire trajectory of the object i is then

$$D_i = \sum_{k=2}^{n-1} d_i^k \tag{15.36}$$

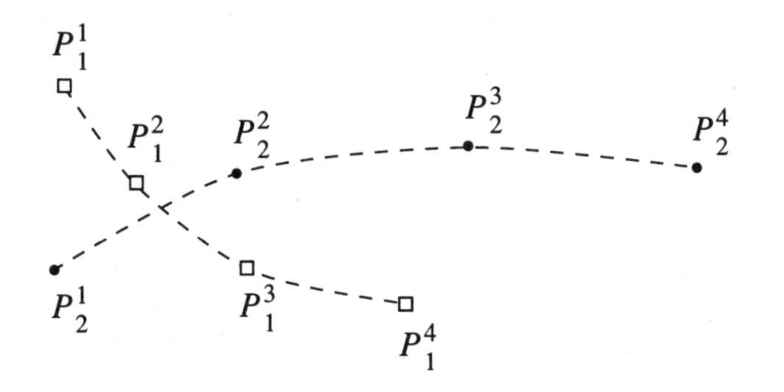

Figure 15.12: *The trajectories of two objects moving simultaneously and independently.*

Similarly, for m trajectories of m moving objects in the image sequence, the overall trajectory deviation D can be determined as

$$D = \sum_{i=1}^{m} D_i \tag{15.37}$$

With the overall trajectory deviation defined in this way, the multiple object trajectory tracking can be solved by minimizing the overall trajectory deviation D.

Path coherence function

So far, only an intuitive definition of the path coherence function Φ has been given. In agreement with motion assumptions, if the image acquisition frequency is high enough, the direction and velocity changes in consecutive images should be smooth. Then the path coherence function can be defined as

$$\begin{aligned}
\Phi(P_i^{k-1}, P_i^k, P_i^{k+1}) &= w_1(1-\cos\theta) + w_2\left(1 - 2\frac{\sqrt{s_k s_{k+1}}}{s_k + s_{k+1}}\right) \\
&= w_1\left(1 - \frac{|\overline{\mathbf{x}_i^{k-1}\mathbf{x}_i^k} \cdot \overline{\mathbf{x}_i^k\mathbf{x}_i^{k+1}}|}{||\overline{\mathbf{x}_i^{k-1}\mathbf{x}_i^k}||\,||\overline{\mathbf{x}_i^k\mathbf{x}_i^{k+1}}||}\right) \\
&\quad + w_2\left(1 - 2\frac{\sqrt{||\overline{\mathbf{x}_i^{k-1}\mathbf{x}_i^k}||\,||\overline{\mathbf{x}_i^k\mathbf{x}_i^{k+1}}||}}{||\overline{\mathbf{x}_i^{k-1}\mathbf{x}_i^k}||\,||\overline{\mathbf{x}_i^k\mathbf{x}_i^{k+1}}||}\right)
\end{aligned} \tag{15.38}$$

where the angle θ and distances s_k, s_{k+1} are given by Figure 15.13. The weights w_1, w_2 reflect the importance of direction coherence and velocity coherence.

Occlusion

When simultaneously tracking several objects with independent motion, object occlusion is almost guaranteed to occur. Consequently, some objects may partially or completely disappear in some image frames which can result in errors in object trajectory. If minimization of the overall trajectory deviation D [equation (15.37)] is performed using the given path coherence function, it is assumed that the same number of objects (object points) is detected

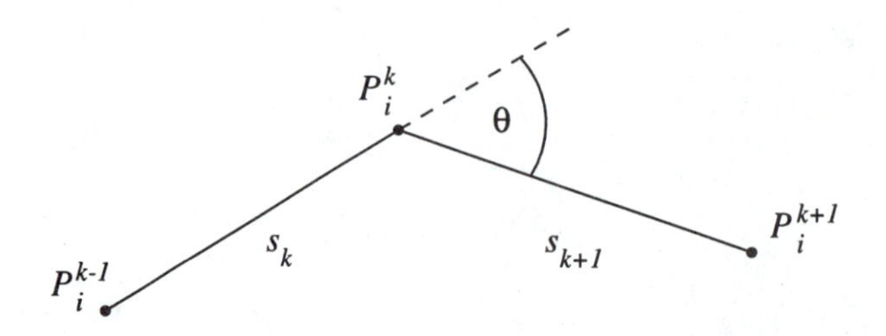

Figure 15.13: *Path coherence function—definition of the angle θ and distances s_k, s_{k+1}.*

in each image of the sequence and that the detected object points consistently represent the same objects (object points) in every image. Clearly, this is not the case if occlusion occurs.

To overcome the occlusion problem, additional local trajectory constraints must be considered and trajectories must be allowed to be incomplete if necessary. Incompleteness may reflect occlusion, appearance or disappearance of an object, or missing object points due to changed object aspect resulting from motion or simply due to poor object detection. Thus, additional motion assumptions that were not reflected in the definition of the path coherence function Φ, e.g., maximum velocity, must be incorporated. An algorithm called the *greedy exchange* was presented in [Sethi and Jain 87] which finds the maximum set of complete or partially complete trajectories and minimizes the sum of local smoothness deviations for all identified trajectories. Local smoothness deviation is constrained not to exceed a preset maximum $\Phi_{\max}$ and the displacement between any two successive trajectory points X_i^k, X_i^{k+1} must be less than a preset threshold $d_{\max}$. To deal efficiently with incomplete trajectories, **phantom points** are introduced and used as substitutes for the missing trajectory points. These hypothetical points allow each potential trajectory to be treated as complete and permit consistent application of the optimization function. Details of the algorithm and example results can be found in [Sethi and Jain 87, Jain et al. 95].

A conceptually similar method that minimizes a proximal uniformity cost function (reflecting the assumption that a small distance is usually traveled in a short time and a smooth trajectory is followed) was presented in [Rangarajan and Shah 91]. A two-stage algorithm was given in [Cheng and Aggarwal 90], in which the first stage performs a forward search that extends trajectories up to the current frame and the second stage is a rule-based backward correcting algorithm that corrects wrong correspondences introduced in the last few frames.

Spatio-temporal approaches to analysis of image sequences with multiple independently moving objects represent another alternative to motion analysis. Several of them based on optical flow computation were mentioned in Section 15.2.3. A minimum description length (MDL) approach to motion analysis of long image sequences was introduced in [Gu et al. 96]. The method first constructs a family of motion models, each model corresponding to some meaningful motion type—translation, rotation, their combination, etc. Using the motion description length, the principle of progressive perception from extension in time, and optimal modeling of a limited period of observations, the objects in the image sequences are segmented to determine when objects change their type of motion or when a new part of an object appears. If the motion information in two consecutive frames is ambiguous, it is resolved by minimizing the motion description length in a long image sequence. Examples and applications for stationary and moving observers are given in [Gu et al. 96].

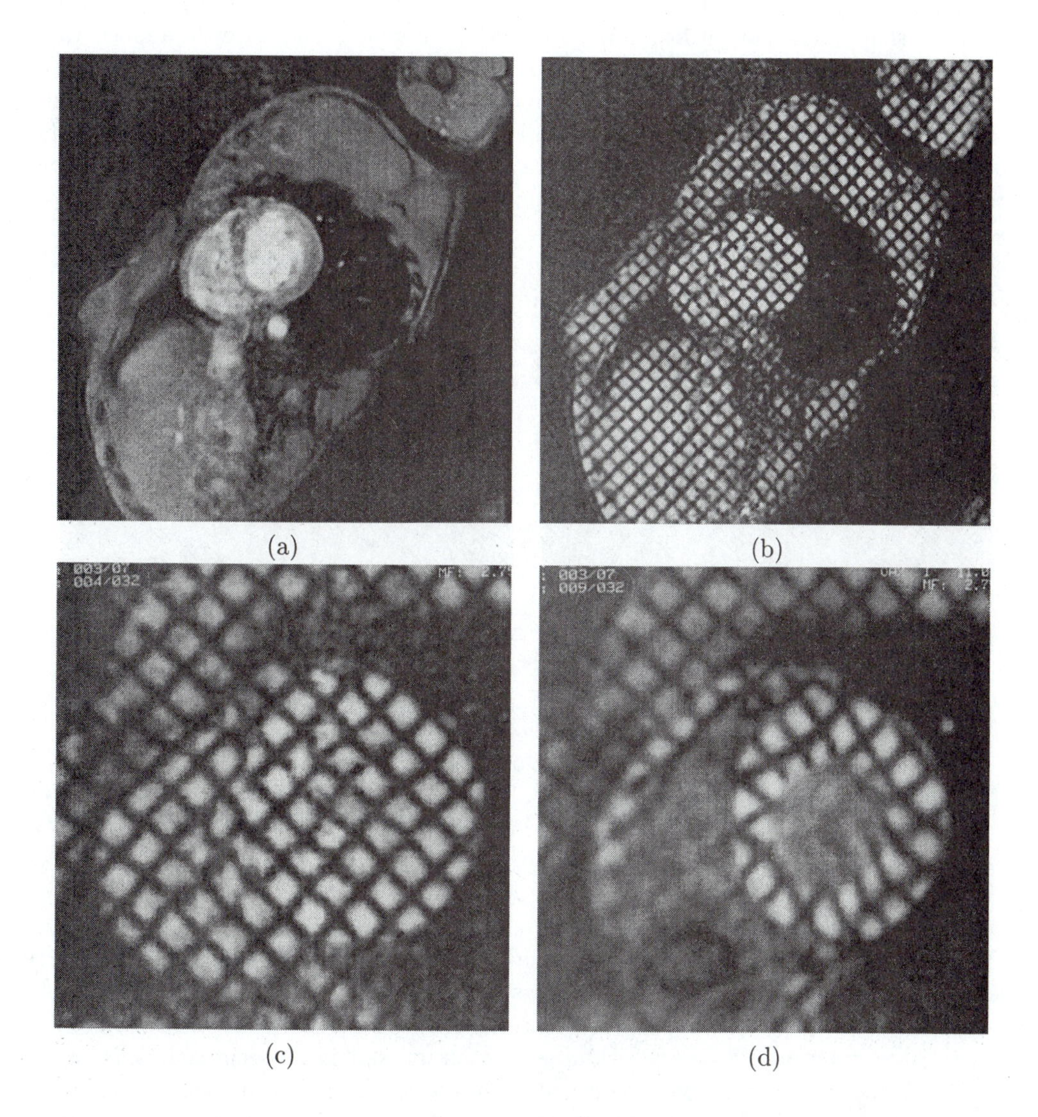

Figure 15.14: *Magnetic resonance image of the heart: (a) original chest image, diastole; (b) chest image with magnetic resonance markers, diastole; (c) image of the heart with markers, diastole; (d) image with markers, systole. Courtesy D. Fisher, S. M. Collins, The University of Iowa.*

A different method of interest-point correspondence and trajectory detection has been used in the analysis of cardiac wall motion from magnetic resonance images [Fisher 90, Fisher et al. 91], where rigid body motion assumptions could not be used since the human heart changes its shape over the cardiac cycle. Interest points were magnetically applied to the heart muscle using a special magnetic resonance pulse sequence known as SPAMM (spatial modulation of magnetization). This results in an image with a rectangular grid of markers, see Figure 15.14; heart motion is clearly visible on images if markers are applied. The first step of the motion analysis algorithm is a precise automatic detection of markers. Using a correlation technique (Section 5.4), the exact position of markers is determined (possibly at sub-pixel resolution), see Figure 15.15.

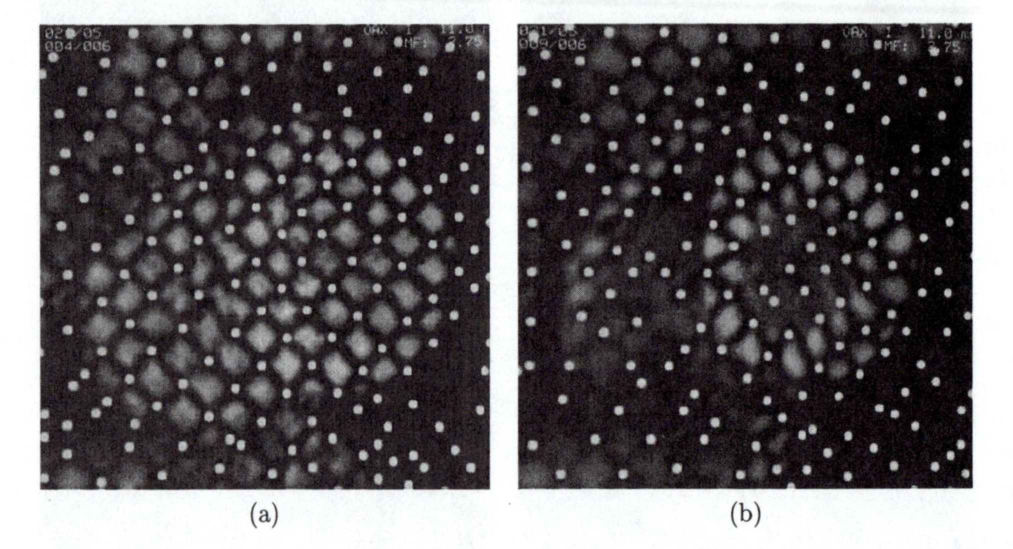

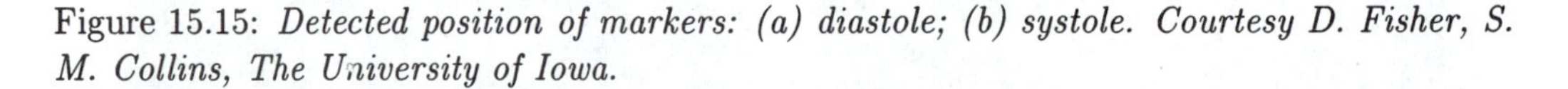

Figure 15.15: *Detected position of markers: (a) diastole; (b) systole. Courtesy D. Fisher, S. M. Collins, The University of Iowa.*

To track the marker locations, specific knowledge about small relative motion of marker positions in consecutive frames is used. Markers are considered as nodes of a two-dimensional graph, and dynamic programming is used to determine optimal trajectories (see Section 5.2.5). The optimality criterion is based on distance between markers in consecutive images, on the quality of marker detection, and on consistency of motion direction in consecutive images. Marker quality evaluation results from the marker detection correlation process. Successor nodes are determined by requiring that the trajectory length between successors be less than some specified constant. Identified and tracked markers are illustrated in Figure 15.16, and a resulting velocity field is shown in Figure 15.17.

Another approach to motion analysis of tagged MR heart images based on left-ventricle boundary detection and matching tag templates in expected directions can be found in [Guttmann and Prince 90, Theotokis and Prince 96]. Deformable models can also be applied to analyze 3D motion using tagged MR heart images, as described in [Young and Axel 92]. In [Kambhamettu and Goldgof 92], estimation of point correspondences on a surface

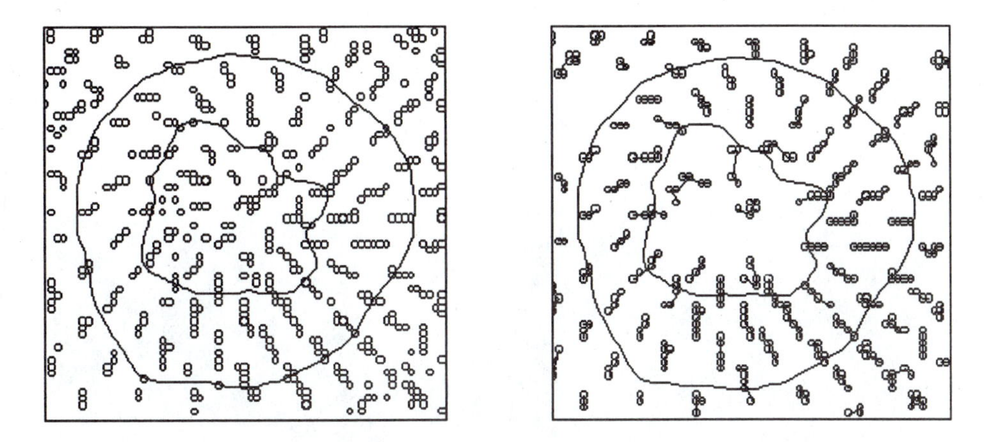

Figure 15.16: *Velocity field: Identified markers (left) and tracked markers (right). Note that the dynamic programming has removed most of the spurious nodes that occur in the center of the cavity.*

undergoing non-rigid motion is based on changes in Gaussian curvature.

Many motion-related applications can be found in remote sensing, especially in meteorological satellite data analysis. An example of a **meteotrend** analysis of the cloud motion in both horizontal and vertical directions, and prediction of cloud motion and cloud types, is given in Figures 15.18 and 15.19. A survey of many other approaches to motion correspondence, trajectory parametrization, representation of relative motion and motion events, overview of useful region-based features, matching and classification approaches, and approaches to motion recognition including cyclic motion, lipreading, gesture interpretation, motion verb recognition, and temporal textures classification, is given in [Cedras and Shah 95].

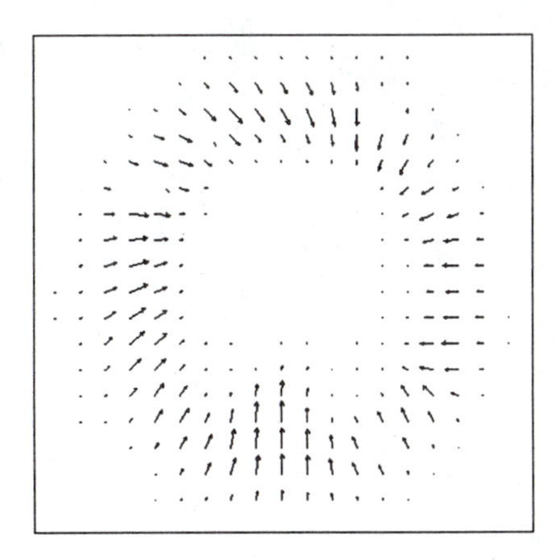

Figure 15.17: *Velocity field derived from the information in Figure 15.16. Courtesy D. Fisher, S. M. Collins, The University of Iowa.*

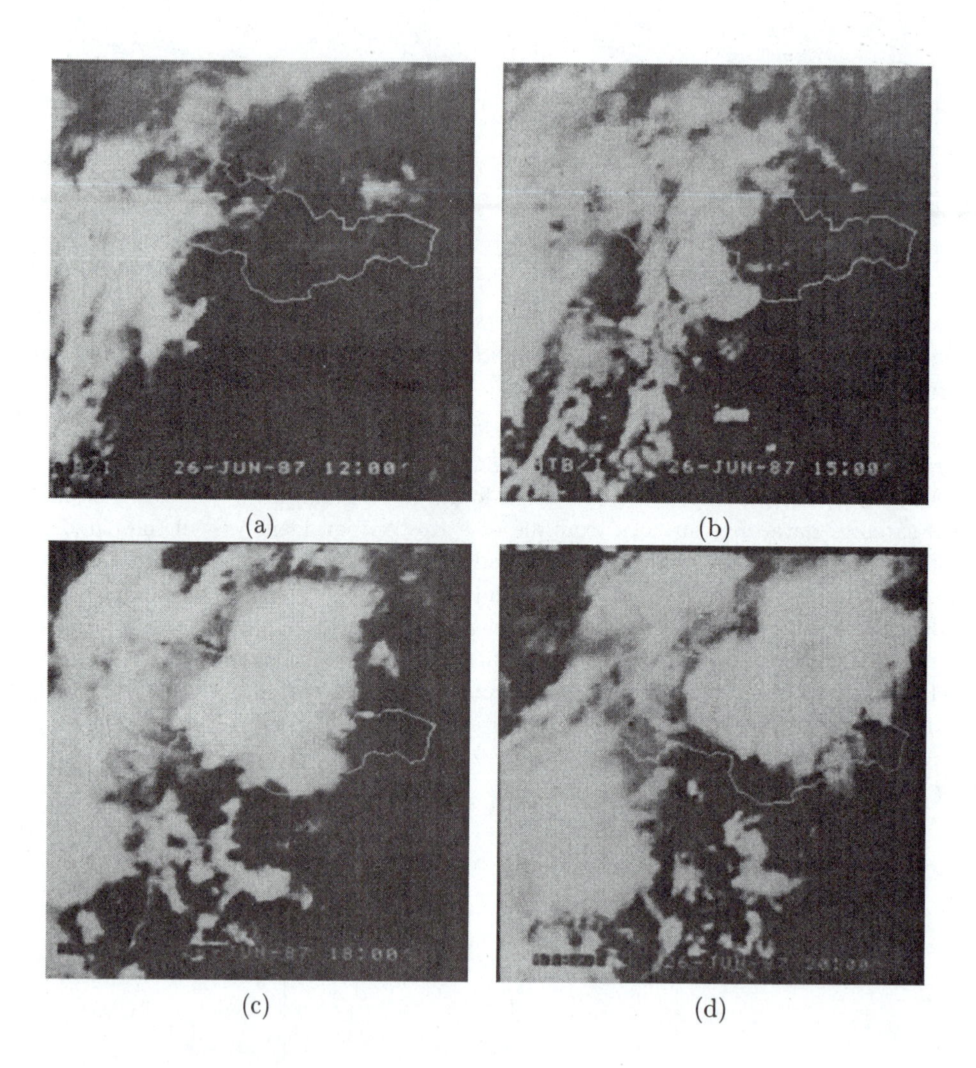

Figure 15.18: *Meteosat image sequence. Image data acquired on June 26, 1987, at (a) noon, (b) 3 p.m., (c) 6 p.m., (d) 8 p.m. Courtesy L. Vlcak, D. Podhorsky, Slovak Hydrometeorological Institute, Bratislava, Slovakia.*

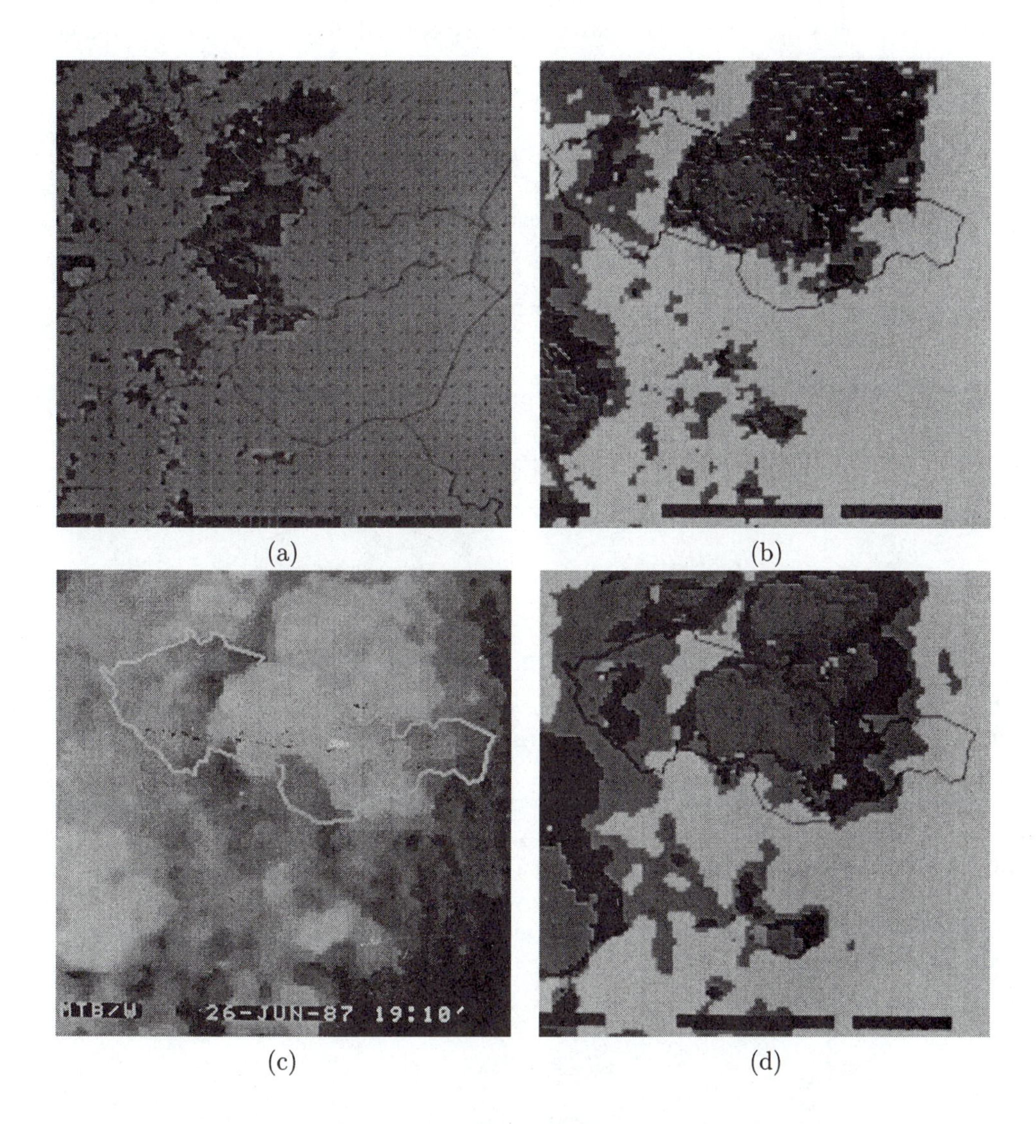

Figure 15.19: *Meteotrend: (a) cloud motion analysis in horizontal and vertical directions, vertical speed coded in color; (b) cloud type classification; (c) cloud cover (motion) prediction; (d) cloud type prediction. A color version of (a), (b) and (d) may be seen in the color inset. Courtesy L. Vlcak, D. Podhorsky, Slovak Hydrometeorological Institute Bratislava, Slovakia.*

15.4 Kalman filters

Frequently, we may have a model of an object in a scene that we see through noise; thus we *model* $\mathbf{x}$ and *observe* $\mathbf{z}$, where $\mathbf{x}$ and $\mathbf{z}$ are feature vectors, possibly not of the same dimension. In the circumstance of observing through an image sequence, in particular motion, we may be able to use the models and observations in two ways.

- Multiple observations $\mathbf{z}_1, \mathbf{z}_2, \ldots$ should permit an improved estimate of the underlying model $\mathbf{x}$. It is possible that this model evolves in time, in which case $\mathbf{z}_k$ will give an estimate of $\mathbf{x}_k$; provided we have a clear understanding of how $\mathbf{x}_k$ changes with k, it should still be possible to use the $\mathbf{z}_k$ to estimate this more complex model.
- The estimate of $\mathbf{x}$ at time k may also provide a prediction for the observation $\mathbf{x}_{k+1}$, and thereby for $\mathbf{z}_{k+1}$.

This suggests a feedback mechanism, or predictor controller (Figure 15.20), whereby we observe $\mathbf{z}_k$, estimate $\mathbf{x}_k$, predict $\mathbf{x}_{k+1}$ thereby predict $\mathbf{z}_{k+1}$, observe $\mathbf{z}_{k+1}$ *taking advantage of the prediction*, and then update our estimate of $\mathbf{x}_{k+1}$.

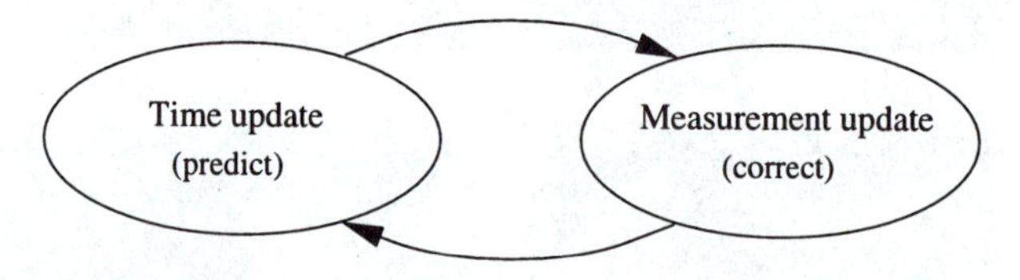

Figure 15.20: *The Kalman filter iterative cycle; the time update predicts events at the next step, and the measurement update adjusts estimates in the light of observation.*

For a particular class of model and observation, this scheme has been realized as the **Kalman filter** [Kalman 60], which is widely used in motion vision applications. Specifically, we assume that the system is linear, that observations of it are linear functions of the underlying state, and that noise, both in the system and in measurement, is white and Gaussian (slight variations and extensions of this model may be found in more specialized texts). Formally, we have the model

$$\begin{aligned} \mathbf{x}_{k+1} &= A_k \mathbf{x}_k + \mathbf{w}_k \\ \mathbf{z}_k &= H_k \mathbf{x}_k + \mathbf{v}_k \end{aligned} \tag{15.39}$$

The matrices A_k describe the evolution of the underlying model state, while $\mathbf{w}_k$ is zero mean Gaussian noise. We assume that $\mathbf{w}_k$ has covariance Q_k:

$$\begin{aligned} Q_k &= E[\mathbf{w}_k \mathbf{w}_k^T] \\ (Q_k)_{ij} &= E(w_k^i w_k^j) \end{aligned}$$

where w_k^i denotes the ith component of vector $\mathbf{w}_k$. The matrices H_k are the measurement matrices, describing how the observations are related to the model; $\mathbf{v}_k$ is another zero mean Gaussian noise factor, with covariance R_k.

Given $\mathbf{x}_{k-1}$ (or an estimate $\hat{\mathbf{x}}_{k-1}$) we can now compute from equation (15.39) an a priori estimate $\mathbf{x}_k = A_{k-1}\mathbf{x}_{k-1}$. Conventionally, this is referred to as $\hat{\mathbf{x}}_k^-$ to indicate (via $\hat{}$) that it is an estimate, and (via $^-$) that it is 'before' observation. Correspondingly, we can define $\hat{\mathbf{x}}_k^+$ as the updated estimate computed 'after' observation—it is this computation that the Kalman filter provides.

Associated with each estimate are errors,

$$\begin{aligned} \mathbf{e}_k^- &= \mathbf{x}_k - \hat{\mathbf{x}}_k^- \\ \mathbf{e}_k^+ &= \mathbf{x}_k - \hat{\mathbf{x}}_k^+ \end{aligned} \tag{15.40}$$

with corresponding covariances P_k^- and P_k^+; note that these errors are caused by $\mathbf{w}_k$ and the error in the estimate.

The Kalman filter operates by examining the residual

$$\mathbf{z}_k - H_k \hat{\mathbf{x}}_k^-$$

to which $\mathbf{e}_k^-$ and the noise $\mathbf{v}_k$ contribute. In the absence of the noise, and if the estimate is perfect, this is zero. The approach is to seek a matrix K_k, the **Kalman gain** matrix, to update $\hat{\mathbf{x}}_k^-$:

$$\hat{\mathbf{x}}_k^+ = \hat{\mathbf{x}}_k^- + K_k(\mathbf{z}_k - H_k\hat{\mathbf{x}}_k^-) \tag{15.41}$$

This matrix K_k is sought to minimize the a posteriori covariance P_k^+; this is done in a least-squares sense, meaning that we are seeking an optimal linear filter. The approach is to substitute equation (15.41) into equation (15.40), calculate P_k^+ from equation (15.40), differentiate with respect to K_k, and set to zero. This leads to the equation

$$K_k = P_k^- H_k^T (H_k P_k^- H_k^T + R_k)^{-1} \tag{15.42}$$

wherein

$$P_k^- = A_k P_{k-1}^+ A_k^T + Q_{k-1} \tag{15.43}$$

Elementary manipulation also provides the relation

$$P_k^+ = (I - K_k H_k) P_k^- \tag{15.44}$$

The full details of this derivation are not relevant here, and are easily accessible in a range of other texts (e.g., [Gelb 74, Maybeck 79]). It is only necessary to observe that, if an application displays the appropriate conditions of linearity, equations (15.42)–(15.44) provide an algorithm for computing the gain matrix, since all its components are observable.

15.4.1 Example

The Kalman filter is non-trivial to comprehend and use. To illustrate it, we present here a trivial example [Gelb 74]. Suppose we have a constant scalar x, observed through uncorrelated Gaussian noise of zero mean and variance r. In this case we have $A_k = I = 1$ and $H_k = I = 1$:

$$\begin{aligned} x_{k+1} &= x_k \\ z_k &= x_k + v_k \end{aligned}$$

where v_k is normally distributed about 0 with variance r. Immediately, we see from equation (15.42) that

$$K_k = \frac{p_k^-}{p_k^- + r} \tag{15.45}$$

We can deduce covariance relations from equations (15.43) and (15.44);

$$\begin{aligned} p_{k+1}^- &= p_k^+ \\ p_{k+1}^+ &= (1 - K_{k+1})p_{k+1}^- \\ &= p_{k+1}^- \frac{r}{p_{k+1}^- + r} \\ &= p_k^+ \frac{r}{p_k^+ + r} \end{aligned} \tag{15.46}$$

Equation (15.46) provides a recurrence relation; writing $p_0 = p_0^+$, we can deduce

$$p_k^+ = \frac{rp_0}{kp_0 + r}$$

which, substituted into equation (15.45), gives

$$K_k = \frac{p_0}{r + kp_0}$$

and so, from equation (15.41),

$$\hat{x}_k^+ = \hat{x}_k^- + \frac{p_0}{r + kp_0}(z_k - \hat{x}_k^-)$$

which tells us the intuitively obvious—as k grows, new measurements provide very little new information.

Section 16.4 presents a more complex 3D vision application that uses the filter for more productive purposes.

15.5 Summary

- **Motion analysis**
  - Motion analysis is dealing with three main groups of motion-related problems:
    * Motion detection
    * Moving object detection and location
    * Derivation of 3D object properties
  - A two-dimensional representation of a (generally) three-dimensional motion is called a **motion field** wherein each point is assigned a **velocity vector** corresponding to the motion direction, velocity, and distance from an observer at an appropriate image location.
  - **Optical flow** represents one approach to motion field construction, in which motion direction and motion velocity are determined at possibly all image points.

  - **Feature point correspondence** is another method for motion field construction. Velocity vectors are determined only for corresponding feature points.
  - Object **motion parameters** can be derived from computed motion field vectors.
  - **Motion assumptions** can help to localize moving objects. Frequently used assumptions include:
    * Maximum velocity
    * Small acceleration
    * Common motion
    * Mutual correspondence

- **Differential motion analysis**
  - Subtraction of images acquired at different instants in time makes motion detection possible, assuming a stationary camera position and constant illumination.
  - There are many problems associated with this approach, and results of subtraction are highly dependent on an object–background contrast.
  - A **cumulative difference image** improves performance of differential motion analysis. It provides information about motion direction and other time-related motion properties, and about slow motion and small object motion.
  - Detecting **moving edges** helps further overcome the limitations of differential motion analysis methods. By combining the spatial and temporal image gradients, the differential analysis can be reliably used for detection of slow-moving edges as well as detection of weak edges that move with higher speed.

- **Optical flow**
  - Optical flow reflects the image changes due to motion during a time interval dt, which must be short enough to guarantee small inter-frame motion changes.
  - The optical flow field is the velocity field that represents the three-dimensional motion of object points across a two-dimensional image.
  - Optical flow computation is based on two assumptions:
    * The observed brightness of any object point is constant over time.
    * Nearby points in the image plane move in a similar manner (the **velocity smoothness** constraint).
  - Optical flow computation will be in error if the constant brightness and velocity smoothness assumptions are violated. In real imagery, their violation is quite common. Typically, the optical flow changes dramatically in highly textured regions, around moving boundaries, at depth discontinuities, etc. Resulting errors propagate across the entire optical flow solution.
  - Global error propagation is the biggest problem of global optical flow computation schemes, and local optical flow estimation helps overcome the difficulties.
  - Optical flow analysis does not result in motion trajectories; instead, more general motion properties are detected that can significantly increase the reliability of complex motion analysis. Parameters that are detected include:

    * Mutual object velocity
    * Focus of expansion (FOE) determination
    * Distance (depth) determination
    * Collision prediction

- **Motion analysis based on correspondence of interest points**
  - This method finds significant points (**interest points, feature points**) in all images of the sequence—points least similar to their surroundings, representing object corners, borders, or any other characteristic features in an image that can be tracked over time.
  - Point detection is followed by a matching procedure, which looks for correspondences between these points in time.
  - The process results in a sparse velocity field.
  - Motion detection based on correspondence works even for relatively long inter-frame time intervals.
  - **Tracking** of object motion is a common application that can be solved using motion correspondence.
  - If several independently moving objects are tracked, the solution methods often rely on motion constraints and minimize a **path coherence function** that represents a measure of agreement between the derived object trajectory and the motion constraints.
- **Kalman filtering**
  - Kalman filtering is an approach that is frequently used in dynamic estimation and represents a powerful tool when used for motion analysis.
  - Kalman filtering requires the system to be linear, with observations of it to be linear functions of the underlying state. Noise, both in the system and in measurement, is assumed to be white and Gaussian.
  - While the assumptions are often unrealistic if applied to image sequences, they represent a convenient choice.

15.6 Exercises

Short-answer questions

1. Describe the differences between motion detection and moving-object detection.
2. Name the object motion assumptions, and explain their rationale.
3. Explain how cumulative difference images can be used in motion analysis.
4. What is the aperture problem? What are its consequences? How can this problem be overcome?
5. Name the two basic assumptions used for optical flow computation. Are the assumptions realistic? What problems arise if they are violated?

6. Describe two approaches that increase optical flow computation robustness when the optical flow assumptions are not valid.

7. Specify how optical flow can be used to determine:

   (a) Mutual velocity of an observer and an object

   (b) The focus of expansion

   (c) Distance of a moving object from the observer

   (d) Possible collision of the object with an observer and time to collision

8. Explain the image frame rate requirements for motion analysis using optical flow and motion analysis based on correspondence of interest points.

9. Explain the concept of motion analysis based on correspondence of interest points. Why is the correspondence problem difficult?

10. Determine properties of a path coherence function and explain how it can be used for object tracking.

11. Explain how occlusion may be handled in object tracking. Why are phantom points needed?

12. Explain the usage of Kalman filtering in motion analysis.

Problems

1. Using a static camera, create image sequences consisting of ten image frames (use the frame rate appropriate for the moving object velocity) depicting one, three, and five moving objects on a non-homogeneous background (the sequence used in Figure 15.2 may serve as an example). Alternatively, create computer-generated image sequences. (Also, the World Wide Web may serve as a source for such image sequences.) The image sequence(s) will be used in the problems below.

2. Develop a program for motion detection in image sequences created in Problem 15.1. If the approach is based on differential image analysis, pay special attention to automated threshold determination.

3. Develop a program for motion analysis using cumulative difference images. Determine the trajectory of the moving objects. Use the image sequences created in Problem 15.1.

4. Develop a program for motion analysis using moving edges, and apply it to the image sequences created in Problem 15.1.

5. Create a sequence depicting slow motion and analyze it using programs developed in Problems 15.2–15.4. Compare performance of the three approaches.

6. At time $t_0 = 0$, a point object is located at real-world co-ordinates $x_0, y_0, z_0 = (30, 60, 10)$ and is moving with a constant speed $(u, v, w) = (-5, -10, -1)$ towards the observer. Assuming unit focal distance of the optical system:

   (a) Determine the location of the object at the image co-ordinates (x', y') at time t_0.

   (b) Determine the image co-ordinates of the focus of expansion.

   (c) Determine the time of the object collision with the observer.

7. Develop a function (subroutine) for detection of interest points.

8. Develop a program for motion analysis using correspondence of interest points (use interest points identified in Problem 15.7). Generate frame-to-frame velocity fields, not necessarily using the subsequent frames. Apply it to image sequences with one, three, and five moving objects and assess the results.

9. Develop an image tracking program using path coherence, and use it to generate object motion trajectories.

15.7 References

[Abdel-Malek et al. 90] A Abdel-Malek, O Hasekioglu, and J Bloomer. Image segmentation via motion vector estimates. In *Medical Imaging IV: Image Processing,* Newport Beach, CA, pages 366–371, SPIE, Bellingham, WA, 1990.

[Adiv 89] G Adiv. Inherent ambiguities in recovering 3D motion and structure from a noisy flow field. *IEEE Transactions on Pattern Analysis and Machine Intelligence,* 11(5):477–489, 1989.

[Aggarwal and Martin 88] J K Aggarwal and W Martin. *Motion Understanding.* Kluwer, Boston, 1988.

[Aggarwal and Nandhakumar 88] J K Aggarwal and N Nandhakumar. On the computation of motion from sequences of images—a review. *Proceedings of the IEEE,* 76(8):917–935, 1988.

[Albus and Hong 90] J S Albus and T H Hong. Motion, depth, and image flow. In *Proceedings of the 1990 IEEE International Conference on Robotics and Automation,* Cincinnati, OH, pages 1161–1170, IEEE, Los Alamitos, CA, 1990.

[Anandan 87] P Anandan. A unified perspective on computational techniques for the measurement of visual motion. In *1st International Conference on Computer Vision,* London, England, pages 219–230, IEEE, Piscataway, NJ, 1987.

[Anandan 88] P Anandan. Motion detection and analysis. State of the art and some requirements from robotics. In *Robot Control 1988 (SYROCO '88)—Selected Papers from the 2nd IFAC Symposium,* pages 347–352, Pergamon Press, Karlsruhe, Germany, 1988.

[Arnspang 93] J Arnspang. Motion constraint equations based on constant image irradiance. *Image and Vision Computing,* 11:577–587, 1993.

[Barnard 79] S T Barnard. The image correspondence problem. PhD thesis, University of Minnesota, 1979.

[Barnard and Thompson 80] S T Barnard and W B Thompson. Disparity analysis of images. *IEEE Transactions on Pattern Analysis and Machine Intelligence,* 2(4):333–340, 1980.

[Barron et al. 92a] J L Barron, D J Fleet, S S Beauchemin, and T A Burkitt. Performance of optical flow techniques. In *Proceedings, 1992 Computer Vision and Pattern Recognition,* Champaign, IL, pages 236–242, IEEE, Los Alamitos, CA, 1992.

[Barron et al. 92b] J L Barron, D J Fleet, S S Beauchemin, and T A Burkitt. Performance of optical flow techniques. Technical Report TR 299, Department of Computer Science, University of Western Ontario, London, Ontario, Canada, 1992.

[Bober and Kittler 94] M Bober and J Kittler. Estimation of complex multimodal motion: An approach based on robust statistics and Hough transform. *Image and Vision Computing,* 12:661–668, 1994.

[Brady 87] J M Brady. Seeds of perception. In *Proceedings of the Third Alvey Vision Conference,* Cambridge, England, pages 259–267. University of Cambridge, 1987.

[Burt et al. 83] P J Burt, C Yen, and X Xu. Multi-resolution flow through motion analysis. In *CVPR '83: Computer Society Conference on Computer Vision and Pattern Recognition,* Washington, DC, pages 246–252, IEEE, Los Alamitos, CA, 1983.

[Cafforio and Rocca 83] C Cafforio and F Rocca. The differential method for image motion estimation. In T S Huang, editor, *Image Sequence Processing and Dynamic Scene Analysis,* pages 104–124, Springer Verlag, Berlin, 1983.

[Cedras and Shah 95] C Cedras and M Shah. Motion-based recognition: A survey. *Image and Vision Computing,* 13:129–154, 1995.

[Cheng and Aggarwal 90] C L Cheng and J K Aggarwal. A two-stage approach to the correspondence problem via forward-searching and backward-correcting. In *10th International Conference on Pattern Recognition,* Atlantic City, NJ, pages 173–177, IEEE, Los Alamitos, CA, 1990.

[Clocksin 80] W F Clocksin. Perception of surface slant and edge labels from optical flow—a computational approach. *Perception,* 9:253–269, 1980.

[Costabile et al. 85] M F Costabile, C Guerra, and G G Pieroni. Matching shapes: A case study in time-varying images. *Computer Vision, Graphics, and Image Processing,* 29:296–310, 1985.

[Davis et al. 83] L S Davis, Z Wu, and H Sun. Contour based motion estimation. *Computer Vision, Graphics, and Image Processing,* 23:246–252, 1983.

[Dreschler and Nagel 82] L S Dreschler and H H Nagel. Volumetric model and 3D trajectory of a moving car derived from monocular TV frame sequences. *Computer Graphics and Image Processing,* 20:199–228, 1982.

[Enkelmann 88] W Enkelmann. Investigations of multigrid algorithms for the estimation of optical flow fields in image sequences. *Computer Vision, Graphics, and Image Processing,* 43:150–177, 1988.

[Enkelmann 91] W Enkelmann. Obstacle detection by evaluation of optical flow fields from image sequences. *Image and Vision Computing,* 9(3):160–168, 1991.

[Fisher 90] D J Fisher. Automatic tracking of cardiac wall motion using magnetic resonance markers. PhD thesis, University of Iowa, 1990.

[Fisher et al. 91] D J Fisher, J C Ehrhardt, and S M Collins. Automated detection of noninvasive magnetic resonance markers. In *Computers in Cardiology,* Chicago, IL, pages 493–496, IEEE, Los Alamitos, CA, 1991.

[Fleet 92] D J Fleet. *Measurement of Image Velocity.* Kluwer, Norwell, MA, 1992.

[Fleet and Jepson 90] D J Fleet and A D Jepson. Computation of component image velocity from local phase information. *International Journal of Computer Vision,* 5:77–105, 1990.

[Fogel 88] S V Fogel. A nonlinear approach to the motion correspondence problem. In *2nd International Conference on Computer Vision,* Tarpon Springs, FL, pages 619–628, IEEE, Piscataway, NJ, 1988.

[Fogel 89] S V Fogel. Implementation of a nonlinear approach to the motion correspondence problem. In *Proceedings, Workshop on Visual Motion,* Irvine, CA, pages 87–98, IEEE, Piscataway, NJ, 1989.

[Freeman and Adelson 91] W T Freeman and E H Adelson. The design and use of steerable filters. *IEEE Transactions on Pattern Analysis and Machine Intelligence,* 13:891–906, 1991.

[Fua and Leclerc 94] P Fua and Y G Leclerc. Registration without correspondences. In *CVPR '94: Computer Society Conference on Computer Vision and Pattern Recognition,* Seattle, WA, pages 121–128, IEEE, Los Alamitos, CA, 1994.

[Gelb 74] A Gelb, editor. *Applied Optimal Estimation.* MIT Press, Cambridge, MA, 1974.

[Girosi et al. 89] F Girosi, A Verri, and V Torre. Constraints for the computation of optical flow. In *Proceedings: Workshop on Visual Motion,* Washington, DC, pages 116–124, IEEE, Piscataway, NJ, 1989.

[Glazer 84] F Glazer. Multilevel relaxation in low level computer vision. In Rosenfeld A, editor, *Multiresolution Image Processing and Analysis*, pages 312–330. Springer Verlag, Berlin, 1984.

[Gong 87] S G Gong. Parallel computation of visual motion. Master's thesis, Oxford University, England, 1987.

[Gong 88] S G Gong. Improved local flow. In *Proceedings of the Fourth Alvey Vision Conference,* Manchester, England, pages 129–134. University of Manchester, 1988.

[Gong 89] S G Gong. Curve motion constraint equation and its application. In *Proceedings, Workshop on Visual Motion,* Irvine, CA, pages 73–80, IEEE, Piscataway, NJ, 1989.

[Gu et al. 96] H Gu, Y Shirai, and M Asada. MDL-based segmentation and motion modeling in a long image sequence of scene with multiple independently moving objects. *IEEE Transactions on Pattern Analysis and Machine Intelligence*, 18:58–64, 1996.

[Guttmann and Prince 90] M A Guttmann and J L Prince. Image analysis methods for tagged MRI cardiac studies. In *Medical Imaging IV: Image Processing,* Newport Beach, CA, pages 168–175, SPIE, Bellingham, WA, 1990.

[Hampel et al. 86] F R Hampel, E Ronchetti, P Rousseeuw, and W A Stahel. *Robust Statistics: The Approach Based on Influence Functions.* Wiley, Chichester, England, 1986.

[Haralick and Watson 81] R M Haralick and L Watson. A facet model for image data. *Computer Graphics and Image Processing*, 15:113–129, 1981.

[Heeger 87] D J Heeger. Model for the extraction of image flow. *Journal of the Optical Society of America*, 4:1455–1471, 1987.

[Heel 90] J Heel. Dynamic motion vision. *Robotics*, 6(3):297–314, 1990.

[Heitz and Bouthemy 93] F Heitz and P Bouthemy. Multimodal estimation of discontinuous optical flow using Markov random fields. *IEEE Transactions on Pattern Analysis and Machine Intelligence*, 15:1217–1232, 1993.

[Hildreth 83] E C Hildreth. Computations underlining the measurement of visual motion. *Artificial Intelligence*, 23(3):309–354, 1983.

[Horn 86] B K P Horn. *Robot Vision.* MIT Press, Cambridge, MA, 1986.

[Horn and Schunk 81] B K P Horn and B Schunk. Determining optical flow. *Artificial Intelligence*, 17:185–204, 1981.

[Hu and Ahuja 93] X Hu and N Ahuja. Motion and structure estimation using long sequence motion models. *Image and Vision Computing*, 11:549–569, 1993.

[Huang 83] T S Huang, editor. *Image Sequence Processing and Dynamic Scene Analysis.* Springer Verlag, Berlin, 1983.

[Huang and Chen 95] C L Huang and Y T Chen. Motion estimation method using a 3D steerable filter. *Image and Vision Computing*, 13:21–32, 1995.

[Hummel and Sundareswaran 93] R Hummel and Sundareswaran. Motion parameter estimation from global flow field data. *IEEE Transactions on Pattern Analysis and Machine Intelligence*, 15:459–476, 1993.

[Jain 81] R Jain. Dynamic scene analysis using pixel–based processes. *Computer*, 14(8):12–18, 1981.

[Jain 83] R Jain. Direct computation of the focus of expansion. *IEEE Transactions on Pattern Analysis and Machine Intelligence*, 5(1):58–64, 1983.

[Jain 84] R Jain. Difference and accumulative difference pictures in dynamic scene analysis. *Image and Vision Computing*, 2(2):99–108, 1984.

[Jain et al. 79] R Jain, W N Martin, and J K Aggarwal. Segmentation through the detection of changes due to motion. *Computer Graphics and Image Processing*, 11:13–34, 1979.

[Jain et al. 87] R Jain, S L Bartlett, and N O'Brien. Motion stereo using ego–motion complex logarithmic mapping. *IEEE Transactions on Pattern Analysis and Machine Intelligence*, 9(3):356–369, 1987.

[Jain et al. 95] R Jain, R Kasturi, and B G Schunck. *Machine Vision.* McGraw-Hill, New York, 1995.

[Kalman 60] R E Kalman. A new approach to linear filtering and prediction problems. *Transactions of the ASME—Journal of Basic Engineering*, 82:35–45, March 1960.

[Kambhamettu and Goldgof 92] C Kambhamettu and D B Goldgof. Point correspondence recovery in non-rigid motion. In *Proceedings, IEEE Conference on Computer Vision and Pattern Recognition,* Champaign, IL, pages 222–227, IEEE, Los Alamitos, CA, 1992.

[Kanatani 85] K I Kanatani. Detecting the motion of a planar surface by line and surface integrals. *Computer Vision, Graphics, and Image Processing*, 29:13–22, 1985.

[Kearney and Thompson 88] J K Kearney and W B Thompson. Bounding constraint propagation for optical flow estimation. In J K Aggarwal and W Martin, editors, *Motion Understanding.* Kluwer, Boston, 1988.

[Kearney et al. 87] J K Kearney, W B Thompson, and D L Boley. Optical flow estimation—an error analysis of gradient based methods with local optimization. *IEEE Transactions on Pattern Analysis and Machine Intelligence*, 9(2):229–244, 1987.

[Koenderink 86] J J Koenderink. Optic flow. *Vision Research*, 26(1):161–180, 1986.

[Koenderink and Doorn 75] J J Koenderink and A J Doorn. Invariant properties of the motion parallax field due to the movement of rigid bodies relative to an observer. *Optica Acta*, 22(9):773–791, 1975.

[Kottke and Sun 94] D P Kottke and Y Sun. Motion estimation via cluster matching. *IEEE Transactions on Pattern Analysis and Machine Intelligence*, 16:1128–1132, 1994.

[Kumar et al. 96] A Kumar, A R Tannenbaum, and G J Balas. Optical flow: A curve evolution approach. *IEEE Transactions on Image Processing*, 5:598–610, 1996.

[Lee et al. 89] D Lee, A Papageorgiou, and G W Wasilkowski. Computing optical flow. In *Proceedings, Workshop on Visual Motion,* Irvine, CA, pages 99–106, IEEE, Piscataway, NJ, 1989.

[Lucas and Kanade 81] B Lucas and T Kanade. An iterative image registration technique with an application to stereo vision. In *DARPA Image Understanding Workshop,* Washington, DC, pages 121–130, DARPA, Los Altos, CA, 1981.

[Marr 82] D Marr. *Vision—A Computational Investigation into the Human Representation and Processing of Visual Information.* Freeman, San Francisco, 1982.

[Maybeck 79] P S Maybeck. *Stochastic Models, Estimation, and Control*, volume 1. Academic Press, New York–London, 1979.

[Mehrotra and Nichani 90] R Mehrotra and S Nichani. Corner detection. *Pattern Recognition Letters*, 23(11):1223–1233, 1990.

[Murray and Buxton 89] D W Murray and B F Buxton. Scene segmentation from visual motion using global optimization. *IEEE Transactions on Pattern Analysis and Machine Intelligence*, 9(2):200–228, 1989.

[Murray and Buxton 90] D W Murray and B F Buxton. *Experiments in the Machine Interpretation of Visual Motion*. MIT Press, Cambridge, MA, 1990.

[Mutch and Thompson 84] K M Mutch and W B Thompson. Hierarchical estimation of spatial properties from motion. In A Rosenfeld, editor, *Multiresolution Image Processing and Analysis*, pages 343–354, Springer Verlag, Berlin, 1984.

[Mutch and Thomson 85] K M Mutch and W B Thomson. Analysis of accretion and deletion at boundaries in dynamic scenes. *IEEE Transactions on Pattern Analysis and Machine Intelligence*, 7(2):133–137, 1985.

[Nagel 83] H H Nagel. Displacement vectors derived from second order intensity variations. *Computer Vision, Graphics, and Image Processing*, 21:85–117, 1983.

[Nagel 86] H H Nagel. An investigation of smoothness constraints for the estimation of displacement vector fields from image sequences. *IEEE Transactions on Pattern Analysis and Machine Intelligence*, 8(5):565–593, 1986.

[Nagel 87] H H Nagel. On the estimation of optical flow: Relations between different approaches and some new results. *Artificial Intelligence*, 33:299–324, 1987.

[Nagel and Enkelmann 84] H H Nagel and W Enkelmann. Towards the estimation of displacement vector fields by oriented smoothness constraint. In *7th International Conference on Pattern Recognition*, Montreal, Canada, pages 6–8, IEEE, Piscataway, NJ, 1984.

[Negahdaripour and Ganesan 92] S Negahdaripour and V Ganesan. Simple direct computation of the FOE with confidence measures. In *Proceedings, 1992 Computer Vision and Pattern Recognition*, Champaign, IL, pages 228–233, IEEE, Los Alamitos, CA, 1992.

[Negahdaripour and Horn 89] S Negahdaripour and B K P Horn. Direct method for locating the focus of expansion. *Computer Vision, Graphics, and Image Processing*, 46(3):303–326, 1989.

[Nesi 93] P Nesi. Variational approach to optical flow estimation managing discontinuities. *Image and Vision Computing*, 11:419–439, 1993.

[Paquin and Dubois 83] R Paquin and E Dubois. A spatio-temporal gradient method for estimating the displacement field in time-varying imagery. *Computer Vision, Graphics, and Image Processing*, 21:205–221, 1983.

[Radig 84] B Radig. Image sequence analysis using relational structure. *Pattern Recognition*, 17(1):161–168, 1984.

[Rangarajan and Shah 91] K Rangarajan and M Shah. Establishing motion correspondence. *CVGIP - Image Understanding*, 54:56–73, 1991.

[Rangarajan et al. 89] K Rangarajan, M Shah, and D van Brackle. Optimal corner detector. *Computer Vision, Graphics, and Image Processing*, 48(2):230–245, 1989.

[Ringach and Baram 94] D L Ringach and Y Baram. A diffusion mechanism for obstacle detection from size-change information. *IEEE Transactions on Pattern Analysis and Machine Intelligence*, 16:76–80, 1994.

[Rong et al. 89] J H Rong, J L Coatrieux, and R Collorec. Combining motion estimation and segmentation in digital subtracted angiograms analysis. In *Sixth Multidimensional Signal Processing Workshop,* Pacific Grove, CA, page 44, IEEE, Piscataway, NJ, 1989.

[Sandini and Tistarelli 86] G Sandini and M Tistarelli. Analysis of camera motion through image sequences. In V Cappelini and R Marconi, editors, *Advances in Image Processing and Pattern Recognition*, pages 100–106. North Holland, Amsterdam, 1986.

[Schalkoff 87] R J Schalkoff. Dynamic imagery modelling and motion estimation using weak formulations. *IEEE Transactions on Pattern Analysis and Machine Intelligence*, 9(4):578–584, 1987.

[Schalkoff 89] R J Schalkoff. *Digital Image Processing and Computer Vision.* Wiley, New York, 1989.

[Schunck 89] B G Schunck. Robust estimation of image flow. In *Sensor Fusion II: Human and Machine Strategies,* Philadelphia, PA, pages 116–127, SPIE, Bellingham, WA, 1989.

[Scott 88] G L Scott. *Local and Global Interpretation of Moving Images.* Pitman–Morgan Kaufmann, London–San Mateo, CA, 1988.

[Sethi and Jain 87] I K Sethi and R Jain. Finding trajectories of feature points in a monocular image sequence. *IEEE Transactions on Pattern Analysis and Machine Intelligence*, 9(1):56–73, 1987.

[Shah and Jain 84] M A Shah and R Jain. Detecting time–varying corners. In *7th International Conference on Pattern Recognition,* Montreal, Canada, pages 42–48, IEEE, Piscataway, NJ, 1984.

[Shahraray and Brown 88] B Shahraray and M K Brown. Robust depth estimation from optical flow. In *2nd International Conference on Computer Vision,* Tarpon Springs, FL, pages 641–650, IEEE, Piscataway, NJ, 1988.

[Shapiro et al. 95] V Shapiro, I Backalov, and V Kavardjikov. Motion analysis via interframe point correspondence establishment. *Image and Vision Computing*, 13:111–118, 1995.

[Snyder 89] M A Snyder. The precision of 3D parameters in correspondence-based techniques: The case of uniform translational motion in a rigid environment. *IEEE Transactions on Pattern Analysis and Machine Intelligence*, 11(5):523–528, 1989.

[Song and Leahy 91] S M Song and R Leahy. Computation of 3D velocity fields from 3D cine CT images of a human heart. *IEEE Transactions on Medical Imaging*, 10(3):295–306, 1991.

[Subbarao 88] M Subbarao. *Interpretation of Visual Motion: A Computational Study.* Pitman–Morgan Kaufmann, London–San Mateo, CA, 1988.

[Subbarao 90] M Subbarao. Bounds on time-to-collision and rotational component from first-order derivatives of image flow. *Computer Vision, Graphics, and Image Processing*, 50(3):329–341, 1990.

[Szeliski and Coughlan 94] R Szeliski and J Coughlan. Hierarchical spline-based image registration. In *CVPR '94: Computer Society Conference on Computer Vision and Pattern Recognition,* Seattle, WA, pages 194–201, IEEE, Los Alamitos, CA, 1994.

[Theotokis and Prince 96] S A Theotokis and J L Prince. Experiments in multiresolution motion estimation for multifrequency tagged cardiac MR images. In *Proceedings of the IEEE International Conference on Image Processing,* Lausanne, Switzerland, pages III:299–302, IEEE, Piscataway, NJ, 1996.

[Thompson and Barnard 81] W B Thompson and S T Barnard. Lower level estimation and interpretation of visual motion. *Computer*, 14(8):20–28, 1981.

[Thompson and Pong 90] W B Thompson and T C Pong. Detecting moving objects. *International Journal of Computer Vision*, 4(1):39–57, 1990.

[Thompson et al. 84] W B Thompson, K M Mutch, and V A Berzins. Analyzing object motion based on optical flow. In *7th International Conference on Pattern Recognition,* Montreal, Canada, pages 791–194, IEEE, Los Alamitos, CA, 1984.

[Thompson et al. 85] W B Thompson, K M Mutch, and V A Berzins. Dynamic occlusion analysis in optical flow fields. *IEEE Transactions on Pattern Analysis and Machine Intelligence*, 7(4):374–383, 1985.

[Thompson et al. 93] W B Thompson, P Lechleider, and E R Stuck. Detecting moving objects using the rigidity constraint. *IEEE Transactions on Pattern Analysis and Machine Intelligence*, 15:162–166, 1993.

[Tsai and Huang 81] R Y Tsai and T S Huang. Estimating three-dimensional motion parameters of a rigid planar patch. In *Proceedings of PRIP Conference,* Dallas, TX, pages 94–97, IEEE, Piscataway, NJ, 1981.

[Tsai and Huang 82] R Y Tsai and T S Huang. Uniqueness and estimation of three-dimensional motion parameters of rigid objects with curved surfaces. In *Proceedings of PRIP Conference,* Las Vegas, pages 112–118, IEEE, Los Alamitos, CA, 1982.

[Tsao et al. 97] A T Tsao, T P Hung, C S Fuh, and Y S Chen. Ego-motion estimation using optical flow fields observed from multiple cameras. In *Computer Vision and Pattern Recognition*, pages 457–462, IEEE Computer Society, Los Alamitos, CA, 1997.

[Ullman 79] S Ullman. *The Interpretation of Visual Motion.* MIT Press, Cambridge, MA, 1979.

[Vega-Riveros and Jabbour 89] J F Vega-Riveros and K Jabbour. Review of motion analysis techniques. *IEE Proceedings, Part I: Communications, Speech and Vision*, 136(6):397–404, 1989.

[Verri et al. 89] A Verri, F Girosi, and V Torre. Mathematical properties of the 2D motion field: From singular points to motion parameters. In *Proceedings: Workshop on Visual Motion,* Washington, DC, pages 190–200, IEEE, Piscataway, NJ, 1989.

[Vesecky 88] J F Vesecky. Observation of sea–ice dynamics using synthetic aperture radar images automated analysis. *IEEE Transactions on Geoscience and Remote Sensing*, 26(1):38–48, 1988.

[Watson and Ahumada 85] A B Watson and A J Ahumada. Model of human-model sensing. *Journal of the Optical Society of America*, 2:322–342, 1985.

[Waxman and Wohn 85] A M Waxman and K Wohn. Contour evolution, neighbourhood deformation, and global image flow. *International Journal of Robotics Research*, 4(3):95–108, 1985.

[Waxman et al. 87] A M Waxman, K Behrooz, and S Muralidhara. Closed-form solutions to image flow equations for 3D structure and motion. *International Journal of Computer Vision*, 1:239–258, 1987.

[Webb and Aggarwal 82] J A Webb and J K Aggarwal. Structure from motion of rigid and jointed objects. *Artificial Intelligence*, 19:107–130, 1982.

[Weng et al. 89] J Weng, T S Huang, and N Ahuja. Motion and structure from two perspective views: Algorithms, error analysis, and error estimation. *IEEE Transactions on Pattern Analysis and Machine Intelligence*, 11(5):451–476, 1989.

[Wu and Kittler 90] S F Wu and J Kittler. General motion estimation and segmentation. In *Visual Communications and Image Processing '90,* Lausanne, Switzerland, pages 1198–1209, SPIE, Bellingham, WA, 1990.

[WVM 89] *Proceedings, Workshop on Visual Motion,* Irvine, CA, Piscataway, NJ, 1989. IEEE.

[Yachida et al. 80] M Yachida, M Ikeda, and S Tsuji. A plan guided analysis of cineangiograms for measurment of dynamic behavior of heart wall. *IEEE Transactions on Pattern Analysis and Machine Intelligence,* 2(6):537–543, 1980.

[Young and Axel 92] A Young and L Axel. Non-rigid heart wall motion using MR tagging. In *Proceedings, 1992 Computer Vision and Pattern Recognition,* Champaign, IL, pages 399–404, IEEE, Los Alamitos, CA, 1992.

[Zheng and Tsuji 90] J Y Zheng and S Tsuji. From anorthoscope perception to dynamic vision. In *Proceedings of the 1990 IEEE International Conference on Robotics and Automation,* Cincinnati, OH, pages 1154–1160, IEEE, Los Alamitos, CA, 1990.

[Zhuang et al. 88] X Zhuang, R M Haralick, and Y Zhao. From depth and optical flow to rigid body motion. In *CVPR '88: Computer Society Conference on Computer Vision and Pattern Recognition,* Ann Arbor, MI, pages 393–397, IEEE, Los Alamitos, CA, 1988.

Chapter 16

Case studies

This chapter presents four case studies that illustrate many of the concepts introduced in this book. No one example falls entirely within the bounds of one chapter and so, while we have attempted to present material in increasing order of complexity, this has not always been possible.

In brief, the topics we illustrate here are

An optical music recognition (OMR) system: (Section 16.1) Image binarization and deskewing; segmentation and region labeling; projection histograms; clustering; rule-based reconstruction; syntactic-and semantic-based feedback

Automated image analysis in cardiology: (Section 16.2) Image acquisition techniques; edge detection; geometric transforms; border detection using graph searching; cost function design

Automated identification of airway trees: (Section 16.3) Semi-thresholding; X-ray CT imaging; image pre-processing; edge-based region growing; rule-based image interpretation; three-dimensional visualization

Passive surveillance: (Section 16.4) Image subtraction; morphological noise reduction; boundary extraction and spline approximation; PDMs and ASMs; Kalman filtering

These examples have been selected to demonstrate a wide range of algorithms and approaches; none of these is 'finished' in the sense of being a complete answer, since they are all examples of live, ongoing research which are the subject of continuing work.

16.1 An optical music recognition system

Optical music recognition takes musical scores (usually a sequence of *staffs*, each usually composed of five *stave* lines) and, after scanning, translates the written music into an internal computer-usable form, of which there are many (e.g., MIDI [Loy 85] and DARMS [Erickson 75]). The similarities of this task to optical character recognition (OCR) are clear inasmuch as the latter translates a 2D representation that we are accustomed to reading into digital form, but the differences are critical. These include:

Figure 16.1: *A section of music, as read by a scanner. Courtesy K. C. Ng, School of Computer Studies, University of Leeds.*

- Unlike characters, symbols come in a variety of sizes, sometimes compound (e.g., beamed groups).
- The information stream is not wholly sequential—events may be simultaneous. This leads to partial superposition of information.
- Variable semantics—a symbol's meaning may depend on its neighbors and context.
- Interconnection and inconsistent spacing.

In addition, we have all the customary problems of scanning documents—ink flooding, symbol breaking, misalignment etc. A section of score appears in Figure 16.1, illustrating some of these difficulties.

This problem has received a lot of attention over many years [Blostein and Baird 92, Selfridge-Field 94] but, despite several partial successes, no complete solution has been developed. We outline here an approach due to Ng [Ng 95] which has overcome many of the problems of its predecessors, and which is a good illustration of how a 2D, binary problem may quickly lead into difficult and complex matters.

Images are first scanned and binarized by a straightforward application of Algorithm 5.2 (see Figure 16.2). Music layout is heavily dependent on the five line staves and the vertical *bar* lines that subdivide them, and the correct alignment of the data is important to all that follows. Possible skew is detected by correlating a long, thin ribbon of black with the image (Section 5.4)—a best fit then provides the best estimate of skew at that point. The exercise is repeated many times and a voting mechanism derives the best overall skew estimate, after which the image is rotated to compensate in a manner described in Section 4.2.2. All the staves are then located, allowing the fundamental geometric unit of the page (the inter-stave line distance) to be computed—this unit provides the scale of all other symbols.

Given the geometry, it is possible to remove the stave lines selectively (although this is non-trivial), leaving the symbols. It is important to remove only black pixels that are parts of stave lines and not parts of a superimposed feature; this can be done by deriving good estimates of the 'likely' thickness of these lines in isolation—if they appear thicker, it is safe to assume another feature is nearby. After this stage, surviving black pixels are sometimes musical primitives (e.g., notes), but more often they are compound or interconnected (see Figure 16.3). A feature of Ng's approach is to decompose all symbols to their graphic (rather than musical) base [Ng and Boyle 92]—thus a note is composed of a notehead *and* a stem, although a musician would see these two as part of the same primitive.

Figure 16.2: *The sample, binarized and deskewed. Courtesy K. C. Ng, School of Computer Studies, University of Leeds.*

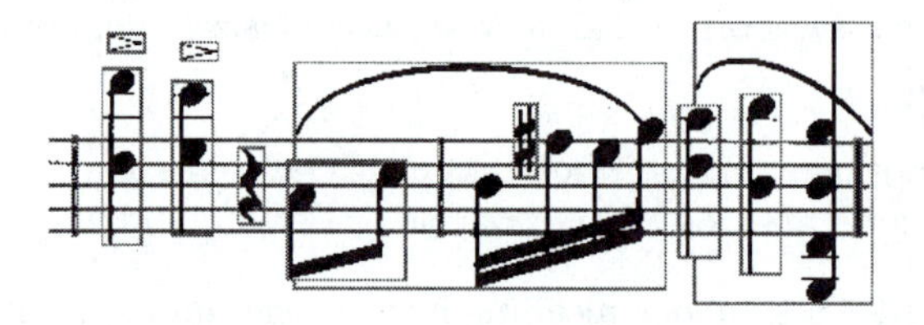

Figure 16.3: *Compound symbols within their bounding boxes. Courtesy K. C. Ng, School of Computer Studies, University of Leeds.*

In its current form, the system can process a significant (and usable) number of musical symbols—notes, rests, accidentals, clefs, bars, and dots. A training set of these (drawn from multiple publishers) is constructed by measuring only the height and width of these symbols' bounding boxes. A cluster-based (Section 7.2.4) nearest-neighbor (minimum distance) classifier is trained; empirically the clustering is seen to be very good, and classification unambiguous; unknown symbols are offered to it and regarded as identified if they fall 'close' to a cluster—this measurement is formalized by examining the spread of each cluster and using this to derive a measure of confidence that an unknown point does or does not belong [Ng and Boyle 96].

In the event of no classification being made, the symbol is partitioned by breaking it with either a horizontal or vertical cut at a point suggested by the vertical and horizontal projection histograms p_h and p_v (Section 6.3.1) of the symbol. This approach in fact uses an adaptation of the projection developed for on OCR application that delivers very good 'cut points' [Kahan et al. 87]—instead of looking for minima of the projection $p_h(i)$ or $p_v(j)$ (thin points of the symbol), we search for maxima of the functions

$$\frac{|p_h''(i)|}{p_h(i)} \quad \text{and} \quad \frac{|p_v''(j)|}{p_v(j)}$$

thereby locating minima of the projection that are simultaneous with large values of its second derivative (see Figure 16.4). This iterative procedure successfully locates reasonable break points in the pixel arrays; early iterations often locate connecting beams or slurs (long curved lines), which may be identified by their thicknesses. These are features of variable length which may be curved or straight—they are traced to their extremities, their locations remembered and then erased. This often requires projection of lines through other features such as stems (because of superposition), for which the curved lines are modeled locally as quadratics.

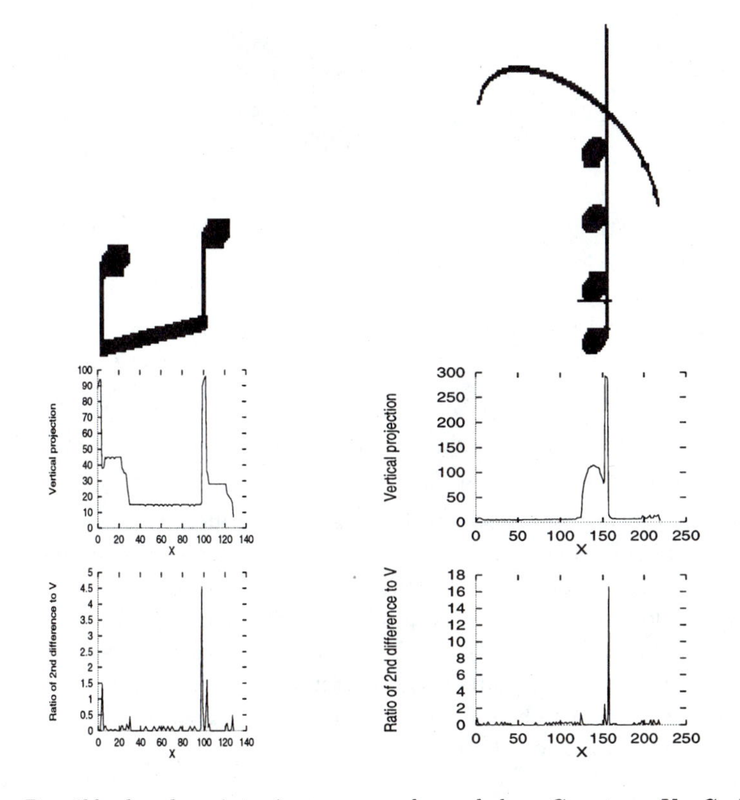

Figure 16.4: *Possible break points in compound symbols. Courtesy K. C. Ng, School of Computer Studies, University of Leeds.*

This iterative subdivision continues until either identification or failure, signaled by the feature being too small—in nearly all circumstances an identification is made, usually accurately. An illustration showing a part of Figure 16.1 is given in Figure 16.5. In the case of features that have *pitch* (e.g., noteheads and accidentals), this is revealed by their geometric position on the staff. This approach is successful because it is simple and robust; the use of easily observed features such as height and width is in contrast to many more elaborate schemes that are frequently undone by their own complexity—interconnectivity, whether intended or accidental, serves to undermine attempts to recognize 'whole' features.

The output of this classification is not perfect, but is a moderately reliable identification of the primitives we seek—a rule-based reconstruction then takes place to complete the representation [Ng and Boyle 96]; see Figure 16.6. At this point, beams, dots and flags on note stems become relevant as they betray the note durations. Two particular musical features are accessible via the image at this stage that are of importance [Ng et al. 95].

1. Timing: Music is divided into *bars* of equal duration—the dividing lines are among the easiest things to locate in the score. Knowledge of the *time signature* gives enormous

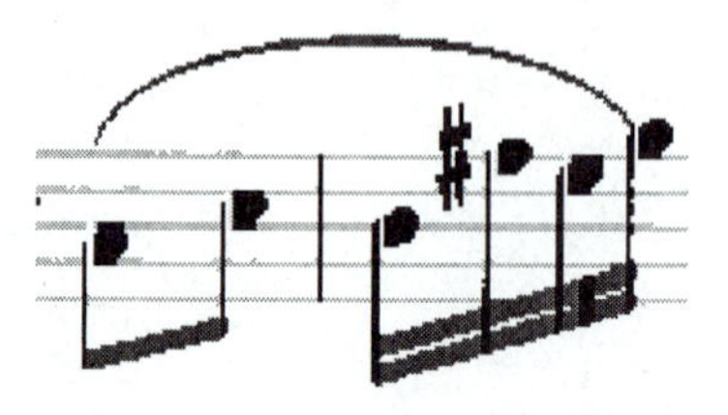

Figure 16.5: *Subsegmentation of primitives. Courtesy K. C. Ng, School of Computer Studies, University of Leeds.*

assistance in verifying note identifications, since it constrains how many notes of which type may appear in a given place; this knowledge resolves many ambiguities.

2. Key: The musical key of the piece is a good high-level clue to what is likely in certain instances—in particular, it can dictate which 'accidentals' (sharp and flat signs) may or may not appear in certain places.

Ng has developed algorithms based on the preliminary identifications that permit automatic identification of both time and key signatures. This is done by examination of a suitably large sample of the music under the assumption that the mis-identifications are few enough to permit statistical features of the music to emerge. Specifically, relative note frequency is a good indicator of key, and time signature is betrayed by a combination of (estimated) duration, together with particular common note patterns to distinguish fine-grained differences (for example, between 3/4 and 6/8 time). These judgments are of course statistical, and small samples (or unusual music) would defeat this phase.

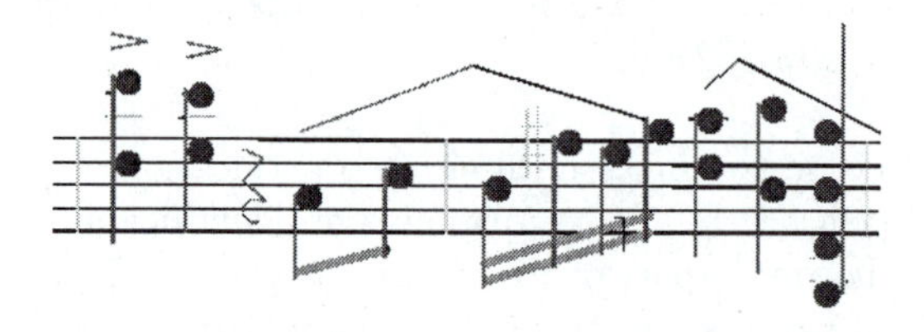

Figure 16.6: *The resulting reconstruction. Courtesy K. C. Ng, School of Computer Studies, University of Leeds.*

This application is a good example of various very low-level techniques coming together to solve (part of) what is known to be a difficult problem. The key to it turns out to be, in addition to a sound understanding of the image processing, a full understanding of the domain (musical notation) and which parts of it may be brought to bear on the pattern recognition problem at hand. Thus, the time and signature identification are critical to verification of identification, and feedback for error correction, and the significant success demonstrated would probably not have been accessible to someone attempting the same job without the musical expertise—this moral is important to bear in mind; an incomplete understanding of the pictures cannot be used to build a complete *automatic* understanding of them.

16.2 Automated image analysis in cardiology

Each year, millions of people suffer myocardial infarction as a result of narrowing of a coronary artery; the death rate from coronary artery disease in the United States alone exceeds 500,000 per year. Assessment of coronary arterial geometry is central to the diagnosis of coronary artery disease, to decisions about pharmacological therapy or coronary artery re-vascularization with transluminal coronary angioplasty or coronary artery bypass surgery, to judgments about the short- and long-term outcome of re-vascularization procedures, and to assessment of patient prognosis. While each of the existing methods for assessing disease in vivo, such as coronary angiography, fluoroscopy, fiberoptic angioscopy, ultra-fast computed tomography, magnetic resonance angiography, epicardial ultrasonography, and intra-vascular ultrasound, has potential advantages [Collins and Skorton 86, Marcus et al. 91, de Feyter et al. 95], only coronary angiography and intra-vascular ultrasound have found widespread clinical use.

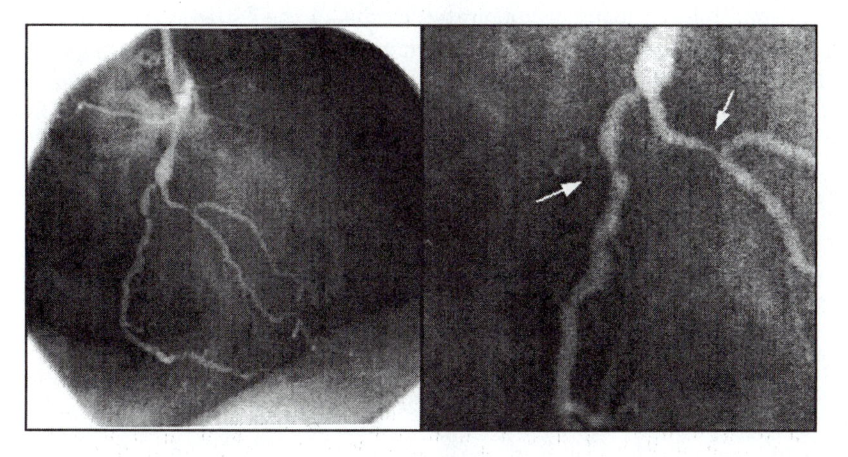

Figure 16.7: *Coronary angiograms. Left: Angiogram of the coronary tree. Right: Stenotic vessel segments are depicted by arrows in an enlargement.*

Coronary angiography produces projection images or silhouettes of the vascular lumen from the absorption of X-rays by radio-opaque dye injected into the coronary tree (Figure 16.7). Conclusions about the presence and significance of atherosclerosis are inferred from visual assessment of the reduction in vessel lumen diameter with respect to the lumen diameter of a nearby vessel segment that is presumed to be normal. Angiography has maintained a pivotal role in the evaluation and treatment of patients with coronary disease and millions of coronary angiograms are performed every year around the world, the majority of which are interpreted visually. Visual assessment of the severity of coronary disease is associated with very substantial inter- and intra-observer variability. Moreover, it does not allow accurate assessment of the physiological significance of coronary obstructions. These shortcomings have prompted clinical investigators to call for the use of quantitative coronary angiography for assessing coronary obstruction severity in patient care. A substantial amount of research has been directed at the development of semi-automated and automated methods for defining coronary arterial borders and calculating indices of lesion severity and physiological significance. As a result, automated approaches were developed to identify vessel borders and calculate indices of stenosis severity from coronary angiograms. In Section 16.2.1, one of the most recent methods for robust coronary border detection is presented; other densitometric and geometric approaches to computer analysis of images from coronary angiography may be found in [Sonka and Collins 97].

With intra-vascular ultrasound, a high-frequency (20–50 MHz) ultrasound source rotates near the tip of a catheter inserted in the arterial lumen [Bom et al. 95]. Either the piezoelectric crystal generating the ultrasound beam or a mirror deflecting the ultrasound beam is rotated. Alternatively, a stationary multi-element crystal array encircling the catheter may be employed. In the former case, the ultrasound element or the mirror is rotated by use of a long flexible shaft driven by an external motor. The lengths of the shaft and arterial path tortuosity may cause non-uniform rotation, creating imaging artifacts. In the latest prototypes, the external motor has been replaced by a micro-motor in the catheter tip. In multi-element ultrasound arrays, no mechanical rotation is involved and the signal is processed and multiplexed using an integrated circuit in the catheter tip.

Atherosclerosis is characterized by an accumulation of plaque material in the arterial wall; Figure 16.8 shows in schematic form the cross-sectional anatomy of a diseased coronary artery. Intra-vascular ultrasound yields tomographic two-dimensional cross-sectional images

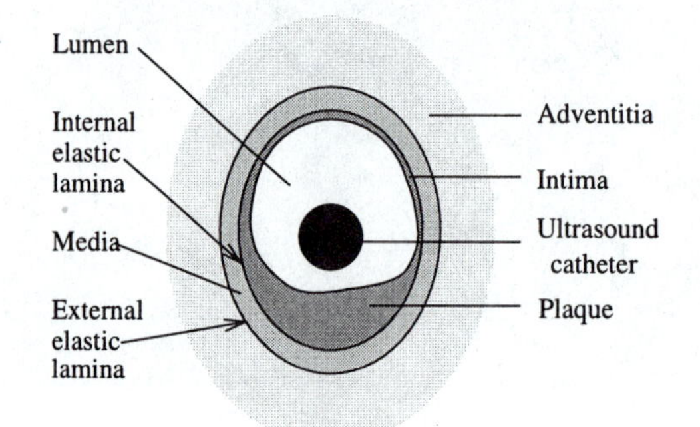

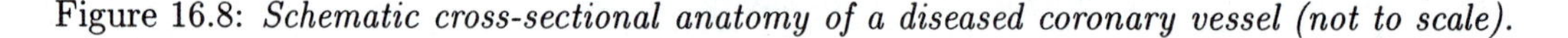
Figure 16.8: *Schematic cross-sectional anatomy of a diseased coronary vessel (not to scale).*

that directly demonstrate arterial lumen, plaque, and wall morphology (Figure 16.9). In adults, normal coronary vessel walls (upper portion of Figure 16.8) have a thin intimal layer bounded by the internal elastic lamina. The media, composed of smooth muscle cells and a reticular collagen network, is located between the internal and external elastic laminae and is of nearly constant thickness, typically 0.1–0.3 mm. Outside the external elastic lamina is fibrous and fatty adventitia tissue.

From the images obtained by intra-vascular ultrasound, morphometric measurements of lumen and plaque can be performed directly. There is substantial enthusiasm in the clinical community for using this emerging imaging method to overcome the shortcomings of angiography and for guiding and evaluating catheter-based interventions. Intra-vascular ultrasound facilitates determination of plaque dimensions and composition, including studies of mechanisms of plaque dilatation, recoil, and dissection after catheter-based intervention. Its role in interventional planning and after-intervention quality control is unique. To date, quantitative intra-vascular ultrasound studies of coronary atherosclerosis have nearly always relied on

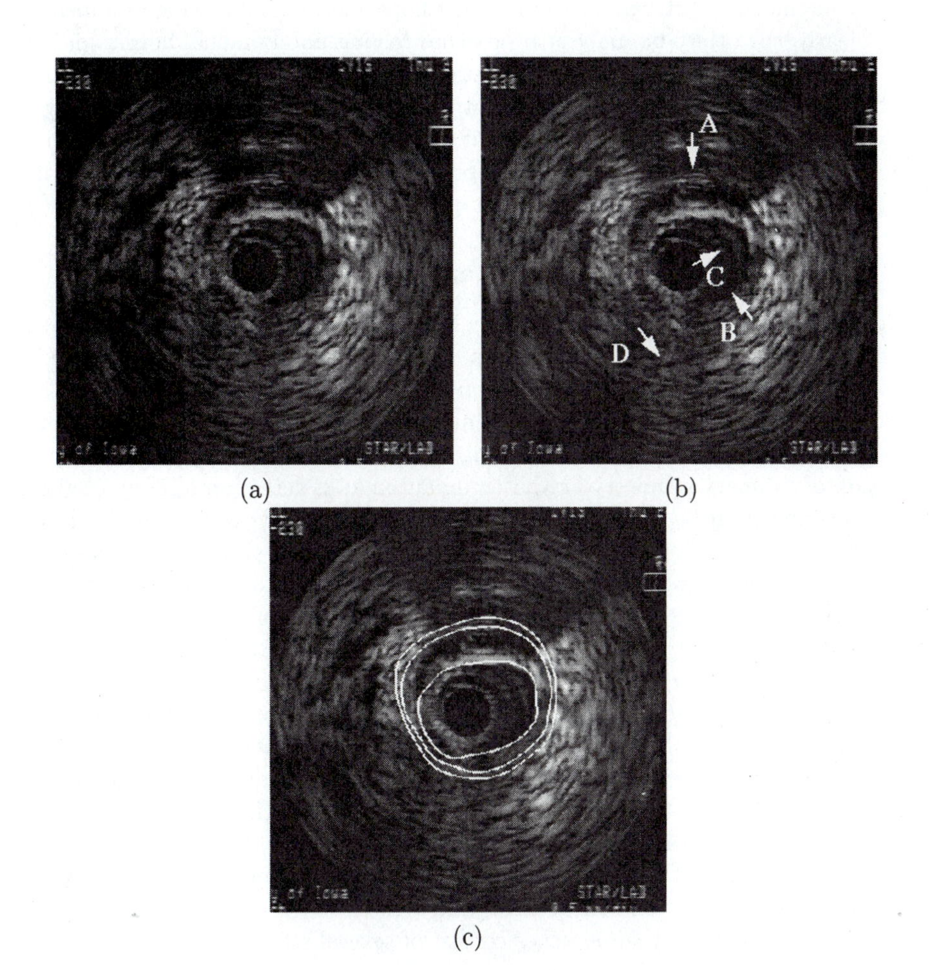

Figure 16.9: *Intra-vascular ultrasound image of a diseased coronary vessel. (a) Original image. (b) Original image showing circumferential plaque that thickens at the 10- to 2-o'clock position. Double-echo response from the internal and external wall borders with intervening sonolucent media is seen at the 12-o'clock position (arrow A) and also at 10-o'clock and 5-o'clock. The interface between lumen and plaque is depicted by arrow B and is visible at 9- to 6-o'clock. The catheter near-field artifact surrounds the catheter in the middle of the image and obscures the plaque–lumen border from 6- to 9-o'clock. Arrow C shows a near-field artifact. Two wide arrows point to regions of signal attenuation caused by fibrotic plaque that is highly reflectant. The dark sonolucent region from 10- to 2-o'clock located inside the plaque is due to echo dropout distal to the fibro-calcific plaque at the lumen–plaque interface. Arrow D shows the strut and guidewire artifact. (c) Positions of external elastic lamina, internal elastic lamina, and coronary lumen defined by an experienced observer.*

manual measurements of arterial structure dimensions and manual tracing of lumen and wall borders. Clearly, the utility of analysis approaches relying upon manual border identification is limited by the need for observers with substantial experience and by the tedious nature of manual tracing which is prone to inter- and intra-observer variability. One of the few existing automated methods for three-dimensional intra-vascular ultrasound image analysis is presented in Section 16.2.2; more detailed descriptions of intra-vascular ultrasound applications and analysis methods can be found in [Sonka et al. 97a].

16.2.1 Robust analysis of coronary angiograms

Automated methods for evaluation of coronary lesion severity from angiograms can be divided into geometric and densitometric approaches. Densitometric approaches are based on the density of contrast material in the vessel lumen and allow estimation of lumen cross-sectional area from a single angiographic view [Mancini et al. 87, Johnson et al. 88, Nichols et al. 84, Wiesel et al. 86, Simmons et al. 88, Whiting et al. 91]. Conventional geometric approaches to evaluation of coronary geometry are based on automated detection of individual coronary lumen borders and on using data from two or more angiographic projections to reconstruct a three-dimensional representation of coronary lesion size and shape. An example of this latter approach is the Brown-Dodge method of quantitative coronary arteriography [Brown et al. 77]. A great variety of methods for automated detection of coronary borders have been introduced [Ellis et al. 86, Wong et al. 86, Le Free et al. 86, Mancini et al. 87, Eichel et al. 88, Fleagle et al. 89, Reiber et al. 91, Buchi et al. 90]. In general, these methods detect coronary borders by identifying image pixels with large gradients. Although this approach works well with high-quality images, it ignores global information that is very useful in detecting local border position in images of intermediate or poor quality. Reiber et al. were the first to report a well-validated coronary border detection technique that identified a border that in a global sense was optimal [Reiber et al. 84]. Computer identification of optimal coronary borders is typically based on 2D graph searching or dynamic programming principles introduced in Sections 5.2.4, 5.2.5, and 8.1.5.

Coronary border detection approaches consist of several steps.

1. *Observer identification of the vessel segment of interest.*

2. *Application of an edge operator* to derive an edge image (Section 4.3.2). The edge operators used are often quite large, to help in detection of low-contrast edges in noisy angiography images. To determine border location accurately and to facilitate correct vessel diameter measurements, combinations of first- and second-derivative edge detectors are common. For example, in studies in which the angiograms were digitized to yield pixel sizes of approximately 0.05 mm, a weighted combination of a Sobel-like 11×11 operator (Section 4.3.2) approximating the first derivative of image intensity and a 21×21 Marr-Hildreth operator (Section 4.3.3) approximating the second derivative of image intensity were used [Fleagle et al. 89, Sonka et al. 94]. The mask sizes and relative weights were determined empirically from phantom studies. Alternatively, adaptive edge operators may be used to enable accurate determination of small vessel diameters [Sonka et al. 97b, Sonka et al. 93a].

3. *Geometric warping* to straighten the edge image (Figure 5.24, Sections 4.2 and 5.2.4).

4. *Construction of a directed graph and optimal border identification* (Figure 5.21, Sections 5.2.4, 5.5.1). In conventional approaches in which individual (left and right) coronary borders are identified independently, a 2D graph is constructed. If both borders are identified simultaneously as discussed below, a 3D graph is constructed and searched.

5. *Mapping of the borders which have been detected back into the original image space* using the inverse of the straightening geometric transform of step 3 and *calculation of a diameter function.*

Although conventional methods based on individual identification of globally optimal borders in 2D graphs were seen to be accurate in uncomplicated images (e.g., Figure 5.26), experience suggests that even the most robust conventional techniques often fail to identify acceptable borders when routinely applied to unselected images in a clinical setting. Images with poor contrast, high noise, branching vessels, or nearby or overlapping structures present particular problems for automated techniques. Figure 16.10a gives an example of such a complex angiogram. Conventional automated methods for coronary border detection fail in part because they identify the left and right borders independently. Clearly, there is information contained in the position of one border that might be useful in identifying the position of the other border. Consequently, a method for simultaneous detection of coronary border pairs was developed to overcome most of these limitations [Sonka et al. 93b, Sonka et al. 95b, Sonka and Collins 97].

The algorithm for 3D graph construction and simultaneous border detection was presented in Section 5.5.1. The cost function used was of the form presented in equation (5.42), and the heuristic graph searching algorithm was used for optimal border detection. To understand better the cost function that was specifically designed for coronary angiography border detection, we describe its most important parameters: Local edge directions were considered, multi-resolution and multi-stage control strategies were used as discussed in Section 8.1.5, a symmetric vessel model (Figure 5.58) was used with parameters $\alpha = 1.8$, $\beta = 2.2$ in the first stage and $\alpha = 1.2$, $\beta = 1.4$ in the second stage [Sonka et al. 95b]. One of the implementational challenges was to achieve reasonable processing times compared to the conventional method. Because of the substantial memory and processing requirements of the 3D graph searching algorithm, special attention was paid to careful design of data structures. For instance, the *open_list* data structure was a triply linked, height-balanced binary search tree [Reingold and Hansen 83]. As nodes were generated and placed on the open list, *open_list* records were dynamically created. Each record contained fields for the partial path cost to the node, the (x, y, z) co-ordinates of the corresponding node, a balance factor describing the relative heights of the left and right subtrees of the record, and pointers to the three adjacent records (i.e., parent, left and right sibling records) in the binary tree. The balance factor was used to keep the binary search tree balanced in order to maintain optimum access times to the open list. Records were inserted into the *open_list* structure so that they remained sorted according to partial path costs.

In medical image analysis applications, it is of primary importance to validate developed methods in a large number of clinical images in comparison to an independent standard. The independent standard is manually defined by one or more expert observers or may be determined using another previously validated method. For coronary border detection, the

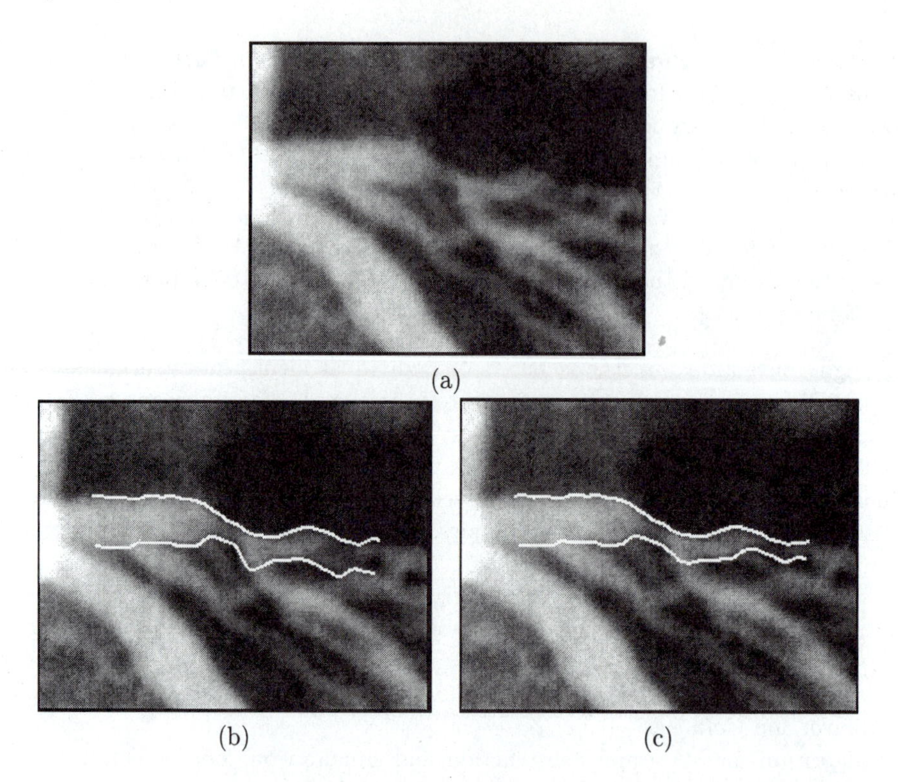

Figure 16.10: *Border identification in a coronary artery with a closely adjacent vessel: (a) original image; (b) conventional border detection incorrectly identified the vessel borders in the right end of the traced vessel—an adjacent vessel border was detected instead; (c) simultaneous border detection correctly detected both coronary borders.*

main performance characteristics are the border detection reliability and the accuracy of minimum lumen diameter measurements. To evaluate the reliability of the simultaneous border detection algorithm, the frequency with which the method failed to yield reasonable borders in unselected clinical angiograms was assessed. The failure rate of the method was compared to the failure rate of a conventional border detection method as assessed by an expert observer. Minimum lumen diameters were compared to diameters derived from manual tracing and editing.

Figures 16.10, 8.2, and 8.4d show examples of coronary borders detected with the conventional and simultaneous border detection methods in a vessel segment with closely adjacent and branching vessels. Conventional border detection failed to identify acceptable borders of the vessel of interest, while the simultaneous border detection algorithm yielded accurate borders. Figure 16.11 summarizes the results of evaluation of the reliability of simultaneous and conventional border detection in 439 clinical angiograms. Conventional coronary border detection failed to yield acceptable borders in 88/439 or 20% of images. Simultaneous border detection failed in only 17/439 or 4% of images. This represents a very substantial reduction in the failure rate, due to the simultaneous character of border detection and the incorpo-

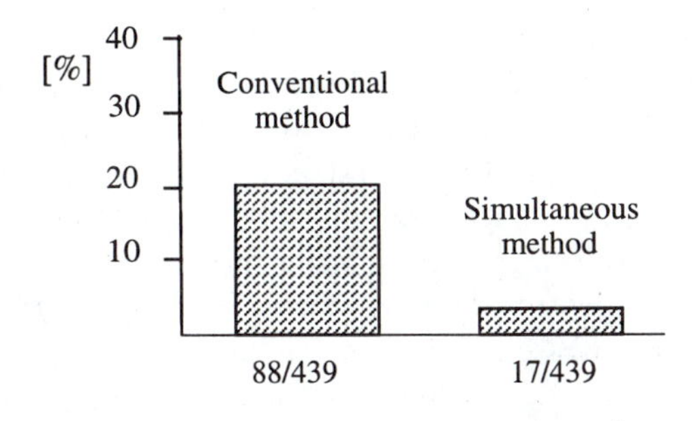

Figure 16.11: *Failure rates of coronary border detection in complex images. Conventional border detection (left) and simultaneous border detection (right).*

rated a priori knowledge about coronary vessels and their angiographic border properties. In uncomplicated images for which the previously validated method was shown to be accurate, minimum lumen diameters derived from the simultaneous and conventional border detection methods were highly correlated ($r = 0.97$), showing that the reduction in the failure rate was not achieved at the expense of decreased accuracy.

16.2.2 Knowledge-based analysis of intra-vascular ultrasound

Several approaches exist to distinguish among vessel lumen, plaque, and surrounding tissues to identify vessel lumen in intra-vascular ultrasound images (Figure 16.9) and to produce three-dimensional renderings of vessel geometry [Sonka et al. 97a]. Usually, these methods are based on image intensity thresholding. Vessel walls detected by these methods typically include plaque, intima, media, and adventitia (Figure 16.8), with no or limited ability to distinguish between them. Since intra-vascular ultrasound images are noisy, much effort is currently being devoted to development of blood backscatter-based techniques to clean up the noise from the imaged vessel lumen [Bom et al. 95, Pasterkamp et al. 93, Evans and Nixon 96]. Such pre-processing may substantially improve performance of threshold-based techniques.

Alternatively, border identification approaches that are based on variations of optimal border detection are used [Dhawale et al. 93, Li et al. 93b, von Birgelen et al. 95, Sonka et al. 95c, Sonka et al. 95a]. The method for automated segmentation of intra-vascular ultrasound images that is discussed below uses global image information and heuristic graph searching (Section 5.2.4) to identify wall and plaque borders [Sonka et al. 95c]. A priori knowledge about coronary artery anatomy and ultrasound image characteristics was incorporated into the border detection cost function. Figure 16.9 clearly demonstrates that intra-vascular ultrasound images are noisy and contain catheter and other artifacts. Without contextual information from image frames adjacent in space and time, single-frame intra-vascular images are difficult to analyze even for the most experienced humans. To be successful, automated methods need to incorporate substantial a priori knowledge about ultrasound imaging physics and arterial anatomy. The following method incorporates a priori information of a variety of types into the border detection process. In particular, to identify the position of the wall borders, the search is not for a connected series of pixels with large edge gradients but for an expected double-echo edge pattern based on the differences in acoustic impedance between the elastic laminae and the surrounding tissues.

At this point, some fundamentals about ultrasound imaging are needed. The interaction of ultrasound with coronary arterial structures gives rise to typical image patterns that may be used to identify arterial wall and intramural structures within the ultrasound image (Figure 16.9). Intra-vascular ultrasound images contain both large-amplitude signals from specular reflectors at the interface between structures with differing acoustic impedances and lower-amplitude signals or diffusely scattered signals arising from regions of nearly uniform acoustic properties. In intra-vascular ultrasound images, the lumen is typically a dark echo-free region adjacent to the imaging catheter. The lumen–intima or lumen–plaque interfaces constitute a large acoustic impedance mismatch and produce specular reflections. The internal and external elastic laminae enclose the sonolucent media and produce acoustic impedance mismatches. Importantly, in ultrasound images, the location of the acoustic interface (border) is not depicted by the largest amplitude of the echo response. Since the ultrasound image visualizes ultrasound signal reflections, the correct border position corresponds to the location of the *leading edge* of the ultrasound echo signal. This fact is also used for accurate identification of the coronary wall and plaque border positions in the knowledge-based method described here.

After digitization of intra-vascular ultrasound images yielding a pixel size of about 0.03 mm, the border detection method consists of the following steps.

1. *Interactive definition of the region of interest.* Due to the image ambiguities shown in Figure 16.9, a region of interest must be identified prior to automated border detection. If a single (2D) image frame is analyzed, an ellipse enclosing the vessel (lumen, plaque, media, and a portion of the adventitia) is manually defined by an operator. The ellipse defines the outer limit of the region of interest. The inner limit is identified by a closed smoothed polygon connecting several points identified by the operator inside the vessel lumen near the blood–plaque interface. If an image sequence is processed, the manually identified region of interest is only defined for the first image frame and results from previous frames are used to estimate the regions of interest in the subsequent frames.

2. *Edge detection and co-ordinate transformation of the edge image.* Similarly to other applications of border detection based on graph searching (Sections 5.2.4–5.5.1, 16.2.1), the edge image is produced, straightened, and a 2D graph is constructed. Since the ultrasound echo features are local, small edge detectors (5×5 or 7×7) are utilized.

3. *Identification of vessel wall and plaque borders.* Using a 2D graph searching approach, optimal borders representing the internal and external laminae and the plaque–lumen interface are sequentially identified. The outline of the region of interest serves as a shape model during the external wall border detection. The identified external wall border then serves as a model for the internal wall border detection.

4. *Mapping of all three borders back into the original image space* using the inverse of the geometric transform used for edge image co-ordinate transform in step 2 and *calculation of quantitative measures of arterial morphology.*

As always, the key to identifying accurate borders is to define an appropriate cost function for nodes in the searched graph. Since the properties of border pixels forming the external and internal vessel wall borders and plaque–lumen border differ, separate cost functions were developed for detection of each border. In contrast to the usual applications of graph searching-based border detection, where costs are inversely related to edge strength, the cost functions in this ultrasound border detection method are related to a specific expected edge pattern and incorporate a priori knowledge about cross-sectional arterial anatomy.

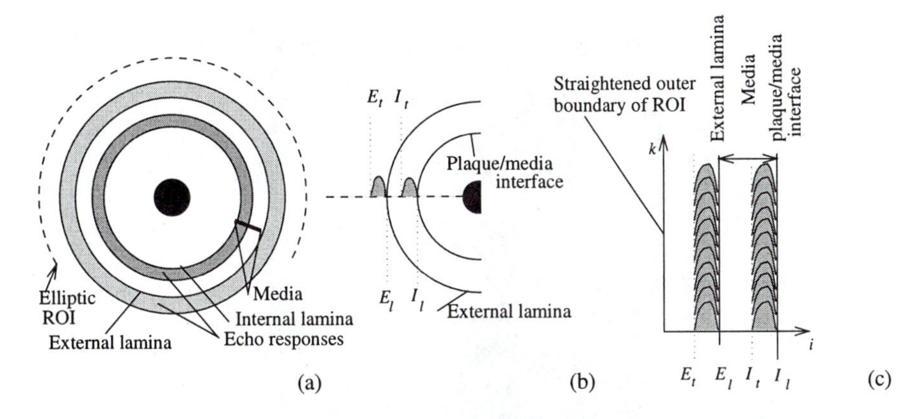

Figure 16.12: *Knowledge-based coronary wall and plaque border detection: (a) ultrasound responses from internal and external wall borders form a double-echo pattern; (b) idealized echo responses along a radial profile; (c) leading and trailing edges of echo responses from internal and external wall borders in the resampled image.*

Two types of a priori information are used to calculate the cost function for the external elastic lamina. First, the outer boundary of the region of interest is used as a rough model for the shape of the external lamina. The second type of a priori information incorporated in the cost function is the expected double-echo pattern arising from the interfaces of the media and the surrounding external and internal elastic laminae. The double-echo pattern is depicted by arrow A in the middle panel of Figure 16.9, coronary anatomy terminology is shown in Figure 16.8, and manually identified borders of coronary wall and plaque are shown in the right panel of Figure 16.9. Costs associated with each node are designed to reflect the degree of match between the local edge configuration and the expected double-echo edge pattern. The expected edge configuration is shown in Figures 16.12a,b. E_l is the leading edge of the ultrasound echo response and represents the location of the external elastic lamina. E_t represents the trailing edge of the echo response to the external lamina. I_l and I_t represent the leading and trailing edges of the echo response depicting the internal elastic lamina. However, the leading edge I_l may be superimposed by the plaque echo and therefore does not contribute to the pattern cost. Considering anatomical knowledge about media thickness, the combination of three strong edges I_t, E_l, and E_t located approximately 0.1–0.3 mm apart and having appropriate edge directions is given the highest likelihood of indicating the position of the external elastic lamina. As such, the identification of the external elastic lamina is based on searching for edge triplets which form a contiguous circumferential pattern in the original ultrasound image. The position of the external lamina is defined as the location of the echo leading edge E_l (Figure 16.12c).

The internal wall border is identified inside of and in proximity to the external lamina border. The cost assigned to a node depends upon its edge strength, its edge direction, and its position with respect to the external lamina border. Information about media thickness is used as a priori knowledge. When searching for the internal lamina border, the previously detected external lamina border is used as a model. The model in this case forms a narrow region of support in which the internal border is most likely to be located. The plaque–lumen border is identified using a standard edge strength-based cost function.

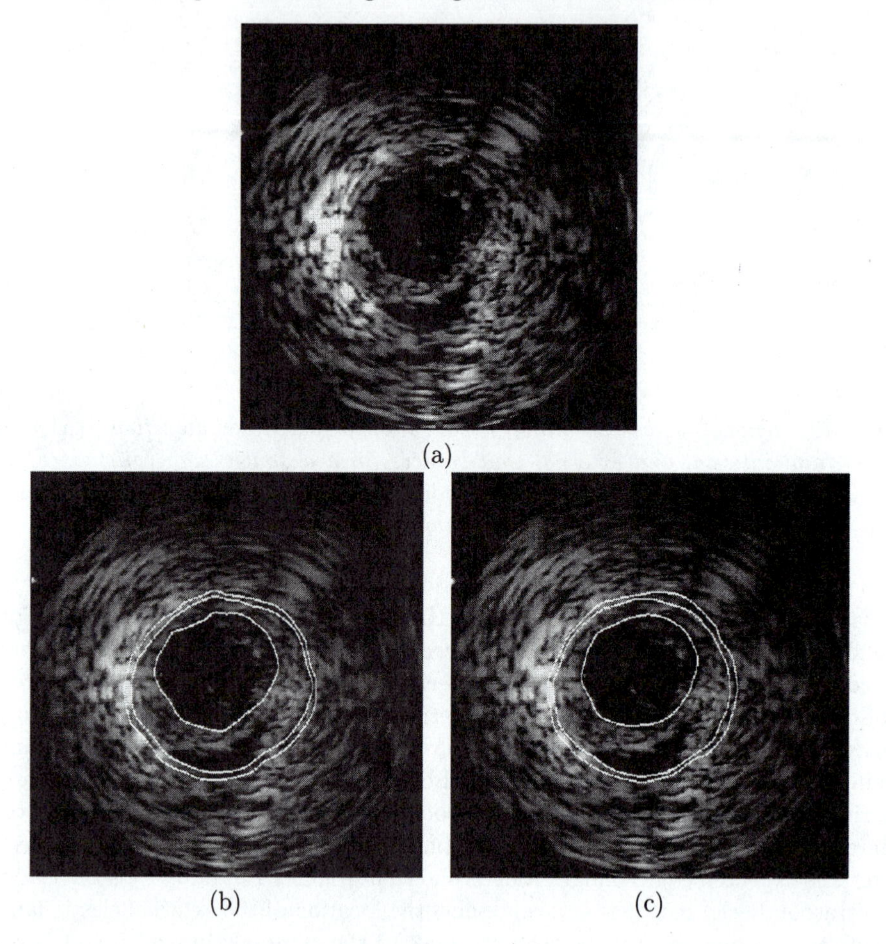

(a)

(b) (c)

Figure 16.13: *Intra-vascular ultrasound border detection in a diseased coronary vessel using the knowledge-based method: (a) original image showing plaque thickening at the 2- to 9-o'clock position; (b) computer-determined internal and external wall and plaque borders; (c) manually identified internal and external wall and plaque borders.*

To assess the performance of the intra-vascular ultrasound border detection method, automatically detected wall and plaque borders were compared with carefully determined manually identified borders that served as an independent standard. To evaluate the accuracy of the border detection method objectively, quantitative measures such as lumen and plaque area obtained from computer-detected borders were compared to measures obtained

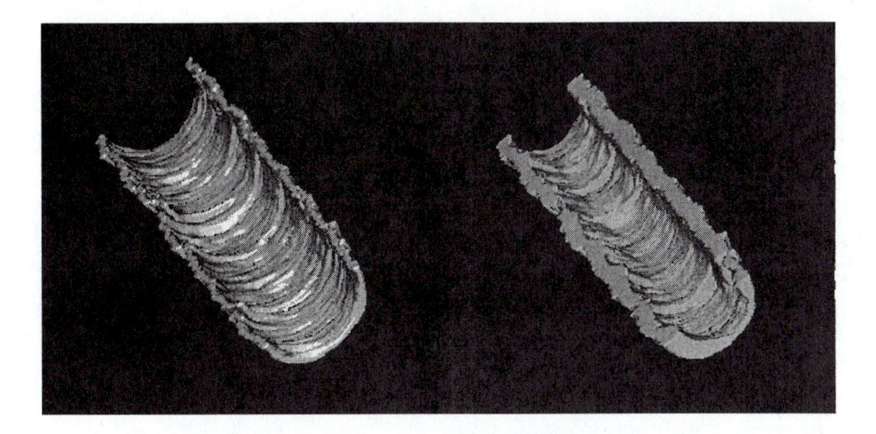

Figure 16.14: *If the ultrasound catheter is pulled back along the vessel, the automated border detection in the image sequence yields a three-dimensional image. Examples show surface rendering of automatically determined coronary wall (left) and plaque (right). The 3D pull-back IVUS data were acquired in vivo after coronary atherectomy. The periodic wall/plaque pattern is caused by cardiac motion.*

from observer-defined borders. To assess border positioning accuracy, computer-detected and observer-defined borders were directly compared by calculating border positioning errors.

Figures 16.13 and 16.14 show examples of single-frame and pull-back sequence analysis. As can be seen, the computer-determined borders closely approximate the borders identified by the expert observer. In the large number of analyzed intra-vascular ultrasound images, computer-determined area and volume measurements correlated well with the independent standard ($r > 0.95$), and the border positioning errors were small (root-mean-square error of vessel wall border position was 0.08 ± 0.02 mm). To appreciate the small size of these border position errors, one needs to remember that the maximal obtainable accuracy is of the order of one wavelength of the transmitted ultrasound [Li et al. 93a]. At a frequency of 30 MHz, the theoretical limiting accuracy is 0.05 mm, as the velocity of sound in blood is approximately 1500 m/s. Thus, the wall and plaque border detection was highly accurate.

In designing the intra-vascular ultrasound border detection method, the aim was to mimic the strategies employed by experienced ultrasonographers when visually identifying arterial borders. When the local distribution of image intensities do not allow unambiguous border identification, expert observers rely upon a priori knowledge they have accumulated about intra-vascular ultrasound image characteristics. Their final decision about border locations represents a balance between the clues provided by local image patterns and their expectations about overall image characteristics. It is of course possible that the two will at times conflict and it is thus necessary to make a trade-off between local border detection accuracy and overall analysis reliability. The same sort of trade-off was made in designing the overall scheme and cost functions used in the automated border detection method. Overall, the two applications given above clearly demonstrate the power of knowledge-based global optimization in low-quality border detection problems.

16.3 Automated identification of airway trees

X-ray computed tomography (CT) is an excellent tool for cross-sectional and three-dimensional imaging of the human body in vivo. X-ray CT has a unique position within the list of imaging modalities in respect to evaluation of lung function and anatomy. X-ray CT does not suffer the artifacts that are present in magnetic resonance (MR) images which relate to air–water interfaces, and in-plane resolution can be brought down to well under a millimeter with the image gray-scale (Hounsfield units) being essentially linearly related to tissue density. Among others, CT plays an irreplaceable role in diagnosis and treatment of pulmonary diseases since it facilitates sensitive and reliable assessing of alterations in regional lung structure and function. Other available techniques, such as pulmonary function tests, that are based on global lung volume measures are often too insensitive to assess peripheral or heterogeneous abnormalities in structure and function.

Considering image-based techniques, the chest radiograph has been the cornerstone of imaging in patients suffering from a pulmonary disease [Chrispin and Norman 74]. However, its image quality and the associated diagnostic value are limited due to the integral character of projectional images. Thus, descriptive abnormalities on chest radiographs are often vague and subject to inter-observer variability. Computed tomography and particularly high-resolution CT eliminates superposition of structures and has advanced the ability of the radiologist to assess regional abnormalities. Nevertheless, cardiac motion and breathing artifacts represent fundamental problems associated with the use of conventional CT scanning when imaging the beating heart, bellowing lung, or flowing blood, since the length of time required to complete a scan is large in comparison to the duration of the physiological motion of interest. To avoid such problems, image acquisition may have to be gated with respect to the ECG signal or breathing.

One of the important X-ray CT imaging tasks is visualization and quantitative analysis of intra-thoracic airway trees. Such analysis is crucial for assessment of a variety of diseases including cystic fibrosis, emphysema, local airway narrowings, and others. High-resolution CT evaluation of the lung fields in patients with pulmonary disease, namely, patients suffering from cystic fibrosis, is superior to the chest radiograph. Bhalla et al. [Bhalla et al. 91] have shown that the chest radiograph is unable to detect the presence of bronchiectasis (irreversible dilatation of the bronchial tree) in 45% of segments where it is demonstrated on high-resolution CT. Even more important, the plain chest X-ray misses 92% of the mucous plugging that is visible with high-resolution CT.

Identification of airway trees from three-dimensional CT image data sets is a very difficult problem, and development of methods for airway tree detection suffers from several natural limitations of the image data quality inherent in CT scanning geometry and physics. Currently, high-resolution CT scanning via the electron beam CT (EBCT, also called ultra-fast CT) offers fast image acquisition (50 msec per slice) with a best in-slice resolution of 0.4 mm/pixel. However, if the airway is not perpendicular to the scanning plane, effective image resolution may be lower due to a partial volume effect. With a much lower resolution along the z axis resulting from the lowest available slice thickness of 1.5 mm, many small objects such as pulmonary airways or vessels become undetectable on a CT image. The second limitation is linked to the in vivo character of CT imaging. While airway tree detection in isolated lung specimens may be achieved by simple threshold-based three-dimensional region growing [Kiatoka and Yumoto 90, Ney et al. 89, Wood et al. 95], in vivo image data quality can be substantially influenced by breathing and heart motion artifacts, which make image analysis of such CT data sets difficult if not impossible. As a result of in vivo imaging limitations, the airway tree is often disconnected in several locations. Airway gray-levels change substantially with decreased airway diameter because of partial volume effects, limited resolution between scans, and motion artifacts. Therefore, airway segmentation methods based solely on three-dimensional connectivity and absolute gray-level properties of airways will predictably fail on in vivo image data. To facilitate computerized analysis of thoracic CT data sets in vivo, complicated CT acquisition protocols must be followed—CT image acquisition must be ECG gated to minimize heart motion artifacts and volume controlled to minimize breathing motion artifacts. With such care, airways as small as or smaller than 1 mm can be visualized (but not quantified) and airway dimensions can be assessed for airways as small as 2 mm in diameter.

Following this form of scanning protocol, high-quality three-dimensional sequences of cross-sectional CT images can be obtained, as demonstrated in Figure 16.15. The airways (marked by arrows) can be manually traced in individual cross-sectional images and the airway tree may be constructed by three-dimensional stacking of the two-dimensional tracings (see the left panel of Figure 16.18). Such manual identification of the airway tree is extremely time consuming and very tedious. Manual identification of an entire airway tree resulting from expert observer identification of airways in 40 cross-sectional CT images requires approximately 9 hours. With 1.5 mm z-axis resolution, human volumetric CT data sets may contain much more than 100 slices. Clearly, manual identification of airway trees in routine clinical care is not feasible. Here, an automated method for intra-thoracic airway tree detection is briefly described; details can be found in [Sonka et al. 96, Park et al. 96].

A sequence of steps must be performed to achieve automated airway tree detection.

1. *Segmentation of lungs* from the volumetric CT data set. In the CT images, lung tissue and air are represented by the lower end of the CT gray-level range and gray-level thresholding can be used for lung segmentation (Section 5.1). However, the exact threshold often varies within the data set and from patient to patient. Nevertheless, fully automated lung segmentation methods can be developed that adaptively adjust the segmentation threshold [Everhart et al. 94].

2. *Definition of a primary airway tree* as that portion of the tree which consists of major tree branches. The major airways fall within a tight gray-value range in CT images and differ significantly from the airway walls, thus exhibiting high-contrast borders. In the major pulmonary tree branches, gray-level-based techniques work well using three-dimensional seeded region growing [Udupa 82, Hoffman et al. 92] (Section 5.3). To minimize the presence of non-tree structures in the primary tree, the gray-level threshold used for region growing must be quite conservative. As a result, the primary tree does not contain small-diameter airways.

3. *Preprocessing* of individual CT slices. The goal of the pre-processing step is to help identify all potential locations of airways and vessels in individual CT slices. Airway appearance is enhanced using a scale-sensitive 13×13 Marr-Hildreth operator–hat transform (Figure 4.9). Since the hat transform kernel size is designed to enhance small- and

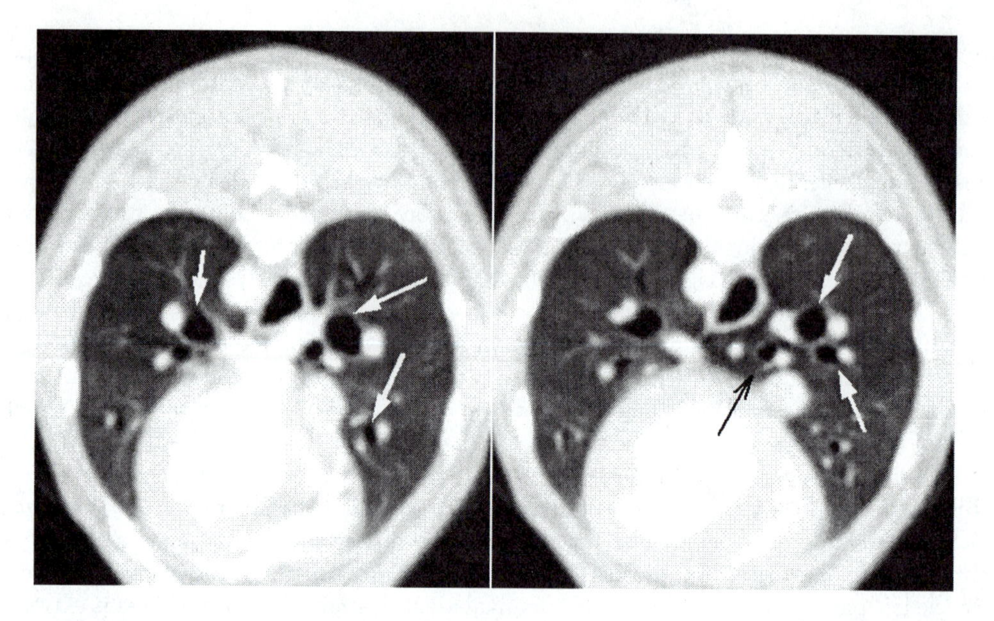

Figure 16.15: *Electron beam CT slices depicting two 3 mm-thick cross sections of canine lungs positioned 6 mm apart (slices n and n+2 shown). Several airways are marked by arrows. Note the airway branching marked by white arrows in the right panel. Bright regions correspond to vessels. Dogs were scanned in the prone position.*

medium-diameter airways and vessels, large-diameter airways forming the primary tree are not enhanced (see left panel of Figure 16.16) and must be superimposed over the hat-transformed images. Similarly, large-diameter vessels are identified using simple gray-level threshold criteria. The pre-processed image is segmented using edge-based region growing [Cornelis et al. 92, De Becker et al. 92]. A large number of regions is formed; the regions with potential correspondence to airways tend to be dark and vessel regions bright. Regions with intermediate gray-levels are assigned a *background* label, meaning that they correspond neither to airways nor vessels. The image regions that have gray-levels darker than those of background are labeled as *candidate airways*, the regions brighter than background are labeled as *candidate vessels*. While the *background* label is final, the *candidate* regions receive final labels of *background, vessel,* or *airway* after further processing described in the next step. Figure 16.16 shows the intermediate results of pre-processing.

4. *Vessel and airway detection in individual image slices.* While vessels appear brighter than the lung tissue and can therefore be easily identified, gray-level properties of airways vary substantially and additional contextual information is needed for airway detection. Anatomical knowledge about the pairwise appearance of vessels and airways and knowledge about existence of airway walls surrounding the airways may be utilized. Vessels are identified first using a priori information regarding realistic sizes and expected gray-level properties. Airways are identified considering the pairwise vessel–airway appearance and employing already labeled *candidate vessel* and *candidate airway*

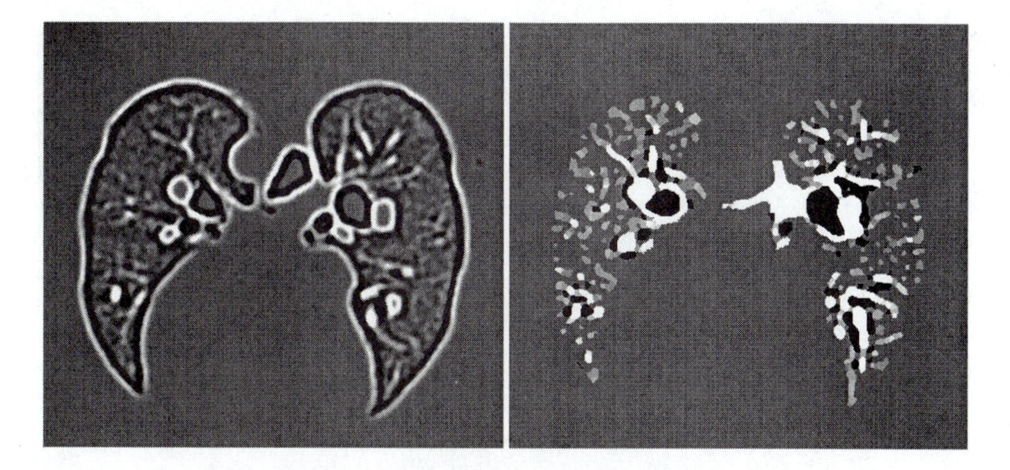

Figure 16.16: *Single-slice processing of CT lung images (see the left panel of Figure 16.15 for the original image). Left: Result of the hat transform to enhance small-diameter vessels and airways. Right: Preliminary region labeling; candidate airways are black, background is gray, and candidate vessels are brighter than background. Note that similar preliminary labeling cannot be achieved by direct application of gray-level-based segmentation to the original image no matter how carefully the airway and vessel thresholds are selected.*

regions to provide the needed contextual information. As a result of this step, each region is labeled as *background, vessel,* or *airway.*

5. *Airway tree construction* using three-dimensional connectivity. Regions identified in the previous labeling steps in individual CT slices may be stacked in the three-dimensional space to represent the three-dimensional airway tree. As a result, airway regions from adjacent CT slices that correspond to the same airway branch are three-dimensionally connected and form a three-dimensional object. Difficulties may arise caused by errors in single-slice airway detection, and by the lack of three-dimensional connectivity both real (due to airway occlusion or severe narrowing) or artifactual (due to the non-zero slice thickness). To minimize the number of false (extra) airways, only regions labeled as *airways* that are exhibiting good 3D connectivity are used to form the final 3D airway tree.

The *background/vessel/airway* labeling process performed in step 4 requires more detailed description. The rules defining region labeling are applied in three sequential passes. In the first pass, the regions are assigned final labels *vessel, background,* or *airway.* As an example, the rule used to determine airways is the following:

$$\begin{aligned}\text{region} = \text{airway} \quad IF \quad & [(\text{region\_gray} \leq \text{max\_airway\_gray}) \\ & AND \ (\text{region } \textit{is-adjacent-to} \text{ vessel}) \\ & AND \ (\text{region } NOT \ \textit{is-adjacent-to} \text{ big\_vessel})]\end{aligned}$$

In the second and third passes, labeling confidence is associated with each *vessel* and *airway* label, respectively. Two confidence levels are defined for the *vessel* label, and four confidence levels for the *airway* label. The label assignment and confidence rating are performed in a region adjacency graph (Section 3.2.3). The region adjacency graph also carries information about original CT gray-level, shape, and size properties of each image region. Obviously, adjacency to a vessel with a higher labeling confidence represents higher labeling confidence for the airway. Regions are more likely to represent airways if they are very dark on the absolute scale or if they are the darkest of all their respective neighboring regions. Additionally, airway labeling confidence depends greatly on the contextual requirement of airways being surrounded by airway walls. Therefore, airway regions which are not surrounded by a reliably identified wall are assigned low airway labeling confidences in the third rule-application pass. An example rule for an airway labeled with the highest confidence follows:

IF [airway *is-adjacent-to* (confidence(vessel) == 1)]
AND (airway *is-darkest-of-all-neighbors*)
AND (airway *is-surrounded-by-a-wall*)
THEN confidence(airway) = 1.0

In the complete set of rules [Sonka et al. 96], several regional and inter-regional relations were used. The Boolean binary relation *is-adjacent-to* referred to adjacency of two regions and was assessed in the region adjacency graph. The inter-regional relation *is-brightest-of-all-neighbors*, (*is-darkest-of-all-neighbors*) required comparison of a region's brightness with the maximum (minimum) *region_brightness* of all adjacent regions and was easily evaluated in the region adjacency graph. The Boolean regional property *is-very-dark* was satisfied if candidate airway region's brightness was less than the value of the parameter *dark_airway_gray*. An airway is usually surrounded by an airway wall (*is-surrounded-by-a-wall*) that can be identified as a bright ring encompassing the airway (Figure 16.17). To determine whether or not the candidate airway is surrounded by an airway wall, existence of a sequence of dark-bright-dark gray-level changes along radial rays emanating from the region centroid is examined as indicated in Figure 16.17. Because of the adjacent vessels, the bright ring of approximately constant width does not always completely surround the airway. Therefore, the wall evidence is collected along each ray directed from the center of gravity of a candidate airway through all its perimeter points.

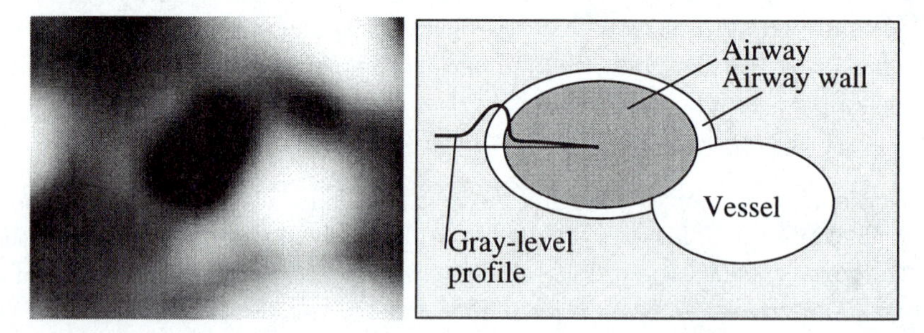

Figure 16.17: *Airway wall detection analysis principles. The original location of the airway shown in the left panel was in Figure 16.15 depicted by a black arrow.*

Performance of the rule-based airway tree detection method was tested in canine EBCT data sets. The parameters for the airway detection method were determined from measurements of canine lung CT images and served as a priori knowledge about canine airway morphology and CT scanning conditions. To evaluate the reliability of our airway tree detection method, the frequency was assessed with which the method wrongly identified or failed to identify airways in individual EBCT image slices as compared to the airway identification provided by an expert observer. By comparing the airway detection accuracy of the rule-based method to that of a more conventional airway detection method based on three-dimensional seeded region growing as described in [Wood et al. 95], the rule-based method significantly out-performed the conventional tree detection. Figure 16.18 gives an example of the manually and automatically determined trees. The result of manual analysis demonstrates that the automated tree detection quality is comparable to that achievable by manual tracing. For comparison, the tree determined using the conventional method based on seeded region growing is given in the middle panel to demonstrate the performance improvement due to the application of knowledge and context in the airway detection process.

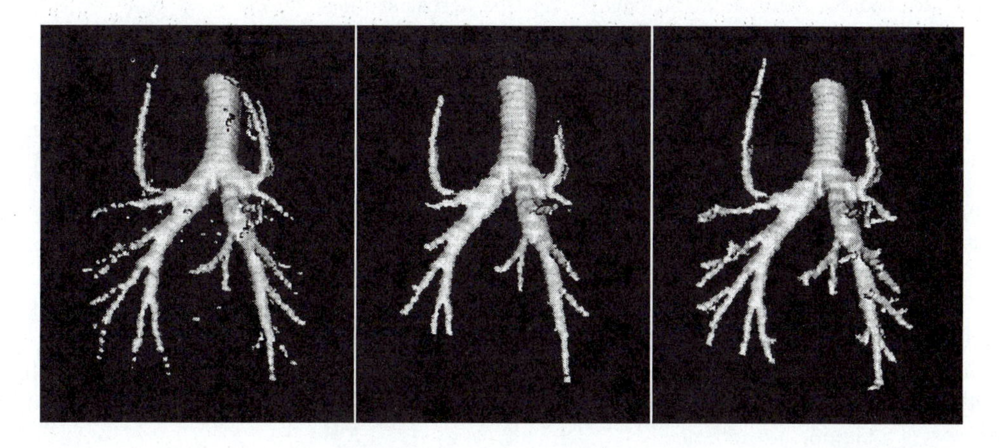

Figure 16.18: *Three-dimensional rendering of airway trees. Left: Manually identified canine airway tree. Manual detection is extremely time consuming and tedious. The tree required 7 hours of tracing and an additional 2 hours of manual editing. Middle: Airway tree detected using the conventional approach based on seeded 3D region growing. Right: Airway tree determined automatically using the rule-based approach.*

While the rule-based method significantly out-performed the conventional seeded region growing approach and the percentages of correctly determined airways were high, there were many extra airways present after the single-slice rule-based labeling that were only partially removed in the 3D tree construction step. Additionally, setting of the method's parameters was based on a trial-and-error strategy and the method was quite sensitive to the parameter setup. As is typical for rule-based systems, it was necessary to dichotomize crisply rules at (often) artificial boundaries. Therefore, an alternative approach based on fuzzy logic labeling (Section 7.1) was developed in which binary rules were substituted with fuzzy rules [Park et al. 96]. Using essentially identical a priori knowledge, it was demonstrated that fuzzy logic significantly decreased the number of extra airways. Still, definition of the fuzzy logic membership function was manual and required time-consuming trial-and-error parameter setup. Consequently, an automated training [Ishibuchi et al. 95] was developed in which fuzzy labeling system parameters were determined during a training stage in which airway examples and counter-examples were used [Park et al. 98]. The resulting fuzzy logic-based system offers ease of automated training. Simultaneously, its performance is comparable to that of the manually trained fuzzy logic system.

Correct quantitative assessment of geometry and structure of pulmonary airway trees requires to perform the measurements in three-dimensional space. Manual three-dimensional analysis of high-resolution computed tomography (HRCT) pulmonary data sets is tedious and impractical in routine evaluation of pulmonary disorders. Accurate identification and quantitation of the airways from high-resolution CT image data is expected to contribute significantly to development and evaluation of early detection techniques and new therapies for pulmonary disorders. It will also facilitate quantitative measurements of individual airway branches, combination of structural and perfusion measurements, and tracking of the full airway tree of the same patient over time to evaluate changes on a regional basis via the ability to find the same location within the tree across studies.

16.4 Passive surveillance

Surveillance refers to the task of observing a scene, often for lengthy periods, in search of particular objects or particular behavior. This task has many applications; foremost among them is security (monitoring for undesirable behavior such as theft or vandalism), but increasing numbers of others in areas such as agriculture also exist [McFarlane and Schofield 95, Sergeant et al. 96, Sumpter et al. 97]. Historically, closed-circuit TV (CCTV) surveillance has been mundane and labor intensive, involving personnel scanning multiple screens, but the advent of reasonably priced fast hardware means that automatic surveillance is becoming a realistic task to attempt in real time. Several attempts at this are underway [Rohr 93, Blake et al. 93, Rohr 94]; we summarize here work completed by Baumberg [Baumberg 95] which represents a significant advance.

Figure 16.19: *A pedestrian scene. Courtesy A. M. Baumberg, School of Computer Studies, University of Leeds.*

For simplicity, consider a scene being monitored by a single stationary camera; a frame from such a scene is illustrated in Figure 16.19. Given a sequence of such scenes, we might hope that any significant change between them may be due to objects of interest in motion—in this example pedestrians, but other applications may pick out vehicles, animals, or some other object. In reality, changes in weather conditions, slight movement of objects such as trees, camera judder etc. can cause there to be many changes from frame to frame—it is possible, however, to maintain a good background estimate by computing the median of the last few frames (perhaps over the last 5 minutes); McFarlane gives an efficient way of doing this without storing all the relevant information [McFarlane and Schofield 95]. Computing image differences frame by frame with such a dynamically maintained reference provides a number of 'blobs' which, when thresholded for size, usually provide a silhouette of the objects of interest. This procedure is similar to that described in Section 15.1, except that the difference image retains the absolute intensity difference (i.e., is not binary). This difference is blurred (see Section 4.3.3) and then thresholded to reduce the effects of noise. Various quantization and remaining noise effects leave boundaries imperfect, but a simple sequence of morphological operations is usually enough to produce usable shapes; a sequence of dilations and erosions (Sections 11.3.1 and 11.3.2) is sufficient to fill in 'gaps' that may appear in shapes. Figure 16.20 illustrates the sort of thing we may see at this stage.

Figure 16.20: *Silhouette extracted from the video sequence. Courtesy A. M. Baumberg, School of Computer Studies, University of Leeds.*

A lengthy sequence of input will generate a large number of such silhouettes which we may then attempt to analyze for shape characteristics [Baumberg and Hogg 94a]. The approach is to approximate the boundary of the region with a cubic B-spline (Section 6.2.5). First, a standard reference point on the boundary is needed; for this application, this is quite straightforward by determining the principal axis of the shape (that which minimizes the sum of squared distances to that axis) and selecting the uppermost intersection with the boundary. For scenes involving humans in upright positions, this usually works well in identifying the 'top' of the figure. Using this as an origin, equally spaced points around the boundary are used to derive a B-spline representation with 40 control points—Figure 16.21 illustrates this.

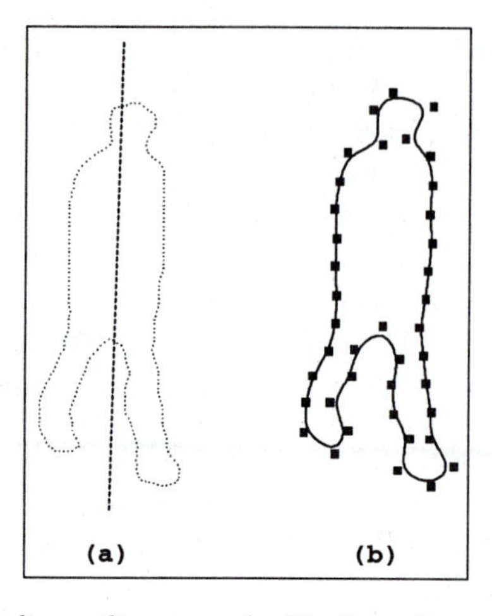

Figure 16.21: *Extracting a spline. Courtesy A. M. Baumberg, School of Computer Studies, University of Leeds.*

Given a training set of such shapes, it is now straightforward to analyze them using the point distribution model method described in Section 8.3. Each shape has 80 parameters (from 40 control points) $(x_1, y_1, x_2, y_2, \ldots, x_{40}, y_{40})$, but it transpires that the first 18 eigenvalues account for virtually all the variation detected. Thus we are representing a shape as a vector **b**,

$$\mathbf{x} = \overline{\mathbf{x}} + P\mathbf{b} \tag{16.1}$$

where $\mathbf{x}$ is the 80-dimensional vector defining the spline, $\overline{\mathbf{x}}$ is the mean shape, P is an 80×18 matrix, and $\mathbf{b}$ is an 18-dimensional vector parameterizing the shape $\mathbf{x}$.

Variations in the first two modes are illustrated in Figure 16.22. Notice that the first

Figure 16.22: *Varying the first and second modes by $\pm 1.5\sigma$. Courtesy A. M. Baumberg, School of Computer Studies, University of Leeds.*

mode captures the difference between silhouettes in which one and two legs are visible, while the second clearly illustrates a 'left-right' swaying of a moving figure.

An immediate application of this analysis is to clean errors in silhouettes; we can take a noisy boundary $\mathbf{x}$, determine the PDM representation $\mathbf{b}$, project this into the closest point $\tilde{\mathbf{b}}$ within the model space defined by a (clean) training set, and map this back to a spline defined by a vector $\tilde{\mathbf{x}}$ which is normally cleaner than the original—an example of the effects of this procedure is given in Figure 16.23. Strictly, what we are doing is constraining how far the eigen co-ordinates may vary, and requiring that $\tilde{\mathbf{b}}$ lie in an 18D hyper-ellipsoid.

A primary aim of this modeling is to assist in tracking, that is, to detect an object (pedestrian) and follow its trajectory through the scene. Given the PDM representation, we

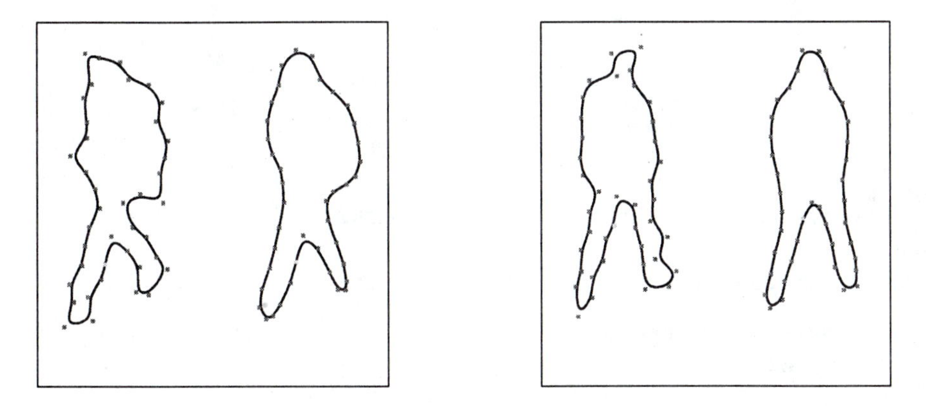

Figure 16.23: *Filtering noisy splines. Courtesy A. M. Baumberg, School of Computer Studies, University of Leeds.*

note that a spline appears in the scene after due translation, rotation, and scaling; if the current offset is (o_x, o_y), the scale is s and the rotation is θ, we may model the boundary by

$$Q = \begin{pmatrix} s\cos\theta & -s\sin\theta \\ s\sin\theta & s\cos\theta \end{pmatrix} \tag{16.2}$$

$$\begin{pmatrix} X_i \\ Y_i \end{pmatrix} = Q\begin{pmatrix} x_i \\ y_i \end{pmatrix} + \begin{pmatrix} o_x \\ o_y \end{pmatrix}$$

where x_i, y_i are given by equation (16.1). Then if we write $\mathbf{o} = (o_x, o_y, o_x, o_y, \ldots, o_x, o_y,)$ (40 times), and

$$\mathcal{Q} = \begin{pmatrix} Q & \ldots & 0 \\ \vdots & \ldots & \vdots \\ 0 & \ldots & Q \end{pmatrix} \tag{16.3}$$

(an 80×80 matrix), then the shape vector $\mathbf{X}$ is related to the state $\mathbf{b}$ by the equation

$$\mathbf{X} = \mathcal{Q}(P\mathbf{b} + \overline{\mathbf{x}}) + \mathbf{o} \tag{16.4}$$

When a new object is detected, it will not be clear what the best estimates of its scale, trajectory, or model parameters are, but given suitable assumptions we might initialize these parameters and then iterate them during successive frames using a Kalman filter (Section 15.4) to converge on a good estimate [Baumberg and Hogg 94b]. We suppose that the object to be initialized has bounding box given by lower left co-ordinates (x_l, y_l) and upper right (x_r, y_r), and that the mean height of figures in a training set is h_m. Rewriting equation (16.2) as

$$\begin{pmatrix} s\cos\theta & -s\sin\theta \\ s\sin\theta & s\cos\theta \end{pmatrix} = \begin{pmatrix} a_x & -a_y \\ a_y & a_x \end{pmatrix}$$

we can then initialize the figure as the mean shape, $\hat{\mathbf{b}}^0 = \mathbf{0}$, and

$$\hat{a}_x^0 = \frac{y_r - y_l}{h_m}$$

$$\begin{aligned}\hat{a}_y^0 &= 0\\ \hat{o}_x^0 &= \frac{x_l + x_r}{2}\\ \hat{o}_y^0 &= \frac{y_l + y_r}{2}\end{aligned}$$

so the figure is scaled to its bounding box, aligned vertically, with the origin at the center of the box.

A reasonable model for the evolution of these parameters is to assume that:

1. A stochastic model for the shape is more stable than an assumption of uniform change, so we assume

$$\mathbf{b}^k = \mathbf{b}^{k-1} + \mathbf{w}^{k-1}$$

where $\mathbf{w}$ is a zero mean, normally distributed noise term, $w_i^k \sim N(0, \sigma_i)$. The eigen analysis gives the variance of b_i over the training set as λ_i, and σ_i is initialized as $\sigma_i = \kappa\lambda_i$, where characteristically $\kappa = 0.05$—thus the shape estimate is allowed to vary within an ellipsoid that is a subset of that defined by the training set. The eigen decomposition assumes that $E(b_i b_j) = 0$.

2. The object is moving uniformly in 2D subject to additive noise,

$$\frac{d}{dt}\begin{pmatrix} o_x \\ \dot{o}_x \end{pmatrix} = \begin{pmatrix} \dot{o}_x \\ 0 \end{pmatrix} + \begin{pmatrix} v_x \\ w_x \end{pmatrix}$$

for noise terms v_x and w_x (and similarly for o_y). This gives the frame update equation

$$\begin{pmatrix} o_x^{k+1} \\ \dot{o}_x{}^{k+1} \end{pmatrix} = \begin{pmatrix} 1 & \Delta t \\ 0 & \Delta t \end{pmatrix}\begin{pmatrix} o_x^k \\ \dot{o}_x{}^k \end{pmatrix} + \begin{pmatrix} v_x \\ w_x \end{pmatrix} \tag{16.5}$$

(where Δt is the time step between frames). v_x and w_x are noise terms, with $v_x \sim N(0, q_v)$, $w_x \sim N(0, q_w)$.

3. The alignment parameters a_x, a_y are constant, subject to noise

$$\begin{pmatrix} a_x^{k+1} \\ a_y^{k+1} \end{pmatrix} = \begin{pmatrix} a_x^k \\ a_y^k \end{pmatrix} + \begin{pmatrix} w_{ax} \\ w_{ax} \end{pmatrix} \tag{16.6}$$

where $w_{ax}, w_{ax} \sim N(0, q_a)$.

The Kalman filter proceeds by estimating origin, alignment, and shape independently of each other. Given estimates of the parameters, we predict the next state $\hat{\mathbf{X}}^-$ from equation (16.4) and use this to make an observation from the image $\mathbf{z}$.

To update the x-origin co-ordinate we need to consider the state $(\hat{o}_x, \hat{\dot{o}}_x)$. Many measurements of the origin are available from the expression

$$\mathbf{z} - Q(P\hat{\mathbf{b}} + \overline{\mathbf{x}})$$

These measurements, together with the update relation equation (16.5) and the noise variance properties, provide the necessary ingredients for the application of the filter, which can then be used to provide the best estimate of the origin in the next frame.

To update the alignment parameters with the observation $\mathbf{z}$, we use the measurement model

$$\mathbf{z} - \hat{\mathbf{o}} = H \begin{pmatrix} a_x \\ a_y \end{pmatrix}$$

where H is a $N \times 2$ measurement matrix defined by manipulation of equation (16.4). Again, updating estimates for the alignment is now a straightforward application of the theory.

Likewise, equation (16.4) provides a measurement model for the 18 shape parameters; it can be shown that each of these may be extracted independently of the others, permitting a model of the form

$$\mathbf{z} - \hat{\mathbf{X}} = \mathbf{h}_i(b_i - \hat{b_i})$$

to be constructed, where $\mathbf{h}_i$ is an $N \times 1$ measurement matrix.

The Kalman filter is key to providing real-time performance in this application. Estimates of where the silhouette ought to be permit very localized search for edges in new frames, obviating the need for operations over the whole image. This operation is performed by determining the edge normal at the predicted position, and searching along it for the position of maximum contrast with the reference background. If no likely point is found (the contrast is low), 'no observation' is recorded and the associated predictions are not updated. This is particularly powerful since it permits the figure to suffer partial occlusion and still be tracked—this is illustrated by scenes with artificially introduced occlusion in Figure 16.24.

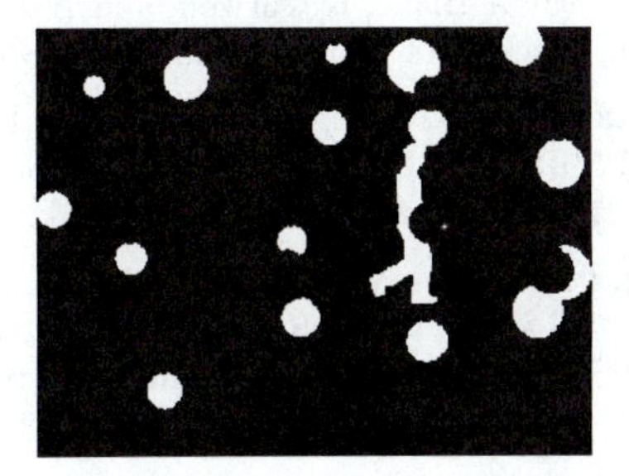

Figure 16.24: *Tracking through occlusion. Courtesy A. M. Baumberg, School of Computer Studies, University of Leeds.*

This has been an abbreviated overview of a large development, and for full details the reader is referred to the original text [Baumberg 95]. In fact, the work goes further than described in many respects; in particular it should be noted that

- The tracking system generalizes to a moving camera, allowing much greater scope. This is done by dispensing with the reference image approach and using a primitive edge detector in the areas indicated by the Kalman filter—the limited area that needs to be covered makes this a viable real-time approach. Initialization still (currently) requires the static camera assumption in order to extract blobs representing moving figures.

- The periodic nature of a walker's movement can also be exploited to produce a spatio-temporal approach that assists in tracking by reducing the size of search windows in future frames [Baumberg and Hogg 96]. This is done by making predictions of the shape expected based on learned characteristics of the periodic motion. This also permits the tracker to survive incidences of complete occlusion (for example, walking behind a tree), since the prediction has a clear idea of what the silhouette should look like even if it is absent from a sequence of frames.

16.5 References

[Baumberg 95] A M Baumberg. Learning deformable models for tracking human motion. PhD thesis, School of Computer Studies, University of Leeds, Leeds, UK, 1995.

[Baumberg and Hogg 94a] A M Baumberg and D C Hogg. An efficient method of contour tracking using active shape models. In *Proceedings of the IEEE Workshop on Motion on Non-rigid and Articulated Objects*, pages 194–199, Texas, November 1994.

[Baumberg and Hogg 94b] A M Baumberg and D C Hogg. Learning flexible models from image sequences. In J-O Eklundh, editor, *3rd European Conference on Computer Vision,* Stockholm, Sweden, pages 299–308, Springer Verlag, Berlin, 1994.

[Baumberg and Hogg 96] A M Baumberg and D C Hogg. Learning spatiotemporal models from examples. *Image and Vision Computing,* 14(8):525–532, 1996.

[Bhalla et al. 91] M Bhalla, N Turcios, V Aponte, V Jenkins, M Leitman, D McCauley, and D Naidick. Cystic fibrosis: Scoring system with thin section CT. *Radiology,* 179:783–788, 1991.

[Blake et al. 93] A Blake, R Curwen, and A Zisserman. A framework for spatio-temporal control in the tracking of visual contours. *International Journal of Computer Vision,* 11:127–145, 1993.

[Blostein and Baird 92] D Blostein and H S Baird. A critical survey of music image analysis. In H S Baird, H Bunke, and K Yamamoto, editors, *Structured Document Image Analysis,* pages 405–434. Springer Verlag, New York, 1992.

[Bom et al. 95] N Bom, W Li, A F W van der Steen, C L de Korte, E J Gussenhoven, C von Birgelen, C T Lancee, and J R T C Roelandt. Intravascular ultrasound: Technical update 1995. In P J de Feyter, C Di Mario, and P W Serruys, editors, *Quantitative Coronary Imaging,* pages 89–106. Barjesteh, Meeuwes & Co., Rotterdam, 1995.

[Brown et al. 77] B G Brown, E Bolson, M Frimer, and H Dodge. Quantitative coronary arteriography estimation of dimensions, hemodynamic resistance, and atheroma mass of coronary artery lesions using the arteriogram and digital computation. *Circulation,* 55:329—337, 1977.

[Buchi et al. 90] M Buchi, O M Hess, R L Kirkeeide, T Suter, M Muser, H P Osenberg, P Niederer, M Anliker, K L Gould, and H P Krayenbuhl. Validation of a new automatic system for biplane coronary arteriography. *International Journal of Cardiac Imaging,* 5:93–103, 1990.

[Chrispin and Norman 74] A R Chrispin and A P Norman. The systematic evaluation of the chest radiograph in cystic fibrosis. *Pediatric Radiology,* 2:101–106, 1974.

[Collins and Skorton 86] S M Collins and D J Skorton. *Cardiac Imaging and Image Processing.* McGraw-Hill, New York, 1986.

[Cornelis et al. 92] J Cornelis, J De Becker, M Bister, C Vanhove, G Demonceau, and A Cornelis. Techniques for cardiac image segmentation. In *Proceedings of the 14th IEEE EMBS Conference, Vol. 14,* Paris, France, pages 1906–1908, IEEE, Piscataway, NJ, 1992.

[De Becker et al. 92] J De Becker, M Bister, N Langloh, C Vanhove, G Demonceau, and J Cornelis. A split-and-merge algorithm for the segmentation of 2-D, 3-D, 4-D cardiac images. In *Proceedings of the IEEE Satellite Symposium on 3D Advanced Image Processing in Medicine,* Rennes, France, pages 185–189, IEEE, Piscataway, NJ, 1992.

[de Feyter et al. 95] P J de Feyter, C Di Mario, and P W Serruys. *Quantitative Coronary Imaging.* Barjesteh, Meeuwes & Co., Rotterdam, 1995.

[Dhawale et al. 93] P J Dhawale, Q Rasheed, N Griffin, D L Wilson, and J McB Hodgson. Intracoronary ultrasound plaque volume quantification. In *Computers in Cardiology*, pages 121–124, IEEE, Los Alamitos, CA, 1993.

[Eichel et al. 88] P H Eichel, E J Delp, K Koral, and A J Buda. Method for a fully automatic definition of coronary arterial edges from cineangiograms. *IEEE Transactions on Medical Imaging*, 7:313–320, 1988.

[Ellis et al. 86] S Ellis, W Sanders, C Goulet, R Miller, K C Cain, J Lesperanz, M G Bourassa, and E L Alderman. Optimal detection of the progression of coronary artery disease: Comparison of methods suitable for risk factor intervention trials. *Circulation*, 74:1235–1242, 1986.

[Erickson 75] R F Erickson. The DARMS project: A status report. *Computers and the Humantities*, 9:292–298, 1975.

[Evans and Nixon 96] A Evans and M S Nixon. Biased motion-adaptive temporal filtering for speckle reduction in echocardiography. *IEEE Transactions on Medical Imaging*, 15:39–50, 1996.

[Everhart et al. 94] J Everhart, M Cannon, J Newell, and D Lynch. Image segmentation applied to CT examination of Lymphangioleimyomatosis (LAM). In *Medical Imaging 1994—Image Processing, Proceedings SPIE*, volume 2167, pages 87–95, SPIE, Bellingham WA, 1994.

[Fleagle et al. 89] S R Fleagle, M R Johnson, C J Wilbricht, D J Skorton, R F Wilson, C W White, M L Marcus, and S M Collins. Automated analysis of coronary arterial morphology in cineangiograms: Geometric and physiologic validation in humans. *IEEE Transactions on Medical Imaging*, 8(4):387–400, 1989.

[Hoffman et al. 92] E A Hoffman, D Gnanaprakasam, K B Gupta, J D Hoford, S D Kugelmass, and R S Kulawiec. VIDA: An environment for multi-dimensional image display and analysis. In *Proceedings SPIE*, volume 1660, pages 694–711, SPIE, Bellingham WA, 1992.

[Ishibuchi et al. 95] H Ishibuchi, K Nozaki, N Yamamoto, and H Tanaka. Selecting fuzzy if-then rules for classification problems using genetic algorithms. *IEEE Transactions on Fuzzy Systems*, 3:260–270, 1995.

[Johnson et al. 88] M R Johnson, D J Skorton, E E Ericksen, S R Fleagle, R F Wilson, L F Hiratzka, C W White, M L Marcus, and S M Collins. Videodensitometric analysis of coronary stenoses: In vivo geometric and physiologic validation in humans. *Investigative Radiology*, 23:891–898, 1988.

[Kahan et al. 87] S Kahan, T Pavlidis, and H Baird. On the recognition of printed characters of any font and size. *IEEE Transactions on Pattern Analysis and Machine Intelligence*, PAMI-9(2):274–288, 1987.

[Kiatoka and Yumoto 90] H Kiatoka and T Yumoto. Three-dimensional CT of the bronchial tree. A trial using an inflated fixed lung specimen. *Investigative Radiology*, 25:813–817, 1990.

[Le Free et al. 86] M T Le Free, S B Simon, G B J Mancini, and R A Vogel. Digital angiographic assessment of coronary arterial geometric diameter and videodensitometric cross-sectional area. In *Proceedings SPIE*, volume 626, pages 334–341, SPIE, Bellingham, WA, 1986.

[Li et al. 93a] S Li, W N McDicken, and P R Hoskins. Blood vessel diameter measurement by ultrasound. *Physiological Measurements*, 14:291–297, 1993.

[Li et al. 93b] W Li, J G Bosch, Y Zhong, H van Urk, E J Gussenhoven, F Mastik, F van Egmond, H Rijstenborgh, J H C Reiber, and N Bom. Image segmentation and 3D reconstruction of intravascular ultrasound images. *Acoustic Imaging*, 20:489–496, 1993.

[Loy 85] G Loy. Musicians make a standard: The MIDI phenomenon. In C Roads, editor, *The Music Machine; Selected Readings from Computer Music Journal*, pages 181–198. MIT Press, Cambridge, MA, 1985.

[Mancini et al. 87] G B J Mancini, S B Simon, M J McGillem, M T Le Free, H Z Friedman, and R A Vogel. Automated quantitative coronary angiography: Morphologic and physiologic validation in vivo of a rapid digital angiographic method. *Circulation*, 75:452–460, 1987.

[Marcus et al. 91] M L Marcus, H R Schelbert, D J Skorton, and G L Wolf. *Cardiac Imaging.* Saunders, Philadelphia, 1991.

[McFarlane and Schofield 95] N J B McFarlane and C P Schofield. Segmentation and tracking of piglets in images. *Machine Vision and Applications*, 8:187–193, 1995.

[Ney et al. 89] D R Ney, J E Kuhlman, R H Hruban, H Ren, G M Hutchins, and E K Fishman. Three-dimensional CT volumetric reconstruction and display of the bronchial tree. *Investigative Radiology*, 25:736–742, 1989.

[Ng 95] K C Ng. Automated computer recognition of music scores. PhD thesis, School of Computer Studies, University of Leeds, Leeds, UK, 1995.

[Ng and Boyle 92] K C Ng and R D Boyle. Segmentation of musical primitives. In D C Hogg and R D Boyle, editors, *Proceedings of the British Machine Vision Conference,* Leeds, UK, pages 472–480, Springer Verlag, London, 1992.

[Ng and Boyle 96] K C Ng and R D Boyle. Recognition and reconstruction of primitives in music scores. *Image and Vision Computing*, 14(1):39–46, 1996.

[Ng et al. 95] K C Ng, R D Boyle, and D Cooper. Automated optical music score recognition and its enhancement using high-level musical knowledge. In *Proceedings XI Colloquium on Musical Informatics* Bologna, Italy, pages 167–172, Associazione di Informatica Musicale Italiana, Bologna, Italy, November 1995.

[Nichols et al. 84] A B Nichols, C F O Gabriele, J J Fenoglio Jr., and P D Esser. Quantification of relative arterial stenosis by cinevideodensitometric analysis of coronary arteriograms. *Circulation*, 69:512–522, 1984.

[Park et al. 96] W Park, E A Hoffman, and M Sonka. Fuzzy logic approach to extraction of intrathoracic airway trees from three-dimensional CT images. In *Image Processing, Proceedings SPIE Vol. 2710*, pages 210–219, SPIE, Bellingham, WA, 1996.

[Park et al. 98] W Park, E A Hoffman, and M Sonka. Segmentation of intrathoracic airway trees: A fuzzy logic approach. *IEEE Transactions on Medical Imaging*, 17(4), 1998. To appear.

[Pasterkamp et al. 93] G Pasterkamp, M S van der Heiden, M J Post, B Ter Haar Romeny, W P T M Mali, and C Borst. Discrimination of the intravascular lumen and dissections in a single 30-MHz ultrasound image: Use of confounding blood backscatter to advantage. *Radiology*, 187:871–872, 1993.

[Reiber et al. 84] J H C Reiber, C J Kooijman, C J Slager, J J Gerbrands, J C H Schuurbiers, A Den Boer, W Wijns, and P W Serryus. Computer assisted analysis of the severity of obstructions from coronary cineangiograms: A methodological review. *Automedica*, 5:219–238, 1984.

[Reiber et al. 91] J H C Reiber, P W Serruys, and J D Barth. Quantitative coronary angiography. In M L Marcus, H R Schelbert, D J Skorton, and G L Wolf, editors, *Cardiac Imaging*, pages 211–280. Saunders, Philadelphia, 1991.

[Reingold and Hansen 83] E M Reingold and W J Hansen. *Data Structures.* Little, Brown, Boston, 1983.

[Rohr 93] K Rohr. Incremental recognition of pedestrians from image sequences. In J K Aggarwal, editor, *Proceedings of Computer Vision and Pattern Recognition 1993*, pages 8–13, IEEE Computer Society Press, New York, 1993.

[Rohr 94] K Rohr. Towards model-based recognition of human movements in image sequences. *Computer Vision, Graphics, and Image Processing*, 59(1):94–115, 1994.

[Selfridge-Field 94] E Selfridge-Field. Optical Recognition of Musical Notation: A Survey of Current Work. In *Computing in Musicology*, volume 9, pages 109–145. Center for Computer Assisted Research in the Humanities (CCARH), Stanford, CA, 1994.

[Sergeant et al. 96] D M Sergeant, J M Forbes, and R D Boyle. Computer vision as a tool for monitoring behaviour. *Occasional Publication of the British Society of Animal Science*, 20:77–79, 1996. BSAS Symposium, Leeds, UK, September.

[Simmons et al. 88] M A Simmons, A D Muskett, R A Kruger, S C Klausner, N A Burton, and J A Nelson. Quantitative digital subtraction coronary angiography using videodensitometry: An in vivo analysis. *Investigative Radiology*, 23:98–106, 1988.

[Sonka and Collins 97] M Sonka and S M Collins. Automated analysis of coronary angiograms. In C T Leondes, editor, *Medical Imaging Techniques and Applications*, pages 147–182. Gordon & Breach, New York, 1997.

[Sonka et al. 93a] M Sonka, G K Reddy, M D Winniford, and S M Collins. Adaptive simultaneous coronary border detection: A method for accurate analysis of small diameter vessels. In *Computers in Cardiology*, pages 109–112, IEEE, Los Alamitos, CA, 1993.

[Sonka et al. 93b] M Sonka, C J Wilbricht, S R Fleagle, S K Tadikonda, M D Winniford, and S M Collins. Simultaneous detection of both coronary borders. *IEEE Transactions on Medical Imaging*, 12(3):588–599, 1993.

[Sonka et al. 94] M Sonka, M D Winniford, X Zhang, and S M Collins. Lumen centerline detection in complex coronary angiograms. *IEEE Transactions on Biomedical Engineering*, 41:520–528, 1994.

[Sonka et al. 95a] M Sonka, W Liang, X Zhang, S DeJong, S M Collins, and C R McKay. Three-dimensional automated segmentation of coronary wall and plaque from intravascular ultrasound pullback sequences. In *Computers in Cardiology 1995*, pages 637–640, IEEE, Los Alamitos, CA, 1995.

[Sonka et al. 95b] M Sonka, M D Winniford, and S M Collins. Robust simultaneous detection of coronary borders in complex images. *IEEE Transactions on Medical Imaging*, 14(1):151–161, 1995.

[Sonka et al. 95c] M Sonka, X Zhang, M Siebes, M S Bissing, S C DeJong, S M Collins, and C R McKay. Segmentation of intravascular ultrasound images: A knowledge-based approach. *IEEE Transactions on Medical Imaging*, 14:719–732, 1995.

[Sonka et al. 96] M Sonka, W Park, and E A Hoffman. Rule-based detection of intrathoracic airway trees. *IEEE Transactions on Medical Imaging*, 15:314–326, 1996.

[Sonka et al. 97a] M Sonka, C R McKay, and C von Birgelen. Computer analysis of intravascular ultrasound images. In C T Leondes, editor, *Medical Imaging Techniques and Applications*, pages 183–226. Gordon & Breach, New York, 1997.

[Sonka et al. 97b] M Sonka, G Reddy, and S M Collins. Adaptive approach to accurate analysis of small diameter vessels in cineangiograms. *IEEE Transactions on Medical Imaging*, 16:87–95, 1997.

[Sumpter et al. 97] N Sumpter, R D Boyle, and R Tillet. Modelling collective animal behaviour using extended PDMs. In A Clark, editor, *Proceedings of the British Machine Vision Conference*, Colchester, UK. BMVA Press, 1997.

[Udupa 82] J K Udupa. Interactive segmentation and boundary surface formation for 3-D digital images. *Computer Graphics and Image Processing*, 18:213–235, 1982.

[von Birgelen et al. 95] C von Birgelen, C Di Mario, F Prati, N Bruining, W Li, P J De Feyter, and J R T C Roelandt. Intracoronary ultrasound: Three-dimensional reconstruction techniques. In C Di Mario, P J De Feyter, and J R T C Roelandt, editors, *Quantitative Coronary Imaging*, pages 181–197. Barjesteh, Meeuwes & Co., Rotterdam, 1995.

[Whiting et al. 91] J S Whiting, J M Pfaff, and N L Eigler. Advantages and limitations of videodensitometry in quantitative coronary angiography. In J H C Reiber and P W Serryus, editors, *Quantitative Coronary Angiography 1991*, pages 55–132. Kluwer Academic Publishers, Dordrecht, The Netherlands, 1991.

[Wiesel et al. 86] J Wiesel, A M Grunwald, C Tobiasz, B Robin, and M M Bodenheimer. Quantitation of absolute area of a coronary arterial stenosis: Experimental validation with a preparation in vivo. *Circulation*, 74:1099–1106, 1986.

[Wong et al. 86] W Wong, R L Kirkeeide, and K L Gould. Computer applications in angiography. In S M Collins and D J Skorton, editors, *Cardiac Imaging and Image Processing*, pages 206–238. McGraw-Hill, New York, 1986.

[Wood et al. 95] S Wood, A Zerhouni, E A Hoffman, and W Mitzner. Measurement of three-dimensional lung tree structures using computed tomography. *Journal of Applied Physiology*, 79:1687–1697, 1995.

Index

Bold text refers to major or defining entries.